Mathematical SETI

Statistics, Signal Processing, Space Missions

Mathematical SETI

Statistics, Signal Processing, Space Missions

Claudio Maccone

Mathematical SETI

Statistics, Signal Processing, Space Missions

Springer

Published in association with
Praxis Publishing
Chichester, UK

PRAXIS

Dr. Claudio Maccone
International Academy of Astronautics and Istituto Nazionale di Astrofisica
Via Martorelli 43 - 10155 Torino (Turin)
Italy
Email at IAA: claudio.maccone@iaamail.org
Email at home: clmaccon@libero.it

Front cover: (Upper right) Some of the 42 six-meter dishes now making up the Allen Telescope Array (ATA) at Hat Creek Radio Observatory in northern California, U.S.A. (courtesy: SETI Institute). (Lower left) Artist concept of the 500-meter Arecibo-like FAST radiotelescope, now under construction in Guizhou Province, southwest China (courtesy: Professor Rendong Nan).

SPRINGER–PRAXIS BOOKS IN ASTRONOMY AND PLANETARY SCIENCES
SUBJECT *ADVISORY EDITOR*: John Mason, M.B.E., B.Sc., M.Sc., Ph.D.

ISBN 978-3-642-27436-7 ISBN 978-3-642-27437-4 (eBook)
DOI 10.1007/978-3-642-27437-4
Springer Heidelberg Dordrecht London New York

Library of Congress Control Number: 2011944195

Cover design: Jim Wilkie
Project management: OPS Ltd., Gt. Yarmouth, Norfolk, U.K.

Printed on acid-free paper

Springer is part of Springer Science+Business Media (www.springer.com)

Contents

Preface

SETI, the modern Search for Extra Terrestrial Intelligence, began in 1959 with the publication of the seminal paper entitled "Searching for Interstellar Communications" by Giuseppe Cocconi (1914–2008) and Philip Morrison (1915–2005) in *Nature*, Vol. 184, Number 4690, pp. 844–846, September 19, 1959. Just one year later, in 1960, Frank Drake started experimental radio SETI with Project Ozma, when he first looked for possible extraterrestrial (ET) signals near the 1.420 gigahertz marker frequency known as "the hydrogen line". Modern radio SETI was thus born, and so it continues today as a result of the enormous progress in electronic equipment and mathematical computing algorithms used to detect ET signals.

But, a further year later, at a meeting on SETI held at the National Radio Astronomy Observatory (NRAO) in Green Bank, West Virginia, Frank Drake gave another outstanding contribution, now called the Drake equation (*http://en.wikipedia.org/wiki/Drake_equation*). This is described in Chapter 1 of this book, together with its extension for the equation to encompass probability and statistics, as discovered by this author in 2007 and first presented in 2008. That analysis occupies the first 11 chapters of this book.

PART I—SETI STATISTICS. This first part of the book comprises 11 chapters.

Chapter 1—The statistical Drake equation. This chapter shows how the classical Drake equation, the product of seven positive numbers, may be replaced by the product of seven positive random variables, called the "statistical Drake equation". This way of doing things is scientifically more robust inasmuch as each input of the classical Drake equation now has an error bar around itself. In other words, the purely numeric inputs of the classical Drake equation now become the mean values of the corresponding random variables, around which a certain standard deviation (to be found experimentally) is subtracted or added, as is customary in serious scientific papers. The mathematical consequences of this transformation are demonstrated and it is proven that the new random variable N for the number of communicating civilizations in the Galaxy must follow the lognormal probability distribution when the number of factors in the Drake equation is allowed to increase at will. This result opens up the possibility of inserting increasingly more factors into the Drake equation letting it become more representative of physical reality: for example, the end of civilization as a result of asteroid impact is absent in the 1961 formulation by Frank Drake probably because it was not until 1980 that the dinosaur demise was accepted by the scientific community as the consequence of an asteroid impact. Chapter 1 also derives another probability distribution (called "Maccone distribution" by Paul Davies and others) yielding the probability density function (pdf) of the distance between any two nearby civilizations in the Galaxy. This is of paramount importance for SETI inasmuch as it explains that we should hardly expect any ET civilization to be located at distances from us smaller than, say, 500 light-years. Thus, the most natural explanation for the apparent failure of 50 years of SETI search (1960–2010) is we did not find them simply because our current radio telescopes do not reach out far enough, being confined to ranges up to 100 or 200 light-years at best.

Chapter 2—Letting Maxima do the calculations. This chapter introduces students and young researchers to the joy of getting rid of hand-written calculations by resorting to the freely downloadable "Maxima" computer algebra code. In other words, the reader will find in the appendixes to the various chapters all those Maxima codes that this author had to write himself in order to prove the many equations given in the book for the first time. This way of doing things for heavily mathematical books like this one is brand new: we are not ashamed to demonstrate the beauty of SETI, astrophysics, and signal processing to our readers, we also teach them how to derive relevant new results by virtue of both Maxima and Mathcad. A couple of examples demonstrate this, in Appendixes 2.A and 2.B we derive the statistical properties of the lognormal distribution, key to the statistical Drake equation as shown in Chapter 1, and, as a demo of the strong capabilities of Maxima in tensor calculus, we derive the closed Einstein universe of 1917 (basic to cosmology), the Friedman equations of 1924, and the ensuing number of protons in the universe, the famous 10^{80} found by Dirac in 1937 (Dirac cosmology).

Chapter 3—How many planets for man and aliens? This chapter introduces the reader to the Dole equation (1964). This equation is mathematically the same as the Drake equation, but applies to the number of habitable planets in the Galaxy, rather than the number of communicating civilizations in the Galaxy. Thus, we extend our study to the statistical Dole equation, rather than the classical Dole equation of 1964, and we reach the conclusion that there should be some 100 million habitable planets for man in the Galaxy, plus a standard deviation of about 200 million. Not bad for future human expansion into the Galaxy, if we survive the many hazards we will have to face in the centuries to come, such as physical adversities and alien opposition. Having found the probability distribution of the distance between two nearby ET civilizations in Chapter 1, so in Chapter 3 we find that the same probability distribution applies to the distance between two nearby habitable planets—after changing the numbers (but not the equations) as necessary, of course.

Chapter 4—Statistical Fermi paradox and Galactic travel. This chapter considers the possible expansion of a certain civilization throughout the Galaxy, whether it be human or ET. The key idea here is that the amount of time required for expansion into space is basically determined by two factors: (1) the speed of spaceships used for hopping from one habitable planet to the next, and (2) the time required to colonize a new planet from scratch and make it a new base for further space voyages. We assume that the first variable (spaceship speed) is essentially deterministic, and so does not require any statistical treatment. However, we also assume that the second variable (colonization time) follows the lognormal distribution, again as a result of the number of unknown factors being so large that it approaches infinity. The Central Limit Theorem of statistics—as used in Chapter 1 to find the lognormal distribution of N and in Chapter 3 for N_{Hab}, respectively—is used here. Having made this assumption, the coral model for the expansion of corals at sea and of a civilization in the Galaxy allows us to determine the probability distribution of the overall time necessary for a given civilization to expand throughout the whole Galaxy. The mathematics gets quite complicated, and only wise use of Maxima allowed us to find the relevant probability distributions. All this is of course a statistical enlargement of the famous Fermi paradox, so far only treated by other authors in naive purely deterministic settings.

Chapter 5—How long does a civilization live? This chapter tries to address the totally unknown value of the last term in the Drake equation: How long might a technological civilization possibly survive? Since nobody knows, for we ourselves are the only example we have, we confine discussion in this chapter to changes in the number N that would be caused by long-lived civilizations vs. short-lived ones. The numerical examples given in this chapter are a statistical extension of the corresponding deterministic values given by Carl Sagan in his book (and TV series) *Cosmos* (1980).

Chapter 6—Life-span modeling by finite *b*-lognormals. This chapter contains profoundly innovative material, regarded by the author as one of the best mathematical models he has devised so far in his 64-year lifetime. The idea is as follows. All living beings are born at a certain time ($t = b =$ birth), they then grow up through adolescence ($t = a =$ adolescence), reach their prime at the peak ($t = p =$ peak), then decline through senility ($t = s =$ senility), and finally die ($t = d =$ death). Is there a probability finite density function that has such a behavior in time? Yes, is our answer and that is the b-lognormal. What is a b-lognormal? It simply is an ordinary (μ, σ) lognormal starting at a certain positive time $t = b > 0$, rather than at $t = 0$. Its equation just requires shifting the origin to a new positive instant $t = b > 0$, which we call a "b-lognormal" since that probability density function does not seem to have a name yet. But, the other four points in time mentioned above do have an immediate mathematical meaning: (1) the adolescence time ($t = a$) is the abscissa on the ascending inflexion point, (2) the peak time ($t = p$) is obviously the abscissa of the maximum, (3) the senility time ($t = s$) is the abscissa of the descending inflexion point, and finally (4) the death time ($t = d$) is the abscissa of the intercept between the tangent to senility and the time axis, and this "mathematical trick" allows us to get rid of the infinite tail on the right, replacing it by an obvious finite point. Such are b-lognormals. Now, Chapter 6 is entirely devoted to discovering new mathematical equations expressing the two unknown parameters (μ, σ) as functions of some of the known inputs such as birth time $(t = b)$ plus two of the four remaining known inputs (a, p, s, d). The author was able to discover some finite equations of this type, and probably still unknown similar equations exist as well. But, what he was able to discover was enough to write Chapters 7 and 8, of key importance to "mathematical history" and "mathematical Darwinian evolution", respectively. Finally, the author also derived an expression for the finite probability density of b-lognormals to renormalize to 1 vs. the ordinary normalization constant of ordinary lognormals. The set of all these new results is a breakthrough in replacing the mountain of words used today to describe Darwinian evolution and mathematical history by a simple set of statistical distributions in agreement with the statistical Drake equation and SETI.

Chapter 7—Historic civilizations as finite *b*-lognormals. We apply the results of Chapter 6 to mathematical history. We compute and compare the finite b-lognormals of eight civilizations whose existence had the greatest effect on the history of the world in the last 3,000 years: Ancient Greece (600 BC–30 BC), Ancient Rome (753 BC–476 AD), Renaissance Italy (1250–1600), Portugal (1419–1974), Spain (1492–1898), France (1524–1962), Britain (1588–1974), and the U.S.A. (1898–2050 tentatively). It will be objected that all these civilizations belong to the so-called Western world: nevertheless, the West has been home to the most advanced civilizations in the last 3,000 years. Whether Asia will replace the West in leading humanity in the future is highly likely, but that is still in the balance right now in 2012. So, these eight b-lognormals were compared on the same plot and a sort of "upper envelope" clearly emerged: it is an exponential curve that, more or less, embraces all b-lognormals as the geometric locus of their peaks! The principal result here is that b-lognormals become increasingly narrow as time elapses (i.e., their peaks get higher and higher), and this reveals progress (i.e., an increasing degree of civilization). To make this result quantitative, rather than just qualitative, we need a new unit measuring the "amount of evolution" reached by a certain civilization at a certain time, just as meters measure lengths, seconds measure time, coulombs measure electric charge, and so on in physics. This new unit of evolution we call the "darwin" and introduce it in the next chapter which deals with Darwinian evolution. We do so because in science "to measure is to understand".

Chapter 8—A mathematical model for evolution and SETI. The "exponential envelope" that was looming in the background of the previous chapter fully comes to light as the link between Darwinian evolution and the family of b-lognormals which is constrained between the exponential and the time axis. In fact, we first define Darwinian evolution simply as the exponential increase in the number of living species on Earth that characterized the last 3.5 billion years of life on Earth. In other words, we assume that 3.5 billion years ago the first and only living organism appeared (RNA?) and draw an exponential curve linking that point to the current (about) 500,000 living species. This exponential curve is then the geometric locus of all maxima of a one-parameter family of *b*-lognormals (the family free parameter is the birth time *b* of any new species) accounting for cladistics (i.e., the modern theory of evolution rigorously based on when a new species appears in the course of evolution, and not on any other naive and simplistic taxonomy claims). In still other words, each new species is an exponential curve, either slightly increasing or slightly decreasing in time, which departs from the "main exponential" (the overall envelope) when that new species originates. As a further new result, we also derive the "NoEv" or "No-Evolution" probability distribution for a given species, namely the pdf applying when a given species does not undergo any change for ages (i.e., when its members are born, grow up, mate, age, and die for millions or billions of years without their number being subject to any significant increase or decrease). Surprisingly, this brand-new probability distribution which stems from our theory is no longer a lognormal or a *b*-lognormal: it is something new, like a static law of evolution, and the fact that the relevant "NoEv" paper was published in a journal like *OLEB* (*Origins of Life and Evolution of Biospheres*) means we are not talking garbage.

Chapter 9—Societal statistics by the statistical Drake equation. This chapter is devoted to a new possibility thrown up by the statistical Drake equation: mathematically deriving new statistical results about previously unknown topics from statistical data we already know about. The unknown topic here is the "societal part" of the Drake equation (i.e., the product of its last three terms $fi \cdot fc \cdot fL$). These three terms yield, respectively: (1) fi the probability that intelligent life (i.e., higher than monkeys) could arise on a planet already teeming with life as has been the case for the historic evolution of humankind since its appearance on Earth some 7 million years ago up until the discovery of radiowaves enabling communications between different ET civilizations in the Galaxy (the existence of radiowaves was first understood mathematically in 1864 by James Clerk Maxwell as sinusoidal solutions to his newly discovered Maxwell equations); (2) fc means the communicative phase of a civilization using radio, lasers, or even neutrinos, which historically started in 1864 for humans and is ongoing right now; (3) fL means the overall life-time of a civilization, from its inception until its death (e.g., as a result of asteroid impact, nearby supernova explosion, a rogue planet or star disrupting the gravitational stability of the stellar system of interest, or even because of nuclear wars among ETs), about which we know nothing at all. Having said this, Chapter 9 suggests we might know something (i.e., a statistical distribution) for the "societal part" $fi \cdot fc \cdot fL$ by rewriting it as the ratio $fi \cdot fc \cdot fL = N/(Ns \cdot fp \cdot ne \cdot fl) = N/N_{\text{Hab}}$. Since the probability distributions of both N and N_{Hab} are known (lognormals of the Drake and Dole equations, respectively), it all boils down to computing the new probability distribution of the ratio between two lognormals, which is not a lognormal but another more general distribution derived by us in Chapter 9.

Chapter 10—Cubics of historical recovery. Carl Sagan in his book (and TV series) *Cosmos* neatly points out the 1,000 years of lost progress by humanity in between the fall of the Western Roman Empire (476 AD) and the recovery of the

Italian Renaissance (about 1400 AD). Well, in Chapter 10 we cast this into a simple (perhaps simplistic) mathematical curve: a cubic (i.e., an algebraic equation of the third degree as a function of time). We show that the numeric values of this cubic match historic progress made in the following fields rather well: (1) astronomy from 1000 BC to 2000 AD, (2) SETI between 1450 and 2000, (3) search for exoplanets between 1950 and 2010, (4) unification of Europe between 1750 and 2010, (5) human life expectancy between 10000 BC and 2000 AD also extrapolated out to 3000 AD and 10000 AD. All these results are offered as simple mathematical models of what appears to be an "historical recovery law" of human civilizations, which might perhaps be extended to other ET civilizations as well, but only after SETI succeeds, of course.

Chapter 11—Exponential evolution in time as geometric Brownian motion. The statistical Drake equation described in Chapter 1 and the following chapters is static (i.e., it does not change in time). It was not until January 8, 2012 that this author came to realize that his static statistical Drake equation actually is a "snapshot" of a very important stochastic process called a "geometric Brownian motion" (GBM), which is more a "movie" than a "snapshot". But GBM is a very, very important stochastic process, probably the most important stochastic process of all: in fact, it was proven back in 1973 to be the key equation in the Black–Scholes model now used everyday in mathematical finance (*http://en.wikipedia.org/wiki/Black%E2%80%93Scholes*). Robert C. Merton was the first to publish a paper expanding the mathematical understanding of the options-pricing model and coined the term "Black–Scholes options-pricing model". Merton and Scholes received the 1997 Nobel Prize in Economics for their work and, though ineligible for the prize because of his death in 1995, Black was mentioned as a contributor by the Swedish academy. Having said this, we prove in Chapter 11 that GBM really is the same as the exponentially increasing number $N(t)$ of communicating civilizations in the Galaxy, subject, however, to uncertainty. In other words: as intelligence and technologies continue to evolve the overall number $N(t)$ of ET civilizations in the Galaxy increase exponentially, but this is subject to the risk of some civilizations suddenly disappearing because of asteroid impacts, nearby supernova explosions, rogue planets or stars disrupting the gravitational stability of the stellar system in question, or even because of nuclear wars among ETs. So, the mean value of $N(t)$ grows exponentially in time as $\langle N(t)\rangle = N_0\, e^{\mu t}$, but $N(t)$ itself is a random process with ups and downs, essentially given by $N(t) = N_0 \cdot e^{\left(\mu - \frac{\sigma^2}{2}\right)} \cdot e^{\sigma B(t)}$, which is what GBN is, $B(t)$ being the standard $(0, 1)$ Brownian motion. So far so good, but after having discovered this, we did more: we found the pdf of the distance stochastic process ("Maccone process"?) given by $D(t) = \dfrac{C}{\sqrt[3]{N(t)}} = \dfrac{\sqrt[3]{6R^2_{\text{Galaxy}}h_{\text{Galaxy}}}}{\sqrt[3]{N(t)}}$. This of course reduces to the "Maccone" distance distribution between any two nearby ET civilizations found in Chapter 1 for the static case, which is also the distance distribution between two nearby habitable planets (with different numbers) as proven in Chapter 3. Thus, in conclusion, we believe that Chapter 11 is the most important chapter in this book as it paves the way to future statistical considerations about ETs and their distances in the Galaxy.

PART II—SPACE MISSIONS TO EXPLOIT GRAVITATIONAL LENSING. This second part of the book comprises five chapters.

Chapter 12—So much gain at 550 AU. Gravitational lensing is the bending of light around massive celestial objects as predicted by Einstein's general theory of relativity. Although Einstein published his predictions in 1915 and in greater detail in 1936, it was not until 1978 that the first "twin-quasar image" (i.e., an image of the same quasar "doubled" by the gravitational lens of an intermediate galaxy) was indeed spotted by astronomers, proving the theory. But, what do we know about single spherical stars acting as lensing objects? Well, in Chapter 12 it is shown that around each star a "focal sphere" exists such that, if the observer is inside that sphere, no lensing effect occurs, but if the observer is outside that sphere, then an enormous magnification of all objects located on the opposite side of the star with respect to the observer's position occurs. In particular, this is true for the radio signals emitted by an ET civilization located so far away from the Sun that no antenna on Earth, however large, would be capable of detecting them. Thus, exploiting the Sun as a gravitational lens has become a must for all precursor interstellar missions reaching at least 550 AU from the Sun in any direction. The distance 550 AU (i.e., 3.17 light-days or 14 times the Sun-to-Pluto distance) is the radius of the focal sphere around the Sun predicted by general relativity. This author has given the name "FOCAL" to any future space mission getting to 550 AU and beyond to take advantage of the huge radio and visual magnifications provided by the gravitational lens of the Sun. These future space missions were described with plenty of technical details in the author's 2009 book entitled *Deep Space Flight and Communications: Exploiting the Sun as a Gravitational Lens*, published by Springer–Praxis. FOCAL is an acronym for "Fast Outgoing Cyclopean Astronomical Lens", which summarizes well what this new space mission is all about.

Chapter 13—FOCAL mission to 1,000 AU as an interstellar precursor. Future FOCAL space missions might have three different targets: (1) getting a magnified image of the black hole (called Sgr A*) located at the center of the Galaxy and of its surroundings. (2) Getting unprecedented details of the nearest triple-star system (Alpha Centauri A, B, and C) in view of the first truly interstellar mission that would probably have that stellar system as its target, simply because it is the nearest to the solar system. (3) Getting unprecedented details about any extrasolar planet that might possibly be discovered in the future to host life, whether in lower forms only or in the higher form of an extraterrestrial civilization. These topics are technically discussed in Chapter 13, keeping in mind that the gravitational lens of the Sun (or of any star) favors space missions getting farther away from their minimal focal sphere inasmuch as the farther they go the more they get rid of the coronal effects of that star, such as the electrons creating divergent lensing effects. This is the reason the real optimal distance that any future FOCAL spacecraft from humanity must reach is about 1,000 AU (i.e., about one light-week), rather than just 550 AU or half a light-week.

Chapter 14—Belt of FOCAL spheres between 550 and 17,000 AU. So far we have discussed the gravitational lens created by the Sun, but what about the gravitational lenses created by the planets? Well, the radius of each focal sphere is proportional to the ratio between the square of the body's physical radius and the body mass. Thus, it is shown in Chapter 14 that an actual belt of eight nearly concentric focal spheres exists between 550 and 17,000 AU, created by the gravitational lensing effect of the Sun and the six more massive planets. In increasing distance from the Sun, there are the focal spheres of the Naked Sun (at 550 AU), Coronal Sun (at 1,000 AU), Jupiter, Neptune, Saturn, Earth, Uranus, and Venus. It follows that any precursor interstellar space mission will have to cross all these focal spheres, taking advantage of each of them in order to magnify the images of celestial objects lying on the other side of the lensing body. In addition, since all planets move around the Sun, planet lenses will be moving lenses with respect to the Sun, thus making sweeping arcs of circles on the celestial sphere as seen from the FOCAL spacecraft. In must be added, however, that the magnification provided by each such planetary lens will be smaller, or much smaller, than the magnification provided by the Sun. Thus, the notion of magnification must be carefully considered. On February 2, 2010, this author received an e-mail from Professor Frank Drake in which he suggested that the Kraus gain used up to then should be replaced by another mathematical formula, called the "Drake gain" by the author of this book, proving much higher numbers for the gain, just as power gain is much higher than voltage gain. This topic is also discussed in Chapter 14, although it seems to the author of this book a matter open for debate (at least as of 2012).

Chapter 15—Galactic Internet by star gravitational lensing. Up to now, we have only considered gravitational lensing made by a single body, whether the Sun or a planet. In Chapter 15 we study the brand-new idea of a "radio bridge" between any two stars, with each star contributing the gravitational lensing focusing power due to its own radius and mass. The combined effect of the two stars would enormously improve radio communications between the two stellar systems inasmuch as the overall gain provided by the two stars would be so large that communications across huge distances would become feasible with modest input powers too. In other words, (in popular terms) the answer to the question "Could we talk to Alpha Centauri A with a cellphone?" would be "Yes, if we exploit the gravitational lenses of both stars." Of course, two FOCAL space missions would now be required to build the radio bridge: one in the solar system reaching the minimal focal distance of 550 AU or beyond, and one in the Alpha Centauri A system reaching the minimal focal distance of 749 AU and beyond. Better still, suppose for a moment that humanity was already capable of sending a FOCAL probe to α Cen A. The answer to the question "What would we do then?" would be "Put it on the other side of α Cen A with respect to the Sun direction and keep the perfect (or nearly perfect) alignment between the four points in space: the FOCAL probe on the opposite side of α Cen A, the center of α Cen A, the center of the Sun, and another FOCAL probe at, say, 600 AU from the Sun in the direction opposite to

α Cen A." Then, a perfect radio bridge exploiting both stars as gravitational lenses would have been constructed, and communications would be feasible at minimal input energy costs, although the required time for all messages to reach the destination would still be 4.37 years, of course. In Chapter 15 we also consider several more radio bridges, like those of the Sun–Barnard Star, the Sun–Sirius, the Sun–Sun_at_Galactic_Center, and even the Sun–Sun_at_Andromeda. In all cases, the radio bridge insures moderate power expense to keep the radio link. This is a wonderful new step ahead, inconceivable before the Sun as a gravitational lens was deeply studied for the first time.

Chapter 16—Extragalactic Internet by black hole gravitational lensing. SETI between galaxies has long been regarded as impossible to achieve because power attenuation across huge intergalactic distances would be fatal to any signal, however energetic. In Chapter 16 we prove that this assumption is incorrect if radio bridges (in the sense described in Chapter 15) between the two black holes located at the center of the two galaxies are exploited. In other words, consider, for instance, the big black hole Sgr A* located at the center of the Milky Way Galactic Bulge, which has a mass of about 4 million Suns, and the similar but 10 times more massive (i.e., a mass of 40 million solar masses) P2 black hole located at the center of the Andromeda Galaxy (M31) at a distance of 2.5 million light-years from the Milky Way. Quite surprisingly, it turns out that the Sgr A*–Andromeda (M31) P2 black hole radio bridge performs better than the "small" Sun–α Cen A radio bridge! This is to say that our calculations show that, to maintain the link between the two black holes at the two galaxies' centers the same transmission power is required as to maintain the link between the Sun and α Cen A. In fact, it is actually much less than that since only about 1/10,000 of the above-mentioned power would be required. This "miracle" is of course due to the huge masses of the two black holes, which more than compensates for the abysmal distance between the Milky Way and Andromeda. Similarly, it is proven in Chapter 16 that other radio bridges between supermassive black holes of other nearby galaxies work equally as well. The most unexpected case is provided by the radio bridge Sgr A*–M67 big black hole, having an estimated mass of 6.6 billion solar masses compensating for the 57 million light-year distance and yielding a performance about 10,000 times better than the "small" Sun–α Cen A radio bridge. All this sounds like science fiction, but it is science fact showing that the black hole at the center of each galaxy is indeed the most important part of the galaxy, where ET civilizations may even be fighting each other to keep control of this powerful supermassive black hole radio station. Human SETI scientists should carefully look at Sgr A* and its surroundings to see if "something strange", such as a "star war", is taking place there right now (i.e., about 25,000 years ago!).

PART III—KLT FOR OPTIMAL SIGNAL PROCESSING. The third part of this book comprises 12 chapters.

Chapter 17—A simple introduction to the KLT and BAM-KLT. The KLT (Karhunen–Loève Transform) is a mathematical algorithm to extract weak signals from thick noise much more powerfully than the Fast Fourier Transform (FFT). In essence, the KLT is the well-known Principal Components Analysis (PCA) applied to both filtering (as just said) and data compression, and it is widely used by scientists and engineers of all kinds, unfortunately under many different names and often without a clear understanding of the mathematics involved inasmuch as the mathematics is often regarded as too difficult to understand by engineers. In Chapter 17, the first of 12 chapters devoted to studying the KLT in different situations, we first provide an easy-to-understand description of the classical KLT (i.e., the KLT as usually taught in university graduate courses—Sections 17.1 through 17.6). The classical KLT may be summarized (in technical language) as a principal axes transformation in the Hilbert space spanned by eigenvectors of the autocorrelation of noise plus a possible embedded signal. Radio SETI, when regarded as the extraction of weak signals from thick noise, is of course the ideal application field of the KLT. Later in the chapter (Sections 17.7 through 17.14) the reader is introduced to the BAM-KLT (an acronym for Bordered Autocorrelation Matrix KLT), investigated and published by this author since 2008 as an advanced section of KLT theory. The basic BAM-KLT goal is to get around the main KLT difficulty of the $O(N^2)$ calculations required to compute the eigenvalues and eigenvectors of an $N \times N$ autocorrelation matrix, which is a much higher computational burden than the $O(N \ln(N))$ calculations required by the FFT. The BAM-KLT, however, is still a fresh new topic, needing more profound investigations. Finally, we would like to point out the recent and important new development that began with the publication of the paper by Arkadiusz Szumski, "Finding the interference: The Karhunen–Loève Transform as an instrument to detect weak RF signals," *InsideGNSS* (Working Papers section), May–June 2011 issue, pp. 56–63. In practice, the young Polish telecommunications engineer Szumski was tasked by the European Space Agency (ESA) with checking the validity of this author's new results about the KLT and the BAM-KLT by writing a numerical code enabling KLT to extract very weak and complicated (chirped, non-stationary, transient, etc.) signals from random noise across both narrow and wide bands and with signal-to-noise ratios (SNRs) like 10^{-3} or smaller (the FFT already fails with SNR = 1). The results of Szumski's simulations have convinced many that the KLT and BAM-KLT are well suited to be applied in GNSS (global navigation satellite systems), where huge investments are being made to improve old-fashioned FFT-based technologies, totally inadequate when compared with the KLT. The future will tell, and this book, originally intended for SETI scientists only, might become palatable to GNSS engineers as well.

Chapter 18—KLT of radio signals from relativistic spaceships in uniform and decelerated motion. By "relativistic spaceship (or probe)" in the following we mean a spacecraft that can reach speeds that are a significant fraction of the speed of light. For such spacecraft, Newtonian physics (nowadays successfully applied to nearly all spacecraft) would no longer hold good, and Einstein's special theory of relativity must be applied instead. No relativistic spaceship yet exists, and so the wealth of results, presented in this and the following chapters about optimal telecommunications between the Earth and a relativistic probe moving either away from Earth or approaching it, might superficially be dismissed. However, we are convinced that, sooner or later, relativistic spacecraft will indeed be built, and so Chapters 18 through 28 are hardly a waste of time for the reader to digest, even if mathematically difficult to understand at first sight. So, let us start by pointing out that the most striking feature of the theory of relativity is of course the existence of two times: (1) "coordinate time" (i.e. the time of guys remaining on Earth, say) and (2) "proper time" (i.e., the time of astronauts living inside the relativistic spacecraft). The key point is that proper time elapses more slowly than coordinate time (i.e., the twin paradox): if one of two twins stays on Earth and the other one flies around in space at relativistic speeds, when he comes back to Earth his brother would probably have long since died, inasmuch as his brother's time had elapsed much faster than the relativistic astronaut's time. With this in mind, the author (in his youth) developed a full mathematical theory of the KLT for time-rescaled Brownian motions for Gaussian stochastic processes (i.e., for the radio noise hampering telecommunications in space between a relativistic spaceship and the Earth). This mathematical theory is described in Chapters 21 ("Brownian motion and its time rescaling") and 22 ("Maccone First KLT Theorem: KLT of all time-rescaled Brownian motions"), as we shall see later. Essentially, the author proved that the KLT eigenfunctions of all time-rescaled Gaussian processes are Bessel functions of the first kind, while the corresponding eigenvalues are the zeros of such Bessel functions and their first derivative. In Chapter 18 these results are applied to two cases of relativistic motions: (1) uniform motion, for which time rescaling is rather obvious since Brownian motion KLT eigenfunctions are just sines, and (2) decelerated motion (i.e., the motion of a spaceship slowing down from high relativistic speed to practically zero speed—as done by alien spaceships in the movie *Independence Day*). Computation of the stochastic integral yielding the total energy of signals received on Earth completes the chapter.

Chapter 19—KLT of radio signals from relativistic spaceships in hyperbolic motion. When we think about interstellar flights to the nearest stars, special relativity shows that the best velocity profile to reach the target with minimal proper time is the so-called "hyperbolic motion": namely, (proper) uniformly accelerated motion from the start to midway and then, after turning the ship by 180 degrees, (proper) uniformly decelerated motion from midway to the target, finally getting there at zero speed. The optimal (proper) uniform acceleration would be $1\,g = 9.8\,\mathrm{m/s^2}$, so as to let the people and equipment on board feel the same weight they have on Earth. At such $1\,g$ (proper) uniform acceleration, the center of the Galaxy would be reached in just 21 years of proper time despite it being 25,000 light-years away from us, but of course the energy requirements for the spacecraft to maintain $1\,g$ (proper) acceleration for 21 years would be huge, which is why the project is unfeasible by today's standards. Less demanding energy requirements would apply of course if the target is the nearest stellar system, Alpha Centauri, just 4.37 light-years away, but, whatever the target may be, telecommunications of some kind will have to be kept between the (proper) uniformly accelerated spacecraft and the Earth. This chapter proves that the KLT eigenfunctions for this hyperbolic motion are Bessel functions of the first kind and order zero. To be precise, this KLT is an asymptotically exact KLT for $t \to \infty$, but in reality the numbers are already acceptable about one year after departure from Earth, thus yielding an applicable result.

Chapter 20—KLT of radio signals from relativistic spaceships in arbitrary motion. The same mathematical techniques already applied in Chapter 19 to hyperbolic motion (i.e., to proper uniformly accelerated motion) are shown in this chapter to be appropriate to derive the KLT for signals emitted by a spacecraft in arbitrary radial motion away from the Earth or towards it (such as an alien spacecraft heading for the Earth would do, as in the *Independence Day* movie of 1996). The calculations are unfortunately long and, at first sight, may appear daunting. But the starting idea is easy: just as was the case for hyperbolic motion, it is possible to exactly integrate the relativistic arbitrary motion radial differential equation. This opens up the possibility of computing the relevant KLT by resorting to the theory of time-rescaled Brownian motions already used in Chapter 18 and described in detail in Chapters 21, 22, and 23. Thus, equations are provided not only for the KLT of the signals received on Earth from a receding or approaching spaceship at an arbitrary speed and acceleration profile, but also for the total amount of energy of the signals emitted by such a spaceship (i.e., for the stochastic integral of the square of the intensity of such radio signals). The time will come when these equations will become important to human expansion (or to human understanding of an alien) into the Galaxy, just as in *Star Trek*.

Chapter 21—Brownian motion and its time rescaling. This is the first of three chapters devoted to detailed mathematical description of the KLT for all time-rescaled Gaussian processes (as mathematicians call them) or time-rescaled Brownian motions (as physicists prefer to call them). This formed a large part of the author's work following his Ph.D. obtained at the Department of Mathematics of the University of London (U.K.) King's College in 1980. The key idea is that if one lets the time variable of ordinary Brownian motion elapse other than in a uniform way (i.e., if one allows time to be stretched, time dilation of special relativity, in an arbitrary fashion), the process still remains Gaussian, and its autocorrelation can be easily expressed in terms of the arbitrary time-rescaling function $f(t)$. Actually, it can easily be proven that the square of the time-rescaling function, $f^2(t)$, equals the well-known radical $\sqrt{1-\frac{v^2(t)}{c^2}}$ of special relativity, and this establishes a firm connection between time-rescaled Brownian motions and special relativity. It is as easy as that.

Chapter 22—Maccone First KLT Theorem: KLT of all time-rescaled Brownian motions. This is the second of three chapters devoted to the detailed mathematical description of the KLT for all time-rescaled Gaussian processes. Since the KLT of ordinary Brownian motion can be proven to be given exactly by sinusoidal functions (i.e., it is basically the same as a Fourier transform), then one may wonder what the KLT of time-rescaled Brownian motions might possibly be. The answer is provided by the "Maccone First KLT Theorem" and is that eigenfuctions are Bessel functions of the first kind suitably time-rescaled, while eigenvalues are the zeros of certain linear combinations of such Bessel functions and their derivatives. It is also possible to find an asymptotic version of this KLT for $t \to \infty$ where things are easier and ready for engineering applications. The mathematical proof of this "Maccone First KLT Theorem" is unfortunately lengthy and hinges on the Sturm–Liouville theory of boundary problems for ordinary differential equations of the second order with non-constant coefficients. The author discovered all this back in the early 1980s, and published these results in the Italian mathematical journal *Bollettino dell'Unione Matematica Italiana* ("Eigenfunctions and energy for time-rescaled Gaussian processes", Bollettino UMI, Series 6, Vol. 3A (1984), pp. 212–219). He was then seeking a university position somewhere, but that paper was regarded as "just a physical application" by pure mathematicians and "useless for particle physics" by physicists of the time. So, the author failed to get a position, and in 1985 he joined the space company Alenia Spazio S.p.A., where the engineers were even less interested in his work than the pure mathematicians and particle physicists. The truth is that nobody cared about the telecommunication theory for relativistic interstellar flights, with the exception of a few bright minds such as Dr. Leslie Shepherd and Dr. Giovanni Vulpetti of the Interstellar Space Exploration Committee (ISEC) of the International Academy of Astronautics (IAA), which this author eventually joined.

Chapter 23—KLT of the $B(t^{2H})$ time-rescaled Brownian motion. This is the third of three chapters devoted to detailed mathematical description of the KLT for all time-rescaled Gaussian processes. It deals with the KLT of a Brownian motion whose time variable does not elapse uniformly, but rather increases like power t^{2H} (time dilation). The H parameter is the Hurst exponent of Mandelbrot since the $B(t^{2H})$ is indeed an H self-similar stochastic process. This author discovered that this process has an exact KLT by solving the relevant integral equation back in 1979, when he was preparing his Ph.D. thesis in Mathematics at the University of London (U.K.) King's College, awarded to him on September 19, 1980. Afterwards, the author returned home to Turin, and finally published his exact KLT of $B(t^{2H})$ in both the *Bollettino dell'Unione Matematica Italiana*, Series 6, Vol. 4-C (1985), pp. 363–378, and *Il Nuovo Cimento* (the Italian physics journal), Series B, Vol. 100 (1987), pp. 329–342. He also found time to study the stochastic integral yielding the total energy of this stochastic process and generalize the Cameron–Martin formula that holds for standard Brownian motion to this process.

Chapter 24—Maccone Second KLT Theorem: KLT of all time-rescaled square Brownian motions. The content of this chapter was long regarded by this author as his best mathematical result ever about the KLT: it was published as "The Karhunen-Loève expansion of the zero-mean square process of a time-rescaled Gaussian process," *Bollettino dell'Unione Matematica Italiana*, Series 7, Vol. 2-A (1988), pp. 221–229. The author still remembers the reviewer's words *con una serie di passaggi ben condotti* ("by virtue of a series of well-conducted steps") it was indeed possible for him to exactly solve the KLT integral equation for the square of any time-rescaled Gaussian process (as described in detail in Chapter 24). The calculations are complicated, and the very few who read that 1988 paper probabaly uttered the usual question "What's the purpose of all this?" This author's paper was regarded as of little use for pure mathematicians or particle physicists, and no use for telecommunication engineers. But, all this simply stems from the fact that no relativistic interstellar flights have been studied, neither back in 1988 nor today. Looked at more generally and more positively, this and related KLT papers could become useful in the study of transient astrophysical phenomena, like supernova explosions, gamma ray bursts, and SETI signals that are just transient because of short-lasting gravitational lensing alignments among stars and/or other bodies. However, Chapter 24 contains a result that might already be of some use today: the exact KLT of the square of standard Brownian motion, which is even simulated numerically in the chapter just to check that everything is right.

Chapter 25—KLT of the $B^2(t^{2H})$ time-rescaled square Brownian motion. This chapter is with respect to Chapter 24 the same as Chapter 23 is with respect to Chapter 22. In other words, in Chapter 25 we prove that the KLT of the time-rescaled stochastic process $B^2(t^{2H})$ is the exact solution of the corresponding KLT integral equation, given by certain Bessel functions of the first kind and, similarly, the eigenvalues are the zeros of similar Bessel functions. The asymptotic version of these results for $t \to \infty$ is also obtained and is particularly simple inasmuch as the Bessel functions then downgrade to ordinary cosines whose time variable is time-rescaled like a power of t. In this chapter we were likewise able to check by numeric simulations that the results are indeed correct.

Chapter 26—Maccone Third KLT Theorem: Asymptotic KLT of GBM. The geometric Brownian motion (GBM) is today the most important stochastic process of all, since it is widely used in the mathematics of finance and, in particular, in the Black–Scholes models. It was not until January 8, 2012 that it dawned on this author that his statistical Drake equation (described in Chapter 1) is just the particular case $t = 1$ and $N_0 = e^{\frac{\sigma^2}{2}}$of a GBM. In other words, all the equations derived in Chapter 1 are static particular cases of much more general corresponding equations of a GBM in which time t is not constrained to any particular numeric value. This was proven in Chapter 11, and the conclusion was that the $N(t)$ stochastic process $(t \geq 0)$, yielding the number of communicating ET civilizations in the Galaxy at any positive time t, is exactly a GBM. We know that every stochastic process—and thus $N(t)$—has a KLT. So, the next natural question arises: What is the KLT of a GBM? The author is unable to answer this question at the moment. However, back in September 2007, he was able to derive the asymptotic KLT (i.e., valid for $t \to \infty$) of the exponential Brownian motion $e^{B(t)}$ which now appears to be the key stochastic (i.e., non-simply deterministic) part of the GBM definition. This asymptotic KLT of a GBM is published here for the first time. The starting point is the surprising result that the time-dependent order $\nu(t)$ of Bessel functions of the first kind described in Chapter 22 is practically equal to 1 for just moderate values of t a few instants after the initial time $t = 0$, and stays as such for ever (i.e., for all higher values of time). Thus, in practice it is possible to simply set $\nu(t) = 1$ as the order of Bessel functions order, and then keep going! The asymptotic KLT of the GBM for $t \to \infty$ is announced—a brand-new result that might have profound impacts on further developments. But, at the moment the author confines himself to giving the new equations.

Chapter 27—A Matlab code for KLT simulations. Mathematical books today should also provide numeric codes to simulate the new equations provided in analytical form. This is the goal of Chapter 27. In 2008, the author and his pupil Dr. Nicolò Antonietti set out to write a Matlab (or Octave) code capable of simulating at least the most important KLT expansions given in this book. The input was usually standard Brownian motion or, alternatively, a user-provided input stochastic process numerically assigned. The output was the KLT displayed in three forms: (1) the full KLT reconstruction based on all instants between 0 and T; (2) the KLT reconstructed by using only the first few eigenvalues and eigenvectors of the autocorrelation matrix, thus yielding—instead of a perfect reconstruction—a smooth curve interpolating the data of the input stochastic process (this we call "an empirical reconstruction" of the input stochastic process where the degree of approximation can be selected at will, of course—in other words, the KLT is also a data compression algorithm); (3) when available, we also provide the exact KLT or, if that is unavailable, we provide the asymptotic KLT for $t \to \infty$ based on the asymptotic expansion of Bessel functions of the first kind for $t \to \infty$.

Chapter 28—KLT applications (by *Stephane Dumas*). The author's new French Canadian pupil, Monsieur Stephane Dumas of Québec, kindly provided for free the whole content of Chapter 28. After an introduction to the many applications of the KLT under different names (PCA = Principal Component Analysis, SVD = Singular Value Decomposition, and so on), Dumas provides the Fortran code that he wrote in order to use the KLT in a number of applications. He also sought to optimize the computing time, which is of course the main obstacle still hampering real-time applications of the KLT instead of the much more customary (and much less profound) FFT. This chapter is of course intended for the applied sciences of all kinds, and not just SETI.

PART IV: THE UNITED NATIONS AND PROTECTION OF THE MOON'S FARSIDE. The fourth part of this book comprises a single chapter.

Chapter 29—The United Nations and protection of the Moon's farside. One of this author's teachers who made a profound impression upon him was the Frenchman, Jean Heidmann (1920–2000) (*http://www.universalis.fr/encyclopedie/jean-heidmann/*). In the 1990s, Heidmann was tasked by the International Academy of Astronautics (IAA) with studying how best the Moon farside could be protected by law against future human polluters of any kind (real estate, industry, military) that could damage the unique environment existing on the Moon's farside, which is free of man-made radio noise. This is more a political problem than an astronautical and scientific problem and needs to be resolved urgently by some international agreement among the main spacefaring nations before it is too late. Just before Heidmann passed away in 2000, he appointed this author as his deputy to lead the relevant IAA team, and so this author took over that difficult task. Although he published several papers on the subject (see, e.g., "Protected Antipode Circle on the farside of the Moon," *Acta Astronautica*, Vol. 63 (2008), pp. 110–118), this was just the beginning of his activity in this field. On June 10, 2010, this author delivered a speech at the United Nations Committee on the Peaceful Uses of Outer Space (COPUOS) in Vienna, Austria, to raise the issue at the United Nations level for the first time. Chapter 29 describes in detail the issues at stake, and how the United Nations could resolve this space problem for the future benefit of all humankind.

PART V: EPILOGUE. The fifth part of this book comprises a single chapter.

Chapter 30—Epilogue: Evolution, progress, and SETI. This book ends with an epilogue about SETI. What if SETI succeeds? It's hard to predict the consequences of the first contact between humankind and ETs. If we look back at past human history, we know that the most important sudden contact between two historic civilizations that had independently developed different levels of technology was the discovery of America by Europeans, which resulted in the violent subjugation of the lower-level civilization by the higher-level one. Yet, those were times when war was common practice to all civilizations on Earth, and so the resulting violence was perhaps inevitable. Perhaps more advanced civilizations than ours might come to put an end to wars and contact would thus be smoother. In any case, as of 2012 we know that an immense number of extrasolar planets do indeed exist in the Galaxy, and so thinking that we are alone is no longer realistic. The best we can do is to keep exploring, as astrobiology does nowadays, and keep "Searching for Life Signatures", as the conferences run by this author since 2008 under the aegis of the IAA are called. Only the future will tell.

Conclusion

This is a textbook for high-level students and research scientists—it is not popular stuff at all. The author feels that a book like this is really needed by the worldwide scientific community since it tries to bridge the gaps among branches of science that are still regarded as independent of each other—just to mention a few: astronomy, evolution of life on Earth and elsewhere, mathematical history (a brand-new discipline?), and of course astronautics (especially, future relativistic interstellar flights). To combine all this in some "unified mathematical description" is something that needed to be done. This book is a step in the right direction toward this new vision of humanity in the Galaxy that future decades and centuries will increasingly develop.

We would like to conclude this preface with the same sentence used by Johannes Kepler (1571–1630) when he disclosed to the world the discovery of his Third Law (1619), which he could only achieve by coupling astronomical observations to the intense use of mathematics. To correctly understand Kepler's sentence, however, one has to remember that, in Kepler's time, Bible reckoning had made scholars believe that the Universe had been created just about 6,000 years earlier. Keeping that in mind, we now report Kepler's sentence:

> *"The die is cast; the book is written, to be read either now or by posterity, I care not which. It may well wait a century for a reader, as God has waited six thousand years for an observer"*

from *Harmonice mundi* (1619), translated in Bill Swainson and Anne H. Soukhanov, *Encarta Book of Quotations* (2000), p. 514.

Claudio Maccone
Torino (Turin), Italy, February 29, 2012

Acknowledgments and dedication

This author is indebted to many individuals for inspiration and help over the 30-year period (1982–2012) of his involvement in the SETI field. To list them all, however, would simply be impossible. Thus, only two colleagues will be publicly acknowledged here because of their influence on the author's thoughts as well as on his career over the last three decades:

(1) Professor Frank Donald Drake, an American researcher, the first experimental SETI scientist ever and discoverer of the Drake equation studied in this book—quite simply this author's most inspiring and respected teacher.
(2) Dr. Jean-Michel Contant, a French researcher, Secretary General of the International Academy of Astronautics (IAA) and constant supporter of SETI activities within the IAA. Indeed, the IAA is responsible for:
    (a) creation of the IAA's SETI Permanent Committee and its two sessions (SETI I: science and technology and SETI II: societal aspects) held yearly at the International Astronautical Congress (IAC);
    (b) publication of the many important SETI papers in the IAA journal *Acta Astronautica* over the years;
    (c) the Moon Farside Protection action undertaken by this author at the United Nations (as described in Chapter 29).

To these two individuals in particular, as well as to all other SETI scientists from all over the world, this book is dedicated.

Claudio Maccone
Technical Director for Scientific Space Exploration
International Academy of Astronautics (IAA)
and Associate, Istituto Nazionale di Astrofisica (INAF)
Torino (Turin), Italy, March 19, 2012

(Left to right) Jean-Michel Contant, Frank Drake, and Claudio Maccone at the Royal Society in London, January 26, 2010.

Foreword by Frank Drake

One of the most important discoveries in the history of astronomy—indeed, in all of science—was made in recent years: this was the detection of thousands of other planetary systems beyond our own. This is the result of the endeavors made by a small group of astronomers diligently observing, night after night, small deviations in the positions of lines in stellar spectra and observing faint occultations of stars by planets as revealed by the Kepler spacecraft. All of this has happened while observing less than 1% of the visible stars in the night sky. The detection technique used by Kepler can only detect about 1% of the planetary systems actually orbiting the stars observed. So these amazing results tell us that there must be millions of planetary systems orbiting the stars we can observe, and simple models put the number of planetary systems in our entire Milky Way Galaxy in the billions. What an exciting result this is, and the age-old question of whether there are other worlds like ours has been answered with a resounding "Yes!"

Equally exciting is the discovery that a large proportion of these systems have planets that may be inhabited by living creatures, just as here on planet Earth. These planets must be the right size and right distance from their star for life as we know it to exist. There are very many of these planets. If there is life and if it has evolved in much the same way as life has on Earth—which seems inevitable—then there must be a large number of planets supporting intelligent life probably with technologies like ours. Such technologies as radio transmitters and powerful lasers may be detectable by instruments that already exist on Earth or that soon may be developed.

All of this strengthens our resolve to extend the scope and speed with which the search for extraterrestrial intelligence, SETI, is undertaken, which has for the last 50 years or so been done with very limited resources. The knowledge that there really is life akin to ours deep in space justifies assigning far greater resources to the search than in the past—resources like telescope capability, telescope observing time, the time of dedicated scientists. To facilitate eventual discovery and to use such resources responsibly calls for the most careful planning.

Thus, it is imperative to analyze our search methods and procedures carefully so that we can create the best possible SETI program from the resources available. A giant step toward quantifying the power of our searches and, in so doing, optimizing our search programs has been made by Claudio Maccone in this book.

The author has created a powerful mathematical framework for quantifying the capability of SETI programs to succeed. This is based on the 50-year-old Drake equation, which can be used to predict the probability that a civilization might be encountered in a given search. In this book the author has applied statistical methods rigorously to application of this equation. He shows how his methods may be applied beneficially to other problems and, in so doing, enhance our understanding of the mathematical approaches he has developed.

Lastly, he points the way to a very powerful method to detect other civilizations from their faint signals, feeble radio transmissions, and dim lights of their cities at night. This is to use our Sun as a gravitational lens to create a giant telescope that concentrates all forms of electromagnetic radiation at a series of foci positioned many hundreds of astronomical units from the Sun. This, in effect, creates a lens that is larger than the diameter of the Sun, more than a million kilometers across, to create brilliant images with image detail of incredible resolution.

Although such a telescope is beyond our technical capabilities at present, it may be probable in the not-too-distant future, perhaps in a hundred years or so. It is important to consider it in detail, as the author does, because many, maybe most, older civilizations than ours will have long since implemented the lens provided by their star. If they use such lenses not only to search for life, but also to amplify signals very powerfully and focus them on distant civilizations, how might these signals appear? This is important because it could well be that such amplified signals will be what we can most easily detect. We need to know what forms they might take so that we are instrumentally and mentally prepared should such a signal be captured. The detection of such a signal would be one of the most monumental discoveries ever and, simultaneously, would demonstrate that many more discoveries are awaiting our attention. The window would be opened to an incredible wealth of beneficial information which would enrich us all.

Frank Drake
SETI Institute, Mountain View, California
April 10, 2012
Emeritus Professor of Astronomy and Astrophysics
University of California at Santa Cruz, U.S.A.

Foreword by Giovanni Bignami

Dr. Claudio Maccone submitted a Proposal to the European Space Agency (ESA) about the FOCAL (Fast Outgoing Cyclopean Astronomical Lens) space mission to 550 AU and beyond in the year 2000, when I was Scientific Director of the Italian Space Agency (ASI). I immediately endorsed Maccone's Proposal acknowledging that ASI would support it if ESA approved it. Unfortunately, ESA failed to select Maccone's Proposal for further study and so his ideas had to be temporarily abandoned, but that did not mean they were bad ideas. In 2009, Dr. Maccone published a whole technical book on the subject, entitled *Deep Space Flight and Communications*, and more interest arose again in 2011 on the occasion of the DARPA-NASA 100 Year Starship Study. Quite simply, any future precursor interstellar mission will have to cross the minimal focal sphere at 550 AU beyond which it is possible to exploit the huge signal magnifications provided by the Sun as a gravitational lens. Thus, Maccone's FOCAL mission is destined to pop up again and again in the future, until favorable political conditions enable some important space agency to make it a reality. In this new book about mathematical SETI, five chapters (Chapter 12 through 16) are devoted to the physics and mathematics of the Sun as a gravitational lens with important updates with respect to the 2009 book by the author. For instance, it is proven that, even for the nearest interstellar mission to Alpha Centauri (4.37 light-years away), the ordinary radio telescopes we have on Earth would be inadequate to maintain the radio link between the Earth and the receding spacecraft unless the huge radio amplification provided by the Sun as a gravitational lens is invoked. In other words, Maccone points out that, before launching a truly interstellar spacecraft to even the nearest star, it would be necessary to send a large space antenna in the opposite direction of the target with respect to the Sun just in order to keep the radio link at moderate powers until the interstellar spacecraft reached Alpha Centauri. But, Maccone goes beyond that. Suppose for a moment that our spacecraft had already reached Alpha Centauri: What would we then do with it? Maccone's suggestion is to put the interstellar probe on the other side of, say,

α Cen A (the most massive of the three stars) and in a direction exactly opposite to the Sun. Then, these two stars and the two FOCAL probes on their outer sides, respectively, would form a huge "radio bridge" capable of reducing the link powers dramatically. In popular terms, Maccone answers the question "Would we be able to maintain the link to α Cen using a cellphone?" by saying "Yes, if we exploited the gravitational lenses of both stars by putting two FOCAL probes on their opposite sides." This notion of a "radio bridge made possible by star gravitational lensing" is then extended to other increasingly distant stars, showing that the links are still feasible even up to thousands of light years and more. Thus, even a sort of "Galactic Internet" could indeed be created, or perhaps it has already been created by more advanced civilizations in the Galaxy, but humans will be unable to join it until their spacecraft reach the minimal focal distance from our own star, the Sun, at 550 AU. Finally, Maccone extends these notions even to extragalactic distances, showing that the notion of a "radio bridge" still holds if one replaces the masses of the stars by the masses of the supermassive black holes located at the center of each galaxy. Thus, for the Milky Way, SgrA* (i.e., the 4,000,000 $M_{\odot}$ central black hole) would be the "magnifying gravitational lens" capable of keeping the Milky Way in touch with other galaxies. Just science fiction? No, just classical physics, but revealed by the new open-minded SETI vision of the universe that many still lack.

Chapters 17 through 28 of this book deal with a completely different mathematical topic: how to improve the ordinary noise-filtering techniques based on the FFT (Fast Fourier Transform) by the much more sophisticated KLT (Karhunen–Loève Transform). The basic idea seems to be that, while the FFT only uses sines and cosines as the set of orthonormal functions in the Hilbert space upon which the series expansion is made, the KLT uses the best set of orthonormal functions changing them according to the input, rather than keeping them fixed to sines and cosines. This optimizes the transform by adapting it to the input, whatever that might be. Unfortunately, KLT mathematics is much less simple than FFT mathematics, and that makes the topic difficult to follow for engineers not particularly familiar with eigenvalues, eigenvectors, and all that. To physicists, on the contrary, the KLT is immediately clear since it is just the ordinary Hilbert space apparatus of quantum physics translated into the language of signal processing. Maccone then finds explicit equations for KLT eigenvalues and eigenfuctions in a number of practical cases of interest to both SETI and astrophysics in general. For instance, the rotational period of the Vela pulsar was impossible to find by the ordinary FFT, but was indeed found by a team from the Asiago Observatory at the European Southern Observatory at La Silla (Chile) using the KLT. More generally, the theory of transient phenomena might greatly benefit by the adoption of the KLT instead of the FFT. It must be said, however, that the computational burden is much higher for the KLT than for the FFT: such as $O(N^2)$ for the KLT as compared with the much easier $(O(N \ln(N))$ for the FFT, which is why many researchers still prefer the FFT just to save computational time and money. Yet, the FFT utterly fails for signal-to-noise ratios (SNRs) lower than one, while the KLT succeeds in retrieving very weak signals in SNRs of even (say) 10^{-3} or smaller.

Thus, for the purpose of research, the KLT is vastly superior to the FFT. Moreover, it also applies to wide bands rather than just to narrow bands to which the FFT is confined. Actual numerical codes to compute the KLT are also provided in this book, making it an ideal tool for research.

But the most profound novelty put forward in this book is the statistical Drake equation and its far-reaching consequences, which are described in the first 11 chapters of the book. The basic idea here is to replace all numerical input variables in the classical Drake equation of 1961 by positive random variables whose probability distribution may be unknown at the outset. Maccone shows that, if the number of such input random variables is made to increase at will (meaning that increasingly more physical factors are taken into account), then the probability distribution of N, the number of communicating civilizations in the Galaxy, approaches the lognormal distribution. This is really a simple consequence of the central limit theorem of statistics, but was not realized as such until 2008, when Maccone first presented it at the SETI I Session of the International Astronautical Congress held in Glasgow, Scotland. After that, Maccone reformulated the Dole equation too, yielding the number of habitable planets in the Galaxy in just the same way as the Drake equation yields the number of ET civilizations. The most important new result of all, however, is the discovery of the probability distribution yielding the distance between any two nearby ET civilizations in the Galaxy. This is no longer a lognormal distribution, but rather a new equation dubbed "Maccone distribution" by Paul Davies (who first understood and popularized it after its discovery by Maccone). This really is at the heart of SETI inasmuch as it predicts that the probability of finding ETs at distances less than about 500 light-years is virtually zero. Since our radio telescopes on Earth cannot detect ET signals arriving from distances greater than 100 or 200 light-years, it is immediately apparent that SETI searches have so far failed to find ETs simply because they do not reach far enough out into the Galaxy! Indeed, this is even more obvious if one considers that the size of the Galaxy is about 100,000 light-years and we are struggling to explore a sphere of 500 to 1,000 light-years around the Sun. Put flippantly, it is as if Christopher Columbus had stopped at the Canary Islands and then "concluded" that America could not possibly exist!

Finally, one last chapter describes Maccone's activity on the United Nations Committee on the Peaceful Uses of Outer Space (COPUOS) in order to protect the farside of the Moon from wild future exploitation by anyone disrespectful of its unique radio-quiet environment. This is of course more politics than science. But, it is high time something was done before private entrepreneurs achieve the capability of flying to the Moon at their own private expense, thus making them believe that they can dispose at will of the Moon just because they have money.

In conclusion, this is a highly innovative research book, opening up new prospects for the future of humanity in space and I can only highly recommend it.

Giovanni F. Bignami
President, COSPAR and INAF (Italian National Institute for Astrophysics)
Milan, Italy, March 2, 2012

Figures

Tables

INPUT TABLES

Abbreviations and acronyms

ALMA	Atacama Large Millimeter/submillimeter Array
ATA	Allen Telescope Array
BAM	Bordered Autocorrelation Method
BER	Bit Error Rate
BPP	Breakthrough Propulsion Physics (program)
CLT	Central Limit Theorem
CFHT	Canada–France–Hawaii Telescope
CLOE	Common Lisp Oriented Environment for PCs by Symbolics
CMB	Cosmic Microwave Background
DSN	Deep Space Network
DOE	Department Of Energy (U.S.)
FFT	Fast Fourier Transform
FT	Fourier Transform
FTL	Faster Than Light
GPL	General Public License
HARPS	High Accuracy Radial Velocity Planet Searcher
HPBW	Half Power Beam Width
KL	Karhunen–Loève
KLT	Karhunen–Loève Transform
IAC	International Astronautical Congress
LAPACK	Linear Algebra PACKage
LOFAR	LOw Frequency ARray
LU	LU factorization (decomposition) of a matrix
MACSYMA	Project MAC's SYmbolic MAnipulator
MIT	Massachusetts Institute of Technology
MKS	Meter Kilogram Second system of units
NESC	National Energy Software Center

NCP	North Celestial Pole
OHP	Observatoire de Haute Provence
PCA	Principal Components Analysis
pdf	probability density function
PA	Position Angle
SETI	Search for ExtraTerrestrial Intelligence
SKA	Square Kilometer Array
SIM	Space Interferometry Mission
SVD	Singular Value Decomposition
TPF	Terrestrial Planet Finder
VLSI	Very Large Scale Integration

Part I

SETI statistics

1

The statistical Drake equation

1.1 INTRODUCTION TO SETI

SETI is an acronym for "Search for ExtraTerrestrial Intelligence".

SETI is a comparatively new branch of scientific research that began only in 1959. The goal of SETI is to ascertain whether alien civilizations exist in the universe, how far from us they exist, and possibly how much more advanced than us they may be.

As of 2012, the only physical tools we know that could help us get in touch with aliens are the electromagnetic waves that an alien civilization could emit and we could detect. This forces us to use the largest radiotelescopes on Earth for SETI research because the higher our collecting area of electromagnetic radiation the higher our sensitivity (i.e., the farther in space we can probe). Yet, even by using the largest radiotelescopes we have on Earth (the 305-meter dish at Arecibo, for instance) we cannot search for aliens beyond, say, a few hundred light-years. This is a very, very small amount of space around us within our Galaxy, the Milky Way, that is about a hundred thousand light-years in diameter. Thus, current SETI can cover only a very tiny fraction of the Galaxy, and it is not surprising that in the past 50 years of SETI searches no extraterrestrial civilization has been discovered. Quite simply, we have not got far enough!

This demands the construction of much more powerful and radically new radiotelescopes. Rather than big and heavy metal dishes, whose mechanical problems hamper SETI research too much, we are now turning to "software radio-telescopes", where a large number of small dishes (ATA = Allen Telescope Array and ALMA = Atacama Large Millimeter/submillimeter Array) or even just of simple dipoles (LOFAR = Low Frequency Array) using state-of-the-art electronics and very high-speed computing can outperform the classical radiotelescopes in many regards. The final dream in this field is the SKA (= Square Kilometer Array), currently being designed and expected to be completed around 2020.

1.2 THE KEY QUESTION: HOW FAR AWAY ARE THEY?

But, still, the key question remains: How far away are they?

Or, more correctly, how far away do we expect the nearest extraterrestrial civilization in the Galaxy to be from the Solar System?

This question was first faced in a scientific manner back in 1961 by the same scientist who also was the first experimental SETI radio astronomer ever: the American Frank Donald Drake. He first considered the shape and size of the Galaxy where we are living: the Milky Way. This is a spiral galaxy measuring some 100,000 light-years in diameter and some 1,600 light-years in the thickness of the Galactic Disk half way from its center. That is:

(1) The diameter of the Galaxy is about 100,000 light-years (abbreviated lt-yr) (i.e., its radius, R_{Galaxy}, is about 50,000 lt-yr).
(2) The thickness of the Galactic Disk half way from its center, h_{Galaxy}, is about 1,600 lt-yr.
(3) The volume of the Galaxy may then be approximated as the volume of the corresponding cylinder, i.e.

$$V_{Galaxy} = \pi R_{Galaxy}^2 h_{Galaxy}. \tag{1.1}$$

Now consider the sphere around us having a radius r. The volume of such a sphere is

$$V_{Our\_Sphere} = \frac{4}{3}\pi \left(\frac{\text{ET\_Distance}}{2}\right)^3. \tag{1.2}$$

In the last equation, we had to divide the distance "ET_Distance" between ourselves and the nearest ET civilization by 2 because we are now going to make the unwarranted assumption that all ET civilizations are equally spaced from each other in the Galaxy! This is a crazy assumption, clearly, and should be replaced by more scientifically grounded assumptions as soon as we know more about our Galactic Neighborhood. At the moment, however, this is the best guess that we can make, and so we shall take it for granted, although we are aware that this is a weak point in the reasoning.

Furthermore, let us denote by N the total number of civilizations now living in the Galaxy, including ourselves. Of course, this number N is unknown. We only know that $N \geq 1$ since one civilization does at least exist! Having thus assumed that ET civilizations are *uniformly spaced in the Galaxy*, we can then write down the proportion:

$$\frac{V_{Galaxy}}{N} = \frac{V_{Our\_Sphere}}{1}. \tag{1.3}$$

That is, upon replacing both (1.1) and (1.2) into (1.3):

$$\frac{\pi R_{Galaxy}^2 h_{Galaxy}}{N} = \frac{\frac{4}{3}\pi \left(\frac{\text{ET\_Distance}}{2}\right)^3}{1}. \tag{1.4}$$

The last equation contains two unknowns, N and ET_Distance, and so we don't know which of them is best to solve for.

However, we may suppose that, by resorting to the (rather uncertain) knowledge that we have about the evolution of the Galaxy through the last 10 billion years or so, we might somehow compute an approximate value for N.

Then, we may solve (1.4) for ET_Distance thus obtaining the (*average*) *distance between any pair of neighboring civilizations in the Galaxy* (*distance law*)

$$\text{ET\_Distance}(N) = \frac{\sqrt[3]{6\, R_{Galaxy}^2\, h_{Galaxy}}}{\sqrt[3]{N}} = \frac{C}{\sqrt[3]{N}}, \tag{1.5}$$

where the positive constant C is defined by

$$C = \sqrt[3]{6\, R_{Galaxy}^2\, h_{Galaxy}} \approx 28{,}845 \text{ light-years}. \tag{1.6}$$

Equations (1.5) and (1.6) are the starting point to understand the origin of the Drake equation that we discuss in detail in Section 1.3.

Let us first complete this section by pointing out three different numerical cases of the distance law (1.5):

(1) We know that we exist, so N may not be smaller than 1 (i.e., $N \geq 1$). Suppose then that we are alone in the Galaxy (i.e., that $N = 1$). Then the distance law (1.5) yields just the constant C (i.e., 28,845 lt-yr) as the distance to the nearest civilization from us. This is about the distance in between ourselves and the center of the Galaxy (i.e., the Galactic Bulge). Thus, this result seems to suggest that, if we do not find any extraterrestrial civilization around us in the outskirts of our Galaxy, then we should look around the Galactic Center first. This is indeed what is happening: many SETI searches are actually pointing antennas towards the Galactic Center, looking for beacons (see, e.g., [1]).
(2) Suppose next that $N = 1{,}000$ (i.e., there are about a thousand extraterrestrial communicating civilizations in the whole Galaxy right now). Then the distance law (1.5) yields an average distance of 2,885 light-years. This is a distance that most radiotelescopes on Earth are not capable of reaching for SETI searches right now: hence the need to build larger radiotelescopes, like ALMA, LOFAR, and the SKA.
(3) Suppose finally that $N = 1{,}000{,}000$ (i.e., there are a million communicating civilizations now in the Galaxy). Then the distance law (1.5) yields an average distance of 288 light-years. This is within the (upper) range of distances that our current radiotelescopes can reach for SETI searches, and is a justification of all SETI searches that have been done so far in the first 50 years of SETI (1960–2010).

In conclusion, interpolating the above three special cases of N, we may say that the distance law (1.5) yields the key diagrams (Figures 1.1 and 1.2) of the average ET

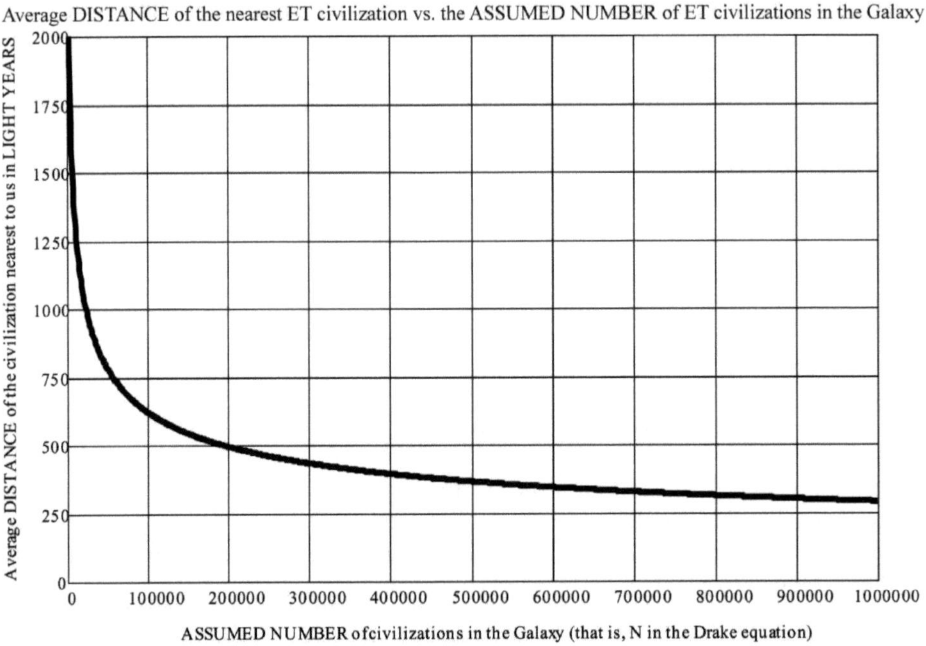

Figure 1.1. *Distance law* (i.e., the average distance—plotted along the vertical axis in light-years—vs. the *number* of communicating civilizations *assumed* to exist in the Galaxy right now).

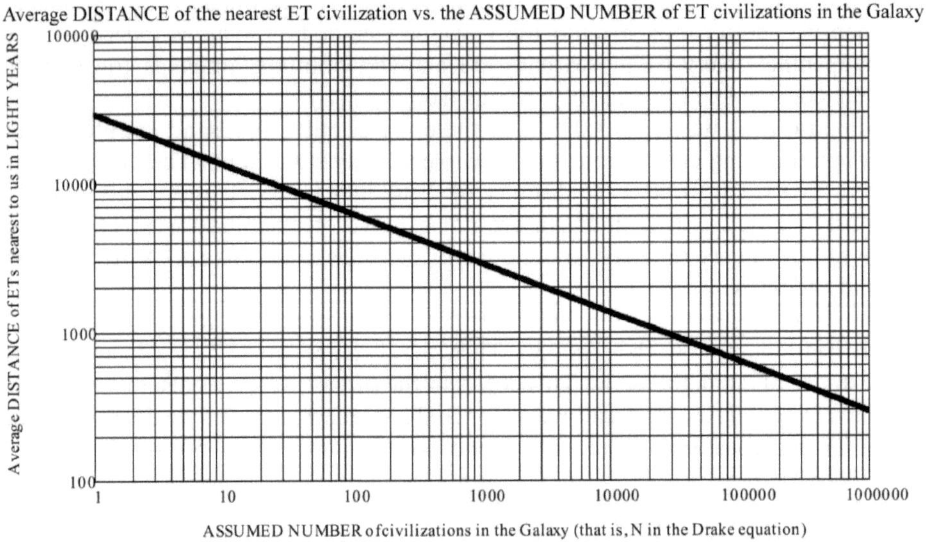

Figure 1.2. Same as Figure 1.1, but in log-log scale, so that the inverse cubic curve is turned into a straight line.

distance vs. the assumed number of communicating civilizations, N, in the Galaxy right now.

1.3 COMPUTING N BY VIRTUE OF THE DRAKE EQUATION (1961)

In the previous section, the problem of finding how close the nearest ET civilization might be was "solved" by reducing it to the computation of N, the total number of extraterrestrial civilizations now existing in the Galaxy. In this section we describe the famous Drake equation, which was proposed back in 1961 by Frank Donald Drake to estimate the numerical value of N. We believe that no better introductory description of the Drake equations exists than that given by Carl Sagan in his 1983 book *Cosmos* [2], in its turn based on the famous TV series *Cosmos*. So, in this section we report Carl Sagan's description of the Drake equation unabridged:

> "But is there anyone out there to talk to? With a third or a half a trillion stars in our Milky Way Galaxy alone, could ours be the only one accompanied by an inhabited planet? How much more likely it is that technical civilizations are a cosmic commonplace, that the Galaxy is pulsing and humming with advanced societies, and, therefore, that the nearest such culture is not so very far away—perhaps transmitting from antennas established on a planet of a naked-eye star just next door. Perhaps when we look up the sky at night, near one of those faint pinpoints of light is a world on which someone quite different from us is then glancing idly at a star we call the Sun and entertaining, for just a moment, an outrageous speculation.
>
> It is very hard to be sure. There may be several impediments to the evolution of a technical civilization. Planets may be rarer than we think. Perhaps the origin of life is not so easy as our laboratory experiments suggest. Perhaps the evolution of advanced life forms is improbable. Or it may be that complex life forms evolve more readily, but intelligence and technical societies require an unlikely set of coincidences—just as the evolution of the human species depended on the demise of the dinosaurs and the ice-age recession of the forests in whose trees our ancestors screeched and dimly wondered. Or perhaps civilizations arise repeatedly, inexorably, on innumerable planets in the Milky Way, but are generally unstable; so all but a tiny fraction are unable to survive their technology and succumb to greed and ignorance, pollution and nuclear war.
>
> It is possible to explore this great issue further and make a crude estimate of N, the number of advanced civilizations in the Galaxy. We define an advanced civilization as one capable of radio astronomy. This is, of course, a parochial if essential definition. There may be countless worlds on which the inhabitants are accomplished linguists or superb poets but indifferent radio astronomers. We will not hear from them. N can be written as the product or multiplication of a number of factors, each a kind of filter, every one of which must be sizable for there to be a large number of civilizations:

Ns, the number of stars in the Milky Way Galaxy;
fp, the fraction of stars that have planetary systems;
ne, the number of planets in a given system that are ecologically suitable for life;
fl, the fraction of otherwise suitable planets on which life actually arises;
fi, the fraction of inhabited planets on which an intelligent form of life evolves;
fc, the fraction of planets inhabited by intelligent beings on which a communicative technical civilization develops; and
fL, the fraction of planetary lifetime graced by a technical civilization.

Written out, the equation reads

$$N = Ns \cdot fp \cdot ne \cdot fl \cdot fi \cdot fc \cdot fL. \tag{1.7}$$

All of the *f*'s are fractions, having values between 0 and 1; they will pare down the large value of *Ns*.

To derive *N* we must estimate each of these quantities. We know a fair amount about the early factors in the equation, the number of stars and planetary systems. We know very little about the later factors, concerning the evolution of intelligence or the lifetime of technical societies. In these cases our estimates will be little better than guesses. I invite you, if you disagree with my estimates below, to make your own choices and see what implications your alternative suggestions have for the number of advanced civilizations in the Galaxy. One of the great virtues of this equation, due to Frank Drake of Cornell, is that it involves subjects ranging from stellar and planetary astronomy to organic chemistry, evolutionary biology, history, politics and abnormal psychology. Much of the Cosmos is in the span of the Drake equation.

We know *Ns*, the number of stars in the Milky Way Galaxy, fairly well, by careful counts of stars in a small but representative region of the sky. It is a few hundred billion; some recent estimates place it at 4×10^{11}. Very few of these stars are of the massive short-lived variety that squander their reserves of thermonuclear fuel. The great majority have lifetimes of billions or more years in which they are shining stably, providing a suitable energy source for the energy and evolution of life on nearby planets.

There is evidence that planets are a frequent accompaniment of star formation: in the satellite systems of Jupiter, Saturn and Uranus, which are like miniature solar systems; in theories of the origin of the planets; in studies of double stars; in observations of accretion disks around stars; and in some preliminary investigations of gravitational perturbations of nearby stars.[1] Many,

[1] Carl Sagan was writing these lines back in the 1970s, when no extrasolar planets had been discovered. The first such discovery occurred in 1995, when Michel Mayor and Didier Queloz, working at the Observatoire de Haute Provence in France, discovered the first extrasolar planet orbiting the nearby star 51 Peg. This first extrasolar planet was hence named 51 Peg B. Many more extrasolar planets have been discovered around nearby stars ever since. As of April 2009, 347 extrasolar planets (exoplanets) are listed in the *Extrasolar Planets Encyclopaedia* [1].

perhaps even most, stars may have planets. We take the fraction of stars that have planets, fp, as roughly equal to $1/3$. Then the total number of planetary systems in the Galaxy would be $Ns\, fp \sim 1.3 \times 10^{11}$ (the symbol $\sim$ means "approximately equal to"). If each system were to have about ten planets, as ours does, the total number of worlds in the Galaxy would be more than a trillion, a vast arena for the cosmic drama.

In our own solar system there are several bodies that may be suitable for life of some sort: the Earth certainly, and perhaps Mars, Titan and Jupiter. Once life originates, it tends to be very adaptable and tenacious. There must be many different environments suitable for life in a given planetary system. But conservatively we choose $ne = 2$. Then the number of planets in the Galaxy suitable for life becomes $Ns\, fp\, ne \sim 3 \times 10^{11}$.

Experiments show that under the most common cosmic conditions the molecular basis of life is readily made, the building blocks of molecules able to make copies of themselves. We are now on less certain grounds; there may, for example, be impediments in the evolution of the genetic code, although I think this is unlikely over billions of years of primeval chemistry. We choose $fl \sim 1/3$, implying a total number of planets in the Milky Way on which life has arisen at least once as $Ns\, fp\, ne\, fl \sim 1 \times 10^{11}$, a hundred billion inhabited worlds. That in itself is a remarkable conclusion. But we are not yet finished.

The choices of fi and fc are more difficult. On the one hand, many individually unlikely steps had to occur in biological evolution and human history for our present intelligence and technology to develop. On the other hand, there must be quite different pathways to an advanced civilization of specified capabilities. Considering the apparent difficulty in the evolution of large organisms, represented by the Cambrian explosion, let us choose $fi \times fc = 1/100$, meaning that only 1 percent of planets on which life arises actually produce a technical civilization. This estimate represents some middle ground among the varying scientific options. Some think that the equivalent of the step from the emergence of trilobites to the domestication of fire goes like a shot in all planetary systems; others think that, even given ten or fifteen billion years, the evolution of a technical civilization is unlikely. This is not a subject on which we can do much experimentation as long as our investigations are limited to a single planet. Multiplying these factors together, we find $Ns\, fp\, ne\, fl\, fi\, fc \sim 1 \times 10^{9}$, a billion planets on which technical civilizations have arisen at least once. But that is very different from saying that there are a billion planets on which technical civilizations now exist. For this we must also estimate fL.

What percentage of the lifetime of a planet is marked by a technical civilization? The Earth has harboured a technical civilization characterized by radio astronomy for only a few decades out of a lifetime of a few billion years. So far, then, for our planet fL is less than $1/10^{8}$, a millionth of a percent. And it is hardly out of the question that we might destroy ourselves tomorrow. Suppose this were a typical case, and the destruction so complete that no other technical civilization—of the human or any other species—were able to emerge in the five or so billion years remaining before the Sun dies. Then $Ns\, fp\, ne\, fl\, fi\, fc\, fL \sim 10$, and, at a given time there would be only a tiny smattering, a handful, a pitiful few technical civilizations in the Galaxy, the steady state number maintained as emerging societies replace those recently self-immolated. The number N might be even as small as 1. If civilizations tend to destroy themselves soon after reaching a technological phase, there might be no one for us to talk with but ourselves. And that we do but poorly. Civilizations would take billions of years of tortuous evolution, and then snuff themselves out in an instant of unforgivable neglect.

But consider the alternative, the prospect that at least some civilizations learn to live with technology; that the contradictions posed by the vagaries of past brain evolution are consciously resolved and do not lead to self-destruction; or that, even if major disturbances occur, they are reveres in the subsequent billions of years of biological evolution. Such societies might live to a prosperous old age, their lifetimes measured perhaps on geological or stellar evolutionary time scales. If 1 percent of civilizations can survive technological adolescence, take the proper fork at this critical historical branch point and achieve maturity, then $fL \sim 1/100$, $N \sim 10^7$, and the number of extant civilizations in the Galaxy is in the millions. Thus, for all our concern about the possible unreliability of our estimates of the early factors in the Drake equation, which involve astronomy, organic chemistry and evolutionary biology, the principal uncertainty comes to economics and politics and what, on Earth, we call human nature. It seems fairly clear that if self-destruction is not the overwhelmingly preponderant fate of galactic civilizations, then the sky is softly humming with messages from the stars.

These estimates are stirring. They suggest that the receipt of a message from space is, even before we decode it, a profoundly hopeful sign. It means that someone has learned to live with high technology; that it is possible to survive technological adolescence. This alone, quite apart from the contents of the message, provides a powerful justification for the search for other civilizations."

1.4 THE DRAKE EQUATION IS OVER-SIMPLIFIED

In the 50 years (1961–2011) that have elapsed since Frank Drake proposed his equation, a number of scientists and writers have tried to discover which of the numerical values of its seven independent variables are more realistically in agreement with our present day knowledge. Thus there is a considerable amount of literature about the Drake equation nowadays, and, as one can easily imagine, the results obtained by the various authors largely differ from one another. In other words, the value of N, which various authors obtained by different assumptions about the astronomy, the biology, and the sociology implied by the Drake equation, may range from a few tens in the pessimist's view to some million or even billions in the optimist's opinion. A lot of uncertainty thus affects our knowledge of N as of 2009. In all cases, however, the final result about N has always been a sheer number (i.e., a positive integer number ranging from 1 to millions or billions). This is precisely the aspect of the Drake equation that the author regarded as "too simplistic" and improved mathematically in his paper #IAC-08-A4.1.4, entitled "The statistical Drake equation". which was presented on October 1, 2008, at the 59th International Astronautical Congress (IAC) held in Glasgow, Scotland, U.K., September 29–October 3, 2008. That paper is attached herewith as Appendix 1.B. Newcomers to SETI and to the Drake equation, however, may find that paper too difficult to be understood mathematically at a first reading. Thus, we shall now explain the content of that paper "by speaking more in lay terms". We thank the reader for his or her attention.

1.5 THE STATISTICAL DRAKE EQUATION

We start with an example.

Consider the first independent variable in the Drake equation (1.7) (i.e., Ns), the number of stars in the Milky Way Galaxy. Astronomers tell us that there should be about 350 million stars in the Galaxy. Of course, nobody has counted (or even seen in the photographic plates) *all* the stars in the Galaxy! There are too many practical difficulties preventing us from doing so: for example, the dust clouds that prevent us from seeing even the Galactic Bulge (i.e., the central region of the Galaxy) in visible light (although we can "see it" at radio frequencies like the famous neutral hydrogen line at 1,420 MHz). So, it doesn't make any sense to state that $Ns = 350 \times 10^6$, or state (even worse) that the number of stars in the Galaxy is, say, 354,233,321 or similar fanciful exact integer numbers. That is just silly and non-scientific. Much more scientific, on the contrary, is to say that the number of stars in the Galaxy is 350 million plus or minus, say, 50 millions (or whatever values the astronomers may regard as more appropiate, since this is just an example to let the reader understand the difficulty).

Thus, it makes sense to *replace* each of the seven independent variables in the Drake equation (1.7) by a *mean value* (350 millions, in the above example) *plus or minus a certain standard deviation* (50 millions, in the above example).

By doing so, we have made a great step ahead: we have abandoned the too-simplistic equation (1.7) and replaced it by something more sophisticated and scientifically more serious: the *statistical* Drake equation. In other words, we have transformed the classical and simplistic Drake equation (1.7) into an advanced statistical tool for the investigation of a host of facts hardly known to us in detail. In still other words:

(1) we replace each independent variable in (1.7) by a *random variable*, labeled D_i (from Drake);
(2) we assume that the *mean value* of each D_i is the same numerical value previously attributed to the corresponding independent variable in (1.7);
(3) but now we also *add a standard deviation* σ_{D_i} on each side of the mean value, which is provided by the knowledge gathered by scientists in each discipline encompassed by each D_i.

Having so done, the next question is: How can we discover the *probability distribution* for each D_i?

For instance, will it be a Gaussian, or what?

This is a difficult question, for nobody knows the probability distribution of the number of stars in the Galaxy, not to mention the probability distribution of the other six variables in the Drake equation (1.7).

There is a brilliant way to get around this difficulty, though.

We start by excluding the Gaussian because each variable in the Drake equation is a *positive* (or, more precisely, a non-negative) random variable, while the Gaussian applies to *real* random variables only. So, the Gaussian is out. Then, one might consider the large class of well-studied and positive probability densities called "gamma distributions", but it is then unclear why one should adopt gamma distributions and not any other. The solution to this apparent conundrum comes from Shannon's Information Theory and a theorem that he proved in 1948: "The probability distribution having maximum entropy (= uncertainty) over any *finite* range of real values is the *uniform* distribution over that range." This is proven in Appendix 1.A of the present chapter.

So, at this point, we assume that each of the seven D_i in (1.7) is a *uniform* random variable, whose mean value and standard deviation is known by scientists working in the respective field (be it astronomy, biology, or sociology). Notice that, for such a uniform distribution, knowledge of the mean value μ_{D_i} and of the standard deviation σ_{D_i} automatically determines the *range* of that random variable in between its lower (called a_i) and upper (called b_i) limits: in fact, these limits are given by the equations

$$\begin{cases} a_i = \mu_{D_i} - \sqrt{3}\,\sigma_{D_i} \\ b_i = \mu_{D_i} + \sqrt{3}\,\sigma_{D_i}. \end{cases} \tag{1.8}$$

The "surprising" factor $\sqrt{3}$ in the above equations comes from the definitions of mean value and standard deviation: please see equations (1.B.12), (1.B.15), and (1.B.17) in Appendix 1.B for the relevant proof. So the uniform distribution of each random variable D_i is perfectly determined by its mean value and standard deviation, and so are all its other properties.

The next problem is the following: OK, since we now know everything about each uniformly distributed D_i, what is the probability distribution of N, given that N is the product (1.7) of all the D_i?

In other words, not only do we want to discover the analytical expression of the probability density function of N, but we also want to relate its mean value μ_N to all mean values μ_{D_i} of the D_i, and its standard deviation σ_N to all standard deviations σ_{D_i} of the D_i.

This is a difficult problem.

It occupied the author's mind for no fewer than 10 years (1997–2007). It is actually an *analytically unsolvable* problem, in that, to the best of the author's knowledge, it is *impossible* to find an analytic expression for any *finite product* of uniform random variables D_i. This result is proven in Sections 1.B.2 through 1.B.3.3 of Appendix 1.B (unfortunately!).

1.6 SOLVING THE STATISTICAL DRAKE EQUATION BY VIRTUE OF THE CENTRAL LIMIT THEOREM (CLT) OF STATISTICS

The solution to the problem of finding the analytical expression for the probability density function of N in the statistical Drake equation was found by the author only in September 2007. The key steps are the following:

(1) Take the natural logs of both sides of the statistical Drake equation (1.7). This changes the product into a sum.
(2) The mean values and standard deviations of the logs of the random variables D_i may all be expressed analytically in terms of the mean values and standard deviations of the D_i.
(3) Recall the Central Limit Theorem (CLT) of statistics, stating that (loosely speaking) if you have a *sum* of independent random variables, each of which is *arbitrarily distributed* (hence, also including uniformly distributed), then, when the number of terms in the sum increases indefinitely (i.e., for a sum of random variables infinitely long) ... the *sum random variable tends to a Gaussian.*
(4) Thus, the natural log of N tends to a Gaussian.
(5) Thus, N tends to the *lognormal distribution.*
(6) The mean value and standard deviations of this lognormal distribution of N may all be expressed analytically in terms of the mean values and standard deviations of the logs of the D_i already found previously.

This result is fundamental

All the relevant equations are summarized in Table 1.1. This table is actually the same as Table 2 of the author's original paper IAC-08-A4.1.4, entitled "The statistical Drake equation" and presented by him at the International Astronautical Congress (IAC) held in Glasgow, U.K., on October 1, 2008. The original paper is reproduced in Appendix 1.B.

To sum up, not only have we found that N approaches the completely known lognormal distribution for an *infinity* of factors in the statistical Drake equation (1.7), but we have also paved the way to further applications by removing the condition that the number of terms in the product (1.7) must be *finite.*

The possibility of *adding any number of factors* to the Drake equation (1.7) was not envisaged, of course, by Frank Drake back in 1961, when "summarizing" the evolution of life in the Galaxy in *seven* simple *steps.* But, today, the number of factors in the Drake equation should be increased: for instance, there is no mention in the original Drake equation of the possibility that asteroidal impacts might destroy life on Earth at any time, and this is because the demise of the dinosaurs at the K/T impact had not yet been understood by scientists in 1961—it took until 1980 for this to be the case!

In practice, we are suggesting *increasing* the number of factors as much as necessary in order to get better and better estimates of N as long as our scientific knowledge increases. This we call the "Data Enrichment Principle" and believe should be the next important goal in the study of the statistical Drake equation.

Table 1.1. Summary of the properties of the lognormal distribution that apply to the random variable N = number of ET communicating civilizations in the Galaxy.

Random variable	N = number of communicating ET civilizations in Galaxy
Probability distribution	Lognormal
Probability density function	$f_N(n) = \frac{1}{n} \cdot \frac{1}{\sqrt{2\pi}\sigma} e^{-\frac{(\ln(n)-\mu)^2}{2\sigma^2}} \quad (n \geq 0)$
Mean value	$\langle N \rangle = e^{\mu} e^{\frac{\sigma^2}{2}}$
Variance	$\sigma_N^2 = e^{2\mu} e^{\sigma^2} \left(e^{\sigma^2} - 1\right)$
Standard deviation	$\sigma_N = e^{\mu} e^{\frac{\sigma^2}{2}} \sqrt{e^{\sigma^2} - 1}$
All the moments (i.e., k-th moment)	$\left\langle N^k \right\rangle = e^{k\mu} e^{k^2 \frac{\sigma^2}{2}}$
Mode (= abscissa of the lognormal peak)	$n_{\text{mode}} \equiv n_{\text{peak}} = e^{\mu} e^{-\sigma^2}$
Value of the mode peak	$f_N(n_{\text{mode}}) = \frac{1}{\sqrt{2\pi}\,\sigma} \cdot e^{-\mu} \cdot e^{\frac{\sigma^2}{2}}$
Median (= 50–50 probability value for N)	Median $= m = e^{\mu}$
Skewness	$\frac{K_3}{(K_2)^{\frac{3}{2}}} = \left(e^{\sigma^2} + 2\right) \sqrt{e^{\sigma^2} - 1}$
Kurtosis	$\frac{K_4}{(K_2)^2} = e^{4\sigma^2} + 2\, e^{3\sigma^2} + 3\, e^{2\sigma^2} - 6$
Expression of μ in terms of the lower (a_i) and upper (b_i) limits of the Drake uniform input random variables D_i	$\mu = \sum_{i=1}^{7} \langle Y_i \rangle = \sum_{i=1}^{7} \frac{b_i[\ln(b_i) - 1] - a_i[\ln(a_i) - 1]}{b_i - a_i}$
Expression of σ^2 in terms of the lower (a_i) and upper (b_i) limits of the Drake uniform input random variables D_i	$\sigma^2 = \sum_{i=1}^{7} \sigma_{Y_i}^2 = \sum_{i=1}^{7} \left(1 - \frac{a_i b_i [\ln(b_i) - \ln(a_i)]^2}{(b_i - a_i)^2}\right)$

Finally, we wish to provide a numerical example explaining how the statistical Drake equation works in practice. This will be done in the next section.

1.7 AN EXAMPLE EXPLAINING THE STATISTICAL DRAKE EQUATION

To understand how things work in practice for the statistical Drake equation, please consider Input Table 1.1. It is made up of three columns:

Input Table 1.1. Input values (i.e., mean values and standard deviations) for the seven Drake uniform random variables D_i. The first column lists the seven input sheer numbers that also become the mean values (middle column). Finally the last column lists the seven input standard deviations. The bottom line is the classical Drake equation (1.7).

$Ns := 350 \cdot 10^9$	$\mu Ns := Ns$	$\sigma Ns := 1 \cdot 10^9$
$fp := \frac{50}{100}$	$\mu fp := fp$	$\sigma fp := \frac{10}{100}$
$ne := 1$	$\mu ne := ne$	$\sigma ne := \frac{1}{\sqrt{3}}$
$fl := \frac{50}{100}$	$\mu fl := fl$	$\sigma fl := \frac{10}{100}$
$fi := \frac{20}{100}$	$\mu fi := fi$	$\sigma fi := \frac{10}{100}$
$fc := \frac{20}{100}$	$\mu fc := fc$	$\sigma fc := \frac{10}{100}$
$fL := \frac{10{,}000}{10^{10}}$	$\mu fL := fL$	$\sigma fL := \frac{1{,}000}{10^{10}}$
$N := Ns \cdot fp \cdot ne \cdot fl \cdot fi \cdot fc \cdot fL$		$N = 3{,}500$

(1) The first column lists the seven input sheer numbers that also become ...
(2) The mean values (middle column).
(3) Finally, the last column lists the seven input standard deviations.

The bottom line is the classical Drake equation (1.7). We see that, for this particular set of seven inputs, the classical Drake equation (i.e., the product of the seven numbers) yields a total of 3,500 communicating extraterrestrial civilizations existing in the Galaxy right now.

The statistical Drake equation, however, provides a much more articulated answer than just the above sheer number $N = 3{,}500$. In fact, a MathCad code written by the author and capable of performing all the numerical calculations required by the statistical Drake equation, for a given set of seven input mean values plus seven input standard deviations, yields for N the lognormal distribution (thin curve) plotted in Figure 1.3. We see immediately that the peak of this thin curve (i.e., the mode) falls at about $n_{\text{mode}} \equiv n_{\text{peak}} = e^{\mu} e^{-\sigma^2} \approx 250$ (this is equation (1.B.98) of Appendix 1.B), while the median (50–50 value splitting the lognormal density in two parts with equal undergoing areas) falls at about $n_{\text{median}} \equiv e^{\mu} \approx 1{,}740$. These seem to be smaller values than $N = 3{,}500$ provided by the classical Drake equations, but it's a wrong impression due to a poor "intuitive" understanding of what statistics

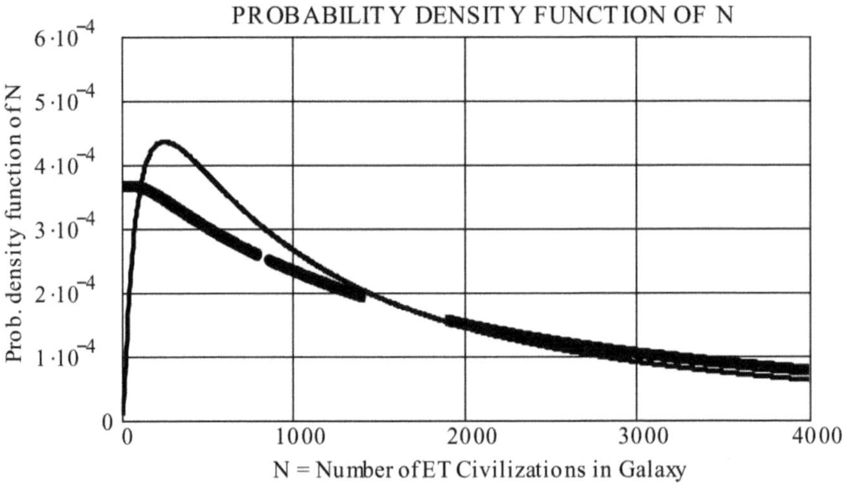

Figure 1.3. Comparing the two probability density functions of the random variable N found: (1) at the end of Section 3.3 in a purely numerical way and without resorting to the CLT at all (thick curve), and (2) analytically by using the CLT and the relevant lognormal approximation (thin curve).

is! In fact, neither the mode nor the median are the "really important" values: the really important value for N is the *mean value*! Now if you look at the thin curve in Figure 1.3 (i.e., the lognormal distribution arising from the Central Limit Theorem), you see that this curve has a *long tail on the right*! In other words, it does *not* immediately go down to nearly zero beyond the peak of the mode. Thus, when you actually compute the mean value, you should not be too surprised to find that it equals ~4,590 communicating civilizations now in the Galaxy. This is the important number, and it is *higher* than the 3,500 provided by the classical Drake equation. Thus, in conclusion, *the statistical extension* of the classical Drake equation that we made *increases our hopes* of finding an extraterrestrial civilization!

Our hopes are increased even more so when we go on to consider the standard deviation associated with the mean value 4,590. In fact, the standard deviation is given by equation (1.B.97) of Appendix 1.B. This yields $\sigma_N = e^{\mu}\, e^{\frac{\sigma^2}{2}} \sqrt{e^{\sigma^2} - 1} = 11{,}195$ and so the expected number of N may actually be even much higher than the 4,590 provided by the mean value alone! The "upper limit of the one-sigma confidence interval", as statisticians call it (i.e., the sum 4,590 + 1,1195 = 15,785), yields a higher number still! Note: please don't tell me that the lower limit of the one-sigma confidence interval (i.e., −6,605) is "negative". It is actually *zero* because the lognormal distribution is *positive* (or, more correctly, non-negative) and so negative values of its support are ruled out from the start!. Finally, the reader should not worry about the thick curve depicted in Figure 1.3: it is just the *numerical* solution of the statistical Drake equation for a *finite* number of seven input factors. Figure 1.3 actually shows that this curve "is well interpolated" by the lognormal distribution (thin curve) (i.e., by the neat analytical expression provided by the Central Limit Theorem for an *infinite* number of factors in the Drake equation). That is, in conclusion, Figure 1.3 visually shows that taking seven factors or an infinity of factors "is almost the same thing" for a value as small as 7.

1.8 FINDING THE PROBABILITY DISTRIBUTION OF THE ET_DISTANCE BY VIRTUE OF THE STATISTICAL DRAKE EQUATION

Having solved the statistical Drake equation by finding the lognormal distribution, we are now in a position to solve the ET_Distance problem by resorting to statistics again, rather than just to the purely deterministic Distance Law (1.5), as we did in Section 1.2. This is "scientifically more serious" than just the purely deterministic Distance Law (1.5) inasmuch as the new statistical Distance Law will yield a *probability density* for the distance, with the relevant mean value and standard deviation, of course. In other words, the Distance Law (1.5) now becomes a random variable itself whose probability distribution, mean value, and standard deviation must be computed by "replacing" (so to say) into (1.5) the fact that N is now known to follow the lognormal distribution. This was also done by the author in September 2007, and is mathematically described in detail in Section 1.B.7 of Appendix 1.B.

So, the important new result is the *probability density for the distance*, which the well-known physicist Paul Davies dubbed "the Maccone distribution" and whose equation reads

$$f_{\mathrm{ET\_Distance}}(r) = \frac{3}{r} \cdot \frac{1}{\sqrt{2\pi}\,\sigma} \cdot e^{-\frac{\left(\ln\left[\frac{6\,R_{Galaxy}^2\,h_{Galaxy}}{r^3}\right]-\mu\right)^2}{2\sigma^2}} \tag{1.9}$$

holding for $r \geq 0$. This is equation (1.B.114) of Appendix 1.B.

Starting from this equation, the author computed the *mean value* of the random variable ET_Distance

$$\langle \mathrm{ET\_Distance} \rangle = C\, e^{-\frac{\mu}{3}}\, e^{\frac{\sigma^2}{18}} \tag{1.10}$$

which is equation (1.B.119) and finally the ET_Distance *standard deviation*

$$\sigma_{\mathrm{ET\_Distance}} = C\, e^{-\frac{\mu}{3}}\, e^{\frac{\sigma^2}{18}} \sqrt{e^{\frac{\sigma^2}{9}} - 1} \tag{1.11}$$

which is equation (1.B.123). Of course, all other descriptive statistical quantities, such as moments, cumulants, etc. can be computed when starting from the probability density (1.9), and the result is Table 1.2, which is Table 1.B.3 of Appendix 1.B.

Finally, we wish to complete this section, as well as this "easy introduction to the statistical Drake equation", by pointing out the numerical values that equations (1.10) and (1.11) yield for Input Table 1.1. They are, respectively:

$$r_{mean\_value} = C\, e^{-\frac{\mu}{3}}\, e^{\frac{\sigma^2}{18}} \approx 2{,}670 \text{ light-years} \tag{1.12}$$

Table 1.2. Summary of the properties of the probability distribution that apply to the random variable ET_Distance yielding the (average) distance between any two neighboring communicating civilizations in the Galaxy.

Random variable	ET_Distance between any two neighboring ET civilizations in the Galaxy assuming they are *uniformly* distributed throughout the whole Galaxy volume
Probability distribution	Unnamed (Paul Davies suggested "Maccone distribution")
Probability density function	$f_{\text{ET\_Distance}}(r) = \frac{3}{r}\cdot\frac{1}{\sqrt{2\pi}\,\sigma}\cdot e^{-\frac{\left(\ln\left[\frac{6\,R_{Galaxy}^2\,h_{Galaxy}}{r^3}\right]-\mu\right)^2}{2\sigma^2}}$
Numerical constant C related to the Milky Way size	$C = \sqrt[3]{6\,R_{Galaxy}^2\,h_{Galaxy}} \approx 28{,}845$ light-years
Mean value	$\langle \text{ET\_Distance}\rangle = C\,e^{-\frac{\mu}{3}}\,e^{\frac{\sigma^2}{18}}$
Variance	$\sigma^2_{\text{ET\_Distance}} = C^2\,e^{-\frac{2}{3}\mu}\,e^{\frac{\sigma^2}{9}}\left(e^{\frac{\sigma^2}{9}}-1\right)$
Standard deviation	$\sigma_{\text{ET\_Distance}} = C\,e^{-\frac{\mu}{3}}\,e^{\frac{\sigma^2}{18}}\sqrt{e^{\frac{\sigma^2}{9}}-1}$
All the moments (i.e., the k-th moment)	$\langle \text{ET\_Distance}^{k}\rangle = C^k\,e^{-k\frac{\mu}{3}}\,e^{k^2\frac{\sigma^2}{18}}$
Mode (= abscissa of the peak)	$r_{\text{mode}} \equiv r_{peak} = C\,e^{-\frac{\mu}{3}}\,e^{-\frac{\sigma^2}{9}}$
Value of the mode peak	Peak value of $f_{\text{ET\_Distance}}(r) = \equiv f_{\text{ET\_Distance}}(r_{\text{mode}}) = \frac{3}{C\sqrt{2\pi}\,\sigma}\cdot e^{\frac{\mu}{3}}\cdot e^{\frac{\sigma^2}{18}}$
Median (= 50–50 probability value for N)	$Median = m = Ce^{-\frac{\mu}{3}}$
Skewness	$\frac{K_3}{(K_2)^{\frac{3}{2}}} = \frac{e^{\frac{2\sigma^2}{9}} + e^{\frac{\sigma^2}{9}} - 2}{\sqrt{e^{\frac{\sigma^2}{9}}-1}}$
Kurtosis	$\frac{K_4}{(K_2)^2} = e^{\frac{4\sigma^2}{9}} + 2\,e^{\frac{\sigma^2}{3}} + 3\,e^{\frac{2\sigma^2}{9}} - 6$
Expression of μ in terms of the lower (a_i) and upper (b_i) limits of the Drake uniform input random variables D_i	$\mu = \sum_{i=1}^{7}\langle Y_i\rangle = \sum_{i=1}^{7}\frac{b_i[\ln(b_i)-1]-a_i[\ln(a_i)-1]}{b_i-a_i}$
Expression of σ^2 in terms of the lower (a_i) and upper (b_i) limits of the Drake uniform input random variables D_i	$\sigma^2 = \sum_{i=1}^{7}\sigma^2_{Y_i} = \sum_{i=1}^{7}\left(1-\frac{a_ib_i[\ln(b_i)-\ln(a_i)]^2}{(b_i-a_i)^2}\right)$

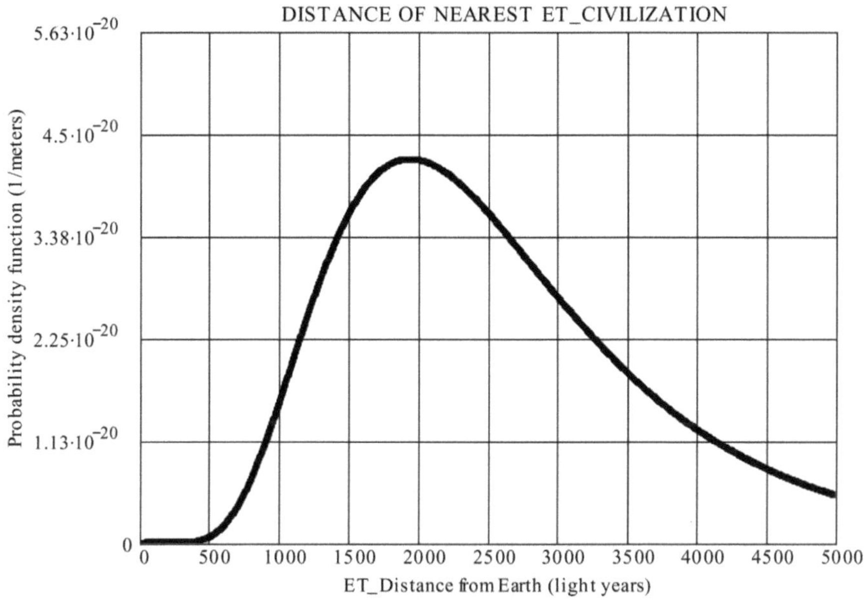

Figure 1.4. This is the probability of finding the nearest extraterrestrial civilization at the distance r from Earth (in light-years) if the values assumed in the Drake equation are those shown in Input Table 1.1. The relevant probability density function $f_{\text{ET\_Distance}}(r)$ is given by equation (1.9). Its mode (peak abscissa) equals 1,933 light-years, but its mean value is higher since the curve has a long tail on the right: the mean value equals in fact 2,670 light-years. Finally, the standard deviation equals 1,309 light-years: *this is good news for SETI*, inasmuch as the nearest ET civilization might lie at just 1 sigma = 2,670 − 1,309 = 1,361 light-years from us.

which is equation (1.B.153), and

$$\sigma_{\text{ET\_Distance}} = C\, e^{-\frac{\mu}{3}} e^{\frac{\sigma^2}{18}} \sqrt{e^{\frac{\sigma^2}{9}} - 1} \approx 1{,}309 \text{ light-years} \tag{1.13}$$

which is equation (1.B.154).

Drawing a graph (Figure 1.4) of the ET_Distance probability density (1.9) helps to clarify matters.

From Figure 1.4 we see that the probability of finding extraterrestrials is practically zero up to a distance of about 500 light-years from Earth. Then it starts increasing with increasing distance from Earth, and reaches its maximum at

$$r_{\text{mode}} \equiv r_{peak} = C\, e^{-\frac{\mu}{3}} e^{-\frac{\sigma^2}{9}} \approx 1{,}933 \text{ light-years} \tag{1.14}$$

This is the *most likely value* of the distance at which we can expect to find the nearest extraterrestrial civilization.

It is not, as we said, the mean value of the probability distribution (1.9) for $f_{\text{ET\_Distance}}(r)$. In fact, the probability density (1.9) has an infinite tail on the right, as clearly shown in Figure 1.4, and hence its mean value must be higher than its peak value. As given by (1.10) and (1.12), its mean value is $r_{mean\_value} = C e^{-\frac{\mu}{3}} e^{\frac{\sigma^2}{18}} \approx 2{,}670$ light-years. This is the *mean* (value of the) *distance* at which we can expect to find extraterrestrials.

After having found the above two distances (1,933 and 2,670 lt-yr, respectively), the next natural question that arises is: "what is the range, back and forth around the mean value of the distance, at which we are most likely to expect to find extraterrestrials?" The answer to this question is given by the notion of standard deviation that we already found to be given by (1.11) and (1.13), $\sigma_{\text{ET\_Distance}} = C e^{-\frac{\mu}{3}} e^{\frac{\sigma^2}{18}} \sqrt{e^{\frac{\sigma^2}{9}} - 1} \approx 1{,}309$ light-years. More precisely, this is the so-called one-sigma (distance) level. Probability theory then shows that the nearest extraterrestrial civilization is expected to be located within this range: that is, within the two distances of $(2{,}670 - 1{,}309) = 1{,}361$ light-years and $(2{,}670 + 1{,}309) = 3{,}979$ light-years, with probability given by the integral of $f_{\text{ET\_Distance}}(r)$ taken in between these two lower and upper limits, that is:

$$\int_{1{,}361 \text{ light-years}}^{3{,}979 \text{ light-years}} f_{\text{ET\_Distance}}(r)\, dr \approx 0.75 = 75\%. \tag{1.15}$$

In plain words: with 75% probability, the nearest extraterrestrial civilization is located in between the distances of 1,361 and 3,979 light-years from us, having assumed the input values to the Drake equation given by Input Table 1.1. If we change these input values, then all the numbers change again, of course.

1.9 THE "DATA ENRICHMENT PRINCIPLE" AS THE BEST CLT CONSEQUENCE UPON THE STATISTICAL DRAKE EQUATION (ANY NUMBER OF FACTORS ALLOWED)

As a fitting climax to all the statistical equations developed so far, let us now state our *Data Enrichment Principle.* It simply states that "the higher the number of factors in the statistical Drake equation, the better."

Put in this simple way, it looks like a new way of saying that the CLT lets the random variable Y approach the normal distribution when the number of terms in the sum (1.4) approaches infinity. And this is the case, indeed. However, our Data Enrichment Principle has more profound methodological consequences that we cannot explain now, but hope to describe more precisely in the future.

1.10 CONCLUSIONS

We have sought to extend the classical Drake equation to let it encompass statistics and probability.

This approach appears to pave the way to future, more profound investigations intended not only to associate "error bars" to each factor in the Drake equation, but especially to increase the number of factors themselves. In fact, this seems to be the only way to incorporate into the Drake equation more and more new scientific information as soon as it becomes available. In the long run, the statistical Drake equation might just become a huge computer code, growing in size and especially in the depth of the scientific information it contained. It would thus be humanity's first "Encyclopedia Galactica".

Unfortunately, to extend the Drake equation to statistics, it was necessary to use a mathematical apparatus that is more sophisticated than just the simple product of seven numbers.

When the author had the honor and privilege to present his results at the SETI Institute on April 11, 2008, in front of an audience that included Frank Drake, he felt he had to add these words: "My apologies, Frank, for disrupting the beautiful simplicity of your equation."

1.11 ACKNOWLEDGMENT

The author is grateful to Drs. Hal Puthoff and Eric Davis of the Institute for Advanced Study at Austin for reading the author's paper #IAC-08-A4.1.4 and suggesting the present publication made up by both the paper itself and its popularized version.

APPENDIX 1.A PROOF OF SHANNON'S 1948 THEOREM STATING THAT THE UNIFORM DISTRIBUTION IS THE "MOST UNCERTAIN" ONE OVER A FINITE RANGE OF VALUES

Information theory was initiated by Claude Shannon (1916–2001) in his two well-known 1948 papers.

In this appendix, we wish to draw attention to a couple of theorems that Shannon proves on pp. 36 and 37 of his work (note that Shannon omits the upper and lower limits of all integrals in the first theorem: they are $-\infty$ and $+\infty$, respectively):

"5. Let $p(x)$ be a one-dimensional distribution. The form of $p(x)$ giving a maximum entropy subject to the condition that the standard deviation of x be fixed at σ is Gaussian. To show this we must maximize

$$H(x) = -\int p(x) \log p(x)\, dx$$

with

$$\sigma^2 = \int p(x) x^2\, dx \quad \text{and} \quad 1 = \int p(x)\, dx$$

as constants. This requires, by the calculus of variations, maximizing

$$\int [-p(x) \log p(x) + \lambda p(x) x^2 + \mu p(x)]\, dx.$$

The condition for this is

$$-1 - \log p(x) + \lambda x^2 + \mu = 0$$

and consequently (adjusting the constants to satisfy the constraints)

$$p(x) = \frac{1}{\sqrt{2\pi}\sigma} e^{-(x^2/2\sigma^2)}$$

and

7. If x is limited to a half line ($p(x) = 0$ for $x \leq 0$) and the first moment of x is fixed at a

$$a = \int_0^\infty p(x) x\, dx,$$

then the maximum entropy occurs when

$$p(x) = \frac{1}{a} e^{-(x/a)}$$

and is equal to $\log ea$."

Now, we wish to point out that there is a third possible case, other than the two given by Shannon. This is the case when the probability density function $p(x)$ is limited to a *finite interval* $a \leq x \leq b$. This is obviously the case with any physical *positive* random variable, such as a distance, or the number N of extraterrestrial communicating civilizations in the Galaxy. In addition, it is easy to prove that for any such finite random variable the maximum entropy distribution is the *uniform* distribution over $a \leq x \leq b$. Shannon did not bother to prove this simple theorem in his 1948 papers since he probably regarded it as too trivial. But we prefer to point out this theorem since, in the language of the statistical Drake equation, it sounds like: "Since we don't know what the probability distribution of any one of the Drake random variables D_i is, it is safer to assume that each of them has the maximum possible entropy over $a_i \leq x \leq b_i$ (i.e., that D_i is *uniformly* distributed there).

The proof of this theorem is along the same lines as for the previous two cases discussed by Shannon:

(1) We start by assuming that $a_i \leq x \leq b_i$.
(2) We then form the linear combination of the entropy integral plus the normalization condition for D_i

$$\delta \int_{a_i}^{b_i} [-p(x) \log p(x) + \lambda\, p(x)]\, dx = 0$$

where λ is a Lagrange multiplier.

(3) Performing the variation, one finds

$$-\log p(x) - 1 + \lambda = 0 \quad \text{that is: } p(x) = e^{\lambda-1}.$$

(4) Applying the normalization condition (constraint) to the last expression for $p(x)$ yields

$$1 = \int_{a_1}^{b_i} p(x)\,dx = \int_{a_1}^{b_i} e^{\lambda-1}\,dx = e^{\lambda-1}\int_{a_1}^{b_i} dx = e^{\lambda-1}(b_i - a_i)$$

which yields

$$e^{\lambda-1} = \frac{1}{b_i - a_i}$$

and finally

$$p(x) = \frac{1}{b_i - a_i} \quad \text{with} \quad a_i \le x \le b_i$$

showing that the maximum entropy probability distribution over any *finite* interval $a_i \le x \le b_i$ is the *uniform* distribution.

APPENDIX 1.B ORIGINAL TEXT OF THE AUTHOR'S #IAC-08-A4.1.4 PAPER ENTITLED "THE STATISTICAL DRAKE EQUATION"

Claudio Maccone at Glasgow, Scotland, on Wednesday, October 1, 2008, presenting his paper IAC-08-A4.1.4 entitled "The statistical Drake equation" at the 59th International Astronautical Congress (IAC).

IAC-08-A4.1.4 THE STATISTICAL DRAKE EQUATION

Claudio Maccone, Co-Vice Chair, SETI Permanent Study Group, International Academy of Astronautics, Via Martorelli, 43 - Torino (Turin) 10155 - Italy,
http://www.maccone.com/* - *E-mail: clmaccon@libero.it

Abstract

We provide the statistical generalization of the Drake equation.

From a simple product of seven positive numbers, the Drake equation is now turned into the product of seven positive random variables. We call this "the statistical Drake equation". The mathematical consequences of this transformation are then derived. The proof of our results is based on the Central Limit Theorem (CLT) of statistics. In loose terms, the CLT states that the sum of any number of independent random variables, each of which may be *arbitrarily* distributed, approaches a Gaussian (i.e., normal) random variable. This is called the Lyapunov Form of the CLT, or the Lindeberg Form of the CLT, depending on the mathematical constraints assumed on the third moments of the various probability distributions. In conclusion, we show that:

(1) The new random variable N, yielding the number of communicating civilizations in the Galaxy, follows the *lognormal* distribution. Then, as a consequence, the mean value of this lognormal distribution is the ordinary N in the Drake equation. The standard deviation, mode, and all the moments of this lognormal N are also found.
(2) The seven factors in the ordinary Drake equation now become seven positive random variables. The probability distribution of each random variable may be *arbitrary*. The CLT in the so-called Lyapunov or Lindeberg forms (both of which do not assume the factors to be identically distributed) allows for that. In other words, the CLT "translates" into our statistical Drake equation by allowing an arbitrary probability distribution for each factor. This is both physically realistic and practically very useful, of course.
(3) An application of our statistical Drake equation then follows. The (average) *distance* between any two neighboring and communicating civilizations in the Galaxy may be shown to be inversely proportional to the cubic root of N. Then, in our approach, this distance becomes a new random variable. We derive the relevant probability density function, apparently previously unknown and dubbed "Maccone distribution" by Paul Davies.
(4) *Data Enrichment Principle.* It should be noticed that *any* positive number of random variables in the statistical Drake equation is compatible with the CLT. So, our generalization allows for many more factors to be added in the future as long as more refined scientific knowledge about each factor becomes known to scientists. This capability to make room for more future factors in the statistical Drake equation we call the "Data Enrichment Principle", and we

regard it as the key to more profound future results in the fields of astrobiology and SETI.

Finally, a practical example is given of how our statistical Drake equation works numerically. We work out in detail the case where each of the seven random variables is uniformly distributed around its own mean value and has a given standard deviation. For instance, the number of stars in the Galaxy is assumed to be uniformly distributed around, say, 350 billion with a standard deviation of, say, 1 billion. Then, the resulting lognormal distribution of N is computed numerically by virtue of a MathCad file that the author has written. This shows that the mean value of the lognormal random variable N is actually of the same order as the classical N given by the ordinary Drake equation, as one might expect from a good statistical generalization.

1.B.1 Introduction

The Drake equation is now a famous result (see [5] for the Wikipedia summary) in the fields of SETI (see [6]) and astrobiology (see [7]). Devised in 1960, the Drake equation was the first scientific attempt to estimate the number N of extraterrestrial civilizations in the Galaxy with which we might come in contact. Frank Drake (see [8]) proposed it as the product of seven factors:

$$N = Ns \cdot fp \cdot ne \cdot fl \cdot fi \cdot fc \cdot fL. \tag{1.B.1}$$

where:

(1) Ns is the estimated number of stars in our Galaxy.
(2) fp is the fraction (= percentage) of such stars that have planets.
(3) ne is the number of "Earth-type" such planets around the given star; in other words, ne is the number of planets, in a given stellar system, on which the chemical conditions exist for life to begin its course: they are "ready for life".
(4) fl is the fraction (= percentage) of such "ready for life" planets on which life actually starts and evolves (but not yet to the "intelligence" level).
(5) fi is the fraction (= percentage) of such "planets with life-forms" that actually evolve until some form of "intelligent civilization" emerges (like the first, historic human civilizations on Earth).
(6) fc is the fraction (= percentage) of such "planets with civilizations" where the civilizations evolve to the point of being able to communicate across interstellar distances with other (at least) similarly evolved civilizations. As far as we know in 2008, this means that they must be aware of the Maxwell equations governing radio waves, as well as of computers and radioastronomy (at least).
(7) fL is the fraction of galactic civilizations alive at the time when we, poor humans, attempt to pick up their radio signals (which they throw out into space just as we have done since 1900, when Marconi started transatlantic transmissions). In other words, fL is the number of civilizations now transmitting and receiving, and this implies an estimate of "how long will a technological civilization survive?" which nobody can make at the moment. Also, are they going to destroy

themselves in a nuclear war, and thus live only a few decades of technological civilization? Or are they slowly becoming wiser, reject war, speak a single language (like English today), and merge into a single "nation", thus living in peace for ages? Or will robots take over one day making "animals made of flesh" disappear for ever (the so-called "post-biological universe")?

No one knows ...

But let us go back to the Drake equation (1.B.1). In the 50 years of its existence, a number of suggestions have been put forward about the different numeric values of its seven factors. Of course, every different set of these seven input numbers yields a different value for N, and we can endlessly play that way. But we claim that these are like ... the play of children!

We claim the classical Drake equation (1.B.1), as we shall call it from now on to distinguish it from our statistical Drake equation to be introduced in the coming sections. Well, the classical Drake equation is scientifically inadequate in one regard at least: it just handles sheer numbers and does not associate an error bar to each of its seven factors. At the very least, we want to associate an error bar to each D_i.

Well, we have thus reached *Step 1* in our improvement of the classical Drake equation: replace each sheer number by a probability distribution!

The flow chart on the opposite page is provided as a guide to this paper.

1.B.2 Step 1: Letting each factor become a random variable

In this paper we adopt the notations of the great book *Probability, Random Variables and Stochastic Processes* by Athanasios Papoulis (1921–2002), now re-published as Papoulis-Pillai [9]. The advantage of this notation is that it makes a neat distinction between probabilistic (or statistical: it's the same thing here) variables, always denoted by capitals, from non-probabilistic (or "deterministic") variables, always denoted by lowercase letters. Adopting the Papoulis notation is also a tribute to him by the author, who was a Fulbright Grantee in the United States with him at the Polytechnic Institute (now Polytechnic University) of New York in the years 1977–1979.

We thus introduce seven new (positive) random variables D_i ("D" from "Drake") defined as

$$\begin{cases} D_1 = Ns \\ D_2 = fp \\ D_3 = ne \\ D_4 = fl \\ D_5 = fi \\ D_6 = fc \\ D_7 = fL \end{cases} \tag{1.B.2}$$

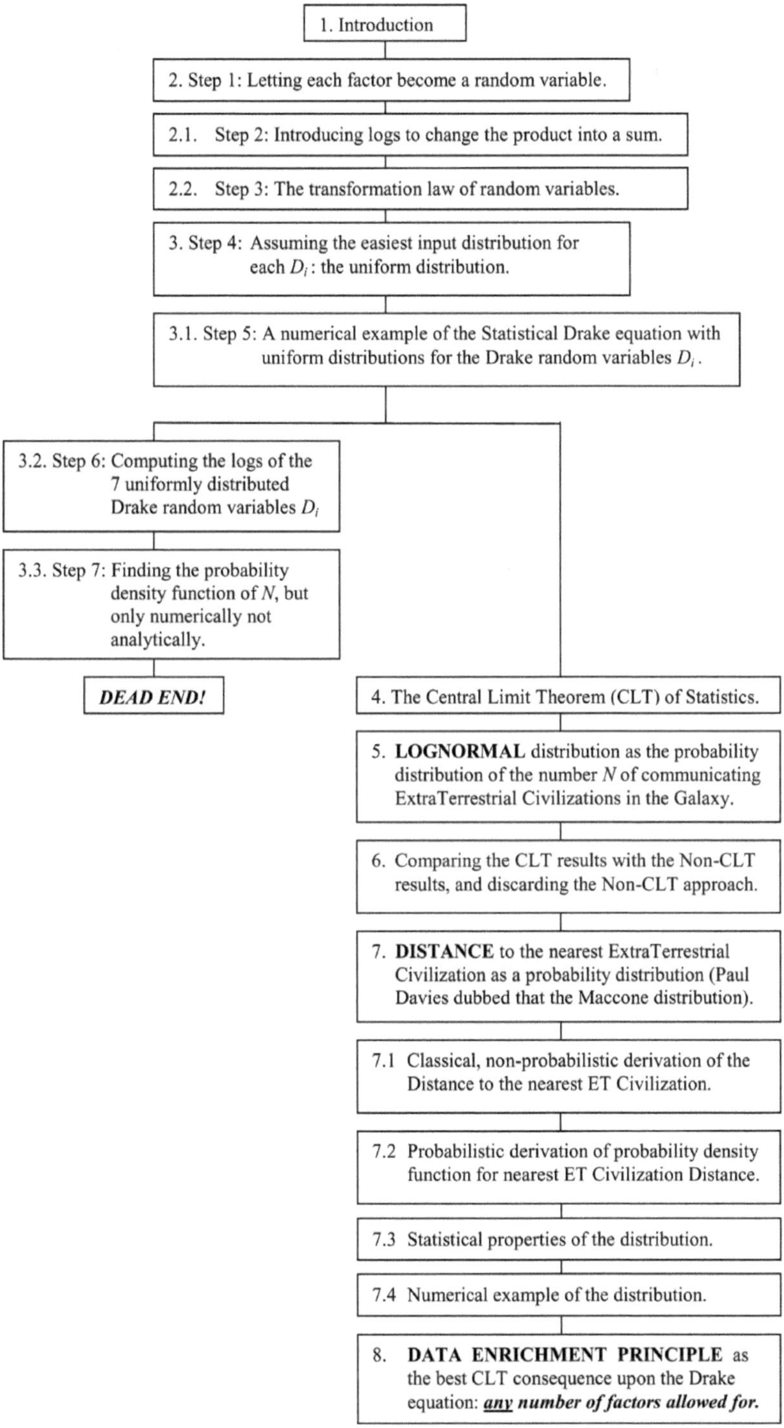
1. Introduction
2. Step 1: Letting each factor become a random variable.
2.1. Step 2: Introducing logs to change the product into a sum.
2.2. Step 3: The transformation law of random variables.
3. Step 4: Assuming the easiest input distribution for each D_i: the uniform distribution.
3.1. Step 5: A numerical example of the Statistical Drake equation with uniform distributions for the Drake random variables D_i.
3.2. Step 6: Computing the logs of the 7 uniformly distributed Drake random variables D_i
3.3. Step 7: Finding the probability density function of N, but only numerically not analytically.
DEAD END!
4. The Central Limit Theorem (CLT) of Statistics.
5. **LOGNORMAL** distribution as the probability distribution of the number N of communicating ExtraTerrestrial Civilizations in the Galaxy.
6. Comparing the CLT results with the Non-CLT results, and discarding the Non-CLT approach.
7. **DISTANCE** to the nearest ExtraTerrestrial Civilization as a probability distribution (Paul Davies dubbed that the Maccone distribution).
7.1 Classical, non-probabilistic derivation of the Distance to the nearest ET Civilization.
7.2 Probabilistic derivation of probability density function for nearest ET Civilization Distance.
7.3 Statistical properties of the distribution.
7.4 Numerical example of the distribution.
8. **DATA ENRICHMENT PRINCIPLE** as the best CLT consequence upon the Drake equation: ***any number of factors allowed for.***

so that our *statistical* Drake equation may be simply rewritten as

$$N = \prod_{i=1}^{7} D_i. \tag{1.B.3}$$

Of course, N now becomes a (positive) random variable too, having its own (positive) mean value and standard deviation. Just as each of the D_i has its own (positive) mean value and standard deviation ...

... the natural question then arises: How are the seven mean values on the right related to the mean value on the left?

... and How are the seven standard deviations on the right related to the standard deviation on the left? Just take the next step ...

Step 2: Introducing logs to change the product into a sum

Products of random variables are not easy to handle in probability theory. It is actually much easier to handle sums of random variables, rather than products, because:

(1) The probability density of the sum of two or more independent random variables is the convolution of the relevant probability densities (don't worry about the equations right now).
(2) The Fourier transform of the convolution is simply the product of the Fourier transforms (again, don't worry about the equations at this point)

So, let us take the natural logs of both sides of the statistical Drake equation (1.B.3) and change it into a sum:

$$\ln(N) = \ln\left(\prod_{i=1}^{7} D_i\right) = \sum_{i=1}^{7} \ln(D_i). \tag{1.B.4}$$

It is now convenient to introduce eight new (positive) random variables defined as follows:

$$\begin{cases} Y = \ln(N) \\ Y_i = \ln(D_i) \quad i = 1, ..., 7. \end{cases} \tag{1.B.5}$$

Upon inversion, the first equation of (1.B.5) yields the important equation that will be used in the sequel

$$N = e^Y. \tag{1.B.6}$$

We are now ready to take *Step 3*.

Step 3: The transformation law of random variables

So far we have not mentioned the problem: "which probability distribution shall we attach to each of the seven (positive) random variables D_i?"

It is not easy to answer this question because we do not have the slightest scientific clue as to which probability distributions fit best to each of the seven points listed in Section 1.B.1.

Yet, at least one trivial error must be avoided: claiming that each of those seven random variables must have a Gaussian (i.e., normal) distribution. In fact, the Gaussian distribution, having the well-known bell-shaped probability density function

$$f_X(x;\mu,\sigma) = \frac{1}{\sqrt{2\pi}\sigma} \cdot e^{-\frac{(x-\mu)^2}{2\,\sigma^2}} \quad (\sigma \geq 0) \tag{1.B.7}$$

has its independent variable y ranging between $-\infty$ and ∞ and so it can apply to a real random variable Y only, and never to positive random variables like those in the statistical Drake equation (1.B.3). Period.

Searching again for probability density functions that represent positive random variables, an obvious choice would be the gamma distributions (see [10]). However, we discarded this choice too for a different reason: keep in mind that, according to (1.B.5), once we have selected a particular type of probability density function (pdf) for the last seven equations (1.B.5), then we must compute the (new and different) pdf of the logs of such random variables. Moreover, the pdf of these logs is certainly no longer gamma type.

It is timely to remind the reader of a certain theorem that is proved in probability courses, but, unfortunately, does not seem to have a specific name. This is the transformation law (as we shall call it, see [9]) which allows us to compute the pdf of a certain new random variable Y that is a known function $Y = g(X)$ of another random variable X having a known pdf. In other words, if the pdf $f_X(x)$ of a certain random variable X is known, then the pdf $f_Y(y)$ of the new random variable Y, related to X by the functional relationship

$$Y = g(X) \tag{1.B.8}$$

can be calculated according to this rule:

(1) First invert the corresponding non-probabilistic equation $y = g(x)$ and denote by $x_i(y)$ the various real roots resulting from the inversion.
(2) Second, note whether these real roots may be either finitely or infinitely many, according to the nature of the function $y = g(x)$.
(3) Third, the probability density function of Y is then given by the (finite or infinite) sum

$$f_Y(y) = \sum_i \frac{f_X(x_i(y))}{|g'(x_i(y))|} \tag{1.B.9}$$

where the summation extends to all roots $x_i(y)$ and $|g'(x_i(y))|$ is the absolute value of the first derivative of $g(x)$ where the i-th root $x_i(y)$ has been replaced instead of x.

Since we must use this transformation law to transfer from the D_i to the $Y_i = \ln(D_i)$, it is clear that we need to start from a D_i pdf that is as simple as possible. The gamma pdf does not respond to this need because the analytic expression of the transformed pdf is very complicated (or, at least, it looked so to the author in the first instance). Also, the gamma distribution has two free parameters in it, and this "complicates" its application to the various meanings of the Drake equation. In conclusion, we discarded gamma distributions and confined ourselves to the simpler uniform distribution instead, as shown in the next section.

1.B.3 Step 4: Assuming the easiest input distribution for each D_i: The uniform distribution

Let us now suppose that each of the seven D_i is distributed *uniformly* in the interval ranging from the lower limit $a_i \geq 0$ to the upper limit $b_i \geq a_i$.

This is the same as saying that the probability density function of each of the seven Drake random variables D_i has the equation

$$f_{\text{uniform\_D}_i}(x) = \frac{1}{b_i - a_i} \quad \text{with } 0 \leq a_i \leq x \leq b_i \tag{1.B.10}$$

as it follows at once from the normalization condition

$$\int_{a_i}^{b_i} f_{\text{uniform\_D}_i}(x)\, dx = 1. \tag{1.B.11}$$

Let us now consider the mean value of such uniform D_i defined by

$$\begin{aligned}\langle \text{uniform\_D}_i \rangle &= \int_{a_i}^{b_i} x\, f_{\text{uniform\_D}_i}(x)\, dx = \frac{1}{b_i - a_i} \int_{a_i}^{b_i} x\, dx \\ &= \frac{1}{b_i - a_i} \left[\frac{x^2}{2} \right]_{a_i}^{b_i} = \frac{b_i^2 - a_i^2}{2\,(b_i - a_i)} = \frac{a_i + b_i}{2}.\end{aligned}$$

In words (as is intuitively obvious): the mean value of the uniform distribution simply is the mean of the lower plus upper limit of the variable range

$$\langle \text{uniform\_D}_i \rangle = \frac{a_i + b_i}{2}. \tag{1.B.12}$$

In order to find the variance of the uniform distribution, we first need finding the second moment

$$\begin{aligned}\langle \text{uniform\_D}_i^2 \rangle &= \int_{a_i}^{b_i} x^2\, f_{\text{uniform\_D}_i}(x)\, dx = \frac{1}{b_i - a_i} \int_{a_i}^{b_i} x^2\, dx = \frac{1}{b_i - a_i} \left[\frac{x^3}{3} \right]_{a_i}^{b_i} \\ &= \frac{b_i^3 - a_i^3}{3\,(b_i - a_i)} = \frac{(b_i - a_i)\,(a_i^2 + a_i b_i + b_i^2)}{3\,(b_i - a_i)} = \frac{a_i^2 + a_i b_i + b_i^2}{3}.\end{aligned}$$

The second moment of the uniform distribution is thus

$$\left\langle \mathrm{uniform\_D_i^2} \right\rangle = \frac{a_i^2 + a_i b_i + b_i^2}{3}. \tag{1.B.13}$$

From (1.B.12) and (1.B.13) we may now derive the variance of the uniform distribution

$$\sigma^2_{\mathrm{uniform\_D_i}} = \left\langle \mathrm{uniform\_D_i^2} \right\rangle - \left\langle \mathrm{uniform\_D_i} \right\rangle^2 = \frac{a_i^2 + a_i b_i + b_i^2}{3} - \frac{(a_i + b_i)^2}{4} = \frac{(b_i - a_i)^2}{12}. \tag{1.B.14}$$

Upon taking the square root of both sides of (1.B.14), we finally obtain the standard deviation of the uniform distribution:

$$\sigma_{\mathrm{uniform\_D_i}} = \frac{b_i - a_i}{2\sqrt{3}}. \tag{1.B.15}$$

We now wish to perform a calculation that is mathematically trivial, but rather unexpected from the intuitive point of view, and very important for our applications to the statistical Drake equation. Just consider the two simultaneous equations (1.B.12) and (1.B.15)

$$\begin{cases} \left\langle \mathrm{uniform\_D_i} \right\rangle = \dfrac{a_i + b_i}{2} \\ \sigma_{\mathrm{uniform\_D_i}} = \dfrac{b_i - a_i}{2\sqrt{3}}. \end{cases} \tag{1.B.16}$$

Upon inverting this trivial linear system, one finds

$$\begin{cases} a_i = \left\langle \mathrm{uniform\_D_i} \right\rangle - \sqrt{3}\,\sigma_{\mathrm{uniform\_D_i}} \\ b_i = \left\langle \mathrm{uniform\_D_i} \right\rangle + \sqrt{3}\,\sigma_{\mathrm{uniform\_D_i}}. \end{cases} \tag{1.B.17}$$

This is of paramount importance for our application of the statistical Drake equation inasmuch as it shows that: if one (scientifically) assigns the mean value and standard deviation of a certain Drake random variable D_i, then the lower and upper limits of the relevant uniform distribution are given by the two equations (1.B.17), respectively.

In other words, there is a factor of $\sqrt{3} = 1.732$ included in the two equations (1.B.17) that is not obvious at all to human intuition, and must indeed be taken into account.

The application of this result to the statistical Drake equation is discussed in the next section.

1.B.3.1 Step 5: A numerical example of the statistical Drake equation with uniform distributions for the Drake random variables D_i

The first variable *Ns* in the classical Drake equation (1.B.1) is the number of stars in our Galaxy. Nobody knows how many there are exactly. Only statistical estimates can be made by astronomers, and they oscillate around, say, a mean value of 350 billion (if this value is indeed correct!). This being the situation, we assume that our uniformly distributed random variable *Ns* has a mean value of 350 billion minus or plus a standard deviation of, say, one billion (we don't care whether this number is scientifically the best estimate as of August 2008: we just want to set up a numerical example of our statistical Drake equation). In other words, we now assume that one has:

$$\begin{cases} \langle \text{uniform\_D}_1 \rangle = 350 \cdot 10^9 \\ \sigma_{\text{uniform\_D}_1} = 1 \cdot 10^9. \end{cases} \tag{1.B.18}$$

Therefore, according to equations (1.B.17) the lower and upper limit of our uniform distribution for the random variable $Ns = D_1$ are, respectively

$$\begin{cases} a_{Ns} = \langle \text{uniform\_D}_1 \rangle - \sqrt{3}\,\sigma_{\text{uniform\_D}_1} = 348.3 \cdot 10^9 \\ b_{Ns} = \langle \text{uniform\_D}_1 \rangle + \sqrt{3}\,\sigma_{\text{uniform\_D}_1} = 351.7 \cdot 10^9. \end{cases} \tag{1.B.19}$$

We proceed similarly for all the other six random variables in the statistical Drake equation (1.B.3).

For instance, we assume that the fraction of stars that have planets is 50% (i.e., 50/100), and this will be the mean value of the random variable $fp = D_2$. We also assume that the relevant standard deviation is 10% (i.e., that $\sigma_{fp} = 10/100$). Therefore, the relevant lower and upper limits for the uniform distribution of $fp = D_2$ turn out to be

$$\begin{cases} a_{fp} = \langle \text{uniform\_D}_2 \rangle - \sqrt{3}\,\sigma_{\text{uniform\_D}_2} = 0.327 \\ b_{fp} = \langle \text{uniform\_D}_2 \rangle + \sqrt{3}\,\sigma_{\text{uniform\_D}_2} = 0.673. \end{cases} \tag{1.B.20}$$

The next Drake random variable is the number *ne* of "Earth-type" planets in a given star system. Taking the Solar System as an example, since only the Earth is truly "Earth-type", the mean value of *ne* is clearly 1, but the standard deviation is not zero if we assume that Mars may also be regarded as Earth type. Since there are thus two Earth-type planets in the Solar System, we must assume a standard deviation of $1/\sqrt{3} = 0.577$ to compensate the $\sqrt{3}$ appearing in (1.B.17) in order to finally yield two "Earth-type" planets (Earth and Mars) for the upper limit of the random variable *ne*. In other words, we assume that

$$\begin{cases} a_{ne} = \langle \text{uniform\_D}_3 \rangle - \sqrt{3}\,\sigma_{\text{uniform\_D}_3} = 0 \\ b_{ne} = \langle \text{uniform\_D}_3 \rangle + \sqrt{3}\,\sigma_{\text{uniform\_D}_3} = 2. \end{cases} \tag{1.B.21}$$

The next four Drake random variables have even more "arbitrarily" assumed values that we simply assume for the sake of making up a numerical example of our statistical Drake equation with uniform entry distributions. So, we really make no

Table 1.B.1. Input values (i.e., mean values and standard deviations) for the seven Drake uniform random variables D_i. The first column lists the seven input sheer numbers that also become the mean values (middle column). Finally, the last column lists the seven input standard deviations. The bottom line is the classical Drake equation (1.B.1).

$Ns := 350 \cdot 10^9$	$\mu Ns := Ns$	$\sigma Ns := 1 \cdot 10^9$
$fp := \frac{50}{100}$	$\mu fp := fp$	$\sigma fp := \frac{10}{100}$
$ne := 1$	$\mu ne := ne$	$\sigma ne := \frac{1}{\sqrt{3}}$
$fl := \frac{50}{100}$	$\mu fl := fl$	$\sigma fl := \frac{10}{100}$
$fi := \frac{20}{100}$	$\mu fi := fi$	$\sigma fi := \frac{10}{100}$
$fc := \frac{20}{100}$	$\mu fc := fc$	$\sigma fc := \frac{10}{100}$
$fL := \frac{10{,}000}{10^{10}}$	$\mu fL := fL$	$\sigma fL := \frac{1{,}000}{10^{10}}$
$N := Ns \cdot fp \cdot ne \cdot fl \cdot fi \cdot fc \cdot fL$		$N = 3{,}500$

assumption about the astronomy, the biology, or the sociology of the Drake equation: we just care about its mathematics.

All our assumed entries are given in Table 1.B.1.

Note that, had we assumed all the standard deviations to equal *zero* in Table 1.B.1, then our statistical Drake equation (1.B.3) would have obviously reduced to the classical Drake equation (1.B.1), and the resulting number of civilizations in the Galaxy would have turned out to be 3,500:

$$N = 3{,}500. \tag{1.B.22}$$

This is the important deterministic number that we will use in the sequel of this paper for comparison with our statistical results on the mean value of N (i.e. $\langle N \rangle$). This will be explained in Sections 1.B.3.3 and 1.B.5.

1.B.3.2 Step 6: Computing the logs of the seven uniformly distributed Drake random variables D_i

Intuitively speaking, the natural log of a uniformly distributed random variable may not be another uniformly distributed random variable! This is obvious from the trivial diagram of $y = \ln(x)$ shown in Figure 1.B.1.

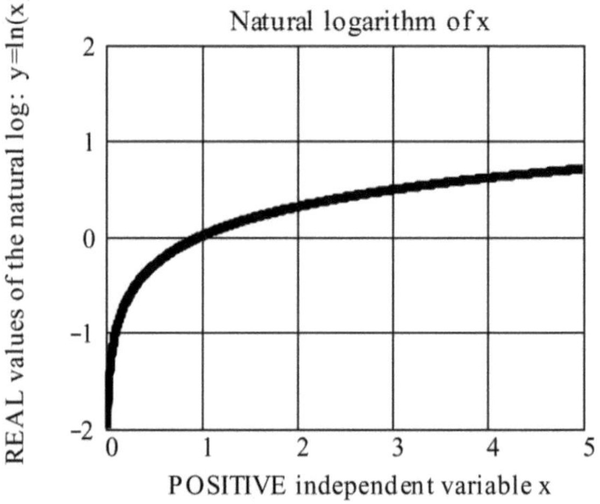

Figure 1.B.1. The simple function $y = \ln(x)$.

So, if we have a uniformly distributed random variable D_i with lower limit a_i and upper limit b_i, the random variable

$$Y_i = \ln(D_i) \quad i = 1, ..., 7 \tag{1.B.23}$$

must have its range limited in between the lower limit $\ln(a_i)$ and the upper limit $\ln(b_i)$. In other words, these are the lower and upper limits of the relevant probability density function $f_{Y_i}(y)$. But what is the actual analytic expression of such a pdf? To find it, we must resort to the general transformation law for random variables, defined by equation (1.B.9). Here we obviously have

$$y = g(x) = \ln(x) \tag{1.B.24}$$

that, upon inversion, yields the single root

$$x_1(y) = x(y) = e^y. \tag{1.B.25}$$

On the other hand, differentiating (1.B.24) one gets

$$g'(x) = \frac{1}{x} \quad \text{and} \quad g'(x_1(y)) = \frac{1}{x_1(y)} = \frac{1}{e^y} \tag{1.B.26}$$

where (1.B.25) has already been used in the last step. By virtue of the uniform probability density function (1.B.10) and of (1.B.26), the general transformation law (1.B.9) finally yields

$$f_Y(y) = \sum_i \frac{f_X(x_i(y))}{|g'(x_i(y))|} = \frac{1}{b_i - a_i} \cdot \frac{1}{\left|\frac{1}{e^y}\right|} = \frac{e^y}{b_i - a_i}. \tag{1.B.27}$$

In other words, the requested pdf of Y_i is

$$f_{Y_i}(y) = \frac{e^y}{b_i - a_i} \quad i = 1, \ldots, 7 \quad \ln(a_i) \le y \le \ln(b_i). \tag{1.B.28}$$

These are the probability density functions of the natural logs of all the uniformly distributed Drake random variables D_i.

This is indeed a positive function of y over the interval $\ln(a_i) \le y \le \ln(b_i)$, as for every pdf, and it is easy to see that its normalization condition is fulfilled:

$$\int_{\ln(a_i)}^{\ln(b)} f_{Y_i}(y)\, dy = \int_{\ln(a_i)}^{\ln(b)} \frac{e^y}{b_i - a_i}\, dy = \frac{e^{\ln(b_i)} - e^{\ln(a_i)}}{b_i - a_i} = 1. \tag{1.B.29}$$

Next we want to find the mean value and standard deviation of Y_i, since these play a crucial role for future developments. The mean value $\langle Y_i \rangle$ is given by

$$\begin{aligned} \langle Y_i \rangle &= \int_{\ln(a_i)}^{\ln(b_i)} y \cdot f_{Y_i}(y)\, dy = \int_{\ln(a_i)}^{\ln(b_i)} \frac{y \cdot e^y}{b_i - a_i}\, dy \\ &= \frac{b_i[\ln(b_i) - 1] - a_i[\ln(a_i) - 1]}{b_i - a_i}. \end{aligned} \tag{1.B.30}$$

This is thus the mean value of the natural log of all the uniformly distributed Drake random variables D_i

$$\langle Y_i \rangle = \langle \ln(D_i) \rangle = \frac{b_i[\ln(b_i) - 1] - a_i[\ln(a_i) - 1]}{b_i - a_i}. \tag{1.B.31}$$

In order to find the variance also, we must first compute the mean value of the square of Y_i, that is

$$\begin{aligned} \langle Y_i^2 \rangle &= \int_{\ln(a_i)}^{\ln(b_i)} y^2 \cdot f_{Y_i}(y)\, dy = \int_{\ln(a_i)}^{\ln(b_i)} \frac{y^2 \cdot e^y}{b_i - a_i}\, dy \\ &= \frac{b_i\left[\ln^2(b_i) - 2\ln(b_i) + 2\right] - a_i\left[\ln^2(a_i) - 2\ln(a_i) + 2\right]}{b_i - a_i}. \end{aligned} \tag{1.B.32}$$

The variance of $Y_i = \ln(D_i)$ is now given by (1.B.32) minus the square of (1.B.31), which, after a few reductions, yield:

$$\sigma_{Y_i}^2 = \sigma_{\ln(D_i)}^2 = 1 - \frac{a_i b_i[\ln(b_i) - \ln(a_i)]^2}{(b_i - a_i)^2}. \tag{1.B.33}$$

Whence the corresponding standard deviation

$$\sigma_{Y_i} = \sigma_{\ln(D_i)} = \sqrt{1 - \frac{a_i b_i[\ln(b_i) - \ln(a_i)]^2}{(b_i - a_i)^2}}. \tag{1.B.34}$$

Let us now turn to another topic: the use of Fourier transforms, which in probability theory are called "characteristic functions". Following again the notation of Papoulis [9] we call the "characteristic function" $\Phi_{Y_i}(\zeta)$ of an assigned probability distribution Y_i the Fourier transform of the relevant probability density function; that is (with $j = \sqrt{-1}$)

$$\Phi_{Y_i}(\zeta) = \int_{-\infty}^{\infty} e^{j\zeta y} f_{Y_i}(y)\, dy. \tag{1.B.35}$$

The use of characteristic functions simplifies things greatly. For instance, the calculation of all moments of a known pdf becomes trivial if the relevant characteristic function is known. The proofs of important theorems of statistics, like the Central Limit Theorem that we will use in Section 1.B.4, are greatly simplified as well. Another important result is that the characteristic function of the sum of a finite number of independent random variables is simply given by the product of the corresponding characteristic functions. This is just the case we are facing in the statistical Drake equation (1.B.3) and so we are now led to find the characteristic function of the random variable Y_i; that is

$$\begin{aligned}\Phi_{Y_i}(\zeta) &= \int_{-\infty}^{\infty} e^{j\zeta y} f_{Y_i}(y)\, dy = \int_{\ln(a_i)}^{\ln(b_i)} e^{j\zeta y} \frac{e^y}{b_i - a_i}\, dy \\ &= \frac{1}{b_i - a_i} \int_{\ln(a_i)}^{\ln(b_i)} e^{(1+j\zeta)y}\, dy = \frac{1}{b_i - a_i} \cdot \frac{1}{1 + i\zeta} \left[e^{(1+j\zeta)y} \right]_{\ln(a_i)}^{\ln(b_i)} \\ &= \frac{e^{(1+j\zeta)\ln(b_i)} - e^{(1+j\zeta)\ln(a_i)}}{(b_i - a_i)\,(1 + j\zeta)} = \frac{b_i^{1+j\zeta} - a_i^{1+j\zeta}}{(b_i - a_i)\,(1 + j\zeta)}.\end{aligned} \tag{1.B.36}$$

Thus, the characteristic function of the natural log of the Drake uniform random variable D_i is given by

$$\Phi_{Y_i}(\zeta) = \frac{b_i^{1+j\zeta} - a_i^{1+j\zeta}}{(b_i - a_i)\,(1 + j\zeta)}. \tag{1.B.37}$$

1.B.3.3. Step 7: Finding the probability density function of N, but only numerically not analytically

Having found the characteristic functions $\Phi_{Y_i}(\zeta)$ of the logs of the seven input random variables D_i, we can now immediately find the characteristic function of the random variable $Y = \ln(N)$ defined by (1.B.5). In fact, by virtue of (1.B.4), the well-known Fourier transform property stating that "the Fourier transform of a convolution is the product of the Fourier transforms", and of (1.B.37), it immediately follows that $\Phi_Y(\zeta)$ equals the product of the seven $\Phi_{Y_i}(\zeta)$:

$$\Phi_Y(\zeta) = \prod_{i=1}^{7} \Phi_{Y_i}(\zeta) = \prod_{i=1}^{7} \frac{b_i^{1+j\zeta} - a_i^{1+j\zeta}}{(b_i - a_i)\,(1 + j\zeta)}. \tag{1.B.38}$$

The next step is to invert this Fourier transform in order to get the probability density function of the random variable $Y = \ln(N)$. In other words, we must compute the following inverse Fourier transform

$$f_Y(y) = \frac{1}{2\pi}\int_{-\infty}^{\infty} e^{-j\zeta y}\Phi_Y(\zeta)\,d\zeta = \frac{1}{2\pi}\int_{-\infty}^{\infty} e^{-j\zeta y}\left[\prod_{i=1}^{7}\Phi_{Y_i}(\zeta)\right]d\zeta$$

$$= \frac{1}{2\pi}\int_{-\infty}^{\infty} e^{-j\zeta y}\left[\prod_{i=1}^{7}\frac{b_i^{1+j\zeta} - a_i^{1+j\zeta}}{(b_i - a_i)\,(1 + j\zeta)}\right]d\zeta. \tag{1.B.39}$$

The author regrets being unable to compute the last integral *analytically*. He had to compute it *numerically* for the particular values of the 14 a_i and b_i that follow from Table 1.B.1 and equations (1.B.17). The result was the probability density function for $Y = \ln(N)$ plotted in Figure 1.B.2.

We are now just one step away from finding the probability density of N, the number of extraterrestrial civilizations in the Galaxy predicted by our statistical Drake equation (1.B.3). The point here is to transfer from the probability density function of Y to that of N, knowing that $Y = \ln(N)$, or alternatively, that $N = \exp(Y)$, as stated by (1.B.6). We must thus resort to the transformation law of random variables (1.B.9) by setting

$$y = g(x) = e^x. \tag{1.B.40}$$

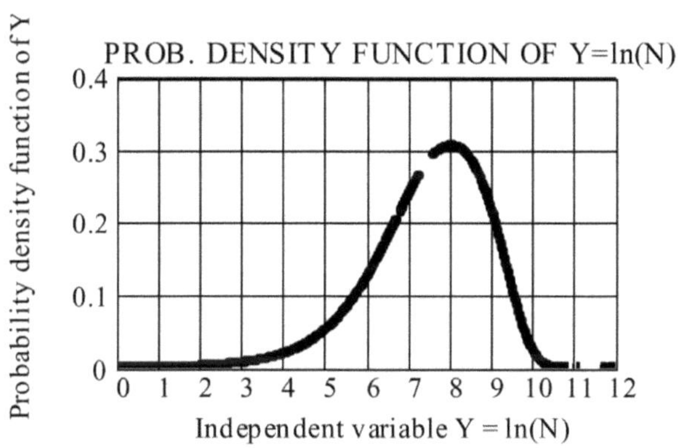

Figure 1.B.2. Probability density function of $Y = \ln(N)$ computed numerically by virtue of the integral (1.B.39). The two "funny gaps" in the curve are due to numeric limitations in the MathCad numeric solver that the author used for this numeric computation.

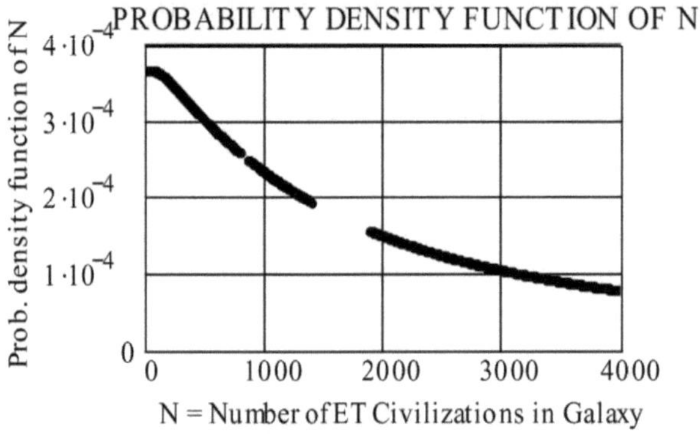

Figure 1.B.3. The numeric (and not analytic) probability density function curve $f_N(y)$ of the number N of extraterrestrial civilizations in the Galaxy according to the statistical Drake equation (1.B.3). We see that the curve peak (i.e., the mode) is very close to low values of N, but the tail on the right is high, meaning that the resulting mean value $\langle N \rangle$ is of the order of thousands.

This, upon inversion, yields the single root

$$x_1(y) = x(y) = \ln(y). \qquad (1.B.41)$$

On the other hand, differentiating (1.B.40) one gets

$$g'(x) = e^x \quad \text{and} \quad g'(x_1(y)) = e^{\ln(y)} = y \qquad (1.B.42)$$

where (1.B.41) has already been used in the last step. The general transformation law (1.B.9) finally yields

$$f_N(y) = \sum_i \frac{f_X(x_i(y))}{|g'(x_i(y))|} = \frac{1}{|y|} f_Y(\ln(y)). \qquad (1.B.43)$$

This probability density function $f_N(y)$ was computed numerically by using (1.B.43) and the numeric curve given by (1.B.39), and the result is shown in Figure 1.B.3.

We now want to compute the mean value $\langle N \rangle$ of the probability density (1.B.43). Clearly, it is given by

$$\langle N \rangle = \int_0^\infty y\, f_N(y)\, dy. \qquad (1.B.44)$$

This integral too was computed numerically, and the result was a perfect match with $N = 3{,}500$ of (1.B.22); that is

$$\langle N \rangle = 3{,}499.99880177509 + 0.000001249146 86\text{i}. \qquad (1.B.45)$$

Note that this result was computed numerically in the complex domain because of the Fourier transforms, and that the real part is virtually 3,500 (as expected) while the imaginary part is virtually zero because of rounding errors. So, this result is excellent, and proves that the theory presented so far is mathematically correct.

Finally, we want to consider the standard deviation. This also had to be computed numerically, resulting in

$$\sigma_N = 3{,}953.42910143389 + 0.000000032800058i. \tag{1.B.46}$$

This standard deviation, higher than the mean value, implies that N might range in between 0 and 7,453.

This completes our study of the probability density function of N if the seven uniform Drake input random variables D_i have the mean values and standard deviations listed in Table 1.B.1.

We conclude that, unfortunately, even under the simplifying assumptions that the D_i be uniformly distributed, it is impossible to solve the full problem analytically, since all calculations beyond equation (1.B.38) had to be performed numerically.

This is no good.

Shall we thus lose faith, and declare "impossible" the task of finding an analytic expression for the probability density function $f_N(y)$?

Rather surprisingly, the answer is "no", and there is indeed a way out of this dead-end, as we shall see in the next section.

1.B.4 The Central Limit Theorem (CLT) of statistics

Indeed, there is a good, approximating analytical expression for $f_N(y)$, and this is the following lognormal probability density function

$$f_N(y, \mu, \sigma) = \frac{1}{y} \cdot \frac{1}{\sqrt{2\pi}\sigma} e^{-\frac{(\ln(y)-\mu)^2}{2\sigma^2}} \quad (y \geq 0). \tag{1.B.47}$$

To understand why, we must resort to what is perhaps the most beautiful theorem of statistics: the Central Limit Theorem (abbreviated CLT). Historically, the CLT was in fact first proven in 1901 by the Russian mathematician Alexandr Lyapunov (1857–1918), and later (1920) by the Finnish mathematician Jarl Waldemar Lindeberg (1876–1932) under weaker conditions. These conditions are certainly fulfilled in the context of the Drake equation because of the "reality" of the astronomy, biology, and sociology involved with it, and we are not going to discuss this point any further here. A good, synthetic description of the Central Limit Theorem (CLT) of statistics is found at the Wikipedia site [11] to which the reader is referred for more details, such as the equations for the Lyapunov and the Lindeberg conditions, making the theorem "rigorously" valid.

Put loosely, the CLT states that, if one has a sum of random variables even *not* identically distributed, this sum tends to a normal distribution when the number of terms making up the sum tends to infinity. Also, the normal distribution mean value is the sum of the mean values of the added random variables, and the normal distribution variance is the sum of the variances of the added random variables.

Let us now write down the equations of the CLT in the form needed to apply it to our statistical Drake equation (1.B.3). The idea is to apply the CLT to the sum of random variables given by (1.B.4) and (1.B.5) whatever their probability distributions. In other words, the CLT applied to the statistical Drake equation (1.B.3) leads immediately to the following three equations:

(1) The sum of the (arbitrarily distributed) independent random variables Y_i makes up the new random variable Y.
(2) The sum of their mean values makes up the new mean value of Y.
(3) The sum of their variances makes up the new variance of Y.

In equations:

$$\begin{cases} Y = \sum_{i=1}^{7} Y_i \\ \langle Y \rangle = \sum_{i=1}^{7} \langle Y_i \rangle \\ \sigma_Y^2 = \sum_{i=1}^{7} \sigma_{Y_i}^2. \end{cases} \tag{1.B.48}$$

This completes our synthetic description of the CLT for sums of random variables.

1.B.5 *The lognormal distribution is the distribution of the number N of extraterrestrial civilizations in the Galaxy*

The CLT may of course be extended to products of random variables upon taking the logs of both sides, just as we did in equation (1.B.3). It then follows that the exponent random variable, like Y in (1.B.6), tends to a normal random variable, and, as a consequence, it follows that the base random variable, like N in (1.B.6), tends to a lognormal random variable.

To understand this fact better in mathematical terms consider again the transformation law (1.B.9) of random variables. The question is: What is the probability density function of the random variable N in equation (1.B.6)? That is: What is the probability density function of the lognormal distribution? To find it, set

$$y = g(x) = e^x. \tag{1.B.49}$$

This, upon inversion, yields the single root

$$x_1(y) = x(y) = \ln(y). \tag{1.B.50}$$

On the other hand, differentiating (1.B.49) one gets

$$g'(x) = e^x \quad \text{and} \quad g'(x_1(y)) = e^{\ln(y)} = y \tag{1.B.51}$$

where (1.B.50) has already been used in the last step. The general transformation law (1.B.9) finally yields

$$f_N(y) = \sum_i \frac{f_X(x_i(y))}{|g'(x_i(y))|} = \frac{1}{|y|} f_Y(\ln(y)). \tag{1.B.52}$$

Therefore, replacing the probability density on the right by virtue of the well-known normal (or Gaussian) distribution given by equation (1.B.7), the lognormal distribution of equation (1.B.47) is found, and the derivation of the lognormal distribution from the normal distribution is proved.

In view of future calculations, it is also useful to point out the so-called "Gaussian integral"; that is:

$$\int_{-\infty}^{\infty} e^{-A\cdot x^2} e^{B\cdot x}\, dx = \sqrt{\frac{\pi}{A}} \cdot e^{\frac{B^2}{4A}}, \quad A > 0, \quad B = \text{real}. \tag{1.B.53}$$

This follows immediately from the normalization condition of the Gaussian (1.B.7); that is

$$\int_{-\infty}^{\infty} \frac{1}{\sqrt{2\pi}\sigma} e^{-\frac{(x-\mu)^2}{2\sigma^2}} dx = 1, \tag{1.B.54}$$

just upon expanding the square at the exponent and making the two replacements (we skip all steps)

$$\begin{cases} A = \dfrac{1}{2\,\sigma^2} > 0\,, \\ B = \dfrac{\mu}{\sigma^2} = \text{real}. \end{cases} \tag{1.B.55}$$

In the sequel of this paper we shall denote the independent variable of the lognormal distribution (1.B.47) by a lowercase letter n to remind the reader that the corresponding random variable N is the positive integer number of extraterrestrial civilizations in the Galaxy. In other words, n will be treated as a positive real number in all calculations to follow because it is a "large" number (i.e., a continuous variable) compared with the only civilization that we know of (i.e., our own). In conclusion, from now on the lognormal probability density function of N will be written as

$$f_N(n) = \frac{1}{n} \cdot \frac{1}{\sqrt{2\pi}\sigma} e^{-\frac{(\ln(n)-\mu)^2}{2\sigma^2}} \quad (n \geq 0). \tag{1.B.56}$$

That said, we now turn to the statistical properties of the lognormal distribution (1.B.55); that is, to the statistical properties that describe the number N of extraterrestrial civilizations in the Galaxy.

Our first goal is to prove an equation yielding all the moments of the lognormal distribution (1.B.56); that is, for every non-negative integer $k = 0, 1, 2, \ldots$ one has

$$\left\langle N^k \right\rangle = e^{k\mu}\, e^{k^2 \cdot \frac{\sigma^2}{2}}. \tag{1.B.57}$$

The relevant proof starts with the definition of the k-th moment

$$\left\langle N^k \right\rangle = \int_0^\infty n^k \cdot f_N(n)\, dn = \int_0^\infty n^k \cdot \frac{1}{n} \cdot \frac{1}{\sqrt{2\pi}\,\sigma} \cdot e^{-\frac{(\ln[n]-\mu)^2}{2\sigma^2}}\, dn$$

One then transforms the above integral by virtue of the substitution

$$\ln[n] = z. \tag{1.B.58}$$

The new integral in z is then seen to reduce to the Gaussian integral (1.B.53) (we skip all steps here) and (1.B.57) follows

$$= e^{k\mu}\, e^{k^2 \cdot \frac{\sigma^2}{2}}.$$

Upon inserting $k = 0$ into (1.B.56), the normalization condition for $f_N(n)$ follows

$$\int_0^\infty f_N(n)\, dn = 1. \tag{1.B.59}$$

Upon inserting $k = 1$ into (1.B.56), the important mean value of the random variable N is found

$$\langle N \rangle = e^{\mu}\, e^{\frac{\sigma^2}{2}}. \tag{1.B.60}$$

Upon inserting $k = 2$ into (1.B.56), the mean value of the square of the random variable N is found

$$\left\langle N^2 \right\rangle = e^{2\mu}\, e^{2\sigma^2}. \tag{1.B.61}$$

The variance of N now follows from the last two formulas:

$$\sigma_N^2 = e^{2\mu}\, e^{\sigma^2}\left(e^{\sigma^2} - 1\right). \tag{1.B.62}$$

The square root of this is the important standard deviation formula for the N random variable

$$\sigma_N = e^{\mu}\, e^{\frac{\sigma^2}{2}} \sqrt{e^{\sigma^2} - 1}. \tag{1.B.63}$$

The third moment is obtained upon inserting $k = 3$ into (1.B.56)

$$\left\langle N^3 \right\rangle = e^{3\mu}\, e^{\frac{9}{2}\sigma^2}. \tag{1.B.64}$$

Finally, upon inserting $k = 4$, the fourth moment of N is found

$$\left\langle N^4 \right\rangle = e^{4\mu}\, e^{8\sigma^2}. \tag{1.B.65}$$

Our next goal is to find the cumulants of N. In principle, we could compute all the cumulants K_i from the generic i-th moment μ_i by virtue of the recursion formula (see [12])

$$K_i = \mu'_i - \sum_{k=1}^{i-1} \binom{i-1}{k-1} K_k \, \mu'_{n-k}. \tag{1.B.66}$$

In practice, however, here we shall confine ourselves to the computation of the first four cumulants simply because only they are required to find the skewness and kurtosis of the distribution. Then, the first four cumulants in terms of the first four moments read:

$$\begin{cases} K_1 = \mu'_1 \\ K_2 = \mu'_2 - K_1^2 \\ K_3 = \mu'_3 - 3\,K_1\,K_2 - K_1^3 \\ K_4 = \mu'_4 - 4\,K_1\,K_3 - 3\,K_2^2 - 6\,K_2\,K_1^2 - K_1^4. \end{cases} \tag{1.B.67}$$

These equations yield, respectively:

$$K_1 = e^{\mu}\, e^{\frac{\sigma^2}{2}} \tag{1.B.68}$$

$$K_2 = e^{2\mu}\, e^{\sigma^2}\left(e^{\sigma^2} - 1\right) \tag{1.B.69}$$

$$K_3 = e^{3\mu}\, e^{\frac{9}{2}\sigma^2} \tag{1.B.70}$$

$$K_4 = e^{4\mu+2\sigma^2}\left(e^{\sigma^2} - 1\right)^3 \left(e^{3\,\sigma^2} + 3\,e^{2\sigma^2} + 6\;e^{\sigma^2} + 6\right). \tag{1.B.71}$$

From these we derive the skewness

$$\frac{K_3}{(K_2)^{\frac{3}{2}}} = \left(e^{\sigma^2} + 2\right)\sqrt{e^{\sigma^2} \quad 1}, \tag{1.B.72}$$

and the kurtosis

$$\frac{K_4}{(K_2)^2} = \; e^{4\,\sigma^2} + 2\;e^{3\,\sigma^2} + 3\;e^{2\,\sigma^2} - 6. \tag{1.B.73}$$

Finally, we want to find the mode of the lognormal probability density function (i.e., the abscissa of its peak). To do so, we must first compute the derivative of the probability density function $f_N(n)$ of equation (1.B.56), and then set it equal to zero. This derivative is actually the derivative of the ratio of two functions of n, as it plainly appears from (1.B.57). Thus, let us set for a moment

$$E(n) = \frac{(\ln[n] - \mu)^2}{2\,\sigma^2} \tag{1.B.74}$$

where "E" stands for "exponent". Upon differentiating this, one gets

$$E'(n) = \frac{1}{2\,\sigma^2} \cdot 2\,(\ln[n] - \mu) \cdot \frac{1}{n}. \tag{1.B.75}$$

But the lognormal probability density function (1.B.56), by virtue of (1.B.74), now reads

$$f_N(n) = \frac{1}{\sqrt{2\pi}\sigma} \cdot \frac{e^{-E(n)}}{n}. \tag{1.B.76}$$

So that its derivative is

$$\frac{df_{\text{ET\_Distance}}{}'(r)}{dr} = \frac{1}{\sqrt{2\pi}\sigma} \cdot \frac{-e^{-E(n)}E'(n) \cdot n - 1 \cdot e^{-E(n)}}{n^2}$$

$$= \frac{1}{\sqrt{2\pi}\sigma} \cdot \frac{-e^{-E(n)}\left[E'(n) \cdot n + 1\right]}{n^2}. \tag{1.B.77}$$

Setting this derivative equal to zero means setting

$$E'(n) \cdot n + 1 = 0. \tag{1.B.78}$$

That is, upon replacing (1.B.75),

$$\frac{1}{\sigma^2} \cdot (\ln[n] - \mu) + 1 = 0. \tag{1.B.79}$$

Rearranging, this becomes

$$\ln[n] - \mu + \sigma^2 = 0 \tag{1.B.80}$$

and finally

$$n_{\text{mode}} \equiv n_{\text{peak}} = e^{\mu}\, e^{-\sigma^2}. \tag{1.B.81}$$

This is the most likely number of extraterrestrial civilizations in the Galaxy.

How likely? To find the value of the probability density function $f_N(n)$ corresponding to this value of the mode, we must obviously insert (1.B.81) into (1.B.56). After a few rearrangements, one then gets

$$f_N(n_{\text{mode}}) = \frac{1}{\sqrt{2\pi}\,\sigma} \cdot e^{-\mu} \cdot e^{\frac{\sigma^2}{2}}. \tag{1.B.82}$$

This is the most likely number of extraterrestrial civilizations in the Galaxy (i.e., it is the peak height in the lognormal probability density function $f_N(n)$).

Next to the mode, the median m [13] is one more statistical number used to characterize any probability distribution. It is defined as the independent variable abscissa m such that a realization of the random variable will take up a value lower than m with 50% probability or a value higher than m with again 50% probability. In other words, the median m splits up our probability density into exactly two equally probable parts. Since the probability of occurrence of the random event equals the area under its density curve (i.e., the definite integral under its density curve) then the median m (of the lognormal distribution, in this case) is defined as the integral upper limit m:

$$\int_0^m f_N(n)\,dn = \int_0^m \frac{1}{n}\cdot\frac{1}{\sqrt{2\pi}\sigma}\,e^{-\frac{(\ln(n)-\mu)^2}{2\sigma^2}} = \frac{1}{2}. \tag{1.B.83}$$

In order to find m, we may *not* differentiate (1.B.83) with respect to m, since the "precise" factor $\frac{1}{2}$ on the right would then disappear into a zero. On the contrary, we may try to perform the obvious substitution

$$z^2 = \frac{(\ln(n)-\mu)^2}{2\sigma^2} \qquad z \geq 0 \tag{1.B.84}$$

into the integral (1.B.83) to reduce it to the following integral defining the error function erf(z)

$$\operatorname{erf}(x) = \frac{2}{\sqrt{\pi}}\int_0^x e^{-z^2}\,dz. \tag{1.B.85}$$

Then, after a few reductions, which we skip for the sake of brevity, the full equation (1.B.83) is turned into

$$\frac{1}{2} + \operatorname{erf}\left(\frac{\ln(m)-\mu}{\sqrt{2}\sigma}\right) = \frac{1}{2} \tag{1.B.86}$$

that is

$$\operatorname{erf}\left(\frac{\ln(m)-\mu}{\sqrt{2}\sigma}\right) = 0. \tag{1.B.87}$$

Since from the definition (1.B.85) one obviously has erf(0) = 0, (1.B.87) becomes

$$\frac{\ln(m)-\mu}{\sqrt{2}\sigma} = 0 \tag{1.B.88}$$

whence finally

$$\text{median} = m = e^{\mu}. \tag{1.B.89}$$

This is the median of the lognormal distribution of N. In other words, this is the number of extraterrestrial civilizations in the Galaxy such that, with 50% probability the actual value of N will be lower than this median, and with 50% probability it will be higher.

In conclusion, we feel it useful to summarize all the equations that we have derived about the random variable N in Table 1.B.2.

We want to complete this section about the lognormal probability density bfunction (1.B.56) by discovering its numeric values for the inputs to the statistical Drake equation (1.B.3) listed in Table 1.B.1.

According to the CLT, the mean value μ to be inserted into the lognormal density (1.B.56) is given, according to the second equation (1.B.48), by the sum of

Table 1.B.2. Summary of the properties of the lognormal distribution that apply to the random variable N = number of ET communicating civilizations in the Galaxy.

Random variable	N = number of communicating ET civilizations in the Galaxy
Probability distribution	Lognormal
Probability density function	$f_N(n) = \frac{1}{n} \cdot \frac{1}{\sqrt{2\pi}\sigma} e^{-\frac{(\ln(n)-\mu)^2}{2\sigma^2}} \quad (n \geq 0)$
Mean value	$\langle N \rangle = e^{\mu} e^{\frac{\sigma^2}{2}}$
Variance	$\sigma_N^2 = e^{2\mu} e^{\sigma^2} \left(e^{\sigma^2} - 1 \right)$
Standard deviation	$\sigma_N = e^{\mu} e^{\frac{\sigma^2}{2}} \sqrt{e^{\sigma^2} - 1}$
All the moments (i.e., the k-th moment)	$\langle N^k \rangle = e^{k\mu} e^{k^2 \cdot \frac{\sigma^2}{2}}$
Mode (= abscissa of the lognormal peak)	$n_{\text{mode}} \equiv n_{\text{peak}} = e^{\mu} e^{-\sigma^2}$
Value of the mode peak	$f_N(n_{\text{mode}}) = \frac{1}{\sqrt{2\pi}\,\sigma} \cdot e^{-\mu} \cdot e^{\frac{\sigma^2}{2}}$
Median (= 50–50 probability value for N)	$\text{Median} = m = e^{\mu}$
Skewness	$\frac{K_3}{(K_2)^{\frac{3}{2}}} = \left(e^{\sigma^2} + 2 \right) \sqrt{e^{\sigma^2} - 1}$
Kurtosis	$\frac{K_4}{(K_2)^2} = e^{4\sigma^2} + 2\, e^{3\sigma^2} + 3\, e^{2\sigma^2} - 6$
Expression of μ in terms of the lower (a_i) and upper (b_i) limits of the Drake uniform input random variables D_i	$\mu = \sum_{i=1}^{7} \langle Y_i \rangle = \sum_{i=1}^{7} \frac{b_i[\ln(b_i) - 1] - a_i[\ln(a_i) - 1]}{b_i - a_i}$
Expression of σ^2 in terms of the lower (a_i) and upper (b_i) limits of the Drake uniform input random variables D_i	$\sigma^2 = \sum_{i=1}^{7} \sigma_{Y_i}^2 = \sum_{i=1}^{7} \left(1 - \frac{a_i b_i [\ln(b_i) - \ln(a_i)]^2}{(b_i - a_i)^2} \right)$

all the mean values $\langle Y_i \rangle$; that is, by virtue of (1.B.31), by:

$$\mu = \sum_{i=1}^{7} \langle Y_i \rangle = \sum_{i=1}^{7} \frac{b_i[\ln(b_i) - 1] - a_i[\ln(a_i) - 1]}{b_i - a_i}. \tag{1.B.90}$$

Upon replacing the 14 a_i and b_i listed in Table 1.B.1 into (1.B.90), the following numeric mean value μ is found

$$\mu \approx 7.462176. \tag{1.B.91}$$

Similarly, to get the numeric variance σ^2 one must resort to the last of equations (1.B.48) and to (1.B.33):

$$\sigma^2 = \sum_{i=1}^{7} \sigma_{Y_i}^2 = \sum_{i=1}^{7} \left(1 - \frac{a_i b_i [\ln(b_i) - \ln(a_i)]^2}{(b_i - a_i)^2}\right) \tag{1.B.92}$$

yielding the following numeric variance σ^2 to be inserted into the lognormal pdf (1.B.56)

$$\sigma^2 \approx 1.938725 \tag{1.B.93}$$

whence the numeric standard deviation σ

$$\sigma \approx 1.392381. \tag{1.B.94}$$

Upon replacing these two numeric values (1.B.84) and (1.B.86) into the lognormal pdf (1.B.56), the latter is perfectly determined. It is plotted in Figure 1.B.4 as the thin curve.

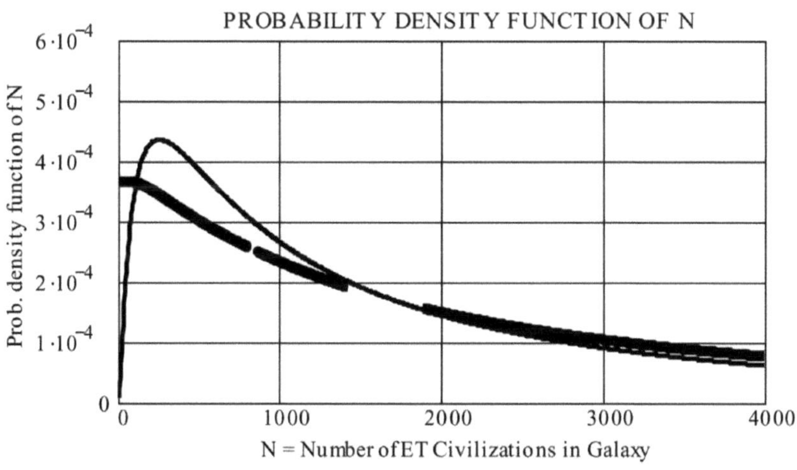

Figure 1.B.4. Comparing the two probability density functions of the random variable N found: (3) at the end of Section 1.B.3.3 in a purely numeric way and without resorting to the CLT at all (thick curve) and (4) analytically by using the CLT and the relevant lognormal approximation (thin curve).

In other words, Figure 1.B.4 shows the lognormal distribution for the number N of extraterrestrial civilizations in the Galaxy derived from the Central Limit Theorem as applied to the Drake equation (with the input data listed in Table 1.B.1).

Let us now point out the most important statistical properties of this lognormal pdf:

(1) *Mean value of N*. This is given by equation (1.B.60) with μ and σ given by (1.B.91) and (1.B.94), respectively:

$$\langle N \rangle = e^{\mu}\, e^{\frac{\sigma^2}{2}} \approx 4589.559. \tag{1.B.95}$$

In other words, there are 4,590 ET civilizations in the Galaxy according the Central Limit Theorem of statistics with the inputs of Table 1.B.1. This number 4,590 is *higher* than the 3,500 foreseen by the classical Drake equation working with sheer numbers only, rather than with probability distributions. Thus equation (1.B.95) *is good news for SETI*, since it shows that the expected number of ETs is *higher* with an adequate statistical treatment than just with the too simple Drake sheer numbers of (1.B.1).

(2) *Variance of N*. The variance of the lognormal distribution is given by (1.B.62) and turns out to be a huge number:

$$\sigma_N^2 = e^{2\mu}\, e^{\sigma^2}\left(e^{\sigma^2} - 1\right) \approx 125{,}328{,}623\,. \tag{1.B.96}$$

(3) *Standard deviation of N*. The standard deviation of the lognormal distribution is given by (1.B.63) and turns out to be:

$$\sigma_N = e^{\mu}\, e^{\frac{\sigma^2}{2}}\sqrt{e^{\sigma^2} - 1} = 11{,}195\,. \tag{1.B.97}$$

Again, this is *good news for SETI*. In fact, such a high standard deviation means that N may range from very low values (0, theoretically, and 1 since humanity exists) up to tens of thousands: 4,590 + 11,195 = 15,785 is (1.B.95) + (1.B.97).

(4) *Mode of N*. The mode (= peak abscissa) of the lognormal distribution of N is given by (1.B.81), and has a surprisingly low numeric value:

$$n_{\text{mode}} \equiv n_{\text{peak}} = e^{\mu}\, e^{-\sigma^2} \approx 250\,. \tag{1.B.98}$$

This is well shown in Figure 1.B.4: the mode peak is very pronounced and close to the origin, but the right tail is high, and this means that the mean value of the distribution is much higher than the mode: 4,590 $\gg$250.

(5) *Median of N*. The median (= 50–50 abscissa, splitting the pdf in two exactly equi-probable parts) of the lognormal distribution of N is given by (1.B.89), and has

the numeric value:

$$n_{\text{median}} \equiv e^{\mu} \approx 1{,}740. \tag{1.B.99}$$

In words, assuming the input values listed in Table 1.B.1, we have exactly a 50% probability that the actual value of N is lower than 1,740, and 50% that it is higher than 1,740.

1.B.6 Comparing the CLT results with the non-CLT results

The time is now ripe to compare the CLT-based results about the lognormal distribution of N just described in Section 1.B.5 with the non-CLT-based results obtained numerically in Section 1.B.3.3.

To do so in a simple, visual way, let us plot two curves on the same diagram:

(1) the numeric curves appearing in Figure 1.B.2 and obtained after laborious Fourier transform calculations in the complex domain; and
(2) the lognormal distribution (1.B.56) with numeric μ and σ given by (1.B.91) and (1.B.94), respectively.

We see that the two curves are virtually coincident for values of N larger than 1,500. This is a consequence of the law of large numbers, of which the CLT is just one of the many facets.

It is much the same case for the natural log of N (i.e., the random variable Y of (1.B.5)), which is plotted in Figure 1.B.5 both in its normal curve version (thin curve) and in its numeric version, obtained via Fourier transforms and already shown in Figure 1.B.2.

The conclusion is simple: from now on we will discard numeric calculations and stick only to the equations derived by virtue of the CLT (i.e., to the lognormal (1.B.56) and its consequences).

1.B.7 Distance of the nearest extraterrestrial civilization as a probability distribution

As an application of the statistical Drake equation developed in the previous sections of this paper, we now want to consider the problem of estimating the distance of the extraterrestrial civilization nearest to us in the Galaxy. In all astrobiology textbooks (see, e.g., [3]) and on several websites, the solution to this problem is reported with only slight differences in the mathematical proofs among the various authors. In the first of the coming two sections (Section 1.B.7.1) we derive the expression for this ET_Distance (as we like to denote it) in the classical, non-probabilistic way: in other words, this is the classical, deterministic derivation. In the second section (1.B.7.2) we provide the probabilistic derivation, arising from our statistical Drake equation, of the corresponding probability density function $f_{\text{ET\_Distance}}(r)$: here r is the distance between us and the nearest ET civilization assumed as the independent variable of its own probability density function. The ensuing sections provide more mathematical

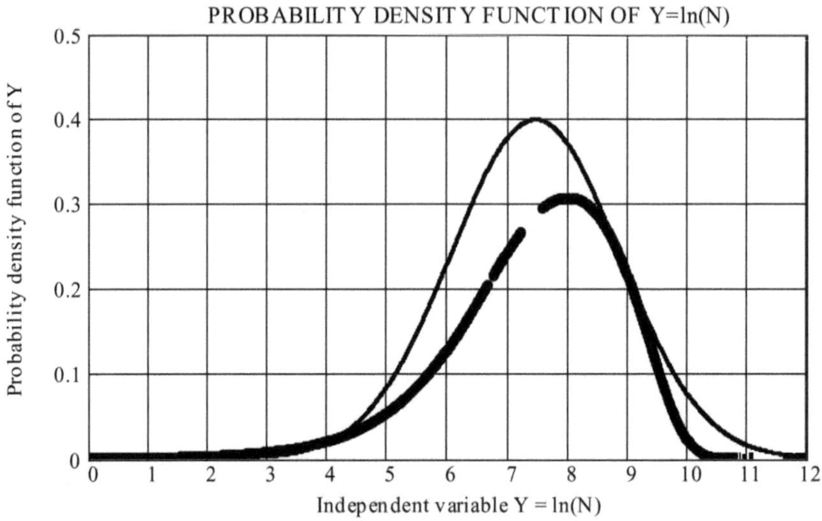

Figure 1.B.5. Comparing the two probability density functions of the random variable $Y = \ln(N)$ found: (1) at the end of Section 1.B.3.3 in a purely numeric way and without resorting to the CLT at all (thick curve) and (2) analytically by using the CLT and the relevant normal (Gaussian) approximation (thin Gaussian curve).

details about this $f_{\mathrm{ET\_Distance}}(r)$ such as its mean value, variance, standard deviation, all central moments, mode, median, cumulants, skewness, and kurtosis.

1.B.7.1 Classical, non-probabilistic derivation of the distance of the nearest ET civilization

Consider the Galactic Disk and assume that:

(4) The diameter of the Galaxy is (about) 100,000 light-years (abbreviated lt-yr) (i.e., its radius, R_{Galaxy}, is about 50,000 lt-yr).
(5) The thickness of the Galactic Disk half way from its center, h_{Galaxy}, is about 16,000 lt-yr.
(6) The volume of the Galaxy may then be approximated as the volume of the corresponding cylinder; that is

$$V_{Galaxy} = \pi R_{Galaxy}^2 h_{Galaxy}. \tag{1.B.100}$$

(7) Now consider the sphere around us having a radius r. The volume of such as sphere is

$$V_{Our\_Sphere} = \frac{4}{3}\pi\left(\frac{\mathrm{ET\_Distance}}{2}\right)^3. \tag{1.B.101}$$

In the last equation, we had to divide the distance ET_Distance between ourselves and the nearest ET civilization by 2 because we are now going to make the unwarranted assumption that all ET civilizations are equally spaced from each other in the Galaxy! This is a crazy assumption, clearly, and should be replaced by more scientifically grounded assumptions as soon as we know more about our Galactic neighborhood. At the moment, however, this is the best guess that we can make, and so we shall take it for granted, although we are aware that this is a weak point in our reasoning.

Having thus assumed that ET civilizations are *uniformly spaced in the Galaxy*, we can write down this proportion:

$$\frac{V_{Galaxy}}{N} = \frac{V_{Our\_Sphere}}{1}. \tag{1.B.102}$$

That is, upon replacing both (1.B.100) and (1.B.101) into (1.B.102):

$$\frac{\pi\, R_{Galaxy}^2\, h_{Galaxy}}{N} = \frac{\frac{4}{3}\pi\left(\frac{\text{ET\_Distance}}{2}\right)^3}{1}. \tag{1.B.103}$$

The only unknown in the last equation is ET_Distance, and so we may solve for it, thus getting the: (*average*) *distance between any pair of neighboring civilizations in the Galaxy*

$$\text{ET\_Distance} = \frac{\sqrt[3]{6\, R_{Galaxy}^2\, h_{Galaxy}}}{\sqrt[3]{N}} = \frac{C}{\sqrt[3]{N}} \tag{1.B.104}$$

where the positive constant C is defined by

$$C = \sqrt[3]{6\ R_{Galaxy}^2\ h_{Galaxy}} \approx 28{,}845 \text{ light-years.} \tag{1.B.105}$$

Equations (1.B.104) and (1.B.105) are the starting point for our first application of the statistical Drake equation, which we discuss in detail in the coming sections of this paper.

1.B.7.2 Probabilistic derivation of the probability density function for ET_Distance

The probability density function (pdf) yielding the distance of the ET civilization nearest to us in the Galaxy and presented in this section was discovered by the author on September 5, 2007. He did not disclose it to other scientists until the SETI meeting run by the famous mathematical physicist and popular science author Paul Davies at the "Beyond" Center of the University of Arizona at Phoenix, over February 5–8, 2008. This meeting was also attended by SETI Institute experts Jill Tarter, Seth Shostak, Doug Vakoch, Tom Pierson, and others. During the author's talk, Paul Davies suggested calling the new probability density function that yields the ET_Distance "the Maccone distribution"; it is derived in this section.

Let us go back to equation (1.B.104). Since N is now a random variable (obeying the lognormal distribution), it follows that the ET_Distance must be a random variable as well. Hence it must have some unknown probability density function which we denote by

$$f_{\mathrm{ET\_Distance}}(r) \tag{1.B.106}$$

where r is the new independent variable of such a probability distribution (it is denoted by r to remind the reader that it expresses the three-dimensional radial distance separating us from the nearest ET civilization in a full spherical symmetry of the space around us).

The question then is: What is the unknown probability distribution (1.B.106) of the ET_Distance? We can answer this question upon making two formal substitutions

$$\begin{cases} N \to x \\ \mathrm{ET\_Distance} \to y \end{cases} \tag{1.B.107}$$

into the transformation law (1.B.8) for random variables. As a consequence, (1.B.104) takes the form

$$y = g(x) = \frac{C}{\sqrt[3]{x}} = C \cdot x^{-\frac{1}{3}}. \tag{1.B.108}$$

In order to find the unknown probability density $f_{\mathrm{ET\_Distance}}(r)$, we now apply rule (1.B.9) to (1.B.108). First, note that (1.B.108), when inverted to yield the various roots $x_i(y)$, yields a single real root only

$$x_1(y) = \frac{C^3}{y^3}. \tag{1.B.109}$$

Then, the summation in (1.B.9) reduces to one term only. Second, differentiating (1.B.108) one finds

$$g'(x) = -\frac{C}{3} \cdot x^{-\frac{4}{3}}. \tag{1.B.110}$$

Thus, the relevant absolute value reads

$$|g'(x)| = \left| -\frac{C}{3} \cdot x^{-\frac{4}{3}} \right| = \frac{C}{3} \cdot x^{-\frac{4}{3}}. \tag{1.B.111}$$

Upon inserting (1.B.111) into (1.B.9), we then find

$$|g'(x_1)| = \frac{C}{3} \cdot x^{-\frac{4}{3}} = \frac{C}{3} \cdot \left[\frac{C^3}{y^3}\right]^{-\frac{4}{3}} = \frac{C}{3} \cdot \left[\frac{C}{y}\right]^{-4} = \frac{y^4}{3\,C^3}. \tag{1.B.112}$$

This is the denominator of (1.B.9). The numerator simply is the lognormal probability density function (1.B.56) where the old independent variable x must now be rewritten in terms of the new independent variable y by virtue of

(1.B.109). By doing so, we finally arrive at the new probability density function $f_Y(y)$

$$f_Y(y) = \frac{3\,C^3}{y^4} \cdot \frac{1}{\frac{C^3}{y^3}} \cdot \frac{1}{\sqrt{2\pi}\,\sigma} \cdot e^{-\frac{\left(\ln\left[\frac{C^3}{y^3}\right]-\mu\right)^2}{2\sigma^2}}.$$

Rearranging and replacing y by r, the final form is:

$$f_{\text{ET\_distance}}(r) = \frac{3}{r} \cdot \frac{1}{\sqrt{2\pi}\,\sigma} \cdot e^{-\frac{\left(\ln\left[\frac{C^3}{r^3}\right]-\mu\right)^2}{2\sigma^2}}. \tag{1.B.113}$$

Now, just replace C in (1.B.113) by virtue of (1.B.105). We have thus discovered the probability density function yielding the probability of finding the nearest extra-terrestrial civilization in the Galaxy in the spherical shell between the distances r and $r + dr$ from Earth:

$$f_{\text{ET\_Distance}}(r) = \frac{3}{r} \cdot \frac{1}{\sqrt{2\pi}\,\sigma} \cdot e^{-\frac{\left(\ln\left[\frac{6\,R_{Galaxy}^2\,h_{Galaxy}}{r^3}\right]-\mu\right)^2}{2\sigma^2}} \tag{1.B.114}$$

holding for $r \geq 0$.

Statistical properties of this distribution

We now want to study this probability distribution in detail. Our next questions are:

(1) What is its mean value?
(2) What are its variance and standard deviation?
(3) What are its moments to any higher order?
(4) What are its cumulants?
(5) What are its skewness and kurtosis?
(6) What are the coordinates of its peak (i.e., the mode (peak abscissa) and its ordinate)?
(7) What is its median?

The first three points in the list are all covered by the following theorem: all the moments of (1.B.113) are given—here k is the generic and non-negative integer exponent (i.e., $k = 0, 1, 2, 3, \ldots \geq 0$)—by

$$\left\langle \text{ET\_Distance}^{\text{k}} \right\rangle = \int_0^\infty r^k \cdot f_{\text{ET\_Distance}}(r)\,dr = \int_0^\infty r^k \cdot \frac{3}{r} \cdot \frac{1}{\sqrt{2\pi}\,\sigma} \cdot e^{-\frac{\left(\ln\left[\frac{C^3}{r^3}\right]-\mu\right)^2}{2\sigma^2}}\,dr$$
$$= C^k\, e^{-k\frac{\mu}{3}}\, e^{k^2 \cdot \frac{\sigma^2}{18}}. \tag{1.B.115}$$

To prove this result, one first transforms the above integral by virtue of the substitution

$$\ln\left[\frac{C^3}{r^3}\right] = z. \tag{1.B.116}$$

Then the new integral in z is then seen to reduce to the known Gaussian integral (1.B.53) and, after several reductions which we skip for the sake of brevity, (1.B.115) follows from (1.B.53). In other words, we have proven that

$$\left\langle \text{ET\_Distance}^k \right\rangle = C^k\, e^{-k\frac{\mu}{3}}\, e^{k^2 \cdot \frac{\sigma^2}{18}}. \tag{1.B.117}$$

Upon inserting $k = 0$ into (1.B.117), the normalization condition for $f_{\text{ET\_Distance}}(r)$ follows

$$\int_0^{\infty} f_{\text{ET\_Distance}}(r)\, dr = 1. \tag{1.B.118}$$

Upon inserting $k = 1$ into (1.B.117), the important mean value of the random variable ET_Distance is found

$$\langle \text{ET\_Distance} \rangle = C\, e^{-\frac{\mu}{3}}\, e^{\frac{\sigma^2}{18}}. \tag{1.B.119}$$

Upon inserting $k = 2$ into (1.B.117), the mean value of the square of the random variable ET_Distance is found

$$\left\langle \text{ET\_Distance}^2 \right\rangle = C^2\, e^{-\frac{2}{3}\mu}\, e^{\frac{2}{9}\sigma^2}. \tag{1.B.120}$$

The variance of ET_Distance now follows from the last two formulas with a few reductions:

$$\sigma^2_{\text{ET\_Distance}} = \left\langle \text{ET\_Distance}^2 \right\rangle - \langle \text{ET\_Distance} \rangle^2 = C^2\, e^{-\frac{2}{3}\mu}\, e^{\frac{\sigma^2}{9}}\left(e^{\frac{\sigma^2}{9}} - 1\right). \tag{1.B.121}$$

So, the variance of ET_Distance is

$$\sigma^2_{\text{ET\_Distance}} = C^2\, e^{-\frac{2}{3}\mu}\, e^{\frac{\sigma^2}{9}}\left(e^{\frac{\sigma^2}{9}} - 1\right). \tag{1.B.122}$$

The square root of this is the important standard deviation of the ET_Distance random variable

$$\sigma_{\text{ET\_Distance}} = C\, e^{-\frac{\mu}{3}}\, e^{\frac{\sigma^2}{18}}\sqrt{e^{\frac{\sigma^2}{9}} - 1}. \tag{1.B.123}$$

The third moment is obtained upon inserting $k = 3$ into (1.B.117)

$$\left\langle \text{ET\_Distance}^3 \right\rangle = C^3\, e^{-\mu}\, e^{\frac{\sigma^2}{2}}. \tag{1.B.124}$$

Finally, upon setting $k = 4$ into (1.B.117), the fourth moment of ET_Distance is found

$$\left\langle \mathrm{ET\_Distance}^4 \right\rangle = C^4\, e^{-\frac{4}{3}\mu}\, e^{\frac{8}{9}\sigma^2}. \tag{1.B.125}$$

Our next goal is to find the cumulants of the ET_Distance. In principle, we could compute all the cumulants K_i from the generic i-th moment μ_i' by virtue of the recursion formula (see [12])

$$K_i = \mu_i' - \sum_{k=1}^{i-1} \binom{i-1}{k-1} K_k\, \mu_{n-k}'. \tag{1.B.126}$$

In practice, however, we shall confine ourselves here to the computation of the first four cumulants because only they are required to find the skewness and kurtosis of the distribution (1.B.113). Then, the first four cumulants in terms of the first four moments read:

$$\begin{cases} K_1 = \mu_1' \\ K_2 = \mu_2' - K_1^2 \\ K_3 = \mu_3' - 3\,K_1\,K_2 - K_1^3 \\ K_4 = \mu_4' - 4\,K_1\,K_3 - 3\,K_2^2 - 6\,K_2\,K_1^2 - K_1^4. \end{cases} \tag{1.B.127}$$

These equations yield, respectively:

$$K_1 = C\, e^{-\frac{\mu}{3}}\, e^{\frac{\sigma^2}{18}} \tag{1.B.128}$$

$$K_2 = C^2\, e^{-\frac{2\mu}{3}}\, e^{\frac{\sigma^2}{9}} \left(e^{\frac{\sigma^2}{9}} - 1 \right) \tag{1.B.129}$$

$$K_3 = C^3\, e^{-\mu} \left(e^{\frac{\sigma^2}{2}} - 3\, e^{\frac{5\sigma^2}{18}} + 2\, e^{\frac{\sigma^2}{6}} \right) \tag{1.B.130}$$

$$K_4 = C^4\, e^{-\frac{4\mu}{3}} \left(e^{\frac{8\sigma^2}{9}} - 4\, e^{\frac{5\sigma^2}{9}} - 3\, e^{\frac{4\sigma^2}{9}} + 12\, e^{\frac{\sigma^2}{3}} - 6\, e^{\frac{2\sigma^2}{9}} \right). \tag{1.B.131}$$

From these we derive the skewness

$$\frac{K_3}{(K_2)^{\frac{3}{2}}} = \frac{e^{\frac{2\sigma^2}{9}} + e^{\frac{\sigma^2}{9}} - 2}{\sqrt{e^{\frac{\sigma^2}{9}} - 1}} \tag{1.B.132}$$

and the kurtosis

$$\frac{K_4}{(K_2)^2} = e^{\frac{4\sigma^2}{9}} + 2\, e^{\frac{\sigma^2}{3}} + 3\, e^{\frac{2\sigma^2}{9}} - 6. \tag{1.B.133}$$

Next we want to find the mode of this distribution (i.e., the abscissa of its peak). To do so, we must first compute the derivative of the probability density function $f_{\mathrm{ET\_Distance}}(r)$ of (1.B.113), and then set it equal to zero. This derivative is actually the derivative of the ratio of two functions of r, as its plainly appears from (1.B.113).

Thus, let us set for a moment

$$E(r) = \frac{\left(\ln\left[\frac{C^3}{r^3}\right] - \mu\right)^2}{2\,\sigma^2}. \tag{1.B.134}$$

where "E" stands for "exponent". Upon differentiating, one gets

$$E'(r) = \frac{1}{2\,\sigma^2} \cdot 2\left(\ln\left[\frac{C^3}{r^3}\right] - \mu\right) \cdot \frac{1}{\frac{C^3}{r^3}} \cdot C^3 \cdot (-3) \cdot r^{-4} = \frac{1}{\sigma^2} \cdot \left(\ln\left[\frac{C^3}{r^3}\right] - \mu\right) \cdot (-3)\frac{1}{r}. \tag{1.B.135}$$

But the probability density function (1.B.113) now reads

$$f_{\mathrm{ET\_Distance}}(r) = \frac{3}{\sqrt{2\pi}\sigma} \cdot \frac{e^{-E(r)}}{r}. \tag{1.B.136}$$

So that its derivative is

$$\frac{df_{\mathrm{ET\_Distance}}{}'(r)}{dr} = \frac{3}{\sqrt{2\pi}\sigma} \cdot \frac{-e^{-E(r)}E'(r) \cdot r - 1 \cdot e^{-E(r)}}{r^2} = \frac{3}{\sqrt{2\pi}\sigma} \cdot \frac{-e^{-E(r)}\left[E'(r) \cdot r + 1\right]}{r^2}. \tag{1.B.137}$$

Setting this derivative equal to zero means setting

$$E'(r) \cdot r + 1 = 0. \tag{1.B.138}$$

That is, upon inserting (1.B.135) into (1.B.138), we get

$$\frac{1}{\sigma^2} \cdot \left(\ln\left[\frac{C^3}{r^3}\right] - \mu\right) \cdot (-3)\frac{1}{r} \cdot r + 1 = 0. \tag{1.B.139}$$

Rearranging, this becomes

$$-3\left(\ln\left[\frac{C^3}{r^3}\right] - \mu\right) \cdot + \sigma^2 = 0 \tag{1.B.140}$$

that is

$$-3\ln\left[\frac{C^3}{r^3}\right] + 3\,\mu + \sigma^2 = 0 \tag{1.B.141}$$

whence

$$\ln\left[\frac{C}{r}\right] = \frac{\mu}{3} + \frac{\sigma^2}{9} \tag{1.B.142}$$

and finally

$$r_{\mathrm{mode}} \equiv r_{\mathrm{peak}} = C\,e^{-\frac{\mu}{3}}e^{-\frac{\sigma^2}{9}}. \tag{1.B.143}$$

This is the most likely ET_Distance from Earth.

How likely? To find the value of the probability density function $f_{\text{ET\_Distance}}(r)$ corresponding to this value of the mode, we must obviously insert (1.B.143) into (1.B.113). After a few rearrangements, which we skip for the sake of brevity, one gets

$$\text{Peak value of } f_{\text{ET\_Distance}}(r) \equiv f_{\text{ET\_Distance}}(r_{\text{mode}}) = \frac{3}{C\sqrt{2\pi}\,\sigma} \cdot e^{\frac{\mu}{3}} \cdot e^{\frac{\sigma^2}{18}}. \quad (1.B.144)$$

This is the peak height in the pdf $f_{\text{ET\_Distance}}(r)$.

Next to the mode, the median m [13] is another statistical number used to characterize any probability distribution. It is defined as the independent variable abscissa m such that a realization of the random variable will take up a value lower than m with 50% probability or a value higher than m again with 50% probability. In other words, the median m splits up our probability density into exactly two equally probable parts. Since the probability of occurrence of the random event equals the area under its density curve (i.e., the definite integral under its density curve) then the median m (of the lognormal distribution, in this case) is defined as the integral upper limit m:

$$\int_0^m f_{\text{ET\_Distance}}(r)\, dr = \frac{1}{2}. \quad (1.B.145)$$

Upon replacing (1.B.113), this becomes

$$\int_0^m \frac{3}{r} \cdot \frac{1}{\sqrt{2\pi}\,\sigma} \cdot e^{-\frac{\left(\ln\left[\frac{C^3}{r^3}\right] - \mu\right)^2}{2\sigma^2}} = \frac{1}{2}. \quad (1.B.146)$$

In order to find m, we may not differentiate (1.B.146) with respect to m, since the "precise" factor $\frac{1}{2}$ on the right would then disappear into a zero. On the contrary, we may try to perform the obvious substitution

$$z^2 = \frac{\left(\ln\left[\frac{C^3}{r^3}\right] - \mu\right)^2}{2\sigma^2} \qquad z \geq 0 \quad (1.B.147)$$

into the integral (1.B.146) to reduce it to the following integral (1.B.85) defining the error function erf(z). Then, after a few reductions, which we leave to the reader as an exercise, the full equation (1.B.145), defining the median, is turned into the corresponding equation involving the error function erf(x) as defined by (1.B.85):

$$\frac{1}{2} + \operatorname{erf}\left(\frac{\ln\left[\frac{C^3}{m^3}\right] - \mu}{\sqrt{2}\,\sigma}\right) = \frac{1}{2} \quad (1.B.148)$$

that is

$$\operatorname{erf}\left(\frac{\ln\left[\frac{C^3}{m^3}\right] - \mu}{\sqrt{2}\,\sigma}\right) = 0. \tag{1.B.149}$$

Since from the definition (1.B.147) one obviously has $\operatorname{erf}(0) = 0$, (1.B.149) yields

$$\frac{\ln\left[\frac{C^3}{m^3}\right] - \mu}{\sqrt{2}\,\sigma} = 0 \tag{1.B.150}$$

whence finally

$$\text{median} = m = C\,e^{-\frac{\mu}{3}}. \tag{1.B.151}$$

This is the median of the lognormal distribution of N. In other words, this is the number of extraterrestrial civilizations in the Galaxy such that, with 50% probability the actual value of N will be lower than this median, and with 50% probability it will be higher.

In conclusion, we feel it useful to summarize all the equations that we derived about the random variable N in Table 1.B.3.

Numerical example of the ET_Distance distribution

In this section we provide a numerical example of the analytic calculations carried out so far.

Consider the Drake equation values reported in Table 1.B.1. The graph of the corresponding probability density function of the nearest ET_Distance, $f_{\text{ET\_Distance}}(r)$, is shown in Figure 1.B.6.

From Figure 1.B.6, we see that the probability of finding extraterrestrials is practically zero up to a distance of about 500 light-years from Earth. Then it starts increasing with increasing distance from Earth, and reaches its maximum at

$$r_{\text{mode}} \equiv r_{peak} = C\,e^{-\frac{\mu}{3}}\,e^{-\frac{\sigma^2}{9}} \approx 1{,}933\,\text{light-years}. \tag{1.B.152}$$

This is the *most likely value* of the distance at which we can expect to find the nearest extraterrestrial civilization.

It is not, however, the mean value of the probability distribution (1.B.113) for $f_{\text{ET\_Distance}}(r)$. In fact, the probability density (1.B.113) has an infinite tail on the right, as clearly shown in Figure 1.B.6, and hence its mean value must be higher than its peak value. As given by (1.B.119), its mean value is

$$r_{mean\_value} = C\,e^{-\frac{\mu}{3}}\,e^{\frac{\sigma^2}{18}} \approx 2{,}670\,\text{light-years}. \tag{1.B.153}$$

Table 1.B.3. Summary of the properties of the probability distribution that applies to the random variable ET_Distance yielding the (average) distance between any two neighboring communicating civilizations in the Galaxy.

Random variable	ET_Distance between any two neighboring ET civilizations in the Galaxy assuming they are *uniformly* distributed throughout the whole Galaxy volume.
Probability distribution	Unnamed (Paul Davies suggested "Maccone distribution")
Probability density function	$f_{\text{ET\_Distance}}(r) = \frac{3}{r} \cdot \frac{1}{\sqrt{2\pi}\,\sigma} \cdot e^{-\frac{\left(\ln\left[\frac{6\,R_{Galaxy}^2\,h_{Galaxy}}{r^3}\right]-\mu\right)^2}{2\sigma^2}}$
Defining the numerical constant C	$C = \sqrt[3]{6\,R_{Galaxy}^2\,h_{Galaxy}} \approx 28{,}845$ light-years
Mean value	$\langle \text{ET\_Distance} \rangle = C\,e^{-\frac{\mu}{3}}\,e^{\frac{\sigma^2}{18}}$
Variance	$\sigma_{\text{ET\_Distance}}^2 = C^2\,e^{-\frac{2}{3}\mu}\,e^{\frac{\sigma^2}{9}}\left(e^{\frac{\sigma^2}{9}}-1\right)$
Standard deviation	$\sigma_{\text{ET\_Distance}} = C\,e^{-\frac{\mu}{3}}\,e^{\frac{\sigma^2}{18}}\sqrt{e^{\frac{\sigma^2}{9}}-1}$
All the moments (i.e., the k-th moment)	$\left\langle \text{ET\_Distance}^k \right\rangle = C^k\,e^{-k\frac{\mu}{3}}\,e^{k^2\cdot\frac{\sigma^2}{18}}$
Mode (= abscissa of the probability density function peak)	$r_{\text{mode}} \equiv r_{peak} = C\,e^{-\frac{\mu}{3}}\,e^{-\frac{\sigma^2}{9}}$
Value of the mode peak	Peak value of $f_{\text{ET\_Distance}}(r) =$ $f_{\text{ET\_Distance}}(r_{\text{mode}}) = \frac{3}{C\sqrt{2\pi}\,\sigma} \cdot e^{\frac{\mu}{3}} \cdot e^{\frac{\sigma^2}{18}}$
Median (= 50–50 probability value for ET_Distance)	Median $= m = Ce^{-\frac{\mu}{3}}$
Skewness	$\frac{K_3}{(K_2)^{\frac{3}{2}}} = \frac{e^{\frac{2\sigma^2}{9}} + e^{\frac{\sigma^2}{9}} - 2}{\sqrt{e^{\frac{\sigma^2}{9}} - 1}}$
Kurtosis	$\frac{K_4}{(K_2)^2} = e^{\frac{4\sigma^2}{9}} + 2\,e^{\frac{\sigma^2}{3}} + 3\,e^{\frac{2\sigma^2}{9}} - 6$
Expression of μ in terms of the lower (a_i) and upper (b_i) limits of the Drake uniform input random variables D_i	$\mu = \sum_{i=1}^{7} \langle Y_i \rangle = \sum_{i=1}^{7} \frac{b_i[\ln(b_i)-1] - a_i[\ln(a_i)-1]}{b_i - a_i}$
Expression of σ^2 in terms of the lower (a_i) and upper (b_i) limits of the Drake uniform input random variables D_i	$\sigma^2 = \sum_{i=1}^{7} \sigma_{Y_i}^2 = \sum_{i=1}^{7} \left(1 - \frac{a_i b_i [\ln(b_i) - \ln(a_i)]^2}{(b_i - a_i)^2}\right)$

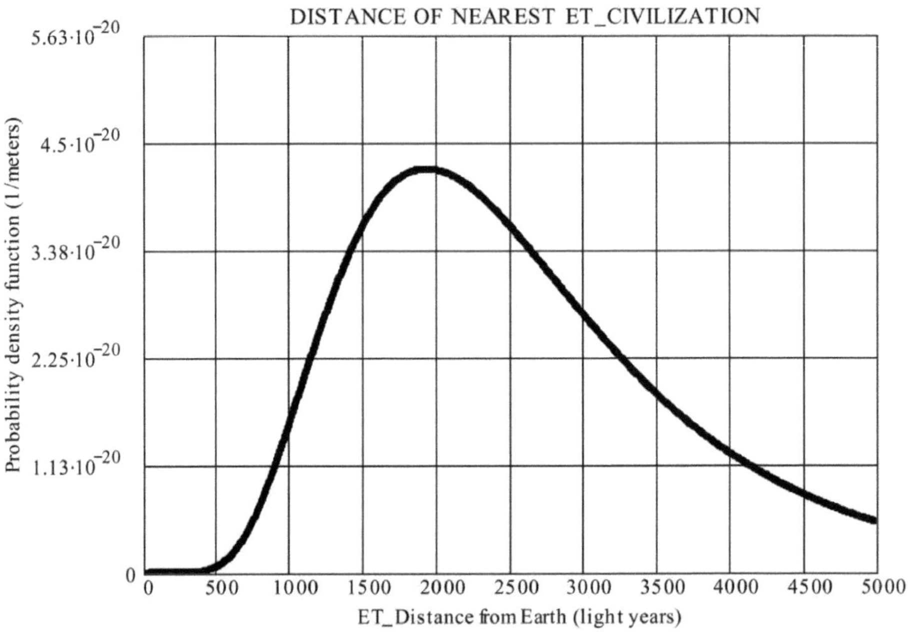

Figure 1.B.6. This is the probability of finding the nearest extraterrestrial civilization at the distance r from Earth (in light-years) if the values assumed in the Drake equation are those shown in Table 1.B.1. The relevant probability density function $f_{\text{ET\_Distance}}(r)$ is given by equation (1.B.113). Its mode (peak abscissa) equals 1,933 light-years, but its mean value is higher since the curve has a high tail on the right: the mean value equals in fact 2,670 light-years. Finally, the standard deviation equals 1,309 light-years: *this is good news for SETI*, inasmuch as the nearest ET civilization might lie at just 1 sigma= 2,670 – 1,309 = 1,361 light-years from us.

This is the *mean* (value of the) *distance* at which we can expect to find extraterrestrials.

After having found the above two distances (1,933 and 2,670 lt-yr, respectively), the next natural question that arises is: What is the range, forth and back around the mean value of the distance, at which we are most likely to expect to find extra-terrestrials? The answer to this question is given by the notion of standard deviation, which we have already found to be given by (1.B.123)

$$\sigma_{\text{ET\_Distance}} = C\, e^{-\frac{\mu}{3}}\, e^{\frac{\sigma^2}{18}} \sqrt{e^{\frac{\sigma^2}{9}} - 1} \approx 1{,}309 \text{ light-years.} \qquad (1.B.154)$$

More precisely, this is the so-called one-sigma (distance) level. Probability theory then shows that the nearest extraterrestrial civilization is expected to be located within this range; that is, within the two distances of $(2{,}670 - 1{,}309) = 1{,}361$ light-years and $(2{,}670 + 1{,}309) = 3{,}979$ light-years, with probability given by the integral of $f_{\text{ET\_Distance}}(r)$ taken in between these two lower and upper limits; that is:

$$\int_{1,361 \text{ light years}}^{3,979 \text{ light years}} f_{\text{ET\_Distance}}(r)\, dr \approx 0.75 = 75\% \tag{1.B.155}$$

In plain words: at the 75% probability, the nearest extraterrestrial civilization is located in between the distances of 1,361 and 3,979 light-years from us, having assumed the input values to the Drake equation given by Table 1.B.1. If we change these input values, then all the numbers change again.

1.B.8 The "Data Enrichment Principle" as the best CLT consequence upon the statistical Drake equation (any number of factors allowed)

As a fitting climax to all the statistical equations developed so far, let us now state our *Data Enrichment Principle*. It simply states that "the higher the number of factors in the statistical Drake equation, the better."

Put in this simple way, it simply looks like a new way of saying that the CLT lets the random variable Y approach the normal distribution when the number of terms in the sum (1.B.4) approaches infinity. And this is indeed the case, However, our Data Enrichment Principle has more profound methodological consequences that we cannot explain now, but hope to describe more precisely in one or more coming papers.

Conclusions

We have sought to extend the classical Drake equation to let it encompass statistics and probability.

This approach appears to pave the way to future, more profound investigations intended not only to associate "error bars" to each factor in the Drake equation, but especially to increase the number of factors themselves. In fact, this seems to be the only way to incorporate into the Drake equation more and more new scientific information as soon as it becomes available. In the long run, the statistical Drake equation might just become a huge computer code, growing in size and especially in the depth of the scientific information it contains. It would thus be humanity's first "Encyclopedia Galactica".

Unfortunately, to extend the Drake equation to statistics, it was necessary to use a mathematical apparatus that is more sophisticated than just the simple product of seven numbers.

When the author had the honor and privilege to present his results at the SETI Institute on April 11, 2008, in front of an audience that included Professor Frank Drake, he felt he had to add these words: "My apologies, Frank, for disrupting the beautiful simplicity of your equation."

Acknowledgments

The author is grateful to Drs. Jill Tarter, Paul Davies, Seth Shostak, Doug Vakoch, Tom Pierson, Carol Oliver, Paul Shuch, and Kathryn Denning for attending his first ever presentation about these topics at the "Beyond" Center of the University of Arizona at Phoenix on February 8, 2008. He also would like to thank Dan Werthimer and his school of young SETI experts for keeping alive the interplay between experimental and theoretical SETI. But the greatest "thanks" goes of course to the teacher of us all: Professor Frank D. Drake, whose equation opened a new way of thinking about the past and future of humans in the Galaxy.

APPENDIX 1.C ORIGINAL POWERPOINT PRESENTATION OF THE AUTHOR'S PAPER #IAC-08-A4.1.4 ENTITLED THE STATISTICAL DRAKE EQUATION

THE STATISTICAL DRAKE EQUATION

Claudio Maccone

Member of the International Academy of Astronautics (IAA)
Co-Vice Chair, SETI Permanent Study Group of the IAA
Home Page : www.maccone.com
E-mail : clmaccon@libero.it

59TH INTERNATIONAL ASTRONAUTICAL CONGRESS GLASGOW, UK, SEPTEMBER 29TH – OCTOBER 3RD, 2008.

The Classical Drake Equation /2

- All Attendees in this meeting do know the meaning of the seven factors in the Drake equation: so, they will not be described here.
- On the contrary, we'd like to stress that the seven factors are just POSITIVE NUMBERS. And that the Drake equation simply is the PRODUCT of these seven positive numbers.
- It is claimed here that Drake's approach is too "simple-minded", since it does NOT yield the ERROR BAR associated to each factor!

ABSTRACT

- We provide the statistical generalization of the Drake equation.
- From a simple product of seven positive numbers, the Drake equation is now turned into the product of seven positive random variables.
- We call this "the Statistical Drake Equation".
- A key advantage is to associate an ERROR BAR to each variable in the classical Drake equation.

The STATISTICAL Drake Equation /1

- If we want to associate an ERROR BAR to each factor of the Drake equation then...
- ... we must regard each factor in the Drake equation as a RANDOM VARIABLE.
- Then the number N of communicating civilizations also becomes a random variable.
- This we call the STATISTICAL DRAKE EQUATION and study for the first time here.

The Classical Drake Equation /1

- In 1961 Frank Drake introduced his famous "Drake equation" described at the web site http://en.wikipedia.org/wiki/Drake_equation. It yields the number N of communicating civilizations in the Galaxy:

$$N = Ns \cdot fp \cdot ne \cdot fl \cdot fi \cdot fc \cdot fL$$

- Frank Donald Drake (b. 1930)

The STATISTICAL Drake Equation /2

- Denoting each random variable by capitals, the STATISTICAL DRAKE EQUATION reads

$$N = \prod_{i=1}^{7} D_i$$

- Where the D sub i ("D from Drake") are the 7 random variables, and N is a random variable too ("to be determined").

The STATISTICAL Drake Equation /3

- The next question is: what is the statistical distribution of each *D sub i*?
- To simplify things, we'll just suppose that each *D sub i* is UNIFORMLY DISTRIBUTED.
- Thus, the MEAN VALUE of each *D sub i* has same numeric value as the corresponding pure number in the Drake equation !
- Each *D sub i* also has an ERROR BAR around the mean value that is equal to the standard deviation of the UNIFORM distribution !

The STATISTICAL Drake Equation /4

- For instance: nobody knows what the number of stars in the Galaxy is ***exactly !*** ☺
- But in our Statistical Drake Equation, we avoid this rough approximation by replacing the unknown number of stars by a UNIFORM distribution centered around the most probable number. Also, it spreads out to the value of the standard deviation of the uniform distribution.
- It can be proven that the latter must be multiplied by a factor of sqrt(3) = 1.7.

The STATISTICAL Drake Equation /5

- This author has completely solved the problem of finding the statistical distribution of *N* given the 7 uniformly distributed random variables *D sub i*. It does not have a name and is rather complicated.
- The mathematical proof is too long to be described here in detail (sorry). It will be published soon.
- Let us further try to GENERALIZE the Statistical Drake Equation to the case where the random variables are OTHER THAN SEVEN. Then...

Conclusion #1
The number of Signaling Civilizations is Log-Normally distributed

- Our Statistical Drake Equation, now Generalized to any number of factors, embodies as a special case the Statistical Drake Equation with just 7 factors.
- The conclusion is that the random variable *N* (the number of communicating ET Civilizations in the Galaxy) is LOG-NORMALLY distributed.
- The classical "old pure-number Drake value" of *N* is now replaced by the MEAN VALUE of such a log-normal distribution.
- But we now also have an ERROR BAR around it !

Generalizing the STATISTICAL Drake Equation to ANY NUMBER OF FACTORS /1

- Consider the statistical equation

$$N = \prod_{i=1}^{any\_number} D_i$$

- This is the generalization of our Statistical Drake Equation to the product of ANY finite NUMBER of positive random variables.
- Is it possible to determine the statistics of *N*?
- Rather surprisingly, the answer is "yes" !

Generalizing the STATISTICAL Drake Equation to ANY NUMBER OF FACTORS /2

- First, you obviously take the natural log of both sides to change the finite product into a finite sum

$$\ln(N) = \sum_{i=1}^{any\_number} \ln(D_i)$$

- Second, to this finite sum one can apply the CENTRAL LIMIT THEOREM OF STATISTICS. It states that, in the limit for an infinite sum, the distribution of the left-hand-side is NORMAL.
- This is true WHATEVER the distributions of the random variables in the sum MAY BE. ***Just Great !***

Generalizing the STATISTICAL Drake Equation to ANY NUMBER OF FACTORS /3

- So, the random variable on the left is NORMAL, i.e.

$$\ln(N)$$

- Thus, the random variable *N* under the log must be LOG-NORMAL and its distribution is determined!
- One must, however, determine the mean value and variance of this log-normal distribution in terms of the mean values and variances of the factor random variables. This is DIFFICULT. But it can be done, for example, by a suitable numeric code that this author wrote in MathCad language.

Conclusion #2
The DISTANCE between ET Civilizations now follows a probability distribution

- In all Astrobiology textbooks (e.g. Bennett & Shostak, 2007 ed., page 404) the average DISTANCE between Signaling Civilizations in the Galaxy is given. This is inversely proportional to the cubic root of the number *N* obtained by the classical Drake equation.
- Now this average distance is a new random variable. This author found the relevant probability distribution. But this is too complicated to be described here.
- The MEAN VALUE of such a new distribution is the old average DISTANCE, but we now also have an ERROR BAR around it !

Example #1
UNIFORM INPUT DISTRIBUTIONS

$Ns := 350 \cdot 10^{9}$	$\mu Ns := Ns$	$\sigma Ns := 1 \cdot 10^{9}$
$fp := \frac{50}{100}$	$\mu fp := fp$	$\sigma fp := \frac{10}{100}$
$ne := 1$	$\mu ne := ne$	$\sigma ne := \frac{1}{\sqrt{3}}$
$fl := \frac{50}{100}$	$\mu fl := fl$	$\sigma fl := \frac{10}{100}$
$fi := \frac{20}{100}$	$\mu fi := fi$	$\sigma fi := \frac{10}{100}$
$fc := \frac{20}{100}$	$\mu fc := fc$	$\sigma fc := \frac{10}{100}$
$fL := \frac{10000}{10^{10}}$	$\mu fL := fL$	$\sigma fL := \frac{1000}{10^{10}}$
$N := Ns \cdot fp \cdot ne \cdot fl \cdot fi \cdot fc \cdot fL$	$fL\_factor := Ns \cdot fp \cdot ne \cdot fl \cdot fi \cdot fc$	$\mu fL = 0.000001$
$N = 3500$	$fL\_factor = 3500000000$	$\sigma fL = 0.0000001$

Example #2
The DISTANCE between ET Civilizations now follows a probability distribution

- In the previous example, the average DISTANCE given by the classic Drake equation is 4093 light years.
- In our statistical approach, the MEAN VALUE of the distance turns out to be 2670 light years...
- ...with an ERROR BAR of ± 1309 light years.
- Thus the MINIMAL STATISTICAL ET DISTANCE is JUST 1361 light years. <u>This is *SETI-ENCOURAGING !*</u>

Example #1
The number of Signaling Civilizations is Log-Normally distributed

- For the above uniform inputs, the log-normal curve

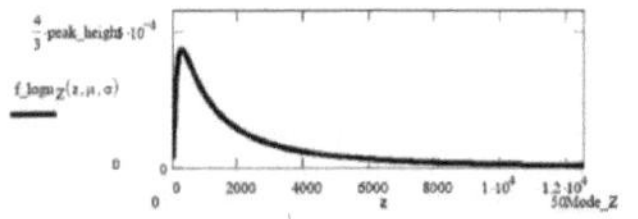

- Yields the number of Signaling Civilizations (mean value) as 4590, while classic Drake eq. yields 3500.
- With a standard deviation of ± 11,195 Civilizations.

Example #2
The DISTANCE between ET Civilizations now follows a probability distribution

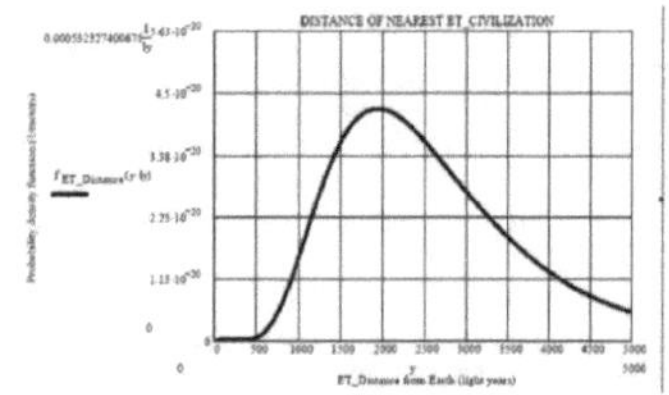

CONCLUSION

- Surprisingly, STATISTICS and the Drake equation MATCH PERFECTLY if one just takes the <u>logs</u> of the Drake equation!
- By doing so, one also invokes the Central Limit Theorem (CLT) of Statistics, enabling ANY number of factors to be considered.
- This is our DATA ENRICHMENT PRINCIPLE that is the key to further future Astrobiology investigations by virtue of our extension.

APPENDIX 1.D STATISTICAL DRAKE EQUATION

Numeric code in Mathcad written by Claudio Maccone in September 2007, updated and completed for this book, *Mathematical SETI*, on February 13, 2012.

Inputs:

$Ns := 350 \cdot 10^9 \qquad \mu Ns := Ns \qquad \sigma Ns := 1 \cdot 10^9$

$fp := \frac{50}{100} \qquad \mu fp := fp \qquad \sigma fp := \frac{10}{100}$

$ne := 1 \qquad \mu ne := ne \qquad \sigma ne := \frac{1}{\sqrt{3}} \qquad \frac{1}{\sqrt{3}} = 5.774 \times 10^{-1}$

$fl := \frac{50}{100} \qquad \mu fl := fl \qquad \sigma fl := \frac{10}{100}$

$fi := \frac{20}{100} \qquad \mu fi := fi \qquad \sigma fi := \frac{10}{100}$

$fc := \frac{20}{100} \qquad \mu fc := fc \qquad \sigma fc := \frac{10}{100}$

$fL := \frac{10000}{10^{10}} \qquad \mu fL := fL \qquad \sigma fL := \frac{1000}{10^{10}}$

$N := Ns \cdot fp \cdot ne \cdot fl \cdot fi \cdot fc \cdot fL \qquad fL\_factor := Ns \cdot fp \cdot ne \cdot fl \cdot fi \cdot fc \qquad \mu fL = 1.000 \times 10^{-6}$

$N = 3.500 \times 10^3 \qquad fL\_factor = 3.500 \times 10^9 \qquad \sigma fL = 1.000 \times 10^{-7}$

Mean value and variance of the logarithm of the number of stars in the Galaxy, ln(Ns)

$aNs := \mu Ns - \sqrt{3} \cdot \sigma Ns \qquad aNs = 3.483 \times 10^{11} \qquad bNs := \mu Ns + \sqrt{3} \cdot \sigma Ns \qquad bNs = 3.517 \times 10^{11}$

$$MeanlnNs := \frac{bNs \cdot \ln(bNs) - bNs - aNs \cdot \ln(aNs) + aNs}{bNs - aNs}$$

$MeanNs = 2.658 \times 10^1$

$$VarlnNs := \frac{2 \cdot aNs \cdot \ln(aNs) \cdot bNs \cdot \ln(bNs) - aNs \cdot bNs \cdot (\ln(bNs))^2 - aNs \cdot bNs \cdot (\ln(aNs))^2 + (bNs - aNs)^2}{(bNs - aNs)^2}$$

$VarlnNs = 8.163 \times 10^{-6}$

Mean value and variance of the logarithm of the fraction of stars that have planets around them, ln(fp)

$afp := \mu fp - \sqrt{3} \cdot \sigma fp \qquad afp = 3.268 \times 10^{-1} bfp := \mu fp + \sqrt{3} \cdot \sigma fp \qquad bfp = 6.732 \times 10^{-1}$

$$Meanlnfp := \frac{bfp \cdot \ln(bfp) - bfp - afp \cdot \ln(afp) + afp}{bfp - afp}$$

$\text{Meanlnfp} = -7.139 \times 10^{-1}$

$$\text{Varlnfp} := \frac{2 \cdot \text{afp} \cdot \ln(\text{afp}) \cdot \text{bfp} \cdot \ln(\text{bfp}) - \text{afp} \cdot \text{bfp} \cdot (\ln(\text{bfp}))^2 - \text{afp} \cdot \text{bfp} \cdot (\ln(\text{afp}))^2 - 2 \cdot \text{afp} \cdot \text{bfp} + \text{afp}^2 + \text{bfp}^2}{(\text{bfp} - \text{afp})^2}$$

$\text{Varlnfp} = 4.241 \times 10^{-2}$

Mean value and variance of the logarithm of the number of Earths per star (i.e., the number of planets per star that are capable of sustaining life, ln(ne))

$\text{ane} := \mu\text{ne} - \sqrt{3} \cdot \sigma\text{ne}$ $\quad \text{ane} = 0.000 \times 10^0$ $\quad \text{bne} := \mu\text{ne} + \sqrt{3} \cdot \sigma\text{ne}$ $\quad \text{bne} = 2.000 \times 10^0$

$$\text{Meanlnne}(\text{ane}, \text{bne}) := \begin{cases} \ln(\text{bne}) - 1 & \text{if ane} = 0 \\ \dfrac{\text{bne} \cdot \ln(\text{bne}) - \text{bne} - \text{ane} \cdot \ln(\text{ane}) + \text{ane}}{\text{bne} - \text{ane}} & \text{otherwise} \end{cases}$$

$\text{Meanlnne}(\text{ane}, \text{bne}) = -3.069 \times 10^{-1}$

$\text{Varlnne} := 1 \qquad \text{Varlnne} = 1.000 \times 10^0$

Mean value and variance of the logarithm of the fraction of planets in ne where life evolves, ln(fl)

$\text{afl} := \mu\text{fl} - \sqrt{3} \cdot \sigma\text{fl}$ $\quad \text{afl} = 3.268 \times 10^{-1} \text{bfl} := \mu\text{fl} + \sqrt{3} \cdot \sigma\text{fl}$ $\quad \text{bfl} = 6.732 \times 10^{-1}$

$$\text{Meanlnfl} := \frac{\text{bfl} \cdot \ln(\text{bfl}) - \text{bfl} - \text{afl} \cdot \ln(\text{afl}) + \text{afl}}{\text{bfl} - \text{afl}}$$

$\text{Meanlnfl} = -7.139 \times 10^{-1}$

$$\text{Varlnfl} := \frac{2 \cdot \text{afl} \cdot \ln(\text{afl}) \cdot \text{bfl} \cdot \ln(\text{bfl}) - \text{afl} \cdot \text{bfl} \cdot (\ln(\text{bfl}))^2 - \text{afl} \cdot \text{bfl} \cdot (\ln(\text{afl}))^2 - 2 \cdot \text{afl} \cdot \text{bfl} \cdot + \text{afl}^2 + \text{bfl}^2}{(\text{bfl} - \text{afl})^2}$$

$\text{Varlnfl} = 4.241 \times 10^{-2}$

Mean value and variance of the logarithm of the fraction of fl where intelligent life evolves, ln(fl)

$\text{afi} := \mu\text{fi} - \sqrt{3} \cdot \sigma\text{fi}$ $\quad \text{afi} = 2.679 \times 10^{-2} \text{bfi} := \mu\text{fi} + \sqrt{3} \cdot \sigma\text{fi}$ $\quad \text{bfi} = 3.732 \times 10^{-1}$

$$\text{Meanlnfi} := \frac{\text{bfi} \cdot \ln(\text{bfi}) - \text{bfi} - \text{afi} \cdot \ln(\text{afi}) + \text{afi}}{\text{bfi} - \text{afi}}$$

$\text{Meanlnfi} = -1.782 \times 10^0$

$$\text{Varlnfi} := \frac{2 \cdot \text{afi} \cdot \ln(\text{afi}) \cdot \text{bfi} \cdot \ln(\text{bfi}) - \text{afi} \cdot \text{bfi} \cdot (\ln(\text{bfi}))^2 - \text{afi} \cdot \text{bfi} \cdot (\ln(\text{afi}))^2 - 2 \cdot \text{afi} \cdot \text{bfi} + \text{afi}^2 + \text{bfi}^2}{(\text{bfi} - \text{afi})^2}$$

$\text{Varlnfi} = 4.219 \times 10^{-1}$

Mean value and variance of the logarithm of the fraction of fi that communicate, ln(fc)

$afc := \mu fc - \sqrt{3} \cdot \sigma fc \qquad afc = 2.679 \times 10^{-2} bfc := \mu fc + \sqrt{3} \cdot \sigma fc \qquad bfc = 3.732 \times 10^{-1}$

$$Meanlnfc := \frac{bfc \cdot \ln(bfc) - bfc - afc \cdot \ln(afc) + afc}{bfc - afc}$$

$Meanlnfc = -1.782 \times 10^{0}$

$$Varlnfc := \frac{2 \cdot afc \cdot \ln(afc) \cdot bfc \cdot \ln(bfc) - afc \cdot bfc \cdot (\ln(bfc))^2 - afc \cdot bfc \cdot (\ln(afc))^2 - 2 \cdot afc \cdot bfc + afc^2 + bfc^2}{(bfc - afc)^2}$$

$Varlnfc = 4.219 \times 10^{-1}$

Mean value and variance of the logarithm of the fraction fL of the planet's life during which the communicating civilizations live, ln(fL)

$afL := \mu fL - \sqrt{3} \cdot \sigma fL \qquad afL = 8.268 \times 10^{-1} bfL := \mu fL + \sqrt{3} \cdot \sigma fL \qquad bfL = 1.173 \times 10^{-6}$

$$MeanlnfL := \frac{bfL \cdot \ln(bfL) - bfL - afL \cdot \ln(afL) + afL}{bfL - afL}$$

$MeanlnfL = -1.382 \times 10^{1}$

$$VarlnfL := \frac{2 \cdot afL \cdot \ln(afL) \cdot bfL \cdot \ln(bfL) - afL \cdot bfL \cdot (\ln(bfL))^2 - afL \cdot bfL \cdot (\ln(afL))^2 - 2 \cdot afL \cdot bfL + afL^2 + bfL^2}{(bfL - afL)^2}$$

$VarlnfL = 1.014 \times 10^{-2}$

Value of the *N* lognormal's μ

$\mu := MeanlnNs + Meanlnfp + Meanlnne(ane, bne) + Meanlnfl + Meanlnfi + Meanlnfc + MeanlnfL$

$\mu = 7.462 \times 10^{0}$

Value of the *N* lognormal's σ

$\sigma := \sqrt{VarlnNs + Varlnfp + Varlnne + Varlnfl + Varlnfi + Varlnfc + VarlnfL}$

$\sigma = 1.392 \times 10^{0} \qquad \sigma^2 = 1.939 \times 10^{0}$

Lognormal of *N* and its statistics

Old value of *N* given by the classical Drake equation:

$N = 3.500 \times 10^{3}$

Mean value of N given by the statistical Drake equation

$\text{MeanValue\_N} := e^{\mu+\frac{\sigma^2}{2}}$ $\text{MeanValue\_N} = 4.590 \times 10^3$

Standard deviation of N given by the statistical Drake equation

$\text{Standard\_Deviation\_N} := e^{\mu+\frac{\sigma^2}{2}} \cdot \sqrt{e^{\sigma^2} - 1}$ $\text{Standard\_Deviation\_N} = 1.120 \times 10^4$

PEAK abscissa of N (mode) given by the statistical Drake equation

$\text{Mode\_N} := e^{\mu-\sigma^2}$ $\text{Mode\_N} = 2.505 \times 10^2$

$\text{Median\_N} := e^{\mu}$ $\text{Median\_N} = 1.741 \times 10^3$

$$\text{f\_lognormal}_N(n, \mu, \sigma) := \frac{1}{n} \cdot \frac{1}{\sqrt{2\pi} \cdot \sigma} \cdot e^{-\frac{(\ln(n)-\mu)^2}{2\sigma^2}}$$

$\text{peak\_height} := \text{f\_lognormal}_N(\text{Mode\_N}, \mu, \sigma)$ $\text{peak\_height} = 4.339 \times 10^{-4}$

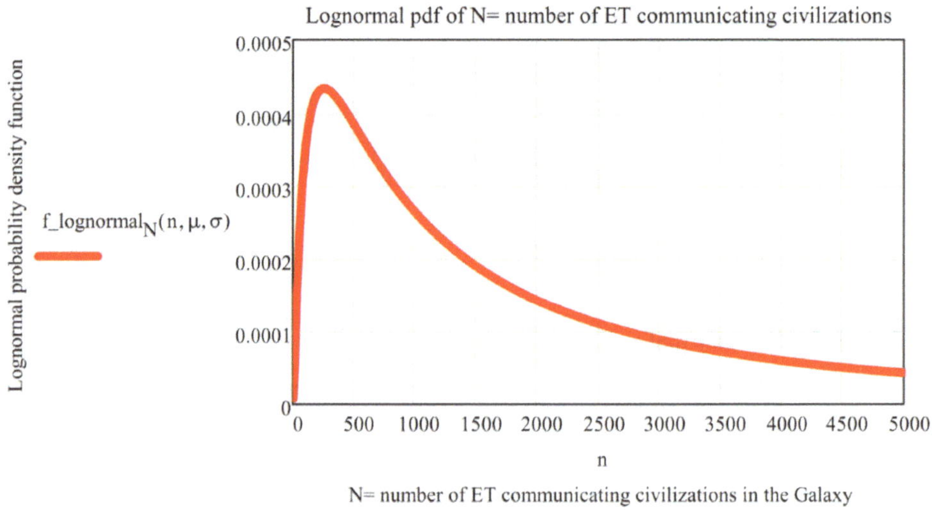

Finding the *distance* between ETs

Defining the light-year (ly), which Mathcad does not know!

$\text{ly} := c \cdot \text{yr}$ $\text{ly} = 9.461 \times 10^{15}\ \text{m}$

Defining the size of the Galaxy

$\text{Radius\_of\_Galaxy} := 50000 \cdot \text{ly}$ $\text{Thickness\_of\_Galaxy\_at\_half\_radius} := 1600 \cdot \text{ly}$

$R_{Galaxy} := \text{Radius\_of\_Galaxy}$ $h_{Galaxy} := \text{Thickness\_of\_Galaxy\_at\_half\_radius}$

Defining the average distance between any two nearby communicating civilizations

$$\text{Tridimensional\_Average\_DISTANCE\_between\_any\_two\_civilizations} := \frac{\sqrt[3]{6 \cdot R_{Galaxy}{}^{2} \cdot h_{Galaxy}}}{\sqrt[3]{N}}$$

$$\text{Tridimensional\_Average\_DISTANCE\_between\_any\_two\_civilizations} = 1.900 \times 10^{3} \cdot \text{ly}$$

$$\text{Av\_DISTANCE(Nciv)} := \frac{\sqrt[3]{6 \cdot R_{Galaxy}{}^{2} \cdot h_{Galaxy}}}{\sqrt[3]{\text{Nciv}}} \qquad \text{Av\_DISTANCE(Ns)} = 4.093 \times 10^{0} \cdot \text{ly}$$

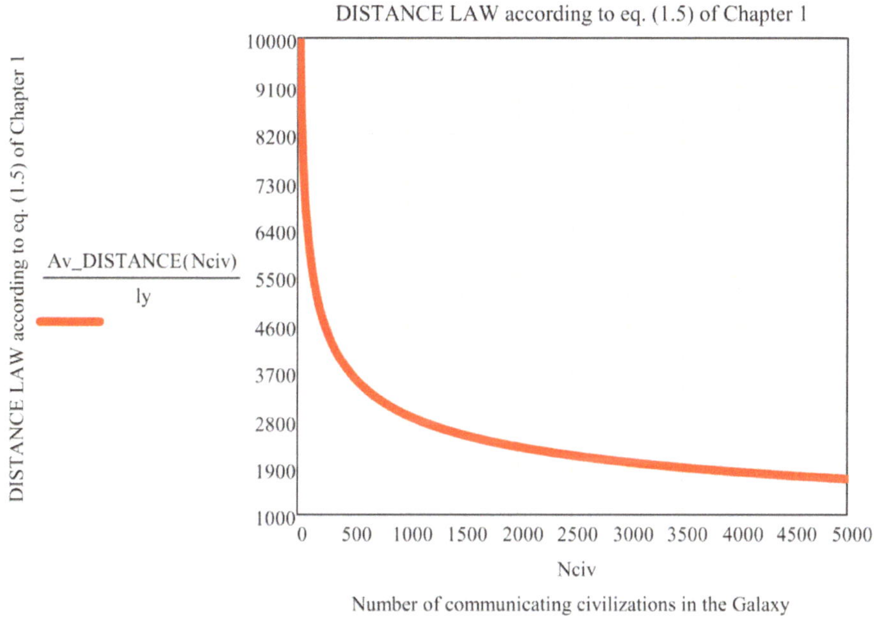

Computing the average distance between any two nearby stars in the Galaxy is necessary to check that the *distance law* of eq. (1.5) holds good for *anything* in the Galaxy *supposed to be uniformly distributed*!—be they stars, or planets, or ET civilizations!

$$\text{Av\_DISTANCE(Ns)} = 4.093 \times 10^{0} \cdot \text{ly}$$

Thus, the average distance between any two nearby stars in the Galaxy is about 4 lt-yr, which is *just the same distance* between the Sun and the nearest star system to us (i.e., the Alpha Cen triple system). This perfect verification of the Distance Law (1.5) is actually possible because the distance of the Solar System from the Galaxy center is about 25,000 lt-yr (i.e., we are just *half way* across the radius of the Galaxy itself!)

***ET distance* as a *random variable* (i.e., as a function of the N random variable) (discovered by this author during the night of September 4 and 5, 2007)**

Defining the C constant as in eq. (1.6)

$$C := \sqrt[3]{6 \cdot R_{Galaxy}{}^{2} \cdot h_{Galaxy}}$$

$C = 2.729 \times 10^{20}\ m \qquad C = 2.884 \times 10^{4} \cdot ly$

ET distance mean value as in eq. (1.B.119)

$$ET\_DISTANCE\_mean\_value := C \cdot e^{\frac{\sigma^2}{18} - \frac{\mu}{3}}$$

$ET\_DISTANCE\_mean\_value = 2.670 \times 10^{3} \cdot ly$

ET distance variance as in eq. (1.B.121)

$Tridimensional\_Average\_DISTANCE\_between\_any\_two\_civilizations = 1.900 \times 10^{3} \cdot ly$

$$ET\_DISTANCE\_variance := C^2 \cdot e^{\frac{\sigma^2}{9} - \frac{2\mu}{3}} \cdot \left(e^{\frac{\sigma^2}{9}} - 1\right)$$

$ET\_DISTANCE\_variance = 1.714 \times 10^{6} \cdot ly^2$

ET distance standard deviation as in eq. (1.B.123)

$$ET\_DISTANCE\_standard\_deviation := C \cdot e^{\frac{\sigma^2}{18} - \frac{\mu}{3}} \cdot \sqrt{e^{\frac{\sigma^2}{9}} - 1}$$

$ET\_DISTANCE\_standard\_deviation = 1.309 \times 10^{3} \cdot ly$

ET maximum distance

$ET\_MAX\_DISTANCE := ET\_DISTANCE\_mean\_value + ET\_DISTANCE\_standard\_deviation$

$ET\_MAX\_DISTANCE = 3.980 \times 10^{3} \cdot ly$

ET minimum distance

$ET\_MIN\_DISTANCE := ET\_DISTANCE\_mean\_value - ET\_DISTANCE\_standard\_deviation$

$ET\_MIN\_DISTANCE = 1.361 \times 10^{3} \cdot ly$

Plotting the ET distance distribution ("Maccone distribution")

Distance probability density function (also called "Maccone distribution"):

$$f_{distance}(d) := \frac{3}{d} \cdot \frac{1}{\sqrt{2 \cdot \pi} \cdot \sigma} \cdot e^{-\frac{\left(\ln\left(\frac{C^3}{d^3}\right) - \mu\right)^2}{2 \cdot \sigma^2}}$$

ET distance peak or mode

$$\text{DISTANCE\_PEAK} := C\, e^{-\frac{\mu}{3}-\frac{\sigma^2}{9}} \qquad \text{DISTANCE\_PEAK} = 1.829\times 10^{19}\cdot \text{m}$$

$$\text{DISTANCE\_PEAK} = 1.933\times 10^{3}\cdot \text{ly}$$

ET distance peak height

$$f_{\text{distance}}\left(C\, e^{-\frac{\mu}{3}-\frac{\sigma^2}{9}}\right) = 3.992\times 10^{-4}\cdot\frac{1}{\text{ly}}$$

$$\text{Peak\_Height} := f_{\text{distance}}\left(C\, e^{-\frac{\mu}{3}-\frac{\sigma^2}{9}}\right) \qquad \text{Peak\_Height} = 4.220\times 10^{-20}\,\frac{1}{\text{m}}$$

$$\text{Mode\_Height} := \frac{3}{C}\cdot\frac{1}{\sqrt{2\cdot\pi}\cdot\sigma}\cdot e^{\frac{\mu}{3}+\frac{\sigma^2}{18}} \qquad \text{Mode\_Height} = 4.220\times 10^{-20}\,\frac{1}{\text{m}}$$

ET *distance* (in light-years) *probability density function*

$$f_{\text{distance\_in\_ly}}(d) := \frac{3}{(d\cdot \text{ly})}\cdot\frac{1}{\sqrt{2\cdot\pi}\cdot\sigma}\cdot e^{\frac{-\left[\ln\left[\frac{C^3}{(d\cdot \text{ly})^3}\right]-\mu\right]^2}{2\cdot\sigma^2}}$$

$$f_{\text{distance\_in\_ly}}(1500) = 3.635\times 10^{-20}\,\frac{1}{\text{m}}$$

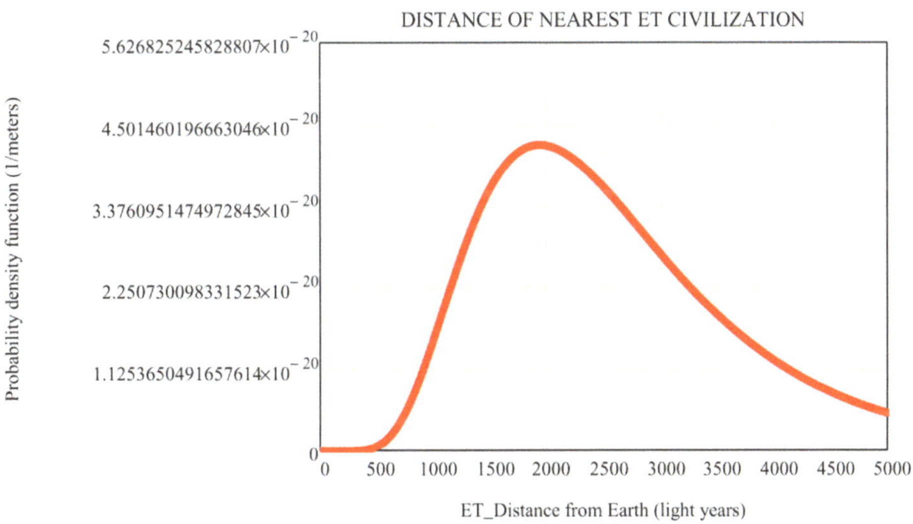

This shows that the probability of finding ETs at distances smaller than 500 light-years is *zero*!

1.12 REFERENCES

[1] Gregory Benford, Jim Benford, and Dominic Benford, "Cost optimized interstellar beacons: SETI", arXiv.org web site (October 22, 2008).
[2] Carl Sagan, *Cosmos*, Random House, New York, 1983. See pp. 298–302 in particular.
[3] Jeffrey Bennet and Seth Shostak, *Life in the Universe*, Second Edition, Pearson/Addison Wesley, San Francisco, 2007, ISBN 0-8053-4753-4. See p. 404 in particular.
[4] Claudio Maccone, "The statistical Drake equation", paper #IAC-08-A4.1.4 presented on October 1, 2008 at the *59th International Astronautical Congress (IAC) held in Glasgow, Scotland, U.K., September 29–October 3, 2008.*
[5] *http://en.wikipedia.org/wiki/Drake_equation*
[6] *http://en.wikipedia.org/wiki/SETI*
[7] *http://en.wikipedia.org/wiki/Astrobiology*
[8] *http://en.wikipedia.org/wiki/Frank_Drake*
[9] Athanasios Papoulis and S. Unnikrishna Pillai, *Probability, Random Variables and Stochastic Processes*, Fourth Edition, Tata McGraw-Hill, New Delhi, 2002, ISBN 0-07-048658-1.
[10] *http://en.wikipedia.org/wiki/Gamma_distribution*
[11] *http://en.wikipedia.org/wiki/Central_limit_theorem*
[12] *http://en.wikipedia.org/wiki/Cumulants*
[13] *http://en.wikipedia.org/wiki/Median*

2

Letting Maxima do the calculations

2.1 A NEW WAY OF PUBLISHING MATHEMATICAL BOOKS

By "mathematical books" we mean books with plenty of written equations. These books are difficult to write and difficult to publish because:

(1) Either the author writes the equations (as "camera-ready" books the equations are not expected to contain misprints).
(2) Or the author submits a file with all equations written by a certain equation editor, but the publisher prefers to have *all equations rewritten anew*. In this case, the rewriting of all equations by virtue of a different equation editor may lead to terrible problems. In fact, the person rewriting the equations does not necessarily understand their meaning and may thus "contribute" his/her own additional misprints, forcing the author to correct them in an endless interchange of emails, criticism, loss of time, etc. thus hampering book publication.
(3) In addition, there is the problem of *proof length*. Some reviewers may or may not like *lengthy* proofs of new mathematical results. They can put pressure on the author to *cut short* all proofs under the false guise of "elegance". As a consequence, students and newcomers to the topics dealt with in the book will then find it impossible to understand these supposedly "elegant" proofs, resulting in them not buying the book quite simply because "nothing is proven in the book" or "the proofs are too difficult to understand".

Having experienced these problems for all his previously published papers and books, the author decided to solve all these problems once and for all by adopting a brand new procedure of writing mathematical books which is described for the first time in this chapter.

The idea is as follows:

(1) The author calculates all his/her mathematical results by virtue of the *free* symbolic manipulator "Maxima". Of course, this implies extra effort for the author of learning the Maxima language, but the subsequent advantages for the publication of all mathematical books absolutely justify this.
(2) The Maxima file containing all inputs (commands telling Maxima what to do) and outputs (the results) is thus produced by the author, who may shorten or enlarge it at will. This we will call the "Input Maxima File".
(3) *Colors* may be used in this file to distinguish inputs (in blue), outputs (in black), and comments in English (in brown), while the input equation numbers are in red and the output equation numbers are in green. Of course, different color conventions might be adopted, but these are the conventions adopted in this book.
(4) This *color file* is then printed by Maxima as a pdf file. The pdf file is then printed in black and white in the book, thus losing the color advantage. But the original color pdf files freely downloadable from a website created by the author, thus allowing the colors to be read again in all their glory! This file we will call the "Output Maxima File".
(5) Both the *Input and Output Maxima Files* will be available from the same site too.
(6) Finally, since Maxima may be downloaded *for free* from the internet, all readers will be capable of *testing* the proofs themselves, as well as "improving them", thus dramatically increasing the number of books sold.

But what is Maxima? Or, more correctly, what is Macsyma, from which Maxima was derived? We briefly answer these questions in the next section.

2.2 A SHORT HISTORY OF "MACSYMA" (1968–PRESENT)

In the early 1960s, the United States was vying with the Soviet Union to be the first country capable of landing humans on the Moon. But the theory of orbits leading to the Moon landing had been calculated *by hand* in the previous 300 years by mathematicians expert in celestial mechanics, and NASA was anxious to have all these calculations rechecked by computers, rather than by hand. This is how Macsyma was born. The following text, describing both the history and the capabilities of Macsyma, was taken by the author from the relevant Wikipedia site (*http://en.wikipedia.org/wiki/Macsyma*).

MACSYMA (Project MAC's SYmbolic MAnipulator) is a computer algebra system that was originally developed between 1968 and 1982 at MIT as part of Project MAC and later marketed commercially. It was the first comprehensive symbolic mathematics system and one of the earliest knowledge-based systems; many of its ideas were later adopted by Mathematica, Maple, and other systems.

Development

The project was initiated in July, 1968 by Carl Engelman, William A. Martin (front end, expression display, polynomial arithmetic), and Joel Moses (simplifier, indefinite integration: heuristic/Risch). Bill Martin was in charge of the project until 1971, and Moses ran it for the next decade. Engelman and his staff left in 1969 to return to the MITRE Corporation. Later major contributors to the core mathematics engine were: Yannis Avgoustis (special functions), David Barton (algsys), Richard Bogen (special functions), Bill Dubuque (limits, Grobner, TriangSys, indefinite integration: Bronstein, power series, number theory, special functions, functional equations, pattern matching, sign queries), Richard Fateman (rational functions, pattern matching, arbitrary precision floating-point), Michael Genesereth (comparison, knowledge database), Jeff Golden (simplifier, language, system), R. W. Gosper (definite summation, special functions, simplification, number theory), Charles Karney (plotting), John Kulp, Ed Lafferty (ODE solution, special functions), Stavros Macrakis (real/imaginary parts, compiler, system), Barry Trager (algebraic integration, factoring, Grobner), Paul Wang (polynomial factorization, limits, definite integration), David Y. Y. Yun, Gail Zacharias (Grobner), and Rich Zippel (power series, polynomial factorization, number theory, combinatorics).

Macsyma was written in Maclisp and was, in some cases, a key motivator for improving that dialect of Lisp in the areas of numerical computing, efficient compilation, and language design. Maclisp itself ran primarily on PDP-6 and PDP-10 computers, but also on the Multics OS and on the Lisp machine architectures. Macsyma was one of the largest, if not the largest, Lisp programs of the time.

Commercialization

In 1979, in response to a request from Richard Fateman, then a professor at UC Berkeley, MIT provided a temporary license for Macsyma code. This was almost immediately brought up on the new VAX-11/780 computer using a dialect of Maclisp, Franz Lisp, written by Fateman's group. MIT reluctantly licensed versions of VAX Macsyma from Berkeley to about 50 universities, starting with Caltech, under condition that the licenses would be revoked when negotiations for a proper license were complete. Indeed this happened when the negotiation with Symbolics (see below) was complete. Symbolics was reluctant to license the VAX product, since VAX constituted competitive hardware to their own Lisp machines and therefore suppressed the VAX software for five years. UC Berkeley also brought up copies of Macsyma on Motorola 68000-based systems, most notably Sun workstations. Symbolics suppressed those as well. At the same time Fateman worked to change the (now revoked) temporary license for Macsyma into something more permanent.

Ultimately, the United States Department of Energy imposed on MIT the requirement that they release a copy into the National Energy Software Center (NESC) library in 1982, while allowing MIT to assert restrictions in terms of high price and no redistribution. This was intended to protect the technology transfer to Symbolics. (Such restrictions were lifted in about 2002.) This so-called DOE

Macsyma had been rewritten at MIT in a pre-Common Lisp dialect called NIL lisp, incompletely implemented on VAX/VMS, an unpopular system in academia, where Berkeley VAX Unix was common. DOE Macsyma formed the basis for the much later open-source Maxima system.

In 1981, Moses and Richard Pavelle, MIT staffers and proponents of applying Macsyma to engineering and science, proposed forming a company to commercialize Macsyma. However, MIT policy prevented MIT personnel from profiting directly from MIT developments. In early 1982, Macsyma was licensed by MIT to Arthur D. Little, Inc., which became the broker for Macsyma and soon licensed Macsyma to Symbolics in late 1982. It appears that Symbolics was not so much interested in the code as in keeping Macsyma out of the software catalog of its competitor in the Lisp machine business, LMI. The business arrangement between Symbolics and Arthur D. Little required a royalty payment to ADL of 15% of Macsyma gross sales. This led to speculation on the desire of MIT and ADL to see Macsyma prosper. The development of Macsyma continued at Symbolics despite the fact that it was seen as a diversion from the sales of Lisp machines, which Symbolics considered to be its main business. Macsyma sales and the leveraged sales of Lisp machines reached 10% of overall sales at Symbolics within two years. Despite resistance from many in Symbolics, Macsyma was released for DEC VAX computers and Sun workstations using Berkeley's Franz Lisp in 1986 and 1987.

However, problems in Symbolics and within the Macsyma group led to the decline of Macsyma. In the first half of 1986, Macsyma revenues were lower than in the first half of 1985, in a growing industry. Wolfram's SMP program and Waterloo Maple were growing at this time, although Macsyma was easily superior to these other packages in symbolic math.

Pavelle ran the Macsyma division at Symbolics until early 1986. In the second half of 1986 Richard Petti became the manager of the Macsyma business. Macsyma cut headcount but expanded its sales force and marketing, and focused its developers more on features that customers asked for. (For example, the Grobner algorithm developed in the 1970s at MIT was not installed in Macsyma copies for sale until 1987.) In 1987, Macsyma annual revenues roughly doubled. Macsyma became more user friendly: documentation and online help were reorganized and expanded; some command names were changed to be more mnemonic. Petti argued to Symbolics management that Macsyma was a "strategic business unit" that should be funded based on its performance and potential, not based on the prospects of the main workstation business. However, Symbolics cut the Macsyma headcount after this period of sharp growth; Symbolics tried to use Macsyma as a tiny cash cow to fund the losses of the workstation business that was 30 times its size.

The greatest product weakness of Macsyma was its numerical analysis. Mastering symbolic math was a Herculean task; but numerical capabilities were critical to get a piece of the much larger engineering and lower end scientific markets. At MIT Macsyma had a link to the IMSL (now Rogue Wave Software) numerical libraries, but this link was severed when Macsyma moved to Symbolics. Lisp developers at Symbolics generally believed that numerical analysis was an old technology that was not important for Lisp applications, so they declined to invest in it. Double precision arithmetic in Macsyma (on a PC version) was about six times slower than Fortran. In addition, Macsyma's matrices were implemented as lists of lists, which cost another factor of perhaps 5–10 in speed for key algorithms. Macsyma did not have many of the basic algorithms of numerical linear algebra, such as LU decomposition.

In 1987–1988, the Macsyma group tried to build a PC Macsyma with Gold Hill Lisp. (Earlier, Symbolics had killed its own project to build a Lisp compiler for standard computers to avoid competing with Lisp machine sales. This was a controversial move that, by some accounts, was undertaken without approval of senior management. Symbolics also declined to cooperate with Sun to make Lisp available on Sun workstations for the same reason.) Gold Hill Lisp was too unstable, and its weak architecture made it impossible for Gold Hill to eliminate the bugs. This was a crucial failure for Macsyma. It meant that Macsyma could not respond on PCs when Mathematica appeared on Apple computers in mid-1988. Macsyma appeared on Windows PCs in August 1989 using the CLOE Lisp from Symbolics. However, the Macsyma staff was too small and lacked the mix of skills needed to add the kind of graphics, notebook interface, and better numerics that Mathematica had.

By 1989, it was clear to Petti that Symbolics would implode due to poor product strategy, and that it would take Macsyma with it. (Some product issues: (1) Symbolics' software was designed for MIT-class software developers without enough concessions to others. (2) Symbolics' world-class software was losing share in the market due to dependence on very high-cost hardware. (3) In moving to VLSI hardware in the mid-1980s, Symbolics converted from 36-bit word size to 40-bit, without justification from the market for the enormous development cost of this change.) Macsyma could not assemble a buy-out team due to lack of cooperation from MIT. After keeping reasonably quiet since 1986 about the product issues, Petti tried in late 1988 to persuade Symbolics to adopt a software-only or board-level strategy; but the fourth president in four years would hear nothing of it. In 1990 Petti left Symbolics for a start-up.

Macsyma, Inc.

Macsyma, Inc. was founded in 1992 by Russell Noftsker (chairman, who had co-founded Symbolics) and Richard Petti (president, who had turned around Macsyma for a time at Symbolics) by raising funds and purchasing the rights to Macsyma from the ailing Symbolics. Although the market was growing fast, Macsyma sales in 1991 and early 1992 were still falling rapidly. Macsyma's market share in symbolic math software had fallen from 70% in 1987 to 1% in 1992. By 1993, market growth had slowed and the market had standardized on Mathematica and Maple. The competitors had development staffs that were four to eight times as large as that of Macsyma, Inc. throughout the 1990s.

In early 1995 the company shipped Macsyma 2.0.5, with many improvements: On Wester's large test of symbolic math, Macsyma 2.0.5 scored 10% better than Maple and 15% better than Mathematica. Although Macsyma 2.0.5 was still very slow at numerics, it had a greatly strengthened portfolio of numerical analysis and linear algebra routines. (In 1996 Macsyma added LAPACK which greatly increased the speed of most numerical linear algebra.) .

The development team, at this time, included Jeff Golden (language, compiler, etc.), Bill Gosper (special functions, summation), Howard Cannon (user interface, optimization), and consultant Bill Dubuque (integration, equation solving, database, optimization). Other developers made major contributions in numerical analysis, graphics, and help systems.

The company could not continue indefinitely to outperform staffs four to eight times as large and reverse the market momentum. Market share did not increase above 2%, because the competitors were entrenched in all key accounts, converting to a new product was very costly in learning time, and market growth declined. Also, starting in 1992 or 1993, Mathsoft engaged in a Pyrrhic strategy of spending $10 million on direct mail at very low prices, which won much of the remaining growth in the symbolic math market, just when Macsyma, Inc. was struggling to rebuild its world-class product.

In 1999, Macsyma was acquired by Tenedos LLC, a holding company that previously had purchased Symbolics. The holding company has not re-released or resold Macsyma, but Macsyma continues to be distributed by Symbolics.

2.3 A SHORT HISTORY OF "MAXIMA" (1982–PRESENT)

Maxima, a descendent of Macsyma, is the freely downloadable symbolic manipulator used by the author to prove many results included in this book.

Again, we resort to the relevant Wikipedia site (*http://en.wikipedia.org/wiki/Maxima_%28software%29*) in order to provide a synthetic description of Maxima and its historic development.

Maxima is a computer algebra system based on a 1982 version of Macsyma. It is written in Common Lisp and runs on all POSIX platforms such as MAC OS X, Unix, BSD, and GNU/Linux as well as under Microsoft Windows. It is free software released under the terms of the GNU General Public License.

History

Maxima is based on a 1982 version of Macsyma, which was developed at MIT with funding from the United States Department of Energy and other government agencies. A version of Macsyma was maintained by Bill Schelter from 1982 until his death in 2001. In 1998 Schelter obtained permission from the Department of Energy to release his version under the GPL. That version, now called Maxima, is maintained by an independent group of users and developers. Maxima does not include any of the many modifications and enhancements made to the commercial version of Macsyma during 1982–1999. Though the core functionality remains similar, code depending on these enhancements may not work on Maxima, and bugs which were fixed in Macsyma may still be present in Maxima, and vice versa.

Features

Maxima includes a complete programming language with ALGOL-like syntax but Lisp-like semantics. It is written in Common Lisp, and can be accessed programmatically and extended, as the underlying Lisp can be called from Maxima. It uses Gnuplot for drawing.

2.4 A FAMOUS EXAMPLE OF MACSYMA CORRECTING HUMAN ERRORS

The French mathematician and astronomer Charles Delaunay (1816–1872, Wikipedia site: *http://en.wikipedia.org/wiki/Charles-Eug%C3%A8ne_Delaunay*) spent 20 years of his life (1847–1867) computing small variations of the lunar orbit due to perturbations of the angle between the orbital planes of the Earth and Moon. The first ten years of his work (1847–1857) Delaunay devoted to computing by hand these effects, and the next ten years (1857–1867) he devoted to checking his results, which he finally published in 1867. Obviously, nobody was interested in spending ten years of his/her life to check the calculations of Delaunay, and so his work was virtually forgotten until World War II, when it was reconsidered with the goal of localizing German U-Boats (submarines), which were using astronomical methods to find their position at sea. After the war, with the advent of the space age, Delaunay's works were studied again to correctly position artificial satellites, and the need to check them mathematically in full became obvious. Macsyma performed this "miracle" in 1973: 20 hours of Macsyma computer calculations were sufficient to check Delaunay's 20 years of work, and it was found that he had made just a single small error, fortunately without important consequences (for more details, see [1]).

2.5 FIRST MAXIMA EXAMPLE: LOGNORMAL PROPERTIES UP TO KURTOSIS

The first, simple example of a Maxima code offered to our readers is the short file "lognormal_properties_up_to_kurtosis" (see Appendix 2.A). It calculates the basic properties of any lognormal distribution, such as:

(1) Definition of lognormal probability density function.
(2) Normalization condition.
(3) Mean value (i.e., first moment).
(4) Second moment.
(5) Variance.
(6) Standard deviation.
(7) Mode (i.e., abscissa of the (only) peak of the lognormal probability density function).

(8) Peak ordinate.
(9) Median (i.e., abscissa dividing the area under the probability density function exactly in two halves).
(10) Third moment.
(11) Fourth moment.
(12) First four cumulants, related to the first four moments by known formulas proven in all statistics courses.
(13) Skewness.
(14) Kurtosis.

These results are of course the same as those listed in Table 1.1, where the lognormal distribution stands for the probability distribution of the number N of communicating civilizations now existing in the Galaxy.

2.6 SECOND MAXIMA EXAMPLE: CORRECTING THE AUTHOR'S WRONG EQUATION FOR THE SKEWNESS OF BOTH THE LOGNORMAL AND MACCONE DISTRIBUTIONS PUBLISHED BETWEEN 2008 AND 2011

The second Maxima file (see Appendix 2.B) deals with the correction of an *error* that the author made in all his papers about the statistical Drake equation published between 2008 and 2010 inclusive. The error is as follows:

(1) The definition of *skewness* of any probability distribution is the ratio of its two *cumulants* K_3 over K_2 raised to the power 3/2 (see the Wikipedia site *http://en.wikipedia.org/wiki/Skewness*)

$$\text{Skewness} = \frac{K_3}{(K_2)^{\frac{3}{2}}}. \tag{2.1}$$

(2) The author in all his papers about the statistical Drake equation published between 2008 and 2011 unfortunately used the *fourth cumulant* K_4 *instead of the third one* K_3, thus inferring an incorrect equation for the skewness of the relevant probability distribution.
(3) In practice, that happened for both the lognormal and the Maccone (distance) probability distribution.
(4) The author finally realized his mistake on April 29, 2011, and he then created this Maxima file yielding the *correct skewness* for both the lognormal distribution:

$$\text{Skewness\_of\_lognormal\_distribution} = \frac{K_3}{(K_2)^{\frac{3}{2}}} = (e^{\sigma^2} + 2)\sqrt{e^{\sigma^2} - 1} \tag{2.2}$$

and for the Maccone (distance) distribution:

$$\text{Skewness\_of\_the\_Maccone\_(distance)\_distribution} = \frac{K_3}{(K_2)^{\frac{3}{2}}} = \frac{e^{\frac{2\sigma^2}{9}} + e^{\frac{\sigma^2}{9}} - 2}{\sqrt{e^{\frac{\sigma^2}{9}} - 1}}. \tag{2.3}$$

The second Maxima file "Incorrect_Skewness_for_lognormal_and_Maccone_distributions_prior_to_April_29_2011" (see Appendix 2.B) just proves that.

2.7 THIRD MAXIMA EXAMPLE: NUMBER OF PROTONS AND NEUTRONS IN EINSTEIN'S STATIC UNIVERSE

Our third Maxima example deals with Einstein's general theory of relativity. This is to show Maxima's great capabilities in handling tensor calculus, Euler–Lagrange equations of a certain variational principle (the Einstein–Hilbert action, in this case), and solving relevant differential equations. However, before reading the file "TENSOR_DEMO_Number_of_protons_and_neutrons_in_Einstein_static_universe" (see Appendix 2.C), the reader must understand how the four components of the input metric tensor (lg, i.e. "lower g sub ik", i.e. the covariant components of the metric tensor in the unknown function $a(t)$) are derived. We do this in this introductory section using the above Maxima example.

If we draw a circle of radius R centered at the origin of two axes, it has the equation $x^2 + y^2 = R^2$. This may be called a one-sphere, meaning that *on* the circle one can only move either forward or backward (i.e., the one-sphere has *one* degree of freedom only). Next comes the two-sphere. It simply is what we popularly call "a sphere", a two-dimensional manifold *on* the sphere (i.e., two degrees of freedom are requested to specify a point *on* the sphere—usually longitude and latitude). This sphere's Cartesian equation is obviously $x^2 + y^2 + z^2 = R^2$ but one must then note that this is the equation of the two-sphere *embedded* in three-dimensional Euclidean space, just as $x^2 + y^2 = R^2$ was the equation of the one-sphere (=circle) *embedded* in the two-dimensional Euclidean space x, y. We are now ready to understand what the three-sphere is. A three-sphere is a three-dimensional sphere *on* the sphere (i.e., it is a three-dimensional sphere *embedded* in a four-dimensional Euclidean space of coordinates x, y, z, w, having called w the new, fourth coordinate of this four-dimensional Euclidean embedding space). Clearly, the Cartesian equation of such a three-sphere is $x^2 + y^2 + z^2 + w^2 = R^2$. In his 1917 seminal paper entitled "Cosmological considerations on the general theory of relativity" (English translation from the German) Einstein defined the "universe" as a three-sphere in which the distance between us and any other galaxy is given by $r^2 = x^2 + y^2 + z^2$. Inserting this definition of "distance" into the three-sphere equation, the latter is turned into:

$$r^2 + w^2 = R^2. \tag{2.4}$$

Let us now introduce a new *parameter* k, allowing us to extend all results from just a three-sphere (i.e., a three-sphere with a *positive* curvature) to a three-dimensional Euclidean space (i.e., the ordinary three-dimensional Euclidean space, which has *no curvature*) and even to a three-dimensional pseudosphere (i.e., a three-sphere with a *negative* radius). The idea is to rewrite (2.4) as follows:

$$r^2 + w^2 = \frac{R^2}{k} \tag{2.5}$$

that obviously reduces to (2.4) if $k = +1$, but it also yields the Euclidean space if $k = 0$ (amounting to a "sphere" with an *infinite radius* $R \to \infty$), and finally the three-pseudosphere if $k = -1$. In other words, one has:

$$k = \begin{cases} +1 & \text{three-sphere (closed universe)} \\ 0 & \text{Euclidean three-dimensional space (open universe)} \\ -1 & \text{three-pseudosphere (open universe).} \end{cases} \tag{2.6}$$

Let us now solve (2.5) for w, taking the positive sign in front of the square root:

$$w = \sqrt{\frac{R^2}{k} - r}. \tag{2.7}$$

Differentiating this with respect to r (r is "the coordinate of the movable galaxy") one gets

$$dw = \frac{-2r\,dr}{2\sqrt{\frac{R^2}{k} - r}} = \frac{-r\,dr}{w} \tag{2.8}$$

yielding

$$dw^2 = \frac{r^2}{w^2}\,dr^2. \tag{2.9}$$

Let us now consider the line element dl^2 (=square of the distance between two infinitesimally closed points) of our three-dimensional spatial manifold *embedded in four-dimensional Euclidean space* with fourth coordinate w. In spherical coordinates (r, θ, ϕ), the three-dimensional part of it equals $dr^2 + r^2\,d\theta^2 + r^2 \sin^2\theta\,d\phi^2$, to which one must add the dw^2 of the fourth coordinate w. In other words, one has

$$dl^2 = (dr^2 + r^2\,d\theta^2 + r^2 \sin^2\theta\,d\phi^2) + dw^2. \tag{2.10}$$

Inserting (2.9) and (2.7) into (2.10) *amounts to projecting the three-sphere onto its equatorial plane* (this is called "orthogonal projection"). In other words, the three-dimensional world in which we humans live is the "projection" of the curved three-sphere embedded in four-dimensional Euclidean space just as the maps of the Earth are a plane projection of the curved two-sphere embedded in Euclidean three-dimensional space. Then, rearranging (2.10) by invoking (2.9) and

(2.7), one gets:

$$dl^2 = (dr^2 + r^2\,d\theta^2 + r^2\sin^2\theta\,d\phi^2) + dw^2 = dr^2 + r^2\,d\theta^2 + r^2\sin^2\theta\,d\phi^2 + \frac{r^2}{w^2}dr^2$$
$$= \left(1 + \frac{r^2}{w^2}\right)dr^2 + (r^2\,d\theta^2 + r^2\sin^2\theta\,d\phi^2) = \left(\frac{w^2 + r^2}{w^2}\right)dr^2 + (r^2\,d\theta^2 + r^2\sin^2\theta\,d\phi^2)$$
$$= \left(\frac{\frac{R^2}{k}}{\frac{R^2}{k} - r^2}\right)dr^2 + (r^2\,d\theta^2 + r^2\sin^2\theta\,d\phi^2) = \left(\frac{1}{1 - \frac{k}{R^2}r^2}\right)dr^2 + (r^2\,d\theta^2 + r^2\sin^2\theta\,d\phi^2)$$
$$= \frac{1}{1 - k\left(\frac{r}{R}\right)^2}dr^2 + (r^2\,d\theta^2 + r^2\sin^2\theta\,d\phi^2) = \frac{dr^2}{1 - k\left(\frac{r}{R}\right)^2} + (r^2\,d\theta^2 + r^2\sin^2\theta\,d\phi^2). \tag{2.11}$$

In other words, the spatial line element of the universe (where R is the universe radius) is

$$dl^2 = \frac{dr^2}{1 - k\left(\frac{r}{R}\right)^2} + (r^2\,d\theta^2 + r^2\sin^2\theta\,d\phi^2). \tag{2.12}$$

We may now *normalize to one* this manifold *by assuming* $R = 1$. This also amounts to saying that "the distance r between us and the moving galaxy is measured in units of the radius of the universe R." The last equation (2.12) is thus turned into

$$dl^2 = \frac{dr^2}{1 - kr^2} + (r^2\,d\theta^2 + r^2\sin^2\theta\,d\phi^2). \tag{2.13}$$

So far, we have *not* allowed the universe to *expand*. This is the mistake that Einstein "made" in 1917 and that he referred to as his "biggest blunder", since he was forced to introduce the cosmological term and constant in his equations to keep his universe static. But in 1922 Alexander Friedman removed this constraint, and in 1927 Edwin Hubble observationally proved that the universe is indeed expanding according to Hubble's Law. Thus, in order to let the universe *expand* (*Big Bang*) or even *contract* (*Big Crunch*), we may *multiply the whole line element* (2.13)—that is, every differential in (2.13)—*by an expansion factor* $a(t)$ that is a function of time t (i.e., coordinate time as measured on Earth). Since all differentials in (2.13) are squared, (2.13) is thus changed into:

$$dl^2 = \frac{a^2(t)\,dr^2}{1 - kr^2} + a^2(t)(r^2\,d\theta^2 + r^2\sin^2\theta\,d\phi^2). \tag{2.14}$$

This is the spatial part of the expanding universe.

What about its temporal part? Well, the line element ds^2 of the four-dimensional curved spacetime of general relativity is notoriously given by (this is the space-like metric that Maxima "prefers" rather than the time-like metric):

$$ds^2 = dl^2 - g_{44}(r, \theta, \phi, t)\, dt^2. \tag{2.15}$$

The physical meaning of $g_{44}(r, \theta, \phi, t)$ is immediately found if one remembers that the light-cone of special relativity has the simple equation $ds^2 = 0$, thus changing (2.15) into

$$g_{44}(r, \theta, \phi, t) = \left(\frac{dl}{dt}\right)^2 = (\text{light speed in the curved manifold})^2. \tag{2.16}$$

Now, we assume that the speed of light in a vacuum is the same all over the universe and equals c. Thus $g_{44}(r, \theta, \phi, t)$ is constant all over the universe (i.e., it does not depend on the coordinates (r, θ, ϕ, t), and equals c^2). Inserting this into (2.15) one finally gets the full spacetime line element of the expanding universe:

$$ds^2 = dl^2 - c^2\, dt^2 = \frac{a^2(t)\, dr^2}{1 - kr^2} + a^2(t)(r^2\, d\theta^2 + r^2 \sin^2\theta\, d\phi^2) - c^2\, dt^2. \tag{2.17}$$

Rewriting this line element as the 4×4 matrix of covariant metric tensor components, g_{ik}, one finally gets:

$$g_{ik} = \text{lg} = \begin{bmatrix} \dfrac{a^2(t)}{1 - kr^2} & 0 & 0 & 0 \\ 0 & a^2(t)r^2 & 0 & 0 \\ 0 & 0 & a^2(t)r^2 \sin^2\theta & 0 \\ 0 & 0 & 0 & -c^2 \end{bmatrix}. \tag{2.18}$$

This is precisely the input covariant metric tensor—equations (%i4) through (%o5)—of our Maxima file: "TENSOR_DEMO_Number_of_protons_and_neutrons_in_Einstein_static_universe".

2.8 REFERENCE

[1] R. Pavelle, M. Rothstein, and J. Fitch, "Computer algebra," *Scientific American*, **245**(6), 102–113 (December 1981).

APPENDIX 2.A "LOGNORMAL_PROPERTIES_UP_TO_KURTOSIS"

LOGNORMAL DISTRIBUTIONS:
their analytic properties derived by virtue of Maxima.
But only up to the fourth cumulant, skewness and kurtosis.
For a general reference, please see the Wikipedia site about
"Log-normal Distributions".

Author: Claudio Maccone (e-mail: clmaccon@libero.it), August 7th, 2011.

```
(%i1) kill(all);
(%o0) done
```

Definition of the lognormal pdf.

```
(%i1) assume(mu>0, sigma>0);
```

(%o1) $[\mu>0,\sigma>0]$

```
(%i2) lgn(t, mu, sigma):=(1/(sqrt(2*%pi)*sigma*(t)))*%e^(-(log(t)-mu)^2/(2*sigma^2));
```

(%o2) $\mathrm{lgn}(t,\mu,\sigma):=\frac{1}{\sqrt{2\pi}\,\sigma\,t}\,\%e^{-\frac{(\log(t)-\mu)^2}{2\sigma^2}}$

Normalization Condition.

```
(%i3) 'integrate(lgn(t, mu, sigma),t,0,inf)=integrate(lgn(t, mu, sigma),t,0,inf);
```

(%o3) $\frac{\int_0^\infty \frac{\%e^{-\frac{(\log(t)-\mu)^2}{2\sigma^2}}}{t}\,dt}{\sqrt{2}\,\sqrt{\pi}\,\sigma}=1$

MEAN VALUE, i.e. first moment.

```
(%i4) 'integrate(t*lgn(t, mu, sigma),t,0,inf)=integrate(t*lgn(t, mu, sigma),t,0,inf);
```

(%o4) $\frac{\int_0^\infty \%e^{-\frac{(\log(t)-\mu)^2}{2\sigma^2}}\,dt}{\sqrt{2}\,\sqrt{\pi}\,\sigma}=\%e^{\frac{\sigma^2+2\mu}{2}}$

```
(%i5) mean_value_eq:mean_value=rhs(%);
```

(%o5) $mean\_value=\%e^{\frac{\sigma^2+2\mu}{2}}$

2nd moment.

(%i6) 'integrate(t^2*lgn(t, mu, sigma),t,0,inf)=integrate(t^2*lgn(t, mu, sigma),t,0,inf);

(%o6) $$\frac{\int_0^{\infty} t\,\%e^{-\frac{(\log(t)-\mu)^2}{2\sigma^2}}\,dt}{\sqrt{2}\sqrt{\pi}\,\sigma}=\%e^{2\sigma^2+2\mu}$$

(%i7) factor(%);

(%o7) $$\frac{\int_0^{\infty} t\,\%e^{-\frac{\log(t)^2}{2\sigma^2}+\frac{\mu\log(t)}{\sigma^2}-\frac{\mu^2}{2\sigma^2}}\,dt}{\sqrt{2}\sqrt{\pi}\,\sigma}=\%e^{2\sigma^2+2\mu}$$

(%i8) second_moment_eq:second_moment=rhs(%);

(%o8) $$second\_moment=\%e^{2\sigma^2+2\mu}$$

VARIANCE.

(%i9) factor(radcan(second_moment_eq-mean_value_eq^2));

(%o9) $$second\_moment-mean\_value^2=\left(\%e^{\sigma^2}-1\right)\%e^{\sigma^2+2\mu}$$

(%i10) variance_eq:variance=rhs(%);

(%o10) $$variance=\left(\%e^{\sigma^2}-1\right)\%e^{\sigma^2+2\mu}$$

STANDARD DEVIATION.

(%i11) sqrt(variance_eq);

(%o11) $$\sqrt{variance}=\sqrt{\%e^{\sigma^2}-1}\;\%e^{\frac{\sigma^2+2\mu}{2}}$$

(%i12) standard_deviation_eq:standard_deviation=rhs(%);

(%o12) $$standard\_deviation=\sqrt{\%e^{\sigma^2}-1}\;\%e^{\frac{\sigma^2+2\mu}{2}}$$

MODE, i.e. PEAK.

(%i13) maximum_cond:diff(lgn(t, mu, sigma), t)=0;

(%o13) $-\dfrac{\%e^{-\frac{(\log(t)-\mu)^2}{2\sigma^2}}(\log(t)-\mu)}{\sqrt{2}\sqrt{\pi}\sigma^3 t^2}-\dfrac{\%e^{-\frac{(\log(t)-\mu)^2}{2\sigma^2}}}{\sqrt{2}\sqrt{\pi}\sigma t^2}=0$

(%i14) peak_time_eq:first(solve(maximum_cond, t));

(%o14) $t=\%e^{\mu-\sigma^2}$

(%i15) mode_time_eq:p=rhs(peak_time_eq);

(%o15) $p=\%e^{\mu-\sigma^2}$

(%i16) finding_the_mode_ord:lgn(rhs(mode_time_eq), mu, sigma);

(%o16) $\dfrac{\%e^{\frac{\sigma^2}{2}-\mu}}{\sqrt{2}\sqrt{\pi}\sigma}$

(%i17) mode_ord_eq:peak_ord=%;

(%o17) $peak\_ord=\dfrac{\%e^{\frac{\sigma^2}{2}-\mu}}{\sqrt{2}\sqrt{\pi}\sigma}$

Median.

(%i18) assume(median>0);

(%o18) $[median>0]$

(%i19) assume(median>1);

(%o19) $[median>1]$

(%i20) median_cond:'integrate(lgn(t, mu, sigma),t,0,median)='integrate(lgn(t, mu, sigma),t,median,inf);

(%o20) $\dfrac{\int_0^{median}\frac{\%e^{-\frac{(\log(t)-\mu)^2}{2\sigma^2}}}{t}dt}{\sqrt{2}\sqrt{\pi}\sigma}=\dfrac{\int_{median}^{\infty}\frac{\%e^{-\frac{(\log(t)-\mu)^2}{2\sigma^2}}}{t}dt}{\sqrt{2}\sqrt{\pi}\sigma}$

(%i21) median_cond:integrate(lgn(t, mu, sigma),t,0,median)=integrate(lgn(t, mu, sigma),t,median,inf);

(%o21) $\dfrac{\frac{\sqrt{\pi}\sigma}{\sqrt{2}}-\frac{\sqrt{\pi}\,\mathrm{erf}\left(\frac{\sqrt{2}\mu-\sqrt{2}\log(median)}{2\sigma}\right)\sigma}{\sqrt{2}}}{\sqrt{2}\sqrt{\pi}\sigma}=\dfrac{\frac{\sqrt{\pi}\,\mathrm{erf}\left(\frac{\sqrt{2}\mu-\sqrt{2}\log(median)}{2\sigma}\right)\sigma}{\sqrt{2}}+\frac{\sqrt{\pi}\sigma}{\sqrt{2}}}{\sqrt{2}\sqrt{\pi}\sigma}$

(%i22) solve(%,median);

(%o22) $[median = \%e^{\mu}]$

(%i23) median_eq:median_time=rhs(first(%));

(%o23) $median\_time = \%e^{\mu}$

3rd Moment.

(%i24) 'integrate(t^3*lgn(t, mu, sigma),t,0,inf)=integrate(t^3*lgn(t, mu, sigma),t,0,inf);

(%o24) $$\frac{\int_0^{\infty} t^2\, \%e^{-\frac{(\log(t)-\mu)^2}{2\sigma^2}}\, dt}{\sqrt{2}\sqrt{\pi}\,\sigma} = \%e^{\frac{9\sigma^2+6\mu}{2}}$$

(%i25) factor(%);

(%o25) $$\frac{\int_0^{\infty} t^2\, \%e^{-\frac{\log(t)^2}{2\sigma^2}+\frac{\mu\log(t)}{\sigma^2}-\frac{\mu^2}{2\sigma^2}}\, dt}{\sqrt{2}\sqrt{\pi}\,\sigma} = \%e^{\frac{9\sigma^2}{2}+3\mu}$$

(%i26) third_moment_eq:third_moment=rhs(%);

(%o26) $$third\_moment = \%e^{\frac{9\sigma^2}{2}+3\mu}$$

4th Moment.

(%i27) 'integrate(t^4*lgn(t, mu, sigma),t,0,inf)=integrate(t^4*lgn(t, mu, sigma),t,0,inf);

(%o27) $$\frac{\int_0^{\infty} t^3\, \%e^{-\frac{(\log(t)-\mu)^2}{2\sigma^2}}\, dt}{\sqrt{2}\sqrt{\pi}\,\sigma} = \%e^{8\sigma^2+4\mu}$$

(%i28) factor(%);

(%o28) $$\frac{\int_0^{\infty} t^3\, \%e^{-\frac{\log(t)^2}{2\sigma^2}+\frac{\mu\log(t)}{\sigma^2}-\frac{\mu^2}{2\sigma^2}}\, dt}{\sqrt{2}\sqrt{\pi}\,\sigma} = \%e^{8\sigma^2+4\mu}$$

(%i29) fourth_moment_eq:fourth_moment=rhs(%);

(%o29) $$fourth\_moment = \%e^{8\sigma^2+4\mu}$$

First four Cumulants.

```
(%i30) def_K1:K[1]=rhs(mean_value_eq);
```

(%o30) $K_1 = \%e^{\frac{\sigma^2+2\mu}{2}}$

```
(%i31) def_K2:K[2]=rhs(second_moment_eq)-K[1]^2;
```

(%o31) $K_2 = \%e^{2\sigma^2+2\mu} - K_1^2$

```
(%i32) def_K3:K[3]=rhs(third_moment_eq)-3*K[1]*K[2]-K[1]^3;
```

(%o32) $K_3 = \%e^{\frac{9\sigma^2}{2}+3\mu} - 3K_1K_2 - K_1^3$

```
(%i33) def_K4:K[4]=rhs(fourth_moment_eq)-4*K[1]*K[3]-3*K[2]^2-6*K[2]*K[1]^2-K[1]^4;
```

(%o33) $K_4 = \%e^{8\sigma^2+4\mu} - 4K_1K_3 - 3K_2^2 - 6K_1^2K_2 - K_1^4$

```
(%i34) def_K2,def_K1;
```

(%o34) $K_2 = \%e^{2\sigma^2+2\mu} - \%e^{\sigma^2+2\mu}$

```
(%i35) K2_eq:factor(radcan(%));
```

(%o35) $K_2 = \left(\%e^{\sigma^2} - 1\right)\%e^{\sigma^2+2\mu}$

```
(%i36) def_K3,def_K2,def_K1;
```

(%o36) $K_3 = \%e^{\frac{9\sigma^2}{2}+3\mu} - 3\,\%e^{\frac{\sigma^2+2\mu}{2}}\left(\%e^{2\sigma^2+2\mu} - \%e^{\sigma^2+2\mu}\right) - \%e^{\frac{3(\sigma^2+2\mu)}{2}}$

```
(%i37) K3_eq:factor(radcan(%));
```

(%o37) $K_3 = \left(\%e^{\sigma^2} - 1\right)^2\left(\%e^{\sigma^2} + 2\right)\%e^{\frac{3\sigma^2}{2}+3\mu}$

```
(%i38) def_K4,def_K1,K2_eq,K3_eq;
```

(%o38) $K_4 = -4\left(\%e^{\sigma^2} - 1\right)^2\left(\%e^{\sigma^2} + 2\right)\%e^{\frac{\sigma^2+2\mu}{2}+\frac{3\sigma^2}{2}+3\mu} + \%e^{8\sigma^2+4\mu} - 6\left(\%e^{\sigma^2} - 1\right)\%e^{2\sigma^2+4\mu} - 3\left(\%e^{\sigma^2} - 1\right)^2\%e^{2(\sigma^2+2\mu)} - \%e^{2(\sigma^2+2\mu)}$

```
(%i39) K4_eq:factor(radcan(%));
```

(%o39) $K_4 = \left(\%e^{\sigma^2} - 1\right)^3\left(\%e^{3\sigma^2} + 3\,\%e^{2\sigma^2} + 6\,\%e^{\sigma^2} + 6\right)\%e^{2\sigma^2+4\mu}$

Skewness.

(%i40) Skewness_eq:radcan(K3_eq/(K2_eq)^(3/2));

(%o40) $\frac{K_3}{K_2^{3/2}}=\sqrt{\%e^{\sigma^2}-1}\left(\%e^{\sigma^2}+2\right)$

Kurtosis.

(%i41) Kurtosis_eq:radcan(K4_eq/(K2_eq)^2);

(%o41) $\frac{K_4}{K_2^2}=\%e^{4\,\sigma^2}+2\,\%e^{3\,\sigma^2}+3\,\%e^{2\,\sigma^2}-6$

APPENDIX 2.B "INCORRECT_SKEWNESS_FOR_LOGNORMAL_AND_MACCONE_DISTRIBUTIONS_PRIOR_TO_APRIL_29_2011"

```
(%i1) kill(all);
(%o0) done
```

Lognormal SKEWNESS, my error corrected April 29, 2011, at 11:20 pm.

My error : I had put K[4] instead of K[2] at the denominator, raised to the power 3/2 !!!!
My apologies. Claudio Maccone.

```
(%i1) def_K1:K[1]=%e^(mu+sigma^2/2);
```

(%o1) $K_1 = \%e^{\frac{\sigma^2}{2}+\mu}$

```
(%i2) def_K2:K[2]=(%e^(2*mu))*(%e^sigma^2)*(%e^sigma^2-1);
```

(%o2) $K_2 = \left(\%e^{\sigma^2}-1\right)\%e^{\sigma^2+2\mu}$

```
(%i3) def_mu_3:mu[3]=%e^(3*mu)*%e^(9*sigma^2/2);
```

(%o3) $\mu_3 = \%e^{\frac{9\sigma^2}{2}+3\mu}$

```
(%i4) def_K3:K[3]=mu[3]-3*K[1]*K[2]-K[1]^3;
```

(%o4) $K_3 = \mu_3 - 3K_1 K_2 - K_1^3$

```
(%i5) def_K3,def_K1,def_K2,def_mu_3;
```

(%o5) $K_3 = \%e^{\frac{9\sigma^2}{2}+3\mu} - 3\left(\%e^{\sigma^2}-1\right)\%e^{\frac{3\sigma^2}{2}+3\mu} - \%e^{3\left(\frac{\sigma^2}{2}+\mu\right)}$

```
(%i6) expression_of_K3:factor(expand(%));
```

(%o6) $K_3 = \left(\%e^{\sigma^2}-1\right)^2\left(\%e^{\sigma^2}+2\right)\%e^{\frac{3\sigma^2}{2}+3\mu}$

```
(%i7) expression_of_K3/(def_K2)^(3/2);
```

(%o7) $\frac{K_3}{K_2^{3/2}} = \sqrt{\%e^{\sigma^2}-1}\left(\%e^{\sigma^2}+2\right)\%e^{-\frac{3(\sigma^2+2\mu)}{2}+\frac{3\sigma^2}{2}+3\mu}$

```
(%i8) LOGNORMAL_SKEWNESS:radcan(%);
```

(%o8) $\frac{K_3}{K_2^{3/2}} = \sqrt{\%e^{\sigma^2}-1}\left(\%e^{\sigma^2}+2\right)$

Checking that the lognormal skewness approaches zero
when the variance approaches zero (as intuitively obviuous).

2_INCORRECT_SKEWNESS_for_lognormal_and_Maccone_distributions_PRIOR_TO_APRIL_29_2011.wxm/ 3

(%i9) 'limit(rhs(LOGNORMAL_SKEWNESS), sigma, 0)=limit(rhs(LOGNORMAL_SKEWNESS), sigma, 0);

(%o9) $\lim_{\sigma \to 0} \sqrt{\%e^{\sigma^2}-1}\left(\%e^{\sigma^2}+2\right)=0$

So, the lognormal pfd has a POSITIVE SKEWNESS.

Maccone SKEWNESS, my error corrected April 29, 2011, at 11:21 pm.

Again, I had put K[4] instead of K[2] at the denominator, raised to the power 3/2 !!!!

(%i10) assume(C>0);

(%o10) $[C>0]$

(%i11) def_mK1:mK[1]=C*(%e^(-mu/3))*%e^(sigma^2/18);

(%o11) $mK_1=\%e^{\frac{\sigma^2}{18}-\frac{\mu}{3}}\,C$

(%i12) def_mK2:mK[2]=C^2*(%e^(-2*mu/3))*%e^(sigma^2/9)*(%e^(sigma^2/9)-1);

(%o12) $mK_2=\%e^{\frac{\sigma^2}{9}-\frac{2\,\mu}{3}}\left(\%e^{\frac{\sigma^2}{9}}-1\right)C^2$

(%i13) def_mmu_3:mmu[3]=C^3*%e^(-mu)*%e^(sigma^2/2);

(%o13) $mmu_3=\%e^{\frac{\sigma^2}{2}-\mu}\,C^3$

(%i14) def_mK3:mK[3]=mmu[3]-3*mK[1]*mK[2]-mK[1]^3;

(%o14) $mK_3=mmu_3-3\,mK_1\,mK_2-mK_1^3$

(%i15) def_mK3,def_mK1,def_mK2,def_mmu_3;

(%o15) $mK_3=-3\,\%e^{\frac{\sigma^2}{6}-\mu}\left(\%e^{\frac{\sigma^2}{9}}-1\right)C^3+\%e^{\frac{\sigma^2}{2}-\mu}\,C^3-\%e^{3\left(\frac{\sigma^2}{18}-\frac{\mu}{3}\right)}\,C^3$

(%i16) expression_of_mK3:factor(expand(%));

(%o16) $mK_3=\%e^{-\mu}\left(\%e^{\frac{\sigma^2}{2}}-3\,\%e^{\frac{5\,\sigma^2}{18}}+2\,\%e^{\frac{\sigma^2}{6}}\right)C^3$

(%i17) expression_of_mK3/(def_mK2)^(3/2);

$$\text{(\%o17)}\ \frac{mK_3}{mK_2^{3/2}} = \frac{\%e^{-\frac{3\left(\frac{\sigma^2}{9}-\frac{2\mu}{3}\right)}{2}-\mu}\left(\%e^{\frac{\sigma^2}{2}} - 3\,\%e^{\frac{5\sigma^2}{18}} + 2\,\%e^{\frac{\sigma^2}{6}}\right)}{\left(\%e^{\frac{\sigma^2}{9}}-1\right)^{3/2}}$$

(%i18) MACCONE_SKEWNESS:radcan(%);

$$\text{(\%o18)}\ \frac{mK_3}{mK_2^{3/2}} = \frac{\%e^{\frac{2\sigma^2}{9}} + \%e^{\frac{\sigma^2}{9}} - 2}{\sqrt{\%e^{\frac{\sigma^2}{9}}-1}}$$

Checking that the Maccone skewness approaches zero
when the variance approaches zero (as intuitively obviuous).

(%i19) 'limit(rhs(MACCONE_SKEWNESS), sigma, 0)=limit(rhs(MACCONE_SKEWNESS), sigma, 0);

$$\text{(\%o19)}\ \lim_{\sigma\to 0} \frac{\%e^{\frac{2\sigma^2}{9}} + \%e^{\frac{\sigma^2}{9}} - 2}{\sqrt{\%e^{\frac{\sigma^2}{9}}-1}} = 0$$

APPENDIX 2.C "TENSOR_DEMO_NUMBER_OF_PROTONS_AND_NEUTRONS_IN_EINSTEIN_STATIC_UNIVERSE"

TENSOR DEMO OF MAXIMA:
computing the total number of protons and neutrons
in Einstein's static universe of 1917.
This is the famous 10^80 used in cosmology!

Author: Claudio Maccone, August 7th, 2011 (fully commented version).
E-mail: clmaccon@libero.it

Spacelike FRIEDMAN METRIC WITH COSMOLOGICAL TERM */

```
(%i103) (date: timedate (), print ("timedate reports current time", date));
timedate reports current time 2012-06-06 13:07:10+01:00
(%o103) 2012-06-06 13:07:10+01:00
```

```
(%i104) kill(all);
remvalue: derivabbrev doesn't appear to be a known variable; just unbind it anyway.
(%o0) done
```

Loading the c-tensor (=component tensor) package of Maxima.

```
(%i1) (if get('ctensor, 'version)=false then load(ctensor) else clear_ctensor(), ratfac:true);
(%o1) true
```

Assuming that all physical constants and parameters are REAL POSITIVE.
This assumption avoids stupid imaginaries from showing up in the calculation.
In other words, Maxima, always ASSUMES all variables and parameters to be REAL (i.e. both positive and negative and zero). Since all PHYSICAL variables and parameters are POSITIVE, you MUST declare so to avoid imaginaries !

```
(%i2) assume(c>0, G>0, M>0, r>0, a>0, %Lambda>0);
(%o2) [c>0,G>0,M>0,r>0,a>0,Λ>0]
```

Defining as 4 the number of dimensions of the continuum (dim: 4).
In other words, Maxima can do tensor calculations in ANY number of dimensions (like in Kaluza-Klein theories). Of course, that would take much more time and memory of your computer.

Also defining the SPHERICAL coordinates for the tri-sphere of the universe.
The "ph" instead of "phi" is a subtle trick to reserve phi for other uses.

Finally, the time dependence a(t) of the tri-sphere radius a(t) is defined also, thus representing the expanding or contracting universe.

(%i3) (dim:4, ct_coords:[r, %theta, ph, t], depends([a], [t]));

(%o3) [a(t)]

Spacelike FRIEDMAN METRIC WITH COSMOLOGICAL TERM in ORTHOGONAL PROJECTION. In other words, the tri-sphere is PROJECTED ONTO ITS EQUATORIAL PLANE !!!
This is the relevant covariant (lg = lower g sub ik) metric tensor.

The k is the space curvature of the tri-sphere normalized to one,
i.e. +1 for closed universes (= spherical geometry),
0 for non-curved (i.e Euclidean) and thus open universes,
-1 for hyperbolic (negative curvature) and thus open universes.

(%i4) lg:matrix([a^2/(1-k*r^2), 0, 0, 0], [0, a^2*r^2, 0, 0], [0, 0, a^2*r^2*sin(%theta)^2, 0], [0, 0, 0, -c^2])$

(%i5) 'lg=lg;

(%o5) $$lg = \begin{pmatrix} \frac{a^2}{1-k\,r^2} & 0 & 0 & 0 \\ 0 & a^2\,r^2 & 0 & 0 \\ 0 & 0 & \sin(\theta)^2\,a^2\,r^2 & 0 \\ 0 & 0 & 0 & -c^2 \end{pmatrix}$$

(%i6) cmetric();

(%o6) done

Determinant of the metric tensor.

(%i7) 'gdet=gdet;

(%o7) $$gdet = \frac{\sin(\theta)^2\,a^6\,c^2\,r^4}{k\,r^2-1}$$

Contravariant metric tensor (ug = upper g super ik).

(%i8) 'ug=ug;

(%o8) $ug=\begin{pmatrix} -\frac{k\,r^2-1}{a^2} & 0 & 0 & 0 \\ 0 & \frac{1}{a^2\,r^2} & 0 & 0 \\ 0 & 0 & \frac{1}{\sin(\theta)^2\,a^2\,r^2} & 0 \\ 0 & 0 & 0 & -\frac{1}{c^2} \end{pmatrix}$

All the non-zero Christoffel symbols of the first kind
(lcs = made up by lower gs only)...
... and second kind
(mcs = mixed Christoffel symbols, i.e. made up by both
lower and upper gs).

(%i9) christof(all);

(%t9) $lcs_{1,1,1} = \frac{a^2 k r}{(1-k r^2)^2}$

(%t10) $lcs_{1,1,4} = -\frac{a\, derivative(a,t,1)}{1-k r^2}$

(%t11) $lcs_{1,2,2} = a^2 r$

(%t12) $lcs_{1,3,3} = \sin(\theta)^2 a^2 r$

(%t13) $lcs_{1,4,1} = \frac{a\, derivative(a,t,1)}{1-k r^2}$

(%t14) $lcs_{2,2,1} = -a^2 r$

(%t15) $lcs_{2,2,4} = -a\, derivative(a,t,1)\, r^2$

(%t16) $lcs_{2,3,3} = \cos(\theta) \sin(\theta)\, a^2 r^2$

(%t17) $lcs_{2,4,2} = a\, derivative(a,t,1)\, r^2$

(%t18) $lcs_{3,3,1} = -\sin(\theta)^2 a^2 r$

(%t19) $lcs_{3,3,2} = -\cos(\theta) \sin(\theta)\, a^2 r^2$

(%t20) $lcs_{3,3,4} = -\sin(\theta)^2 a\, derivative(a,t,1)\, r^2$

(%t21) $lcs_{3,4,3} = \sin(\theta)^2 a\, derivative(a,t,1)\, r^2$

(%t22) $mcs_{1,1,1} = -\frac{k r}{k r^2 - 1}$

(%t23) $mcs_{1,1,4} = -\frac{a\, derivative(a,t,1)}{c^2 (k r^2 - 1)}$

(%t24) $mcs_{1,2,2} = \frac{1}{r}$

(%t25) $mcs_{1,3,3} = \frac{1}{r}$

(%t26) $mcs_{1,4,1} = \frac{derivative(a,t,1)}{a}$

(%t27) $mcs_{2,2,1} = r (k r^2 - 1)$

(%t28) $mcs_{2,2,4} = \frac{a\, derivative(a,t,1)\, r^2}{c^2}$

(%t29) $mcs_{2,3,3} = \frac{\cos(\theta)}{\sin(\theta)}$

(%t30) $mcs_{2,4,2} = \frac{derivative(a,t,1)}{a}$

(%t31) $mcs_{3,3,1} = \sin(\theta)^2 r (k r^2 - 1)$

(%t32) $mcs_{3,3,2} = -\cos(\theta) \sin(\theta)$

(%t33) $mcs_{3,3,4} = \frac{\sin(\theta)^2 a\, derivative(a,t,1)\, r^2}{c^2}$

derivative(a,t,1)

All the non-zero components of the Riemann curvature tensor.
This is the fully covariant Riemann tensor (R sub abcd).

(%i35) lriemann(true);

(%t35) $lriem_{2,2,1,1} = -\frac{a^2\,(c^2\,k+\mathrm{derivative}(a,t,1)^2)\,r^2}{c^2\,(k\,r^2-1)}$

(%t36) $lriem_{3,3,1,1} = -\frac{\sin(\theta)^2\,a^2\,(c^2\,k+\mathrm{derivative}(a,t,1)^2)\,r^2}{c^2\,(k\,r^2-1)}$

(%t37) $lriem_{3,3,2,2} = \frac{\sin(\theta)^2\,a^2\,(c^2\,k+\mathrm{derivative}(a,t,1)^2)\,r^4}{c^2}$

(%t38) $lriem_{4,4,1,1} = \frac{a\,\mathrm{derivative}(a,t,2)}{k\,r^2-1}$

(%t39) $lriem_{4,4,2,2} = -a\,\mathrm{derivative}(a,t,2)\,r^2$

(%t40) $lriem_{4,4,3,3} = -\sin(\theta)^2\,a\,\mathrm{derivative}(a,t,2)\,r^2$

(%o40) done

All the non-zero components of the covariant Ricci tensor (R sub ik).

(%i41) ricci(true);

(%t41) $ric_{1,1} = -\frac{2\,c^2\,k+a\,\mathrm{derivative}(a,t,2)+2\,\mathrm{derivative}(a,t,1)^2}{c^2\,(k\,r^2-1)}$

(%t42) $ric_{2,2} = \frac{(2\,c^2\,k+a\,\mathrm{derivative}(a,t,2)+2\,\mathrm{derivative}(a,t,1)^2)\,r^2}{c^2}$

(%t43) $ric_{3,3} = \frac{\sin(\theta)^2\,(2\,c^2\,k+a\,\mathrm{derivative}(a,t,2)+2\,\mathrm{derivative}(a,t,1)^2)\,r^2}{c^2}$

(%t44) $ric_{4,4} = -\frac{3\,\mathrm{derivative}(a,t,2)}{a}$

(%o44) done

Introducing the Ricci scalar R and declaring its space dependence
upon the distance r from the origin (i.e. the observer = i.e. us)
and the receding galaxy at distance r in the expanding universe).

The Ricci scalar is called "tracer" by Maxima.

This dependence R(r) must be declared here before we define
the Einstein equations as the Euler-Lagrange equations yielded
by the Hilbert variational principle of general relativity.

(%i45) depends(R, r);

(%o45) [R(r)]

(%i46) ricci_scalar:R=factor(expand(tracer));

(%o46) $R = \frac{6\,(c^2\,k + a\,\mathrm{derivative}(a,t,2) + \mathrm{derivative}(a,t,1)^2)}{a^2\,c^2}$

Defining the lagrangian of the Hilbert variational principle of general relativity with the cosmological term also (Lambda is the cosmological constant).

(%i47) lagrangian_for_metric:'((R-2*%Lambda)*sqrt(-gdet))=rhs((ricci_scalar-2*%Lambda)*sqrt(-gdet));

(%o47) $\sqrt{-gdet}\,(R - 2\,\Lambda) = \frac{\%i\,\sin(\theta)\,a^3\,c\left(\frac{6\,(c^2\,k + a\,\mathrm{derivative}(a,t,2) + \mathrm{derivative}(a,t,1)^2)}{a^2\,c^2} - 2\,\Lambda\right)r^2}{\sqrt{k\,r^2 - 1}}$

Loading the Maxima package "optvar"=variational optimization.
This package is able to calculate the Euler-Lagrange differential equations for any given lagrangian.

(%i48) if get('optvar, 'version)=false then load(optvar);

(%o48) *C:/PROGRA~2/MAXIMA~1.1-G/share/maxima/5.25.1/share/calculus/optvar.mac*

Here is the ONLY Euler-Lagrange equation for the above lagrangian.
This is because there is ONLY ONE independent variable: the time t.
Notice that the paramenter r is NOT and independent variable, but rather the function r(t) in the above lagrangian. It represents the increasing distance r(t) between us and the reference expanding galaxy while THE WHOLE UNIVERSE EXPANDS (comoving reference frame).

Note: worry not about the fictitious imaginary %i here below:
it is just a stupid multiplicative constant coming out of the calculations.

(%i49) el(rhs(lagrangian_for_metric), [a], [r]);

(%t49) $0 = \frac{3\,\%i\,\sin(\theta)\,a^2\,c\left(\frac{6\,(c^2\,k + a\,\mathrm{derivative}(a,t,2) + \mathrm{derivative}(a,t,1)^2)}{a^2\,c^2} - 2\,\Lambda\right)r^2}{\sqrt{k\,r^2 - 1}} + \frac{\%i\,\sin(\theta)\,a^3\,c\left(\frac{6\,\mathrm{derivative}(a,t,2)}{a^2\,c^2} - \frac{12\,(c^2\,k + a\,\mathrm{derivative}(a,t,2) + \mathrm{derivative}(a,t,1)^2)}{a^3\,c^2}\right)r^2}{\sqrt{k\,r^2 - 1}}$

(%o49) [%t49]

Solving the above only Einstein equation with respect, for instance, the second derivative of the unknown function a(t) yields the Friedman differential equation with the cosmological term.

(%i50) Friedman_eq_with_Lambda:first(solve(%t49, 'diff(a, t, 2)));

$$(\%o50)\ \mathrm{derivative}(a,t,2)=\frac{-c^2\,k+\Lambda\,a^2\,c^2-\mathrm{derivative}(a,t,1)^2}{2\,a}$$

Setting to zero the cosmological constant Lambda, we get the classical Friedman differential equation.

(%i51) Friedman_eq:subst(0, %Lambda, Friedman_eq_with_Lambda);

$$(\%o51)\ \mathrm{derivative}(a,t,2)=\frac{-c^2\,k-\mathrm{derivative}(a,t,1)^2}{2\,a}$$

Maxima now solves the above Friedman equation for the case of the CLOSED universe (i.e k>0). The result is the famous CYCLOID up two the two undetermined integration constants %k1 and %k2.

(%i52) assume(k>0);

(%o52) [k>0]

(%i53) function_a_for_closed_universe:ode2(Friedman_eq, a, t);

$$(\%o53)\ \left[\frac{\%e^{-2\,\%k1}\left(\%e^{\%k1}\sqrt{a}\,\mathrm{atan}\left(\frac{\%e^{-\frac{\%k1}{2}}\sqrt{1-\%e^{\%k1}\,a\,c^2\,k}}{\sqrt{a}\,c\,\sqrt{k}}\right)+\%e^{\frac{3\,\%k1}{2}}\,a\,c\,\sqrt{k}\,\sqrt{1-\%e^{\%k1}\,a\,c^2\,k}\right)}{\sqrt{a}\,c^3\,k^{3/2}}=t+\%k2,\right.$$
$$\left.-\frac{\%e^{-2\,\%k1}\left(\%e^{\%k1}\sqrt{a}\,\mathrm{atan}\left(\frac{\%e^{-\frac{\%k1}{2}}\sqrt{1-\%e^{\%k1}\,a\,c^2\,k}}{\sqrt{a}\,c\,\sqrt{k}}\right)+\%e^{\frac{3\,\%k1}{2}}\,a\,c\,\sqrt{k}\,\sqrt{1-\%e^{\%k1}\,a\,c^2\,k}\right)}{\sqrt{a}\,c^3\,k^{3/2}}=t+\%k2\right]$$

(%i54) closed_universe_with_k_plus_1:subst(1, k, function_a_for_closed_universe);

$$(\%o54)\ \left[\frac{\%e^{-2\,\%k1}\left(\%e^{\%k1}\sqrt{a}\,\mathrm{atan}\left(\frac{\%e^{-\frac{\%k1}{2}}\sqrt{1-\%e^{\%k1}\,a\,c^2}}{\sqrt{a}\,c}\right)+\%e^{\frac{3\,\%k1}{2}}\,a\,c\,\sqrt{1-\%e^{\%k1}\,a\,c^2}\right)}{\sqrt{a}\,c^3}=t+\%k2,\right.$$
$$\left.-\frac{\%e^{-2\,\%k1}\left(\%e^{\%k1}\sqrt{a}\,\mathrm{atan}\left(\frac{\%e^{-\frac{\%k1}{2}}\sqrt{1-\%e^{\%k1}\,a\,c^2}}{\sqrt{a}\,c}\right)+\%e^{\frac{3\,\%k1}{2}}\,a\,c\,\sqrt{1-\%e^{\%k1}\,a\,c^2}\right)}{\sqrt{a}\,c^3}=t+\%k2\right]$$

Next, the same is done for the OPEN hyperbolic universe.

(%i55) forget(k>0);

(%o55) [k>0]

(%i56) assume(k<0);

(%o56) [k<0]

(%i57) general_integral_for_hyperbolic_universe:ode2(Friedman_eq, a, t);

(%o57) [

$$\frac{\%e^{-2\,\%k1}\left(\%e^{\%k1}\sqrt{a}\,\log\left(\frac{\sqrt{1-\%e^{\%k1}\,a\,c^2\,k}-\%e^{\%k1/2}\sqrt{a}\,c\,\sqrt{-k}}{\sqrt{1-\%e^{\%k1}\,a\,c^2\,k}+\%e^{\%k1/2}\sqrt{a}\,c\,\sqrt{-k}}\right)+2\,\%e^{\frac{3\,\%k1}{2}}\,a\,c\,\sqrt{-k}\,\sqrt{1-\%e^{\%k1}\,a\,c^2\,k}\right)}{2\sqrt{a}\,c^3\sqrt{-k}\,k}=t+\%k2$$

$$,-\frac{\%e^{-2\,\%k1}\left(\%e^{\%k1}\sqrt{a}\,\log\left(\frac{\sqrt{1-\%e^{\%k1}\,a\,c^2\,k}-\%e^{\%k1/2}\sqrt{a}\,c\,\sqrt{-k}}{\sqrt{1-\%e^{\%k1}\,a\,c^2\,k}+\%e^{\%k1/2}\sqrt{a}\,c\,\sqrt{-k}}\right)+2\,\%e^{\frac{3\,\%k1}{2}}\,a\,c\,\sqrt{-k}\,\sqrt{1-\%e^{\%k1}\,a\,c^2\,k}\right)}{2\sqrt{a}\,c^3\sqrt{-k}\,k}=t+\%k2]$$

(%i58) hyperbolic_universe_with_k_minus_1:subst(-1, k, general_integral_for_hyperbolic_universe);

(%o58) [

$$-\frac{\%e^{-2\,\%k1}\left(\%e^{\%k1}\sqrt{a}\,\log\left(\frac{\sqrt{\%e^{\%k1}\,a\,c^2+1}-\%e^{\%k1/2}\sqrt{a}\,c}{\sqrt{\%e^{\%k1}\,a\,c^2+1}+\%e^{\%k1/2}\sqrt{a}\,c}\right)+2\,\%e^{\frac{3\,\%k1}{2}}\,a\,c\,\sqrt{\%e^{\%k1}\,a\,c^2+1}\right)}{2\sqrt{a}\,c^3}=t+\%k2,$$

$$\frac{\%e^{-2\,\%k1}\left(\%e^{\%k1}\sqrt{a}\,\log\left(\frac{\sqrt{\%e^{\%k1}\,a\,c^2+1}-\%e^{\%k1/2}\sqrt{a}\,c}{\sqrt{\%e^{\%k1}\,a\,c^2+1}+\%e^{\%k1/2}\sqrt{a}\,c}\right)+2\,\%e^{\frac{3\,\%k1}{2}}\,a\,c\,\sqrt{\%e^{\%k1}\,a\,c^2+1}\right)}{2\sqrt{a}\,c^3}=t+\%k2]$$

Finally we turn to the case of EUCLIDEAN-SPACE UNIVERSE,
i.e. the Einstein-De Sitter universe of 1932.

(%i59) Friedman_eq_for_Einstein_De_Sitter_universe:subst(0, k, Friedman_eq);

(%o59) $\text{derivative}(a,t,2)=-\frac{\text{derivative}(a,t,1)^2}{2\,a}$

(%i60) Einstein_De_Sitter_universe:ode2(Friedman_eq_for_Einstein_De_Sitter_universe, a, t);

(%o60) $\frac{2\,a^{3/2}}{3\,\%k1}=t+\%k2$

(%i61) Einstein_De_Sitter_Radius_of_the_universe:Einstein_De_Sitter_universe^(2/3);

(%o61) $\frac{2^{2/3}\, a}{3^{2/3}\, \%k1^{2/3}} = (t+\%k2)^{2/3}$

The famous SEMICUBIC PARABOLA (i.e. a grows like t^(2/3)) is thus found.

(%i62) Einstein_De_Sitter_Radius:first(solve(%, a));

(%o62) $a = \frac{3^{2/3}\, \%k1^{2/3}\, (t+\%k2)^{2/3}}{2^{2/3}}$

Let us now go back to the original EINSTEIN STATIC UNIVERSE of 1917.
His paper was entitled "Cosmologische Betrachtungen zur Allgemeine Relativitaetstheorie" and is found in English in the Dover book "The Principle of Relativity".

So we again need the differential equation for a(t) with the cosmological constant Lambda and with POSITIVE curvature k=1, i.e. CLOSED UNIVERSE again.

(%i63) diff_eq_for_Einstein_universe:subst(1, k, Friedman_eq_with_Lambda);

(%o63) $\mathrm{derivative}(a,t,2) = \frac{\Lambda\, a^2\, c^2 - c^2 - \mathrm{derivative}(a,t,1)^2}{2\, a}$

Since the universe we are looking for is STATIC, i.e. its radius does NOT change in time, the time dependence a(t) must be REMOVED.

(%i64) Einstein_universe:remove(a, depends);

(%o64) done

With such a condition, the two time derivatives in the above differential equations both equal zero, and one is just left with the simple result:

(%i65) eq_of_Einstein_universe:ev(diff_eq_for_Einstein_universe, 'diff);

(%o65) $0 = \frac{\Lambda\, a^2\, c^2 - c^2}{2\, a}$

This is thus the famous relationship between the cosmological constant Lambda and the radius of the static universe found by Einstein in 1917:

(%i66) Lambda_for_Einstein_universe:first(solve(eq_of_Einstein_universe, %Lambda));

(%o66) $\Lambda = \frac{1}{a^2}$

So as to say, this is a PURELY GEOMETRIC relationship between the universe static radius a and the cosmological constant Lambda.

But we want more to proceed: we want the relationship between the overall MASS OF THE UNIVERSE, M, i.e. the mass of all galaxies making up the universe, and the SIZE (i.e. the RADIUS) of the universe again. In other words, in Einstein's static universe the attractive gravitational force among the galaxies exacly balances the repulsive force created by the presence of the cosmological constant, and the universe is thus STATIC.

So, let us re-introduce the time dependence of a(t) for a moment, and let us find the covariant Einstein tensor lein (= lower Einstein tensor = G sub ik).

```
(%i67) depends(a, t);
(%o67) [a(t)]

(%i68) leinstein(true);
```

(%t68) $\mathrm{lein}_{1,1} = -\dfrac{c^2 k + 2\, a\, \mathrm{derivative}(a,t,2) + \mathrm{derivative}(a,t,1)^2}{c^2\,(1 - k\, r^2)}$

(%t69) $\mathrm{lein}_{2,2} = -\dfrac{(c^2 k + 2\, a\, \mathrm{derivative}(a,t,2) + \mathrm{derivative}(a,t,1)^2)\, r^2}{c^2}$

(%t70) $\mathrm{lein}_{3,3} = -\dfrac{\sin(\theta)^2\,(c^2 k + 2\, a\, \mathrm{derivative}(a,t,2) + \mathrm{derivative}(a,t,1)^2)\, r^2}{c^2}$

(%t71) $\mathrm{lein}_{4,4} = \dfrac{3\,(c^2 k + \mathrm{derivative}(a,t,1)^2)}{a^2}$

(%o71) done

This clearly is a DIAGONAL second-rank tensor, i.e. a matrix where all elements outside the main diagonal are equal to zero.

The Einstein equations with the cosmological term are found by equating this diagonal Einstein tensor to the DIAGONAL STRESS-ENERGY TENSOR, that we are now going to find.

Understanding the UNIVERSE AS A FLUID MADE UP BY GALAXIES, then the Eulerian theory of fluids yields for the stress-energy tensor of such a fluid, having pressure p>0 and density rho>0, the following MIXED (I.E. WITH ONE COVARIANT AND ONE CONTRAVARIANT INDEX) tensor:

(%i72) assume(p>0, %rho>0);

(%o72) [p>0, ρ>0]

(%i73) MIXED_stress_energy_tensor:matrix([p, 0,0,0], [0, p, 0, 0], [0, 0, p, 0], [0, 0, 0, -c^2*%rho]);

(%o73)
$$\begin{pmatrix} p & 0 & 0 & 0 \\ 0 & p & 0 & 0 \\ 0 & 0 & p & 0 \\ 0 & 0 & 0 & -\rho c^2 \end{pmatrix}$$

To transform this MIXED tensor into a FULLY COVARIANT one, we must COMPOSE the above mixed tensor with the covariant metric tensor, namely we must do the MATRIX PRODUCT (denoted by a dot among the two matrices in Maxima):

(%i74) l_set:MIXED_stress_energy_tensor.lg;

(%o74)
$$\begin{pmatrix} \frac{a^2 p}{1-k r^2} & 0 & 0 & 0 \\ 0 & a^2 p r^2 & 0 & 0 \\ 0 & 0 & \sin(\theta)^2 a^2 p r^2 & 0 \\ 0 & 0 & 0 & \rho c^4 \end{pmatrix}$$

At last we now have both the fully covariant Einstein tensor and the fully covariant stress-energy tensor, and so the Einstein equations are obtained by equalling the Einstein tensor to the stress-energy tensor multiplied by the proportionality constant 8*%pi*G/c^4.

(%i75) for i thru dim do for j thru dim do (einstein_eqs[i, j]:lein[i, j]=(8*%pi*G/c^4)*l_set[i, j]);

(%o75) done

(%i76) for i thru dim do display(einstein_eqs[i, i]);

$$-\frac{c^2\,k+2\,a\,\text{derivative}(a,t,2)+\text{derivative}(a,t,1)^2}{c^2\,(1-k\,r^2)}=\frac{8\,\pi\,a^2\,p\,G}{c^4\,(1-k\,r^2)}$$

$$-\frac{(c^2\,k+2\,a\,\text{derivative}(a,t,2)+\text{derivative}(a,t,1)^2)\,r^2}{c^2}=\frac{8\,\pi\,a^2\,p\,r^2\,G}{c^4}$$

$$-\frac{\sin(\theta)^2\,(c^2\,k+2\,a\,\text{derivative}(a,t,2)+\text{derivative}(a,t,1)^2)\,r^2}{c^2}=\frac{8\,\pi\,\sin(\theta)^2\,a^2\,p\,r^2\,G}{c^4}$$

$$\frac{3\,(c^2\,k+\text{derivative}(a,t,1)^2)}{a^2}=8\,\pi\,\rho\,G$$

(%o76) done

A glance to the above four Einstein differential equations for a(t) shows that the first and second equations really are the same thing.

And so is the third equation either, yielding.

(%i77) einstein_eqs[2,2]/einstein_eqs[3,3];

(%o77) $$\frac{1}{\sin(\theta)^2}=\frac{1}{\sin(\theta)^2}$$

Please notice that Maxima writes sin(theta)^2 rather than sin^2(theta).
This is, in our opinion, a TERRIBLY BAD HABIT, VERY MISLEADING.
But there is nothing we can do against it. So is the code.
More BAD consequences of this BAD NOTATION will be found later when dealing with lognormal probability distributions.

The full solution of the above Einstein equations with the cosmological term would lead us to the Lemaitre universe and related topics, but we now want to confine ourselves to the EINSTEIN STATIC UNIVERSE only.
So, let us remove again the time dependence of a(t):

(%i78) Einstein_universe:remove(a, depends);

(%o78) done

Then the last of the above Einstein equation yields, for the Einstein STATIC universe, the second "physical" relationship between the DENSITY rho of the universe and its radius a:

(%i79) second_eq_for_Einstein_universe:ev(einstein_eqs[4,4], 'diff);

(%o79) $\frac{3\,c^2\,k}{a^2}=8\,\pi\,\rho\,G$

Since we are now interested in the CLOSED universe only, we set k=1:

(%i80) second_eq_for_Einstein_universe:subst(1, k, second_eq_for_Einstein_universe);

(%o80) $\frac{3\,c^2}{a^2}=8\,\pi\,\rho\,G$

(%i81) a_vs_rho:rootscontract(radcan(last(solve(second_eq_for_Einstein_universe, a))));

(%o81) $a=c\,\sqrt{\frac{3}{8\,\pi\,\rho\,G}}$

This is thus the relationship between the radius of the Einstein static universe and its density:

(%i82) Radius_of_Einstein_universe:a[\E]=rhs(a_vs_rho);

(%o82) $a_E=c\,\sqrt{\frac{3}{8\,\pi\,\rho\,G}}$

But now we can match this equation to the previous one yielding the cosmological constant Lambda:

(%i83) Lambda_expression:Lambda_for_Einstein_universe, a_vs_rho;

(%o83) $\Lambda=\frac{8\,\pi\,\rho\,G}{3\,c^2}$

We thus conclude that FOR THE EINSTEIN STATIC UNIVERSE, THE COSMOLOGICAL CONSTANT LAMBDA SUB E EQUALS:

(%i84) Lambda_of_Einstein_universe:%Lambda[\E]=rhs(%);

(%o84) $\Lambda_E=\frac{8\,\pi\,\rho\,G}{3\,c^2}$

Let us now introduce the numerical values of G, c and of the (mean) density of the observable universe, rho = 5 x 10^(-27) kg/meter^3 :

(%i85) assume(m>0, kg>0, sec>0);

(%o85) [m>0,kg>0,sec>0]

```
(%i86) numerical_constants:[G=6.67*10^(-11)*m^3/(kg*sec^2), c=299792458*m/sec, %rho=5*10^(
-27)*kg/m^3];
```

(%o86) $[G=\frac{6.67\ 10^{-11}\ m^3}{kg\ sec^2}, c=\frac{299792458\ m}{sec}, \rho=\frac{kg}{200000000000000000000000000\ m^3}]$

```
(%i87) numeric_Lambda_of_Einstein_universe:Lambda_of_Einstein_universe, numerical_constants;
```

(%o87) $\Lambda_E=\frac{9.8951678318368472\ 10^{-54}\ \pi}{m^2}$

So, for the Einstein static universe, the cosmological constant Lambda equals 3.10 x 10^(-53) /meter^2, the value reported in all textbooks.

```
(%i88) numeric_Lambda_of_Einstein_universe:ev(%, numer);
```

(%o88) $\Lambda_E=\frac{3.1086586566536678\ 10^{-53}}{m^2}$

As for the RADIUS = size of the Einstein static universe, it is given by 1.8 x 10^10 light years, i.e. 18 billion light years:

```
(%i89) Radius_of_Einstein_universe:a[E]=rhs(a_vs_rho);
```

(%o89) $a_E=c\sqrt{\frac{3}{8\ \pi\ \rho\ G}}$

```
(%i90) numeric_Radius_of_Einstein_universe:Radius_of_Einstein_universe,numerical_constants;
```

(%o90) $a_E=\frac{3.1789845531858917\ 10^{26}\ m}{\sqrt{\pi}}$

```
(%i91) numeric_Radius_of_Einstein_universe:ev(numeric_Radius_of_Einstein_universe, numer);
```

(%o91) $a_E=1.7935499711666983\ 10^{26}\ m$

```
(%i92) def_light_year:ly=c*365.25*60*60*24*sec;
```

(%o92) $ly=3.15576\ 10^7\ c\ sec$

```
(%i93) light_year_eq:def_light_year, numerical_constants;
```

(%o93) $ly=9.4607304725808\ 10^{15}\ m$

```
(%i94) numeric_Radius_of_Einstein_universe_in_ly:numeric_Radius_of_Einstein_universe/light_year_eq;
```

(%o94) $\frac{a_E}{ly}=1.8957838153880249\ 10^{10}$

Finally, the VOLUME of the universe for Einstein's static model.
To find it, we must remember that the volume of a tri-sphere is given by
2 %pi^2 radius^3. So, one gets:

```
(%i95) volume_of_Einstein_universe:2*%pi^2*Radius_of_Einstein_universe^3;
```

(%o95) $2\,\pi^2\,a_E^3 = \dfrac{3^{3/2}\sqrt{\pi}\,c^3}{2^{7/2}\,\rho^{3/2}\,G^{3/2}}$

```
(%i96) numeric_volume_of_Einstein_universe:volume_of_Einstein_universe, numerical_constants;
```

(%o96) $2\,\pi^2\,a_E^3 = 1.2365547970638998\;10^{79}\;3^{3/2}\sqrt{\pi}\;m^3$

And the volume equals 1.1 x 10^80 cubic meters:

```
(%i97) numeric_volume_of_Einstein_universe:V[E]=ev(rhs(numeric_volume_of_Einstein_universe), numer);
```

(%o97) $V_E = 1.1388595947073545\;10^{80}\;m^3$

But we now know both the volume and the density of the universe.
So we can compute its MASS:

```
(%i98) mass_of_Einstein_universe:M[E]=%rho*rhs(volume_of_Einstein_universe);
```

(%o98) $M_E = \dfrac{3^{3/2}\sqrt{\pi}\,c^3}{2^{7/2}\sqrt{\rho}\,G^{3/2}}$

```
(%i99) universe_numeric_mass:mass_of_Einstein_universe, numerical_constants;
```

(%o99) $M_E = 6.1827739853194985\;10^{52}\;3^{3/2}\sqrt{\pi}\;kg$

The MASS of the Einstein static universe turns thus out to be 5.7 x 10^53 kilograms:

```
(%i100) universe_numeric_mass:ev(universe_numeric_mass, numer);
```

(%o100) $M_E = 5.6942979735367728\;10^{53}\;kg$

DIVIDING THIS MASS BY THE MASS OF THE PROTON (AND NEARLY EQUAL MASS OF THE NEUTRON) WE FIND THE TOTAL NUMBER OF PROTONS PLUS NEUTRONS IN THE EINSTEIN STATIC UNIVERSE.
This is the famous value 10^80 used by Dirac in 1937 to set up the Dirac cosmological theory, and largely used today for more speculations about dark mass, dark energy, etcetera.

```
(%i101) proton_numeric_mass:M_proton=1.6726231*10^(-27)*kg;
```

(%o101) M_proton $= 1.672623100000001\;10^{-27}\;kg$

(%i102) Number_of_protons_AND_neutrons=rhs(universe_numeric_mass)/(2*rhs(proton_numeric_mass));

(%o102) Number_of_protons_AND_neutrons = 1.7022059463177247 10^{80}

3

How many planets for man and aliens?

3.1 HABITABLE PLANETS FOR MAN

Let us now change topics completely! Rather than looking for ETs in the Galaxy, we now seek habitable planets for man in the Galaxy. How many are there and how far from us is the nearest such habitable planet?

These topics were seriously considered for the first time in 1964 by Stephen H. Dole, then with the Rand Corporation. Unfortunately, the author has been unable to find when and where Dole was born and died, although the author's friend Dr. Laurance Doyle of the SETI Institute is aware of Dole's passing.

Back in 1964, just three years after Frank Drake had made known his now famous Drake equation, Dole fully understood the ramifications of the Drake equation. In his now famous book entitled *Habitable Planets for Man* [8] he used the same mathematical structure as the Drake equation (3.1) to find the *number of habitable planets for man in the Galaxy*. For instance, on p. 82 of his book he gives the same mathematical equation as Drake's, but applies it to habitable planets.

Figure 3.1 reproduces this crucial page of Dole's 1964 book, which nowadays can be downloaded free from the Rand Corporation website. Hereafter we call this equation "the classical Dole equation".

As we can see from Figure 3.1, the classical Dole equation is made up of *ten* factors (instead of *seven* factors as in the Drake equation):

$$N_{Hab} = Ns \cdot Pp \cdot Pi \cdot PD \cdot PM \cdot Pe \cdot PB \cdot PR \cdot PA \cdot PL. \tag{3.1}$$

Here N_{Hab} is the total number of habitable planets for man in the Galaxy, and it is given by the product of the following *ten* input numbers:

(1) Ns is the number of stars in the suitable mass range 0.35 to 1.43 solar masses (this is Dole's assumption of the mass of "habitable stars").
(2) Pp is the probability that a given star has planets in orbit around it.

CHAPTER 5

Probability of Occurrence of Habitable Planets

Having summarized the properties of habitable planets and the astronomical requirements implied by these properties, we can now attempt to estimate the prevalence of such bodies in our Galaxy (the Milky Way); and to do this with any reasonable degree of accuracy (in the spirit of the present study), it is necessary to consider the following factors:

N_s, the prevalence of stars in the suitable mass range, 0.35 to 1.43 solar masses;

P_p, the probability that a given star has planets in orbit about it;

P_i, the probability that the inclination of the planet's equator is correct for its orbital distance;

P_D, the probability that at least one planet orbits within an ecosphere;

P_M, the probability that the planet has a suitable mass, 0.4 to 2.35 Earth masses;

P_e, the probability that the planet's orbital eccentricity is sufficiently low;

P_B, the probability that the presence of a second star has not rendered the planet uninhabitable;

P_R, the probability that the planet's rate of rotation is neither too fast nor too slow;

P_A, the probability that the planet is of the proper age;

P_L, the probability that, all astronomical conditions being proper, life has developed on the planet.

Once values for all of these factors have been established, the estimated number of habitable planets N_{HP} in the Galaxy can be expressed as the product:

$$N_{HP} = N_s P_p P_i P_D P_M P_e P_B P_R P_A P_L.$$

82

Figure 3.1. Reproduction of p. 82 of Stephen H. Dole's book *Habitable Planets for Man*, First Edition published in 1964; it can be downloaded free today from the website of the Rand Corporation.

(3) *Pi* is the probability that the inclination of the planet's equator is correct for its orbital distance.
(4) *PD* is the probability that at least one planet orbits within an ecosphere.
(5) *PM* is the probability that the planet has a suitable mass, 0.4 to 2.35 Earth masses (again, this is Dole's assumption).
(6) *Pe* is the probability that the planet's orbital eccentricity is sufficiently low.
(7) *PB* is the probability that the presence of a second star has not rendered the planet uninhabitable.
(8) *PR* is the probability that the planet's rate of rotation is neither too fast nor too slow.
(9) *PA* is the probability that the planet is of the proper age.
(10) *PL* is the probability that, all astronomical conditions being appropriate, life has developed on the planet.

3.2 THE STATISTICAL DOLE EQUATION

Let us rename the above 10 input variables of the classical Dole equation (3.1) as follows:

$$\begin{cases} D_1 = Ns \\ D_2 = Pp \\ D_3 = Pi \\ D_4 = PD \\ D_5 = PM \\ D_6 = Pe \\ D_7 = PB \\ D_8 = PR \\ D_9 = PA \\ D_{10} = PL \end{cases} \tag{3.2}$$

so that our classical Dole equation may be simply rewritten as

$$N_{Hab} = \prod_{i=1}^{10} D_i. \tag{3.3}$$

We now let (3.3) undergo exactly the same changes that we applied to the classical Drake equation (1.7). In other words:

(1) All the input variables on the right-hand side of (3.3) now become *positive random variables.*
(2) All these random variables are supposed to be *uniformly distributed* with assigned mean values μ_{D_i} and standard deviations σ_{D_i}. It can then be shown that doing so

actually amounts to assigning the lower and upper limits (a_i and b_i, respectively) of each uniform random variable D_i.

(3) As a consequence of these assumptions, the total number of habitable planets in the Galaxy, N_{Hab}, also becomes a *random variable*, which we already know to be *lognormally distributed* from our previous similar work about the Drake equation.

Thus, we may now call (3.3) the *statistical Dole equation*. The notation D_i obviously comes from "Dole", but the lucky coincidence that both Frank Drake's and Stephen Dole's family names both start with the same letter will save us from introducing new notations other than these D_i!

It is true that the classical Drake equation (1.7) and the classical Dole equation (3.1) have a different number of factors (7 and 10, respectively), but ... frankly speaking, who cares? This is perfectly in line with what we did for the Drake equation, and so *the number of factors in both (1.7) and (3.1) is totally irrelevant, thanks to the Central Limit Theorem!*

3.3 THE NUMBER OF HABITABLE PLANETS FOR MAN IN THE GALAXY FOLLOWS THE LOGNORMAL DISTRIBUTION

We now just repeat the same arguments developed for the Drake equation to immediately conclude that: *the total number of habitable planets in the Galaxy follows the lognormal distribution* given in Table 1.1.

3.4 THE DISTANCE BETWEEN ANY TWO NEARBY HOSPITABLE PLANETS FOLLOWS THE MACCONE DISTRIBUTION

Again we now just repeat the same arguments developed for the Drake equation to immediately conclude that *the distance between any two nearby hospitable planets follows the Maccone distribution* given in Table 1.2.

3.5 A NUMERICAL EXAMPLE: SOME 100 MILLION HABITABLE PLANETS EXIST IN THE GALAXY!

We just need to complete this chapter by giving a numerical example of how our statistical Dole equation (3.3) works.

Consider the following Input Table 3.1. This is in principle comparable with Input Table 1.1 for the statistical Drake equation. In fact, the arguments developed by Dole in Chapter 5 of [8] do provide the mean values of each D_i, but only such mean values, and not the relevant standard deviations, of course.

Input Table 3.1. Input values (i.e., mean values and standard deviations) for the 10 Dole uniform random variables D_i. The first column lists the 10 input sheer numbers that also are the mean values (middle column). The last column lists the 10 input standard deviations. The bottom line is the classical Dole equation (3.1). So, the number of habitable planets in the Galaxy given by the classical Dole equation as a sheer number is 35,171,930.

$Ns := 6.448 \cdot 10^8$	$\mu Ns := Ns$	$\sigma Ns := 1 \cdot 10^7$
$Pp := 1.0$	$\mu Pp := Pp$	$\sigma Pp := \frac{10}{100}$
$Pi := 0.81$	$\mu Pi := Pi$	$\sigma Pi := \frac{10}{100}$
$PD := 0.63$	$\mu PD := PD$	$\sigma PD := \frac{10}{100}$
$PM := 0.19$	$\mu PM := PM$	$\sigma PM := \frac{10}{100}$
$Pe := 0.94$	$\mu Pe := Pe$	$\sigma Pe := \frac{10}{100}$
$PB := 0.95$	$\mu PB := PB$	$\sigma PB := \frac{10}{100}$
$PR := 0.9$	$\mu PR := PR$	$\sigma PR := \frac{10}{100}$
$PA := 0.7$	$\mu PA := PA$	$\sigma PA := \frac{10}{100}$
$PL := 1$	$\mu PL := PL$	$\sigma PL := \frac{10}{100}$
$N_{Hab} := Ns \cdot Pp \cdot Pi \cdot PD \cdot PM \cdot Pe \cdot PB \cdot PR \cdot PA \cdot PL$		
$N_{Hab} := 3.5171930508624 \times 10^7$		

However, to set up a working example of the statistical Dole equation, we must also assign the 10 standard deviations that were not given by Dole and are unknown to the author from the current scientific literature about these matters.

So, to keep things simple, the author has assigned the value of 1/10 (i.e., 10%) to each of the 10 standard deviations listed in Input Table 3.1. Input Table 3.1 is now complete.

Having assumed all the values listed in Input Table 3.1 as the input values, a new (unpublished) MathCad code was created by the author for the statistical Dole equation. This code yielded the following results for the input values of Input Table 3.1.

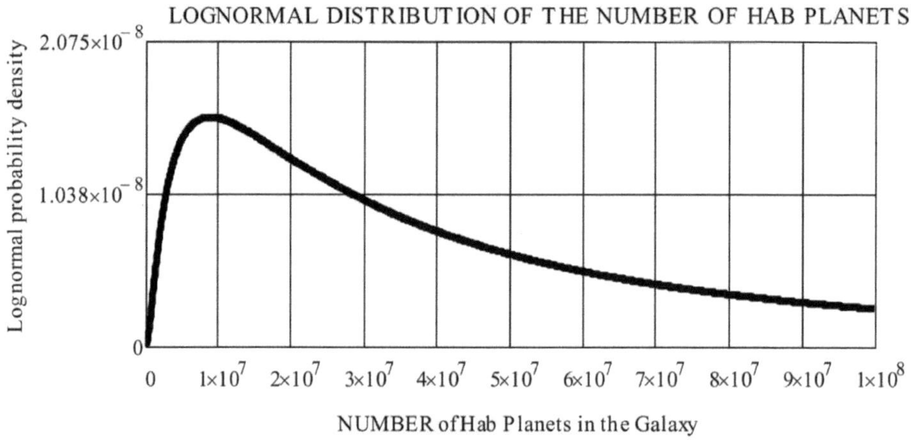

Figure 3.2. The lognormal probability density of the overall *number* of habitable planets in the Galaxy as described in Stephen H. Dole's book *Habitable Planets for Man*, First Edition published in 1964, and implemented by assigning a 10% standard deviation to all the 10 input random variables listed in Input Table 3.1.

First of all, the lognormal probability density for the random variable N_{Hab} is shown in Figure 3.2. We see that the peak (i.e., the mode) corresponds to about 10 million planets, but the tail is rather long.

To quantify these remarks, let us first point out that the author's MathCad code yields the following numerical values for the two parameters μ and σ given by the last two rows in both Tables 1.1 and 1.2:

$$\begin{cases} \mu_{Hab} = 1.76268289631314 \cdot 10^1 \\ \sigma_{Hab} = 1.27010132908265 \cdot 10^0. \end{cases} \tag{3.4}$$

Then, the mean value of the random variable N_{Hab}, given by the fourth row in Table 1.1, is given by

$$\langle N_{Hab} \rangle = e^{\mu_{Hab}} e^{\frac{\sigma_{Hab}^2}{2}} = 1.012 \cdot 10^8 \approx 100 \text{ million}. \tag{3.5}$$

In other words, our statistical (and thus more serious, scientifically speaking) treatment of the Dole equation yields 100 million expected habitable planets in the Galaxy.

This figure is higher than the 35 million given by the classical Dole equation, and much higher than the value of the mode (10 million) shown by the lognormal curve in Figure 3.2.

The last result, stating that there are about 100 million habitable planets in the Galaxy, is of course good news for the future "human conquest of the Galaxy" (if there will ever be one!), since it raises to 100 million the expected number of "Earths" to land on!

But what about the standard deviation around the mean value given in Input

Table 3.1? Table 1.1, row 6, shows that such a standard deviation of the random variable N_{Hab} is given by

$$\sigma_{N_{Hab}} = e^{\mu_{Hab}}\, e^{\frac{\sigma_{Hab}^2}{2}} \sqrt{e^{\sigma_{Hab}^2} - 1} = 2.0 \cdot 10^8 \approx 200 \text{ million.} \tag{3.6}$$

In other words, the standard deviation of the number of habitable planets is 200 million. So, with probability one sigma, we might expect the actual number of habitable planets to rise up to 100 million + 200 million = 300 million.

Finally, the median (50–50 probability of the lognormal distribution shown in Figure 3.1) yields a value of

$$\text{median} = m = e^{\mu_{Hab}} = 4.521 \cdot 10^7 \approx 45 \text{ million.} \tag{3.7}$$

3.6 DISTANCE (MACCONE) DISTRIBUTION OF THE NEAREST HABITABLE PLANET TO US ACCORDING TO THE PREVIOUS NUMERICAL INPUT

Next comes the *distance* distribution of the nearest habitable planet to us (of course, under the easy hypothesis that the distribution of habitable planets in the Galaxy is *uniform*). Well, from the third row of Table 1.2 it follows that the relevant probability density is given by the Maccone distribution, and this is plotted in Figure 3.3.

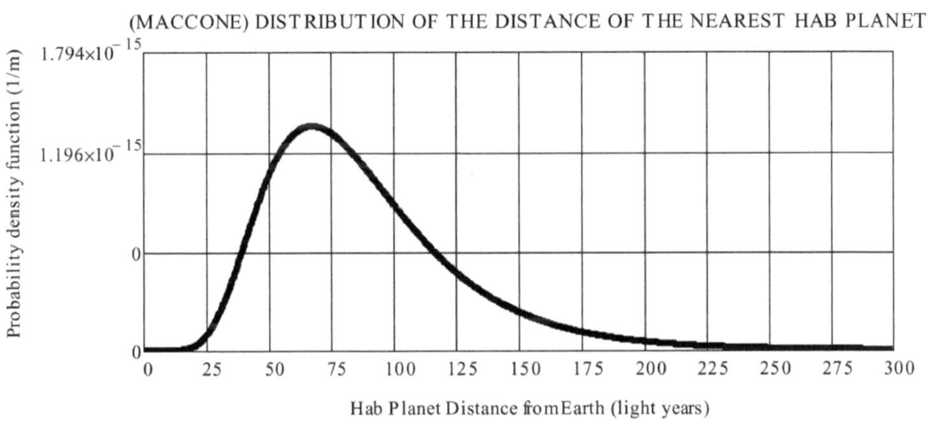

Figure 3.3. The Maccone *probability distribution of the distance of the nearest habitable planet to us in the Galaxy* using data of Input Table 3.1 assumed as inputs to the statistical Dole equation (3.3). A glance at this plot immediately reveals that there is little hope of detecting a habitable planet at distances smaller than 25 light-years from us, since the value of the Maccone distribution is practically zero at such distances. Thus, future interstellar spacecraft designers should keep this lower bound in mind should they wish to land on habitable planets rather than just on any planet. Also, the curve reaches its peak (mode) at about 67 light-years from us, its mode (50–50 probability) at about 80 light-years and, above all, its mean value at 88 light-years from us. The relevant standard deviation turns out to be about 40 light-years, since the distribution tail is rather short.

The mean value of the Maccone distribution is given by the fifth row in Table 1.2 (i.e., for the data given by Input Table 3.1):

$$\langle \text{Hab\_Distance} \rangle = C\, e^{-\frac{\mu_{Hab}}{3}}\, e^{\frac{\sigma_{Hab}^2}{18}} = 8.8 \cdot 10^1 \text{ lt-yr} \approx 88 \text{ lt-yr}. \tag{3.8}$$

The relevant standard deviation is given by the seventh row in Table 3.2, and reads

$$\sigma_{\text{Hab\_Distance}} = C\, e^{-\frac{\mu_{Hab}}{3}}\, e^{\frac{\sigma_{Hab}^2}{18}} \sqrt{e^{\frac{\sigma_{Hab}^2}{9}} - 1} = 3.9 \cdot 10^1 \text{ lt-yr} \approx 40 \text{ lt-yr}. \tag{3.9}$$

Thus, with probability one sigma, it should not be beyond expectations to detect a habitable planet even at, say, just $88 - 40 = 48 \sim 50$ light-years from us.

3.7 COMPARING THE STATISTICAL DOLE AND DRAKE EQUATIONS: NUMBER OF HABITABLE PLANETS VS. NUMBER OF ET CIVILIZATIONS IN THIS GALAXY

It is now appropriate to make a comparison between the number of habitable planets and the number of expected ET civilizations in the Galaxy.

In other words, we want the "get a feeling" for the numbers that we have worked out in this chapter just to see if an intercomparison makes sense.

We can do this by putting on the same table

(1) the mean value and standard deviation of the total number of both habitable planets and ET civilizations, and
(2) the mean value and standard deviation of their respective distances from us (of course, under the hypothesis that both are uniformly scattered throughout the Galaxy).

The result is Table 3.1, which clearly shows how much "more rare" ET civilizations are than habitable planets. Roughly, one has:

$$\frac{\langle N_{Hab} \rangle}{\langle N_{ET} \rangle} = \frac{100 \text{ million}}{4{,}950} \approx 20{,}202 \tag{3.10}$$

so that habitable planets seem to be 20,000 times more frequent than ET civilizations or, put another way, only one ET civilization emerges from 20,000 habitable planets.

As for the distances, the ratio is the other way round:

$$\frac{\langle \text{Hab\_Distance} \rangle}{\langle \text{ET\_Distance} \rangle} = \frac{2{,}670 \text{ lt-yr}}{88 \text{ lt-yr}} \approx 30.340 \tag{3.11}$$

meaning that ETs are, on the average, 30 times farther out than habitable planets. However, it is important to keep in mind that all these results are of course just statistical!

Table 3.1. Comparing the results of the statistical Dole and Drake equation found by inputting Input Table 3.1 and Table 1.1, respectively.

	Statistical Dole equation	*Statistical Drake equation*
Mean value of the *total number* of	Habitable planets in the Galaxy ~100 million	ET civilizations in the Galaxy ~4,590
Standard deviation of the *total number* of	Habitable planets in the Galaxy ~200 million	ET civilizations in the Galaxy ~11,195
Mean value of the *distance* of	Nearest habitable planet ~88 light-years	Nearest ET civilization ~2,670 light-years
Standard deviation of the *distance* of	Nearest habitable planet ~40 light-years	Nearest ET civilization ~1,309 light-years

3.8 SEH, THE "STATISTICAL EQUATION FOR THE HABITABLES" IS JUST THE STATISTICAL DOLE EQUATION

Up to now we have referred to (3.3) as the statistical Dole equation. In view of further improvements in the mathematical analysis of this equation, however, it might be better to rename it "SEH", an acronym standing for "Statistical Equation for the Habitables". This will become clear in future papers by the author, where a number higher than 10 will be the new number of independent, uniform random variables describing equation inputs.

3.9 CONCLUSIONS

We have sought to extend both the classical Drake and Dole equations to let them encompass statistics and probability.

This approach appears to pave the way for future, more profound investigations intended not only to associate "error bars" to each factor in the equation, but especially to increase the number of factors themselves. In fact, this seems to be the only way to incorporate new scientific information in the equations as it becomes available.

In the long run, our statistical equations might just become a huge computer code, growing in size and especially in the depth of scientific information it contains. It would thus become humanity's first "Encyclopedia Galactica". Unfortunately, to extend the Drake equation to statistics, it was necessary to use a mathematical apparatus that is more sophisticated than just the simple product of seven numbers.

APPENDIX 3.A STATISTICAL DOLE EQUATION

Numeric code in Mathcad written by Claudio Maccone on September 4, 2008, updated and completed for this book, *Mathematical SETI*, on February 14, 2012.

Classical Dole equation and numerical inputs assumed by Dole in his book

We show the parsec value to be sure that Mathcad knows about it:

$$\text{parsec} := 30.857 \cdot 10^{15} \cdot \text{m}$$

The density of the number of *habitable* stellar systems Ns in the Galaxy assumed by Dole in his book:

$$\rho_{\text{Ns\_Dole}} := 4.033 \cdot 10^{-4} \cdot \frac{1}{\text{parsec}^3}$$

The volume of the Galaxy assumed by Dole in his book:

$$V_{\text{Galaxy\_Dole}} := 1.6 \cdot 10^{12} \cdot \text{parsec}^3$$

The resulting number of *habitable* stellar systems in the Galaxy according to Dole:

$$\rho_{\text{Ns\_Dole}} \cdot V_{\text{Galaxy\_Dole}} = 6.453 \times 10^8$$

The same as above, but with a smaller precision (one order of magnitude less):

$$\text{Total\_No.\_of\_HAB\_STARS\_Dole} := \text{Total\_DENSITY\_of\_HAB\_STARS\_Dole} \cdot V_{\text{Galaxy\_Dole}}$$

$$\text{Total\_No.\_of\_HAB\_STARS\_Dole} = 6.448 \times 10^8$$

645 million habitable stars in the Galaxy is the result given by Dole in his book, who makes *no use of statistics at all* (i.e., by virtue of the *classical Dole equation*—namely, pure *multiplication* given by equation (3.1)). Now *suppose* that each habitable star system only has *one habitable planet*, as seems to be the case in the Solar System (Mars may have been habitable in the past, but it is *not* now, at least for humans).

The statistical Dole equation and numerical inputs assumed by Dole in his book for mean values, but not for the last column (standard deviations) which the author of this book assumes to be 10% for all 10 inputs, just to fix the ideas

Inputs to the statistical Dole equation (3.3) also given by Input Table 3.1:

$Ns := 6.448 \cdot 10^8 \quad \mu Ns := Ns \quad \sigma Ns := 1 \cdot 10^7$

$Pp := 1.0 \quad \mu Pp := Pp \quad \sigma Pp := \frac{10}{100}$

$Pi := 0.81 \quad \mu Pi := Pi \quad \sigma Pi := \frac{10}{100}$

$PD := 0.63 \quad \mu PD := PD \quad \sigma PD := \frac{10}{100}$

$PM := 0.19 \quad \mu PM := PM \quad \sigma PM := \frac{10}{100}$

$Pe := 0.94 \quad \mu P3 := Pe \quad \sigma Pe := \frac{10}{100}$

$PB := 0.95 \quad \mu PB := PB \quad \sigma PB := \frac{10}{100}$

$PR := 0.9 \quad \mu PR := PR \quad \sigma PR := \frac{10}{100}$

$PA := 0.7 \quad \mu PA := PA \quad \sigma PA := \frac{10}{100}$

$PL := 1 \quad \mu PL := PL \quad \sigma PL := \frac{10}{100}$

$N_{Hab} := Ns \cdot Pp \cdot Pi \cdot PD \cdot PM \cdot Pe \cdot PB \cdot PR \cdot PA \cdot PL \qquad N_{Hab} = 3.517 \times 10^7$

$N_{Hab\_Dole} := 645 \cdot 10^6$

According to the above mean values, the *classical* Dole equation (i.e., pure multiplication) would yield *35 million habitable planets* in the Galaxy. This is contrasted by the *645 million habitable planets* claimed by Dole in his book.

Mean value and variance of the logarithm of the number of habitable stars in the Galaxy, ln(Ns)

These are F, G, K, and M stars, but *not* O, B, and A, since the life of the latter is too short to create planets:

$aNs := \mu Ns - \sqrt{3} \cdot \sigma Ns \qquad aNs = 6.275 \times 10^8$

$bNs := \mu Ns + \sqrt{3} \cdot \sigma Ns \qquad bNs = 6.621 \times 10^8$

$$\text{MlnNs} := \frac{\text{bNs}\cdot(\ln(\text{bNs})-1)-\text{aNs}\cdot(\ln(\text{aNs})-1)}{\text{bNs}-\text{aNs}}$$

$$\text{MlnNs} = 2.028\times10^{1}$$

$$\text{VlnNs} := 1-\frac{\text{aNs}\cdot\text{bNs}\cdot\left(\ln\left(\frac{\text{bNs}}{\text{aNs}}\right)\right)^2}{(\text{bNs}-\text{aNs})^2} \qquad \text{VlnNs} = 2.406\times10^{-4}$$

Mean value and variance of the logarithm of the probability that such an F, G, K, or M star has planets in orbit around it

$$\text{aPp} := \mu\text{Pp}-\sqrt{3}\cdot\sigma\text{Pp} \qquad \text{aPp} = 8.268\times10^{-1}$$

$$\text{bPp} := \mu\text{Pp}+\sqrt{3}\cdot\sigma\text{Pp} \qquad \text{bPp} = 1.173\times10^{0}$$

$$\text{MlnPp} := \frac{\text{bPp}\cdot\ln(\text{bPp})-\text{bPp}-\text{aPp}\cdot\ln(\text{aPp})+\text{aPp}}{\text{bPp}-\text{aPp}}$$

$$\text{MlnPp} = -5.046\times10^{-3}$$

$$\text{VlnPp} := 1-\frac{\text{aPp}\cdot\text{bPp}\cdot\left(\ln\left(\frac{\text{bPp}}{\text{aPp}}\right)\right)^2}{(\text{bPp}-\text{aPp})^2} \qquad \text{VlnPp} = 1.014\times10^{-2}$$

Mean value and variance of the logarithm of the probability that the inclination of the planet's equator is correct for its orbital distance

$$\text{aPi} := \mu\text{Pi}-\sqrt{3}\cdot\sigma\text{Pi} \qquad \text{aPi} = 6.368\times10^{-1}$$

$$\text{bPi} := \mu\text{Pi}+\sqrt{3}\cdot\sigma\text{Pi} \qquad \text{bPi} = 9.832\times10^{-1}$$

$$\text{MlnPi} := \frac{\text{bPp}\cdot\ln(\text{bPp})-\text{bPp}-\text{aPp}\cdot\ln(\text{aPp})+\text{aPp}}{\text{bPp}-\text{aPp}} \qquad \text{MlnPi} = -5.046\times10^{-3}$$

$$\text{VlnPi} := 1 \qquad \text{VlnPi} = 1\times10^{0}$$

Mean value and variance of the logarithm of the probability that at least one planet orbits within an ecosphere

$$\text{aPD} := \mu\text{PD}-\sqrt{3}\cdot\sigma\text{PD} \qquad \text{aPD} = 4.568\times10^{-1}$$

$$\text{bPD} := \mu\text{PD}+\sqrt{3}\cdot\sigma\text{PD} \qquad \text{bPD} = 8.032\times10^{-1}$$

$$\text{MlnPD} := \frac{\text{bPD}\cdot\ln(\text{bPD})-\text{bPD}-\text{aPD}\cdot\ln(\text{aPD})+\text{aPD}}{\text{bPD}-\text{aPD}} \qquad \text{MlnPD} = -4.749\times10^{-1}$$

$$\text{VlnPD} := 1-\frac{\text{aPD}\cdot\text{bPD}\cdot\left(\ln\left(\frac{\text{bPD}}{\text{aPD}}\right)\right)^2}{(\text{bPD}-\text{aPD})^2} \qquad \text{VlnPD} = 2.613\times10^{-2}$$

Mean value and variance of the logarithm of the probability that the planet has a suitable mass = 0.4 to 2.35 Earth masses

$aPM := \mu PM - \sqrt{3} \cdot \sigma PM \qquad aPM = 1.679 \times 10^{-2}$

$bPM := \mu PM + \sqrt{3} \cdot \sigma PM \qquad bPM = 3.632 \times 10^{-1}$

$$MlnPM := \frac{bPM \cdot \ln(bPM) - bPM - aPM \cdot \ln(aPM) + aPM}{bPM - aPM}$$

$MlnPM = -1.864 \times 10^{0}$

$$VlnPM := 1 - \frac{aPM \cdot bPM \cdot \left(\ln\left(\frac{bPM}{aPM}\right)\right)^2}{(bPM - aPM)^2} \qquad VlnPM = 5.197 \times 10^{-1}$$

Mean values and variance of the logarithm of the probability that the planet's orbital eccentricity is sufficiently low

$aPe := \mu Pe - \sqrt{3} \cdot \sigma Pe \qquad aPe = 7.668 \times 10^{-1}$

$bPe := \mu Pe + \sqrt{3} \cdot \sigma Pe \qquad bPe = 1.113 \times 10^{0}$

$$MlnPe := \frac{bPe \cdot \ln(bPe) - bPe - aPe \cdot \ln(aPe) + aPe}{bPe - aPe}$$

$MlnPe = -6.759 \times 10^{-2}$

$$VlnPe := 1 - \frac{aPe \cdot bPe \cdot \left(\ln\left(\frac{bPe}{aPe}\right)\right)^2}{(bPe - aPe)^2} \qquad VlnPe = 1.15 \times 10^{-2}$$

Mean value and variance of the logarithm of the probability that the presence of a second star has not rendered the planet uninhabitable

$aPB := \mu PB - \sqrt{3} \cdot \sigma PB \qquad aPB = 7.768 \times 10^{-1}$

$bPB := \mu PB + \sqrt{3} \cdot \sigma PB \qquad bPB = 1.123 \times 10^{0}$

$$MlnPB := \frac{bPB \cdot \ln(bPB) - bPB - aPB \cdot \ln(aPB) + aPB}{bPB - aPB}$$

$MlnPB = -5.689 \times 10^{-2}$

$$VlnPB := 1 - \frac{aPB \cdot bPB \cdot \left(\ln\left(\frac{bPB}{aPB}\right)\right)^2}{(bPB - aPB)^2} \qquad VlnPB = 1.126 \times 10^{-2}$$

Mean value and variance of the logarithm of the probability that the planet's rate of rotation is neither too fast nor too slow

$aPR := \mu PR - \sqrt{3} \cdot \sigma PR \qquad aPR = 7.268 \times 10^{-1}$

$bPR := \mu PR + \sqrt{3} \cdot \sigma PR \qquad bPR = 1.073 \times 10^{0}$

$$MlnPR := \frac{bPR \cdot \ln(bPR) - bPR - aPR \cdot \ln(aPR) + aPR}{bPR - aPR}$$

$MlnPR = -1.116 \times 10^{-1}$

$$VlnPR := 1 - \frac{aPR \cdot bPR \cdot \left(\ln\left(\frac{bPR}{aPR}\right)\right)^2}{(bPR - aPR)^2} \qquad VlnPR = 1.256 \times 10^{-2}$$

Mean value and variance of the logarithm of the probability that the planet is of the proper age

$aPA := \mu PA - \sqrt{3} \cdot \sigma PA \qquad aPA = 7.668 \times 10^{-1}$

$bPA := \mu PA + \sqrt{3} \cdot \sigma PA \qquad bPA = 1.113 \times 10^{0}$

$$MlnPA := \frac{bPA \cdot \ln(bPA) - bPA - aPA \cdot \ln(aPA) + aPA}{bPA - aPA}$$

$MlnPA = -6.759 \times 10^{-2}$

$$VlnPA := 1 - \frac{aPA \cdot bPA \cdot \left(\ln\left(\frac{bPA}{aPA}\right)\right)^2}{(bPA - aPA)^2} \qquad VlnPA = 1.15 \times 10^{-2}$$

Mean value and variance of the logarithm of the probability that, all astronomical conditions being appropriate, life has developed on the planet

$aPL := \mu PL - \sqrt{3} \cdot \sigma PL \qquad aPL = 8.268 \times 10^{-1}$

$bPL := \mu PL + \sqrt{3} \cdot \sigma PL \qquad bPL = 1.173 \times 10^{0}$

$$MlnPL := \frac{bPL \cdot \ln(bPL) - bPL - aPL \cdot \ln(aPL) + aPL}{bPL - aPL}$$

$MlnPL = -5.046 \times 10^{-3}$

$$VlnPL := 1 - \frac{aPL \cdot bPL \cdot \left(\ln\left(\frac{bPL}{aPL}\right)\right)^2}{(bPL - aPL)^2} \qquad VlnPL = 1.014 \times 10^{-2}$$

Value of the N_{Hab} lognormal's μ

$$\mu := MlnNs + MlnPp + MlnPi + MlnPD + MlnPM + MlnPe + MlnPB + MlnPR + MlnPA + MlnPL$$

$$\mu = 1.763 \times 10^{1}$$

Value of the N_{Hab} lognormal's σ

$$\sigma := \sqrt{VlnNs + VlnPp + VlnPi + VlnPD + VlnPM + VlnPe + VlnPB + VlnPR + VlnPA + VlnPL}$$

$$\sigma = 1.27 \times 10^{0} \qquad \sigma^2 = 1.613 \times 10^{0}$$

Lognormal of N_{Hab} and its statistics

The old value of N_{Hab} given by the classical Dole equation:

$$N_{Hab} = 3.517 \times 10^{7}$$

The value given by Dole in his book:

$$N_{Hab\_Dole} = 6.45 \times 10^{8}$$

Mean value of N_{Hab} given by the statistical Dole equation

$$MeanValue\_N\_Hab := e^{\mu + \frac{\sigma^2}{2}} \qquad MeanValue\_N\_Hab = 1.013 \times 10^{8}$$

That is, *100 million is the mean value of the number of habitable planets in the Galaxy* according to Input Table 3.1 and the statistical Dole equation.

Variance of N_{Hab} given by the statistical Dole equation

$$Variance\_N\_Hab := e^{2\cdot\mu+\sigma^2} \cdot (e^{\sigma^2} - 1) \qquad Variance\_N\_Hab = 4.122 \times 10^{16}$$

Standard deviation of N_{Hab} given by the statistical Dole equation

$$Standard\_Deviation\_N\_Hab := e^{\mu + \frac{\sigma^2}{2}} \cdot \sqrt{e^{\sigma^2} - 1} \qquad Standard\_Deviation\_N\_Hab = 2.03 \times 10^{8}$$

So, the *standard deviation*, around the 100 million habitable planet mean value, is *plus or minus 200 million.*

Maximum number of N_{Hab} given by the statistical Dole equation

$$MeanValue\_N\_Hab + Standard\_Deviation\_N\_Hab = 3.043 \times 10^{8}$$

So, the maximum number of habitable planets in the Galaxy (statistically speaking) turns out to be about *300 million.*

Median number of N_{Hab} given by the statistical Dole equation

Median_N_Hab := e^{μ} Median_N_Hab = 4.521×10^7

The *median* (= 50–50 in probability) number of habitable planets is thus about 45 million.

Mode = peak of N_{Hab} given by the statistical Dole equation

Mode_N_Hab := $e^{\mu-\sigma^2}$ Mode_N_Hab = 9.008×10^6

The *peak abscissa* (= *mode*) of the N_{Hab} lognormal distribution is of course smaller than all the above values: just about 9 million habitable planets in the Galaxy.

Plotting the lognormal distribution of the number of habitable planets according to the statistical Dole equation

$$f_{N\_Hab}(n, \mu, \sigma) := \frac{1}{n} \cdot \frac{1}{\sqrt{2\pi} \cdot \sigma} \cdot e^{-\frac{(\ln(n)-\mu)^2}{2\sigma^2}}$$

peak_height := f_{N_Hab}(Mode_N_Hab, μ, σ) peak_height = 1.556×10^{-8}

Computing the peak height of this lognormal just to plot its peak at best:

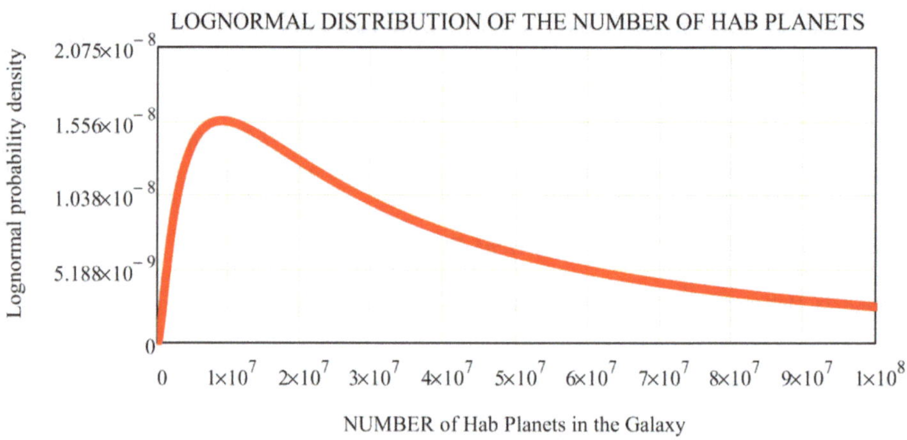

Finding the *distance* between *habitable planets*

Let us define the light-year (ly), which Mathcad does not know:

ly := c · yr ly = 9.461×10^{15} m

Defining the size of the Galaxy

Radius_of_Galaxy := 50000 · ly Thickness_of_Galaxy_at_half_radius := 1600 · ly

R_{Galaxy} := Radius_of_Galaxy h_{Galaxy} := Thickness_of_Galaxy_at_half_radius

Defining the average distance between any two nearby habitable planets

$N_{Hab} = 3.517 \times 10^{7}$

$$Av\_DISTANCE(n) := \frac{\sqrt[3]{6 \cdot R_{Galaxy}{}^{2} \cdot h_{Galaxy}}}{\sqrt[3]{n}}$$

$Av\_DISTANCE(N_{Hab}) = 8.804 \times 10^{1} \cdot ly$

Thus, the average distance between any two nearby habitable planets is given, according to the classical Dole equation, as about 88 light-years. This appears to be reasonable.

Defining the constant C

$$C := \sqrt[3]{6 \cdot R_{Galaxy}{}^{2} \cdot h_{Galaxy}}$$

$C = 2.884 \times 10^{4} \cdot ly \qquad C = 2.729 \times 10^{20}\ m$

Mean value of the average distance between any two nearby habitable planets

Once again, we expect to find the nearest habitable planet to us at about 88 light-years:

$$Mean\_Value\_of\_Hab\_Planet\_DIST := C \cdot e^{\frac{\sigma^2}{18} - \frac{\mu}{3}}$$

$Mean\_Value\_of\_Hab\_Planet\_DIST = 8.856 \times 10^{1} \cdot ly$

Variance of the average distance between any two nearby habitable planets

$$Variance\_of\_Hab\_Planet\_DIST := C^2 \cdot e^{\frac{\sigma^2}{9} - \frac{2\mu}{3}} \cdot \left(e^{\frac{\sigma^2}{9}} - 1\right)$$

$Variance\_of\_Hab\_Planet\_DIST = 1.54 \times 10^{3} \cdot ly^{2}$

Standard deviation of the average distance between any two nearby habitable planets

The corresponding standard deviation of the average distance between any two nearby habitable planets is about 40 light-years:

$$St\_Dev\_of\_Hab\_Planet\_DIST := C \cdot e^{\frac{\sigma^2}{18} - \frac{\mu}{3}} \cdot \sqrt{e^{\frac{\sigma^2}{9}} - 1}$$

$St\_Dev\_of\_Hab\_Planet\_DIST = 3.924 \times 10^{1} \cdot ly$

The maximum average distance between any two nearby habitable planets is about 130 light-years

$$Max\_Stat\_Dist\_Hab\_Planet := Mean\_Value\_of\_Hab\_Planet\_DIST + St\_Dev\_of\_Hab\_Planet\_DIST$$

$Max\_Stat\_Dist\_Hab\_Planet = 1.278 \times 10^{2} \cdot ly$

The minimum average distance between any two nearby habitable planets is about 50 light-years

$$\text{Max\_Stat\_Dist\_Hab\_Planet} := \text{Mean\_Value\_of\_Hab\_Planet\_DIST} - \text{St\_Dev\_of\_Hab\_Planet\_DIST}$$

$$\text{Max\_Stat\_Dist\_Hab\_Planet} = 4.932 \times 10^{1} \cdot \text{ly}$$

The median of the average distance between any two nearby habitable planets

$$\text{Distance\_Median} := C \cdot e^{-\frac{\mu}{3}} \qquad \text{Distance\_Median} = 8.097 \times 10^{1} \cdot \text{ly}$$

Plotting the average distance distribution between any two nearby habitable planets

Maccone distribution for the *average distance* between any two nearby habitable planets:

$$f_{Hab\_Dist}(y) := \frac{3}{y} \cdot \frac{1}{\sqrt{2 \cdot \pi} \cdot \sigma} \cdot e$$

Peak (= mode) abscissa of the average distance between any two nearby habitable planets

$$\text{Distance\_Mode} := C \cdot e^{-\frac{\mu}{3} - \frac{\sigma^2}{9}} \qquad \text{Distance\_Mode} = 6.768 \times 10^{1} \cdot \text{ly}$$

So, the *peak* of the *distance distribution* is at about 67 light-years from the Earth. On the other hand, the *peak height* of the *distance distribution* is given by:

$$\text{Distance\_Peak\_Height} := f_{Hab\_Dist}\left(C \cdot e^{-\frac{\mu}{3} - \frac{\sigma^2}{9}}\right) \qquad \text{Distance\_Peak\_Height} = 1.345 \times 10^{-18} \cdot \frac{1}{m}$$

The *plot* of the *average distance distribution* between any two nearby habitable planets is thus

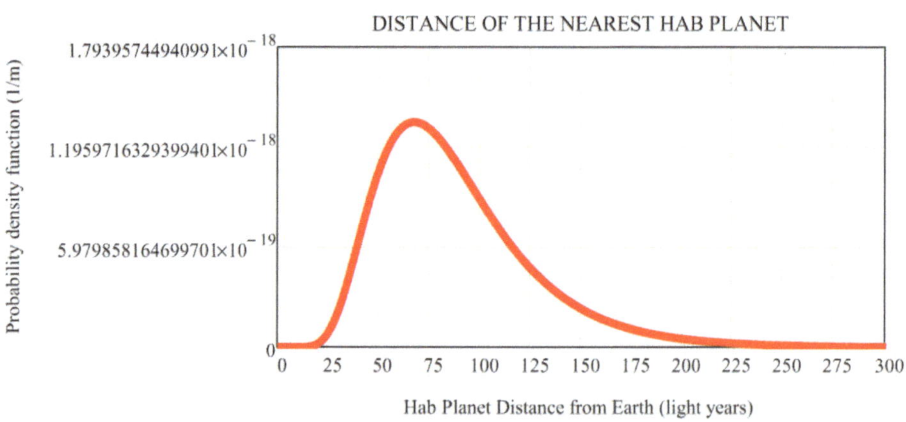

This plot neatly shows that the probability of finding a habitable exoplanet at distances *smaller than about 25 light-years* from Earth is virtually *zero*!

3.10 BIBLIOGRAPHY

[1] Gregory Benford Jim Benford, and Dominic Benford, "Cost optimized interstellar beacons: SETI," arXiv.org website, October 22, 2008.

[2] Jeffrey Bennet and Seth Shostak, *Life in the Universe*, Second Edition, Pearson/Addison Wesley, San Francisco, 2007, see p. 404 in particular.

[3] Carl Sagan, *Cosmos*, Random House, New York, 1983, see pp. 298–302 in particular.

[4] Claudio Maccone, "The statistical Drake equation," paper #IAC-08-A4.1.4 presented on October 1, 2008, at the *59th International Astronautical Congress (IAC) held in Glasgow, Scotland, U.K., September 29–October 3, 2008.*

[5] Claudio Maccone, "The statistical Drake equation," in press, *Acta Astronautica*, 2010, in the *Proceedings of the First IAA Symposium on "Searching for Life Signatures" held at UNESCO, Paris, September 22–26, 2008.*

[6] Stephen H. Dole, *Habitable Planets for Man*, First Edition, 1964, RAND Corporation, see p. 82 in particular (i.e., the beginning of Chapter 5, "Probability of occurrence of habitable planets").

[7] Claudio Maccone, "SETI and SEH (Statistical Equation for Habitables)," paper #IAC-09.A4.1.7 presented at the *60th International Astronautical Congress held at Daejeon, Republic of Korea, October 12–16, 2009*. Later, this paper was published in *Acta Astronautica*, **68**(2011), 63–75.

[8] Stephen H. Dole, *Habitable Planets for Man*, First Edition published in 1964 (downloadable free today from the website of the Rand Corporation).

4

Statistical Fermi paradox and Galactic travels

4.1 THE CLASSICAL CORAL MODEL OF GALACTIC COLONIZATION

The following description of the coral model of Galactic colonization is taken from *Life in the Universe* (Second Edition, 2007), by Jeffrey O. Bennett and G. Seth Shostak [3, see pp. 459 and 476 especially].

Let's start by assuming that another civilization decided to start sending out spacecraft to colonize other habitable planets. How long would it take for this civilization to colonize the entire Galaxy?

The answer clearly depends on the civilization's technological capabilities. For example, if it has the technology to build spacecraft that can travel at speeds close to the speed of light, then it could add colonies throughout the Galaxy fairly quickly, since trips between nearby stars would take only a few years. Perhaps surprisingly, the conclusion is not that much different if we assume much lower speeds.

In fact, consider a civilization that has nuclear rockets such as Project Orion which was developed between 1958 and 1963 (*http://en.wikipedia.org/wiki/Project_Orion_(nuclear_propulsion)* or Project Daedalus developed between 1973 and 1978 (*http://en.wikipedia.org/wiki/Project_Daedalus*): such rockets do not seem that much beyond our technological grasp and they might attain speeds of about 10% of the speed of light (0.1*c*). Given that a typical distance between star system in our region of the Galaxy is about 5 light-years, a nuclear spacecraft traveling at 10% of the speed of light could journey from one star system to the next in about 50 years. This trip would be possible in a human lifetime and might be practical if the colonizers have found ways to hibernate during the voyage or if they have somewhat longer life-spans than we do (either naturally or through medical intervention).

After arriving at a new star system, the colonists establish themselves and set about increasing the population. Once the population has grown sufficiently, these colonists send their own pilgrims into space, adding yet more star systems to the growing civilization. Thus, the process starts at the home star system and the first few colonies are located within just a few light-years. These colonies then lead to other colonies at greater distances, as well as at unexplored locations in between. Growth tends to expand the empire around the edges of the existing empire, much like the growth of coral in the sea. This is the reason this type of colonization model is called "the coral model of Galactic colonization".

The overall result is a gradually expanding region in which all habitable planets are colonized. The colonization rate depends on the speed of spacecraft and the time it takes each colony to start sending its own spacecraft to other stars. For travel at 10% the speed of light and assuming that it takes 150 years before each colony's population grows enough to send out more colonists, the calculations that we shall make in the next section show that the inhabited region of the Galaxy expands outwards from the home world at about 1% of the speed of light. Thus, if the home star is near one edge of the Galactic Disk, so that colonizing the entire Galaxy means inhabiting star systems 100,000 light-years away, the civilization would expand through the entire Galaxy in about 10 million years. The required time would be a few million years less if the home star is in a more central part of the Galaxy.

For an even more conservative estimate, suppose the colonists have rockets that travel at only 1% of the speed of light and that it takes each new colony 5,000 years until it is ready to send out additional colonists. Even in this case, the region occupied by this civilization would grow at a rate of roughly 1/1,000 (0.1%) the speed of light and the entire Galaxy would be colonized in 100 million years. This is still a very short time compared with the time that has been available for civilizations to arise (4.5 billion years for humanity), further deepening the mystery of why we see no evidence that anyone else has done it by now. This is, of course, the well-known Fermi Paradox, first stated by Enrico Fermi (1901–1954) to his colleagues during a lunch discussion at Los Alamos back in 1950 (for a good summary, see the Wikipedia site: *http://en.wikipedia.org/wiki/Fermi_paradox*).

But let us go back to the coral model of Galactic colonization. We now want to cast it as a sound mathematical theory. First of all, let us write down the fact that the overall expansion speed of the empire, v_{exp}, is the ratio of the average distance among any two nearby stars, D, to the sum of twice

(1) the time of actual spaceflight from one star to the next one, t_{flight}, plus
(2) the time t_{col} requested to colonize a planet (i.e., to develop there a civilization until the time is ripe for one more spaceflight jump to the next star).

That is, we assume that the equation holds

$$v_{\text{exp}} = k \frac{D}{t_{flight} + t_{col}}. \tag{4.1}$$

We insert a factor k in front of (4.1) just to take into account the "zigzag" motion of expansion from one star to the next in three-dimensional space. This k is explained by Bennett and Shostak [3, p. 476] in a way that we rephrase as follows. The purely numerical factor k would be equal to 1 only if colonization was always directed straight outward from the home star. In reality, the colonists will sometimes go to uncolonized star systems in other directions: k accounts for this zigzag motion. For zigzag motion in three-dimensional space, we assume that

$$k = \frac{1}{2}. \tag{4.2}$$

Apart from (4.1) and (4.2) another equation is needed to cast our coral expansion model in mathematical form: this is the obvious relationship between the flight time t_{flight} from one star to the next, and the corresponding spaceship (average) speed, v_{ss}, that is

$$t_{flight} = \frac{D}{v_{ss}}. \tag{4.3}$$

Inserting then (4.2) and (4.3) into (4.1), a little rearranging yields

$$v_{\exp}(D, t_{col}, v_{ss}) = k \frac{v_{ss} \cdot D}{D + v_{ss} \cdot t_{col}}. \tag{4.4}$$

This is the expression of the expansion speed of the empire throughout the Galaxy that we want to concentrate on in the coming sections.

An immediate consequence of (4.4) is the Galaxy colonization time, denoted T_{Galaxy} (i.e., the overall time that our expanding empire will need to colonize the whole Galaxy). If a civilization starts conquering the Galaxy from the outskirts (more or less like ours!), the largest possible amount of time is clearly given by

$$T_{Galaxy} = \frac{2\, R_{Galaxy}}{v_{\exp}} = \frac{2\, R_{Galaxy}}{k} \cdot \frac{D + v_{ss} \cdot t_{col}}{v_{ss} \cdot D}. \tag{4.5}$$

As we said, it could take a little less if the conquerors lived near the center of the Galaxy; so, to be conservative, let us take (4.5) for granted and just rewrite it as

$$T_{Galaxy} = \frac{2\, R_{Galaxy}}{k} \cdot \left(\frac{1}{v_{ss}} + \frac{t_{col}}{D} \right)$$

or

$$T_{Galaxy} = \frac{2\, R_{Galaxy}}{k} \cdot \frac{t_{col}}{D} + \frac{2\, R_{Galaxy}}{k \cdot v_{ss}}. \tag{4.6}$$

Let us now introduce two positive constants a and b:

$$\begin{cases} a = \dfrac{2\, R_{Galaxy}}{k} \approx 200{,}000 \text{ light-year} \\ b = \dfrac{2\, R_{Galaxy}}{k \cdot v_{ss}} \approx 20 \text{ million year} \qquad \text{for } v_{ss} \approx 0.01c. \end{cases} \tag{4.7}$$

Then, the whole Galaxy colonization time T_{Galaxy} takes the final form that we will use in the coming sections

$$T_{Galaxy} = a \cdot \frac{t_{col}}{D} + b. \tag{4.8}$$

4.2 THE CLASSICAL FERMI PARADOX (1950)

Let us now consider three different numerical cases of (4.8) and check them against each other.

(1) First, suppose that one has

$$\begin{cases} k = \dfrac{1}{2} \\ v_{ss} = 0.1c \\ D = 5 \text{ lt-yr} \\ t_{col} = 150 \text{ yr.} \end{cases} \tag{4.9}$$

Then, (4.4) and (4.8) yield, respectively:

$$\begin{cases} v_{\exp} = 3{,}747 \ \dfrac{\text{km}}{\text{s}} = 0.0125c \\ T_{Galaxy} \approx 8 \text{ million year.} \end{cases} \tag{4.10}$$

(2) Second, suppose that one has

$$\begin{cases} k = \dfrac{1}{2} \\ v_{ss} = 0.01c \\ D = 5 \text{ lt-yr} \\ t_{col} = 1{,}000 \text{ yr.} \end{cases} \tag{4.11}$$

Then, (4.4) and (4.8) yield, respectively:

$$\begin{cases} v_{\exp} = 500 \ \dfrac{\text{km}}{\text{s}} = 0.001c \\ T_{Galaxy} \approx 60 \text{ million year.} \end{cases} \tag{4.12}$$

(3) Third, assume the *human case*. By this we mean that *habitable planets* are just those planets *habitable by humans*, and not planets of any other kind! Thus, we must apply the classical Dole equation (4.18) of "habitable planets for man" to reach the important conclusion that the average distance between planets habitable by humans is 84 light-years, and *not* just 5 light-years, as in (4.35) and (4.37). In other words, let us assume the inputs:

$$\begin{cases} k = \dfrac{1}{2} \\ v_{ss} = 0.01c \\ D = 84 \text{ lt-yr} \\ t_{col} = 1{,}000 \text{ yr.} \end{cases} \tag{4.13}$$

Then, (4.4) and (4.8) yield, respectively:

$$\begin{cases} v_{\exp} = 1{,}339\ \dfrac{\text{km}}{\text{s}} = 0.004c \\ T_{Galaxy} \approx 22 \text{ million year.} \end{cases} \tag{4.14}$$

So, about 22 million years would be the overall time necessary for humankind to colonize the whole Milky Way (as long as no alien civilization was trying to stop us, of course!) if human spaceships were capable of traveling at 1% of the speed of light and if the average colonization time for every new planet was about 1,000 years.

The basic difference between the humanity expansion model and the two previous models is of course the difference in the average distance among habitable extrasolar planets. Actually, it is interesting to take the limit of (4.5) for distance D as it increases (i.e., $D \to \infty$). This yields

$$\lim_{D\to\infty} T_{Galaxy} = \frac{2\ R_{Galaxy}}{k} \cdot \frac{1}{v_{ss}} \tag{4.15}$$

meaning that spaceship speed v_{ss} plays an increasing role in Galaxy colonization as the average distance D increases. In yet other words, if ETs of a certain "race" can only live on a reduced number of planets, then they must have much faster spaceships to colonize the Galaxy than ETs that can live on a variety of planets! Not a small result at all ... meaning that *adaptable* ETs can more easily conquer the Galaxy.

In conclusion, from the above three examples we see that the time of colonization of the whole Galaxy seems to be of the order of some tens of millions of years: just a blink of the eye compared with the Galaxy age of about 10 billion years; and that is the Fermi paradox, of course. Many papers and books have been written about the Fermi paradox, especially in recent years, but none has been able to solve it so far. In the coming sections, however, we are going to present the *statistical* Fermi paradox.

4.3 THE STATISTICAL CORAL MODEL OF GALACTIC COLONIZATION

The goal of this chapter is to present for the first time the statistical generalization of the classical Fermi paradox described in the previous section.

This is a difficult mathematical job. In fact, consider first the statistical expansion speed (4.4) of the empire, which we rewrite here in the form

$$V_{\exp}(D, T_{col}) = k \frac{v_{ss} \cdot D}{D + v_{ss} \cdot T_{col}}. \tag{4.16}$$

We follow here the convention of denoting all random variables by capitals, in contrast to ordinary (deterministic) variables, which we continue to denote by lowercase letters. Thus, in (4.16) we have three random variables in the game, V_{exp}, D, and T_{col}, while v_{ss} is just a real positive, known parameter. Why do we proceed this way? Because:

(1) The (average) spaceship speed, allowing us to jump from one star to the next, is entirely under human (or ET) control, and so can be regarded just as a sheer number rather than a random variable. This assumption simplifies things greatly from the mathematical point of view and makes the problem mathematically solvable.
(2) On the contrary, the colonization time T_{col} is indeed a random variable, inasmuch as we don't know in advance what difficulties we will have to face in order to colonize a new planet, and so we don't know how long it will actually take to develop facilities on this planet so that we are finally ready for the next jump. Because of the random character of T_{col} we denote it in capitals in (4.16).
(4) Finally, D, the (average) distance in between any two nearby habitable planets follows the Maccone distribution, as we already know. We thus have for its probability density function (pdf) the Maccone distribution

$$f_D(d) = \frac{3}{d} \cdot \frac{1}{\sqrt{2\pi}\,\sigma} \cdot e^{-\frac{\left(\ln\left[\frac{C^3}{d^3}\right]-\mu\right)^2}{2\sigma^2}}. \tag{4.17}$$

But, then, what about the probability density function of the new (positive) random variable T_{col} yielding the amount of time needed to colonize a new planet? Well, we are of course free to choose any density function we wish, but the best choice seems to be a *lognormal* pdf because (as we already know from the Drake and Dole equations) this can be thought of as the multiplicative product whose many random variables are all positive. Thus, we assume that

$$f_{T_{col}}(t) = \frac{1}{\text{yr}} \cdot \frac{\text{yr}}{t} \cdot \frac{1}{\sqrt{2\pi}\,\sigma} \cdot e^{-\frac{\left(\ln\left[\frac{t}{\text{yr}}\right]-\mu\right)^2}{2\sigma^2}}. \tag{4.18}$$

A question now arises: Is any lognormal pdf fully specified if we know in advance both its mean value and its standard deviation? The answer to this question is "yes". In fact, consider the relevant mean value and standard deviation, given by the fourth and sixth line in Table 1.1, respectively, and set up the two simultaneous equations

$$\begin{cases} \langle N \rangle = e^{\mu}\, e^{\frac{\sigma^2}{2}} \\ \sigma_N = e^{\mu}\, e^{\frac{\sigma^2}{2}} \sqrt{e^{\sigma^2} - 1}. \end{cases} \tag{4.19}$$

This system may indeed be inverted by dividing the first equation by the second, solving for σ^2, inserting this σ^2 into the first equation, and finally solving the equation for μ. One thus gets:

$$\begin{cases} \mu = \ln\left(\dfrac{\langle N\rangle^2}{\sqrt{\langle N\rangle^2 + \sigma_N^2}}\right) \\ \sigma = \sqrt{\ln\left(\dfrac{\sigma_N^2}{\langle N\rangle^2} + 1\right)}. \end{cases} \tag{4.20}$$

While that was good for the lognormally distributed random variable N, here we must apply (46) to the lognormally distributed new random variable T_{col}. Thus, $\langle N\rangle$ must be replaced by the mean value of colonization time, μ_{t_col}, and σ_N must be replaced by the standard deviation of colonization time, σ_{t_col}. Thus, for the colonization time, (4.20) becomes

$$\begin{cases} \mu = \ln\left(\dfrac{\mu_{t\_col}{}^2}{\sqrt{\mu_{t\_col}{}^2 + \sigma_{t\_col}{}^2}}\right) \\ \sigma = \sqrt{\ln\left(\dfrac{\sigma_{t\_col}{}^2}{\mu_{t\_col}{}^2} + 1\right)}. \end{cases} \tag{4.21}$$

For instance, suppose that the mean colonization time equals 1,000 years, with a standard deviation of ±500 years. Then inserting these two values (previously divided by yr to make things dimensionally correct) into (4.21) yields

$$\begin{cases} \mu = \ln\left(\dfrac{\mu_{t\_col}{}^2}{\sqrt{\mu_{t\_col}{}^2 + \sigma_{t\_col}{}^2}}\right) = 6.7961835 \\ \sigma = \sqrt{\ln\left(\dfrac{\sigma_{t\_col}{}^2}{\mu_{t\_col}{}^2} + 1\right)} = 0.4723807 \end{cases} \tag{4.22}$$

and the relevant lognormal distribution (4.18) is thus perfectly determined. This lognormal pdf of the time (in years) needed to colonize a new planet if one assumes a mean value of 1,000 years plus or minus a standard deviation of 500 years is plotted in the following Figure 4.1. In our opinion, much of the history on Earth, such as the colonization of America, is similar.

4.4 FINDING THE PROBABILITY DISTRIBUTION OF THE OVERALL TIME NEEDED TO COLONIZE THE WHOLE GALAXY

In this section we face the (difficult) mathematical problem of finding the probability distribution (i.e., the probability density function, or pdf) of the overall time needed

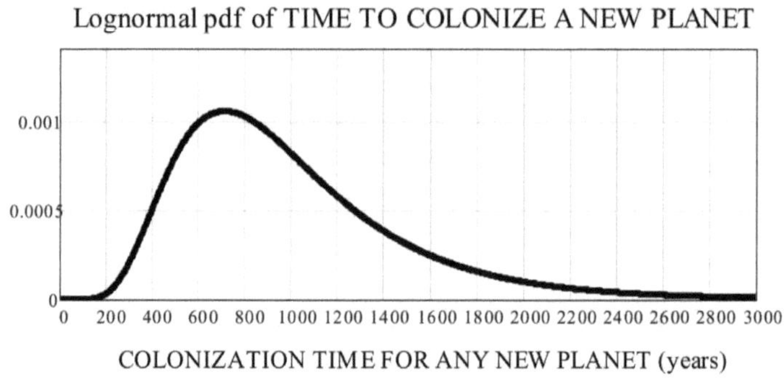

Figure 4.1. The lognormal distribution of the time needed to colonize each new planet in our statistical extension of the coral Galactic expansion model, assuming that, for instance, it takes 1,000 years ±500 years to colonize that planet. Our assumption that this probability distribution must be lognormally distributed is not "arbitrary", but is ensured by the fact that the lognormal distribution can be thought of as the multiplicative product whose many random variables are all positive and unknown (according to the Central Limit Theorem of statistics in its multiplicative version, rather than in its additive version, where the lognormal would be replaced by a Gaussian). The author believes that this very model could also be applied to human history on Earth (like the colonization of the Americas by the Europeans), with the result of creating a new scientific discipline called "mathematical history".

to colonize the Galaxy. This is given by the positive random variable T_{Galaxy} defined by (4.8) where the variable T_{col} is now a positive random variable lognormally distributed as described in the last section, while D is the random variable yielding the average distance between any two nearby "habitable planets for man" (as Steve Dole would have said) and given by the Maccone distribution (4.17). In other words, in this section we are going to find the pdf of the positive random variable

$$T_{Galaxy} = a \cdot \frac{T_{col}}{D} + b \tag{4.23}$$

where all capitals denote random variables while a and b are just the two positive constants defined by (4.7). In other words, apart from the constants a and b, we really must find the pdf of the quotient of the two random variables defined, in loose terms, by the fraction

$$\frac{T_{col}}{D} = \frac{\text{lognormal}}{\text{Maccone}}. \tag{4.24}$$

Standard textbooks on probability theory (e.g., see Papoulis and Pillai, *Probability Random Variables and Stochastic Processes* [6, pp. 186– 187 in particular, equation (6-59)]) tell us that the random variable Z quotient of the two random variables X and Y

$$Z = \frac{X}{Y} \tag{4.25}$$

has its pdf given by the integral

$$f_Z(z) = \int_{-\infty}^{\infty} |y| \cdot f_{XY}(yz, y)\, dy \tag{4.26}$$

where the function $f_{XY}(\ldots,\ldots)$ is the joint pdf of the two random variables X and Y. Now, the two random variables

$$\begin{cases} X = T_{col} \\ Y = D \end{cases} \tag{4.27}$$

are for sure statistically independent of each other, inasmuch as the average time to colonize a planet T_{col} is a "human thing" (or an "alien thing", if referred to an alien civilization), while the average distance D among any two nearby habitable planets is an "astrophysical thing", depending on how the Galaxy formed billion of years ago. Thus, their joint pdf $f_{XY}(\ldots,\ldots)$ simply is the product of the two pdfs—that is, the lognormal one (4.18) and the Maccone one (4.17):

$$f_{T_{col}D}(t,d) = f_{T_{col}}(t) \cdot f_D(d) = \frac{1}{t} \cdot \frac{1}{\sqrt{2\pi}\,\sigma_{t\_col}} \cdot e^{-\frac{\left(\ln\left[\frac{t}{yr}\right] - \mu_{t\_col}\right)^2}{2\sigma_{t\_col}^2}} \cdot \frac{3}{d} \cdot \frac{1}{\sqrt{2\pi}\,\sigma_D} \cdot e^{-\frac{\left(\ln\left[\frac{C^3}{d^3}\right] - \mu_D\right)^2}{2\sigma_D^2}}. \tag{4.28}$$

Rearranging, this becomes

$$f_{T_{col}D}(t,d) = \frac{3}{2\pi\, t\, d\, \sigma_{t\_col}\, \sigma_D} e^{-\frac{\left(\ln\left[\frac{t}{yr}\right] - \mu_{t\_col}\right)^2}{2\sigma_{t\_col}^2}} e^{-\frac{\left(\ln\left[\frac{C^3}{d^3}\right] - \mu_D\right)^2}{2\sigma_D^2}}. \tag{4.29}$$

This is the joint pdf that must be introduced into the integral (4.26). Note, however, that the integral actually ranges from 0 to infinity only, since both t and d do so. The modulus affecting y in (4.26) thus disappears also, and we are just left with the computation of the definite integral

$$f_{\frac{T_{col}}{D}}(z) = \int_0^{\infty} y \cdot f_{T_{col}D}(zy, y)\, dy. \tag{4.30}$$

That is

$$f_{\frac{T_{col}}{D}}(z) = \frac{3}{2\pi\, \sigma_{t\_col}\, \sigma_D} \int_0^{\infty} y \cdot \frac{1}{zy} \cdot \frac{1}{y} \cdot e^{-\frac{\left(\ln\left[\frac{zy}{yr}\right] - \mu_{t\_col}\right)^2}{2\sigma_{t\_col}^2}} \cdot e^{-\frac{\left(\ln\left[\frac{C^3}{y^3}\right] - \mu_D\right)^2}{2\sigma_D^2}}\, dy. \tag{4.31}$$

This is a tough integral to compute. Basically, it can be reduced to the Gauss integral (i.e., to the normalization condition of the ordinary Gaussian or normal curve), but many, many steps are required to perform the integration with respect to y. The author, when faced with its computation, turned to Macsyma, the wonderful computer algebra code which was created at the MIT Artificial Intelligence Laboratory back in the 1960s to let NASA recompute analytically the orbits requested for the

Apollo astronauts to safely reach the Moon and come back. So, Macsyma was able to perform the integration in (57) in a matter of seconds. The outcome was the function of z shown in Figure 4.2. As one can see, this function of z is a complicated mix of exponentials in z through the natural log of z squared, times a power of z at the denominator, times many other constants, like the dimensional yr = year.

$$f_{\frac{T_{col}}{D}}(z) = \frac{3\sqrt{2}\,C^{\frac{3\mu_D+9\mu_{t\_col}}{\sigma_D^2+9\sigma_{t\_col}^2}} \cdot e^{-\frac{\left(\mu_D+3\mu_{t\_col}\right)^2+(3\cdot\log(C\ \mathrm{yr}\ z))^2}{2\left(\sigma_D^2+9\sigma_{t\_col}^2\right)}}}{2\sqrt{\pi}\,\mathrm{yr}^{\frac{3\mu_D+9\mu_{t\_col}-9\log C-9\log z}{\sigma_D^2+9\sigma_{t\_col}^2}} \cdot z^{\frac{\sigma_D^2+9\sigma_{t\_col}^2-3\mu_D-9\mu_{t\_col}+9\log C}{\sigma_D^2+9\sigma_{t\_col}^2}}\sqrt{\sigma_D^2+9\sigma_{t\_col}^2}}. \tag{4.32}$$

Figure 4.2. The probability density function (pdf) (i.e., a function of z) of the quotient of two positive random variables: T_{col}, the time requested to colonize a new planet (in years = yr) over D, the average distance between nearby planets belonging to different stellar system (in units of the constant $C = 28{,}845$ light-years, typical of the Milky Way Galaxy).

Though the pdf (4.32) is difficult to handle by hand, it can easily be handled by Macsyma. Thus, one can prove that it does indeed fulfill the normalization condition

$$\int_0^\infty f_{\frac{T_{col}}{D}}(z)\,dz = 1. \tag{4.33}$$

A similar calculation then shows that the mean value of the quotient of random variables $T_{col/D}$ reads

$$\mu_{\frac{T_{col}}{D}} = \int_0^\infty z\,f_{\frac{T_{col}}{D}}(z)\,dz = e^{\frac{\sigma_D^2+6\mu_D+9\sigma_{t\_col}^2+18\mu_{t\_col}}{18}} \cdot \frac{\mathrm{yr}}{C}. \tag{4.34}$$

The corresponding variance was again found by Macsyma through a similar calculation, and reads

$$\sigma^2_{\frac{T_{col}}{D}} = e^{\frac{\sigma_D^2}{9}+\frac{2}{3}\mu_D+\sigma_{t\_col}^2+2\mu_{t\_col}}\left(e^{\frac{\sigma_D^2}{9}+\sigma_{t\_col}^2}-1\right)\cdot\frac{\mathrm{yr}^2}{C^2}. \tag{4.35}$$

Its square root is thus the relevant standard deviation:

$$\sigma_{\frac{T_{col}}{D}} = e^{\frac{\frac{\sigma_D^2}{9}+\frac{2}{3}\mu_D+\sigma_{t\_col}^2+2\mu_{t\_col}}{2}} \cdot \sqrt{e^{\frac{\sigma_D^2}{9}+\sigma_{t\_col}^2}-1}\cdot\frac{\mathrm{yr}}{C}. \tag{4.36}$$

In order to find the mode of the pdf (4.32) (i.e., the abscissa of its peak), we must discover the first derivative of (58) with respect to z and then set the resulting equation equal to zero. Good old Macsyma did a good job again, and the two results are the two abscissas of the minimum of (4.32), obviously at $z = 0$, and of the maximum (i.e. the peak, or mode) at

$$z_{\text{mode}} = e^{-\frac{\sigma_D^2}{9}+\frac{1}{3}\mu_D-\sigma_{t\_col}^2+\mu_{t\_col}} \cdot \frac{\text{yr}}{C}. \tag{4.37}$$

Finally, we had Macsyma prove that the two inflexion points of (4.32) (i.e., the one before and the one after the peak, or mode) are the two roots of a quadratic algebraic equation in $\log(z)$, found after equaling to zero the second derivative of (4.32) with respect to z. Thus, it is found that the abscissas of two such inflexion points of (4.32) read, respectively

$$z_{\text{inflexion\_1}} = e^{-\frac{\sqrt{\sigma_D^2+9\sigma_{t\_col}^2}\cdot\sqrt{\sigma_D^2+9\sigma_{t\_col}^2+36}+3\sigma_D^2-6\mu_D+27\sigma_{t\_col}^2-18\mu_{t\_col}}{18}} \cdot \frac{\text{yr}}{C}, \tag{4.38}$$

and

$$z_{\text{inflexion\_2}} = e^{\frac{\sqrt{\sigma_D^2+9\sigma_{t\_col}^2}\cdot\sqrt{\sigma_D^2+9\sigma_{t\_col}^2+36}-3\sigma_D^2-6\mu_D-27\sigma_{t\_col}^2-18\mu_{t\_col}}{18}} \cdot \frac{\text{yr}}{C}. \tag{4.39}$$

We shall not derive any further statistical properties of (4.32) here, though a series of lengthy calculations performed by Macsyma would probably enable us to do so.

We only need to find the pdf and basic statistical properties of the random variable T_{Galaxy} defined by (4.8). So, recall that if one has the pdf of a random variable X and wants to find the pdf of the new, linearly transformed random variable $a\,X + b$, where a and b are just constants (i.e., non-random variables), then the two pdfs are related to each other by

$$f_{aX+b}(x) = \frac{1}{|a|} \cdot f_X\left(\frac{x-b}{|a|}\right). \tag{4.40}$$

Now, the constant a defined by (4.7) is positive, and so no absolute value is needed. Thus, (4.8) and (4.40) yield at once

$$f_{T_{Galaxy}}(t) = \frac{1}{a} \cdot f_{\frac{T_{col}}{D}}\left(\frac{t-b}{a}\right). \tag{4.41}$$

We thus conclude that the pdf of the random variable T_{Galaxy} is obtained from (4.23) by letting (4.23) undergo the transformation given by (4.41). In other words, the pdf of the overall time needed to colonize the whole Galaxy is given by (Figure 4.3):

$$f_{T_{Galaxy}}(t) = \frac{1}{a} \cdot \frac{3\sqrt{2}\, C^{\frac{3\mu_D+9\mu_{t\_col}}{\sigma_D^2+9\sigma_{t\_col}^2}} \cdot e^{-\frac{\left(\mu_D+3\mu_{t\_col}\right)^2+\left(3\cdot\log\left(C\ \mathrm{yr}\ \frac{t-b}{a}\right)\right)^2}{2\left(\sigma_D^2+9\sigma_{t\_col}^2\right)}}}{2\sqrt{\pi}\,\mathrm{yr}^{\frac{3\mu_D+9\mu_{t\_col}-9\log C-9\log\left(\frac{t-b}{a}\right)}{\sigma_D^2+9\sigma_{t\_col}^2}} \cdot \left(\frac{t-b}{a}\right)^{\frac{\sigma_D^2+9\sigma_{t\_col}^2-3\mu_D-9\mu_{t\_col}+9\log C}{\sigma_D^2+9\sigma_{t\_col}^2}} \sqrt{\sigma_D^2+9\sigma_{t\_col}^2}}. \tag{4.42}$$

Figure 4.3. The probability density function (i.e., a function of t) of the positive random variables, T_{Galaxy}, the time required to colonize the whole Milky Way Galaxy. Note that this probability distribution only holds for positive values of time t such that they also are larger than the constant b defined by the second equation in (4.7). This is of course requested to avoid imaginaries that would otherwise be brought in by the real power of $(t-b)$ at the denominator. In other words, for values of t ranging between zero and b, the above pdf is understood to be equal to zero.

Again, although the pdf (4.42) is difficult to handle by hand, it can easily be handled by Macsyma. Note that this probability distribution holds only for those positive values of time that are also larger than the constant b defined by the second equation in (4.7). This is of course required to avoid imaginaries that would otherwise be brought in by the real power of $(t-b)$ at the denominator. In other words, for values of t ranging between zero and b, the above pdf is understood to be equal to zero.

Thus, one can prove (by virtue of Macsyma) that it does indeed fulfill the normalization condition

$$\int_0^{\infty} f_{T_{Galaxy}}(t)\, dt = 1. \tag{4.43}$$

We now want to derive the mean value of the time T_{Galaxy} needed to colonize the whole Galaxy. Since the mean value operator is a linear operator, the required mean value is found immediately by letting (4.34) undergo the linear transformation (4.41) and so one gets at once

$$\mu_{T_{Galaxy}} = a\ e^{\frac{\sigma_D^2+6\mu_D+9\sigma_{t\_col}^2+18\mu_{t\_col}}{18}} \cdot \frac{\mathrm{yr}}{C} + b. \tag{4.44}$$

Note that this equation is dimensionally correct since $a = 2\ R_{Galaxy}/k = 4\ R_{Galaxy}$ has the dimension of a length that cancels out length C at the denominator, leaving just a time (in years) as required. Of course, b, given by (4.7), has the dimension of time and depends only on the speed v_{ss} of interstellar flight to hop between stars.

The corresponding variance was again found by Macsyma through a similar calculation, and reads

$$\sigma_{T_{Galaxy}}^2 = a^2\ e^{\frac{\sigma_D^2}{9}+\frac{2}{3}\mu_D+\sigma_{t\_col}^2+2\mu_{t\_col}} \left(e^{\frac{\sigma_D^2}{9}+\sigma_{t\_col}^2} - 1\right) \cdot \frac{\mathrm{yr}^2}{C^2}. \tag{4.45}$$

Its square root is thus the relevant standard deviation:

$$\sigma_{T_{Galaxy}} = a\ e^{\frac{\frac{\sigma_D^2}{9}+\frac{2}{3}\mu_D+\sigma_{t\_col}^2+2\mu_{t\_col}}{2}} \cdot \sqrt{e^{\frac{\sigma_D^2}{9}+\sigma_{t\_col}^2} - 1} \cdot \frac{\text{yr}}{C}. \tag{4.46}$$

The mode of the pdf (4.42) (i.e., the abscissa of its peak) is found by setting to zero the first derivative of (4.42) with respect to z. Macsyma once again did a good job, and the two results are the two abscissas of the minimum of (4.42), obviously at $z = 0$, and of the maximum (i.e., the peak, or mode) at

$$z_{\text{mode\_}T_{\text{Galaxy}}} = a\ e^{-\frac{\sigma_D^2}{9}+\frac{1}{3}\mu_D-\sigma_{t\_col}^2+\mu_{t\_col}} \cdot \frac{\text{yr}}{C} + b. \tag{4.47}$$

Finally, we had Macsyma prove that the two inflexion points of (4.42) (i.e., the one before and the one after the peak, or mode) are the two roots of a quadratic algebraic equation in $\log(z)$ found after equaling to zero the second derivative of (4.42) with respect to z. Thus, it is found that the abscissas of such two inflexion points of (4.42) read, respectively

$$z_{\text{inflexion\_1}} = a\,e^{-\frac{\sqrt{\sigma_D^2+9\sigma_{t\_col}^2}\cdot\sqrt{\sigma_D^2+9\sigma_{t\_col}^2+36}+3\sigma_D^2-6\mu_D+27\sigma_{t\_col}^2-18\mu_{t\_col}}{18}} \cdot \frac{\text{yr}}{C} + b. \tag{4.48}$$

and

$$z_{\text{inflexion\_2}} = a\,e^{\frac{\sqrt{\sigma_D^2+9\sigma_{t\_col}^2}\cdot\sqrt{\sigma_D^2+9\sigma_{t\_col}^2+36}-3\sigma_D^2-6\mu_D-27\sigma_{t\_col}^2-18\mu_{t\_col}}{18}} \cdot \frac{\text{yr}}{C} + b. \tag{4.49}$$

We shall not derive further statistical properties of (4.42) here, though a series of lengthy calculations performed by Macsyma would probably enable us to do so.

4.5 CONCLUSIONS

We extended the classical Fermi paradox to let it encompass statistics and probability. We gave (difficult) statistical equations that are related to both the statistical Drake equation and the statistical Dole equation for habitable planets for man. This provided an analytical theory, but now a good and portable numeric code should be written to let our results be applied to cases of practical interest.

This approach appears to pave the way for future, more profound investigations intended not only to associate "error bars" to each factor in the Drake and Dole equations, but especially to increase the number of factors themselves. In fact, this seems to be the only way to incorporate in these equations new scientific information as soon as it becomes available. As we said, in the long run, our statistical results might just become a huge computer code, growing in size and especially in the depth of the scientific information it contained. It would thus be humanity's first "Encyclopedia Galactica".

Unfortunately, to extend the Drake and Dole equation to statistics, it was necessary to use a mathematical apparatus that was more sophisticated than just the simple product of numbers.

4.6 ACKNOWLEDGMENTS

The author is grateful to Drs. Hal Puthoff and Eric Davis of the Institute for Advanced Study at Austin for reading the author's paper #IAC-08-A4.1.4 and suggesting its publication made up by both the paper itself and its popularized version. Thanks are also due to Professor Frank Drake and Drs. Jill Tarter, Seth Shostak, and Doug Vakoch of the SETI Institute, as well as Dr. H. Paul Shuch, for their appreciation of the statistical work described in this paper. Finally, special thanks go to Marc Millis and Paul Gilster of the Tau Zero Foundation for realizing the importance of our statistical extension of the Drake and Dole equations and helping to have our results popularized within the international scientific community.

4.7 REFERENCES

[1] Gregory Benford, Jim Benford, and Dominic Benford, "Cost optimized interstellar beacons: SETI", arXiv.org web site (October 22, 2008).

[2] Carl Sagan, *Cosmos*, Random House, New York, 1983. See pp. 298–302 in particular.

[3] Jeffrey Bennet and Seth Shostak, *Life in the Universe*, Second Edition, Pearson/Addison Wesley, San Francisco, 2007. See p. 404 in particular.

[4] Claudio Maccone, "The statistical Drake equation," paper #IAC-08-A4.1.4 presented on October 1, 2008 at the *59th International Astronautical Congress (IAC) held in Glasgow, Scotland, U.K., September 29–October 3, 2008.*

[5] Stephen H. Dole, *Habitable Planets for Man*, First Edition, 1964, RAND Corporation. See p. 82 in particular (i.e., the beginning of Chapter 5 entitled "Probability of occurrence of habitable planets").

[6] Athanasios Papoulis and S. Unnikrishna Pillai, *Probability, Random Variables and Stochastic Processes*, Fourth Edition, Tata/McGraw-Hill, New Delhi, 2002. See pp. 186–187 in particular.

[7] Claudio Maccone, "The statistical Fermi paradox," paper #IAC-09.A4.2.8 presented at the *60th International Astronautical Congress held at Daejeon, Republic of Korea, October 12–16, 2009*. Later, this paper was published in the *Journal of the British Interplanetary Society*, **63** (2010), No. 5/6 (May/June 2010), 222–239.

5

How long does a civilization live?

5.1 TWO EXAMPLES OF THE STATISTICAL DRAKE EQUATION RELATED TO THE LIFETIME *fL* OF A TECHNOLOGICAL CIVILIZATION

Nobody knows how long a technological civilization would live on average. In other words, nobody knows fL in either the classical or statistical Drake equation:

$$N = Ns \cdot fp \cdot ne \cdot fl \cdot fi \cdot fc \cdot fL. \tag{5.1}$$

We now provide a further two examples of our statistical Drake equation that are related to the lifetime fL of a civilization.

Consider pp. 299–302 of the famous book *Cosmos* by Carl Sagan [2], where he describes the classical Drake equation (5.1). At this point, Carl Sagan makes a neat distinction between two different numerical values for the seventh factor in the Drake equation, fL.

5.2 EXAMPLE #1: A TECHNOLOGICAL CIVILIZATION LASTING ONLY 45 OF THE 4.5 BILLION YEARS OF OUR EARTH, SUCH AS HUMANITY

The first such numeric value is $fL \sim 10^{-8}$ and applies to us, humans, and our current stage of technological development.

5.3 EXAMPLE #2: A TECHNOLOGICAL CIVILIZATION LASTING 90 MILLION OUT OF 9 BILLION YEARS OF THEIR SUN'S LIFETIME

So much for Carl Sagan's two different numerical examples of the classical Drake equation (5.1), the outcome of which is in all cases sheer numbers.

Input Table 5.1. The first column shows the numerical values of the first six inputs to the classical Drake equation (5.1) for the example provided by Carl Sagan in his book *Cosmos* [2]. Moving from the classical Drake equation (5.1) to our statistical Drake equation described in Section 5.2, the mean values of all six random variables are equal to the corresponding six input numbers given by Sagan. In addition, however, we chose to define the corresponding six input standard deviations at 20% (i.e., one fifth) of the corresponding mean value, to make a complete numerical example of how our statistical Drake equation works.

$Ns = 4 \cdot 10^{11}$	$\mu Ns = 4 \cdot 10^{11}$	$\sigma Ns = \frac{Ns}{5}$
$fp = \frac{1}{3}$	$\mu fp = \frac{1}{3}$	$\sigma fp = \frac{fp}{5}$
$ne = 2$	$\mu ne = 2$	$\sigma ne = \frac{ne}{5}$
$fl = \frac{1}{3}$	$\mu fl = \frac{1}{3}$	$\sigma fl = \frac{fl}{5}$
$fi = \frac{1}{10}$	$\mu fi = \frac{1}{10}$	$\sigma fi = \frac{fi}{5}$
$fc = \frac{1}{10}$	$\mu fc = \frac{1}{10}$	$\sigma fc = \frac{fc}{5}$

Let us now cast all of this into the new language of our statistical Drake equation. The first six input variables, common to both examples, are those listed in Input Table 5.1. There the mean values of the six input random variables are just the same as the six input numbers proposed by Carl Sagan. But we also had to input their standard deviations. We did this for the sake of making an example, by assuming that each standard deviation equals 20% (i.e., one fifth) of the corresponding mean value.

5.4 EXAMPLE #1 REWRITTEN IN THE LANGUAGE OF THE STATISTICAL DRAKE EQUATION

Next we define the seventh input variable, fL, as Carl Sagan did in his first example, again with the same mean value and a 20% standard deviation, thus obtaining Input Table 5.2.

The outputs for this example, obtained by applying all the equations shown in Input Table 5.1, are as follows:

(1) Lognormal probability density for the random variable N (where n is the algebraic real positive independent variable of the corresponding probability density function) yielding the number of technological civilizations in the

Input Table 5.2. Input value of fL for a technological civilization that survives only 45 years of its own technological adolescence (as described by Sagan). Again, the mean value of the input random variable to our statistical Drake equation is the same as the numerical input of the classical Drake equation, but, in addition, we had to input a standard deviation value which we assumed equal to 20% of the corresponding mean value.

$$fL = \frac{4.5 \cdot 10 \text{ yr}}{4.5 \cdot 10^9 \text{ yr}} = 1 \cdot 10^{-8} \quad \mu fL = fL \quad \sigma fL = fL \cdot \frac{20}{100} \tag{5.2}$$

Galaxy if they last 45 years only (and then immolate themselves in a nuclear holocaust):

(2) Mean value of N:

$$\text{Mean\_value\_of\_}N = e^{\mu + \frac{\sigma^2}{2}} = 1.4392 \cdot 10^1 \approx 14. \tag{5.3}$$

So, the mean value of the lognormal distribution of Figure 5.1 is about 14 technological civilizations in the whole Galaxy, while the classical Drake equation (5.1) would yield 8.8 (i.e., nearly 9) technological civilizations. Thus, one more advantage of our statistical Drake equation over the deterministic Drake equation (5.1) is that the statistical treatment is not only more serious scientifically, but it also automatically increases the expected number of civilizations as well as the hopes for a SETI success.

(3) Standard deviation of N:

$$\text{Standard\_deviation\_of\_}N = e^{\mu + \frac{\sigma^2}{2}} \sqrt{e^{\sigma^2} - 1} = 2.2784 \cdot 10^1 \approx 23. \tag{5.4}$$

So, the standard deviation is about 23 technological civilizations, a further increase for SETI hopes!

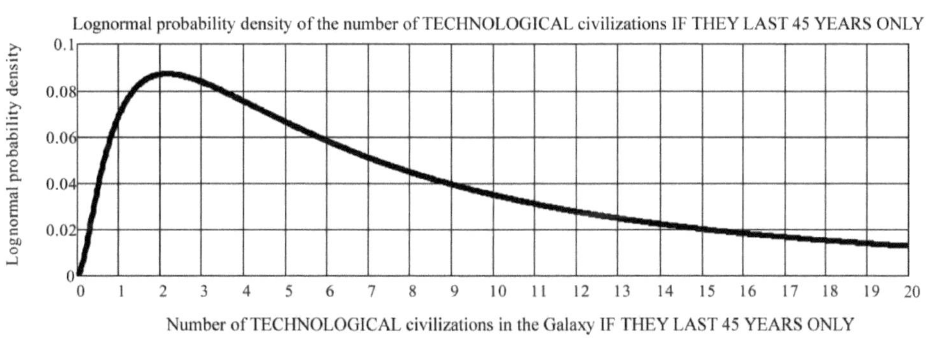

Figure 5.1. Lognormal probability distribution of the number of ET technological civilizations if they last 45 years only.

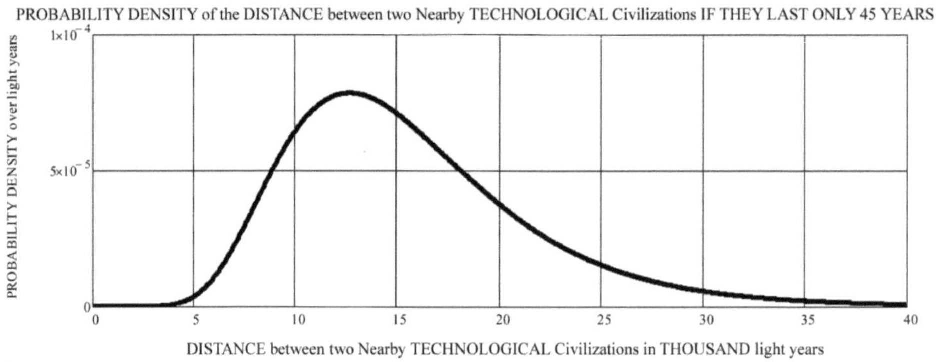

Figure 5.2. Probability distribution of the distance between any two nearby civilizations if they last only 45 years.

Now about the distance in between any two nearby such technological civilizations lasting 45 years only. Using the probability density function written in the third line of Table 1.2 (the "Maccone distribution"), one gets:

(4) Mean value of the distance between any two nearby civilizations:

$$\text{Mean\_distance\_between\_ETs} = C \cdot e^{\frac{\sigma^2}{18} - \frac{\mu}{3}} = 1.5671 \cdot 10^4 \text{ lt-yr} \approx 15{,}671 \text{ lt-yr} \approx 16{,}000 \text{ lt-yr}. \tag{5.5}$$

So, the mean distance between any two nearby civilizations is about 16,000 light-years, while the classical Drake equation (5.1) would yield about 14,000 light-years: "bad news" for SETI. However, the corresponding standard deviation—which, of course, does not exist at all for the classical Drake equation (5.1)—more than compensates for that slightly smaller mean value.

(5) Standard deviation of the distance:

$$\begin{aligned}\text{Standard\_deviation\_of\_the\_distance\_between\_ETs} &= C \cdot e^{\frac{\sigma^2}{18} - \frac{\mu}{3}} \sqrt{e^{\frac{\sigma^2}{9}} - 1} \\ &= 6.0606 \cdot 10^3 \text{ lt-yr} \\ &\approx 6{,}060 \text{ lt-yr} \\ &\approx 6{,}000 \text{ lt-yr}\end{aligned} \tag{5.6}$$

So, the standard deviation of the distance is about 6,000 light-years. If we subtract this value from the mean value above, we get about 9,000 light-years as the minimal expected distance, much lower than the mean value (with no standard deviation) provided by the classical Drake equation (5.1), and thus a further boost to SETI hopes!

5.5 EXAMPLE #2 REWRITTEN IN THE LANGUAGE OF THE STATISTICAL DRAKE EQUATION

Next we define the seventh input variable, fL, as Carl Sagan did in his second example, again with the same mean value and a 20% standard deviation, thus obtaining Input Table 5.3.

Input Table 5.3. Input value of fL for a technological civilization that survives 90 million years (as described by Sagan). Again, the mean value of the random variable input to our statistical Drake equation is the same as the numerical input to the classical Drake equation, but, in addition, we had to input a standard deviation value that we assumed to equal 20% of the corresponding mean value.

$$fL = \frac{9 \cdot 10^7 \text{ yr}}{9 \cdot 10^9 \text{ yr}} = 1 \cdot 10^{-2} \quad \mu fL = fL \quad \sigma fL = fL \cdot \frac{20}{100} \tag{5.7}$$

The outputs for this example, obtained by applying the equations shown in Input Table 5.3, are as follows:

(1) Lognormal probability density for the random variable N (where n is the algebraic real positive independent variable of the corresponding probability density function) yielding the number of technological civilizations in the Galaxy that are likely to have lasted 90 million years.

(2) Mean value of N:

$$\text{Mean\_value\_of\_}N = e^{\mu+\frac{\sigma^2}{2}} = 1.4392 \cdot 10^7 \approx 14 \text{ million.} \tag{5.8}$$

So, the mean value of the lognormal distribution of Figure 5.3 is about 14 million technological civilizations in the whole Galaxy, while the classical Drake equation (7) would yield 8.8 million (i.e., nearly 9 million technological civilizations). Again, another advantage of our statistical Drake equation over the deterministic Drake equation (7) is that the statistical treatment is not only scientifically more serious, but also automatically increases the expected number of civilizations.

(3) Standard deviation of N:

$$\text{Standard\_deviation\_of\_}N = e^{\mu+\frac{\sigma^2}{2}}\sqrt{e^{\sigma^2} - 1} = 2.2784 \cdot 10^7 \approx 23 \text{ million.} \tag{5.9}$$

So, the standard deviation is about 23 million technological civilizations, a further boost to SETI hopes!

Let us now consider the distance between any two nearby such technological civilizations lasting 90 million years. Using the probability density function written

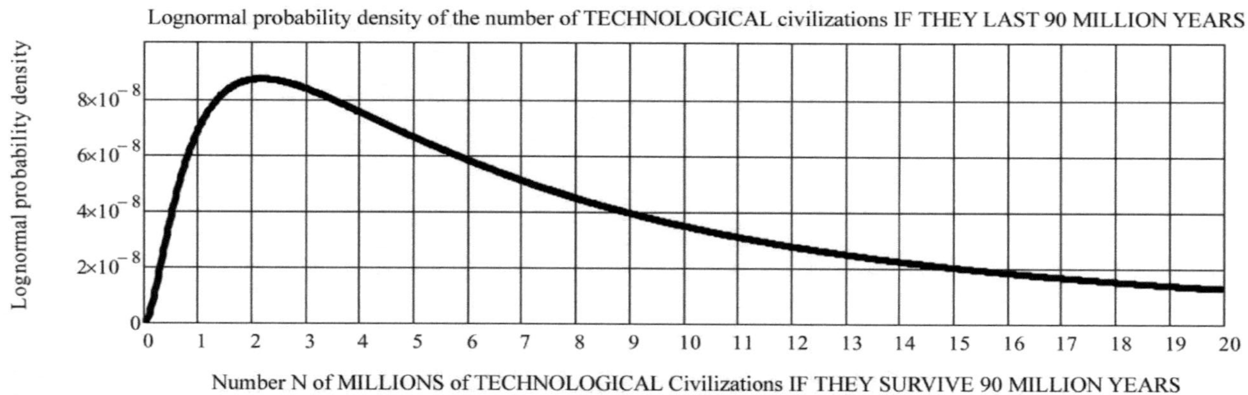

Figure 5.3. Lognormal probability distribution of the number of ET technological civilizations that are likely to have lasted 90 million years.

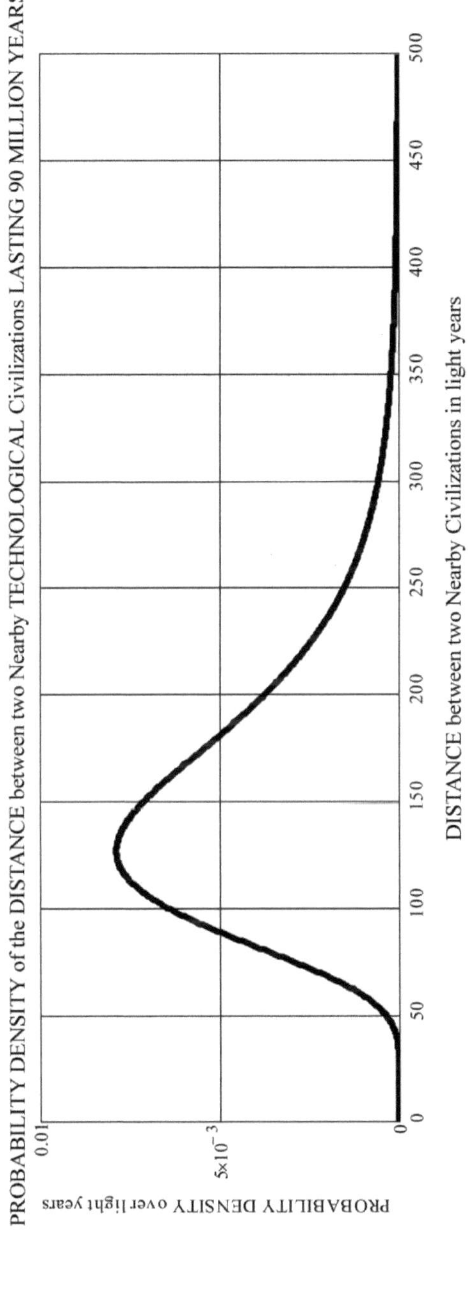

Figure 5.4. Distribution of the distance between any two nearby civilizations lasting 90 million years.

in the third line of Table 1.2, one gets:

(1) Mean value of the distance between any two nearby civilizations lasting 90 million years:

$$\text{Mean\_distance\_between\_ETs} = C \cdot e^{\frac{\sigma^2}{18}-\frac{\mu}{3}} = 1.5671 \cdot 10^2 \text{ lt-yr} \approx 157 \text{ lt-yr}. \tag{5.10}$$

So, the mean distance between any two nearby 90-million-year-old civilizations is just about 157 light-years, while the classical Drake equation (5.1) would yield about 140 light-years: slightly bad news for SETI. However, the corresponding standard deviation—which, of course, does not exist at all for the classical Drake equation (5.1)—more than compensates for that slightly smaller mean value.

(2) Standard deviation of the distance:

$$\text{Standard\_deviation\_of\_the\_distance\_between\_ETs} = C \cdot e^{\frac{\sigma^2}{18}-\frac{\mu}{3}}\sqrt{e^{\frac{\sigma^2}{9}} - 1} = 6.0606 \cdot 10^1 \text{ lt-yr} \approx 60 \text{ lt-yr}. \tag{5.11}$$

So, the standard deviation of the distance is about 60 light-years. If we subtract this value from the mean value above, we get about 97 light-years as the minimal expected distance, still lower than the mean value (with no standard deviation) provided by the classical Drake equation (5.1), and thus a further boost to SETI hopes!

5.6 CONCLUSIONS

We have sought to extend the classical Drake equation to let it encompass statistics and probability.

This approach appears to pave the way for future, more profound investigations intended not only to associate "error bars" to each factor in the Drake equation, but especially to increase the number of factors themselves. In fact, this seems to be the only way to incorporate new scientific information into the Drake equation as soon as it becomes available. In the long run, the statistical Drake equation might just become a huge computer code, growing in size and especially in the depth of the scientific information it contained. It would thus be humanity's first "Encyclopedia Galactica".

5.7 ACKNOWLEDGMENTS

The author is grateful to Drs. Jill Tarter, Paul Davies, Seth Shostak, Doug Vakoch, Tom Pierson, Carol Oliver, Paul Shuch, and Kathryn Denning for attending his first ever presentation on topics at the "Beyond" Center of the University of Arizona at Phoenix on February 8, 2008. He would also like to thank Dan Werthimer and his School of SETI young experts for keeping alive the interplay between experimental and theoretical SETI. But the greatest "thanks" goes, of course, to the teacher of us all: Professor Frank D. Drake, whose equation opened a new way of thinking about the past and the future of humans in the Galaxy.

5.8 REFERENCES

[1] Claudio Maccone, "The statistical Drake equation," paper #IAC-08-A4.1.4 presented on October 1, 2008, at the *59th International Astronautical Congress (IAC) in Glasgow, Scotland, U.K., September 29–October 3, 2008.*

[2] Carl Sagan, *Cosmos*, Random House, New York, 1983. See pp. 298–302 in particular.

[3] Claudio Maccone, "The living Drake equation of the Tau Zero Foundation", a paper presented at the *Sixth IAA Symposium on "Missions to the Outer Solar System and Beyond", Aosta, Italy, July 6–9, 2009* and later published in *Acta Astronautica*, **68** (2011), 582–590.

6

Life-span modeling by finite *b*-lognormals

6.1 INTRODUCING *b*-LOGNORMALS AS OUR MODEL FOR ALL LIFE-SPANS

It is well known that the probability density function (pdf) $f_{aX+b}(x)$ of the random variable $aX + b$, where a and b are arbitrary real constants with respect to the independent variable x, is related to the pdf $f_X(x)$ of the random variable X by the "linear transformation formula for random variables" that reads:

$$f_{aX+b}(x) = \frac{1}{|a|} f_X\left(\frac{x-b}{|a|}\right). \tag{6.1}$$

In the particular case $a = 1$ this formula yields:

$$f_{X+b}(x) = f_X(x - b). \tag{6.2}$$

Thus, one may define the new pdf "lognormal starting at $b > 0$", or simply "b-lognormal", as the pdf obtained by applying (6.2) to the ordinary lognormal starting at zero; that is:

$$\begin{cases} \text{b\_lognormal\_pdf}(x) = 0 & \text{for } 0 \le x \le b \\ \text{b\_lognormal\_pdf}(x) = 0 & \text{for } x = b \\ \text{b\_lognormal\_pdf}(x) = \dfrac{1}{\sqrt{2\pi}\,\sigma(x-b)} e^{-\frac{(\ln(x-b)-\mu)^2}{2\sigma^2}} & \text{for } x \ge b. \end{cases} \tag{6.3}$$

This chapter is entirely devoted to "discovering" the mathematical properties of b-lognormals.

In fact, we shall later use them as the standard mathematical model to symbolize the life-span of any living being, be this living being a cell, a human, a society, a civilization, or even an ET civilization.

On the one hand, all such life-spans are of course finite in time: namely, they are born at a certain instant $t = b$ (b standing for "birth") and they die at a later instant $t = d$ (d standing for "death"), with $d > b$.

On the other hand, b-lognormals like (6.3) are infinite in time and span from $t = b$ to $t = +\infty$; so one might immediately wonder how could (6.3) possibly represent a finite life-span. Well, the answer to such a question will be given later in this chapter, when we introduce the notion of "death instant" $t = d$ as the intersection point between the tangent to (6.3) in its descending inflexion point and the time axis. We will come back to this point later in Section 6.7.

At the moment, we content ourselves with studying the mathematical properties of the b-lognormal pdf (6.3) and this we will do in a highly innovative way for a mathematical book like this one. In fact, the mathematical proof of each of the theorems proven in this chapter will not be demonstrated line by line in the text. Rather, the reader will find them in the Maxima file "*b*_lognormals_inflexion_points_and_DEATH_time.wmx", which is reprinted in Appendix 6.A. Thus, we shall simply state the equation numbers in that Maxima code proving a certain result about b-lognormals, and the interested reader will then find the relevant proof by reading the corresponding Maxima command lines ("i" = input lines) and output lines ("o" = output lines). This way of proving mathematical results simplifies things greatly compared with the usual lengthy proofs!

6.2 ADOLESCENCE ($t = a$) AND SENILITY ($t = s$) POINTS OF *b*-LOGNORMALS (IN OTHER WORDS, THEIR TWO INFLEXION POINTS)

Let us first understand why every lognormal (and so every b-lognormal) has two and only two inflexion points. Well, if one takes the second derivative of the lognormal and equates it to zero, the resulting equation is a second-degree algebraic equation in $\ln(t)$ for the lognormal, and in $\ln(t - b)$ for the b-lognormal. For the b-lognormal, these are given, respectively, by

$$\text{Adolescence point} = a = e^{-\frac{\sigma\sqrt{\sigma^2+4}}{2} - \frac{3\sigma^2}{2} + \mu} + b \tag{6.4}$$

and

$$\text{Senility point} = s = e^{\frac{\sigma\sqrt{\sigma^2+4}}{2} - \frac{3\sigma^2}{2} + \mu} + b. \tag{6.5}$$

The names are obvious: the adolescence point comes after birth and before the peak, while the senility point comes after the peak and before death (i.e., the intersection of the straight line tangent to the b-lognormal at the senility point with the time axis, as we shall precisely define in Section 6.7).

Let us now consider the two ordinates of the b-lognormal, corresponding to the adolescence and senility points, respectively. They are:

$$b\text{-lognormal ordinate at adolescence} = \frac{e^{-\frac{\sigma\sqrt{\sigma^2+4}}{4}} e^{-\mu+\frac{\sigma^2}{4}-\frac{1}{2}}}{\sqrt{2\pi}\,\sigma} \tag{6.6}$$

and

$$b\text{-lognormal ordinate at senility} = \frac{e^{\frac{\sigma\sqrt{\sigma^2+4}}{4}} e^{-\mu+\frac{\sigma^2}{4}-\frac{1}{2}}}{\sqrt{2\pi}\,\sigma}. \tag{6.7}$$

Note that both these ordinates (6.6) and (6.7) are independent of the birth time b, as it must be inasmuch as the position of the birth instant along the time axis is totally irrelevant and arbitrary with respect to ordinates.

Proof. See lines (%i8) through (%o20) of the file in Appendix 6.A.

6.3 FINDING THE b-LOGNORMAL FROM THE ORDINATES OF ITS TWO INFLEXION POINTS PLUS EITHER a OR s

The following sections are devoted to the problem of finding the b-lognormal (i.e., finding the expressions of μ and σ from other numerically known variables). For instance, in this section we show how to find of μ and σ from the values of the ordinates of both the b-lognormal's inflexion points plus one more variable, an abscissa that we show may be either a or s. In general, it is not obvious which known variables will lead to two equations exactly solvable for μ and σ. The only way to know is to check whether a given couple of equations may or may not be solved for μ and σ exactly (i.e., without any numeric approximation). We have tried many combinations and we have found some good ones, but we cannot exclude that more exact formulas might exist yielding both μ and σ exactly.

An interesting first result along these lines follows from (6.6) and (6.7): it is possible to solve these two simultaneous equations with respect to σ exactly, finding:

$$\sigma = \sqrt{2}\sqrt{\sqrt{\left[\ln\left(\frac{b\_\text{lognormal}\_\text{at}\_s}{b\_\text{lognormal}\_\text{at}\_a}\right)\right]^2+1}-1}. \tag{6.8}$$

Then, two possible choices for μ exist: either we solve (6.4) for μ and get the solution:

$$\begin{cases} \sigma = \sqrt{2}\sqrt{\sqrt{\left[\ln\left(\frac{b\_\text{lognormal}\_\text{at}\_s}{b\_\text{lognormal}\_\text{at}\_a}\right)\right]^2+1}-1} \\ \mu = \dfrac{\sigma\sqrt{\sigma^2+4}+3\,\sigma^2+2\,\ln(a-b)}{2} \end{cases} \tag{6.9}$$

or we solve (6.5) for μ and get the solution:

$$\begin{cases} \sigma = \sqrt{2}\sqrt{\sqrt{\left[\ln\left(\frac{b\_lognormal\_at\_s}{b\_lognormal\_at\_a}\right)\right]^2+1}-1} \\ \mu = -\frac{\sigma\sqrt{\sigma^2+4}-3\,\sigma^2-2\,\ln(s-b)}{2}. \end{cases} \tag{6.10}$$

Proof. See lines (%i21) through (%o26) of the file in Appendix 6.A.

In summary, the above two results allow one to determine the *b*-lognormal exactly if the ordinates at its two inflexion points are known, plus either the adolescence abscissa a is known or the senility abscissa s is known.

All these three quantities, however, may be rather hard to measure empirically, since it is hard to draw a pdf empirically. The ordinates, in particular, are hard to measure. We shall thus now concentrate on finding pairs of simultaneous equations yielding σ and μ exactly starting from abscissae only.

6.4 FINDING THE *b*-LOGNORMAL FROM THE BIRTH (*b*), ADOLESCENCE (*a*), AND SENILITY (*s*) ABSCISSAE

Let us first clearly point out that, up to now, only four important abscissae have been introduced:

$$\begin{cases} b = \text{Birth time, where the } b\text{-lognormal starts (this is always assumed to be known).} \\ a = \text{Adolescence time, where the } b\text{-lognormal has its ascending inflexion point.} \\ p = \text{Peak time (i.e., abscissa of the maximum, also called the mode).} \\ s = \text{Senility time, where the } b\text{-lognormal has its descending inflexion point.} \end{cases} \tag{6.11}$$

A fifth, important abscissa will be the death time (d), but this will only be introduced in Section 6.7.

At the moment we only wish to concentrate on the various combinations of the above four abscissae which may possibly lead to equations exactly solvable for σ and μ.

The first such case is offered by the two equations

$$\begin{cases} \sigma = \sqrt{2}\sqrt{\sqrt{\left[\ln\left(\frac{\sqrt{s-b}}{\sqrt{a-b}}\right)\right]^2+1}-1} \\ \mu = \frac{3\,\sigma^2}{2}+\frac{\ln((a-b)(s-b))}{2}. \end{cases} \tag{6.12}$$

Alternatively, one may rewrite the first equation in a slightly simpler form; that is:

$$\begin{cases} \sigma = \sqrt{\sqrt{\left[\ln\left(\frac{s-b}{a-b}\right)\right]^2 + 4} - 2} \\ \mu = \frac{3\,\sigma^2}{2} + \frac{\ln((a-b)(s-b))}{2}. \end{cases} \tag{6.13}$$

These two pairs (6.12) and (6.13) of equations completely solve the problem of determining the b-lognormal when the birth, adolescence, and senility times are all assigned.

Proof. See lines (%i24) thru (%o35) of the file in Appendix 6.A.

6.5 FINDING (APPROXIMATELY) THE *b*-LOGNORMAL FROM ITS PEAK ONLY: THAT IS, GIVEN ITS ABSCISSA *p* AND ORDINATE *P*

Two really important formulas for future developments are the two coordinates of the b-lognormal peak.

The abscissa p of the b-lognormal peak (also called the mode m) is given by

$$\text{peak\_abscissa} = \text{mode\_abscissa} = p = +b\,e^{\mu-\sigma^2} \tag{6.14}$$

and the corresponding ordinate P is given by

$$\text{peak\_ordinate} = \text{mode\_ordinate} = P = \frac{e^{\frac{\sigma^2}{2}-\mu}}{\sqrt{2\pi}\sigma}. \tag{6.15}$$

The last equation will assume a very important role in the sequel of this book, meaning in practice "how advanced a living being or a civilization may be at the time of its zenith" (as we shall see in the next chapter). In this section, however, we want to face the practical problem of determining the b-lognormal from p and P only.

Unfortunately, it is not possible to determine the b-lognormal (i.e., to find σ and μ) exactly from the knowledge of both p and P only. In fact, multiplying (6.14) by (6.15) allows μ to be eliminated but the resulting equation in σ may not be solved for σ exactly, since it reads:

$$(p-b)P = \frac{e^{-\frac{\sigma^2}{2}}}{\sqrt{2\pi}\sigma}. \tag{6.16}$$

One may try, however, to solve (6.16) for sigma numerically. In fact, assuming $0 < \sigma < 1$ (as appears to be the case in most practical applications) one may expand the right-hand side of (1.16) into a convergent Taylor series in σ around the point $\sigma = 0$ and up to the order σ^2, so that (6.16) is turned into its approximated form

$$(p-b)P = \frac{1-\frac{\sigma^2}{2}}{\sqrt{2\pi}\sigma} \quad \text{if } 0 < \sigma < 1. \tag{6.17}$$

Rearranging, this yields the quadratic equation in σ

$$\sigma^2 + 2\sqrt{2\pi}(p-b)P\,\sigma - 2 = 0. \tag{6.18}$$

Solving for σ one then gets

$$\sigma = -\sqrt{2\pi}(p-b)P \pm \sqrt{2}\sqrt{\pi(p-b)^2P^2+1} \tag{6.19}$$

from which the resulting σ with a negative sign must obviously be discarded. This is our approximated expression for σ in terms of b, p, and P (i.e., in terms of the peak abscissa and ordinate only). From this, we finally infer μ from (6.14); that is

$$\mu = \sigma^2 + \ln(p-b). \tag{6.20}$$

6.6 FINDING THE *b*-LOGNORMAL FROM THE BIRTH (*b*), ADOLESCENCE (*a*), AND PEAK (*p*) ABSCISSAE

One more case exists where σ and μ may be found exactly (i.e., without any approximation) from three abscissae: this is the case when the birth (b), adolescence (a), and peak time (p) are assigned. The pair of resolving equations for this case read:

$$\begin{cases} \sigma = \dfrac{\ln\left(\dfrac{p-b}{a-b}\right)}{\sqrt{\ln\left(\dfrac{p-b}{a-b}\right)+1}} \\[2ex] \mu = \dfrac{\sigma\sqrt{\sigma^2+4}+3\,\sigma^2+2\ln(a-b)}{2}. \end{cases} \tag{6.21}$$

Proof. See lines (%i36) through (%o44) of the file in Appendix 6.A.

6.7 FINITE *b*-LOGNORMALS: DEFINING THE DEATH TIME (*d*)

The b-lognormal extends up to $t = +\infty$ and this is in sharp contrast with the fact that every living being sooner or later dies at the finite time d such that $b < d < \infty$. We thus must somehow define this finite death time in order to make b-lognormals become a realistic mathematical model for the life and death of every living being.

We solved this problem by defining the death time $t = d$ as the intercept point between the time axis and the straight line tangent to the b-lognormal at its descending inflexion point $t = s$ (i.e., the tangent line to the lognormal curve at senility).

And, from now on, we shall call *finite b-lognormal* any such truncated b-lognormal, ending just at $t = d$.

This section is devoted to calculation of the equation yielding the d point in terms of the b-lognormal's μ and σ, and the whole procedure is described at lines %i45 through %o56 of the attached file "b-lognormals_inflexion_points_and_DEATH_time.wxm" in Appendix 6.B.

Let us start by recalling the simple formula yielding the equation of the straight line having an angular coefficient m and tangent to the curve $y(t)$ at the point having the coordinates (t_0, y_0):

$$y - y_0 = m(t - t_0). \tag{6.22}$$

Then, the value of y_0 clearly is the value of the b-lognormal at its senility time, given by (6.7); that is, rearranging:

$$y_0 = \frac{e^{\frac{\sigma\sqrt{\sigma^2+4}+\sigma^2-4\mu-2}{4}}}{\sqrt{2\pi}\,\sigma}. \tag{6.23}$$

On the other hand, the abscissa of the senility time $t = s$ is given by (6.5); that is

$$t_0 = e^{\frac{\sigma\sqrt{\sigma^2+4}}{2}-\frac{3\sigma^2}{2}+\mu} + b. \tag{6.24}$$

Finally, we must find the expression of the angular coefficient m at the senility time, and this involves finding the b-lognormal's derivative at senility. Maxima does not have any problem finding this, and lines %i48 and %o48 show that one gets, after some rearranging,

$$m = -\frac{\sqrt{2}\,e^{\frac{7\sigma^2}{4}}\left(\sqrt{\sigma^2+4}-\sigma\right)e^{-\frac{\sigma\sqrt{\sigma^2+4}+8\mu+2}{4}}}{4\sqrt{\pi}\,\sigma^2}. \tag{6.25}$$

Inserting then (6.23), (6.24), and (6.25) into (6.22) one gets the equation of the desired straight line tangent to the b-lognormal at senility:

$$y - \frac{e^{\frac{\sigma\sqrt{\sigma^2+4}+\sigma^2-4\mu-2}{4}}}{\sqrt{2\pi}\,\sigma} = -\frac{\sqrt{2}\,e^{\frac{7\sigma^2}{4}}\left(\sqrt{\sigma^2+4}-\sigma\right)e^{-\frac{\sigma\sqrt{\sigma^2+4}+8\mu+2}{4}}}{4\sqrt{\pi}\,\sigma^2}\left(t - e^{\frac{\sigma\sqrt{\sigma^2+4}}{2}-\frac{3\sigma^2}{2}+\mu} - b\right). \tag{6.26}$$

In order to find the abscissa of the death point $t = d$, we just need to insert $y = 0$ into the above equation (6.26) and solve for the resulting t. Maxima yields at first a rather complicated result (%o52). However, keeping in mind that the term in b must obviously appear "alone" in the final equation since the b-lognormal is just an ordinary lognormal shifted to make it start at b, the way to simplify (%o52) becomes obvious, and the final result is simply

$$d = \frac{\left(\sqrt{\sigma^2+4}+\sigma\right)^2 e^{\frac{\sigma\sqrt{\sigma^2+4}}{2} - \frac{3\sigma^2}{2} + \mu}}{4} + b. \tag{6.27}$$

This is the "death time" of all living beings.

6.8 FINDING THE *b*-LOGNORMAL GIVEN THE TIMES OF BIRTH ($t = b$), SENILITY ($t = s$), AND DEATH ($t = d$)

Let us again clearly point out that, up to now, five important abscissae have been introduced:

$$\begin{cases} b = \text{Birth time, where the } b\text{-lognormal starts (this is always assumed to be known).} \\ a = \text{Adolescence time, where the } b\text{-lognormal has its ascending inflexion point.} \\ p = \text{Peak time (i.e., abscissa of the maximum, also called the mode).} \\ s = \text{Senility time, where the } b\text{-lognormal has its descending inflexion point.} \\ d = \text{Death time, where the } b\text{-lognormal's tangent at senility intercepts the time axis.} \end{cases} \tag{6.28}$$

The question that arises now is: Having introduced the death time (d), do some equations exist enabling one to determine the b-lognormal in terms of the birth time b, the death time d, and some of the remaining three parameters a, p, and s?

The author could only discover one such pair of equations, yielding μ and σ exactly. At the moment the author cannot rule out that more such equation pairs may possibly exist, but he has been unable to discover either of them. So also for approximated (i.e., obtained by the Taylor series) similar equations, which up to the moment he has not had the time to investigate.

So, the only known example of exact equations determining the b-lognormal in terms of birth, senility, and death is provided by the two equations

$$\begin{cases} \mu = \ln(s-b) + \dfrac{2s^2 - (3d+b)s + d^2 + b\,d}{(d-b)s - b\,d + b^2} \\ \sigma = \dfrac{d-s}{\sqrt{d-b}\sqrt{s-b}}. \end{cases} \tag{6.29}$$

These will be extensively used in Chapter 7.

Proof. See equations (%i57) through (%o72) of the file in Appendix 6.A.

6.9 FINDING THE DEATH TIME (d) FROM BIRTH (b), PEAK (p), AND SENILITY (s)

Many people (though not everybody) would like to know when they will die given their birth time, peak time, and senility time.

Well, such a problem is solved exactly (i.e., without numeric approximations) by the equation

$$d = \frac{s + b\,\ln\left(\frac{p-b}{s-b}\right)}{\ln\left(\frac{p-b}{s-b}\right) + 1}. \tag{6.30}$$

Proof. See equations (%i73) through (%o77) of the file in Appendix 6.A.

6.10 FINDING THE PROBABILITY OF THE LENGTH OF SOMEONE'S LIFE; THAT IS, THE AREA UNDER THE RELEVANT FINITE *b*-LOGNORMAL

A basic fact of elementary probability is that the area between two points t_1 and t_2 under any assigned probability density function in the independent variable t represents the probability that the event following that probability distribution will occur between the times t_1 and t_2. This is of course true for any b-lognormal, in that the area in between b and infinity equals 1 (probability of a sure event) and so a fraction of this area = 1 represents the relevant probability of occurrence.

However, this is not the case for the finite b-lognormal defined in Section 6.7, inasmuch as the area under such a finite probability density function is not equal to 1, but to a certain positive number smaller than 1. This begs the next question: How can we renormalize the finite b-lognormal so that the area under it becomes equal to 1? The answer is easy: we must compute the area under the finite b-lognormal between birth b and death d and then divide the finite b-lognormal by such a constant, thus renormalizing it to 1.

Let us thus compute the area under the finite b-lognormal. This area is clearly made up of two parts:

(1) The area under the b-lognormal between b and s (i.e., the integral of the b-lognormal between b and s).
(2) The area under the tangent-at-s straight line between s and d (i.e., the area of the triangle between s and d).

Maxima has no difficulty in computing the integral of the b-lognormal between b and s, and the result is:

$$b\text{-lognormal\_probability\_between\_birth\_and\_senility} = \frac{\operatorname{erf}\left(\frac{\sqrt{\sigma^2+4} - 3\,\sigma}{2^{\frac{3}{2}}}\right)}{2} + \frac{1}{2} \tag{6.31}$$

in which the "error function" erf(x) (see, e.g., the Wikipedia site *http://en.wikipedia.org/wiki/Error_function*) is defined by the integral

$$\operatorname{erf}(x) = \frac{2}{\sqrt{\pi}} \int_0^x e^{-t^2}\, dt. \tag{6.32}$$

Note that the probability given by (6.31) is always higher than 50% (i.e., higher than $\frac{1}{2}$) and that it is independent of μ. The last fact is not surprising, since μ is just the mean value around which the normal distribution is centered and so it may be anywhere along the x-axis; in other words, the normalization constant of the normal (i.e., $\sqrt{2\pi}\,\sigma$) must be independent of μ, and so it is; the b-lognormal, in turn, is nothing more than e raised to the normal and so again the b-lognormal normalization constant must be independent of μ as well. This is thus the case for (6.31).

Proof. See lines (%i78) and (%o79) of the file in Appendix 6.A.

As for the triangle area, one may of course compute it either directly from the triangle formula or as the integral under the straight line between s and d, and in both cases the result is

finite_b-lognormal_probability_between_senility_and_death

$$= \frac{\left(\sqrt{\sigma^2+4} + \sigma\right) e^{\frac{3\,\sigma\sqrt{\sigma^2+4}}{4} - \frac{5\,\sigma^2}{4} - \frac{1}{2}}}{2^{\frac{5}{2}}\sqrt{\pi}}. \tag{6.33}$$

Proof. See lines (%i80) through (%o86) of the file in Appendix 6.A.

The overall finite b-lognormal area is thus the sum of (6.31) and (6.33). This result thus reads:

Probability_between_birth_and_death

$$= \frac{\operatorname{erf}\left(\frac{\sqrt{\sigma^2+4} - 3\,\sigma}{2^{\frac{3}{2}}}\right)}{2} + \frac{\left(\sqrt{\sigma^2+4} + \sigma\right) e^{\frac{3\,\sigma\sqrt{\sigma^2+4}}{4} - \frac{5\,\sigma^2}{2} - \frac{1}{2}}}{2^{\frac{5}{2}}\sqrt{\pi}} + \frac{1}{2}. \tag{6.34}$$

Now, this result must be a positive number smaller than 1, since we know the b-lognormal to be normalized to 1 and (6.34) represents an area that is just a fraction of the whole area under the b-lognormal from 0 to ∞. For instance, if $\sigma = 0.373$, then (6.34) yields a value of 0.892 for the probability between birth and death (i.e., a probability of 89.2%). In geometric terms, this statement is the same as saying the area under the finite b-lognormal equals 89.2% of the area under the full, infinite b-lognormal with the same value of σ, no matter what the value of μ might possibly be.

APPENDIX 6.A "B_LOGNORMALS_INFLEXION_POINTS_AND_DEATH_TIME.WMX"

"FINITE" b-LOGNORMALS as a new mathematical model
for the lifetime of a living being:
a cell, a human, a society, a civilization, whether human or ET.

Part 1:
Elementary properties of b-lognormals up to kurtosis.

Author: Claudio Maccone (e-mail: clmaccon@libero.it), June 23th, 2011.

```
(%i1) kill(all);
(%o0) done
```

Definition of the lognormal pdf starting at b>0.

```
(%i1) assume(b>0, t>b, mu>0, sigma>0);
(%o1) [b>0,t>b,μ>0,σ>0]
(%i2) b_lgn(t, mu, sigma, b):=(1/(sqrt(2*%pi)*sigma*(t-b)))*%e^(-(log(t-b)-mu)^2/(2*sigma^2));
```

(%o2) $$\mathrm{b\_lgn}(t,\mu,\sigma,b):=\frac{1}{\sqrt{2\pi}\,\sigma\,(t-b)}\,\%e^{-\frac{(\log(t-b)-\mu)^2}{2\sigma^2}}$$

Normalization Condition.

```
(%i3) 'integrate(b_lgn(t, mu, sigma, b),t,b,inf)=integrate(b_lgn(t, mu, sigma, b),t,b,inf);
```

(%o3) $$\frac{\int_b^{\infty}\frac{\%e^{-\frac{(\log(t-b)-\mu)^2}{2\sigma^2}}}{t-b}\,dt}{\sqrt{2}\sqrt{\pi}\,\sigma}=1$$

MEAN VALUE.

```
(%i4) 'integrate(t*b_lgn(t, mu, sigma, b),t,b,inf)=integrate(t*b_lgn(t, mu, sigma, b),t,b,inf);
```

(%o4) $$\frac{\int_b^{\infty}\frac{t\,\%e^{-\frac{(\log(t-b)-\mu)^2}{2\sigma^2}}}{t-b}\,dt}{\sqrt{2}\sqrt{\pi}\,\sigma}=\frac{\sqrt{2}\sqrt{\pi}\,\sigma\,\%e^{\frac{\sigma^2}{2}+\mu}+\sqrt{2}\sqrt{\pi}\,b\,\sigma}{\sqrt{2}\sqrt{\pi}\,\sigma}$$

(%i5) distrib(%);

(%o5) $$\frac{\int_b^\infty \frac{t\,\%e^{-\frac{(\log(t-b)-\mu)^2}{2\sigma^2}}}{t-b}\,dt}{\sqrt{2}\sqrt{\pi}\,\sigma}=\%e^{\frac{\sigma^2}{2}+\mu}+b$$

(%i6) mean_value_eq:mean_value=rhs(%);

(%o6) $$mean\_value=\%e^{\frac{\sigma^2}{2}+\mu}+b$$

2nd moment.

(%i7) 'integrate(t^2*b_lgn(t, mu, sigma, b),t,b,inf)=integrate(t^2*b_lgn(t, mu, sigma, b),t,b,inf);

(%o7) $$\frac{\int_b^\infty \frac{t^2\,\%e^{-\frac{(\log(t-b)-\mu)^2}{2\sigma^2}}}{t-b}\,dt}{\sqrt{2}\sqrt{\pi}\,\sigma}=\frac{\sqrt{2}\sqrt{\pi}\,\sigma\,\%e^{2\sigma^2+2\mu}+2^{3/2}\sqrt{\pi}\,b\,\sigma\,\%e^{\frac{\sigma^2}{2}+\mu}+\sqrt{2}\sqrt{\pi}\,b^2\sigma}{\sqrt{2}\sqrt{\pi}\,\sigma}$$

(%i8) factor(%);

(%o8) $$\frac{\int_b^\infty \frac{t^2\,\%e^{-\frac{\log(t-b)^2}{2\sigma^2}+\frac{\mu\log(t-b)}{\sigma^2}-\frac{\mu^2}{2\sigma^2}}}{t-b}\,dt}{\sqrt{2}\sqrt{\pi}\,\sigma}=\%e^{2\sigma^2+2\mu}+2\,b\,\%e^{\frac{\sigma^2}{2}+\mu}+b^2$$

(%i9) second_moment_eq:second_moment=rhs(%);

(%o9) $$second\_moment=\%e^{2\sigma^2+2\mu}+2\,b\,\%e^{\frac{\sigma^2}{2}+\mu}+b^2$$

VARIANCE.

(%i10) factor(radcan(second_moment_eq-mean_value_eq^2));

(%o10) $$second\_moment-mean\_value^2=\left(\%e^{\sigma^2}-1\right)\%e^{\sigma^2+2\mu}$$

(%i11) variance_eq:variance=rhs(%);

(%o11) $$variance=\left(\%e^{\sigma^2}-1\right)\%e^{\sigma^2+2\mu}$$

STANDARD DEVIATION.

(%i12) sqrt(variance_eq);

(%o12) $\sqrt{variance} = \sqrt{\%e^{\sigma^2} - 1}\ \%e^{\frac{\sigma^2 + 2\mu}{2}}$

(%i13) standard_deviation_eq:standard_deviation=rhs(%);

(%o13) $standard\_deviation = \sqrt{\%e^{\sigma^2} - 1}\ \%e^{\frac{\sigma^2 + 2\mu}{2}}$

MODE, i.e. PEAK.

(%i14) maximum_cond:diff(b_lgn(t, mu, sigma, b), t)=0;

(%o14) $-\frac{\%e^{-\frac{(\log(t-b)-\mu)^2}{2\sigma^2}}(\log(t-b)-\mu)}{\sqrt{2}\sqrt{\pi}\,\sigma^3\,(t-b)^2} - \frac{\%e^{-\frac{(\log(t-b)-\mu)^2}{2\sigma^2}}}{\sqrt{2}\sqrt{\pi}\,\sigma\,(t-b)^2} = 0$

(%i15) peak_time_eq:first(solve(maximum_cond, t));

(%o15) $t = \%e^{\mu - \sigma^2} + b$

(%i16) mode_time_eq:p=rhs(peak_time_eq);

(%o16) $p = \%e^{\mu - \sigma^2} + b$

(%i17) finding_the_mode_ord:b_lgn(rhs(mode_time_eq), mu, sigma, b);

(%o17) $\frac{\%e^{\frac{\sigma^2}{2} - \mu}}{\sqrt{2}\sqrt{\pi}\,\sigma}$

(%i18) mode_ord_eq:peak_ord=%;

(%o18) $peak\_ord = \frac{\%e^{\frac{\sigma^2}{2} - \mu}}{\sqrt{2}\sqrt{\pi}\,\sigma}$

Median.

(%i19) assume(median>b);

(%o19) $[median > b]$

(%i20) assume(median>(b+1));

(%o20) $[median - b - 1 > 0]$

```
(%i21) median_cond:'integrate(b_lgn(t, mu, sigma, b),t,b,median)='integrate(b_lgn(t, mu, sigma, b),t,median,inf);
```

$$(\%o21)\ \frac{\int_b^{median} \frac{\%e^{-\frac{(\log(t-b)-\mu)^2}{2\sigma^2}}}{t-b}\,dt}{\sqrt{2}\sqrt{\pi}\,\sigma} = \frac{\int_{median}^{\infty} \frac{\%e^{-\frac{(\log(t-b)-\mu)^2}{2\sigma^2}}}{t-b}\,dt}{\sqrt{2}\sqrt{\pi}\,\sigma}$$

```
(%i22) median_cond:integrate(b_lgn(t, mu, sigma, b),t,b,median)=integrate(b_lgn(t, mu, sigma, b),t,median,inf);
```

$$(\%o22)\ \frac{\frac{\sqrt{\pi}\,\sigma}{\sqrt{2}} - \frac{\sqrt{\pi}\,\mathrm{erf}\left(\frac{\sqrt{2}\,\mu-\sqrt{2}\,\log(median-b)}{2\sigma}\right)\sigma}{\sqrt{2}}}{\sqrt{2}\sqrt{\pi}\,\sigma} = \frac{\frac{\sqrt{\pi}\,\mathrm{erf}\left(\frac{\sqrt{2}\,\mu-\sqrt{2}\,\log(median-b)}{2\sigma}\right)\sigma}{\sqrt{2}} + \frac{\sqrt{\pi}\,\sigma}{\sqrt{2}}}{\sqrt{2}\sqrt{\pi}\,\sigma}$$

```
(%i23) solve(%,median);
```

$$(\%o23)\ [median = \%e^{\mu} + b]$$

```
(%i24) median_eq:median_time=rhs(first(%));
```

$$(\%o24)\ median\_time = \%e^{\mu} + b$$

3rd Moment.

```
(%i25) 'integrate(t^3*b_lgn(t, mu, sigma, b),t,b,inf)=integrate(t^3*b_lgn(t, mu, sigma, b),t,b,inf);
```

$$(\%o25)\ \frac{\int_b^{\infty} \frac{t^3\,\%e^{-\frac{(\log(t-b)-\mu)^2}{2\sigma^2}}}{t-b}\,dt}{\sqrt{2}\sqrt{\pi}\,\sigma} = \frac{\sqrt{2}\sqrt{\pi}\,\sigma\,\%e^{\frac{9\sigma^2}{2}+3\mu} + 3\sqrt{2}\sqrt{\pi}\,b\,\sigma\,\%e^{2\sigma^2+2\mu} + 3\sqrt{2}\sqrt{\pi}\,b^2\,\sigma\,\%e^{\frac{\sigma^2}{2}+\mu} + \sqrt{2}\sqrt{\pi}\,b^3\,\sigma}{\sqrt{2}\sqrt{\pi}\,\sigma}$$

```
(%i26) factor(%);
```

$$(\%o26)\ \frac{\int_b^{\infty} \frac{t^3\,\%e^{-\frac{\log(t-b)^2}{2\sigma^2}+\frac{\mu\log(t-b)}{\sigma^2}-\frac{\mu^2}{2\sigma^2}}}{t-b}\,dt}{\sqrt{2}\sqrt{\pi}\,\sigma} = \%e^{\frac{9\sigma^2}{2}+3\mu} + 3\,b\,\%e^{2\sigma^2+2\mu} + 3\,b^2\,\%e^{\frac{\sigma^2}{2}+\mu} + b^3$$

```
(%i27) third_moment_eq:third_moment=rhs(%);
```

$$(\%o27)\ third\_moment = \%e^{\frac{9\sigma^2}{2}+3\mu} + 3\,b\,\%e^{2\sigma^2+2\mu} + 3\,b^2\,\%e^{\frac{\sigma^2}{2}+\mu} + b^3$$

4th Moment.

(%i28) 'integrate(t^4*b_lgn(t, mu, sigma, b),t,b,inf)=integrate(t^4*b_lgn(t, mu, sigma, b),t,b,inf);

$$(\%o28)\ \frac{\int_b^{\infty} \frac{t^4\ \%e^{-\frac{(\log(t-b)-\mu)^2}{2\sigma^2}}}{t-b}dt}{\sqrt{2}\sqrt{\pi}\sigma} = (\sqrt{2}\sqrt{\pi}\sigma\ \%e^{8\sigma^2+4\mu}+2^{5/2}\sqrt{\pi}\,b\,\sigma\ \%e^{\frac{9\sigma^2}{2}+3\mu}+3\,2^{3/2}\sqrt{\pi}\,b^2\sigma\ \%e^{2\sigma^2+2\mu}+2^{5/2}\sqrt{\pi}\,b^3\sigma\ \%e^{\frac{\sigma^2}{2}+\mu}+\sqrt{2}\sqrt{\pi}\,b^4\sigma)/(\sqrt{2}\sqrt{\pi}\sigma)$$

(%i29) factor(%);

$$(\%o29)\ \frac{\int_b^{\infty} \frac{t^4\ \%e^{-\frac{\log(t-b)^2}{2\sigma^2}+\frac{\mu\log(t-b)}{\sigma^2}-\frac{\mu^2}{2\sigma^2}}}{t-b}dt}{\sqrt{2}\sqrt{\pi}\sigma} = \%e^{8\sigma^2+4\mu}+4\,b\ \%e^{\frac{9\sigma^2}{2}+3\mu}+6\,b^2\ \%e^{2\sigma^2+2\mu}+4\,b^3\ \%e^{\frac{\sigma^2}{2}+\mu}+b^4$$

(%i30) fourth_moment_eq:fourth_moment=rhs(%);

$$(\%o30)\ fourth\_moment = \%e^{8\sigma^2+4\mu}+4\,b\ \%e^{\frac{9\sigma^2}{2}+3\mu}+6\,b^2\ \%e^{2\sigma^2+2\mu}+4\,b^3\ \%e^{\frac{\sigma^2}{2}+\mu}+b^4$$

First four Cumulants.

(%i31) def_K1:K[1]=rhs(mean_value_eq);

$$(\%o31)\ K_1 = \%e^{\frac{\sigma^2}{2}+\mu}+b$$

(%i32) def_K2:K[2]=rhs(second_moment_eq)-K[1]^2;

$$(\%o32)\ K_2 = \%e^{2\sigma^2+2\mu}+2\,b\ \%e^{\frac{\sigma^2}{2}+\mu}+b^2-K_1^2$$

(%i33) def_K3:K[3]=rhs(third_moment_eq)-3*K[1]*K[2]-K[1]^3;

$$(\%o33)\ K_3 = \%e^{\frac{9\sigma^2}{2}+3\mu}+3\,b\ \%e^{2\sigma^2+2\mu}+3\,b^2\ \%e^{\frac{\sigma^2}{2}+\mu}+b^3-3\,K_1\,K_2-K_1^3$$

(%i34) def_K4:K[4]=rhs(fourth_moment_eq)-4*K[1]*K[3]-3*K[2]^2-6*K[2]*K[1]^2-K[1]^4;

(%o34) $K_4=\%e^{8\sigma^2+4\mu}+4\,b\,\%e^{\frac{9\sigma^2}{2}+3\mu}+6\,b^2\,\%e^{2\sigma^2+2\mu}+4\,b^3\,\%e^{\frac{\sigma^2}{2}+\mu}+b^4-4\,K_1\,K_3-3\,K_2^2-6\,K_1^2\,K_2-K_1^4$

(%i35) def_K2,def_K1;

(%o35) $K_2=\%e^{2\sigma^2+2\mu}-\left(\%e^{\frac{\sigma^2}{2}+\mu}+b\right)^2+2\,b\,\%e^{\frac{\sigma^2}{2}+\mu}+b^2$

(%i36) K2_eq:factor(radcan(%));

(%o36) $K_2=\left(\%e^{\sigma^2}-1\right)\%e^{\sigma^2+2\mu}$

(%i37) def_K3,def_K2,def_K1;

(%o37) $K_3=\%e^{\frac{9\sigma^2}{2}+3\mu}-3\left(\%e^{\frac{\sigma^2}{2}+\mu}+b\right)\left(\%e^{2\sigma^2+2\mu}-\left(\%e^{\frac{\sigma^2}{2}+\mu}+b\right)^2+2\,b\,\%e^{\frac{\sigma^2}{2}+\mu}+b^2\right)+3\,b\,\%e^{2\sigma^2+2\mu}-\left(\%e^{\frac{\sigma^2}{2}+\mu}+b\right)^3+3\,b^2\,\%e^{\frac{\sigma^2}{2}+\mu}+b^3$

(%i38) K3_eq:factor(radcan(%));

(%o38) $K_3=\left(\%e^{\sigma^2}-1\right)^2\left(\%e^{\sigma^2}+2\right)\%e^{\frac{3\sigma^2}{2}+3\mu}$

(%i39) def_K4,def_K1,K2_eq,K3_eq;

(%o39) $K_4=\%e^{8\sigma^2+4\mu}+4\,b\,\%e^{\frac{9\sigma^2}{2}+3\mu}+6\,b^2\,\%e^{2\sigma^2+2\mu}-4\left(\%e^{\frac{\sigma^2}{2}+\mu}+b\right)\left(\%e^{\sigma^2}-1\right)^2\left(\%e^{\sigma^2}+2\right)\%e^{\frac{3\sigma^2}{2}+3\mu}-3\left(\%e^{\sigma^2}-1\right)^2\%e^{2(\sigma^2+2\mu)}-6\left(\%e^{\frac{\sigma^2}{2}+\mu}+b\right)^2\left(\%e^{\sigma^2}-1\right)\%e^{\sigma^2+2\mu}-\left(\%e^{\frac{\sigma^2}{2}+\mu}+b\right)^4+4\,b^3\,\%e^{\frac{\sigma^2}{2}+\mu}+b^4$

(%i40) K4_eq:factor(radcan(%));

(%o40) $K_4=\left(\%e^{\sigma^2}-1\right)^3\left(\%e^{3\sigma^2}+3\,\%e^{2\sigma^2}+6\,\%e^{\sigma^2}+6\right)\%e^{2\sigma^2+4\mu}$

Skewness.

(%i41) Skewness_eq:radcan(K3_eq/(K2_eq)^(3/2));

(%o41) $\frac{K_3}{K_2^{3/2}}=\sqrt{\%e^{\sigma^2}-1}\left(\%e^{\sigma^2}+2\right)$

Kurtosis.

(%i42) Kurtosis_eq:radcan(K4_eq/(K2_eq)^2);

(%o42) $\frac{K_4}{K_2^2} = \%e^{4\sigma^2} + 2\,\%e^{3\sigma^2} + 3\,\%e^{2\sigma^2} - 6$

APPENDIX 6.B

"FINITE" LOGNORMALS as a new mathematical model
for the lifetime of a living being:
a cell, a human, a society, a civilization, whether human or ET.

Part 2: Inflexion points, Death time, and overall model Probability.

Author: Claudio Maccone (e-mail: clmaccon@libero.it), March 10th, 2011.

(%i1) kill(all);

(%o0) *done*

Definition of the lognormal pdf starting at b>0.

(%i1) assume(b>0, t>b, mu>0, sigma>0);

(%o1) $[b>0, t>b, \mu>0, \sigma>0]$

(%i2) b_lgn(t, mu, sigma, b):=(1/(sqrt(2*%pi)*sigma*(t-b)))*%e^(-(log(t-b)-mu)^2/(2*sigma^2));

(%o2) $\mathrm{b\_lgn}(t,\mu,\sigma,b) := \frac{1}{\sqrt{2\pi}\,\sigma\,(t-b)}\,\%e^{-\frac{(\log(t-b)-\mu)^2}{2\sigma^2}}$

MODE, i.e. PEAK.

(%i3) maximum_cond:diff(b_lgn(t, mu, sigma, b), t)=0;

(%o3) $-\frac{\%e^{-\frac{(\log(t-b)-\mu)^2}{2\sigma^2}}\,(\log(t-b)-\mu)}{\sqrt{2}\sqrt{\pi}\,\sigma^3\,(t-b)^2} - \frac{\%e^{-\frac{(\log(t-b)-\mu)^2}{2\sigma^2}}}{\sqrt{2}\sqrt{\pi}\,\sigma\,(t-b)^2} = 0$

(%i4) peak_time_eq:first(solve(maximum_cond, t));

(%o4) $t = \%e^{\mu-\sigma^2} + b$

(%i5) mode_time_eq:p=rhs(peak_time_eq);

(%o5) $p = \%e^{\mu-\sigma^2} + b$

(%i6) finding_the_mode_ord:b_lgn(rhs(mode_time_eq), mu, sigma, b);

(%o6) $\frac{\%e^{\frac{\sigma^2}{2}-\mu}}{\sqrt{2}\sqrt{\pi}\,\sigma}$

(%i7) mode_ord_eq:peak_ord=%;

(%o7) $peak\_ord = \dfrac{\%e^{\frac{\sigma^2}{2}-\mu}}{\sqrt{2}\,\sqrt{\pi}\,\sigma}$

TWO INFLEXION POINTS: t=a (adolescence) and t=s (senility).

(%i8) inflexion_cond:diff(b_lgn(t, mu, sigma, b), t,2)=0;

(%o8) $\dfrac{\%e^{-\frac{(\log(t-b)-\mu)^2}{2\sigma^2}}(\log(t-b)-\mu)^2}{\sqrt{2}\,\sqrt{\pi}\,\sigma^5\,(t-b)^3}+\dfrac{\sqrt{2}\,\%e^{-\frac{(\log(t-b)-\mu)^2}{2\sigma^2}}(\log(t-b)-\mu)}{\sqrt{\pi}\,\sigma^3\,(t-b)^3}+\dfrac{\%e^{-\frac{(\log(t-b)-\mu)^2}{2\sigma^2}}(\log(t-b)-\mu)}{\sqrt{2}\,\sqrt{\pi}\,\sigma^3\,(t-b)^3}+\dfrac{\sqrt{2}\,\%e^{-\frac{(\log(t-b)-\mu)^2}{2\sigma^2}}}{\sqrt{\pi}\,\sigma\,(t-b)^3}-\dfrac{\%e^{-\frac{(\log(t-b)-\mu)^2}{2\sigma^2}}}{\sqrt{2}\,\sqrt{\pi}\,\sigma^3\,(t-b)^3}=0$

(%i9) inflexion_times:solve(inflexion_cond, t);

(%o9) $[t=\%e^{-\frac{\sigma\sqrt{\sigma^2+4}}{2}-\frac{3\sigma^2}{2}+\mu}+b, t=\%e^{\frac{\sigma\sqrt{\sigma^2+4}}{2}-\frac{3\sigma^2}{2}+\mu}+b]$

(%i10) asc_infl_time:first(inflexion_times);

(%o10) $t=\%e^{-\frac{\sigma\sqrt{\sigma^2+4}}{2}-\frac{3\sigma^2}{2}+\mu}+b$

(%i11) assume(a>0);

(%o11) $[a>0]$

(%i12) assume(a>b);

(%o12) $[a>b]$

(%i13) a_point_eq:a=rhs(asc_infl_time);

(%o13) $a=\%e^{-\frac{\sigma\sqrt{\sigma^2+4}}{2}-\frac{3\sigma^2}{2}+\mu}+b$

(%i14) ord_of_asc_infl:b_lognormal_at_a=radcan(subst(rhs(first(inflexion_times)), t, b_lgn(t, mu, sigma, b)));

(%o14) $b\_lognormal\_at\_a=\dfrac{\%e^{-\frac{\sigma\sqrt{\sigma^2+4}-\sigma^2+4\mu+2}{4}}}{\sqrt{2}\,\sqrt{\pi}\,\sigma}$

(%i15) desc_infl_time:last(inflexion_times);

(%o15) $t=\%e^{\frac{\sigma\sqrt{\sigma^2+4}}{2}-\frac{3\sigma^2}{2}+\mu}+b$

(%i16) assume(s>0);

(%o16) $[s>0]$

(%i17) assume(s>b);

(%o17) $[s>b]$

(%i18) assume(s>a);

(%o18) $[s>a]$

(%i19) s_point_eq:s=rhs(desc_infl_time);

(%o19) $s=\%e^{\frac{\sigma\sqrt{\sigma^2+4}}{2}-\frac{3\sigma^2}{2}+\mu}+b$

(%i20) ord_of_desc_infl:b_lognormal_at_s=radcan(subst(rhs(last(inflexion_times)), t, b_lgn(t, mu, sigma, b)));

(%o20) $b\_lognormal\_at\_s=\frac{\%e^{\frac{\sigma\sqrt{\sigma^2+4}+\sigma^2-4\mu-2}{4}}}{\sqrt{2}\sqrt{\pi}\,\sigma}$

Finding mu and sigma WHEN the two ORDINATES
of the b-lognormal at the AT THE TWO INFLEXION POINTS
are assigned and the ADOLESCENCE time (t=a) is assigned.

(%i21) expression_in_sigma_only:log(radcan(ord_of_desc_infl/ord_of_asc_infl));

(%o21) $\log\left(\frac{b\_lognormal\_at\_s}{b\_lognormal\_at\_a}\right)=\frac{\sigma\sqrt{\sigma^2+4}}{2}$

(%i22) sigma_from_the_two_inflection_ordinates:second(solve(expression_in_sigma_only^2, sigma));

(%o22) $\sigma=\sqrt{2}\sqrt{\sqrt{\log\left(\frac{b\_lognormal\_at\_s}{b\_lognormal\_at\_a}\right)^2+1}-1}$

Important NOTE about an "error" in Maxima's NOTATION.

Unfortunately, Maxima's notation is WRONG! (and should be corrected!).
In fact, Maxima writes, for instance, sin(x^2) instead of sin^2(x).
So, Maxima writes sin(x^2)+cos(x^2)=1 instead of the correct
sin^2(x)+cos^2(x)=1. This is a SERIOUS MISPRINT!

Such a stupid misprint did NOT occur in the good old MACSYMA
(i.e. in the "commercial" Macsyma) but it does occur in Maxima.

So, in the last equation, the square applies TO THE WHOLE LOG, RATHER
THAN JUST TO THE LOG ARGUMENT!!! Be aware of that, please.

```
(%i23) mu_from_asc_infl_time:first(solve(a_point_eq, mu));
```

(%o23) $\mu = \frac{\sigma\sqrt{\sigma^2+4}+3\sigma^2+2\log(a-b)}{2}$

Finding mu and sigma WHEN the two ORDINATES
of the b-lognormal at the AT THE TWO INFLEXION POINTS
are assigned and the SENILITY time (t=s) is assigned.

```
(%i24) expression_in_sigma_only:log(radcan(ord_of_desc_infl/ord_of_asc_infl));
```

(%o24) $\log\left(\frac{b\_lognormal\_at\_s}{b\_lognormal\_at\_a}\right) = \frac{\sigma\sqrt{\sigma^2+4}}{2}$

```
(%i25) sigma_from_the_two_inflection_ordinates:second(solve(expression_in_sigma_only^2, sigma));
```

(%o25) $\sigma = \sqrt{2}\sqrt{\sqrt{\log\left(\frac{b\_lognormal\_at\_s}{b\_lognormal\_at\_a}\right)^2+1}-1}$

```
(%i26) mu_from_desc_infl_time:first(solve(s_point_eq, mu));
```

(%o26) $\mu = -\frac{\sigma\sqrt{\sigma^2+4}-3\sigma^2-2\log(s-b)}{2}$

FINDING mu and sigma WHEN the
ADOLESCENCE AND SENILITY times, a and s, are assigned.
That is, our life-lognormal model is completely determined
if a and s are assigned.

```
(%i27) a_point_eq;
```

(%o27) $a = \%e^{-\frac{\sigma\sqrt{\sigma^2+4}}{2}-\frac{3\sigma^2}{2}+\mu}+b$

(%i28) s_point_eq;

(%o28) $s = \%e^{\frac{\sigma\sqrt{\sigma^2+4}}{2} - \frac{3\sigma^2}{2} + \mu} + b$

(%i29) first_eq:(a_point_eq-b)*(s_point_eq-b);

(%o29) $(a-b)(s-b) = \%e^{2\mu - 3\sigma^2}$

(%i30) mu_vs_a_and_s:distrib(factor(first(solve(first_eq, mu))));

(%o30) $\mu = \frac{3\sigma^2}{2} + \frac{\log(-(b-a)(s-b))}{2}$

(%i31) to_sigma_eq_1:a_point_eq, mu_vs_a_and_s;

(%o31) $a = \%e^{\frac{\log(-(b-a)(s-b))}{2} - \frac{\sigma\sqrt{\sigma^2+4}}{2}} + b$

(%i32) to_sigma_eq_2:log(radcan(expand(radcan(to_sigma_eq_1-b))/(sqrt(b-a)*sqrt(b-s))));

(%o32) $\log\left(-\frac{\sqrt{b-a}}{\sqrt{b-s}}\right) = -\frac{\sigma\sqrt{\sigma^2+4}}{2}$

(%i33) four_expressions_for_sigma:solve(to_sigma_eq_2^2,sigma);

(%o33) $[\sigma = -\sqrt{2}\sqrt{\sqrt{\log\left(-\frac{\sqrt{b-a}}{\sqrt{b-s}}\right)^2+1}-1}, \sigma = \sqrt{2}\sqrt{\sqrt{\log\left(-\frac{\sqrt{b-a}}{\sqrt{b-s}}\right)^2+1}-1}, \sigma = -\sqrt{2}\,\%i\sqrt{\sqrt{\log\left(-\frac{\sqrt{b-a}}{\sqrt{b-s}}\right)^2+1}+1}, \sigma = \sqrt{2}\,\%i\sqrt{\sqrt{\log\left(-\frac{\sqrt{b-a}}{\sqrt{b-s}}\right)^2+1}+1}]$

(%i34) sigma_vs_a_and_s:radcan(second(four_expressions_for_sigma));

(%o34) $\sigma = \sqrt{\sqrt{4\log(-\sqrt{b-s})^2 - 4\log(b-a)\log(-\sqrt{b-s}) + \log(b-a)^2 + 4} - 2}$

(%i35) final_sigma:sigma=sqrt(2)*sqrt(sqrt(((log(sqrt(s-b)/sqrt(a-b)))^2)+1)-1);

(%o35) $\sigma = \sqrt{2}\sqrt{\sqrt{\log\left(\frac{\sqrt{s-b}}{\sqrt{a-b}}\right)^2+1}-1}$

mu and sigma FROM ADOLESCENCE AND PEAK.

(%i36) eq_1:log(a_point_eq-b);

(%o36) $$\log(a-b)=-\frac{\sigma\sqrt{\sigma^2+4}}{2}-\frac{3\,\sigma^2}{2}+\mu$$

(%i37) eq_2:log(mode_time_eq-b);

(%o37) $$\log(p-b)=\mu-\sigma^2$$

(%i38) eq_3:eq_1-eq_2;

(%o38) $$\log(a-b)-\log(p-b)=-\frac{\sigma\sqrt{\sigma^2+4}}{2}-\frac{\sigma^2}{2}$$

(%i39) eq_4:expand(eq_3^2);

(%o39) $$\log(p-b)^2-2\log(a-b)\log(p-b)+\log(a-b)^2=\frac{\sigma^3\sqrt{\sigma^2+4}}{2}+\frac{\sigma^4}{2}+\sigma^2$$

(%i40) eq_5:distrib((eq_4-sigma^4/2-sigma^2)/sigma^3);

(%o40) $$-\frac{\sigma}{2}-\frac{1}{\sigma}+\frac{\log(p-b)^2}{\sigma^3}-\frac{2\log(a-b)\log(p-b)}{\sigma^3}+\frac{\log(a-b)^2}{\sigma^3}=\frac{\sqrt{\sigma^2+4}}{2}$$

(%i41) eq_6:ratsimp(sigma^6*expand(eq_5^2));

(%o41) $$(\sigma^8+4\,\sigma^6+(-4\log(p-b)^2+8\log(a-b)\log(p-b)-4\log(a-b)^2+4)\,\sigma^4+(-8\log(p-b)^2+16\log(a-b)\log(p-b)-8\log(a-b)^2)\,\sigma^2+4\log(p-b)^4-16\log(a-b)\log(p-b)^3+24\log(a-b)^2\log(p-b)^2-16\log(a-b)^3\log(p-b)+4\log(a-b)^4)/4=\frac{\sigma^8+4\,\sigma^6}{4}$$

(%i42) eq_7:radcan(logcontract(solve(eq_6, sigma)));

(%o42) $$[\sigma=-\frac{\log(b-p)-\log(b-a)}{\sqrt{-\log(b-p)+\log(b-a)+1}},\sigma=\frac{\log(b-p)-\log(b-a)}{\sqrt{-\log(b-p)+\log(b-a)+1}},\sigma=-\frac{\log(b-p)-\log(b-a)}{\sqrt{\log(b-p)-\log(b-a)+1}},\sigma=\frac{\log(b-p)-\log(b-a)}{\sqrt{\log(b-p)-\log(b-a)+1}}]$$

(%i43) sigma_vs_b_p_and_a:logcontract(last(eq_7));

(%o43) $$\sigma=\frac{\log\left(-\frac{p-b}{b-a}\right)}{\sqrt{\log\left(-\frac{p-b}{b-a}\right)+1}}$$

(%i44) mu_vs_b_p_and_a:eq_8:first(solve(eq_1,mu));

(%o44) $$\mu=\frac{\sigma\sqrt{\sigma^2+4}+3\,\sigma^2+2\log(a-b)}{2}$$

INTRODUCING DEATH,
defined as the time of interception between the
TANGENT STRAIGHT LINE
TO THE b-LOGNORMAL AT SENILITY
and the time axis.

```
(%i45) tangent_eq:(y-y0)=m*(t-t0);
```

(%o45) $y-y0=m\,(t-t0)$

```
(%i46) finding_y0:y0=radcan(b_lgn(rhs(s_point_eq), mu, sigma, b));
```

(%o46) $$y0=\frac{\%e^{\frac{\sigma\sqrt{\sigma^2+4}+\sigma^2-4\mu-2}{4}}}{\sqrt{2}\sqrt{\pi}\,\sigma}$$

```
(%i47) finding_t0:t0=rhs(s_point_eq);
```

(%o47) $$t0=\%e^{\frac{\sigma\sqrt{\sigma^2+4}}{2}-\frac{3\sigma^2}{2}+\mu}+b$$

```
(%i48) finding_m:m=radcan(subst(rhs(finding_t0), t, diff(b_lgn(t, mu, sigma, b), t)));
```

(%o48) $$m=-\frac{\left(\sqrt{2}\sqrt{\sigma^2+4}\,\%e^{\frac{7\sigma^2}{4}}-\sqrt{2}\,\sigma\,\%e^{\frac{7\sigma^2}{4}}\right)\%e^{-\frac{\sigma\sqrt{\sigma^2+4}+8\mu+2}{4}}}{4\sqrt{\pi}\,\sigma^2}$$

```
(%i49) tangent_straight_line_eq:tangent_eq,finding_y0,finding_t0,finding_m;
```

(%o49) $$y-\frac{\%e^{\frac{\sigma\sqrt{\sigma^2+4}+\sigma^2-4\mu-2}{4}}}{\sqrt{2}\sqrt{\pi}\,\sigma}=-\frac{\left(\sqrt{2}\sqrt{\sigma^2+4}\,\%e^{\frac{7\sigma^2}{4}}-\sqrt{2}\,\sigma\,\%e^{\frac{7\sigma^2}{4}}\right)\%e^{-\frac{\sigma\sqrt{\sigma^2+4}+8\mu+2}{4}}\left(t-\%e^{\frac{\sigma\sqrt{\sigma^2+4}}{2}-\frac{3\sigma^2}{2}+\mu}-b\right)}{4\sqrt{\pi}\,\sigma^2}$$

```
(%i50) tangent_at_s:radcan(first(solve(tangent_straight_line_eq, y)));
```

(%o50) $$y=-\left(\%e^{-\frac{\sigma\sqrt{\sigma^2+4}+8\mu+4}{4}}\left(\sqrt{\%e}\left(-\sigma\,\%e^{\frac{7\sigma^2}{4}}\,t-\sigma\,\%e^{\frac{2\sigma\sqrt{\sigma^2+4}+\sigma^2+4\mu}{4}}+b\,\sigma\,\%e^{\frac{7\sigma^2}{4}}\right)+\sqrt{\%e}\sqrt{\sigma^2+4}\left(\%e^{\frac{7\sigma^2}{4}}\,t-\%e^{\frac{2\sigma\sqrt{\sigma^2+4}+\sigma^2+4\mu}{4}}-b\,\%e^{\frac{7\sigma^2}{4}}\right)\right)\right)/\left(2^{3/2}\sqrt{\pi}\,\sigma^2\right)$$

(%i51) finding_death_point:solve(subst(0, y, tangent_at_s), t);

(%o51) $$[t=\frac{\%e^{-\frac{7\sigma^2}{4}}\left(\sqrt{\sigma^2+4}\,\%e^{\frac{\sigma\sqrt{\sigma^2+4}}{2}+\frac{\sigma^2}{4}+\mu}+\sigma\,\%e^{\frac{\sigma\sqrt{\sigma^2+4}}{2}+\frac{\sigma^2}{4}+\mu}+(b\sqrt{\sigma^2+4}-b\,\sigma)\,\%e^{\frac{7\sigma^2}{4}}\right)}{\sqrt{\sigma^2+4}-\sigma}]$$

(%i52) death_point_eq:death_point=rhs(first(finding_death_point));

(%o52) $$death\_point=\frac{\%e^{-\frac{7\sigma^2}{4}}\left(\sqrt{\sigma^2+4}\,\%e^{\frac{\sigma\sqrt{\sigma^2+4}}{2}+\frac{\sigma^2}{4}+\mu}+\sigma\,\%e^{\frac{\sigma\sqrt{\sigma^2+4}}{2}+\frac{\sigma^2}{4}+\mu}+(b\sqrt{\sigma^2+4}-b\,\sigma)\,\%e^{\frac{7\sigma^2}{4}}\right)}{\sqrt{\sigma^2+4}-\sigma}$$

(%i53) death_time_eq:d=rhs(death_point_eq);

(%o53) $$d=\frac{\%e^{-\frac{7\sigma^2}{4}}\left(\sqrt{\sigma^2+4}\,\%e^{\frac{\sigma\sqrt{\sigma^2+4}}{2}+\frac{\sigma^2}{4}+\mu}+\sigma\,\%e^{\frac{\sigma\sqrt{\sigma^2+4}}{2}+\frac{\sigma^2}{4}+\mu}+(b\sqrt{\sigma^2+4}-b\,\sigma)\,\%e^{\frac{7\sigma^2}{4}}\right)}{\sqrt{\sigma^2+4}-\sigma}$$

(%i54) def_dd:dd=%e^(-7*sigma^2/4)*%e^((sigma*sqrt(sigma^2+4))/2+sigma^2/4+mu)*(sqrt(sigma^2+4)+sigm

(%o54) $$dd=\frac{(\sqrt{\sigma^2+4}+\sigma)^2\,\%e^{\frac{\sigma\sqrt{\sigma^2+4}}{2}-\frac{3\sigma^2}{2}+\mu}}{4}+b$$

(%i55) radcan(rhs(def_dd)-rhs(death_time_eq));

(%o55) 0

(%i56) death_time:d=rhs(def_dd);

(%o56) $$d=\frac{(\sqrt{\sigma^2+4}+\sigma)^2\,\%e^{\frac{\sigma\sqrt{\sigma^2+4}}{2}-\frac{3\sigma^2}{2}+\mu}}{4}+b$$

Most IMPORTANT !

FINDING THE b-LOGNORMAL
WITH A GIVEN BIRTH (t=b),
SENILITY (t=s),
AND DEATH (t=d).

In other words, the problem of determing the b-lognormal of a living organism when its birth time b, its senility time s and its death time d are known, is now completely solved.

This is done in two parts: PART 1 and PART 2 hereafter.

PART 1: finding sigma as a function of the birth, b, senility, s and death, d.

(%i57) towards_an_eq_in_sigma:4*(death_time-b);

(%o57) $4\,(d-b)=\left(\sqrt{\sigma^2+4}+\sigma\right)^2 \%e^{\frac{\sigma\sqrt{\sigma^2+4}}{2}-\frac{3\,\sigma^2}{2}+\mu}$

(%i58) s_point_eq-b;

(%o58) $s-b=\%e^{\frac{\sigma\sqrt{\sigma^2+4}}{2}-\frac{3\,\sigma^2}{2}+\mu}$

(%i59) towards_an_eq_in_sigma/%;

(%o59) $\frac{4\,(d-b)}{s-b}=\left(\sqrt{\sigma^2+4}+\sigma\right)^2$

(%i60) sqrt(%);

(%o60) $\frac{2\sqrt{d-b}}{\sqrt{s-b}}=\sqrt{\sigma^2+4}+\sigma$

(%i61) %-sigma;

(%o61) $\frac{2\sqrt{d-b}}{\sqrt{s-b}}-\sigma=\sqrt{\sigma^2+4}$

(%i62) %^2;

(%o62) $\left(\frac{2\sqrt{d-b}}{\sqrt{s-b}}-\sigma\right)^2=\sigma^2+4$

(%i63) expand(%);

(%o63) $\sigma^2-\frac{4\sqrt{d-b}\,\sigma}{\sqrt{s-b}}+\frac{4\,d}{s-b}-\frac{4\,b}{s-b}=\sigma^2+4$

(%i64) sigma_vs_b_d_and_s:radcan(first(solve(%, sigma)));

(%o64) $\sigma=-\frac{s-d}{\sqrt{d-b}\,\sqrt{s-b}}$

PART 2: Finding mu a a function of birth, b, senility s and death, d.

(%i65) b_vs_d_eq:first(solve(death_time,b));

(%o65) $b=-\frac{(\sigma\sqrt{\sigma^2+4}+\sigma^2+2)\,\%e^{\frac{\sigma\sqrt{\sigma^2+4}}{2}-\frac{3\sigma^2}{2}+\mu}-2d}{2}$

(%i66) mu_vs_b_and_d_eq:solve(death_time, mu);

(%o66) $[\mu=-\frac{-2\log\left(\frac{2d}{\sigma\sqrt{\sigma^2+4}+\sigma^2+2}-\frac{2b}{\sigma\sqrt{\sigma^2+4}+\sigma^2+2}\right)+\sigma\sqrt{\sigma^2+4}-3\sigma^2}{2}]$

(%i67) mu_vs_b_and_d:distrib(first(mu_vs_b_and_d_eq));

(%o67) $\mu=\log\left(\frac{2d}{\sigma\sqrt{\sigma^2+4}+\sigma^2+2}-\frac{2b}{\sigma\sqrt{\sigma^2+4}+\sigma^2+2}\right)-\frac{\sigma\sqrt{\sigma^2+4}}{2}+\frac{3\sigma^2}{2}$

(%i68) mu_vs_b_d_s:mu_vs_b_and_d,sigma_vs_b_d_and_s;

(%o68) $\mu=\log\left(\frac{2d}{\frac{(s-d)\sqrt{\frac{(s-d)^2}{(d-b)\,(s-b)}+4}}{\sqrt{d-b}\sqrt{s-b}}+\frac{(s-d)^2}{(d-b)\,(s-b)}+2}-\frac{2b}{\frac{(s-d)\sqrt{\frac{(s-d)^2}{(d-b)\,(s-b)}+4}}{\sqrt{d-b}\sqrt{s-b}}+\frac{(s-d)^2}{(d-b)\,(s-b)}+2}\right)+\frac{(s-d)\sqrt{\frac{(s-d)^2}{(d-b)\,(s-b)}+4}}{2\sqrt{d-b}\sqrt{s-b}}+\frac{3(s-d)^2}{2(d-b)\,(s-b)}$

(%i69) mu_vs_b_s_and_d_eq:distrib(radcan(mu_vs_b_d_s));

(%o69) $\mu=\log(s-b)+\frac{2s^2}{(d-b)\,s-b\,d+b^2}+\frac{(-3d-b)\,s}{(d-b)\,s-b\,d+b^2}+\frac{d^2}{(d-b)\,s-b\,d+b^2}+\frac{b\,d}{(d-b)\,s-b\,d+b^2}$

(%i70) (2*s^2)/((d-b)*s-b*d+b^2)+((-3*d-b)*s)/((d-b)*s-b*d+b^2)+d^2/((d-b)*s-b*d+b^2)+(b*d)/((d-b)*s-b*d+b

(%o70) $\frac{2s^2}{(d-b)\,s-b\,d+b^2}+\frac{(-3d-b)\,s}{(d-b)\,s-b\,d+b^2}+\frac{d^2}{(d-b)\,s-b\,d+b^2}+\frac{b\,d}{(d-b)\,s-b\,d+b^2}$

(%i71) radcan(%);

(%o71) $\frac{2s^2+(-3d-b)\,s+d^2+b\,d}{(d-b)\,s-b\,d+b^2}$

(%i72) mu_vs_b_s_and_d:mu=log(s-b)+(2*s^2+(-3*d-b)*s+d^2+b*d)/((d-b)*s-b*d+b^2);

(%o72) $\mu=\log(s-b)+\frac{2s^2+(-3d-b)\,s+d^2+b\,d}{(d-b)\,s-b\,d+b^2}$

Finding the DEATH TIME from the known
PEAK and SENILITY times.

(%i73) invoking_p:mode_time_eq-b;

(%o73) $p-b=\%e^{\mu-\sigma^2}$

(%i74) eq_in_p_s_d:invoking_p,sigma_vs_b_d_and_s;

(%o74) $p-b=\%e^{\mu-\frac{(s-d)^2}{(d-b)\,(s-b)}}$

(%i75) expression_of_p_minus_b:eq_in_p_s_d,mu_vs_b_s_and_d;

(%o75) $p-b=(s-b)\,\%e^{\frac{2\,s^2+(-3\,d-b)\,s+d^2+b\,d}{(d-b)\,s-b\,d+b^2}-\frac{(s-d)^2}{(d-b)\,(s-b)}}$

(%i76) simplified_expression_of_p_minus_b:radcan(expression_of_p_minus_b);

(%o76) $p-b=(s-b)\,\%e^{\frac{s-d}{d-b}}$

(%i77) death_from_peak_and_senility:first(solve(simplified_expression_of_p_minus_b, d));

(%o77) $d=\dfrac{s+b\,\log\left(\frac{b}{b-s}-\frac{p}{b-s}\right)}{\log\left(\frac{b}{b-s}-\frac{p}{b-s}\right)+1}$

Computing the PROBABILITY by which
this "living organism" model is true, i.e.
computing the AREA under the finite_b_lognormal
and the time axis, from birth (t=b) to death (t=d).

In the practice, this means computing the INTEGRAL
under the finite_b_lognormal curve from b to s,
and then ADDING to this the area of the TRIANGLE
under the tangent-at-s straight line and the time axis,
from the descending inflexion point t=s up to the
death time t=d.

(%i78) 'integrate(b_lgn(t,mu,sigma,b), t, b, rhs(s_point_eq))=integrate(b_lgn(t,mu,sigma,b), t, b, rhs(s_point_eq));

(%o78)
$$\frac{\displaystyle\int_b^{\%e^{\frac{\sigma\sqrt{\sigma^2+4}}{2}-\frac{3\sigma^2}{2}+\mu}+b} \frac{\%e^{-\frac{(\log(t-b)-\mu)^2}{2\sigma^2}}}{t-b}\,dt}{\sqrt{2}\sqrt{\pi}\,\sigma} = \frac{\frac{\sqrt{\pi}\,\sigma\,\mathrm{erf}\left(\frac{\sqrt{2}\sqrt{\sigma^2+4}-3\sqrt{2}\,\sigma}{4}\right)}{\sqrt{2}}+\frac{\sqrt{\pi}\,\sigma}{\sqrt{2}}}{\sqrt{2}\sqrt{\pi}\,\sigma}$$

(%i79) prob_from_b_to_s:prob_from_b_to_s=distrib(rhs(%));

(%o79)
$$prob\_from\_b\_to\_s = \frac{\mathrm{erf}\left(\frac{\sqrt{2}\sqrt{\sigma^2+4}-3\sqrt{2}\,\sigma}{4}\right)}{2}+\frac{1}{2}$$

(%i80) triangle_area_:triangle_area=(rhs(death_time)-rhs(s_point_eq))*(rhs(ord_of_desc_infl))/2;

(%o80)
$$triangle\_area = \frac{\left(\frac{(\sqrt{\sigma^2+4}+\sigma)^2\,\%e^{\frac{\sigma\sqrt{\sigma^2+4}}{2}-\frac{3\sigma^2}{2}+\mu}}{4}-\%e^{\frac{\sigma\sqrt{\sigma^2+4}}{2}-\frac{3\sigma^2}{2}+\mu}\right)\%e^{\frac{\sigma\sqrt{\sigma^2+4}+\sigma^2-4\mu-2}{4}}}{2^{3/2}\sqrt{\pi}\,\sigma}$$

(%i81) death_time;

(%o81)
$$d = \frac{(\sqrt{\sigma^2+4}+\sigma)^2\,\%e^{\frac{\sigma\sqrt{\sigma^2+4}}{2}-\frac{3\sigma^2}{2}+\mu}}{4}+b$$

(%i82) s_point_eq;

(%o82)
$$s = \%e^{\frac{\sigma\sqrt{\sigma^2+4}}{2}-\frac{3\sigma^2}{2}+\mu}+b$$

(%i83) prob_from_b_to_d_eq:prob_from_b_to_d=expand(rhs(prob_from_b_to_s)+rhs(triangle_area_));

(%o83)
$$prob\_from\_b\_to\_d = \frac{\mathrm{erf}\left(\frac{\sqrt{\sigma^2+4}}{2^{3/2}}-\frac{3\sigma}{2^{3/2}}\right)}{2}+\frac{\sqrt{\sigma^2+4}\,\%e^{\frac{3\sigma\sqrt{\sigma^2+4}}{4}-\frac{5\sigma^2}{4}-\frac{1}{2}}}{2^{5/2}\sqrt{\pi}}+\frac{\sigma\,\%e^{\frac{3\sigma\sqrt{\sigma^2+4}}{4}-\frac{5\sigma^2}{4}-\frac{1}{2}}}{2^{5/2}\sqrt{\pi}}+\frac{1}{2}$$

(%i84) ev(subst(0.373,sigma,%),numer);

(%o84) $prob\_from\_b\_to\_d = 0.89266088792049$

So, the total probability from birth to death is HIGHER than 1/2 = 50%, and depends upon sigma only.
Please don't be cheated by the minus sign coming from the triangle area: the area is of course POSITIVE, and the minus sign in front of it is just a Maxima way of writing things (other negative terms at the numerator make the whole fraction POSITIVE).

We now provide a CONCISE FORMULA for the whole above probability expression.
That is:

(%i85) concise__prob_from_b_to_d:concise_version=erf((sqrt(sigma^2+4)-3*sigma)/(2^(3/2)))/2+((sqrt(sigma^2-

$$(\%o85)\ concise\_version=\frac{\mathrm{erf}\left(\frac{\sqrt{\sigma^2+4}-3\sigma}{2^{3/2}}\right)}{2}+\frac{(\sqrt{\sigma^2+4}+\sigma)\,\%e^{\frac{3\sigma\sqrt{\sigma^2+4}}{4}-\frac{5\sigma^2}{4}-\frac{1}{2}}}{2^{5/2}\sqrt{\pi}}+\frac{1}{2}$$

(%i86) radcan(prob_from_b_to_d_eq-concise__prob_from_b_to_d)=0;

$$(\%o86)\ prob\_from\_b\_to\_d-concise\_version=\frac{\mathrm{erf}\left(\frac{\sqrt{2}\sqrt{\sigma^2+4}-3\sqrt{2}\sigma}{4}\right)-\mathrm{erf}\left(\frac{\sqrt{\sigma^2+4}-3\sigma}{2^{3/2}}\right)}{2}=0$$

(%i87) final__prob_from_b_to_d:prob_from_b_to_d=erf((sqrt(sigma^2+4)-3*sigma)/2^(3/2))/2+((sqrt(sigma^2+4

$$(\%o87)\ prob\_from\_b\_to\_d=\frac{\mathrm{erf}\left(\frac{\sqrt{\sigma^2+4}-3\sigma}{2^{3/2}}\right)}{2}+\frac{(\sqrt{\sigma^2+4}+\sigma)\,\%e^{\frac{3\sigma\sqrt{\sigma^2+4}}{4}-\frac{5\sigma^2}{4}-\frac{1}{2}}}{2^{5/2}\sqrt{\pi}}+\frac{1}{2}$$

Example: suppose that sigma=0.373.
Then the total probability is given by 89%.
In fact, one has:

(%i88) ev(subst(0.373,sigma,final__prob_from_b_to_d),numer);

$$(\%o88)\ prob\_from\_b\_to\_d=0.89266088792049$$

7

Civilizations as finite *b*-lognormals: Mathematical history

7.1 INTRODUCTION

Centuries of human history on Earth should have taught us something.

Basically, civilizations are born, fight against each other, and "die", *merging*, however, with newer civilizations.

To cast all this in terms of mathematical equations is hard. The reason nobody has done so is because the task is so daunting. Indeed, no course on "Mathematical History" is taught at any university in the world.

In this chapter we will have a stab at this. Our idea is simple: any civilization is born, reaches a peak, then declines ... just like a *b-lognormal*!

7.2 EXAMPLES OF HISTORIC CIVILIZATIONS AS FINITE *b*-LOGNORMALS

We now offer eight examples of such a view; that is, historic development of the civilizations of:

(1) Ancient Greece
(2) Ancient Rome
(3) Renaissance Italy
(4) Portuguese Empire
(5) Spanish Empire
(6) French Empire
(7) British Empire
(8) American (U.S.A.) Empire.

Other historic empires (e.g., the Dutch, German, Russian, Chinese, and Japanese, not to mention the Aztec and Inca Empires, or ancient ones, like the Egyptian,

Persian, Parthian, or the medieval Mongol Empire) should certainly be added to such a picture, but we regret we don't have the time to carry out studies of these in this chapter. Such historic–mathematical studies will be left to be made at a later stage of development of this new research field, which we call "Mathematical History": a mathematical view based on *b*-lognormal probability distributions.

To summarize this chapter's content, we define for each of the eight civilizations listed above:

(a) Birth b (i.e., the year that civilization was supposed to be "born", even if only approximately).
(b) Senility s (i.e., the year of an historic event that marked the beginning of the decline of that civilization).
(c) Death d (i.e., the year another historic event marked the "official passing away" of that civilization from history).

Then, consider the two equations (6.29):

$$\begin{cases} \mu = \ln(s-b) + \dfrac{2s^2 - (3d+b)s + d^2 + b\,d}{(d-b)s - b\,d + b^2} \\ \sigma = \dfrac{d-s}{\sqrt{d-b}\sqrt{s-b}} \end{cases} \tag{7.1}$$

For each civilization, these two equations allow us to compute both μ and σ in terms of the three assigned numbers $(b, \quad s, \quad d)$. As a consequence, the time that a given civilization peaked is found immediately from equation (%o15) in Appendix 6.A, that is

$$\text{peak\_time} = \text{abscissa\_of\_the\_maximum} = p = b + e^{\mu-\sigma^2} \tag{7.2}$$

Also, we can then write the equation of the corresponding *b*-lognormal immediately:

$$b\_\text{lognormal}(t,\mu,\sigma,b) = \begin{cases} = 0 & \text{for } 0 \le t \le b \\ = \dfrac{1}{\sqrt{2\pi}\sigma(t-b)} e^{-\frac{(\log(t-b)-\mu)^2}{2\sigma^2}} & \text{for } b \le t \le \infty \end{cases} \tag{7.3}$$

The plot of this function of time gives a clear picture of the historic development of that civilization.

Inserting the peak time (7.2) into (7.3), the peak ordinate of the civilization is found (i.e., "how civilized that civilization was at its peak") and this is explicitly given by equation (%o18) of Appendix 6.A:

$$\text{peak\_ordinate} = \frac{e^{\frac{\sigma^2}{2}-\mu}}{\sqrt{2\pi}\,\sigma} \tag{7.4}$$

Next we must define the equation of the straight line tangent to the *b*-lognormal at senility (i.e., at the descending inflexion point). This tangent straight line equation is given by equation (%o50) of Appendix 6.A, which we reproduce hereafter as an image just for the sake of completeness:

$$y - \frac{\%e^{\frac{\sigma\sqrt{\sigma^2+4}+\sigma^2-4\mu-2}{4}}}{\sqrt{2}\sqrt{\pi}\sigma} = -\frac{\left(\sqrt{2}\sqrt{\sigma^2+4}\,\%e^{\frac{7\sigma^2}{4}} - \sqrt{2}\sigma\,\%e^{\frac{7\sigma^2}{4}}\right)\%e^{-\frac{\sigma\sqrt{\sigma^2+4}+8\mu+2}{4}}\left(t - \%e^{\frac{\sigma\sqrt{\sigma^2+4}}{2}-\frac{3\sigma^2}{2}+\mu} - b\right)}{4\sqrt{\pi}\sigma^2} \tag{7.5}$$

In the Mathcad file making up Appendix 7.A, we had to introduce ways to abbreviate this long equation (7.5) in order to let it fit the page width. In particular, for each civilization, we had to introduce two constants K and L summarizing the coefficients in (7.5): just a tedious detail of no importance for the understanding of history, but we wanted to make the reader aware of it.

Finally, the plot of each civilization's b-lognormal can be drawn, as well as the tangent straight line at senility, intercepting the time axis at death. This is what we call the *finite* b-lognormal.

7.3 PLOTTING ALL *b*-LOGNORMALS TOGETHER AND FINDING THE TRENDS

Having drawn the b-lognormal plot for each civilization we wish to study, the time is ripe to plot all of them *together* and see what the trends are. This is done in the Mathcad file of Appendix 7.A with the result shown hereafter as Figure 7.1.

We immediately notice some trends:

(1) The first two civilizations in time (Greece and Rome) are separated from the six modern ones by a large 1,000-year gap. This is, of course, the Middle Ages (i.e., the Dark Ages that hampered the development of Western civilization for about 1,000 years). As Carl Sagan said, "the millennium gap in the middle of the diagram represents a poignant lost opportunity for the human species" [1].

(2) While the first two civilizations of Greece and Rome lasted more than 600 years each, all modern civilizations lasted much less: 500 years at most, but usually less or indeed much less.

(3) Since b-lognormals are probability density functions, the area under each b-lognormal must be the same (i.e., just 1—normalization condition). Thus, the shorter a civilization lasts the higher its peak must be! This is obvious from Figure 7.1: Greece and Rome lasted so long and their peak was so much smaller than the British or the American peak!

(4) In other words, our theory accounts for the higher level of more recent historic civilizations in a natural fashion, with no need to introduce further free parameters. Not a small result, we think.

(5) All these remarks are relevant to Appendix 7.A and the figures therein, starting with Figure 7.1.

7.3.1 All Western civilizations (Greek to U.S.) as finite *b*-lognormals

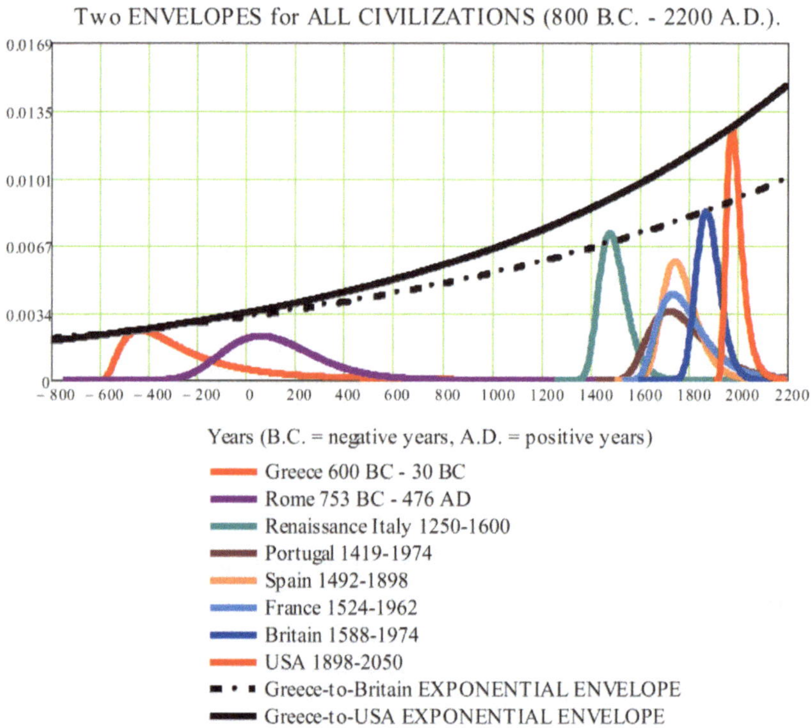

Figure 7.1. The *b*-lognormals of eight civilizations in human history and two exponential envelopes for them.

7.4 FINDING THE EXPONENTIAL ENVELOPES OF ALL *b*-LOGNORMALS

Let us now find the exponential envelopes of all civilizations studied so far.

We are going to erroneously use the term "envelope" when it is not strictly so from a mathematical point of view. See, for instance, the Wikipedia site *http://en.wikipedia.org/wiki/Envelope_(mathematics)* where a delightful calculus theorem shows that, if a family of curves in the $(x, \quad y)$ plane is characterized by the presence of a parameter λ (i.e., each curve of the family corresponds to a different numeric value of λ), then the equation of the family written in the form

$$f(x, \quad y, \quad \lambda) = 0 \tag{7.6}$$

and its partial derivative with respect to λ set to zero

$$\frac{\partial f(x, \quad y, \quad \lambda)}{\partial \lambda} = 0 \tag{7.7}$$

make up a system of two simultaneous equations. If λ can be eliminated from these two equations, then the resulting function of x and y only is the envelope of the family of curves, meaning that each curve of the family is tangent to the envelope but different from it.

Returning to our family of b-lognormals, we are not going to use the word "envelope" in the above strict mathematical sense, but rather we define our "exponential envelope", or briefly "envelope", as the exponential curve of equation

$$E(t) = A\,e^{Bt} \tag{7.8}$$

This increasing exponential curve is supposed to depart from the peak of the lowest b-lognormal and reach the peak of the highest b-lognormal. According to this definition, we first need to write down the two equations requiring the exponential curve $E(t) = A\,e^{Bt}$ to intercept both points of coordinates $(p_1, \quad P_1)$ and $(p_2, \quad P_2)$

$$\begin{cases} P_1 = A\,e^{Bp_1} \\ P_2 = A\,e^{Bp_2} \end{cases} \tag{7.9}$$

Subtracting the log of the second equation from the log of the first one, the constant A disappears and one gets

$$\ln P_2 - \ln P_1 = B(p_2 - p_1) \tag{7.10}$$

whence

$$B = \frac{\ln \dfrac{P_2}{P_1}}{p_2 - p_1} = \ln\left[\left(\frac{P_2}{P_1}\right)^{\frac{1}{p_2-p_1}}\right]. \tag{7.11}$$

Taking the exponential of the last equation multiplied by p_1, one gets

$$e^{Bp_1} = e^{p_1 \ln\left[\left(\frac{P_2}{P_1}\right)^{\frac{1}{p_2-p_1}}\right]} = e^{\ln\left[\left(\frac{P_2}{P_1}\right)^{\frac{p_1}{p_2-p_1}}\right]} = \left(\frac{P_2}{P_1}\right)^{\frac{p_1}{p_2-p_1}}. \tag{7.12}$$

Inserting this into the upper equation (7.9), and solving for A, one gets

$$A = \frac{P_1}{e^{Bp_1}} = \frac{P_1}{\left(\dfrac{P_2}{P_1}\right)^{\frac{p_1}{p_2-p_1}}}. \tag{7.13}$$

This equation and (7.11) completely solve the problem of determining the exponential envelope $E(t) = A\,e^{Bt}$ (i.e., finding its two constants A and B that pass through two different points $(p_1, \quad P_1)$ and $(p_2, \quad P_2)$):

$$\begin{cases} A = \dfrac{P_1}{\left(\dfrac{P_2}{P_1}\right)^{\frac{p_1}{p_2-p_1}}} \\ B = \ln\left[\left(\dfrac{P_2}{P_1}\right)^{\frac{1}{p_2-p_1}}\right]. \end{cases} \tag{7.14}$$

The reader should notice that the equations we develop in Section 8.2 for the exponential increase of the number of species in Darwinian evolution are little more than (7.14) given by

$$\begin{cases} p_1 = t_1 = -3.5 \cdot 10^9 \text{ year} \\ P_1 = 1 \\ p_2 = 0 \\ P_2 = 5 \cdot 10^7. \end{cases} \quad (7.15)$$

In fact, inserting (7.15) into (7.14), the latter becomes

$$\begin{cases} A = \dfrac{P_1}{\left(\dfrac{P_2}{P_1}\right)^{\frac{p_1}{p_2-p_1}}} = \dfrac{1}{(P_2)^{\frac{p_1}{0-p_1}}} = \dfrac{1}{P_2^{-1}} = P_2 = 5 \cdot 10^7 \\ B = \ln\left[\left(\dfrac{P_2}{P_1}\right)^{\frac{1}{p_2-p_1}}\right] = \ln\left[P_2^{\frac{1}{-p_1}}\right] = \dfrac{1}{-p_1} \ln P_2 = \dfrac{\ln(5 \cdot 10^7)}{-(-3.5 \cdot 10^9 \text{year})} = \dfrac{1.605 \cdot 10^{-16}}{\text{sec}}. \end{cases} \quad (7.16)$$

These are little more than equations (8.3) and (8.8), respectively, and show the A and B numerical values typical of Darwinian evolution on Earth over the last 3.5 billion years, assuming that today $5 \cdot 10^7$ living species exist on Earth. Please see Chapter 8 for more details.

7.5 *b*-LOGNORMALS OF FUTURE CIVILIZATIONS

So much for the past: What about the future? What are the *b*-lognormals of ET civilizations in this Galaxy?

Nobody knows, of course; and nobody will know until SETI succeeds.

Science fiction fans, however, might take pleasure in casting the *Star Trek* timeline into the mathematical language of *b*-lognormals. To this end, a good starting point seems to be the Wikipedia site dealing with the timeline of *Star Trek* (*http://en.wikipedia.org/wiki/Timeline_of_Star_Trek*). *The Star Trek Chronology* [3] is also interesting.

7.6 REFERENCES

[1] The interested reader must get a copy of this wonderful book: *Interstellar Migration and the Human Experience*, by Ben R. Finney and Eric M. Jones, University of California Press, 1986—354 (*http://books.google.com/books/about/Interstellar_Migration_and_the_Human_Exp.html?hl=ja&id=iKnaLbRtQasC*). This book is indeed "revolutionary", inasmuch it looks at the history of many past civilizations with the glasses of the new science of SETI. A lot can be learned from this book; moreover, it is free of mathematics.

[2] Carl Sagan, *Cosmos*, Random House, 1980. See in particular p. 335 and the caption to the diagram there.
[3] Michael and Denise Okuda, *The Star Trek Chronology*, 1996, available from Amazon.com.

APPENDIX 7.A CIVILIZATIONS AS FINITE *b*-LOGNORMALS

Ancient Greece (600 BC–30 BC) as a finite *b*-lognormal

bG := − 600	sG := − 323	dG := − 30
Birth Greek expansion by ship along all Mediterranean coasts is in full swing. Pythagoras' (570–495 BC) theorem is proved.	*Senility* Alexander the Great dies in Babylon after expanding the Greek civilization from Greece to Egypt and the Indus River. The Hellenistic civilization starts in 323 BC and ends with Cleopatra in 30 BC.	*Death* Cleopatra commits suicide in 30 BC and with her the last Hellenistic kingdom (Egypt) ends. The Romans take over along all the Mediterranean coasts.

Deriving the values of the b-lognormal's μ and σ from just the above three blocks of data: b, s, and d

Note that we will prefix all equations by "sd" to make it clear that all data in this appendix are derived from *s* and *d* only:

sdsG := − 323 sddG := − 30

The two relevant solving equations are %o72 and %o64 of the Maxima file "Printing 2 b-lognormal inflexion points and Death time" in Chapter 6. These two equations are, respectively, listed hereafter:

$$\mu = \log(s-b) + \frac{2s^2 + (-3d-b)s + d^2 + bd}{(d-b)s - bd + b^2}$$

$$\text{sd}\mu\text{G} := \ln(\text{sdsG} - \text{bG}) + \frac{2\cdot \text{sG}^2 + (-3\cdot \text{sddG} - \text{bG})\cdot \text{sG} + \text{sddG}^2 + \text{bG}\cdot \text{sddG}}{(\text{sddG} - \text{bG})\cdot \text{sG} - \text{bG}\cdot \text{sddG} + \text{bG}^2}$$

The value of μ for the *b*-lognormal of the Greek civilization, as derived from the two values of senility and death:

sdμG = 5.654

$$\sigma = -\frac{s-d}{\sqrt{d-b}\sqrt{s-b}}$$

$$\text{sd}\sigma\text{G} := \frac{\text{sddG} - \text{sG}}{\sqrt{\text{sddG} - \text{bG}}\cdot\sqrt{\text{sG} - \text{bG}}}$$

The value of σ for the Greek civilization as derived from the two values of senility and death:

sdσG = 0.737

Let us now consider the peak of this *b*-lognormal. The peak is reached in 434 BC. This was indeed the "golden age" of Athens and Greece, the period during which Pericles led Athens, from 461 to 429 BC, which is sometimes known as the "Age of Pericles". The Parthenon was completed in 438 BC, although further decorations of the Parthenon continued until 432 BC. Major artists and philosophers (e.g., Socrates, 469–399 BC) lived about that time; so the peak is right:

$\text{sdpG} := e^{\text{sd}\mu\text{G} - \text{sd}\sigma\text{G}^2} + \text{bG}$ sdpG = −434.333

(time of the peak of the Greek civilization—Age of Pericles).

Let us define the equation of the *b*-lognormal having the above μ and σ. This *b*-lognormal is the solid line in the graph below:

$$\text{sdblognormalG}(t, \text{sd}\mu\text{G}, \text{sd}\sigma\text{G}, \text{bG}) := \frac{1}{\sqrt{2 \cdot \pi} \cdot \text{sd}\sigma\text{G} \cdot (t - \text{bG})} \cdot e^{-\frac{(\ln(t-\text{bG}) - \text{sd}\mu\text{G})^2}{2 \cdot \text{sd}\sigma\text{G}^2}}$$

$$\text{sdblognormalG}(\text{sdpG}, \text{sd}\mu\text{G}, \text{sd}\sigma\text{G}, \text{bG}) = 2.488 \times 10^{-3}$$

(ordinate = height of the peak of the Greek civilization):

$$p_1 := \text{sdpG} \qquad p_1 = -434.333$$

$$P_1 := \text{sdblognormalG}(\text{sdpG}, \text{sd}\mu\text{G}, \text{sd}\sigma\text{G}, \text{bG}) \qquad P_1 = 2.488 \times 10^{-3}$$

Defining the straight line tangent to senility and intercepting the time axis at death

Let us define the equation of the straight line tangent to the above *b*-lognormal at its descending inflexion point $t = s$. Since this equation hardly fits within the page width, we are forced to introduce two constants *K* and *L* first:

$$\text{sdKG} := e^{\frac{\text{sd}\sigma\text{G} \cdot \sqrt{\text{sd}\sigma\text{G}^2 + 4} + \text{sd}\sigma\text{G}^2 - 4 \cdot \text{sd}\mu\text{G} - 2}{4}} \qquad \text{sdLG} := e^{-\frac{\text{sd}\sigma\text{G} \cdot \sqrt{\text{sd}\sigma\text{G}^2 + 4} + 8 \cdot \text{sd}\mu\text{G} + 2}{4}}$$

Now we can define the equation of the straight line tangent to the *b*-lognormal at $t = s$. Its intercept with the time axis marks the *death* time of Cleopatra in 30 BC. This straight line is represented by the dashed straight line in the graph below:

$$\text{sdyG}(t) := \frac{\text{sdKG}}{\sqrt{2 \cdot \pi} \cdot \text{sd}\sigma\text{G}} - \frac{\left(\sqrt{2} \cdot \sqrt{\text{sd}\sigma\text{G}^2 + 4}\, e^{\frac{7 \cdot \text{sd}\sigma\text{G}^2}{4}} - \sqrt{2} \cdot \text{sd}\sigma\text{G} \cdot e^{\frac{7 \cdot \text{sd}\sigma\text{G}^2}{4}}\right) \cdot \text{sdLG} \cdot \left(t - e^{\frac{\text{sd}\sigma\text{G} \cdot \sqrt{\text{sd}\sigma\text{G}^2 + 4}}{2} - \frac{3\text{sd}\sigma\text{G}^2}{2} + \text{sd}\mu\text{G}} - \text{bG}\right)}{4 \cdot \sqrt{\pi} \cdot \text{sd}\sigma\text{G}^2}$$

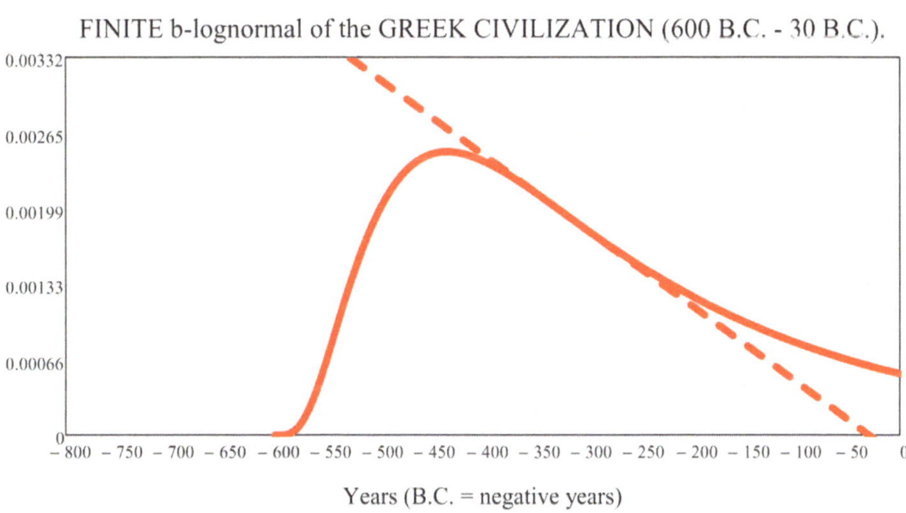

As a mathematical consequence of this *b*-lognormal, we may also derive the adolescence time for the Greek civilization. This turns out to be 542 BC:

$$\text{sdaG} := e^{-\frac{\text{sd}\sigma\text{G} \cdot \sqrt{\text{sd}\sigma\text{G}^2 + 4}}{2} - \frac{3 \cdot \text{sd}\sigma\text{G}^2}{2} + \text{sd}\mu\text{G}} + \text{bG} \qquad \text{sdaG} = -542.475$$

Ancient Rome (753 BC–476 AD) as a finite *b*-lognormal

bR := −753

Birth
Rome is founded by Romulus (according to legend) on April 21, 753 BC

sR := 235

Senility (begins in AD 235)
Emperor Alexander Severus is killed, Rome is no longer the true capital, and the crisis of the 3rd century pervades the empire.

dR := 476

Death
The official end of the Western Roman Empire occurs when no successor to Romulus Augustulus is appointed after September, 4, AD 476.

Deriving the values of the b-lognormal's μ and σ from just the above three blocks of data: b, s, and d

Note that we will prefix all equations by "sd" to make it clear that all data in this appendix are derived from *s* and *d* only:

sdsR := 235 sddR := 476

The two relevant solving equations are %o72 and %o64 of the Maxima file "Printing 2 b-lognormal inflexion points and Death time" in Chapter 6. These two equations are, respectively, listed hereafter:

$$\mu = \log(s-b) + \frac{2s^2 + (-3d-b)s + d^2 + bd}{(d-b)s - bd + b^2}$$

$$\mathrm{sd\mu R} := \ln(\mathrm{sdsR} - \mathrm{bR}) + \frac{2\cdot \mathrm{sR}^2 + (-3\cdot \mathrm{sddR} - \mathrm{bR})\cdot \mathrm{sR} + \mathrm{sddR}^2 + \mathrm{bR}\cdot \mathrm{sddR}}{(\mathrm{sddR} - \mathrm{bR})\cdot \mathrm{sR} - \mathrm{bR}\cdot \mathrm{sddR} + \mathrm{bR}^2}$$

The value of σ for the *b*-lognormal of the civilization of Rome as derived from the two values of senility and death:

$\mathrm{sd\mu R} = 6.747$

$$\sigma = -\frac{s-d}{\sqrt{d-b}\sqrt{s-b}}$$

$$\mathrm{sd\sigma R} := \frac{\mathrm{sddR} - \mathrm{sR}}{\sqrt{\mathrm{sddR} - \mathrm{bR}}\cdot \sqrt{\mathrm{sR} - \mathrm{bR}}}$$

The value of σ for the civilization of Rome as derived from the two values of senility and death:

$\mathrm{sd\sigma R} = 0.219$

Finding the abscissa of the peak of this *b*-lognormal. The peak is reached in 59 AD (i.e., during the reign of Nero, 54–68 AD). If one regards Christianity as the highest legacy left by the Roman Empire, these years were indeed the peak of the Roman Empire, inasmuch as St. Peter and St. Paul were still free to preach Christianity in Rome just a few years before Nero began persecuting people in 64 AD (i.e., just after the (in)famous fire of Rome):

$\mathrm{sdpR} := e^{\mathrm{sd\mu R} - \mathrm{sd\sigma R}^2} + \mathrm{bR}$ $\mathrm{sdpR} = 59.071$

(time of the peak of the Roman civilization—Christianity starts).

Let us define the equation of the *b*-lognormal having the above μ and σ. This *b*-lognormal is the solid line in the graph below:

$$\mathrm{sdblognormalR}(t, \mathrm{sd\mu R}, \mathrm{sd\sigma R}, \mathrm{bR}) := \frac{1}{\sqrt{2\cdot\pi}\cdot \mathrm{sd\sigma R}\cdot (t - \mathrm{bR})}\cdot e^{-\frac{(\ln(t-\mathrm{bR}) - \mathrm{sd\mu R})^2}{2\cdot \mathrm{sd\sigma R}^2}}$$

$\text{sdblognormalR}(\text{sdpR}, \text{sd}\mu\text{R}, \text{sd}\sigma\text{R}, \text{bR}) = 2.193 \times 10^{-3}$

(the ordinate is the height of the peak of the Roman civilization). This value of 2.193×10^{-3} is *slightly smaller* than the corresponding value for the Greeks: 2.488×10^{-3}. This fact may be interpreted as saying the Romans were *less civilized* than the Greeks. This indeed was the case: *Graecia capta ferum victorem coepit*, says Horace (=captured Greece indeed captured its ferocious conqueror (Rome) *by culture*).

Defining the straight line tangent to senility and intercepting the time axis at death

Let us define the equation of the straight line tangent to the above *b*-lognormal at its descending inflexion point $t = s$. Since this equation hardly fits within the page width, we are forced to introduce two constants K and L first:

$$\text{sdKR} := e^{\frac{\text{sd}\sigma\text{R}\cdot\sqrt{\text{sd}\sigma\text{R}^2+4}+\text{sd}\sigma\text{R}^2-4\cdot\text{sd}\mu\text{R}-2}{4}} \qquad \text{sdLR} := e^{-\frac{\text{sd}\sigma\text{R}\cdot\sqrt{\text{sd}\sigma\text{R}^2+4}+8\cdot\text{sd}\mu\text{R}+2}{4}}$$

Now we can define the equation of the straight line tangent to the *b*-lognormal at $t = s$. Its intercept with the time axis marks the *death* time of the Western Roman Empire in AD 476. This straight line is represented by the dashed straight line in the graph below:

$$\text{sdyR}(t) := \frac{\text{sdKR}}{\sqrt{2\cdot\pi}\cdot\text{sd}\sigma\text{R}} - \frac{\left(\sqrt{2}\cdot\sqrt{\text{sd}\sigma\text{R}^2+4}\,e^{\frac{7\cdot\text{sd}\sigma\text{R}^2}{4}} - \sqrt{2}\cdot\text{sd}\sigma\text{R}\cdot e^{\frac{7\cdot\text{sd}\sigma\text{R}^2}{4}}\right)\cdot\text{sdLR}\cdot\left(t - e^{\frac{\text{sd}\sigma\text{R}\cdot\sqrt{\text{sd}\sigma\text{R}^2+4}}{2}-\frac{3\text{sd}\sigma\text{R}^2}{2}+\text{sd}\mu\text{R}} - \text{bR}\right)}{4\cdot\sqrt{\pi}\cdot\text{sd}\sigma\text{R}^2}$$

FINITE b-lognormal of the CIVILIZATION OF ROME (753 B.C. - 476 A.D).

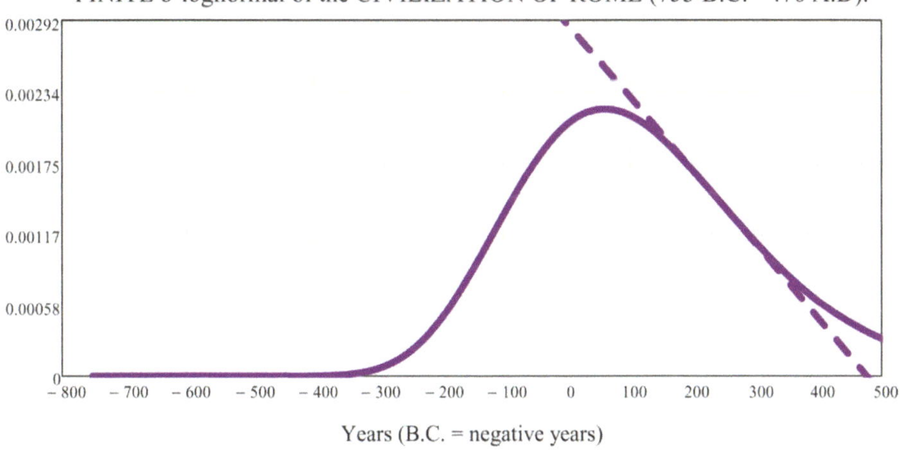

Years (B.C. = negative years)

As a mathematical consequence of this *b*-lognormal, we may also derive the adolescence time for the Roman civilization. This turns out to be 116 BC. This was the peak time of the Roman Republic: Marius (157–86 BC) was a successful Roman general and military reformer. His improvements to the structure and organization of the Roman legion were profound and effective, paving the way to later transformation of the Roman Republic into the Roman Empire:

$$\text{sdaR} := e^{-\frac{\text{sd}\sigma\text{R}\cdot\sqrt{\text{sd}\sigma\text{R}^2+4}}{2}-\frac{3\cdot\text{sd}\sigma\text{R}^2}{2}+\text{sd}\mu\text{R}} + \text{bR} \qquad \text{sdaR} = -116.706$$

The Italian Renaissance (1250–1600) as a finite *b*-lognormal

bI := 1250	sI := 1527	dI := 1660
Birth Frederick II, the last medieval emperor, dies. After him, the "free" Italian towns "breathe" the Renaissance freedom.	*Senility* Starts with the sack of Rome by German Landsknecht soldiers. The Italian Renaissance fades, replaced by a tough Roman Catholic counter reformation and Spanish rule.	*Death* Giordano Bruno is burned at the stake in Rome on February 17, 1600, by order of the Inquisition. Science in Italy is defended by Galileo Galilei but opposed by the strong power of the Catholic Church.

Deriving the values of the b-lognormal's μ and σ from just the above three blocks of data: b, s, and d

Note that we will prefix all equations by "sd" to make it clear that all data in this appendix are derived from *s* and *d* only:

sdsI := 1527   sddI := 1600

The two relevant solving equations are %o72 and %o64 of the Maxima file "Printing 2 b-lognormal inflexion points and Death time" in Chapter 6. These two equations are, respectively, listed hereafter:

$$\mu = \log(s-b) + \frac{2s^2 + (-3d-b)s + d^2 + bd}{(d-b)s - bd + b^2}$$

$$\mathrm{sd\mu I} := \ln(\mathrm{sdsI} - \mathrm{bI}) + \frac{2\cdot \mathrm{sI}^2 + (-3\cdot \mathrm{sddI} - \mathrm{bI})\cdot \mathrm{sI} + \mathrm{sddI}^2 + \mathrm{bI}\cdot \mathrm{sddI}}{(\mathrm{sddI}-\mathrm{bI})\cdot \mathrm{sI} - \mathrm{bI}\cdot \mathrm{sddI} + \mathrm{bI}^2}$$

The value of μ for the *b*-lognormal of the Italian Renaissance as derived from the two values of senility and death:

$\mathrm{sd\mu I} = 5.47$

$$\sigma = -\frac{s-d}{\sqrt{d-b}\sqrt{s-b}}$$

$$\mathrm{sd\sigma I} := \frac{\mathrm{sddI} - \mathrm{sI}}{\sqrt{\mathrm{sddI} - \mathrm{bI}}\cdot\sqrt{\mathrm{sI} - \mathrm{bI}}}$$

The values of σ for the *b*-lognormal of the Italian Renaissance as derived from the two values of senility and death:

$\mathrm{sd\sigma I} = 0.234$

The abscissa of the peak of this *b*-lognormal shows the peak being reached in 1475. Leonardo da Vinci is now aged 23, Cristoforo Colombo (Christopher Columbus) is aged 24, and Michelangelo Buonarroti is born on March 6 that year. The Italian Renaissance is in full swing. Copernicus, aged 2 in 1475, will later go and study in Italy at the universities of Ferrara and Bologna, only to find out that … *all* astronomy had to be rewritten from scratch!:

$\mathrm{sdpI} := e^{\mathrm{sd\mu I} - \mathrm{sd\sigma I}^2} + \mathrm{bI}$ $\mathrm{sdpI} = 1.475 \times 10^3$

Let us define the equation of the *b*-lognormal having the above μ and σ. This *b*-lognormal is the solid line in the graph below:

$$\text{sdblognormalI}(t, \text{sd}\mu\text{I}, \text{sd}\sigma\text{I}, \text{bI}) := \frac{1}{\sqrt{2\cdot\pi}\cdot \text{sd}\sigma\text{I}\cdot(t-\text{bI})}\cdot e^{-\frac{(\ln(t-\text{bI})-\text{sd}\mu\text{I})^2}{2\cdot \text{sd}\sigma\text{I}^2}}$$

$$\text{sdblognormalG}(\text{sdpI}, \text{sd}\mu\text{I}, \text{sd}\sigma\text{I}, \text{bI}) = 7.363\times 10^{-3}$$

(the ordinate is the height of the peak of the civilization of the Italian Renaissance. This equals 7.363×10^{-3} and is much *higher* than the two corresponding peaks of the Greeks (2.488×10^{-3}) and the Romans (2.193×10^{-3}). Just a coincidence! *Not at all!* It is actually the sign of a *more advanced, better civilization*!

Defining the straight line tangent to senility and intercepting the time axis at death

Let us define the equation of the straight line tangent to the above *b*-lognormal at its descending inflexion point $t = s$. Since this equation hardly fits within the page width, we are forced to introduce two constants K and L first:

$$\text{sdKI} := e^{\frac{\text{sd}\sigma\text{I}\cdot\sqrt{\text{sd}\sigma\text{I}^2+4}+\text{sd}\sigma\text{I}^2-4\cdot\text{sd}\mu\text{I}-2}{4}} \qquad \text{sdLI} := e^{-\frac{\text{sd}\sigma\text{I}\cdot\sqrt{\text{sd}\sigma\text{I}^2+4}+8\cdot\text{sd}\mu\text{I}+2}{4}}$$

Now we can define the equation of the straight line tangent to the *b*-lognormal at $t = s$. Its intercept with the time axis marks the *death* time of the Italian Renaissance in 1600. This straight line is represented by the dashed straight line in the graph below:

$$\text{sdyI}(t) := \frac{\text{sdKI}}{\sqrt{2\cdot\pi}\cdot\text{sd}\sigma\text{I}} - \frac{\left(\sqrt{2}\cdot\sqrt{\text{sd}\sigma\text{I}^2+4}\,e^{\frac{7\cdot\text{sd}\sigma\text{I}^2}{4}} - \sqrt{2}\cdot\text{sd}\sigma\text{I}\cdot e^{\frac{7\cdot\text{sd}\sigma\text{I}^2}{4}}\right)\cdot\text{sdLI}\cdot\left(t - e^{\frac{\text{sd}\sigma\text{I}\cdot\sqrt{\text{sd}\sigma\text{I}^2+4}}{2}-\frac{3\text{sd}\sigma\text{I}^2}{2}+\text{sd}\mu\text{I}} - \text{bI}\right)}{4\cdot\sqrt{\pi}\cdot\text{sd}\sigma\text{I}^2}$$

FINITE b-lognormal of the ITALIAN RENAISSANCE CIVILIZATION (1250 - 1600).

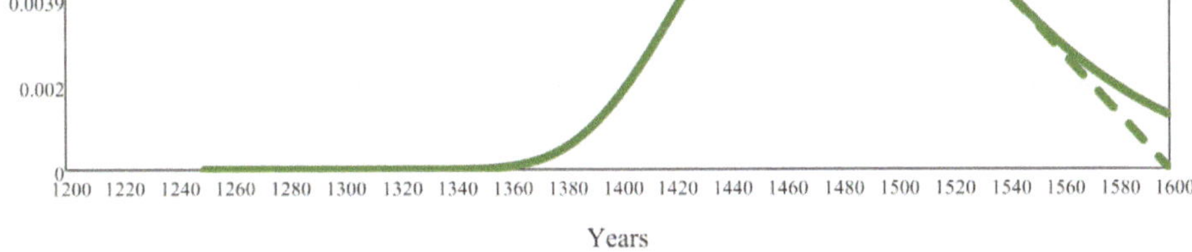

As a mathematical consequence of this *b*-lognormal, we may also derive the adolescence time for the Italian Renaissance civilization. This turns out to be 1423. This was the time of Filippo Brunelleschi (1377–April 15, 1446), one of the foremost architects and engineers of the Italian Renaissance. He is perhaps most famous for inventing linear perspective and designing the dome of Florence Cathedral, but his accomplishments also include bronze artwork, architecture (churches and chapels, fortifications, a hospital, etc.), mathematics, engineering (hydraulic machinery, clockwork mechanisms, theatrical machinery, etc.), and even ship design:

$$\text{sdaI} := e^{-\frac{\text{sd}\sigma\text{I}\cdot\sqrt{\text{sd}\sigma\text{I}^2+4}}{2}-\frac{3\cdot\text{sd}\sigma\text{I}^2}{2}+\text{sd}\mu\text{I}} + \text{bI} \qquad \text{sdaI} = 1.423\times 10^3$$

The Portuguese Empire (1419–1999) as a finite *b*-lognormal

bP := 1419	sP := 1822	dP := 1999
Birth Portuguese ships discover the Atlantic island of Madeira. Portuguese sea expansion starts.	*Senility* Brazil, the most important and richest of all Portuguese colonies, proclaims its independence from Portugal as the Empire of Brazil (September 7, 1822) under Emperor Dom Pedro I.	*Death* Macau, the last Portuguese colony, is handed over to China on December 20, 1999. The first Portuguese ships had arrived in Macau in 1535.

Deriving the values of the b-lognormal's μ and σ from just the above three blocks of data: b, s, and d

Note that we will prefix all equations by "sd" to make it clear that all data in this appendix are derived from *s* and *d* only:

sdsP := 1822   sddP := 1999

The two relevant solving equations are %o72 and %o64 of the Maxima file "Printing 2 b-lognormal inflexion points and Death time" in Chapter 6. These two equations are, respectively, listed hereafter:

$$\mu = \log(s-b) + \frac{2s^2 + (-3d - b)s + d^2 + bd}{(d-b)s - bd + b^2}$$

$$\mathrm{sd\mu P} := \ln(\mathrm{sdsP} - \mathrm{bP}) + \frac{2 \cdot \mathrm{sP}^2 + (-3 \cdot \mathrm{sddP} - \mathrm{bP}) \cdot \mathrm{sP} + \mathrm{sddP}^2 + \mathrm{bP} \cdot \mathrm{sddP}}{(\mathrm{sddP} - \mathrm{bP}) \cdot \mathrm{sP} - \mathrm{bP} \cdot \mathrm{sddP} + \mathrm{bP}^2}$$

The value of μ for the *b*-lognormal of the Portuguese Empire as derived from the two values of senility and death:

$\mathrm{sd\mu P} = 5.828$

$$\sigma = -\frac{s-d}{\sqrt{d-b}\sqrt{s-b}}$$

$$\mathrm{sd\sigma P} := \frac{\mathrm{sddP} - \mathrm{sP}}{\sqrt{\mathrm{sddP} - \mathrm{bP}} \cdot \sqrt{\mathrm{sP} - \mathrm{bP}}}$$

The values of σ for the *b*-lognormal of the Portuguese Empire as derived from the two values of senility and death:

$\mathrm{sd\sigma P} = 0.366$

The abscissa of the peak of this *b*-lognormal shows the peak to be 1716. This does not seem to be a time of special excellence for the Portuguese Empire except in a negative sense: the slave trade from Africa (especially Angola) to Brazil by Portuguese ships was at its peak, carrying an estimated 30,000 black slaves every year.

$\mathrm{sdpP} := e^{\mathrm{sd\mu P} - \mathrm{sd\sigma P}^2} + \mathrm{bP}$ $\mathrm{sdpP} = 1.716 \times 10^3$

(abscissa of the Portuguese peak).

Let us define the equation of the b-lognormal having the above μ and σ. This b-lognormal is the solid line in the graph below:

$$\text{sdblognormalP}(t, \text{sd}\mu\text{P}, \text{sd}\sigma\text{P}, \text{bP}) := \frac{1}{\sqrt{2\cdot\pi}\cdot\text{sd}\sigma\text{P}\cdot(t-\text{bP})}\cdot e^{-\frac{(\ln(t-\text{bP})-\text{sd}\mu\text{P})^2}{2\cdot\text{sd}\sigma\text{P}^2}}$$

$$\text{sdblognormalP}(\text{sdpP}, \text{sd}\mu\text{P}, \text{sd}\sigma\text{P}, \text{bP}) = 3.431\times 10^{-3}$$

(ordinate of the Portuguese peak). This is smaller than the Italian Renaissance peak (i.e., less civilized when you take slavery into account).

Defining the straight line tangent to senility and intercepting the time axis at death

Let us define the equation of the straight line tangent to the above b-lognormal at its descending inflexion point $t = s$. Since this equation hardly fits within the page width, we are forced to introduce two constants K and L first:

$$\text{sdKP} := e^{\frac{\text{sd}\sigma\text{P}\cdot\sqrt{\text{sd}\sigma\text{P}^2+4}+\text{sd}\sigma\text{P}^2-4\cdot\text{sd}\mu\text{P}-2}{4}} \qquad \text{sdLP} := e^{-\frac{\text{sd}\sigma\text{P}\cdot\sqrt{\text{sd}\sigma\text{P}^2+4}+8\cdot\text{sd}\mu\text{P}+2}{4}}$$

We can now define the equation of the straight line tangent to the b-lognormal at $t = s$. Its intercept with the time axis marks the *death* time of the Portuguese Empire in 1999 (when Portugal handed Macau back to China). This straight line is represented by the dashed straight line in the graph below:

$$\text{sdyP}(t) := \frac{\text{sdKP}}{\sqrt{2\cdot\pi}\cdot\text{sd}\sigma\text{P}} - \frac{\left(\sqrt{2}\cdot\sqrt{\text{sd}\sigma\text{P}^2+4}\,e^{\frac{7\cdot\text{sd}\sigma\text{P}^2}{4}} - \sqrt{2}\cdot\text{sd}\sigma\text{P}\cdot e^{\frac{7\cdot\text{sd}\sigma\text{P}^2}{4}}\right)\cdot\text{sdLP}\cdot\left(t - e^{\frac{\text{sd}\sigma\text{P}\cdot\sqrt{\text{sd}\sigma\text{P}^2+4}}{2}-\frac{3\text{sd}\sigma\text{P}^2}{2}+\text{sd}\mu\text{P}} - \text{bP}\right)}{4\cdot\sqrt{\pi}\cdot\text{sd}\sigma\text{P}^2}$$

FINITE b-lognormal of the PORTUGUESE CIVILIZATION (1419 - 1999).

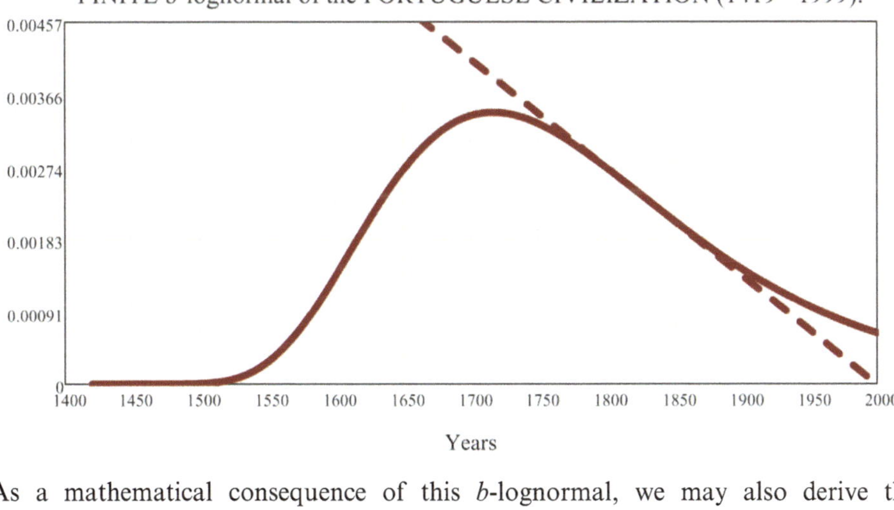

As a mathematical consequence of this b-lognormal, we may also derive the adolescence time for the Portuguese civilization. This turns out to be 1610:

$$\text{sdaP} := e^{-\frac{\text{sd}\sigma\text{P}\cdot\sqrt{\text{sd}\sigma\text{P}^2+4}}{2}-\frac{3\cdot\text{sd}\sigma\text{P}^2}{2}+\text{sd}\mu\text{P}} + \text{bP} \qquad \text{sdaP} = 1.61\times 10^3$$

The Spanish Empire (1492–1898) as a finite *b*-lognormal

bS := 1492 sS := 1805 dS := 1898

Birth
Christopher Columbus discovers America on October 12.

Senility (begins in 1805)
The Spanish fleet (as well as the French fleet) is destroyed by the British at the Battle of Trafalgar, October 21, 1805.

Death
The United States conquer the last Spanish colonies (Cuba, Puerto Rico, and the Philippines), thus ending the Spanish Colonial Empire forever.

Deriving the values of the b-lognormal's μ and σ from just the above three blocks of data: b, s, and d

Note that we will prefix all equations by "sd" to make it clear that all data in this appendix are derived from *s* and *d* only:

sdsS := 1805 sddS := 1898

The two relevant solving equations are %o72 and %o64 of the Maxima file "Printing 2 b-lognormal inflexion points and Death time" in Chapter 6. These two equations are, respectively, listed hereafter:

$$\mu = \log(s-b) + \frac{2s^2 + (-3d - b)s + d^2 + bd}{(d-b)s - bd + b^2}$$

$$\mathrm{sd\mu S} := \ln(\mathrm{sdsS} - \mathrm{bS}) + \frac{2 \cdot \mathrm{sS}^2 + (-3 \cdot \mathrm{sddS} - \mathrm{bS}) \cdot \mathrm{sS} + \mathrm{sddS}^2 + \mathrm{bS} \cdot \mathrm{sddS}}{(\mathrm{sddS} - \mathrm{bS}) \cdot \mathrm{sS} - \mathrm{bS} \cdot \mathrm{sddS} + \mathrm{bS}^2}$$

The value of μ for the *b*-lognormal of the Spanish Empire as derived from the two values of senility and death:

$\mathrm{sd\mu S} = 6.747$

$$\sigma = -\frac{s-d}{\sqrt{d-b}\sqrt{s-b}}$$

$$\mathrm{sd\sigma S} := \frac{\mathrm{sddS} - \mathrm{sS}}{\sqrt{\mathrm{sddS} - \mathrm{bS}} \cdot \sqrt{\mathrm{sS} - \mathrm{bS}}}$$

The value of σ for the *b*-lognormal of the Spanish Empire as derived from the two values of senility and death:

$\mathrm{sd\sigma S} = 0.219$

The abscissa of the peak of this *b*-lognormal shows the peak to be 1741. If one regards the Spanish conquest of the Indians of California (*Alta California* in Spanish) as the most enduring legacy of the Spanish Empire, the years around 1741 were indeed the peak:

$\mathrm{sdpS} := e^{\mathrm{sd\mu S} - \mathrm{sd\sigma S}^2} + \mathrm{bS} \qquad \mathrm{sdpS} = 1.741 \times 10^3$

(abscissa of the peak of the Spanish Empire—1741).

Let us define the equation of the b-lognormal having the above μ and σ. This b-lognormal is the solid line in the graph below:

$$\mathrm{sdblognormalS}(t, \mathrm{sd\mu S}, \mathrm{sd\sigma S}, \mathrm{bS}) := \frac{1}{\sqrt{2 \cdot \pi} \cdot \mathrm{sd\sigma S} \cdot (t - \mathrm{bS})} \cdot e^{-\frac{(\ln(t-\mathrm{bS}) - \mathrm{sd\mu S})^2}{2 \cdot \mathrm{sd\sigma S}^2}}$$

$$\mathrm{sdblognormalS}(\mathrm{sdpS}, \mathrm{sd\mu S}, \mathrm{sd\sigma S}, \mathrm{bS}) = 5.938 \times 10^{-3}$$

(ordinate of the peak of the Spanish Empire). This is higher than the Portuguese ordinate.

Defining the straight line tangent to senility and intercepting the time axis at death

Let us define the equation of the straight line tangent to the above b-lognormal at its descending inflexion point $t = s$. Since this equation hardly fits within the page width, we are forced to introduce two constants K and L first:

$$\mathrm{sdKS} := e^{\frac{\mathrm{sd\sigma S} \cdot \sqrt{\mathrm{sd\sigma S}^2 + 4} + \mathrm{sd\sigma S}^2 - 4 \cdot \mathrm{sd\mu S} - 2}{4}} \qquad \mathrm{sdLS} := e^{-\frac{\mathrm{sd\sigma S} \cdot \sqrt{\mathrm{sd\sigma S}^2 + 4} + 8 \cdot \mathrm{sd\mu S} + 2}{4}}$$

We can now define the equation of the straight line tangent to the b-lognormal at $t = s$. Its intercept with the time axis marks the *death* time of the Spanish Empire in 1898. This straight line is represented by the dashed straight line in the graph below:

$$\mathrm{sdyS}(t) := \frac{\mathrm{sdKS}}{\sqrt{2 \cdot \pi} \cdot \mathrm{sd\sigma S}} - \frac{\left(\sqrt{2} \cdot \sqrt{\mathrm{sd\sigma S}^2 + 4}\, e^{\frac{7 \cdot \mathrm{sd\sigma S}^2}{4}} - \sqrt{2} \cdot \mathrm{sd\sigma S} \cdot e^{\frac{7 \cdot \mathrm{sd\sigma S}^2}{4}}\right) \cdot \mathrm{sdLS} \cdot \left(t - e^{\frac{\mathrm{sd\sigma S} \cdot \sqrt{\mathrm{sd\sigma S}^2 + 4}}{2} - \frac{3 \mathrm{sd\sigma S}^2}{2} + \mathrm{sd\mu S}} - \mathrm{bS}\right)}{4 \cdot \sqrt{\pi} \cdot \mathrm{sd\sigma S}^2}$$

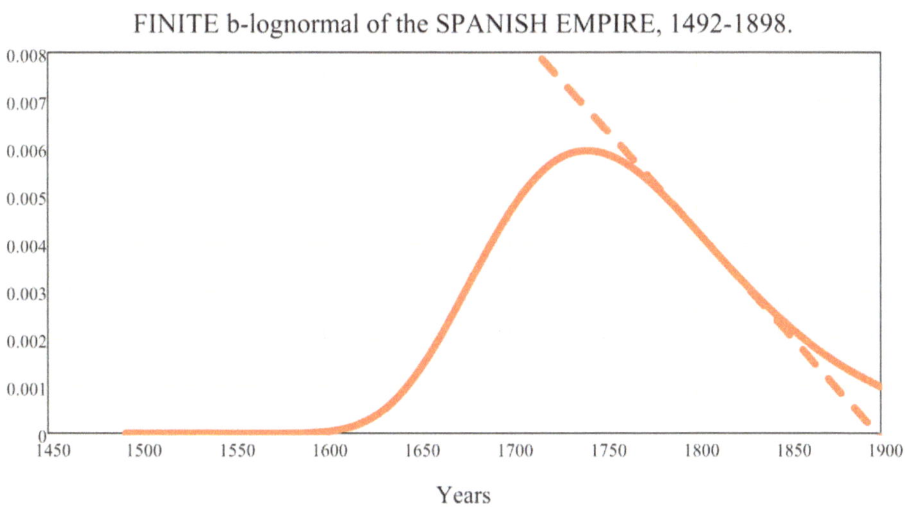

As a mathematical consequence of this b-lognormal, we may also derive the adolescence time for the Spanish Empire. This turns out to equal 1677, a time when Spain had virtually conquered the whole of South America (except for Brazil) and the Philippines, but not Alta California yet. This happened between 1759 and 1776, when San Francisco was founded (i.e., shortly after the peak of the above b-lognormals in 1741):

$$\mathrm{sdaS} := e^{-\frac{\mathrm{sd\sigma S} \cdot \sqrt{\mathrm{sd\sigma S}^2 + 4}}{2} - \frac{3 \cdot \mathrm{sd\sigma S}^2}{2} + \mathrm{sd\mu S}} + \mathrm{bS} \qquad \mathrm{sdaS} = -1.677 \times 10^3$$

The French Empire (1524–1962) as a finite *b*-lognormal

bF := 1524	sF := 1815	dF := 1962
Birth Francis I starts French expansion across the Atlantic at Verrazano in New York Bay.	*Senility* (begins in 1815) With Napoleon's defeat by the British and Prussians at Waterloo, France is forced back to a Bourbon monarchy.	*Death* The French colonial army in Algeria is defeated. Today, smaller colonies in the Atlantic and Pacific are regarded as part of Metropolitan France.

Deriving the values of the b-lognormal's μ and σ from just the above three blocks of data: b, s, and d

Note that we will prefix all equations by "sd" to make it clear that all data in this appendix are derived from *s* and *d* only:

sdsF := 1815 sddF := 1962

The two relevant solving equations are %o72 and %o64 of the Maxima file "Printing 2 b-lognormal inflexion points and Death time" in Chapter 6. These two equations are, respectively, listed hereafter:

$$\mu = \log(s - b) + \frac{2s^2 + (-3d - b)s + d^2 + bd}{(d - b)s - bd + b^2}$$

$$\text{sd}\mu\text{F} := \ln(\text{sdsF} - \text{bF}) + \frac{2 \cdot \text{sF}^2 + (-3 \cdot \text{sddF} - \text{bF}) \cdot \text{sF} + \text{sddF}^2 + \text{bF} \cdot \text{sddF}}{(\text{sddF} - \text{bF}) \cdot \text{sF} - \text{bF} \cdot \text{sddF} + \text{bF}^2}$$

The value of μ for the *b*-lognormal of the French Empire as derived from the two values of senility and death:

$\text{sd}\mu\text{F} = 5.507$

$$\sigma = -\frac{s - d}{\sqrt{d - b}\sqrt{s - b}}$$

$$\text{sd}\sigma\text{F} := \frac{\text{sddF} - \text{sF}}{\sqrt{\text{sddF} - \text{bF} \cdot \sqrt{\text{sF} - \text{bF}}}}$$

The values of σ for the French Empire as derived from the two values of senility and death:

$\text{sd}\sigma\text{F} = 0.219$

As a mathematical consequence of this *b*-lognormal, we may also derive the peak time for the French Colonial Empire. This turns out to be 1732, a time when France had virtually conquered the lands around the St. Lawrence River in Canada and around the Mississippi in the current United States. In addition, France had conquered much of central and southern India under Dupleix. But it soon lost everything to the British: Dupleix was forced by the French government (seeking peace) to leave India in 1754, and in Canada the last French defending commander Louis de Montcalm was defeated and died in 1763, obliging France to accept British rule in Canada.

$\text{sdpF} := e^{\text{sd}\mu\text{F} - \text{sd}\sigma\text{F}^2} + \text{bF}$ $\text{sdpF} = 1.732 \times 10^3$

(abscissa of the peak of the French Empire—1732).

Let us define the equation of the b-lognormal having the above μ and σ. This b-lognormal is the solid line in the graph below.

$$\mathrm{sdblognormalF}(t, \mathrm{sd\mu F}, \mathrm{sd\sigma F}, \mathrm{bF}) := \frac{1}{\sqrt{2\cdot\pi}\cdot \mathrm{sd\sigma F}\cdot(t-\mathrm{bF})}\cdot e^{-\frac{(\ln(t-\mathrm{bF})-\mathrm{sd\mu F})^2}{2\cdot \mathrm{sd\sigma F}^2}}$$

$$\mathrm{sdblognormalF}(\mathrm{sdpF}, \mathrm{sd\mu F}, \mathrm{sd\sigma F}, \mathrm{bF}) = 4.279\times 10^{-3}$$

(ordinate of the peak of the French Empire).

Defining the straight line tangent to senility and intercepting the time axis at death

Let us define the equation of the straight line tangent to the above b-lognormal at its descending inflexion point $t = s$. Since this equation hardly fits within the page width, we are forced to introduce two constants K and L first:

$$\mathrm{sdKF} := e^{\frac{\mathrm{sd\sigma F}\cdot\sqrt{\mathrm{sd\sigma F}^2+4}+\mathrm{sd\sigma F}^2-4\cdot \mathrm{sd\mu F}-2}{4}} \qquad \mathrm{sdLF} := e^{-\frac{\mathrm{sd\sigma F}\cdot\sqrt{\mathrm{sd\sigma F}^2+4}+8\cdot \mathrm{sd\mu F}+2}{4}}$$

We can now define the equation of the straight line tangent to the b-lognormal at $t = s$. Its intercept with the time axis marks the *death* time of the French Colonial Empire in 1962. This straight line is represented by the dashed straight line in the graph below:

$$\mathrm{sdyF}(t) := \frac{\mathrm{sdKF}}{\sqrt{2\cdot\pi}\cdot \mathrm{sd\sigma F}} - \frac{\left(\sqrt{2}\cdot\sqrt{\mathrm{sd\sigma F}^2+4}\,e^{\frac{7\cdot \mathrm{sd\sigma F}^2}{4}} - \sqrt{2}\cdot \mathrm{sd\sigma F}\cdot e^{\frac{7\cdot \mathrm{sd\sigma F}^2}{4}}\right)\cdot \mathrm{sdLF}\cdot\left(t - e^{\frac{\mathrm{sd\sigma F}\cdot\sqrt{\mathrm{sd\sigma F}^2+4}}{2}-\frac{3\mathrm{sd\sigma F}^2}{2}+\mathrm{sd\mu F}} - \mathrm{bF}\right)}{4\cdot\sqrt{\pi}\cdot \mathrm{sd\sigma F}^2}$$

FINITE b-lognormal of the FRENCH COLONIAL EMPIRE, 1525-1962.

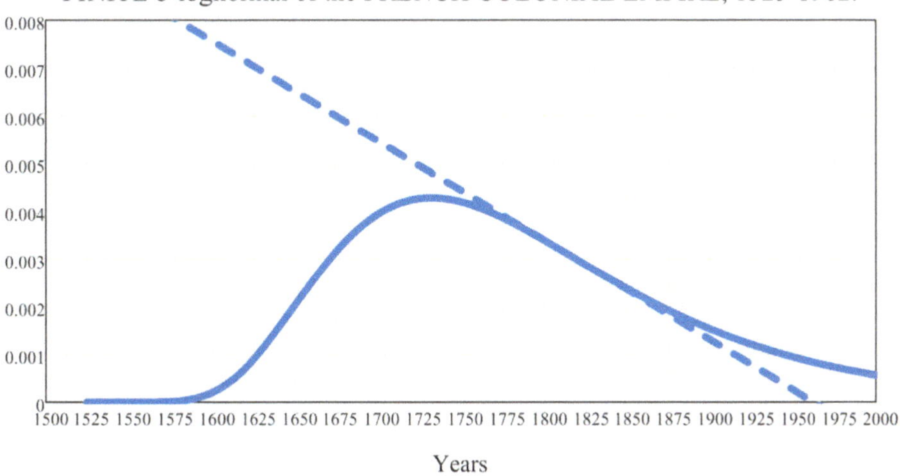

This b-lognormal also lets us compute the adolescence time of the French Colonial Empire. This is 1650, a time when the colonies of France were indeed growing under the guidance of Louis XIV:

$$\mathrm{sdaF} := e^{-\frac{\mathrm{sd\sigma F}\cdot\sqrt{\mathrm{sd\sigma F}^2+4}}{2}-\frac{3\cdot \mathrm{sd\sigma F}^2}{2}+\mathrm{sd\mu F}} + \mathrm{bF} \qquad \mathrm{sdaF} = 1.65\times 10^3$$

The British Empire (1588–1973) as a finite *b*-lognormal

bB := 1588	sB := 1914	dB := 1973
Birth The English fleet of Queen Elizabeth I defeats the Spanish Armada on August 8.	*Senility* Starts in 1914 with the outbreak of World War I. Britain wins the war but at a high cost. So also in World War II, with the loss of the empire.	*Death* Edward Heath, Prime Minister, brings Britain into the European Community, seeking for Britain "a European future after an imperial past".

Deriving the values of the b-lognormal's μ and σ from just the above three blocks of data: b, s, and d

Note that we will prefix all equations by "sd" to make it clear that all data in this appendix are derived from *s* and *d* only:

sdsB := 1914 sddB := 1973

The two relevant solving equations are %o72 and %o64 of the Maxima file "Printing 2 b-lognormal inflexion points and Death time" in Chapter 6. These two equations are, respectively, listed hereafter:

$$\mu = \log(s-b) + \frac{2s^2 + (-3d-b)s + d^2 + bd}{(d-b)s - bd + b^2}$$

$$\mathrm{sd\mu B} := \ln(\mathrm{sdsB} - \mathrm{bB}) + \frac{2 \cdot \mathrm{sB}^2 + (-3 \cdot \mathrm{sddB} - \mathrm{bB}) \cdot \mathrm{sB} + \mathrm{sddB}^2 + \mathrm{bB} \cdot \mathrm{sddB}}{(\mathrm{sddB} - \mathrm{bB}) \cdot \mathrm{sB} - \mathrm{bB} \cdot \mathrm{sddB} + \mathrm{bB}^2}$$

The value of μ for the *b*-lognormal of the British Empire as derived from the two values of senility and death:

$\mathrm{sd\mu B} = 5.661$

$$\sigma = -\frac{s-d}{\sqrt{d-b}\sqrt{s-b}}$$

$$\mathrm{sd\sigma B} := \frac{\mathrm{sddB} - \mathrm{sB}}{\sqrt{\mathrm{sddB} - \mathrm{bB}} \cdot \sqrt{\mathrm{sB} - \mathrm{bB}}}$$

The value of σ for the British Empire as derived from the two values of senility and death:

$\mathrm{sd\sigma B} = 0.167$

The abscissa of the peak of this *b*-lognormal shows the peak to be 1868. At the peak of the Victorian age, James Clerk Maxwell gives the world his Maxwell equations (i.e., the domain of electricity). In 1859 Charles Darwin shakes all religious beliefs by proving that humans come from monkeys. In 1898 Joseph John Thompson discovers the electron, beginning electronics. In 1911 Ernest Rutherford proves that the atom is a "miniature planetary system" accounting for the Mendeleev Table and all of chemistry. Britain is indeed "the best country" for science ... :

$\mathrm{sdpB} := e^{\mathrm{sd\mu B} - \mathrm{sd\sigma B}^2} + \mathrm{bB}$ $\mathrm{sdpB} = 1.868 \times 10^3$ $\mathrm{p}_2 := \mathrm{sdpB}$ $\mathrm{p}_2 = 1.868 \times 10^3$

(abscissa of the peak of the British Empire, ready for later use in the *enveloping British exponential* E_British(t)).

Let us define the equation of the *b*-lognormal having the above μ and σ. This *b*-lognormal is represented by the solid line in the graph below.

$$\mathrm{sdblognormalB}(t, \mathrm{sd\mu B}, \mathrm{sd\sigma B}, \mathrm{bB}) := \frac{1}{\sqrt{2\cdot\pi}\cdot \mathrm{sd\sigma B}\cdot(t-\mathrm{bB})}\cdot e^{-\frac{(\ln(t-\mathrm{bB})-\mathrm{sd\mu B})^2}{2\cdot \mathrm{sd\sigma B}^2}}$$

$$\mathrm{sdblognormalB}(\mathrm{sdpB}, \mathrm{sd\mu B}, \mathrm{sd\sigma B}, \mathrm{bB}) = 8.447\times 10^{-3}$$

(ordinate of the peak of the British Empire). This is much *higher* than the *b*-lognormal for any other European country. This seems to justify the British saying "British is best". Indeed, the international language of the world today is English:

$$P_2 := \mathrm{sdblognormalB}(p_2, \mathrm{sd\mu B}, \mathrm{sd\sigma B}, \mathrm{bB}) \qquad P_2 = 8.447\times 10^{-3}$$

(ordinate of the peak of the British Empire, ready for later use in the *enveloping British exponential* E_British(t)).

Defining the straight line tangent to senility and intercepting the time axis at death

Let us define the equation of the straight line tangent to the above *b*-lognormal at its descending inflexion point $t = s$. Since this equation hardly fits within the page width, we are forced to introduce two constants K and L first:

$$\mathrm{sdKB} := e^{\frac{\mathrm{sd\sigma B}\cdot\sqrt{\mathrm{sd\sigma B}^2+4}+\mathrm{sd\sigma B}^2-4\cdot \mathrm{sd\mu B}-2}{4}} \qquad \mathrm{sdLB} := e^{-\frac{\mathrm{sd\sigma B}\cdot\sqrt{\mathrm{sd\sigma B}^2+4}+8\cdot \mathrm{sd\mu B}+2}{4}}$$

We can now define the equation of the straight line tangent to the *b*-lognormal at $t = s$. Its intercept with the time axis marks the *death* time of the British Empire in 1973. This straight line is represented by the dashed straight line in the graph below:

$$\mathrm{sdyB}(t) := \frac{\mathrm{sdKB}}{\sqrt{2\cdot\pi}\cdot \mathrm{sd\sigma B}} - \frac{\left(\sqrt{2}\cdot\sqrt{\mathrm{sd\sigma B}^2+4}\,e^{\frac{7\cdot \mathrm{sd\sigma B}^2}{4}} - \sqrt{2}\cdot \mathrm{sd\sigma B}\cdot e^{\frac{7\cdot \mathrm{sd\sigma B}^2}{4}}\right)\cdot \mathrm{sdLB}\cdot\left(t - e^{\frac{\mathrm{sd\sigma B}\cdot\sqrt{\mathrm{sd\sigma B}^2+4}}{2}-\frac{3\mathrm{sd\sigma B}^2}{2}+\mathrm{sd\mu B}} - \mathrm{bB}\right)}{4\cdot\sqrt{\pi}\cdot \mathrm{sd\sigma B}^2}$$

FINITE b-lognormal of the BRITISH EMPIRE, 1588-1973.

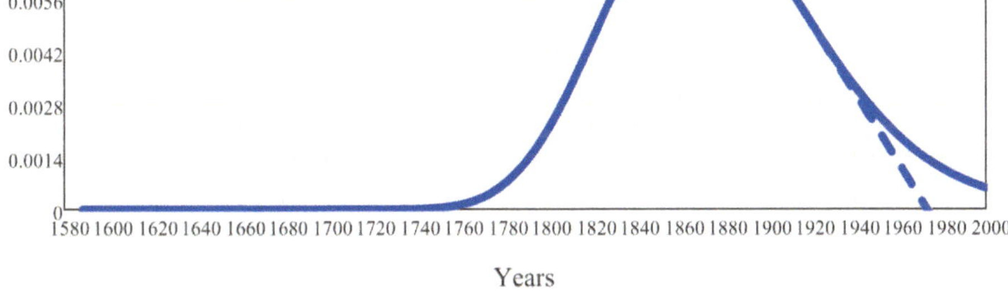

As a mathematical consequence of this *b*-lognormal, we may also derive the adolescence time for the British Empire. It turns out to be 1821:

$$\mathrm{sdaB} := e^{-\frac{\mathrm{sd\sigma B}\cdot\sqrt{\mathrm{sd\sigma B}^2+4}}{2}-\frac{3\cdot \mathrm{sd\sigma B}^2}{2}+\mathrm{sd\mu B}} + \mathrm{bB} \qquad \mathrm{sdaB} = 1.821\times 10^3$$

The American Empire (1898–?2050) as a finite *b*-lognormal

bA := 1898	sA := 2001	dA := 2050
Birth The U.S.A. defeats Spain and gets Cuba, Puerto Rico, and the Philippines. Birth of American imperialism.	*Senility* 9/11 changes the course of history, starting the decline of the U.S.A. as the world's leader. The financial crisis of 2008 adds more troubles.	*Death* As of 2012 the U.S.A. is still the leading country in the world. But, it is widely believed that 2050 or thereabouts will mark the time the U.S.A. will be surpassed by China.

Deriving the values of the b-lognormal's μ and σ from just the above three blocks of data: b, s, and d

Note that we will prefix all equations by "sd" to make it clear that all data in this appendix are derived from *s* and *d* only:

sdsA := 2001 sddA := 2050

The two relevant solving equations are %o72 and %o64 of the Maxima file "Printing 2 b-lognormal inflexion points and Death time" in Chapter 6. These two equations are, respectively, listed hereafter:

$$\mu = \log(s-b) + \frac{2s^2 + (-3d-b)s + d^2 + bd}{(d-b)s - bd + b^2}$$

$$\mathrm{sd\mu A} := \ln(\mathrm{sdsA} - \mathrm{bA}) + \frac{2\cdot \mathrm{sA}^2 + (-3\cdot \mathrm{sddA} - \mathrm{bA})\cdot \mathrm{sA} + \mathrm{sddA}^2 + \mathrm{bA}\cdot \mathrm{sddA}}{(\mathrm{sddA} - \mathrm{bA})\cdot \mathrm{sA} - \mathrm{bA}\cdot \mathrm{sddA} + \mathrm{bA}^2}$$

The value of μ for the *b*-lognormal of the American Empire as derived from the two values of senility and death:

$\mathrm{sd\mu A} = 4.466$

$$\sigma = -\frac{s-d}{\sqrt{d-b}\sqrt{s-b}}$$

$$\mathrm{sd\sigma A} := \frac{\mathrm{sddA} - \mathrm{sA}}{\sqrt{\mathrm{sddA} - \mathrm{bA}}\cdot\sqrt{\mathrm{sA} - \mathrm{bA}}}$$

The value of σ for the *b*-lognormal of the American Empire as derived from the two values of senility and death:

$\mathrm{sd\sigma A} = 0.392$

The United States not only truly won both world wars, but after 1945 they acquired absolute world dominance in all fields of *science* and *technology*. In this they were also helped by speaking the same language as the previous world-leading empire, the British Empire. The American *b*-lognormal *peak* is 1973, and that is certainly true: they landed on the Moon, they had absolute dominance in computers and nuclear weapons, and overtook the other superpower of the day, the Soviet Union, in 1992. But, 9/11 proves that they are now (2012) losing ground. Will China replace them?

$\mathrm{sdpA} := e^{\mathrm{sd\mu A} - \mathrm{sd\sigma A}^2} + \mathrm{bA}$ $\quad \mathrm{sdpA} = 1.973\times 10^3$ $\quad p_3 := \mathrm{sdpA}$ $\quad p_3 = 1.973\times 10^3$

Let us define the equation of the *b*-lognormal having the above μ and σ. This *b*-lognormal is represented by the solid line in the graph below:

$$\mathrm{sdblognormalA}(t, \mathrm{sd\mu A}, \mathrm{sd\sigma A}, \mathrm{bA}) := \frac{1}{\sqrt{2\cdot\pi}\cdot \mathrm{sd\sigma A}\cdot(t-\mathrm{bA})}\cdot e^{-\frac{(\ln(t-\mathrm{bA})-\mathrm{sd\mu A})^2}{2\cdot \mathrm{sd\sigma A}^2}}$$

$$\mathrm{sdblognormalA}(\mathrm{sdpA}, \mathrm{sd\mu A}, \mathrm{sd\sigma A}, \mathrm{bA}) = 0.013$$

$$P_3 := \mathrm{sdblognormalA}(\mathrm{sdpA}, \mathrm{sd\mu A}, \mathrm{sd\sigma A}, \mathrm{bA}) \qquad P_3 = 0.013$$

(ordinate of the peak of the American Empire, ready for later use in the *enveloping American exponential* E_USA(t)).

Defining the straight line tangent to senility and intercepting the time axis at death

Let us define the equation of the straight line tangent to the above b-lognormal at its descending inflexion point $t = s$. Since this equation hardly fits within the page width, we are forced to introduce two constants K and L first:

$$\mathrm{sdKA} := e^{\frac{\mathrm{sd\sigma A}\cdot\sqrt{\mathrm{sd\sigma A}^2+4}+\mathrm{sd\sigma A}^2-4\cdot \mathrm{sd\mu A}-2}{4}} \qquad \mathrm{sdLA} := e^{-\frac{\mathrm{sd\sigma A}\cdot\sqrt{\mathrm{sd\sigma A}^2+4}+8\cdot \mathrm{sd\mu A}+2}{4}}$$

We can now define the equation of the straight line tangent to the b-lognormal at $t = s$. Its intercept with the time axis marks the *assumed death* time of the American Empire in 2050. This straight line is represented by the dashed straight line in the graph below:

$$\mathrm{sdyA}(t) := \frac{\mathrm{sdKA}}{\sqrt{2\cdot\pi}\cdot \mathrm{sd\sigma A}} - \frac{\left(\sqrt{2}\cdot\sqrt{\mathrm{sd\sigma A}^2+4}\,e^{\frac{7\cdot \mathrm{sd\sigma A}^2}{4}} - \sqrt{2}\cdot \mathrm{sd\sigma A}\cdot e^{\frac{7\cdot \mathrm{sd\sigma A}^2}{4}}\right)\cdot \mathrm{sdLA}\cdot\left(t - e^{\frac{\mathrm{sd\sigma A}\cdot\sqrt{\mathrm{sd\sigma A}^2+4}}{2}-\frac{3\mathrm{sd\sigma A}^2}{2}+\mathrm{sd\mu A}} - \mathrm{bA}\right)}{4\cdot\sqrt{\pi}\cdot \mathrm{sd\sigma A}^2}$$

FINITE b-lognormal of the AMERICAN EMPIRE, 1898 - 2050.

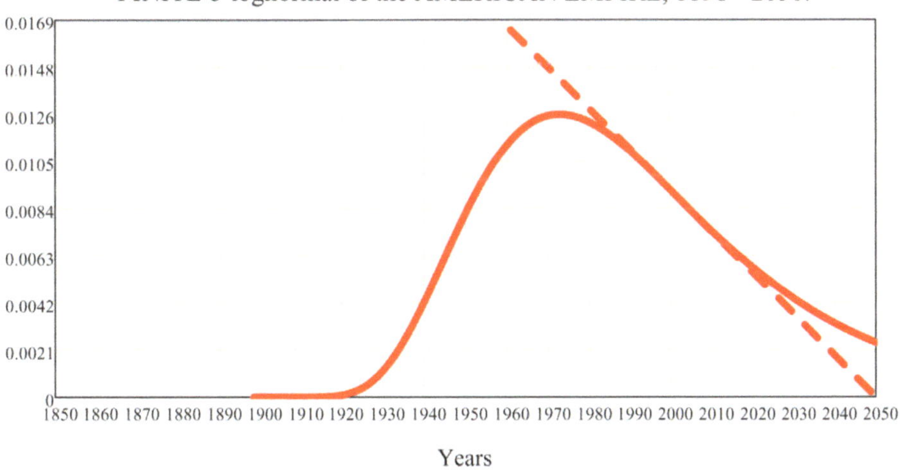

As a mathematical consequence of this b-lognormal, we may also derive the adolescence time for the American Empire. It turns out to be 1944: Americans were "adolescents" when they embarked on their own period of worldwide domination:

$$\mathrm{sdaA} := e^{-\frac{\mathrm{sd\sigma A}\cdot\sqrt{\mathrm{sd\sigma A}^2+4}}{2}-\frac{3\cdot \mathrm{sd\sigma A}^2}{2}+\mathrm{sd\mu A}} + \mathrm{bA} \qquad \mathrm{sdaA} = 1.944\times 10^3$$

All civilizations together as finite *b*-lognormals

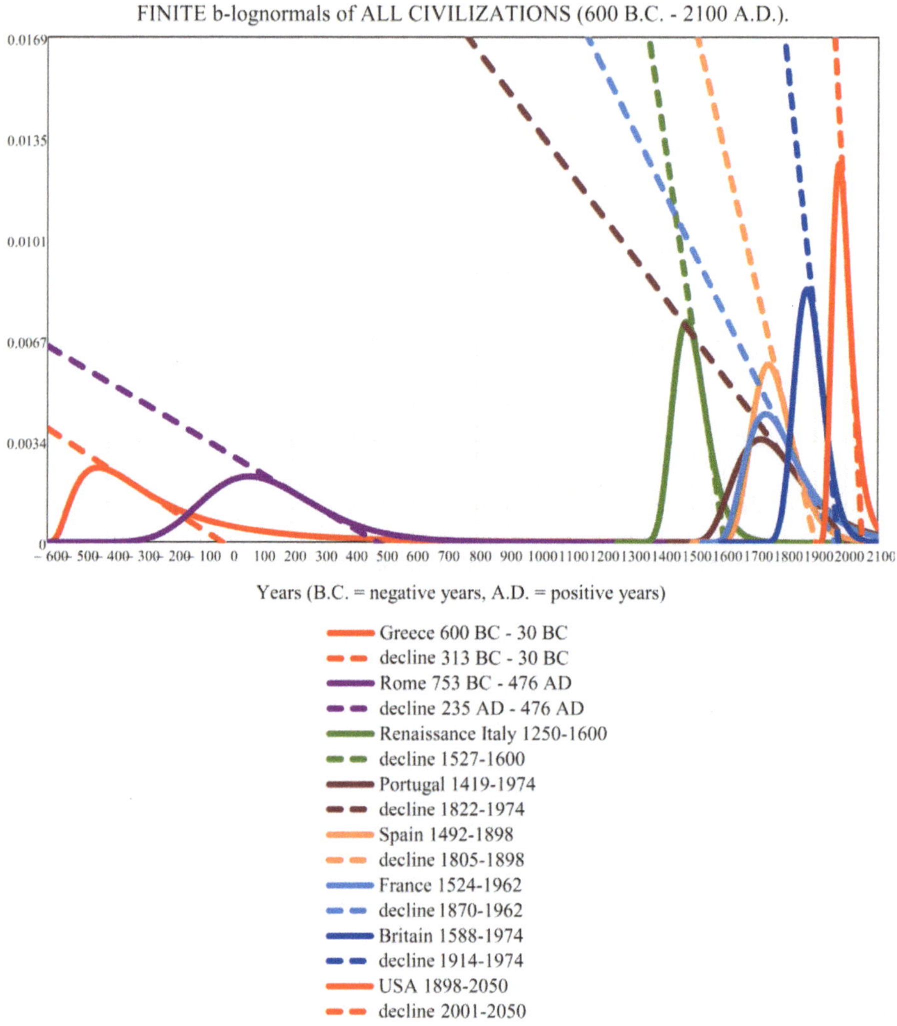

All modern Western European civilizations (Renaissance Italy to Britain) as finite *b*-lognormals

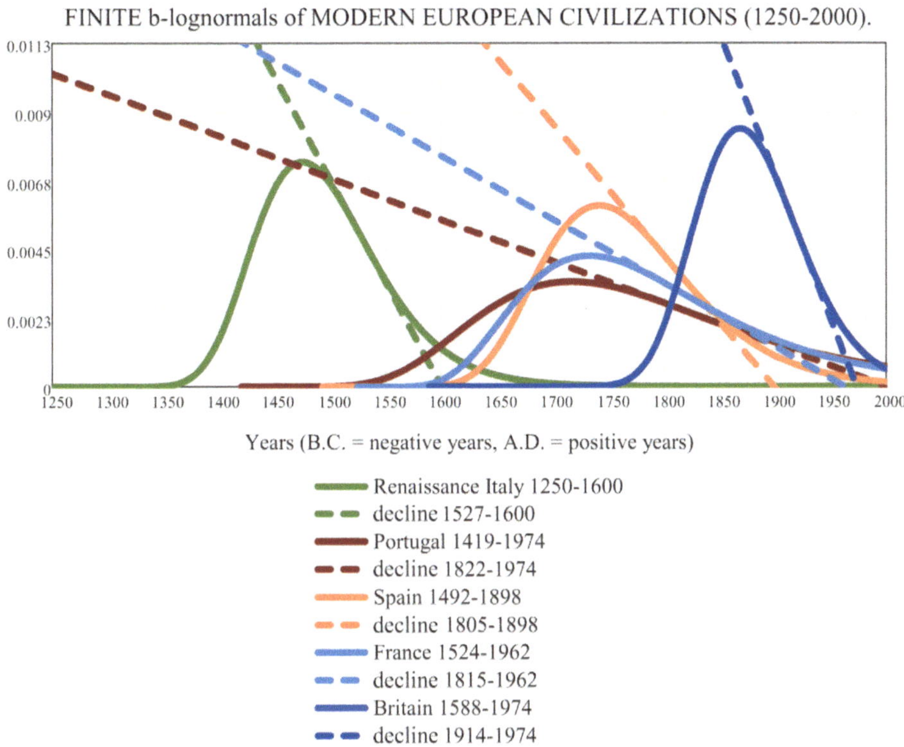

Defining the exponential envelope of all European civilizations together

Let (p_1, P_1) be the coordinates of the *peak* of the *Greek* civilization. We want our exponential envelope $A * e^{(B*t)}$ to pass through this point. To do so it must fulfill the equation $P_1 = A * e^{(B*p_1)}$.

Let (p_2, P_2) be the coordinates of the *peak* of the *British* civilization (the "highest" among all European civilizations). We want our exponential envelope $A * e^{(B*t)}$ to pass through this point. To do so it must fulfill the equation $P_2 = A * e^{(B*p_2)}$.

Then, upon solving the last two equations for *A* and *B*, one can prove that *A* and *B* are given by, respectively:

$$\text{A\_Britain} := \frac{P_1}{\left(\frac{P_2}{P_1}\right)^{\frac{p_1}{p_2-p_1}}} \quad \text{A\_Britain} = 3.134 \times 10^{-3} \quad \text{B\_Britain} := \ln\left[\left(\frac{P_2}{P_1}\right)^{\frac{1}{p_2-p_1}}\right] \quad \text{B\_Britain} = 5.309 \times 10^{-4}$$

Let us now define the *exponential* passing through the two peaks of the Greek and British civilizations, given by:

$$\text{E\_Britain}(t) := \text{A\_Britain} \cdot e^{\text{B\_Britain} \cdot t}$$

Is this exponential somehow similar to the *envelope* of all European *b*-lognormals? Look at the next plot.

Defining the exponential envelope of all European civilizations plus the United States

$$\text{A\_USA} := \frac{P_1}{\left(\frac{P_3}{P_1}\right)^{\frac{p_1}{p_3-p_1}}} \quad \text{A\_USA} = 3.337 \times 10^{-3} \quad \text{B\_USA} := \ln\left[\left(\frac{P_3}{P_1}\right)^{\frac{1}{p_3-p_1}}\right] \quad \text{B\_USA} = 6.754 \times 10^{-4}$$

$$\text{E\_USA}(t) := \text{A\_USA} \cdot e^{\text{B\_USA} \cdot t}$$

All Western European civilizations (Greece to Britain) as finite *b*-lognormals

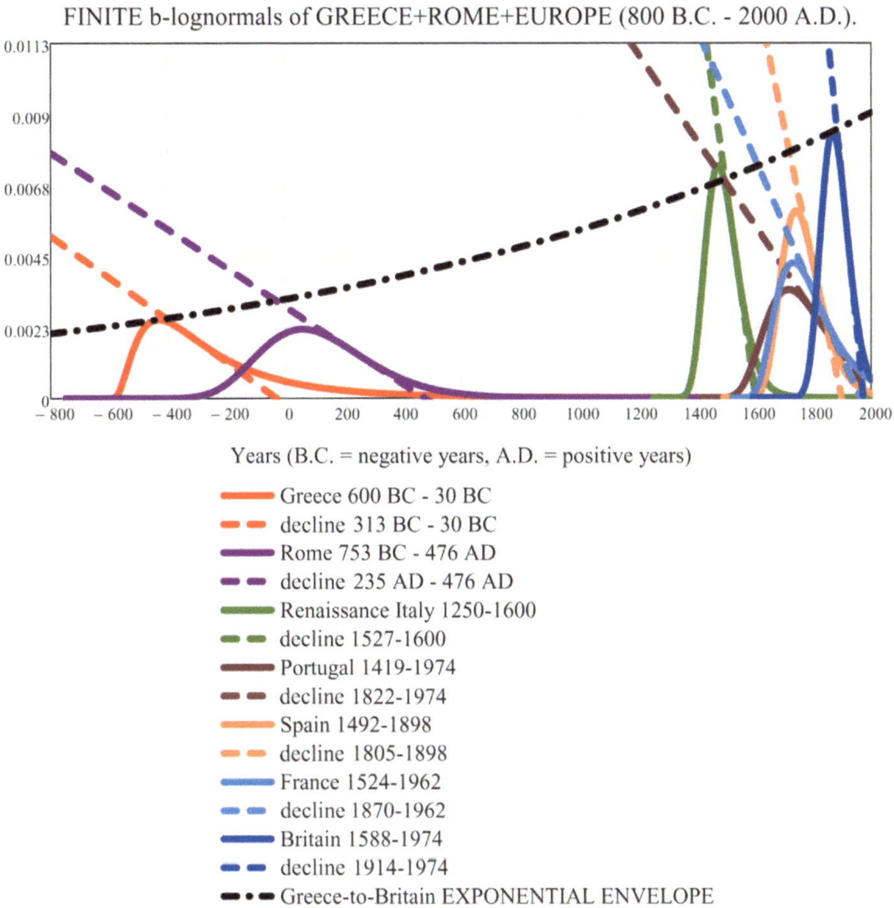

All modern Western European civilizations (Renaissance Italy to Britain) as finite *b*-lognormals

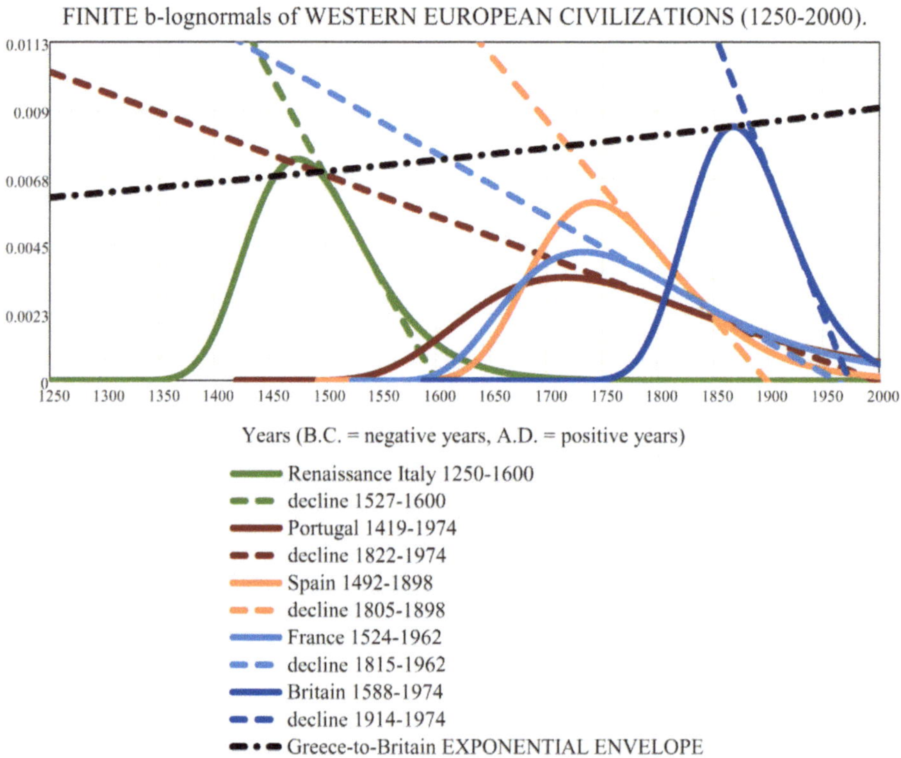

All Western civilizations (Greece to U.S.A.) as finite *b*-lognormals

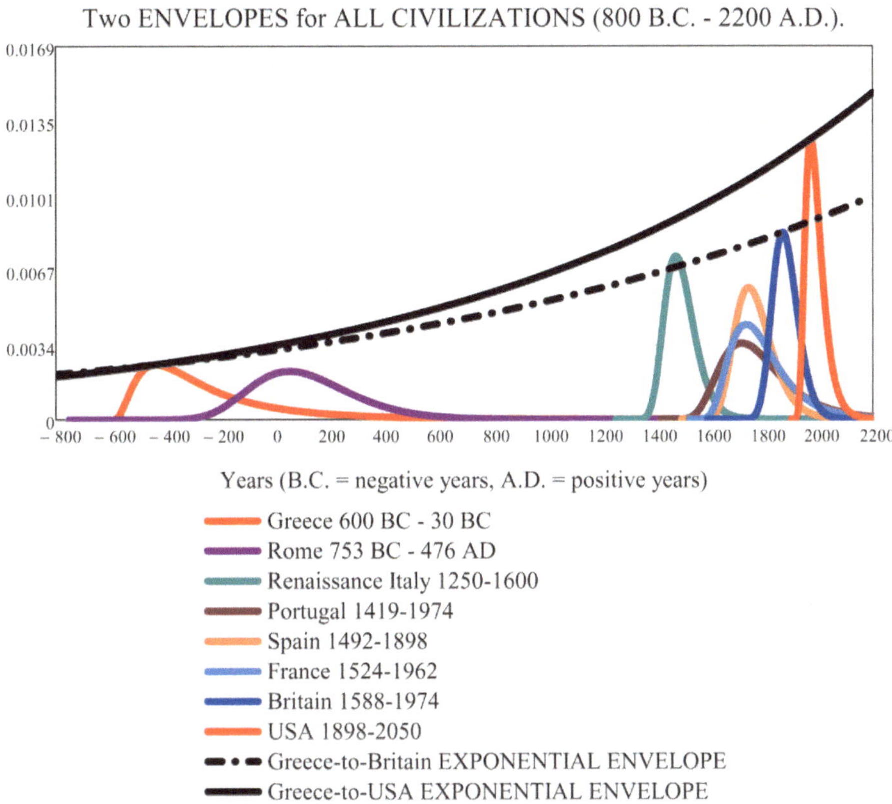

8

A mathematical model for evolution and SETI

8.1 INTRODUCTION: EVOLUTION OF LIFE ON EARTH AND THE STATISTICAL DRAKE EQUATION

Darwinian evolution theory may be regarded as a part of SETI theory in that the factor *fl* in the Drake equation represents the fraction of planets suitable for life on which life actually arises.

In this chapter we first recall our statistical generalization of the Drake equation where the factor *fl* is shown to follow the lognormal probability distribution. This lognormal distribution is a consequence of the Central Limit Theorem (CLT) of statistics, which states that the product of a number of independent random variables whose probability densities are unknown and independent of each other approaches the lognormal distribution when the number of factors increases to infinity.

In addition, we show that the exponential growth of the number of species typical of Darwinian evolution may be regarded as the geometric locus of the peaks of a one-parameter family of lognormal distributions (*b*-lognormals) constrained between the time axis and the exponential growth curve.

Finally, since each *b*-lognormal distribution in the family may in turn be regarded as the product of a large number (actually "an infinity") of independent lognormal probability distributions, the mathematical way is paved to further cast Darwinian evolution into a mathematical theory in agreement with both its typical exponential growth in the number of living species and the statistical Drake equation.

8.2 EVOLUTION AS EXPONENTIAL INCREASE IN THE NUMBER OF LIVING SPECIES

Let us now consider Darwinian evolution. To assume that the number of species increased exponentially over the 3.5-billion-year evolutionary time span is certainly a gross oversimplification of the real situation, as proven, for instance, by Rohde and Muller [1]. However, we will assume this exponential increase in the number of living species in time for the purpose of casting the theory into a mathematically simple and fruitful form. Later in the chapter we will improve on this.

In other words, we assume that 3.5 billion years ago there was only one living species on Earth, whereas now there may be (say) 50 million living species or more (see, e.g., *http://en.wikipedia.org/wiki/Species*). Note that the actual number of species currently living on Earth does not really matter as a number for us: we just want to stress the exponential character of the growth of species. Thus, we shall assume that the number of living species on Earth increases in time as $E(t)$ (standing for "exponential in time"):

$$E(t) = A\, e^{B\,t} \tag{8.1}$$

where A and B are two positive constants that we will soon determine numerically. Let us now adopt the convention that the current epoch corresponds to the origin of the time axis (i.e., to the instant $t = 0$). This means that all the past epochs of Darwinian evolution correspond to negative times, whereas the future ahead of us (including finding ETs) corresponds to positive times. Setting $t = 0$ in (8.1), we immediately find

$$E(0) = A \tag{8.2}$$

proving that the constant A equals the number of living species on Earth right now. We shall assume

$$A = 50 \text{ million species} = 5 \cdot 10^7 \text{ species.} \tag{8.3}$$

To determine the constant B numerically as well, consider the two values of the exponential (8.1) at two different instants t_1 and t_2, with $t_1 < t_2$; that is

$$\begin{cases} E(t_1) = A\, e^{B\,t_1} \\ E(t_2) = A\, e^{B\,t_2}. \end{cases} \tag{8.4}$$

Dividing the lower equation by the upper one, A disappears and we are left with an equation in B only:

$$\frac{E(t_2)}{E(t_1)} = e^{B(t_2 - t_1)}. \tag{8.5}$$

Solving this for B yields

$$B = \frac{\ln(E(t_2)) - \ln(E(t_1))}{t_2 - t_1}. \tag{8.6}$$

We may now impose the initial condition stating that 3.5 billion years ago there was just one species on Earth, the first one (whether this was RNA is unimportant in the present simple mathematical formulation):

$$\begin{cases} t_1 = -3.5 \cdot 10^9 \text{ years} \\ E(t_1) = 1 \quad \text{whence } \ln(E(t_1)) = \ln(1) = 0. \end{cases} \tag{8.7}$$

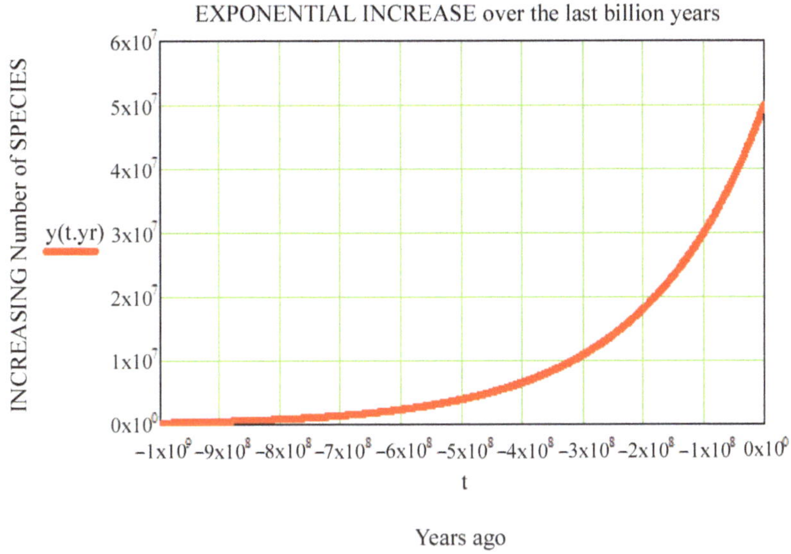

Figure 8.1. Darwinian exponential curve representing the growing number of species on Earth up to now.

The final condition is of course that today ($t_2 = 0$) the number of species equals A given by (8.3). Upon inserting both (8.3) and (8.7) into (8.6), the latter becomes:

$$B = -\frac{\ln(E(t_2))}{t_1} = -\frac{\ln(5 \cdot 10^7)}{-3.5 \cdot 10^9 \text{ year}} = \frac{1.605 \cdot 10^{-16}}{\text{second}}. \tag{8.8}$$

Having thus determined the numerical values of both A and B, the exponential in (8.1) is thus fully specified. This curve is plotted in Figure 8.1 over just the last billion years, rather than over the full range between -3.5 billion years and now.

8.3 INTRODUCING THE DARWIN (d) UNIT, MEASURING THE AMOUNT OF EVOLUTION THAT A GIVEN SPECIES REACHED

In all sciences "to measure is to understand". In physics and chemistry this is done by virtue of units such as the meter, second, kilogram, coulomb, etc. So, it appears useful to introduce a new unit measuring the degree of evolution that a certain species has reached at a certain time t of Darwinian evolution, and the obvious name for such a new unit is the "darwin", denoted by a lowercase "d". For instance, if we adopt the exponential evolution curve described in the previous section, we might say that the dominant species on Earth right now (humans) has reached an evolution level of 50 million darwins.

How many darwins might an alien civilization have already reached? Certainly more than 50 million (i.e., more than 50 Md), but we will not check this out until SETI has its first success.

We are not going to discuss further this notion of measuring the "amount of evolution" since we are aware that endless discussions will result. But it is clear to us that such a new measuring unit (and ways to measure it for different species) will sooner or later have to be introduced to make evolution a fully quantitative science.

8.4 DARWINIAN EXPONENTIAL AS THE ENVELOPE OF ALL *b*-LOGNORMALS, EACH REPRESENTING A DIFFERENT SPECIES TRIGGERED BY EVOLUTION AT THE TIME $t = b > 0$ (CLADISTICS)

How is it possible to match the Darwinian exponential curve with the lognormals appearing in the statistical Drake equation? Our answer to such a question is "by letting the Darwinian exponential become the *envelope* of the *b*-lognormals representing the cladistic branches (i.e., the new species that were produced by evolution at different times as evolution unfolded).

Let us now have a look at Figure 8.2.

The envelope shown in Figure 8.2 is *not* really an envelope in the strictly mathematical sense as explained in calculus textbooks. However, it is "nearly the same thing in practice" because it actually is the geometric *locus of the peaks* of all *b*-lognormals. We shall now explain this in detail.

First of all, let us write down the equation of the *b*-lognormal, which is the lognormal starting at any positive instant $t = b > 0$ (while ordinary lognormals all

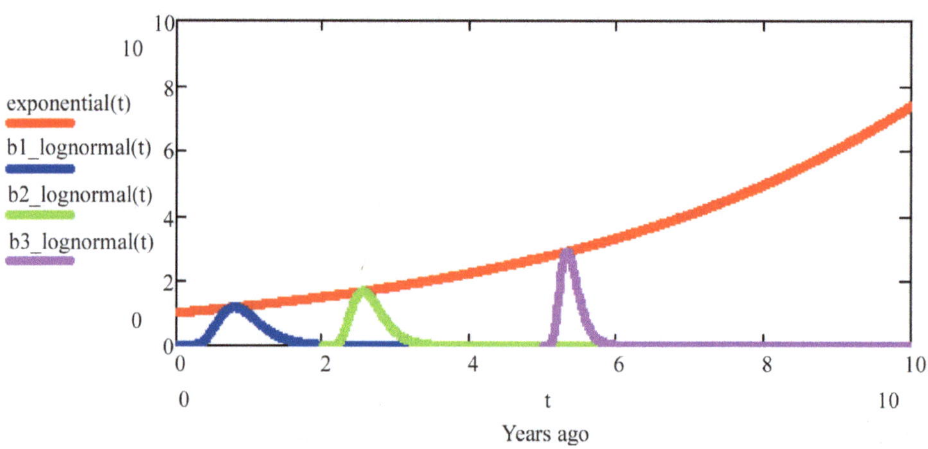

Figure 8.2. Darwinian exponential as the *envelope* of *b*-lognormals. Each *b*-lognormal is a lognormal starting at a time ($t = b =$ birth time) larger than zero and represents a different species "born" at time b of Darwinian evolution.

start just at zero):

$$\begin{cases} b\_\text{lognormal}(t, \mu, \sigma, b) = \dfrac{1}{\sqrt{2\pi}\sigma(t-b)} e^{-\frac{(\ln(t-b)-\mu)^2}{2\sigma^2}} \\ \text{holding for } t > b \text{ and up to } t = \infty. \end{cases} \tag{8.9}$$

Note that its *peak* falls at the abscissa p and ordinate P given, respectively (see the eighth and ninth line in Table 1.1), by:

$$\begin{cases} p = b + e^{\mu-\sigma^2} = b\_\text{lognormal}\_\text{peak}\_\text{abscissa}, \\ P = \dfrac{e^{\frac{\sigma^2}{2}-\mu}}{\sqrt{2\pi}\sigma} = b\_\text{lognormal}\_\text{peak}\_\text{ordinate}. \end{cases} \tag{8.10}$$

Can we *match* the second equation (8.10) with the Darwinian exponential (8.1)? Yes, if we set at time $t = p$:

$$\begin{cases} A = \dfrac{1}{\sqrt{2\pi}\sigma} \\ e^{B\,p} = e^{\frac{\sigma^2}{2}-\mu} \end{cases} \quad \text{that is} \quad \begin{cases} A = \dfrac{1}{\sqrt{2\pi}\sigma} \\ B\,p = \dfrac{\sigma^2}{2} - \mu. \end{cases} \tag{8.11}$$

The last system of two equations may then be inverted (i.e., exactly solved with respect to μ and σ):

$$\begin{cases} \sigma = \dfrac{1}{\sqrt{2\pi}A} \\ \mu = -B\,p + \dfrac{1}{4\pi A^2} \end{cases} \tag{8.12}$$

showing that each b-lognormal in Figure 8.2 (i.e., its μ and σ) is perfectly determined by the Darwinian exponential (i.e., by A and B) plus a precise value of the peak time p. In other words, this is a one-parameter (the parameter is p) family of curves that are all constrained between the time axis and the Darwinian exponential.

Clearly, as long as one moves to higher values of p, the peaks of these curves become increasingly narrower and higher. For instance, Figure 8.3 shows the two b-lognormals corresponding to the two largest mass extinctions on Earth, which occurred about 250 and 64 million years ago, respectively (end of the Paleozoic and Mesozoic eras, respectively).

8.5 CLADOGRAM BRANCHES CONSTRUCTED OF INCREASING, DECREASING, OR STABLE (HORIZONTAL) EXPONENTIAL ARCHES

It is now possible to understand how cladograms shape up in our mathematical theory of evolution: they depart from the time axis at the birth time (b) of the new species and then either:

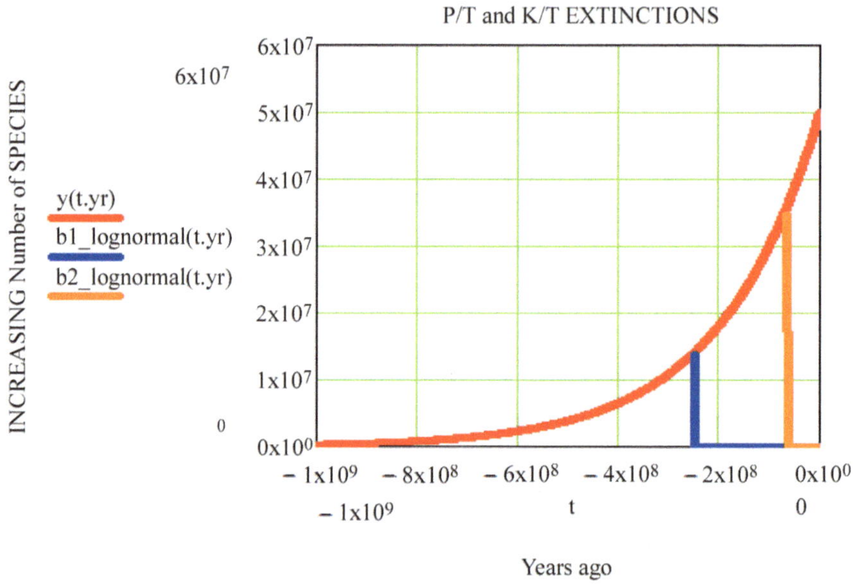

Figure 8.3. Darwinian exponential as the *envelope* of two important *b*-lognormals: those positioned at the P/T and K/T mass extinctions, ending the Primary (or Paleozoic) era and the Secondary (or Mesozoic) era, respectively.

(1) *Increase* if the *b*-lognormal of the *i*-th new species has

$$\begin{cases} A_i = \dfrac{1}{\sqrt{2\pi}\sigma_i} \\ B_i = \dfrac{\dfrac{\sigma_i^2}{2} - \mu_i}{p_i} > 0 \qquad \text{that is } \dfrac{\sigma_i^2}{2} > \mu_i. \end{cases} \tag{8.13}$$

(2) *Decrease* if the same *b*-lognormal has

$$\begin{cases} A_i = \dfrac{1}{\sqrt{2\pi}\sigma_i} \\ B_i = \dfrac{\dfrac{\sigma_i^2}{2} - \mu_i}{p_i} < 0 \qquad \text{that is } \dfrac{\sigma_i^2}{2} < \mu_i. \end{cases} \tag{8.14}$$

(3) *Remain constant* (i.e., rather than exponential arches we have horizontal segments) for all time values for which the *i*-th *b*-lognormal is characterized by:

$$\begin{cases} A_i = \dfrac{1}{\sqrt{2\pi}\sigma_i} \\ B_i = 0 \qquad \text{that is } \dfrac{\sigma_i^2}{2} = \mu_i. \end{cases} \tag{8.15}$$

This case really is the most "routine" one, inasmuch as the given species neither increases nor decreases in time, but rather over multiple generations individuals are born, mate, and die, and the cycle continues endlessly. We call this a *stationary* species. Mathematically, the surprise is that a *stationary* species is no longer described by b-lognormals, but rather by the new probability density found by inserting the last equation (8.15) into (8.9), with the result that (8.9) becomes the *new stationary* pdf:

$$f_{\text{NoEv}}(t,\sigma,b) = \frac{1}{\sqrt{2\pi}\sigma\sqrt{t-b}}\, e^{-\frac{(\ln(t-b))^2}{2\,\sigma^2}}\, e^{-\frac{\sigma^2}{8}}. \tag{8.16}$$

In plain words, this is the pdf for species that undergo *no evolution* at all! Clearly, more words and examples would be needed to better clarify our theory, but we have no space for that here. Table 8.1 shows the main statistical properties of this new NoEv probability density function. They were evaluated by the author using a suitable Maxima code (Maxima is the symbolic manipulator described at *http://maxima.sourceforge.net/*), which is given in Appendix 8.A.

Table 8.1. Summary of the statistical properties of the new random variable NoEv given by equation (8.16) and representing the *stationary life* of a new species born at time b and undergoing *no evolution* thereafter.

Random variable	NoEv = NoEvolution probability = *stationary life*
Probability distribution	(no name yet)
Probability density function	$f_{\text{NoEv}}(t,\,\sigma,\,b) = \frac{1}{\sqrt{2\pi}\sigma\sqrt{t-b}}\, e^{-\frac{(\ln(t-b))^2}{2\sigma^2}-\frac{\sigma^2}{8}}$ $(t \geq b)$
Mean value	$\langle \text{NoEv} \rangle = b + e^{\sigma^2}$
Variance	$\sigma_{\text{NoEv}}^2 = e^{2\sigma^2}\left(e^{\sigma^2} - 1\right)$
Standard deviation	$\sigma_{\text{NoEv}} = e^{\sigma^2}\sqrt{e^{\sigma^2} - 1}$
Mode (= abscissa of the NoEv peak)	$t_{\text{mode}} \equiv t_{\text{peak}} = b + e^{-\frac{\sigma^2}{2}}$
Value of the mode meak (= ordinate of the NoEv peak)	$f_{\text{NoEv}}(t_{\text{mode}}) = \frac{1}{\sqrt{2\pi}\,\sigma}$
Median (= 50–50 probability value for NoEv)	$\text{Median} = m = b + e^{\frac{\sigma^2}{2}}$
Skewness	$\frac{K_3}{(K_2)^{\frac{3}{2}}} = \left(e^{\sigma^2} + 2\right)\sqrt{e^{\sigma^2} - 1}$
Kurtosis	$\frac{K_4}{(K_2)^2} = e^{4\sigma^2} + 2\,e^{3\sigma^2} + 3\,e^{2\sigma^2} - 6$

8.6 KLT FILTERING IN THE HILBERT SPACE AND DARWINIAN SELECTION ARE THE SAME THING IN OUR THEORY

When considering future developments of our mathematical theory of Darwinian evolution, we recall that the KLT (standing for Karhunen–Loève Transform) is a principal axis transformation in the Hilbert space spanned by the eigenfunctions of the autocorrelation of noise with a possible signal in it. Put this way, the KLT may look hard to understand [2, 3]. But we wish to describe in everyday words that it amounts to the well-known Darwinian selection process. In fact, consider a Euclidean space with a large number N of dimensions. A point there means giving N coordinates. We assume each coordinate to be a function of the body that humans have in common with other animals, but other animals may *or may not* (because they are too primordial) have in common with humans. Then, the axis representing humans in this N-space has the largest variance of the set of points around it because humans have *all* functions. Monkeys have *nearly* the same number of functions as humans but in practice they have *fewer* of them. Thus, the monkey axis in the N-space has the *second largest variance* around it. In the mathematical jargon of the KLT this is rephrased by saying that humans are the *dominant = first eigenvalue* in the KLT of the N-space, whereas monkeys are the *second eigenvalue*, and so on for lower species, which are really almost "noise" (i.e., rubbish) when compared with humans.

Filtering involves extracting a tiny signal by virtue of the KLT from thick noise (the KLT works so much better than the trivial FFT used by engineers all over the world, but that is another story, see [2]).

So, just as Darwinian evolution *filtered humans out of a lot of noise* (i.e., other lower-level living organisms), so the KLT applied to the above large N-dimensional space may *describe mathematically the selection* carried on by Darwinian evolution across 3.5 billion years.

But that requires another chapter at least, or, better, the new book entitled *Mathematical SETI* that the author is now writing.

8.7 CONCLUSION

Evolution, as has occurred on Earth over the last 3.5 billion years, is just one chapter of the larger book encompassed by the Drake equation, which covers the time span of 10 billion years or so.

In this chapter we sought to outline a unified and simple mathematical vision of both evolution and SETI.

Our vision is based on the lognormal probability distribution characterizing N in the statistical Drake equation.

We have shown that the envelope of such lognormal distributions changing in time (b-lognormals) may account for the exponential increase in the number of species that have lived on Earth over 3.5 billion years.

APPENDIX 8.A "NOEV_04.WXM"

LIFE as a STATIONARY FINITE NO-EVOLUTION pdf.

This is a mathematical model for the ORDINARY NoEv time of a living being:
BIRTH, REPRODUCTION AND DEATH:
JUST SURVIVAL OF THE SPECIES, AND NO PROGRESS AT ALL.
for a cell, a plant, an animal, a human.
But NOT for an evolving society or civilization, whether human or ET.

Author: Claudio Maccone (e-mail: clmaccon@libero.it), August 16th, 2011.

```
(%i98) kill(all);
(%o0) done
```

Let us recall the b-lognormal pdf.

```
(%i1) assume(b>0, t>b, mu>0, sigma>0);
```

(%o1) $[b>0, t>b, \mu>0, \sigma>0]$

```
(%i2) b_lgn(t, mu, sigma, b):=(1/(sqrt(2*%pi)*sigma*(t-b)))*%e^(-(log(t-b)-mu)^2/(2*sigma^2));
```

(%o2) $\mathrm{b\_lgn}(t,\mu,\sigma,b):=\frac{1}{\sqrt{2\pi}\,\sigma\,(t-b)}\,\%e^{\frac{-(\log(t-b)-\mu)^2}{2\sigma^2}}$

Defining the NoEv of an individual (a cell, a plant, an animal, a human)
as DERIVED from the b-lognormal pdf upon replacing mu=(sigma^2)/2.

```
(%i3) NoEv:factor(radcan(subst(sigma^2/2, mu, b_lgn(t, mu, sigma, b))));
```

(%o3) $\frac{\%e^{-\frac{\log(t-b)^2}{2\sigma^2}-\frac{\sigma^2}{8}}}{\sqrt{2}\,\sqrt{\pi}\,\sigma\,\sqrt{t-b}}$

```
(%i4) NoEv(t, sigma, b):=%e^(-log(t-b)^2/(2*sigma^2)-sigma^2/8)/(sqrt(2)*sqrt(%pi)*sigma*sqrt(t-b));
```

(%o4) $\mathrm{NoEv}(t,\sigma,b):=\frac{\%e^{\frac{-\log(t-b)^2}{2\sigma^2}-\frac{\sigma^2}{8}}}{\sqrt{2}\,\sqrt{\pi}\,\sigma\,\sqrt{t-b}}$

Normalization Condition.

(%i5) 'integrate(NoEv(t, sigma, b),t,b,inf)=integrate(NoEv(t, sigma, b),t,b,inf);

$$\text{(\%o5)}\quad \frac{\int_b^{\infty} \frac{\%e^{-\frac{\log(t-b)^2}{2\sigma^2}-\frac{\sigma^2}{8}}}{\sqrt{t-b}}\,dt}{\sqrt{2}\sqrt{\pi}\,\sigma}=1$$

MEAN VALUE.

(%i6) 'integrate(t*NoEv(t, sigma, b),t,b,inf)=integrate(t*NoEv(t, sigma, b),t,b,inf);

$$\text{(\%o6)}\quad \frac{\int_b^{\infty} \frac{t\,\%e^{-\frac{\log(t-b)^2}{2\sigma^2}-\frac{\sigma^2}{8}}}{\sqrt{t-b}}\,dt}{\sqrt{2}\sqrt{\pi}\,\sigma}=\frac{\sqrt{2}\sqrt{\pi}\,\sigma\,\%e^{\sigma^2}+\sqrt{2}\sqrt{\pi}\,b\,\sigma}{\sqrt{2}\sqrt{\pi}\,\sigma}$$

(%i7) distrib(%);

$$\text{(\%o7)}\quad \frac{\int_b^{\infty} \frac{t\,\%e^{-\frac{\log(t-b)^2}{2\sigma^2}-\frac{\sigma^2}{8}}}{\sqrt{t-b}}\,dt}{\sqrt{2}\sqrt{\pi}\,\sigma}=\%e^{\sigma^2}+b$$

(%i8) mean_value_eq:mean_value=rhs(%);

$$\text{(\%o8)}\quad mean\_value=\%e^{\sigma^2}+b$$

2nd moment.

(%i9) 'integrate(t^2*NoEv(t, sigma, b),t,b,inf)=integrate(t^2*NoEv(t, sigma, b),t,b,inf);

$$\text{(\%o9)}\quad \frac{\int_b^{\infty} \frac{t^2\,\%e^{-\frac{\log(t-b)^2}{2\sigma^2}-\frac{\sigma^2}{8}}}{\sqrt{t-b}}\,dt}{\sqrt{2}\sqrt{\pi}\,\sigma}=\frac{\sqrt{2}\sqrt{\pi}\,\sigma\,\%e^{3\sigma^2}+2^{3/2}\sqrt{\pi}\,b\,\sigma\,\%e^{\sigma^2}+\sqrt{2}\sqrt{\pi}\,b^2\,\sigma}{\sqrt{2}\sqrt{\pi}\,\sigma}$$

(%i10) factor(%);

$$\text{(\%o10)}\quad \frac{\int_b^{\infty} \frac{t^2\,\%e^{-\frac{\log(t-b)^2}{2\sigma^2}-\frac{\sigma^2}{8}}}{\sqrt{t-b}}\,dt}{\sqrt{2}\sqrt{\pi}\,\sigma}=\%e^{3\sigma^2}+2\,b\,\%e^{\sigma^2}+b^2$$

(%i11) second_moment_eq:second_moment=rhs(%);

(%o11) $second\_moment = \%e^{3\,\sigma^2} + 2\,b\,\%e^{\sigma^2} + b^2$

VARIANCE.

(%i12) factor(radcan(second_moment_eq-mean_value_eq^2));

(%o12) $second\_moment - mean\_value^2 = \left(\%e^{\sigma^2} - 1\right) \%e^{2\,\sigma^2}$

(%i13) variance_eq:variance=rhs(%);

(%o13) $variance = \left(\%e^{\sigma^2} - 1\right) \%e^{2\,\sigma^2}$

STANDARD DEVIATION.

(%i14) sqrt(variance_eq);

(%o14) $\sqrt{variance} = \sqrt{\%e^{\sigma^2} - 1}\ \%e^{\sigma^2}$

(%i15) standard_deviation_eq:standard_deviation=rhs(%);

(%o15) $standard\_deviation = \sqrt{\%e^{\sigma^2} - 1}\ \%e^{\sigma^2}$

LIFE PEAK (= NoEv mode, in the probability jargon!)

(%i16) maximum_cond:diff(NoEv(t, sigma, b), t)=0;

(%o16) $$-\frac{\%e^{-\frac{\log(t-b)^2}{2\,\sigma^2} - \frac{\sigma^2}{8}} \log(t-b)}{\sqrt{2}\,\sqrt{\pi}\,\sigma^3\,(t-b)^{3/2}} - \frac{\%e^{-\frac{\log(t-b)^2}{2\,\sigma^2} - \frac{\sigma^2}{8}}}{2^{3/2}\,\sqrt{\pi}\,\sigma\,(t-b)^{3/2}} = 0$$

(%i17) peak_time_eq:expand(first(solve(maximum_cond, t)));

(%o17) $t = \%e^{-\frac{\sigma^2}{2}} + b$

(%i18) mode_time_eq:peaktime=rhs(peak_time_eq);

(%o18) $peaktime = \%e^{-\frac{\sigma^2}{2}} + b$

(%i19) sigma_from_peaktime:first(solve(mode_time_eq,sigma^2));

(%o19) $\sigma^2 = 2\,\log\left(-\frac{1}{b - peaktime}\right)$

(%i20) sigma_from_peaktime,Claudio;

(%o20) $\sigma^2 = 2\log\left(-\frac{1}{b-peaktime}\right)$

(%i21) ev(%,numer);

(%o21) $\sigma^2 = 2\log\left(-\frac{1}{b-peaktime}\right)$

(%i22) finding_the_mode_ord:NoEv(rhs(mode_time_eq), sigma, b);

(%o22) $\frac{1}{\sqrt{2}\sqrt{\pi}\,\sigma}$

(%i23) mode_ord_eq:peak_ord=%;

(%o23) $peak\_ord = \frac{1}{\sqrt{2}\sqrt{\pi}\,\sigma}$

TWO INFLEXION POINTS in one's NoEv:
t=a (adolescence) and t=s (senility).

(%i24) inflexion_cond:diff(NoEv(t, sigma, b), t,2)=0;

(%o24) $\frac{\%e^{-\frac{\log(t-b)^2}{2\sigma^2}-\frac{\sigma^2}{8}}\log(t-b)^2}{\sqrt{2}\sqrt{\pi}\,\sigma^5\,(t-b)^{5/2}} + \frac{\sqrt{2}\,\%e^{-\frac{\log(t-b)^2}{2\sigma^2}-\frac{\sigma^2}{8}}\log(t-b)}{\sqrt{\pi}\,\sigma^3\,(t-b)^{5/2}} + \frac{3\,\%e^{-\frac{\log(t-b)^2}{2\sigma^2}-\frac{\sigma^2}{8}}}{2^{5/2}\sqrt{\pi}\,\sigma\,(t-b)^{5/2}} - \frac{\%e^{-\frac{\log(t-b)^2}{2\sigma^2}-\frac{\sigma^2}{8}}}{\sqrt{2}\sqrt{\pi}\,\sigma^3\,(t-b)^{5/2}} = 0$

(%i25) inflexion_times:solve(inflexion_cond, t);

(%o25) $[t = \%e^{-\frac{\sigma\sqrt{\sigma^2+4}}{2}-\sigma^2} + b,\ t = \%e^{\frac{\sigma\sqrt{\sigma^2+4}}{2}-\sigma^2} + b]$

(%i26) asc_infl_time:first(inflexion_times);

(%o26) $t = \%e^{-\frac{\sigma\sqrt{\sigma^2+4}}{2}-\sigma^2} + b$

(%i27) assume(adolescence>0);

(%o27) $[adolescence > 0]$

(%i28) assume(adolescence>b);

(%o28) $[adolescence > b]$

(%i29) a_point_eq:adolescence=rhs(asc_infl_time);

(%o29) $adolescence = \%e^{-\frac{\sigma \sqrt{\sigma^2+4}}{2} - \sigma^2} + b$

(%i30) ord_of_asc_infl:NoEv_ordinate_at_adolescence=radcan(subst(rhs(first(inflexion_times)), t, NoEv(t, sigma

(%o30) $NoEv\_ordinate\_at\_adolescence = \frac{\%e^{-\frac{\sigma \sqrt{\sigma^2+4} + \sigma^2 + 2}{4}}}{\sqrt{2}\sqrt{\pi}\,\sigma}$

(%i31) desc_infl_time:second(inflexion_times);

(%o31) $t = \%e^{\frac{\sigma \sqrt{\sigma^2+4}}{2} - \sigma^2} + b$

(%i32) assume(senility>0);

(%o32) $[senility > 0]$

(%i33) assume(senility>b);

(%o33) $[senility > b]$

(%i34) assume(senility>adolescence);

(%o34) $[senility > adolescence]$

(%i35) s_point_eq:senility=rhs(desc_infl_time);

(%o35) $senility = \%e^{\frac{\sigma \sqrt{\sigma^2+4}}{2} - \sigma^2} + b$

(%i36) ord_of_desc_infl:NoEv_ordinate_at_senility=radcan(subst(rhs(second(inflexion_times)), t, NoEv(t, sigma,

(%o36) $NoEv\_ordinate\_at\_senility = \frac{\%e^{\frac{\sigma \sqrt{\sigma^2+4} - \sigma^2 - 2}{4}}}{\sqrt{2}\sqrt{\pi}\,\sigma}$

Finding sigma WHEN the two ORDINATES of the b-lognormal at the AT THE TWO INFLEXION POINTS are assigned and the ADOLESCENCE time (t=a) is assigned.

(%i37) expression_in_sigma_only:log(radcan(ord_of_desc_infl/ord_of_asc_infl));

(%o37) $\log\left(\frac{NoEv\_ordinate\_at\_senility}{NoEv\_ordinate\_at\_adolescence}\right) = \frac{\sigma \sqrt{\sigma^2+4}}{2}$

NoEv_04.wxm 6 / 1

(%i38) sigma_from_the_two_inflection_ordinates:second(solve(expression_in_sigma_only^2, sigma));

(%o38) $\sigma = \sqrt{2}\sqrt{\sqrt{\log\left(\frac{NoEv\_ordinate\_at\_senility}{NoEv\_ordinate\_at\_adolescence}\right)^2+1}-1}$

Important NOTE about an "error" in Maxima's NOTATION.

Unfortunately, Maxima's notation is WRONG! (and should be corrected!).
In fact, Maxima writes, for instance, sin(x^2) instead of sin^2(x).
So, Maxima writes sin(x^2)+cos(x^2)=1 instead of the correct sin^2(x)+cos^2(x)=1. This is a SERIOUS MISPRINT!

Such a stupid misprint did NOT occur in the good old MACSYMA (i.e. in the "commercial" Macsyma) but it does occur in Maxima.

So, in the last equation, the square applies TO THE WHOLE LOG, RATHER THAN JUST TO THE LOG ARGUMENT!!! Be aware of that, please.

FINDING sigma WHEN the
ADOLESCENCE AND SENILITY times, a and s, are assigned.
That is, our NoEv-lognormal model is completely determined if a and s are assigned.

(%i39) a_point_eq;

(%o39) $adolescence = \%e^{-\frac{\sigma\sqrt{\sigma^2+4}}{2}-\sigma^2}+b$

(%i40) s_point_eq;

(%o40) $senility = \%e^{\frac{\sigma\sqrt{\sigma^2+4}}{2}-\sigma^2}+b$

(%i41) first_eq:(a_point_eq-b)*(s_point_eq-b);

(%o41) $(adolescence-b)\,(senility-b) = \%e^{-2\,\sigma^2}$

(%i42) four_expressions_for_sigma:solve(first_eq,sigma);

(%o42) $[\sigma = -\sqrt{\log\left(-\frac{1}{\sqrt{adolescence-b}\sqrt{senility-b}}\right)}, \sigma = \sqrt{\log\left(-\frac{1}{\sqrt{adolescence-b}\sqrt{senility-b}}\right)}, \sigma = -\sqrt{\log\left(\frac{1}{\sqrt{adolescence-b}\sqrt{senility-b}}\right)}, \sigma = \sqrt{\log\left(\frac{1}{\sqrt{adolescence-b}\sqrt{senility-b}}\right)}]$

(%i43) final_sigma:last(four_expressions_for_sigma);

(%o43) $\sigma=\sqrt{\log\left(\frac{1}{\sqrt{adolescence-b}\,\sqrt{senility-b}}\right)}$

(%i44) final_sigma,Claudio;

(%o44) $\sigma=\sqrt{\log\left(\frac{1}{\sqrt{adolescence-b}\,\sqrt{senility-b}}\right)}$

(%i45) ev(%,numer);

(%o45) $\sigma=\log\left(\frac{1}{(adolescence-b)^{0.5}\,(senility-b)^{0.5}}\right)^{0.5}$

sigma FROM ADOLESCENCE AND PEAK.

(%i46) eq_1:log(a_point_eq-b);

(%o46) $\log(adolescence-b)=-\frac{\sigma\sqrt{\sigma^2+4}}{2}-\sigma^2$

(%i47) eq_2:log(mode_time_eq-b);

(%o47) $\log(peaktime-b)=-\frac{\sigma^2}{2}$

(%i48) eq_3:logcontract(-eq_1+2*eq_2)^2;

(%o48) $\log\left(-\frac{(peaktime-b)^2}{b-adolescence}\right)^2=\frac{\sigma^2\,(\sigma^2+4)}{4}$

(%i49) assume(LHS>0);
(%o49) $[LHS>0]$

(%i50) def_LHS:LHS=lhs(eq_3);

(%o50) $LHS=\log\left(-\frac{(peaktime-b)^2}{b-adolescence}\right)^2$

(%i51) eq_4:LHS=rhs(eq_3);

(%o51) $LHS=\frac{\sigma^2\,(\sigma^2+4)}{4}$

(%i52) eq_5:solve(eq_4, sigma);

(%o52) $[\sigma=-\sqrt{2}\,\sqrt{\sqrt{LHS+1}-1},\sigma=\sqrt{2}\,\sqrt{\sqrt{LHS+1}-1},\sigma=-\sqrt{2}\,\%i\,\sqrt{\sqrt{LHS+1}+1},\sigma=\sqrt{2}\,\%i\,\sqrt{\sqrt{LHS+1}+1}]$

(%i53) eq_6:eq_5,def_LHS;

(%o53) $[\sigma=-\sqrt{2}\sqrt{\sqrt{\log\left(-\frac{(peaktime-b)^2}{b-adolescence}\right)^2+1}-1},\sigma=\sqrt{2}\sqrt{\sqrt{\log\left(-\frac{(peaktime-b)^2}{b-adolescence}\right)^2+1}-1},\sigma=-\sqrt{2}\,\%i\sqrt{\sqrt{\log\left(-\frac{(peaktime-b)^2}{b-adolescence}\right)^2+1}+1},\sigma=\sqrt{2}\,\%i\sqrt{\sqrt{\log\left(-\frac{(peaktime-b)^2}{b-adolescence}\right)^2+1}+1}]$

(%i54) final_sigma_from_a_and_p:second(eq_6);

(%o54) $\sigma=\sqrt{2}\sqrt{\sqrt{\log\left(-\frac{(peaktime-b)^2}{b-adolescence}\right)^2+1}-1}$

(%i55) final_sigma_from_a_and_p,Claudio;

(%o55) $\sigma=\sqrt{2}\sqrt{\sqrt{\log\left(-\frac{(peaktime-b)^2}{b-adolescence}\right)^2+1}-1}$

(%i56) ev(%,numer);

(%o56) $\sigma=1.414213562373095\left(\left(\log\left(-\frac{(peaktime-b)^2}{b-adolescence}\right)^2+1\right)^{0.5}-1\right)^{0.5}$

INTRODUCING DEATH,
defined as the time of interception between the
TANGENT STRAIGHT LINE
TO THE STATIONARY b-LOGNORMAL AT SENILITY
and the time axis.

(%i57) tangent_eq:(y-y0)=m*(t-t0);

(%o57) $y-y0=m\,(t-t0)$

(%i58) finding_y0:y0=radcan(NoEv(rhs(s_point_eq), sigma, b));

(%o58) $y0=\frac{\%e^{\frac{\sigma\sqrt{\sigma^2+4}-\sigma^2-2}{4}}}{\sqrt{2}\sqrt{\pi}\,\sigma}$

(%i59) finding_t0:t0=rhs(s_point_eq);

(%o59) $t0=\%e^{\frac{\sigma\sqrt{\sigma^2+4}}{2}-\sigma^2}+b$

(%i60) finding_m:m=radcan(subst(rhs(finding_t0), t, diff(NoEv(t, sigma, b), t)));

$$(\%o60)\ m=-\frac{\left(\sqrt{2}\sqrt{\sigma^2+4}\ \%e^{\frac{3\sigma^2}{4}}-\sqrt{2}\,\sigma\,\%e^{\frac{3\sigma^2}{4}}\right)\%e^{-\frac{\sigma\sqrt{\sigma^2+4}+2}{4}}}{4\sqrt{\pi}\,\sigma^2}$$

(%i61) tangent_straight_line_eq:tangent_eq,finding_y0,finding_t0,finding_m;

$$(\%o61)\ y-\frac{\%e^{\frac{\sigma\sqrt{\sigma^2+4}-\sigma^2-2}{4}}}{\sqrt{2}\sqrt{\pi}\,\sigma}=-\frac{\left(\sqrt{2}\sqrt{\sigma^2+4}\ \%e^{\frac{3\sigma^2}{4}}-\sqrt{2}\,\sigma\,\%e^{\frac{3\sigma^2}{4}}\right)\%e^{-\frac{\sigma\sqrt{\sigma^2+4}+2}{4}}\left(t-\%e^{\frac{\sigma\sqrt{\sigma^2+4}}{2}-\sigma^2}-b\right)}{4\sqrt{\pi}\,\sigma^2}$$

(%i62) tangent_at_s:radcan(first(solve(tangent_straight_line_eq, y)));

$$(\%o62)\ y=-$$

$$\frac{\%e^{-\frac{\sigma\sqrt{\sigma^2+4}+\sigma^2+4}{4}}\left(\sqrt{\%e}\left(-\sigma\,\%e^{\sigma^2}t-\sigma\,\%e^{\frac{\sigma\sqrt{\sigma^2+4}}{2}}+b\,\sigma\,\%e^{\sigma^2}\right)+\sqrt{\%e}\sqrt{\sigma^2+4}\left(\%e^{\sigma^2}t-\%e^{\frac{\sigma\sqrt{\sigma^2+4}}{2}}-b\,\%e^{\sigma^2}\right)\right)}{2^{3/2}\sqrt{\pi}\,\sigma^2}$$

(%i63) finding_death_point:solve(subst(0, y, tangent_at_s), t);

$$(\%o63)\ \left[t=\frac{\%e^{-\sigma^2}\left(\left(\sqrt{\sigma^2+4}+\sigma\right)\%e^{\frac{\sigma\sqrt{\sigma^2+4}}{2}}+\left(b\sqrt{\sigma^2+4}-b\,\sigma\right)\%e^{\sigma^2}\right)}{\sqrt{\sigma^2+4}-\sigma}\right]$$

(%i64) death_point_eq:stationary_death_point=rhs(first(finding_death_point));

$$(\%o64)\ stationary\_death\_point=\frac{\%e^{-\sigma^2}\left(\left(\sqrt{\sigma^2+4}+\sigma\right)\%e^{\frac{\sigma\sqrt{\sigma^2+4}}{2}}+\left(b\sqrt{\sigma^2+4}-b\,\sigma\right)\%e^{\sigma^2}\right)}{\sqrt{\sigma^2+4}-\sigma}$$

(%i65) death_time_eq:stationary_d=rhs(death_point_eq);

$$(\%o65)\ stationary\_d=\frac{\%e^{-\sigma^2}\left(\left(\sqrt{\sigma^2+4}+\sigma\right)\%e^{\frac{\sigma\sqrt{\sigma^2+4}}{2}}+\left(b\sqrt{\sigma^2+4}-b\,\sigma\right)\%e^{\sigma^2}\right)}{\sqrt{\sigma^2+4}-\sigma}$$

(%i66) def_dd:stationary_dd=%e^(-7*sigma^2/4)*%e^((sigma*sqrt(sigma^2+4))/2+3*sigma^2/4)*(sqrt(sigma^2+

$$(\%o66)\ stationary\_dd=\frac{\left(\sqrt{\sigma^2+4}+\sigma\right)^2\%e^{\frac{\sigma\sqrt{\sigma^2+4}}{2}-\sigma^2}}{4}+b$$

(%i67) radcan(rhs(def_dd)-rhs(death_time_eq));

(%o67) 0

(%i68) death_time:deathtime=rhs(def_dd);

(%o68) $deathtime=\frac{\left(\sqrt{\sigma^2+4}+\sigma\right)^2 \%e^{\frac{\sigma\sqrt{\sigma^2+4}}{2}-\sigma^2}}{4}+b$

APPROXIMATED sigma from BIRTH and DEATH only.

We assume that 0<sigma<1 and expand the last equation
in a Taylor series with respect to sigma, around sigma=0
and to the second order in sigma.

(%i69) approximated_quadratic_in_sigma:taylor(death_time-b, sigma, 0, 2);

(%o69)/T/ $(-b+deathtime)+\ldots=1+2\,\sigma+\sigma^2+\ldots$

(%i70) approximated_sigma_from_birth_and_death:last(solve(approximated_quadratic_in_sigma, sigma));

(%o70) $\sigma=\sqrt{deathtime-b}-1$

Most IMPORTANT !

FINDING THE b-LOGNORMAL
WITH A GIVEN BIRTH (t=b),
SENILITY (t=s),
AND stationary DEATHTIME (t=deathtime).

In other words, the problem of determing the b-lognormal
of a living organism when its birth time b, its senility time s
and its death time d are known, is now completely solved.
For organisms that are born, reproduce and die, but make no
evolutionary progress.

(%i71) assume(deathtime>senility);

(%o71) $[deathtime>senility]$

(%i72) towards_an_eq_in_sigma:4*(death_time-b);

(%o72) $4\,(deathtime-b)=\left(\sqrt{\sigma^2+4}+\sigma\right)^2 \%e^{\frac{\sigma\sqrt{\sigma^2+4}}{2}-\sigma^2}$

(%i73) s_point_eq-b;

(%o73) $senility-b=\%e^{\frac{\sigma\sqrt{\sigma^2+4}}{2}-\sigma^2}$

(%i74) towards_an_eq_in_sigma/%;

(%o74) $\frac{4\,(deathtime-b)}{senility-b}=\left(\sqrt{\sigma^2+4}+\sigma\right)^2$

(%i75) sqrt(%);

(%o75) $\frac{2\sqrt{deathtime-b}}{\sqrt{senility-b}}=\sqrt{\sigma^2+4}+\sigma$

(%i76) %-sigma;

(%o76) $\frac{2\sqrt{deathtime-b}}{\sqrt{senility-b}}-\sigma=\sqrt{\sigma^2+4}$

(%i77) %^2;

(%o77) $\left(\frac{2\sqrt{deathtime-b}}{\sqrt{senility-b}}-\sigma\right)^2=\sigma^2+4$

(%i78) expand(%);

(%o78) $\sigma^2-\frac{4\sqrt{deathtime-b}\,\sigma}{\sqrt{senility-b}}+\frac{4\,deathtime}{senility-b}-\frac{4\,b}{senility-b}=\sigma^2+4$

(%i79) sigma_vs_b_d_and_s:radcan(first(solve(%, sigma)));

(%o79) $\sigma=-\frac{senility-deathtime}{\sqrt{deathtime-b}\sqrt{senility-b}}$

Finding the DEATH TIME from the known PEAK and SENILITY times.

(%i80) invoking_p:mode_time_eq-b;

(%o80) $peaktime-b=\%e^{-\frac{\sigma^2}{2}}$

(%i81) eq_in_p_s_d:invoking_p,sigma_vs_b_d_and_s;

(%o81) $peaktime-b=\%e^{-\frac{(senility-deathtime)^2}{2\,(deathtime-b)\,(senility-b)}}$

(%i82) expression_of_p_minus_b:eq_in_p_s_d,mu_vs_b_s_and_d;

(%o82) $peaktime-b=\%e^{-\frac{(senility-deathtime)^2}{2\,(deathtime-b)\,(senility-b)}}$

(%i83) simplified_expression_of_p_minus_b:radcan(expression_of_p_minus_b);

$$(\%o83)\ peaktime - b = \%e^{-\frac{senility^2 - 2\ deathtime\ senility + deathtime^2}{(2\ deathtime - 2\ b)\ senility - 2\ b\ deathtime + 2\ b^2}}$$

(%i84) death_from_peak_and_senility:first(solve(simplified_expression_of_p_minus_b, deathtime));

$$(\%o84)\ deathtime = (1 - \log(peaktime - b))\ senility + \sqrt{\log(peaktime - b)^2 - 2\log(peaktime - b)\ (b - senility)} + b\log(peaktime - b)$$

Computing the PROBABILITY by which
this "living organism" model is true, i.e.
computing the AREA under the finite_b_lognormal
and the time axis, from birth (t=b) to death (t=d).

In the practice, this means computing the INTEGRAL
under the finite_b_lognormal curve from b to s,
and then ADDING to this the area of the TRIANGLE
under the tangent-at-s straight line and the time axis,
from the descending inflexion point t=s up to the
death time t=d.

(%i85) 'integrate(NoEv(t,sigma,b), t, b, rhs(s_point_eq))=integrate(NoEv(t,sigma,b), t, b, rhs(s_point_eq));

$$(\%o85)\ \frac{\displaystyle\int_b^{\%e^{\frac{\sigma\sqrt{\sigma^2+4}}{2}-\sigma^2}+b} \frac{\%e^{-\frac{\log(t-b)^2}{2\sigma^2}-\frac{\sigma^2}{8}}}{\sqrt{t-b}}\,dt}{\sqrt{2}\sqrt{\pi}\,\sigma} = \frac{\dfrac{\sqrt{\pi}\,\sigma\,\mathrm{erf}\left(\dfrac{\sqrt{2}\sqrt{\sigma^2+4}-3\sqrt{2}\,\sigma}{4}\right)}{\sqrt{2}}+\dfrac{\sqrt{\pi}\,\sigma}{\sqrt{2}}}{\sqrt{2}\sqrt{\pi}\,\sigma}$$

(%i86) prob_from_b_to_s:prob_from_b_to_s=distrib(rhs(%));

$$(\%o86)\ prob\_from\_b\_to\_s = \frac{\mathrm{erf}\left(\dfrac{\sqrt{2}\sqrt{\sigma^2+4}-3\sqrt{2}\,\sigma}{4}\right)}{2}+\frac{1}{2}$$

(%i87) triangle_area_:triangle_area=(rhs(death_time)-rhs(s_point_eq))*(rhs(ord_of_desc_infl))/2;

$$(\%o87)\ triangle\_area = \frac{\left(\dfrac{\left(\sqrt{\sigma^2+4}+\sigma\right)^2\%e^{\frac{\sigma\sqrt{\sigma^2+4}}{2}-\sigma^2}}{4} - \%e^{\frac{\sigma\sqrt{\sigma^2+4}}{2}-\sigma^2}\right)\%e^{\frac{\sigma\sqrt{\sigma^2+4}-\sigma^2-2}{4}}}{2^{3/2}\sqrt{\pi}\,\sigma}$$

(%i88) death_time;

(%o88) $deathtime = \frac{(\sqrt{\sigma^2+4}+\sigma)^2 \, \%e^{\frac{\sigma\sqrt{\sigma^2+4}}{2} - \sigma^2}}{4} + b$

(%i89) s_point_eq;

(%o89) $senility = \%e^{\frac{\sigma\sqrt{\sigma^2+4}}{2} - \sigma^2} + b$

(%i90) prob_from_b_to_d_eq:prob_from_b_to_d=expand(rhs(prob_from_b_to_s)+rhs(triangle_area_));

(%o90) $prob\_from\_b\_to\_d = \frac{\mathrm{erf}\left(\frac{\sqrt{\sigma^2+4}}{2^{3/2}} - \frac{3\sigma}{2^{3/2}}\right)}{2} + \frac{\sqrt{\sigma^2+4}\, \%e^{\frac{3\sigma\sqrt{\sigma^2+4}}{4} - \frac{5\sigma^2}{4} - \frac{1}{2}}}{2^{5/2}\sqrt{\pi}} + \frac{\sigma\, \%e^{\frac{3\sigma\sqrt{\sigma^2+4}}{4} - \frac{5\sigma^2}{4} - \frac{1}{2}}}{2^{5/2}\sqrt{\pi}} + \frac{1}{2}$

(%i91) ev(subst(0.373,sigma,%),numer);

(%o91) $prob\_from\_b\_to\_d = 0.89266088792049$

So, the total probability from birth to death is HIGHER than 1/2 = 50%, and depends upon sigma only.
Please don't be cheated by the minus sign coming from the triangle area: the area is of course POSITIVE, and the minus sign in front of it is just a Maxima way of writing things (other negative terms at the numerator make the whole fraction POSITIVE).

We now provide a CONCISE FORMULA for the whole above probability expression.
That is:

(%i92) concise__prob_from_b_to_d:concise_version=erf((sqrt(sigma^2+4)-3*sigma)/(2^(3/2)))/2+((sqrt(sigma^2-

(%o92) $concise\_version = \frac{\mathrm{erf}\left(\frac{\sqrt{\sigma^2+4}-3\sigma}{2^{3/2}}\right)}{2} + \frac{(\sqrt{\sigma^2+4}+\sigma)\, \%e^{\frac{3\sigma\sqrt{\sigma^2+4}}{4} - \frac{5\sigma^2}{4} - \frac{1}{2}}}{2^{5/2}\sqrt{\pi}} + \frac{1}{2}$

(%i93) radcan(prob_from_b_to_d_eq-concise__prob_from_b_to_d)=0;

(%o93) $prob\_from\_b\_to\_d - concise\_version = \frac{\mathrm{erf}\left(\frac{\sqrt{2}\sqrt{\sigma^2+4}-3\sqrt{2}\,\sigma}{4}\right) - \mathrm{erf}\left(\frac{\sqrt{\sigma^2+4}-3\sigma}{2^{3/2}}\right)}{2} = 0$

(%i94) final__prob_from_b_to_d:prob_from_b_to_d=erf((sqrt(sigma^2+4)-3*sigma)/2^(3/2))/2+((sqrt(sigma^2+4

(%o94) $prob\_from\_b\_to\_d = \frac{\mathrm{erf}\left(\frac{\sqrt{\sigma^2+4}-3\,\sigma}{2^{3/2}}\right)}{2} + \frac{\left(\sqrt{\sigma^2+4}+\sigma\right)\%e^{\frac{3\,\sigma\sqrt{\sigma^2+4}}{4}-\frac{5\,\sigma^2}{4}-\frac{1}{2}}}{2^{5/2}\sqrt{\pi}} + \frac{1}{2}$

Example: suppose that sigma=0.373.
Then the total probability is given by 89%.
In fact, one has:

(%i95) ev(subst(0.373,sigma,final__prob_from_b_to_d),numer);

(%o95) *prob_from_b_to_d* =0.89266088792049

8.8 REFERENCES

[1] Robert A. Rohde and Richard A. Muller, "Cycles in fossil diversity," *Nature*, **434** (2005), 208–210.
[2] Claudio Maccone, "The KLT (Karhunen–Loève Transform) to extend SETI searches to broad-band and extremely feeble signals," *Acta Astronautica*, **67** (2010), 1427–1439.
[3] A. Szumski, "Finding the interference: The Karhunen–Loève Transform as an instrument to detect weak RF signals," *InsideGNSS* (Working Papers), May/June 2011, 56–63.
[4] C. Maccone, "A mathematical model for evolution and SETI," *Origin of Life and Evolutionary Biospheres (OLEB)*, available online December 3, 2011.

8.9 FURTHER READING

Balazs, B. (1988) "The Galactic belt of intelligent life", in G. Marx (Ed.), *Bioastronomy: The Next Steps*, Kluwer Academic, pp. 61–66.
Burchell, M. J. (2006), "W(h)ither the Drake equation?" *International Journal of Astrobiology*, **5**, 243–250.
Čirkovič, M. M. (2004), "On the temporal aspect of the Drake equation and SETI," *Astrobiology*, **4**, 225–231.
Čirkovič, M. M. and Bradbury, R. J. (2006),"Galactic gradients, postbiological evolution and the apparent failure of SETI," *New Astronomy*, **11**, 628–639.
Gonzalez, G. (2005), "Habitable zones in the Universe," *Origin of Life and Evolution of the Biosphere*, **35**, 555–606.
Gonzalez, G., Brownlee, D., and Ward, P. (2001), "The Galactic habitable zone: Galactic chemical evolution," *Icarus*, **152**, 185–200.
Ksanfomality, L. V. (2004), "The Drake equation may need new factors based on peculiarities of planets of Sun-like stars," in *Planetary Systems in the Universe: Proceedings of IAU Symposium #202*, p. 458.
Lineweaver, C. H., Fenner, Y., and Gibson, B. K. (2004), "The Galactic habitable zone and the age distribution of complex life in the Milky Way," *Science*, **303**, 59–62.
Maccone, C. (2008), "The statistical Drake equation," paper #IAC-08-A4.1.4 presented on October 1, 2008, at the *59th International Astronautical Congress (IAC) held in Glasgow, Scotland, U.K., September 29–October 3, 2008.*
Maccone, C. (2010), "The statistical Drake equation," *Acta Astronautica*, **67**, 1366–1383.
Maccone, C. (2010), "The statistical Fermi paradox," *Journal of the British Interplanetary Society*, **63**, 222–239.
Maccone, C. (2011), "SETI and SEH (Statistical Equation for Habitables)," *Acta Astronautica*, **68**, 63–75.
Marochnik, L. S. and Mukhin, L. M. (1988), "Belt of life in the Galaxy", in G. Marx (Ed.), *Bioastronomy: The Next Steps*, Kluwer Academic, pp. 49–59.
Wallenhorst, S. G. (1981), "The Drake equation reexamined," *QJRAS*, **22**, 380.
Walters, C., Hoover, R. A., and Kotra, R. K. (1980), "Interstellar colonization: A new parameter for the Drake equation?" *Icarus*, **41**, 193–197.

9

Societal statistics by the statistical Drake equation

9.1 INTRODUCTION: BREAKING DOWN THE DRAKE EQUATION SO AS TO ISOLATE ITS SOCIETAL PART

As we know, the Drake equation, first proposed by Frank D. Drake in 1961, is the foundational equation of SETI. It yields an estimate of the number N of extraterrestrial communicating civilizations in the Galaxy given by the product

$$N = Ns\, fp\, ne\, fl\, fi\, fc\, fL \tag{9.1}$$

where: Ns is the number of stars in the Milky Way Galaxy;
fp is the fraction of stars that have planetary systems;
ne is the number of planets in a given system that are ecologically suitable for life;
fl is the fraction of otherwise suitable planets on which life actually arises;
fi is the fraction of inhabited planets on which an intelligent form of life evolves;
fc is the fraction of planets inhabited by intelligent beings on which a communicative technical civilization develops; and
fL is the fraction of planetary lifetime graced by a technical civilization.

The first three terms may be called "the astrophysical terms" in the Drake equation since their numerical value is provided by astrophysical considerations. The fourth term, fl, may be called "the origin-of-life term" and entails biology. The last three terms may be called "the societal terms" inasmuch as their respective numerical values are provided by anthropology, telecommunication science, and "futuristic science", respectively.

In this chapter we seek to provide a statistical estimate of the three societal terms in the Drake equation and base our calculations on the statistical Drake equation first proposed by the author at the 2008 IAC. In that paper the author extended the simple seven-factor product so as to embody statistics. He proved that, no matter which probability distribution may be assigned to each factor, if the number of factors tends to infinity, then the random variable N follows the lognormal distribution (Central Limit Theorem of statistics).

The author also proved at the 2009 IAC that the Dole equation

$$N_{Hab} = Ns \cdot Pp \cdot Pi \cdot PD \cdot PM \cdot Pe \cdot PB \cdot PR \cdot PA \cdot PL, \tag{9.2}$$

yielding the number of habitable planets for man in the Galaxy, has the same mathematical structure as the Drake equation. So the number of habitable planets follows the lognormal distribution as well.

But the Dole equation is described by the first *four* factors of the Drake equation. Thus, we may "divide" the seven-factor Drake equation by the four-factor Dole equation to get the probability distribution of the last three-factor Drake equation (i.e., the probability distribution of the *societal terms only*). These we study in detail in this chapter, achieving new statistical results about the *societal aspects of SETI.*

9.2 THE PROBABILITY DISTRIBUTION OF THE RATIO OF TWO LOGNORMALLY DISTRIBUTED RANDOM VARIABLES

This section is purely mathematical, meaning that we shall only look at and solve the mathematical problem of finding the probability distribution of the ratio of two lognormally distributed random variables. Why we are doing so will become apparent in the next section.

So, let us consider a first lognormally distributed random variable X, as usual characterized by its own two parameters μ_X and σ_X, and thus having the pdf:

$$f_X(x) = \frac{1}{x} \cdot \frac{1}{\sqrt{2\pi}\sigma_X} e^{-\frac{(\ln(x)-\mu_X)^2}{2\sigma_X^2}} \quad (x \geq 0). \tag{9.3}$$

Apart from this, we also consider another arbitrary and different lognormally distributed random variable Y having the pdf:

$$f_Y(y) = \frac{1}{y} \cdot \frac{1}{\sqrt{2\pi}\sigma_Y} e^{-\frac{(\ln(y)-\mu_Y)^2}{2\sigma_Y^2}} \quad (y \geq 0). \tag{9.4}$$

The problem we solve in this section is to answer the question: What is the pdf of the third positive random variable Z given by the ratio of the two previous positive and lognormally distributed X and Y?

$$Z = \frac{X}{Y} = \frac{\text{a certain lognormal distribution}}{\text{another lognormal distribution}}. \tag{9.5}$$

Standard textbooks about probability theory (e.g., the book by Papoulis and Pillai, *Probability Random Variables and Stochastic Processes*, [8], pp. 186–187 and equation (6-59) in particular) tell us that the random variable Z has its pdf given by the integral

$$f_Z(z) = \int_{-\infty}^{\infty} |y| \cdot f_{XY}(yz, y)\, dy \tag{9.6}$$

where the function $f_{XY}(..., \ ...)$ is the joint pdf of the two random variables X and Y. Now, if we assume the two random variables X and Y to be statistically independent of each other, inasmuch as the physics tells us this is the case, then their joint pdf $f_{XY}(..., \ ...)$ simply is the product of the two pdfs (i.e., the two lognormal distributions (9.3) and (9.4)). That is:

$$\begin{aligned} f_{X,Y}(x,y) = f_X(x) \cdot f_Y(y) &= \frac{1}{x} \cdot \frac{1}{\sqrt{2\pi}\sigma_X}\, e^{-\frac{(\ln(x)-\mu_X)^2}{2\sigma_X^2}} \cdot \frac{1}{y} \cdot \frac{1}{\sqrt{2\pi}\sigma_Y}\, e^{-\frac{(\ln(y)-\mu_Y)^2}{2\sigma_Y^2}} \\ &= \frac{1}{2\pi\, \sigma_X \sigma_Y\, x\, y}\, e^{-\left[\frac{(\ln(x)-\mu_X)^2}{2\sigma_X^2} + \frac{(\ln(y)-\mu_Y)^2}{2\sigma_Y^2}\right]}. \end{aligned} \tag{9.7}$$

This is the joint pdf that must be introduced into the integral (9.6). Note, however, that the integral actually ranges from 0 to infinity only, since both x and y do so. The modulus affecting y in (9.6) thus also disappears, and we are just left with the computation of the definite integral

$$f_Z(z) = f_{\frac{X}{Y}}(z) = \int_0^{\infty} y \cdot f_{X,Y}(yz, \ y)\, dy. \tag{9.8}$$

That is, inserting the last expression (9.7) under the integral sign in (9.6)

$$\begin{aligned} f_Z(z) &= \frac{1}{2\pi\, \sigma_X \sigma_Y} \int_0^{\infty} y \cdot \frac{1}{y^2\, z}\, e^{-\left[\frac{(\ln(yz)-\mu_X)^2}{2\sigma_X^2} + \frac{(\ln(y)-\mu_Y)^2}{2\sigma_Y^2}\right]} dy \\ &= \frac{1}{2\pi\, \sigma_X \sigma_Y z} \int_0^{\infty} \frac{e^{-\left[\frac{(\ln(yz)-\mu_X)^2}{2\sigma_X^2} + \frac{(\ln(y)-\mu_Y)^2}{2\sigma_Y^2}\right]}}{y}\, dy. \end{aligned} \tag{9.9}$$

This is a tough integral to compute analytically. Basically, it can be reduced to the Gauss integral (i.e., to the normalization condition of the ordinary Gaussian curve), but many, many steps are required to perform the integration with respect to y. The author, when faced with this analytical computation, turned to Macsyma (concisely described at *http://en.wikipedia.org/wiki/Macsyma*). Macsyma is a wonderful computer algebra code that was created at the MIT Artificial Intelligence Laboratory back in the 1960s to let NASA recompute analytically the orbits requested for the Apollo astronauts to safely reach the Moon and return. Macsyma was able to perform the analytical integration in (9.9) in a matter of seconds. The outcome is the function of z given by:

$$f_Z(z) = z^{\frac{\mu_X-\mu_Y}{\sigma_X^2+\sigma_Y^2}-1} \cdot \frac{1}{\sqrt{2\pi}\sqrt{\sigma_X^2+\sigma_Y^2}}\, e^{-\frac{(\mu_X-\mu_Y)^2+(\ln(z))^2}{2\left(\sigma_X^2+\sigma_Y^2\right)}}. \tag{9.10}$$

As can be seen, this function of z is a complicated mix of exponentials in z through the natural log of z squared, times a power of z, times many other constants. Also, the pdf (9.10) does not have any specific name. Hence, the author thought of calling it LOL, an acronym for Lognormal-Over-Lognormal pdf, as indeed it is.

Although the LOL pdf (9.10) is difficult for humans to handle by hand, it can easily be handled by Macsyma! Thus, one can quickly prove that it indeed fulfills the normalization condition of any pdf,

$$\int_0^\infty f_Z(z)\, dz = 1. \tag{9.11}$$

A similar calculation then shows that the mean value of the LOL random variable Z reads

$$\mu_Z = \langle Z \rangle = \int_0^\infty z\, f_Z(z)\, dz = e^{\mu_X-\mu_Y}\, e^{\frac{\sigma_X^2+\sigma_Y^2}{2}}. \tag{9.12}$$

The corresponding LOL variance was again found by Macsyma through a similar calculation, and reads

$$\sigma_Z^2 = e^{2(\mu_X-\mu_Y)+\sigma_X^2+\sigma_y^2}\left(e^{\sigma_X^2+\sigma_y^2} - 1\right). \tag{9.13}$$

Its square root is thus the relevant LOL standard deviation:

$$\sigma_Z = e^{(\mu_X-\mu_Y)+\frac{\sigma_X^2+\sigma_y^2}{2}}\sqrt{e^{\sigma_X^2+\sigma_y^2} - 1}. \tag{9.14}$$

In order to find the mode of the pdf (9.10) (i.e., the abscissa of its peak), we must discover the first derivative of (9.10) with respect to z and then set the resulting equation equal to zero. Good old Macsyma did a fine job once again, and the two results are the two abscissas of the minimum of (9.10), obviously at $z = 0$, and of the maximum (i.e., the peak, or mode) at

$$z_{\text{mode}} \equiv z_{\text{peak}} = e^{\mu_X-\mu_Y}\, e^{-\left(\sigma_X^2+\sigma_Y^2\right)}. \tag{9.15}$$

Inserting this into (9.10) and rearranging, we find the LOL pdf peak value:

$$f_Z(z_{\text{mode}}) = \frac{1}{\sqrt{2\pi}\,\sqrt{\sigma_X^2+\sigma_Y^2}} \cdot e^{-(\mu_X-\mu_Y)} \cdot e^{\frac{\sigma_X^2+\sigma_Y^2}{2}}. \tag{9.16}$$

To summarize, all the statistical results about the LOL random variable Z that we derived in this section are listed in Table 9.1.

Table 9.1. Summary of the properties of the LOL (lognormal-over-lognormal) distribution that applies to the random variable $Z = \frac{X}{Y} = \frac{N}{N_{Hab}} = \frac{Ns \cdot fp \cdot ne \cdot fl \cdot fi \cdot fc \cdot fL}{Ns \cdot fp \cdot ne \cdot fl} = fi \cdot fc \cdot fL =$ societal part of the statistical Drake equation.

Random variable X	$X = N =$ number of communicating ET civilizations in the Galaxy = left-hand-side of the Drake equation
Probability distribution	Lognormal
Probability density function	$f_X(x) = \frac{1}{x} \cdot \frac{1}{\sqrt{2\pi}\sigma_X} e^{-\frac{(\ln(x)-\mu_X)^2}{2\sigma_X^2}} \quad (x \geq 0)$
Random variable Y	$Y = N_{Hab}$ = number of habitable planets for man in the Galaxy = left-hand-side of the Dole equation
Probability distribution	Lognormal
Probability density function	$f_Y(y) = \frac{1}{y} \cdot \frac{1}{\sqrt{2\pi}\sigma_Y} e^{-\frac{(\ln(y)-\mu_Y)^2}{2\sigma_Y^2}} \quad (y \geq 0)$
Random variable Z	$Z = \frac{X}{Y} = \frac{N}{N_{Hab}} = \frac{Ns \cdot fp \cdot ne \cdot fl \cdot fi \cdot fc \cdot fL}{Ns \cdot fp \cdot ne \cdot fl} = fi \cdot fc \cdot fL$
Probability distribution	No name, but let us call it LOL = Lognormal-Over-Lognormal
Probability density function	$f_Z(z) = z^{\frac{\mu_X-\mu_Y}{\sigma_X^2+\sigma_Y^2}-1} \cdot \frac{1}{\sqrt{2\pi}\sqrt{\sigma_X^2+\sigma_Y^2}} e^{-\frac{(\mu_X-\mu_Y)^2+(\ln(z))^2}{2\left(\sigma_X^2+\sigma_Y^2\right)}} \quad (z \geq 0)$
Mean value	$\langle Z \rangle = e^{\mu_X-\mu_Y} e^{\frac{\sigma_X^2+\sigma_Y^2}{2}}$
Variance	$\sigma_Z^2 = e^{2(\mu_X-\mu_Y)+\sigma_X^2+\sigma_y^2}\left(e^{\sigma_X^2+\sigma_y^2}-1\right)$
Standard deviation	$\sigma_Z = e^{(\mu_X-\mu_Y)+\frac{\sigma_X^2+\sigma_y^2}{2}}\sqrt{e^{\sigma_X^2+\sigma_y^2}-1}$
All the moments (i.e., the k-th moment)	$\langle Z^k \rangle = e^{k(\mu_X-\mu_Y)} e^{k^2 \cdot \frac{\sigma_X^2+\sigma_Y^2}{2}}$
Mode (= abscissa of the lognormal peak)	$z_{\text{mode}} \equiv z_{\text{peak}} = e^{\mu_X-\mu_Y} e^{-(\sigma_X^2+\sigma_Y^2)}$
Value of the mode peak	$f_Z(z_{\text{mode}}) = \frac{1}{\sqrt{2\pi}\sqrt{\sigma_X^2+\sigma_Y^2}} \cdot e^{-(\mu_X-\mu_Y)} \cdot e^{\frac{\sigma_X^2+\sigma_Y^2}{2}}$

9.3 BREAKING THE DRAKE EQUATION DOWN INTO THE DOLE EQUATION TIMES THE SOCIETAL PART

We ar now in a position to achieve the main result of the present chapter, namely estimating the societal part of the Drake equation by virtue of both the full Drake equation and the Dole equation. The idea is as follows.

The Dole equation yields the number of habitable planets for man in the Galaxy. This number we regard as the product of the first four terms of the Drake equation, namely

$$N_{Hab} = Ns \cdot fp \cdot ne \cdot fl. \tag{9.17}$$

We regard this "new definition" of the Dole equation as acceptable inasmuch as for man to live on a certain planet not only must that planet exist, have liquid water, and have an atmosphere that is breathable, but also animals and fishes (= food) akin to those on Earth must be there. Thus, the *fl* term of the Drake equation must be included in the Dole equation inasmuch as it too symbolizes all sorts of living creatures (from trilobites to monkeys) that we know developed during evolution on Earth but may not be regarded as "intelligent" in the way that *Homo sapiens* is.

Having accepted (9.17) as the new definition of the Dole equation as a subset of the Drake equation (9.1), we may then break the Drake equation (9.1) up into two parts: the Dole equation (9.2) times the remaining last three terms of the Drake equation (9.1); that is

$$N = N_{Hab} \cdot fi \cdot fc \cdot fL. \tag{9.18}$$

From this partition it obviously follows that

$$fi \cdot fc \cdot fL = \frac{N}{N_{Hab}} = \frac{\text{Drake lognormal}}{\text{Dole lognormal}}. \tag{9.19}$$

Here we have made a "discovery"; that is: the societal part of the Drake equation (namely, the product of the three random variables $fi \cdot fc \cdot fL$), has a probability distribution given by the ratio of the two lognormal distributions of Drake (at the numerator) over Dole (at the denominator). This is the key new result presented in this chapter.

We may now use the mathematics developed in the previous section (Section 9.2) to immediately conclude that the pdf of the societal part of the Drake equation is given by (9.10), namely:

$$f_{fi \cdot fc \cdot fL}(z) = z^{\frac{\mu_X - \mu_Y}{\sigma_X^2 + \sigma_Y^2} - 1} \cdot \frac{1}{\sqrt{2\pi}\sqrt{\sigma_X^2 + \sigma_Y^2}} e^{-\frac{(\mu_X - \mu_Y)^2 + (\ln(z))^2}{2\left(\sigma_X^2 + \sigma_Y^2\right)}}. \tag{9.20}$$

The key result here is that we have identified the following three random variables

$$\begin{cases} N = X \\ N_{Hab} = Y \\ \dfrac{N}{N_{Hab}} = Z. \end{cases} \tag{9.21}$$

This of course implies identification of the relevant means and standard deviations as well; that is, it implies the following four equations:

$$\begin{cases} \mu_X = \mu_N = \langle N \rangle \\ \sigma_X = \sigma_N \\ \mu_Y = \mu_{Hab} = \langle N_{Hab} \rangle \\ \sigma_Y = \sigma_{N_{Hab}}. \end{cases} \tag{9.22}$$

Inserting (9.22) into (9.20), the latter is turned into its final form:

$$f_{fi \cdot fc \cdot fL}(z) = \frac{z^{\frac{\mu_N - \mu_{N_{Hab}}}{\sigma_N^2 + \sigma_{Hab}^2} - 1}}{\sqrt{2\pi}\sqrt{\sigma_N^2 + \sigma_{Hab}^2}} e^{-\frac{(\mu_N - \mu_{Hab})^2 + (\ln(z))^2}{2\left(\sigma_N^2 + \sigma_{Hab}^2\right)}}. \tag{9.23}$$

This is the probability density function followed by the societal statistics of human-like beings in the Galaxy. Is it a lognormal distribution? Of course not. But it reduces to a lognormal distribution in the special case

$$\mu_N = \mu_{Hab}. \tag{9.24}$$

This is the case when all habitable planets are indeed inhabited by humans (statistically speaking), so that the two mean values coincide. In fact, inserting (9.24) into (9.23), one finds that (9.23) reduces in this special case to:

$$f_{fi \cdot fc \cdot fL\_with\_\mu_N = \mu_{Hab}}(z) = \frac{z^{-1}}{\sqrt{2\pi}\sqrt{\sigma_N^2 + \sigma_{Hab}^2}} e^{-\frac{(\ln(z))^2}{2\left(\sigma_N^2 + \sigma_{Hab}^2\right)}}, \tag{9.25}$$

which is indeed a lognormal distribution with zero mean value and variance equal to $\left(\sigma_N^2 + \sigma_{Hab}^2\right)$.

9.4 CONCLUSIONS

We have written this chapter for the mathematical delight of showing how "playing mathematically" with both the Drake and Dole equations one may find new and unexpected results concerning the societal part of the Drake equation—something that, notoriously, nobody knows anything about.

This chapter is just a beginning. For instance: how could one modify the Drake equation by inserting a special further term dealing with the end of life caused by an asteroid impacting a living planet? Such a question deserves much more mathematical research work.

9.5 REFERENCES

[1] Wikipedia 2011 site about the Drake equation: *http://en.wikipedia.org/wiki/Drake_equation*

[2] Carl Sagan, *Cosmos*, Random House, New York, 1983. See pp. 298–302 in particular.

[3] Claudio Maccone, "The statistical Drake equation," paper #IAC-08-A4.1.4 presented on October 1, 2008, at the *59th International Astronautical Congress (IAC) held in Glasgow, Scotland, U.K., September 29–October 3, 2008.*

[4] Claudio Maccone, "The statistical Drake equation," *Acta Astronautica*, **67** (2010), 1366–1383. This was a special *Acta Astronautica* volume edited by John Elliott which collected all the papers presented at the *First IAA Symposium on "Searching for Life Signatures", UNESCO, Paris, September 22–26, 2008.*

[5] Claudio Maccone, "The statistical Drake equation and A. M. Lyapunov's theorem: Problems in the search for extraterrestrial intelligence, Part I," *International Scientific Journal "Actual Problems of Aviation and Aerospace Systems"*, **1**(32), volume 16, 2011, 38–63. This is essentially a Russian translation of [4].

[6] Claude E. Shannon, "A mathematical theory of communication," *Bell System Technical Journal*, **27**, 379–423, 623–656, 1948.

[7] Stephen H. Dole, *Habitable Planets for Man*, First Edition, 1964, RAND Corporation. See p. 82 in particular (i.e., the beginning of Chapter 5 entitled "Probability of occurrence of habitable planets").

[8] Athanasios Papoulis and S. Unnikrishna Pillai, *Probability, Random Variables and Stochastic Processes*, Fourth Edition, Tata/McGraw-Hill, New Delhi, 2002. See pp. 186–187 in particular .

[9] Claudio Maccone, "Societal statistics by virtue of the statistical Drake equation," paper #IAC-10.A4.2.8, presented at the *61st International Astronautical Congress (IAC), Prague, Czech Republic, September 27–October 1, 2010*, later published in *Acta Astronautica*, available online November 21, 2011.

10

Cubics of historical recovery

10.1 INTRODUCTION

Will you remember the famous men,
Who had to fall to rise again?
So take a deep breath,
Pick yourself up,
Dust yourself off,
Start all over again.

(*Pick Yourself Up* from the Fred Astaire and Ginger Rogers movie *Swing Time* of 1936, lyrics by Dorothy Fields and music by Jerome Kern: *http://lyricsplayground.com/alpha/songs/p/pickyourselfup.shtml*)

These words seem to summarize well a trend that often happened in history: an individual, or a civilization, rises from obscurity, reaches a peak, then falls to a minimum, but finally rises again and at such a high speed that even all previous achievements are dwarfed.

Given the above trend, or "law", this chapter is an attempt to cast it into a simple mathematical model built upon an algebraic equation of the third degree (a "cubic"). In the author's view, this cubic might be capable of summarizing known things, then providing some deeper insight into the chain of events, so as to yield some predictions about the future.

An example of the law we wish to investigate mathematically is provided by the 2000-year-old history of astronomy. The Ancient Greeks in the person of Aristarchus (floruit circa 290 BC at the peak of Greek astronomy) were the first to realize that the Earth revolves around the Sun. Greek astronomy, however, was lost during the time of the Roman Empire and later the Dark Ages, and it was not until the Renaissance that it was rediscovered and published by Copernicus (1543 AD).

This chapter tries to describe that string of events by virtue of a simple algebraic cubic curve, in the hope that it might be the starting point for deeper mathematical investigations.

Moreover, this chapter is intended to be a tribute to Carl Sagan (1934–1996), with whom the author had the honor and privilege to be acquainted on the eve of the opening of the NASA SETI program, which began October 11, 1992, in Barstow, California. Carl Sagan's seminal book *Cosmos* (based on his 13-part television series [1]) inspired the author to develop the mathematical model developed in this chapter. In particular, the figure on p. 335 in Part XIII of *Cosmos* (titled "Who Speaks for Earth?"), was the impetus for the author to consider "casting into some simple mathematical formula":

(1) the achievements of ancient Greek astronomy in the time of Aristarchus (290 BC);
(2) the Dark Ages of the Middle Ages; and
(3) finally the Copernican revolution (1543) and its aftermath, up to the Moon landing and, beyond, into the future.

The simplest formula the author could conceive was simply a cubic fit.

The fit appears to be somewhat remarkable: the resulting inflection point of the cubic indeed corresponds to a turning point in history. The years (around 321 AD) when the Roman Emperor Constantine openly accepted the change in religion from Paganism to Christianity. Not a small change!

Another good fit of our model occurs in the midst of the Dark Ages, predicted by our cubic model to occur around the year 932 AD. Actually, this was the time when the dissolution of the former Empire of Charlemagne reached its peak. More notable coincidences are pointed out in Section 10.3.

10.2 HISTORY OF ASTRONOMY AS A CUBIC

To a mathematician, the history of astronomy is charming (apart from the charm of the topic itself!) as it clearly shows a peak (Aristarchus, 290 BC), then a fall (Dark Ages), and then a rise again (Copernicus, 1543, and beyond). It is natural to think of a simple mathematical curve that would adapt to these three key features. The easiest fitting curve appears to be a cubic (i.e., an algebraic equation of the third degree) in which time t is obviously the independent variable:

$$Cubic(t) = a_3 \cdot t^3 + a_2 \cdot t^2 + a_1 \cdot t + a_0. \tag{10.1}$$

Our task is then to find "intelligent formulas" for the four coefficients a_3, a_2, a_1, and a_0.

This may be done in different ways. In the first three sections of this chapter, we confine ourselves to describing a procedure that appears to us to fit the history of astronomy the best:

(1) we require the cubic's maximum to occur at the time when Aristarchus first claimed (about 290 BC) the Solar System to have all its planets revolving around the Sun; and

(2) we require that Copernicus rediscovered Aristarchus' results in the year 1543, the year of both his death and of the publication of his book *De Revolutionibus Orbium Coelestium.*

In the next section, we cast the above two requirements into equations and prove that they are sufficient to let us infer, mathematically at least, the most important dates in the history of astronomy and, indeed, of civilization, as outlined by Carl Sagan in [1].

10.3 CASTING ARISTARCHUS AND COPERNICUS INTO EQUATIONS

Let us first agree on a convention that we would like to adopt: on the horizontal axis of time we denote dates in the Christian era by positive numbers, while dates before Christ are denoted by negative numbers.

In so assuming, we have really requested our cubic to pass through two points:

(1) The Aristarchus point

$$Maximum\_time \equiv Mt = -290 \tag{10.2}$$

at which the value of the cubic equals a certain maximum value My supposed to be known:

$$Cubic(Mt) \;= My. \tag{10.3}$$

Equations (10.1), (10.2), and (10.3) then amount to

$$a_3 \cdot Mt^3 + a_2 \cdot Mt^2 + a_1 \cdot Mt + a_0 = My. \tag{10.4}$$

This is a first linear algebraic equation relating the four unknown coefficients a_3, a_2, a_1, and a_0 and the known My.

(2) The Copernicus point

$$Recovery\_time\_Rt = 1543 \tag{10.5}$$

at which the cubic equals another value Ry supposed to be known:

$$Cubic(Rt) \;= Ry. \tag{10.6}$$

This equation, when inserted into (10.1), yields

$$a_3 \cdot Rt^3 + a_2 \cdot Rt^2 + a_1 \cdot Rt + a_0 = Ry. \tag{10.7}$$

This is a second linear algebraic equation among the four unknown coefficients a_3, a_2, a_1, and a_0 and the known quantity, Ry.

There is, however, a conceptual difference between the Aristarchus and the Copernicus points. While the Copernicus point simply belongs to the cubic, the Aristarchus point actually is a maximum of the cubic (the capital M in the maths stands for "maximum" and is used simply to distinguish it from the "minimum", written in lowercase). Now, a maximum may exist only if the cubic's first derivative equals zero at the maximum itself. Thus, by differentiating the cubic (10.1) with respect to time, one gets

$$\frac{d\ Cubic\ (t)}{dt} = 3 \cdot a_3 \cdot t^2 + 2 \cdot a_2 \cdot t + a_1 = 0 \tag{10.8}$$

in which time t must be replaced by the Aristarchus time $Mt = -290$, thus yielding:

$$3 \cdot a_3 \cdot (Mt)^2 + 2 \cdot a_2 \cdot (Mt) + a_1 = 0. \tag{10.9}$$

This is a third linear algebraic equation among the four unknown coefficients a_3, a_2, a_1, and a_0 and the known Mt.

Summarizing, so far we have found the three linear algebraic equations (10.4), (10.7), and (10.9) in the four unknown coefficients a_3, a_2, a_1, and a_0. Clearly, our problem of fitting the cubic to the history of astronomy would be completely solved were we to find one more independent linear algebraic equation relating the four unknown coefficients a_3, a_2, a_1, and a_0.

At this point, we get a pleasant surprise: it only suffices to know the first three coefficients of the cubic, a_3, a_2, and a_1, in order to determine the cubic's behavior in time and apply it to the history of astronomy. In other words, we don't need to know the precise value of a_0 (called the "known term" in the language of equations) in order to fit the cubic to the history of astronomy. To see why, note that a_0 is the distance along the vertical axis between the origin and the cubic intercept with the vertical axis itself. So, changing a_0 simply amounts to pulling the whole cubic up or down without changing in any way the relative positions of its maximum, its minimum, or the inflection point located in between them. A simpler way of saying this is if one multiplies or divides all four coefficients a_3, a_2, a_1, and a_0 by the same number, then the scale of the vertical axis is changed, but not the relative positions of the maximum, minimum, and inflection point.

Let us now determine the positions of the maximum, the minimum and the inflection point along the time axis:

(1) Epoch of the Maximum. This is the lower valued root of the second-degree algebraic equation (10.9) and is given by

$$t_{Maximum} = -\frac{a_2}{3\ a_3} - \frac{\sqrt{a_2^2 - 3\ a_1\ a_3}}{3\ a_3}. \tag{10.10}$$

(2) Epoch of the minimum. This is the higher valued root of the second-degree algebraic equation (10.9) and is given by

$$t_{minimum} = -\frac{a_2}{3\ a_3} + \frac{\sqrt{a_2^2 - 3\ a_1\ a_3}}{3\ a_3}. \tag{10.11}$$

(3) Epoch of the inflection point. This is the zero of the second derivative of the cubic (10.1), which is the (only) zero of the first-degree equation

$$\frac{d^2 \; Cubic \; (t)}{dt^2} = 6 \cdot a_3 \cdot t + 2 \cdot a_2 = 0. \tag{10.12}$$

Solving this for t, its zero turns out to be

$$t_{\text{inflection}} = -\frac{a_2}{3 \; a_3}. \tag{10.13}$$

(4) Symmetry of the cubic with respect to its own inflection point. Every cubic is an "odd function" (i.e., it is symmetrical with respect to its own inflection point). A glance at the plot reveals this to be obvious, but could be rigorously proven by transporting the origin to the inflection point and then noting that the new cubic equation does not change by replacing positive values with negative ones along both the horizontal and vertical axis. We shall not waste time by proving that here, but we do point out that equations (10.10), (10.11), and (10.13) clearly reveal that the inflection point is located halfway between the maximum and the minimum, with the distance on both sides given by

$$\frac{t_{mimimum} - t_{Maximum}}{2} = \frac{\sqrt{a_2^2 - 3 \; a_1 \; a_3}}{3 \; a_3}. \tag{10.14}$$

(5) Epochs of the cubic's zeros. The cubic equation (10.1) has, in general, three complex zeros given by the formulas of Cardan, lengthy expressions involving cubic and square roots that are given in analysis textbooks (the details of which we do not report here). Depending on the actual values of the four coefficients a_3, a_2, a_1, and a_0, only one of these roots may however be real and two of them complex conjugate. Further, there is as a special subcase of the last case when the two complex-conjugate zeros might have a zero imaginary part (i.e., in practice there can be two coinciding real zeros plus a third, real and different zero). This case will be of particular interest to us in the next section.

10.4 THE SCIENTIFICALLY "TRUE OR FALSE" CONVENTION

Let us now adopt the following second convention: on the vertical axis, negative numbers correspond to times when "people regarded as true what nowadays is regarded by science as false." Then, in astronomy, the cubic curve that we are seeking must have negative values for times shorter than −290 (times before Aristarchus) and after Aristarchus but before Copernicus (dark ages of astronomy), while the cubic must be positive after Copernicus.

Let us look carefully at the graph shown in Figure 10.1. This is our way to "show" or "summarize" the history of astronomy by virtue of the cubic (10.1). A glance at this graph reveals that:

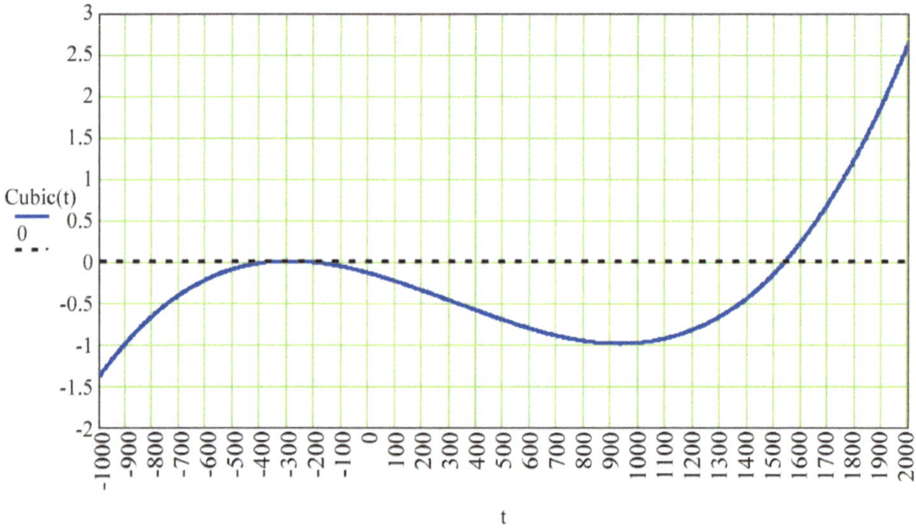

Figure 10.1. History of astronomy as a cubic.

(1) Before Aristarchus all values of the cubic are negative, meaning that the truth (e.g., the Earth's 24-hour rotation and its revolution around the Sun in just one year) had not yet been understood. Aristarchus was the first to understand this around the year 290 BC. According to St. Andrew's University (*http://www.gap.dcs.st-and.ac.uk/~history/Mathematicians/Aristarchus.html*), Aristarchus lived in between 310 and 230 BC and, according to the site *http://www.gap.dcs.st-and.ac.uk/~history/Chronology/500BC_1AD.html#290*, he put forward his revolutionary theory around 290 BC. Unfortunately, Aristarchus was not understood by his fellow countrymen. The *Encyclopedia Britannica* reports (*http://www.britannica.com/eb/article?eu=9551*) that Cleanthes the Stoic declared that Aristarchus ought to be indicted for impiety. So, Aristarchus' greatest discovery hardly had any immediate influence of the development of astronomy, and the whole world had to wait for 1,833 more years for the truth to come out. Along our cubic curve, Aristarchus thus represents the "single point" with coordinates $(Mt, 0)$. This is precisely the case where the cubic has two coinciding real roots and this root is a zero as well, as we pointed out at the end of Section 10.3. In Jesus' time (0–33 AD), the decline of the Greek civilization (negative values and negative derivative of the cubic) had begun already. This is historically correct, inasmuch as the Romans (from about 0 AD through 476 AD) were in essence culturally inferior to the Greeks.

(2) After Aristarchus and before Copernicus (1543 AD) all the cubic's values are negative again. These were the dark ages of Astronomy, and, indeed, the Dark Ages of civilization too. Our graph shows the midst of the Dark Ages was reached around the year 932, when Western Europe was lying in a state of virtually complete feudal anarchy. In addition, Vikings, Magyars, and Muslims

were freely raiding Western Europe. A slow recovery from this havoc only began after 955 AD, when the Holy Roman Emperor Otto I (i.e., German Emperor) defeated the Magyars at Lechfeld. The year 1000 AD is usually regarded as "when the rebirth of Western civilization began."

(3) Note that after 1000 AD the recovery of the West proceeded at a much faster pace than that of its earlier decline. Actually, this pace was twice as fast! This fact is faithfully mirrored by our cubic. In fact, on the one hand, the decline time span (the amount of time between the maximum and the minimum) is given by (10.14) multiplied by 2; that is

$$Decline\_Time\_Span = t_{minimum} - t_{Maximum} = 2\,\frac{\sqrt{a_2^2 - 3\,a_1\,a_3}}{3\,a_3}. \qquad (10.15)$$

On the other hand, the time span (which we call the "recovery time span") in between the minimum at 932 AD and the zero at Copernicus (1543) is given by

$$Recovery\_Time\_Span = Rt - t_{\text{minimum}} = \frac{\sqrt{a_2^2 - 3\,a_1\,a_3}}{3\,a_3}. \qquad (10.16)$$

A full mathematical proof of (10.16) would fill pages and pages of radicals (i.e., Cardan's formulas) (*http://mpec.sc.mahidol.ac.th/numer/FORMULAE. HTM*). See Appendix 10.B for the relevant Macsyma file. But the final result is simple and neat:

$$Recovery\_Time\_Span = \frac{Decline\_Time\_Span}{2}. \qquad (10.17)$$

(4) After Copernicus, of course, all values of the cubic are positive. Giordano Bruno in 1582 was the first not only to fully accept the Copernican structure of the Solar System, but also to understand that the Sun is a just star like many, and that other living beings might well inhabit the planets orbiting other stars (i.e., the true historical beginning of SETI). Bruno was burned at the stake on February 17, 1600 by the Inquisition. But the truth could no longer be hidden. After Kepler, Galileo, and Newton, science started to free itself from religious assumptions of any kind.

10.5 WHAT IS THE REAL MEANING OF THE CUBIC'S VERTICAL AXIS?

So far, we have carefully avoided talking about the cubic's vertical axis. Our second convention, where the vertical axis is mentioned, is not affected by the previous discussion, since the latter simply involves the sign of the positive vs. negative values of the vertical axis. So, it is natural to ask: "What is plotted along the vertical axis of the cubic?"

This question may be given different answers according to the research field that each author wants to investigate. For instance, geography can be related to history if one plots along the vertical axis the size of explored land vs. time. In other words, we may plot the distance from Greece (or from Western Europe, assumed as "the civilization center") actually explored by Western Europeans century after century: this distance then stretches to America in 1492 and to the Moon in 1969. In mathematical terms, the last remark means that we introduce one more instant in time, hereafter called "extra time" and denoted by *Et*, at which the value of y is known (and denoted yEt). For instance, if the extra time is assumed to be the Moon landing, i.e.

$$Extra\_time \equiv Et = 1969 \tag{10.18}$$

the distance along the vertical axis is then the distance of the Moon from the Earth, i.e.

$$y\_at\_Extra\_time \equiv yE = 384{,}000 \text{ km}. \tag{10.19}$$

Clearly, on this scale of distances, the "center of civilization" may now be assumed to be the Earth as a whole, rather then just Greece (as in Aristarchus' time) or Western Europe (as in Copernicus' time), and so our theory implicitly takes today's "globalization" into account as well.

Mathematically, upon inserting equations (10.18) and (10.19) into the cubic (10.1), one gets

$$a_3 \cdot Et^3 + a_2 \cdot Et^2 + a_1 \cdot Et + a_0 = yE. \tag{10.20}$$

This is a fourth linear algebraic equation among the four unknown coefficients a_3, a_2, a_1, and a_0. By adding this equation (10.20) to the three previous equations (10.4), (10.7), and (10.9), we thus get a system of four simultaneous algebraic equations in the four unknown coefficients a_3, a_2, a_1, and a_0, which solves the problem of finding the cubic completely. The actual expressions of a_3, a_2, a_1, and a_0 are found in the Maxima file "1_cubic_results_used_in_text.wxm", which makes up Appendix 10.A.

10.6 DETERMINING THE CUBIC IN TERMS OF *Mt* AND *Rt* ONLY

We think that there are at least three possible ways of choosing coefficients a_3, a_2, a_1, and a_0. These three alternatives are:

Problem 1. Normalizing the cubic to -1 (i.e., in falsity units), as will be done in Section 10.7, followed by two more examples.

Problem 2. Assigning the values of the Extra time *Et* and of the relevant *yEt*, as in (10.17) and (10.18), as proven in Appendix 10.A.

Problem 3. Assuming that the cubic actually is the autocorrelation of a certain stochastic process and, consequently, finding the process's spectrum (this is left for a forthcoming paper).

To solve Problem 1, we must solve the four simultaneous linear algebraic equations (10.9), (10.4), (10.7), and (10.20), respectively, in the four unknown coefficients a_3,

a_2, a_1, and a_0. The relevant coefficient matrix is

$$Coefficient\_Matrix = \begin{bmatrix} 0 & 1 & 2Mt & 3Mt^2 \\ 1 & Mt & Mt^2 & Mt^3 \\ 1 & Rt & Rt^2 & Rt^3 \\ 1 & Et & Et^2 & Et^3 \end{bmatrix} \tag{10.21}$$

and the relevant vector of the known terms is

$$Known\_Term\_Vector = \begin{bmatrix} 0 \\ My \\ Ry \\ Ey \end{bmatrix}. \tag{10.22}$$

In the solution, we assume that the three inequalities are fulfilled

$$Rt > Mt, \quad Et > Rt, \quad Ey > 0. \tag{10.23}$$

Then, the determinant of the coefficient matrix (10.21) looks very much like a Cauchy–Vandermonde determinant and, indeed, after factorization it turns out to equal

$$(Mt - Rt)^2(Rt - Et)\ (Rt - Mt)^2. \tag{10.24}$$

This remark explains why the solution we are seeking, which is formally given by

$$\begin{bmatrix} a_0 \\ a_1 \\ a_2 \\ a_3 \end{bmatrix} = Coefficient\_Matrix^{-1} \cdot Known\_Term\_Vector. \tag{10.25}$$

embodies, after the trivial but lengthy calculations described in the Maxima file in Appendix 10.A, the determinant (10.24) at all four denominators. Indeed, the solution is given by

$$\begin{bmatrix} a_0 \\ a_1 \\ a_2 \\ a_3 \end{bmatrix} = \begin{bmatrix} \text{first expression in Appendix 10.A} \\ \text{second expression in Appendix 10.A} \\ \text{third expression in Appendix 10.A} \\ \text{fourth expression in Appendix 10.A} \end{bmatrix}. \tag{10.26}$$

Appendix 10.A gives the actual expressions of the four unknown coefficients a_3, a_2, a_1, and a_0 as functions of the assumed-to-be-known Mt, Rt, Et, and Ey. Again, all calculations are trivial, but too lengthy do be done by hand. So, we did them using the "Maxima symbolic manipulator code", created by MIT for NASA back in the 1960s. Despite being so old, Maxima has at least one big advantage: virtually all bugs have now been fixed.

Next we must give the four equations (10.26), reported in Appendix 10.A, for the special assumption

$$My = 0 \quad Ry = 0 \tag{10.27}$$

in agreement with our second convention. Thus, under the assumption (10.27), the cubic's coefficients boil down to

$$\begin{bmatrix} a_0 = \dfrac{Ey\,Mt^2\,Rt}{(Mt-Et)^2\,(Rt-Et)} \\[2ex] a_1 = -\dfrac{Ey\,Mt\,(2Rt+Mt)}{(Mt-Et)^2\,(Rt-Et)} \\[2ex] a_2 = \dfrac{Ey\,(Rt+2\,Mt)}{(Mt-Et)^2\,(Rt-Et)} \\[2ex] a_3 = -\dfrac{Ey}{(Mt-Et)^2\,(Rt-Et)} \end{bmatrix}. \tag{10.28}$$

One can check the validity of (10.28), for instance, by inserting it into the maximum condition (10.11) and then getting, as expected,

$$t_{Maximum} = Mt. \tag{10.29}$$

More interesting results are found by inserting (10.28) into the expression of the time of the minimum (10.12), which simply becomes

$$t_{minimum} = \tfrac{1}{3}Mt + \tfrac{2}{3}Rt. \tag{10.30}$$

The corresponding y-axis value of the minimum is found by inserting (10.30) into the cubic (10.1), and reads

$$y_{minimum} = \frac{4\,Ey\,(Rt-Mt)^3}{27\,(Mt-Rt)^2(Rt-Et)}. \tag{10.31}$$

Similarly, upon inserting (10.28) into the time of the inflection point (10.13), the latter is expressed in terms of Mt and Rt only

$$t_{inflection} = \tfrac{2}{3}Mt + \tfrac{1}{3}Rt \tag{10.32}$$

with the corresponding y value given by

$$y_{inflection} = \frac{2\,Ey\,(Rt-Mt)^3}{27\,(Mt-Rt)^2(Rt-Et)}. \tag{10.33}$$

Along the y-axis, from (10.31) and (10.33) it follows that

$$y_{minimum} = 2\,y_{inflection}. \tag{10.34}$$

The decline time span, defined by (10.15), is re-expressed in terms of Mt and Rt, as

$$Decline\_Time\_Span = \tfrac{2}{3}Rt - \tfrac{2}{3}Mt, \tag{10.35}$$

while the recovery time span of (10.16) now becomes

$$Recovery\_Time\_Span = \tfrac{1}{3}Rt - \tfrac{1}{3}Mt. \tag{10.36}$$

Once again, the ratio of (10.36) to (10.35) equals 2:

$$Recovery\_Time\_Span = \frac{Decline\_Time\_Span}{2}. \tag{10.37}$$

Next, we found it useful to define the pre-Maximum time, denoted $t_{preMaximum}$, as the time before the Maximum at which the y value is as low as the y value of the minimum will later be. In the language of the history of astronomy, the pre-maximum time is a time before Aristarchus at which the knowledge of astronomy was as abysmal as it would later be only in the very midst of the Dark Ages. In mathematical language, the pre-Maximum time is defined as the smallest real root, $t_{preMaximum}$, of the algebraic equation of the third degree

$$y_{\text{minimum}} = a_3 \cdot t_{preMaximum}^3 + a_2 \cdot t_{preMaximum}^2 + a_1 \cdot t_{preMaximum} + a_0. \tag{10.38}$$

The analytic expression for $t_{preMaximum}$ is very long, since it is given by Cardan formulas, and will not be reported here. We just mention that, after many lengthy radical reductions, the pre-Maximum time, once expressed in terms of Mt and Rt only, turns out to be given by

$$t_{preMaximum} = \tfrac{4}{3}Mt - \tfrac{1}{3}Rt. \tag{10.39}$$

10.7 NORMALIZING THE CUBIC TO −1 (i.e., in recession or falsity units)

In this section we normalize the cubic in units of recession (i.e., in falsity units); in other words, we request the distance between the ordinates of the Maximum and minimum to equal one. Since in (10.28) the ordinate of the Maximum is set to zero, one must then request that the ordinate of the minimum equal −1; that is

$$y_{minimum} = -1. \tag{10.40}$$

This we call the "normalization condition of the cubic to −1." It solves, in some sense, the problem of understanding "what the vertical axis means" by saying that the vertical axis is measured in units of recession, or in units of falsity.

In order to find the new expressions of the four cubic coefficients a_3, a_2, a_1, and a_0 for this normalized-to-−1 cubic, we first equate the normalization-to-−1 condition (10.40) to the older expression (10.31) of the minimum ordinate

$$-1 = \frac{4\,Ey\,(Rt - Mt)^3}{27\,(Mt - Et)^2\,(Rt - Et)}. \tag{10.41}$$

This is a new equation enabling us to get rid of the extra-time ordinate Ey previously used to fix the vertical axis scale. Thus, solving (10.41) for Ey, we get

$$Ey = -\frac{27\,(Mt - Et)^2\,(Rt - Et)}{4\,(Rt - Mt)^3}. \tag{10.42}$$

Upon inserting (10.42) into each one of the four equations (10.28), they are turned into

$$\begin{bmatrix} a_0 = -\dfrac{27\,Mt^2\,Rt}{4\,(Rt - Mt)^3} \\ a_1 = \dfrac{27\,Mt\,(2Rt + Mt)}{4\,(Rt - Mt)^3} \\ a_2 = -\dfrac{27\,(Rt + 2\,Mt)}{4\,(Rt - Mt)^3} \\ a_3 = \dfrac{27}{4\,(Rt - Mt)^3} \end{bmatrix}, \tag{10.43}$$

which are the four coefficients of the cubic normalized to −1. These equations were used (without proof) in all the author's published papers on cubics.

10.8 HISTORY OF ASTRONOMY NORMALIZED IN FALSITY UNITS (i.e., units at the very depth of the Dark Ages)

An immediate application of (10.43) is of course the cubic of astronomy now normalized to −1. This cubic is defined by the two conditions: the Aristarchus point and the Copernicus point, respectively,

$$Mt = -290\,, \quad Rt = 1543. \tag{10.44}$$

No further condition is requested inasmuch as the vertical axis is now normalized in units of falsity (1 unit of falsity = the false ideas about astronomy they had in the midst of the Dark Ages, i.e. around 932 AD). Numerical calculations then yield:

(a) Transition from Paganism to Christianity

$$t_{\text{inflection}} = 321. \tag{10.45}$$

(b) Very midst of the Dark Ages

$$t_{minimum} = 932. \tag{10.46}$$

(c) How far back in ancient Greek history one has to go to find as many false ideas about astronomy as they had in the midst of the Dark Ages

$$t_{preMaximum} = -901. \tag{10.47}$$

(d) How many years it took for astronomy to reach the worst level of misunderstanding after the truth had first been discovered by Aristarchus

$$Decline\_Time\_Span = 1{,}222. \tag{10.48}$$

(e) How many years it took to rediscover the truth about astronomy since the very depth of the Dark Ages

$$Recovery\_Time\_Span = 611. \tag{10.49}$$

(f) Actual coefficients of the normalized-to-−1 cubic of astronomy, obtained by inserting (10.44) into (10.43):

$$\begin{bmatrix} a_0 = -0.14223 \\ a_1 = -8.88692 * 10^{-4} \\ a_2 = -1.05546 * 10^{-6} \\ a_3 = 1.09602 * 10^{-9} \end{bmatrix}. \tag{10.50}$$

(g) We would like to complete this section by making two historical remarks. The first one is summarized by the Latin sentence *Graecia capta ferum victorem coepit* (i.e., "the captured Greeks indeed captured [culturally] their own ferocious conquerors [the Romans]"). So said the Roman poet Horace (65 BC–8 BC), and he was right! He meant that the Greeks were always culturally superior to the Romans, even close to Jesus' time, when Horace lived. Our cubic simply quantifies this statement: the Greeks were already 17% culturally superior to the Romans at Jesus Christ's and Horace's time, and much more later on," as one can find by applying the two equations

$$t_{Jesus\_death} = 33 \quad cubic\_for\_Astronomy = -0.17266. \tag{10.51}$$

(h) The second remark is that the Copernicus point might actually correspond to the year 1514 rather than 1543. This is because, according to the University of St. Andrew (*http://www-gap.dcs.st-and.ac.uk/~history/Mathematicians/Copernicus.html*) Copernicus actually put forward his revolutionary ideas for the first time in 1514, rather than in 1543, in a handwritten (not printed) work called *Little Commentary*. Figure 10.2 compares these two cubics. It would be interesting to develop a "theory of errors" in this regard, but space prohibits us from doing so now.

10.9 HISTORY OF SETI AS A CUBIC

Let us now change the scenario completely. In this section we claim that even the history of SETI (the modern Search for ExtraTerrestrial Intelligence) can basically be cast as a cubic shape. SETI, in fact, had its true historic beginnings back in 1582, when the Italian scholar Giordano Bruno (1548–1600), in his work *De l'Infinito, Universo e Mondi* ("About Infinity, the Universe and the Worlds"), demonstrated that:

(1) The Copernican System was the right one.
(2) Our Sun is just one star of many. It looks much brighter only because it is much closer to us.

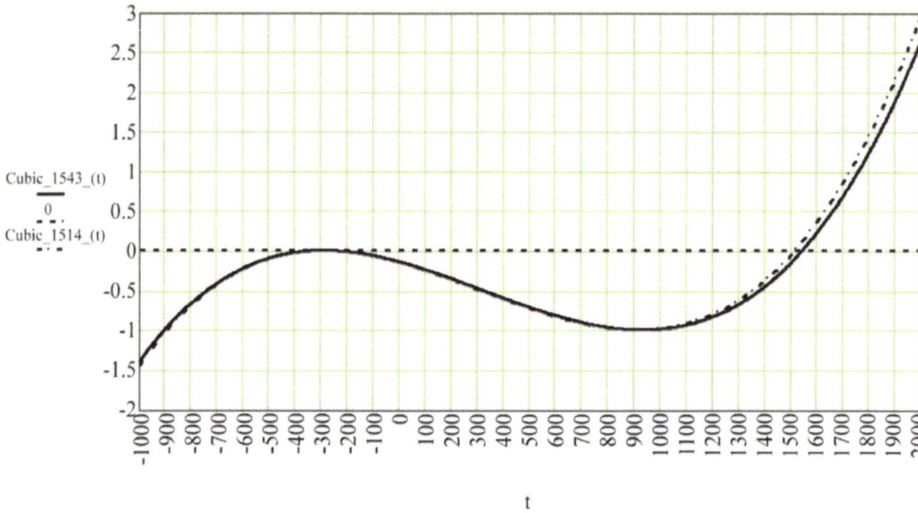

Figure 10.2. Two cubics for the two different Copernicus points: the 1514 (dash-dot) one and the 1543 one.

(3) Just as we know planets orbit our Sun, so we might infer that other planets orbit other stars (a fact not proven scientifically until 1995, when Michel Mayor and Didier Queloz of Geneva discovered the first extrasolar planet around the star 51 Pegasi[1]).
(4) Just as our planet is inhabited by humans and animals, so other planets might host living extraterrestrials.

Because of these great conceptual achievements by Bruno, we claim that the maximum of the cubic of SETI is given by

(1) The Giordano Bruno point:

$$Maximum\_time \equiv Mt = 1582. \tag{10.52}$$

Bruno was the Aristarchus of SETI. He was burned at the stake by order of the Roman Catholic Inquisition on February 17, 1600 in Rome (in Campo dei Fiori, where his memorial now stands). At least in part, he died at the stake because his doctrines were such a blow to the ruling Roman Catholic establishment of the time. Ironically, the name of the Pope who had Bruno burnt at the stake was Clemens VIII—in Latin, *clemens* means "merciful"!

(2) The Cocconi–Morrison point :

$$Recovery\_time \equiv \mathrm{Rt} = 1959. \tag{10.53}$$

Giuseppe (misspelled Guiseppe in many English textbooks) Cocconi (1914–2008)

[1] They made the discovery at OHP in France.

was an Italian physicist and member of the research group assembled by Enrico Fermi in Rome in the 1920s and 1930s and later was director of the CERN proton synchrotron in Geneva. Phil Morrison (1915–2005) was a physics professor at MIT. In 1959, Cocconi and Morrison published a seminal paper proposing the potential of microwaves in the search for interstellar communications. Their paper marked the beginning of modern, experimental SETI, and led to Frank Drake in 1960 making the first search for possible intelligent signals around 1.420 GHz from two nearby stars (ε Eridani and τ Ceti) .

By virtue of the two conditions (10.52) and (10.53), the four equations (10.28) thus yield a new cubic that we call the cubic of SETI (it is plotted in Figure 10.3).

To see that this cubic makes sense, consider (10.11) which yields

$$t_{minimum} = 1833. \tag{10.54}$$

This was right in the middle of the darkest days of the Restoration in Europe, namely the 33 years that elapsed between the fall of Napoleon (1815) and the European Revolution of 1848. (*Note*: We do not regard the Parisian revolution of July 1830 as very significant. In fact, it only brought changes to the history of France—not to the history of Europe as a whole, as the subsequent revolution of 1848 did). As a result of Giordano Bruno's death in 1600 and continued strong persecution by the Roman Catholic Church and most other national churches against individuals making such claims as "we are not alone in the universe" people had little choice but to keep silent about their scientific beliefs. In any event, the technological know-how necessary to prove the existence of ETs was not available until about 1950.

Indeed, the decline time span, given by (10.15), for SETI is, in years,

$$Decline\_Time\_Span = 251 \tag{10.55}$$

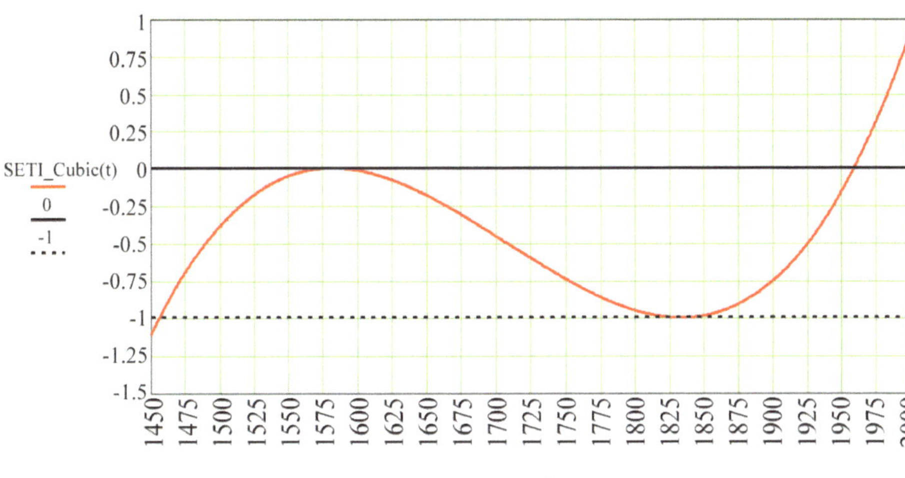

Figure 10.3. History of SETI as a cubic—as usual, normalized to −1 (i.e., in falsity units).

the SETI inflection point, by (10.5), falls at

$$t_{inflection} = 1707 \tag{10.56}$$

and, finally, the recovery time span, by (10.16), is

$$Recovery\_Time\_Span = 125. \tag{10.57}$$

The pre-maximum time falls at

$$t_{preMaximum} = 1456. \tag{10.58}$$

(corresponding to the time of the Renaissance and a new conception of the universe), with actual coefficients of the SETI cubic

$$\begin{bmatrix} a_0 = -617.6282 \\ a_1 = 1.0961 \\ a_2 = -6.45363 * 10^{-4} \\ a_3 = 1.25974 * 10^{-7} \end{bmatrix}. \tag{10.59}$$

10.10 EXTRAPOLATING THE TWO CUBICS TO THE INVENTION OF THE WARP DRIVE (2067 AD?)

So far, we have only fitted our two cubics to human history (i.e., the past few centuries). This is interpolation. Let us now try to have a look at our future; to do this we need to extrapolate. To this end, let us plot both the cubics of astronomy and SETI on the same plot (Figure 10.4). Note that only "the right branch" of the cubic of astronomy appears in this plot because the time span only starts at 1450 AD

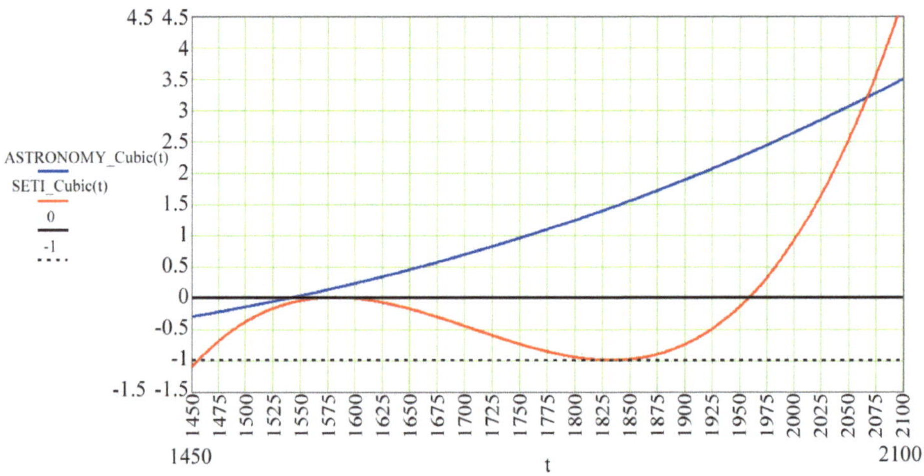

Figure 10.4. Combining the two cubics of astronomy and SETI and finding their intercept.

(as was the case in Figure 10.3). Moreover, the plot now extends up to the end of the present century (i.e., to 2100 AD).

It is immediately apparent that, after the year in which Cocconi and Morrison's work was published (1959), the SETI cubic is much steeper than the astronomy cubic. In other words, SETI is developing much faster (in importance) than the rest of traditional astronomy. The prediction is that SETI will become even more important than astronomy itself after 2067 AD (see the abscissa of the intercept between the two cubics in Figure 10.4).

Does this prediction of our theory make sense? Well, nobody can predict the future, of course. But if we are to believe the *Star Trek* TV series ... well the year 2063 will be extremely significant! It will be the year when Zephram Cochrane (born in 2030) invents the warp drive and pilots Earth's first faster-than-light space flight (see [6]).

Science fiction only? May be not. In the years 1996–2002, NASA's Breakthrough Propulsion Physics program (BPP) (*http://www.grc.nasa.gov/WWW/bpp/*) was the first NASA forum to let experts in relativity and quantum field theory study the possibilities of faster-than-light (FTL) space flight—not to mention other photon-state teleportation experiments that seem to be challenging special relativity itself!

But let us go back to Figure 10.4 and take a few final, bold steps of extrapolation into the future. Let us assume that a variable on the vertical axis represents the "distance from Earth reached by humans at various epochs". Under this assumption, the ordinate corresponding to the year 1969 must then be the Moon distance (i.e., 384,000 km). Having calibrated the two cubics in this fashion, when will humans reach the nearest star, α Centauri, at the distance of 4.29 light-years? Here are some answers:

(1) According to the (slow) cubic of astronomy, humans will reach α Centauri in 612,810 years.
(2) According to the (steeper – faster) cubic of SETI, humans will reach α Centauri in 55,803 years.
(3) But these predictions are rather discouraging! More realistically, then, we may replace the Moon distance of 1969 by the Voyager 1 distance of 2004 (i.e., we consider robotic probes instead of manned flights) and recalibrate the vertical axis accordingly. In this case the results are more encouraging.
(4) According to the (slow) cubic of astronomy, robotic probes will reach α Centauri in 18,220 years.
(5) According to the (steeper = faster) cubic of SETI, robotic probes will reach α Centauri in 4,377 years.
(6) We are now reaching the ultimate conclusion. We now claim that only result (5) is the right one! In other words, we claim that human robotic probes will reach α Centauri in 4,377 years, or in about 4,000 years from now. The reason for our claim is that the cubic of SETI is so steep that it shows interest in SETI will overcome general astronomy as far as the general public is concerned at least from the year 2060 onward. It is to some extent already the case now (2011): many people believe in UFOs, thus revealing their need for contact with alien

civilizations. What is more, since 2005 both NASA and ESA have been planning space missions tasked with finding other Earths outside the Solar System. These two missions (robotic high-tech telescopes to be located at Lagrangian point L2 of the Earth–Sun system) are TPF (Terrestrial Planet Finder, unfortunately canceled in 2011) and Darwin (*http://en.wikipedia.org/wiki/Darwin_%28 spacecraft%29*). Should other Earths be found, it will be very difficult for humans to refrain from sending robotic probes to them. Thus, we may well conclude that the evolution of human knowledge on Earth has reached an unprecedented peak, one that is actually paving the way to human expansion into space.

10.11 EXTRASOLAR PLANETS SINCE 1995

Just as SETI officially started in 1960, so the search for extrasolar planets officially started on October 6, 1995. This was when Michel Mayor and Didier Queloz of the Geneva Observatory announced the (later confirmed by others) discovery of the first extrasolar planet: 51 Pegasi B.

In the last 16 years (1995–2011) progress in this field has been enormous; more than 1,500 extrasolar planets are now known. For a good summary visit Wikipedia (*http://en.wikipedia.org/wiki/Extrasolar_planet*) or the Extrasolar Planets Encyclopedia maintained by Jean Schneider of the Paris Observatory (*http://vo.obspm.fr/exoplanetes/encyclo/encycl.html*). However, the size of discovered planets was initially rather large ("hot Jupiters") but, with technological advances, it has become smaller and smaller (apparently up to 7.5 Earth masses)—so we are now getting closer to detecting new Earths (the aspiration of SETI).

Despite TPF being canceled in 2011, two space missions are currently being planned to detect Earth-sized planets:

(1) NASA-JPL Space Interferometry Mission (SIM). Launch expected in or about 2014 (*http://en.wikipedia.org/wiki/Space_Interferometry_Mission*).
(2) ESA's Darwin, launch expected in or about 2015 (*http://en.wikipedia.org/wiki/Darwin_%28ESA%29*).

Much more could be said about extrasolar planets, but we prefer to close this section by telling the story of the Dutch American astronomer Piet (Peter) van de Kamp, which we'll use later in this chapter as a test case for our mathematical cubic model of the development of human abilities.

The unfortunate story of Peter van de Kamp was masterfully told by Alan Boss in his excellent and popular book about the discovery of extrasolar planets, *Looking for Earths* (with the subtitle "The Race to Find New Solar Systems") [7]. On April 18, 1963, van de Kamp announced the discovery of the first extrasolar planet around Barnard's star, the fastest moving star in the sky that had been discovered by astronomer Edward E. Barnard in 1916 and lies only 5.96 light-years away from the Sun. Van de Kamp claimed that he had detected a wobble in this star's motions suggesting that a planet with a mass only 60% greater than that of Jupiter was orbiting around it. Ten years elapsed without anyone questioning van de Kamp's claims. However, by 1973 astronomers George Gatewood and Heinrich Eichhorn had remeasured the motion of Barnard's Star and found no wobble. Later still, it became clear that the wobble that van de Kamp had seen was actually due to an ill-placed lens in the telescope he had used for his observations: when observed by other telescopes, Barnard's star showed no wobble, and so van de Kamp had *not* discovered the first extrasolar planet after all!

10.12 HISTORY OF EXTRASOLAR PLANET DISCOVERIES IN OUR CUBIC MODEL

The sad story of van de Kamp's announced and failed discovery of the first extrasolar planet in 1963 will now be used together with the true story of Mayor and Queloz's actual discovery of the first extrasolar planet (1995). In other words, we fit the years 1963 and 1995 into our cubic model as:

(1) The van de Kamp point:

$$Maximum\_time \equiv Mt = 1963. \tag{10.60}$$

(2) The Mayor–Queloz point:

$$Recovery\_time \equiv Rt = 1995. \tag{10.61}$$

By virtue of the two conditions (10.60) and (10.61), the four equations (10.28) thus yield a new cubic that we call the "cubic of exoplanets"; it is plotted in Figure 10.5.

10.13 MATCHING (THE TWO CUBICS OF) SETI AND EXOPLANETS!

We have thus found two cubics: the cubic of SETI (shown in Figure 10.3) and the cubic of exoplanets (shown in Figure 10.5). Where do they intercept? The answer to this question is provided by Figure 10.5. Please note that the two cubics do not intercept in the 1960s, since the cubic of SETI crosses the horizontal time axis in 1959 (the Cocconi–Morrison point) whereas the cubic of exoplanets is only tangential to the time axis in 1963 (the van de Kamp point). So, the only true interception occurs much later (actually around May 1998), as one might find by solving the resulting algebraic equation of the third degree, which we won't describe here in detail for the sake of brevity.

Comparison of the two cubics of SETI and exoplanets leads to the following straightforward conclusion: the two scientific fields *merged* in May 1998. In other words, we have found the answer to the question: When are SETI and extrasolar planet search going to merge? In yet other words: SETI and the search for exoplanets are little more than the two sides of the same coin!

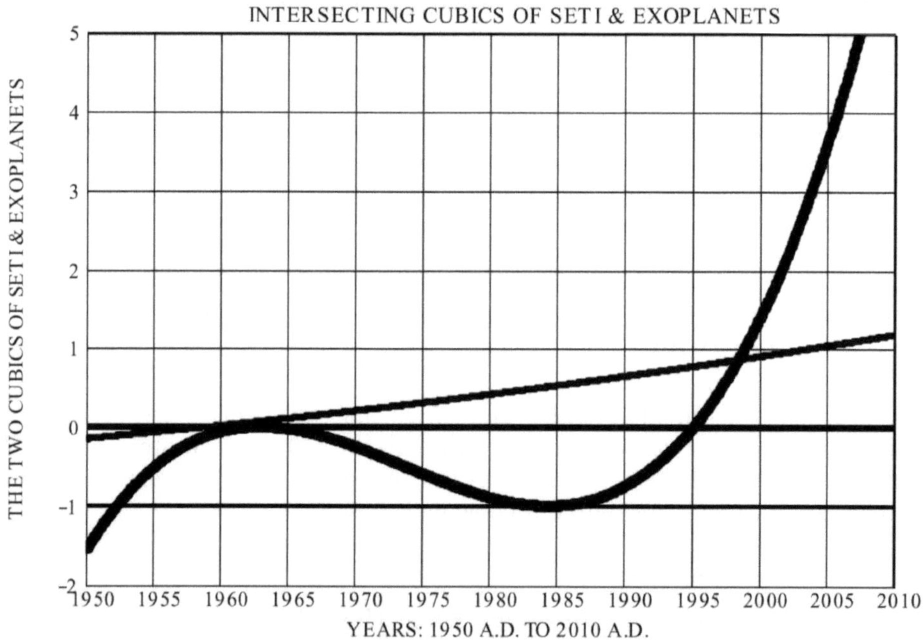

Figure 10.5. The two cubics of SETI (thinner non-horizontal increasing curve) and of exoplanets (thicker curve, lower than the other one until May 1998) intercept each other about May 1998.

This sends a clear message to all SETI radioastronomers: *search all the known exoplanets for SETI signals right now!* These SETI searches may not be on the hydrogen line only! *Do all possible SETI searches on the known exoplanets: next to a "silent" Hot Jupiter there might be a little Earth*, just like in the Solar System! Who knows?

10.14 HISTORY OF THE UNIFICATION OF EUROPE AS A CUBIC

The author does not have sufficient knowledge of the history of the United States to provide an example of "cubic behavior" taken from American history. However, he believes he has enough knowledge of the history of Europe to produce the example of cubic behavior provided in this section.

The introduction of the unified European currency (the euro) on January 1, 2002, is usually regarded as a landmark event ensuring that the unification of Europe will sooner or later be achieved. In the past, the unification of Europe has been a goal, despite being forcibly imposed on the peoples of Europe than freely accepted by them. So (disregarding British opinions!) we consider

(1) The Napoleon point

$$Maximum\_time \equiv Mt = 1812. \tag{10.62}$$

We think that Napoleon went closer than anyone else to unifying Europe. The peak of Napoleon's power occurred in the earlier part of 1812. We won't analyze how justified or unjustified his wars were. We simply take for granted that his time meant the spread of the ideology behind the French Revolution throughout continental Europe—considered by many as a good thing. So, we assume 1812 to be the maximum of our cubic.

(2) The euro (currency) point. The euro, the single currency partly introduced to unify Europe, despite being in disarray at the time of publication, became effective on January 1, 2002. Clearly this date, much more than sound bites, may be regarded as the most important achievement reached so far towards unifying Europe. So, we set

$$Recovery\_time \equiv Rt = 2002. \tag{10.63}$$

The cubic of European unification is now set up. We want to analyze the mathematical consequences of this cubic. First, from (10.39), the pre-Maximum time turns out to be

$$t_{preMaximum} = 1748. \tag{10.64}$$

This was the time of the peace of Aachen, signed at the end of the war of Austrian Succession. More importantly, the years around 1750 may be regarded as the beginning of the Enlightenment in Europe. The Enlightenment continued developing until the outbreak of the French Revolution in 1789. Moreover, it was 1748 when Leonhard Euler published the first modern and extensive textbook on calculus (*Institutiones Calculi Differentialis*) embodying, for instance, the exponential form of complex numbers. So, the years around 1748 do appear significant as the beginning of a new age of free thought—a clear break from the past.

From (10.62), (10.63), and (10.32), we also infer that the inflection point in the history towards a unified Europe occurred around the year

$$t_{inflection} = 1875. \tag{10.65}$$

This was indeed the (dark) peak time of imperial nationalism all over the world—the British Empire, German Empire, Russian Empire, Japanese Empire, U.S. expansion in the Pacific, and so on—leading to World War One in 1914. It was also the time of blind nationalism, so much so that even the brightest scientists found it natural to publish their scientific results only in their own national language, thus hampering the progress of science itself.

But the most frightening (and realistic) prediction of our cubic occurs at the minimum. Because of (10.62), (10.63), and (10.30), the minimum is found to occur in the year

$$t_{minimum} = 1938. \tag{10.66}$$

This was the start of the darkest age of them all in Europe: World War Two, the Nazi extermination camps, and the Holocaust. No more comments needed.

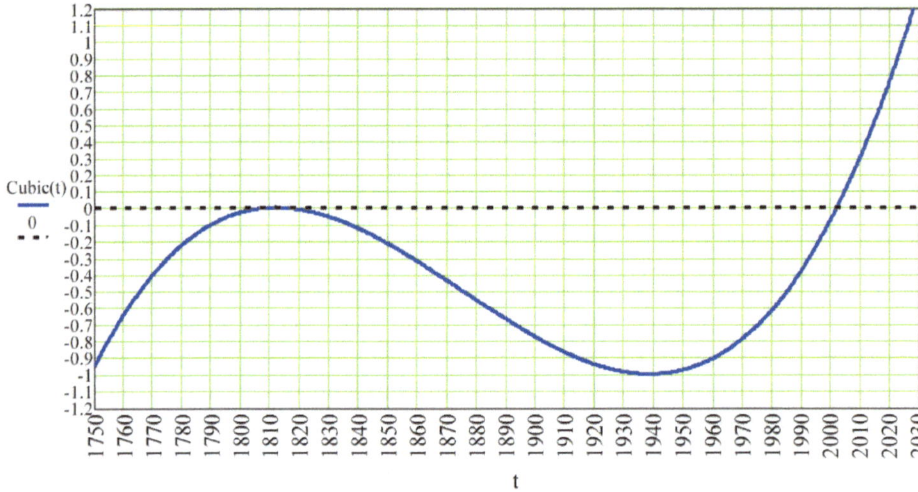

Figure 10.6. History of the unification of Europe as a cubic.

It was not until after 1945 that signs of moral and political resurgence show up in Europe again, as well as "dreams" about the unification of Europe. From (10.62), (10.63), and (10.36), our cubic shows that it took 63 years for Europe to emerge from the dark

$$Recovery\_Time\_Span = 63. \tag{10.67}$$

This time span, however, was but half the time it had taken Europe to sink into the barbaric state that led to the two world wars

$$Decline\_Time\_Span = 126. \tag{10.68}$$

as (10.36) and (10.37) show.

In conclusion, by checking the value of the cubic for the year 2025, that is

$$\begin{cases} t_{2025} = 2025 \\ cubic\_for\_United\_Europe = 1.0269 \end{cases} \tag{10.69}$$

we infer that only around 2025 will Europe be as united as it was disunited at the time of the two world wars!

The four numerical coefficients of our cubic model for the history of the unification of Europe are:

$$\begin{bmatrix} a_0 = -6.4688 * 10^3 \\ a_1 = 10.37112 \\ a_2 = -5.53659 * 10^{-3} \\ a_3 = 9.84108 * 10^{-7} \end{bmatrix}. \tag{10.70}$$

Figure 10.6 shows the cubic of the unification of Europe, having the coefficients given by (10.70).

10.15 HUMAN LIFE EXPECTANCY CUBIC

Life expectancy is the expected (in the statistical sense) number of years of life remaining at a given age. From the SETI perspective, we should add "for humans", since it may have quite different values "for aliens". In addition, the life expectancy of humans was clearly much shorter in the past than it is nowadays. Of course, average living conditions were much worse in the past than they are nowadays. For instance, Wikipedia (*http://en.wikipedia.org/wiki/Life_expectancy*) reports that the current average world life expectancy is 67.2 years at birth. The same source, however, clearly shows that life expectancy at birth was much lower in the past, reaching an all-time-low minimum of 20 years during the Neolithic (see Table 10.1). In other words, during the Upper Paleolithic (roughly 10,000 years ago) the average life expectancy was about 33 years, then it dropped to 20 years in the Neolithic (roughly 3333 BC) and then, with the advent of the Greek and Roman societies, it went up again to about 33 years or so. Thus, we conventionally assume for the numeric times of the Upper Paleolithic, Neolithic, and

Table 10.1. Life expectancy in history (from *http://en.wikipedia.org/wiki/Life_expectancy*).

Era	*Life expectancy at birth* (years)	*Comment*
Upper Paleolithic	33	At age 15, life expectancy an additional 39 years (total age 54)
Neolithic	20	
Bronze Age and Iron Age	26	
Classical Greece	33	
Classical Rome	33	At age 15, life expectancy an additional 37 years (total age 52)
Pre-Columbian North America	25–30	
Medieval Islamic Caliphate	35+	
Medieval Britain	30	At age 21, life expectancy an additional 43 years (total age 64)
Early Modern Britain	25–40	
Early 20th Century	31	
Current world average	67.2	2010 estimate

Greco-Roman, respectively:

$$\begin{cases} Paleolithic\_time = Mt = -10{,}000, \\ Neolithic\_time = t_{minimum} = -3333, \\ Greco\text{-}Roman\_time = Christ\_birth = Rt = 0. \end{cases} \tag{10.71}$$

Armed with such data, the author wondered whether they could fit his cubic model. The answer is "yes", as long as *linear transformation* is carried out on the new life expectancy cubic $L(t)$ so as to let it:

(1) completely lie above the time axis; and
(2) stretch the cubic along the vertical axis so as to let it fit the data listed in Table 10.1.

Let us write these two conditions mathematically. We are now dealing with two cubics:

(1) The old basic cubic

$$C(t) = a_3\,t^3 + a_2\,t^2 + a_1\,t + a_0 \tag{10.72}$$

namely the same as equation (10.1) and

(2) The new life expectancy cubic $L(t)$, related to (10.72) by the linear transformation

$$L(t) = a\,C(t) + b \tag{10.73}$$

where a and b are two constants with respect to t that will be determined immediately. To find them, let us look at Figure 10.7 and let us see what happens at the time origin $t = 0$, corresponding to the time of Jesus' birth (in our calendar), namely let us consider the equation

$$L(0) = a\,C(0) + b. \tag{10.74}$$

Since the cubic $C(t)$ intersects the time axis right at the origin, namely one has $C(0) = 0$, the known term a_0 in its equation must equal zero:

$$C(0) = a_0 = 0. \tag{10.75}$$

Inserting (10.75) into (10.74), the latter yields b at once:

$$L(0) = b = 33. \tag{10.76}$$

In order to find a, we recall that the minimum of $C(t)$ equals -1 (by definition). Thus, we are led to write (10.73) at the time of the minimum:

$$L(t_{\text{minimum}}) = a\,C(t_{\text{minimum}}) + b. \tag{10.77}$$

Since $L(t_{\text{minimum}}) = 20$, (10.77) at the minimum yields

$$20 = a\,(-1) + b. \tag{10.78}$$

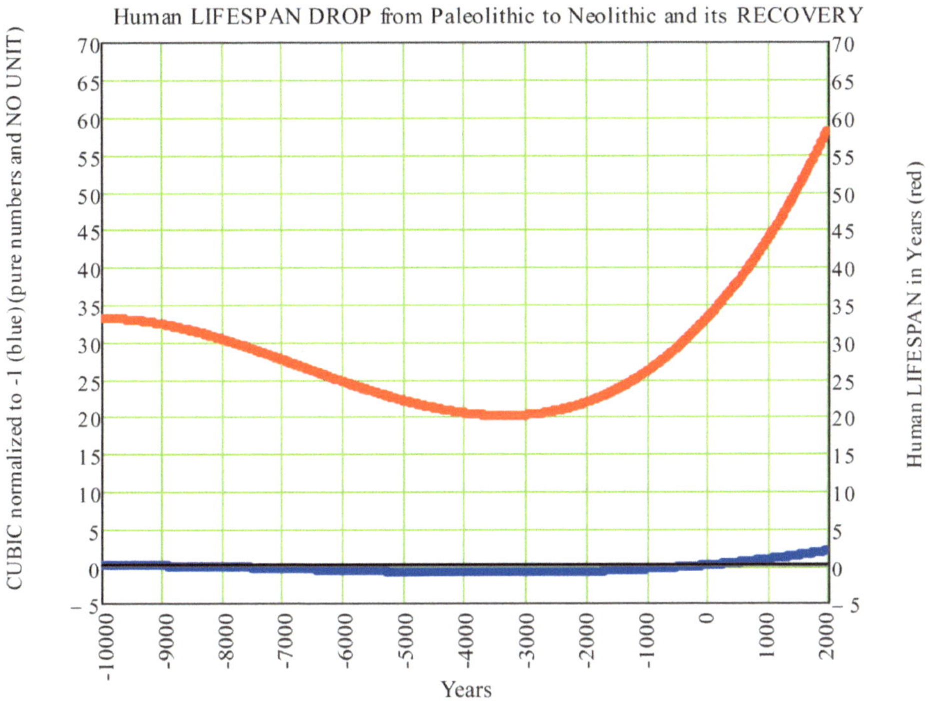

Figure 10.7. Human life expectancy (upper red cubic) from the Paleolithic (–10,000) to the Neolithic (–3,333), to Greco-Roman times (Christ's birth time = 0), and finally to the present age (2000 AD).

This, by virtue of (10.76), finally yields for a the value:

$$a = 33 - 20 = 13. \tag{10.79}$$

The life expectancy cubic $L(t)$ is thus fully determined in terms of the old cubic $C(t)$, and the two curves are plotted on the same graph in Figure 10.7.

We have thus successfully applied our cubic recovery model, plus a simple linear transformation on the cubic, to describe life expectancy *decrease* in our remote past, from the Paleolithic (–10,000) to Greco-Roman times (0 AD).

Let us now continue this investigation into the Middle Ages and up to modern times by keeping the same equations and seeing what emerges. Thus, Figure 10.8 shows in greater detail what was already shown on the extreme right-hand side of Figure 10.7, namely the life expectancy *increase* over the last 2,000 years. The conclusion is that at present (about 2000 AD) the average life expectancy is about 60 years, as expected, showing that our cubic model is at least roughly correct.

Finally, our curiosity pushes us to look into the future: for instance, what will human life expectancy be in a thousand years from now? Well, our cubic model predicts something like 80 years (as shown in Figure 10.9).

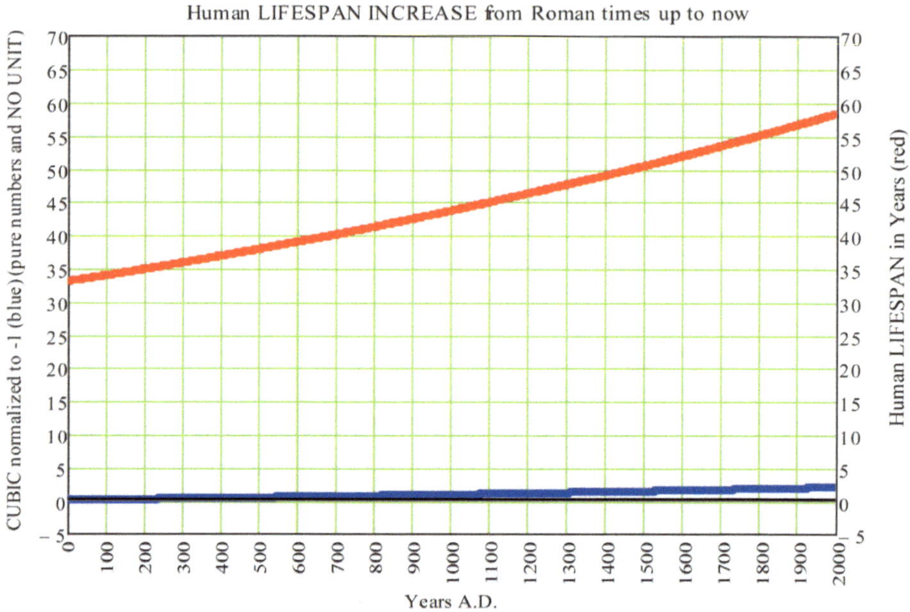

Figure 10.8. Life expectancy from Greco-Roman times (0) to the present (2000 AD).

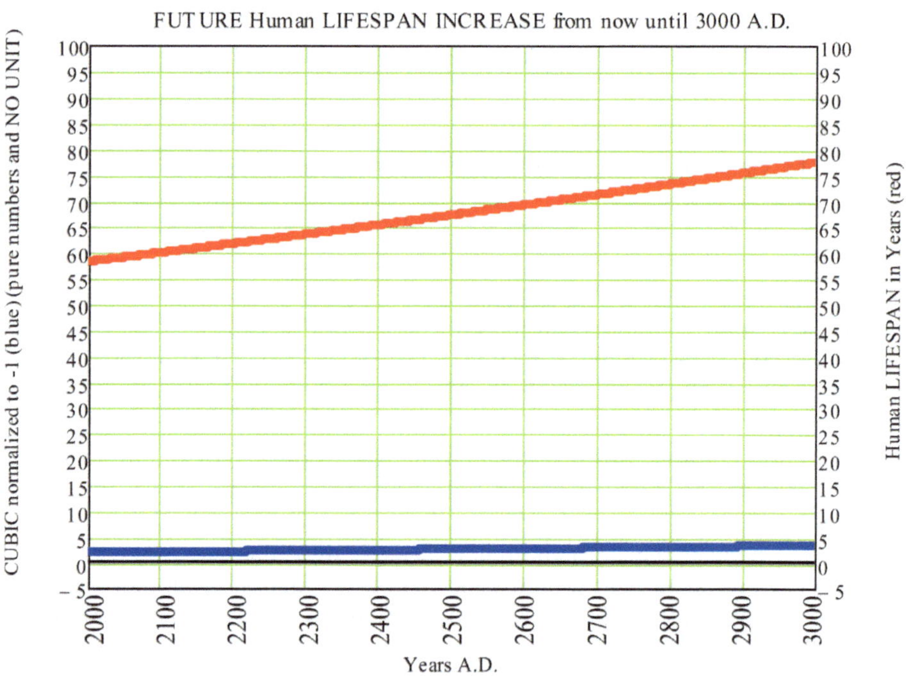

Figure 10.9. *Future* life expectancy from the present (2000 AD) to 1,000 years from now.

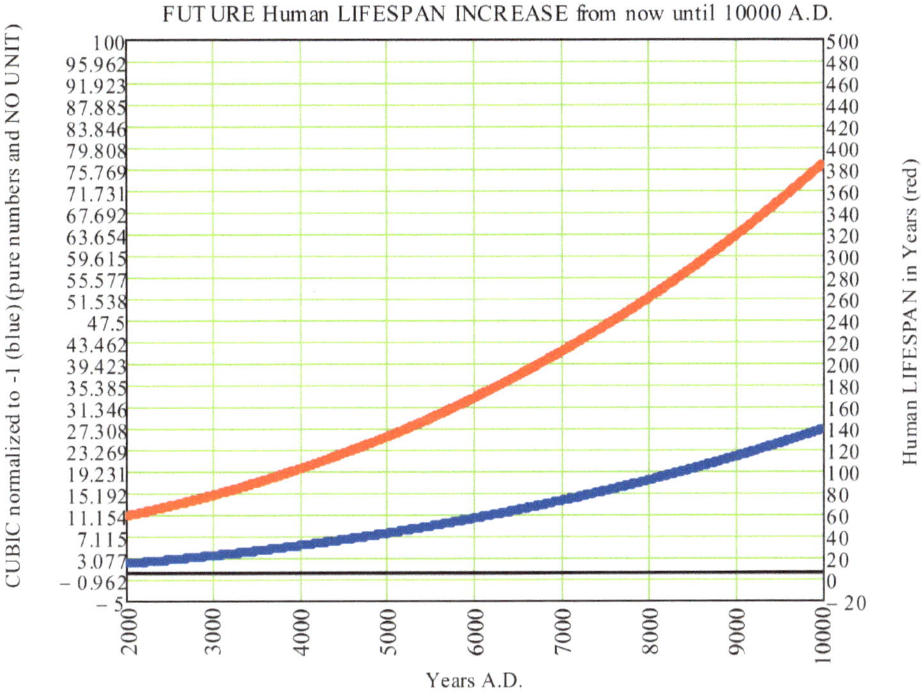

Figure 10.10. *Future* life expectancy from the present (2000 AD) to 10,000 years from now.

Even crazier is our prediction of human life expectancy in 8,000 years from now (i.e., in 10,000 AD): the result is something like a 400-year life-span (as shown in Figure 10.10). This brings to mind the film *Star Wars*, where some of the beings claim to be 900 years old or so.

However, these far future considerations also call into question the fitness of our cubic models to reality. In fact, there seem to be several different paths along which our cubic model might be extended. First, instead of just a polynomial of the third degree in time t, one might decide to consider a polynomial of a higher degree, like a fourth (a "quartic") or even fifth (a "quintic") degree. Well, the author confesses to having tried this several times already but the calculations got so involved that even Maxima could not handle them. Let us not forget that the exact solution of an algebraic equation has to be written down exactly (i.e., with no numeric approximation) only using radicals up to the fourth degree (as was already known in 1545 by the Italian mathematicians Ferro, Tartaglia, Cardan, Ferrari, and Bombelli who discovered the solution of both the cubic and quartic). As for the quintic, the solution involves elliptic functions, as was shown by Charles Hermite in the 1890s. So, let us content ourselves with our "simple" cubics and perhaps try non-polynomial approaches to model drops and recoveries in life expectancy. But this is a goal for further research and will not be considered here.

10.16 IS THE "FORCE" BEHIND PROGRESS INCREASING LINEARLY IN TIME?

We have found a cubic of time. But which differential equation does this cubic fulfill? If you differentiate the cubic (10.1) twice with respect to time, then you find:

$$\frac{d^2\ Cubic\ (t)}{dt^2} = 6 \cdot a_3 \cdot t + 2 \cdot a_2. \tag{10.80}$$

From this equation, one might naively infer that "the force behind progress increases linearly in time." This conclusion is too simplistic, though.

A more serious mathematical model would regard the cubic (10.1) as the (deterministic) mean value, $Cubic\ (t) = \langle X(t) \rangle$, of some unknown stochastic process $X(t)$, which one would then try to determine by advanced mathematical techniques. We leave such an endeavor, however, for future study.

10.17 CONCLUSION

We feel we have just scratched the (mathematical) surface of some much more profound mathematical theory. This theory would embrace human history, the geography of Earth and of space, and sociological features in a single mathematical envelope.

10.18 ACKNOWLEDGMENT

The author is grateful to Dr. Steve Dick, Former Director of the NASA History Office and NASA Chief Historian, for reading and commenting on the manuscript of the first, basic version of this chapter.

10.19 REFERENCES

[1] C. Sagan, *Cosmos*, Random House, New York, 1980. See the unnumbered figure on p. 335, which inspired much of this chapter.

[2] C. Maccone, "A mathematical 'cubic law of recovery', Part 1: Applications to history of astronomy, SETI and modern Europe," *Frontier Perspectives*, **13**(2), Fall/Winter 2004, 22–33.

[3] C. Maccone, "Past and future of astronomy and SETI cast in maths," paper #IAC.05.A4.2.11 presented at the *56th Interntional Astronautical (IAC) Congress, Fukuoka, Japan, October 16–21, 2005*.

[4] C. Maccone, "Past and future of astronomy and SETI cast in maths," *Journal of the British Interplanetary Society*, **59** (2006), 283–289.

[5] C. Maccone, "SETI, extrasolar planets search and interstellar flight: When are they going to merge?" *Acta Astronautica*, **64**, 2009, 724734.

[6] M. Okuda and D. Okuda, *Star Trek Chronology: The History of the Future*, Pocket Books, New York, 1996.

[7] A. Boss, *Looking for Earths: The Race to Find New Solar Systems*, Wiley, New York, 1998.

APPENDIX 10.A

Solution of the four simultaneous linear algebraic equations (10.9), (10.4), (10.7), and (10.20) in the four unknown coefficients a_1, a_2, a_1, and a_0.

Maxima proof of the four equations yielding the CUBIC COEFFICIENTS according to the various possible cases described in the text.

Author: Claudio Maccone E-mail: clmaccon@libero.it

Commented Maxima file completed on August 15th, 2011.

```
(%i36) kill(all);
(%o0) done
```

Eq. 8: the derivative of the cubic at the Maximum point Mt equals zero.

```
(%i1) eq_8:3*a[3]*Mt^2+2*a[2]*Mt+a[1]=0;
```

(%o1) $3\, a_3\, Mt^2 + 2\, a_2\, Mt + a_1 = 0$

Eq. 4: the ordinate of the cubic Maximum is My, supposed to be known.

```
(%i2) eq_4:a[3]*Mt^3+a[2]*Mt^2+a[1]*Mt+a[0]=My;
```

(%o2) $a_3\, Mt^3 + a_2\, Mt^2 + a_1\, Mt + a_0 = My$

Eq. 7: the ordinate of the cubic Recovery point is Ry, supposed to be known.

```
(%i3) eq_7:a[3]*Rt^3+a[2]*Rt^2+a[1]*Rt+a[0]=Ry;
```

(%o3) $a_3\, Rt^3 + a_2\, Rt^2 + a_1\, Rt + a_0 = Ry$

Eq. 20: the ordinate of a generic extra-point (i.e. a point that is neithet the Maximum nor the Recovery Point) is Ey, supposed to be known.

```
(%i4) eq_20:a[3]*Et^3+a[2]*Et^2+a[1]*Et+a[0]=Ey;
```

(%o4) $a_3\, Et^3 + a_2\, Et^2 + a_1\, Et + a_0 = Ey$

Setting up the list of the four equations (8), (4), (7) and (20).

```
(%i5) four_linear_equations:[eq_8, eq_4, eq_7, eq_20];
```

(%o5) $[3\, a_3\, Mt^2 + 2\, a_2\, Mt + a_1 = 0, a_3\, Mt^3 + a_2\, Mt^2 + a_1\, Mt + a_0 = My, a_3\, Rt^3 + a_2\, Rt^2 + a_1\, Rt + a_0 = Ry, a_3\, Et^3 + a_2\, Et^2 + a_1\, Et + a_0 = Ey]$

Defining the list of the four unknowns: a sub 3, a sub 2, a sub 1 and a sub 0.

(%i6) four_unknowns:[a[3], a[2], a[1], a[0]];

(%o6) $[a_3, a_2, a_1, a_0]$

Maxima's command "linsolve" is powerful: it immediately solve systems of linear algebraic equations. Here the above four linear equations are immediately solved for the four unknown cubic coefficients a sub 3, a sub 2, a sub 1 and a sub 0.

The following four equations are exactly those given in Appendix 1, and we like to call them the coefficients og the GENERAL CUBIC, meaning that three points (Maximum, Recovery and Extra-point) are given arbitrarily and the cubic follows.

In other words, the list of the four cubic coefficients is the following:

(%i7) four_cubic_coefficients:factor(linsolve(four_linear_equations, four_unknowns));

(%o7) $$[a_3 = \frac{Mt^2\,Ry - 2\,Et\,Mt\,Ry + Et^2\,Ry + My\,Rt^2 - Ey\,Rt^2 - 2\,Mt\,My\,Rt + 2\,Ey\,Mt\,Rt + 2\,Et\,Mt\,My - Et^2\,My - Ey\,Mt^2}{(Mt-Et)^2\,(Rt-Et)\,(Rt-Mt)^2}, a_2 = -\frac{2\,Mt^3\,Ry - 3\,Et\,Mt^2\,Ry + Et^3\,Ry + My\,Rt^3 - Ey\,Rt^3 - 3\,Mt^2\,My\,Rt + 3\,Ey\,Mt^2\,Rt + 3\,Et\,Mt^2\,My - Et^3\,My - 2\,Ey\,Mt^3}{(Mt-Et)^2\,(Rt-Et)\,(Rt-Mt)^2},$$

$$a_1 = (Mt\,(Mt^3\,Ry - 3\,Et^2\,Mt\,Ry + 2\,Et^3\,Ry + 2\,My\,Rt^3 - 2\,Ey\,Rt^3 - 3\,Mt\,My\,Rt^2 + 3\,Ey\,Mt\,Rt^2 + 3\,Et^2\,Mt\,My - 2\,Et^3\,My - Ey\,Mt^3))/((Mt-Et)^2\,(Rt-Et)\,(Rt-Mt)^2), a_0 = -(Et\,Mt^4\,Ry - 2\,Et^2\,Mt^3\,Ry + Et^3\,Mt^2\,Ry + 2\,Et\,Mt\,My\,Rt^3 - Et^2\,My\,Rt^3 - Ey\,Mt^2\,Rt^3 - 3\,Et\,Mt^2\,My\,Rt^2 + Et^3\,My\,Rt^2 + 2\,Ey\,Mt^3\,Rt^2 + 3\,Et^2\,Mt^2\,My\,Rt - 2\,Et^3\,Mt\,My\,Rt - Ey\,Mt^4\,Rt)/((Mt-Et)^2\,(Rt-Et)\,(Rt-Mt)^2)]$$

We could stop our proof of the four coefficients of the GENERIC CUBIC at this point.

But we prefer to show also how Maxima handles MATRICES.
And so we start again from the coefficient matrix, namely eq. (21).

(%i8) coefficient_matrix:Coefficient_Matrix=matrix([0, 1, 2*Mt, 3*Mt^2],[1, Mt, Mt^2, Mt^3],[1, Rt, Rt^2, Rt^

(%o8) $$Coefficient\_Matrix = \begin{bmatrix} 0 & 1 & 2\,Mt & 3\,Mt^2 \\ 1 & Mt & Mt^2 & Mt^3 \\ 1 & Rt & Rt^2 & Rt^3 \\ 1 & Et & Et^2 & Et^3 \end{bmatrix}$$

Maxima has no problem to immediately find the INVERSE of this matrix:

(%i9) inverse_of_coefficient_matrix:factor(coefficient_matrix^^-1);

(%o9) $$Coefficient\_Matrix^{-1} = \begin{bmatrix} \frac{Et\,Mt\,Rt}{(Mt-Et)\,(Rt-Mt)} & -\frac{Et\,Rt\,(2\,Mt\,Rt-Et\,Rt-3\,Mt^2+2\,Et\,Mt)}{(Mt-Et)^2\,(Rt-Mt)^2} & -\frac{Et\,Mt^2}{(Rt-Et)\,(Rt-Mt)^2} & \frac{Mt^2\,Rt}{(Mt-Et)^2\,(Rt-Et)} \\ -\frac{Mt\,Rt+Et\,Rt+Et\,Mt}{(Mt-Et)\,(Rt-Mt)} & \frac{Mt\,(2\,Rt^2-3\,Mt\,Rt+2\,Et\,Rt-3\,Et\,Mt+2\,Et^2)}{(Mt-Et)^2\,(Rt-Mt)^2} & \frac{Mt\,(Mt+2\,Et)}{(Rt-Et)\,(Rt-Mt)^2} & -\frac{Mt\,(2\,Rt+Mt)}{(Mt-Et)^2\,(Rt-Et)} \\ \frac{Rt+Mt+Et}{(Mt-Et)\,(Rt-Mt)} & -\frac{Rt^2+Et\,Rt-3\,Mt^2+Et^2}{(Mt-Et)^2\,(Rt-Mt)^2} & -\frac{2\,Mt+Et}{(Rt-Et)\,(Rt-Mt)^2} & \frac{Rt+2\,Mt}{(Mt-Et)^2\,(Rt-Et)} \\ -\frac{1}{(Mt-Et)\,(Rt-Mt)} & \frac{Rt-2\,Mt+Et}{(Mt-Et)^2\,(Rt-Mt)^2} & \frac{1}{(Rt-Et)\,(Rt-Mt)^2} & -\frac{1}{(Mt-Et)^2\,(Rt-Et)} \end{bmatrix}$$

The known term vector, namely eq. (22), is the column vector:

(%i10) known_term_vector:Known_Term_Vector=transpose([0, My, Ry, Ey]);

(%o10) $$Known\_Term\_Vector = \begin{bmatrix} 0 \\ My \\ Ry \\ Ey \end{bmatrix}$$

Thus, because of eq. (25), the four coefficients of the generci cubic may be found again this way.

(%i11) transpose([a[0], a[1], a[2], a[3]])=(rhs(factor(coefficient_matrix^^-1))).(rhs(known_term_vector));

(%o11) $$\begin{bmatrix} a_0 \\ a_1 \\ a_2 \\ a_3 \end{bmatrix} = \begin{bmatrix} -\frac{Et\,Mt^2\,Ry}{(Rt-Et)\,(Rt-Mt)^2}+\frac{Ey\,Mt^2\,Rt}{(Mt-Et)^2\,(Rt-Et)}-\frac{Et\,My\,Rt\,(2\,Mt\,Rt-Et\,Rt-3\,Mt^2+2\,Et\,Mt)}{(Mt-Et)^2\,(Rt-Mt)^2} \\ \frac{Mt\,(Mt+2\,Et)\,Ry}{(Rt-Et)\,(Rt-Mt)^2}+\frac{Mt\,My\,(2\,Rt^2-3\,Mt\,Rt+2\,Et\,Rt-3\,Et\,Mt+2\,Et^2)}{(Mt-Et)^2\,(Rt-Mt)^2}-\frac{Ey\,Mt\,(2\,Rt+Mt)}{(Mt-Et)^2\,(Rt-Et)} \\ -\frac{(2\,Mt+Et)\,Ry}{(Rt-Et)\,(Rt-Mt)^2}-\frac{My\,(Rt^2+Et\,Rt-3\,Mt^2+Et^2)}{(Mt-Et)^2\,(Rt-Mt)^2}+\frac{Ey\,(Rt+2\,Mt)}{(Mt-Et)^2\,(Rt-Et)} \\ \frac{Ry}{(Rt-Et)\,(Rt-Mt)^2}+\frac{My\,(Rt-2\,Mt+Et)}{(Mt-Et)^2\,(Rt-Mt)^2}-\frac{Ey}{(Mt-Et)^2\,(Rt-Et)} \end{bmatrix}$$

(%i12) four_coefficients:factor(%);

(%o12) $$\begin{bmatrix} a_0 \\ a_1 \\ a_2 \\ a_3 \end{bmatrix} =$$

$$\begin{bmatrix} -\dfrac{Et\ Mt^4\ Ry - 2\ Et^2\ Mt^3\ Ry + Et^3\ Mt^2\ Ry + 2\ Et\ Mt\ My\ Rt^3 - Et^2\ My\ Rt^3 - Ey\ Mt^2\ Rt^3 - 3\ Et\ Mt^2\ My\ Rt^2 + Et^3\ My\ Rt^2 + 2\ Ey\ Mt^3\ Rt^2 + 3\ Et^2\ Mt^2\ My}{(Mt - Et)^2\ (Rt - Et)\ (Rt - Mt)^2} \\ \dfrac{Mt\ (Mt^3\ Ry - 3\ Et^2\ Mt\ Ry + 2\ Et^3\ Ry + 2\ My\ Rt^3 - 2\ Ey\ Rt^3 - 3\ Mt\ My\ Rt^2 + 3\ Ey\ Mt\ Rt^2 + 3\ Et^2\ Mt\ My - 2\ Et^3\ My - Ey\ M}{(Mt - Et)^2\ (Rt - Et)\ (Rt - Mt)^2} \\ -\dfrac{2\ Mt^3\ Ry - 3\ Et\ Mt^2\ Ry + Et^3\ Ry + My\ Rt^3 - Ey\ Rt^3 - 3\ Mt^2\ My\ Rt + 3\ Ey\ Mt^2\ Rt + 3\ Et\ Mt^2\ My - Et^3\ My - 2\ Ey\ Mt^3}{(Mt - Et)^2\ (Rt - Et)\ (Rt - Mt)^2} \\ \dfrac{Mt^2\ Ry - 2\ Et\ Mt\ Ry + Et^2\ Ry + My\ Rt^2 - Ey\ Rt^2 - 2\ Mt\ My\ Rt + 2\ Ey\ Mt\ Rt + 2\ Et\ Mt\ My - Et^2\ My - Ey\ Mt^2}{(Mt - Et)^2\ (Rt - Et)\ (Rt - Mt)^2} \end{bmatrix}$$

We now proceed to determine the particular cubic NORMALIZED TO -1,
i.e. measured in UNITS OF FALSITY, as described in the text.

First, let us write the four cubic coefficients as four separate equations.

(%i13) a_sub_3:first(four_cubic_coefficients);

(%o13) $$a_3 = \frac{Mt^2\ Ry - 2\ Et\ Mt\ Ry + Et^2\ Ry + My\ Rt^2 - Ey\ Rt^2 - 2\ Mt\ My\ Rt + 2\ Ey\ Mt\ Rt + 2\ Et\ Mt\ My - Et^2\ My - Ey\ Mt^2}{(Mt - Et)^2\ (Rt - Et)\ (Rt - Mt)^2}$$

(%i14) a_sub_2:second(four_cubic_coefficients);

(%o14) $$a_2 = -\frac{2\ Mt^3\ Ry - 3\ Et\ Mt^2\ Ry + Et^3\ Ry + My\ Rt^3 - Ey\ Rt^3 - 3\ Mt^2\ My\ Rt + 3\ Ey\ Mt^2\ Rt + 3\ Et\ Mt^2\ My - Et^3\ My - 2\ Ey\ Mt^3}{(Mt - Et)^2\ (Rt - Et)\ (Rt - Mt)^2}$$

(%i15) a_sub_1:third(four_cubic_coefficients);

(%o15) $a_1 = (Mt\ (Mt^3\ Ry - 3\ Et^2\ Mt\ Ry + 2\ Et^3\ Ry + 2\ My\ Rt^3 - 2\ Ey\ Rt^3 - 3\ Mt\ My\ Rt^2 + 3\ Ey\ Mt\ Rt^2 + 3$
$Et^2\ Mt\ My - 2\ Et^3\ My - Ey\ Mt^3))\ /\ ((Mt - Et)^2\ (Rt - Et)\ (Rt - Mt)^2)$

(%i16) a_sub_zero:fourth(four_cubic_coefficients);

(%o16) $a_0 = -(Et\ Mt^4\ Ry - 2\ Et^2\ Mt^3\ Ry + Et^3\ Mt^2\ Ry + 2\ Et\ Mt\ My\ Rt^3 - Et^2\ My\ Rt^3 - Ey\ Mt^2\ Rt^3 - 3\ Et$
$Mt^2\ My\ Rt^2 + Et^3\ My\ Rt^2 + 2\ Ey\ Mt^3\ Rt^2 + 3\ Et^2\ Mt^2\ My\ Rt - 2\ Et^3\ Mt\ My\ Rt - Ey\ Mt^4\ Rt)\ /\ ((Mt - Et)^2$
$(Rt - Et)\ (Rt - Mt)^2)$

Then let us invoke the condition that both Mt and Rt lie on the time axis.

(%i17) time_axis_thru_Mt_and_Rt_condition:[My=0, Ry=0];

(%o17) $[My=0, Ry=0]$

The four cubic coefficients then boil down to the easier form:

(%i18) time_axis_thru_Mt_and_Rt:four_cubic_coefficients,time_axis_thru_Mt_and_Rt_condition;

(%o18) $$[a_3=\frac{-Ey\,Rt^2+2\,Ey\,Mt\,Rt-Ey\,Mt^2}{(Mt-Et)^2\,(Rt-Et)\,(Rt-Mt)^2}, a_2=-\frac{-Ey\,Rt^3+3\,Ey\,Mt^2\,Rt-2\,Ey\,Mt^3}{(Mt-Et)^2\,(Rt-Et)\,(Rt-Mt)^2}, a_1=\frac{Mt\,(-2\,Ey\,Rt^3+3\,Ey\,Mt\,Rt^2-Ey\,Mt^3)}{(Mt-Et)^2\,(Rt-Et)\,(Rt-Mt)^2}, a_0=-\frac{-Ey\,Mt^2\,Rt^3+2\,Ey\,Mt^3\,Rt^2-Ey\,Mt^4\,Rt}{(Mt-Et)^2\,(Rt-Et)\,(Rt-Mt)^2}]$$

(%i19) zero_Mt_and_Rt_cubic_coefficients:factor(time_axis_thru_Mt_and_Rt);

(%o19) $$[a_3=-\frac{Ey}{(Mt-Et)^2\,(Rt-Et)}, a_2=\frac{Ey\,(Rt+2\,Mt)}{(Mt-Et)^2\,(Rt-Et)}, a_1=-\frac{Ey\,Mt\,(2\,Rt+Mt)}{(Mt-Et)^2\,(Rt-Et)}, a_0=\frac{Ey\,Mt^2\,Rt}{(Mt-Et)^2\,(Rt-Et)}]$$

Since we want to normalize to cubic to -1, we first need the expression of the ordinate of the GENEARIC CUBIC minimum point. This is found by setting to zero the first derivative of the cubic passing thru Mt and Rt.

(%i20) cubic:a[3]*t^3+a[2]*t^2+a[1]*t+a[0];

(%o20) $a_3\,t^3+a_2\,t^2+a_1\,t+a_0$

(%i21) cubic_with_zero_My_and_Ry:cubic,zero_Mt_and_Rt_cubic_coefficients;

(%o21) $$-\frac{Ey\,t^3}{(Mt-Et)^2\,(Rt-Et)}+\frac{Ey\,(Rt+2\,Mt)\,t^2}{(Mt-Et)^2\,(Rt-Et)}-\frac{Ey\,Mt\,(2\,Rt+Mt)\,t}{(Mt-Et)^2\,(Rt-Et)}+\frac{Ey\,Mt^2\,Rt}{(Mt-Et)^2\,(Rt-Et)}$$

(%i22) maximum_and_minimum_condition:diff(cubic_with_zero_My_and_Ry, t)=0;

(%o22) $$-\frac{3\,Ey\,t^2}{(Mt-Et)^2\,(Rt-Et)}+\frac{2\,Ey\,(Rt+2\,Mt)\,t}{(Mt-Et)^2\,(Rt-Et)}-\frac{Ey\,Mt\,(2\,Rt+Mt)}{(Mt-Et)^2\,(Rt-Et)}=0$$

We thus find again that the maximum is at time Mt (of course), but now we also find the time (i.e. the abscissa) of the minimum.

(%i23) abscissae_of_maximum_and_minimum:solve(maximum_and_minimum_condition,t);

(%o23) $[t=\frac{2\,Rt+Mt}{3}, t=Mt]$

(%i24) abscissa_of_minimum:first(abscissae_of_maximum_and_minimum);

(%o24) $t=\frac{2\,Rt+Mt}{3}$

The ordinate of the minimum is now found by replacing the last abscissa into the cubic, and simplyfying what comes out.

(%i25) ordinate_of_cubic_at_minimum:cubic_with_zero_My_and_Ry,abscissa_of_minimum;

(%o25) $-\frac{Ey\,(2\,Rt+Mt)^3}{27\,(Mt-Et)^2\,(Rt-Et)}-\frac{Ey\,Mt\,(2\,Rt+Mt)^2}{3\,(Mt-Et)^2\,(Rt-Et)}+\frac{Ey\,(Rt+2\,Mt)\,(2\,Rt+Mt)^2}{9\,(Mt-Et)^2\,(Rt-Et)}+\frac{Ey\,Mt^2\,Rt}{(Mt-Et)^2\,(Rt-Et)}$

(%i26) simplified_ordinate_of_cubic_at_minimum:factor(ordinate_of_cubic_at_minimum);

(%o26) $\frac{4\,Ey\,(Rt-Mt)^3}{27\,(Mt-Et)^2\,(Rt-Et)}$

We now NORMALIZE THE CUBIC TO MINUS ONE upon imposing that the last expression equals minus one.

(%i27) normalizing_the_cubic_to_minus_one:-1=simplified_ordinate_of_cubic_at_minimum;

(%o27) $-1=\frac{4\,Ey\,(Rt-Mt)^3}{27\,(Mt-Et)^2\,(Rt-Et)}$

Solving for Ey (the ordinate of the extra time that we want to eliminate) yields

(%i28) solving_for_Ey:first(solve(normalizing_the_cubic_to_minus_one,Ey));

(%o28) $Ey=-\frac{(27\,Mt^2-54\,Et\,Mt+27\,Et^2)\,Rt-27\,Et\,Mt^2+54\,Et^2\,Mt-27\,Et^3}{4\,Rt^3-12\,Mt\,Rt^2+12\,Mt^2\,Rt-4\,Mt^3}$

that, once simplyfied, becomes:

(%i29) Ey_expression_for_cubic_normnalized_to_minus_one:factor(solving_for_Ey);

(%o29) $Ey=-\frac{27\,(Mt-Et)^2\,(Rt-Et)}{4\,(Rt-Mt)^3}$

The COEFFICIENTS OF THE CUBIC NORMALIZED TO -1
are now immediately found replaring the last expression for Ey into the former cubic coefficients and simplifying. That is:

(%i30) zero_Mt_and_Rt_cubic_coefficients,Ey_expression_for_cubic_normnalized_to_minus_one;

(%o30) $[a_3=\frac{27}{4\,(Rt-Mt)^3},a_2=-\frac{27\,(Rt+2\,Mt)}{4\,(Rt-Mt)^3},a_1=\frac{27\,Mt\,(2\,Rt+Mt)}{4\,(Rt-Mt)^3},a_0=-\frac{27\,Mt^2\,Rt}{4\,(Rt-Mt)^3}]$

(%i31) four_coefficients_of_cubic_normalized_to_minus_one:%;

(%o31) $[a_3=\frac{27}{4\,(Rt-Mt)^3},a_2=-\frac{27\,(Rt+2\,Mt)}{4\,(Rt-Mt)^3},a_1=\frac{27\,Mt\,(2\,Rt+Mt)}{4\,(Rt-Mt)^3},a_0=-\frac{27\,Mt^2\,Rt}{4\,(Rt-Mt)^3}]$

These are the four equations that the author used in all his papers about the Cubics, without proving and re-proving them in every paper since it would have taken too many pages.

Incidentally, Maxima knows about the Cardan formulae yielding a cubic's zeros. For instance, let us write down the cubic equation set to zero:

(%i32) cubic_to_solve_by_the_Cardan_formulae:cubic=0;

(%o32) $a_3\,t^3+a_2\,t^2+a_1\,t+a_0=0$

Then, the Cardan formulae (first published by Cardan, i.e. Girolamo Cardano (1501-1576) of Milan in his book "Ars Magna" i.e. "the great art", i.e. algebra, in the year 1545) yield:

(%i33) Cardan_formulae:solve(cubic_to_solve_by_the_Cardan_formulae,t);

(%o33)
$$\left[t=\left(-\frac{\sqrt{3}\,\%i}{2}-\frac{1}{2}\right)\left(\frac{\sqrt{27\,a_0^2\,a_3^2+(4\,a_1^3-18\,a_0\,a_1\,a_2)\,a_3+4\,a_0\,a_2^3-a_1^2\,a_2^2}}{2\,3^{3/2}\,a_3^2}-\frac{27\,a_0\,a_3^2-9\,a_1\,a_2\,a_3+2\,a_2^3}{54\,a_3^3}\right)^{1/3}-\frac{\left(\frac{\sqrt{3}\,\%i}{2}-\frac{1}{2}\right)(3\,a_1\,a_3-a_2^2)}{9\,a_3^2\left(\frac{\sqrt{27\,a_0^2\,a_3^2+(4\,a_1^3-18\,a_0\,a_1\,a_2)\,a_3+4\,a_0\,a_2^3-a_1^2\,a_2^2}}{2\,3^{3/2}\,a_3^2}-\frac{27\,a_0\,a_3^2-9\,a_1\,a_2\,a_3+2\,a_2^3}{54\,a_3^3}\right)^{1/3}}-\frac{a_2}{3\,a_3},\right.$$

$$t=\left(\frac{\sqrt{3}\,\%i}{2}-\frac{1}{2}\right)\left(\frac{\sqrt{27\,a_0^2\,a_3^2+(4\,a_1^3-18\,a_0\,a_1\,a_2)\,a_3+4\,a_0\,a_2^3-a_1^2\,a_2^2}}{2\,3^{3/2}\,a_3^2}-\frac{27\,a_0\,a_3^2-9\,a_1\,a_2\,a_3+2\,a_2^3}{54\,a_3^3}\right)^{1/3}-\frac{\left(-\frac{\sqrt{3}\,\%i}{2}-\frac{1}{2}\right)(3\,a_1\,a_3-a_2^2)}{9\,a_3^2\left(\frac{\sqrt{27\,a_0^2\,a_3^2+(4\,a_1^3-18\,a_0\,a_1\,a_2)\,a_3+4\,a_0\,a_2^3-a_1^2\,a_2^2}}{2\,3^{3/2}\,a_3^2}-\frac{27\,a_0\,a_3^2-9\,a_1\,a_2\,a_3+2\,a_2^3}{54\,a_3^3}\right)^{1/3}}-\frac{a_2}{3\,a_3},$$

$$\left.t=\left(\frac{\sqrt{27\,a_0^2\,a_3^2+(4\,a_1^3-18\,a_0\,a_1\,a_2)\,a_3+4\,a_0\,a_2^3-a_1^2\,a_2^2}}{2\,3^{3/2}\,a_3^2}-\frac{27\,a_0\,a_3^2-9\,a_1\,a_2\,a_3+2\,a_2^3}{54\,a_3^3}\right)^{1/3}-\frac{3\,a_1\,a_3-a_2^2}{9\,a_3^2\left(\frac{\sqrt{27\,a_0^2\,a_3^2+(4\,a_1^3-18\,a_0\,a_1\,a_2)\,a_3+4\,a_0\,a_2^3-a_1^2\,a_2^2}}{2\,3^{3/2}\,a_3^2}-\frac{27\,a_0\,a_3^2-9\,a_1\,a_2\,a_3+2\,a_2^3}{54\,a_3^3}\right)^{1/3}}-\frac{a_2}{3\,a_3}\right]$$

If we restrict ourselves to the cubic normalized to -1, then, invoking the relevant four coefficients, the above lot looks even more awful:

(%i34) Cardan_for_normalized_cubic:Cardan_formulae,four_coefficients_of_cubic_normalized_to_minus_one;

$$\text{(\%o34) } [t=\left(-\frac{\sqrt{3}\,\%i}{2}-\frac{1}{2}\right)((8\,(Rt-Mt)^6\,sqrt(\frac{27\left(\frac{19683\,Mt^3\,(2\,Rt+Mt)^3}{16\,(Rt-Mt)^9}-\frac{177147\,Mt^3\,Rt\,(Rt+2\,Mt)\,(2\,Rt+Mt)}{32\,(Rt-Mt)^9}\right)}{4\,(Rt-Mt)^3}$$

$$-\frac{531441\,Mt^2\,(Rt+2\,Mt)^2\,(2\,Rt+Mt)^2}{256\,(Rt-Mt)^{12}}+\frac{531441\,Mt^2\,Rt\,(Rt+2\,Mt)^3}{64\,(Rt-Mt)^{12}}+\frac{14348907\,Mt^4\,Rt^2}{256\,(Rt-Mt)^{12}}))/3^{15/2}-$$

$$\frac{32\,(Rt-Mt)^9\left(-\frac{19683\,(Rt+2\,Mt)^3}{32\,(Rt-Mt)^9}-\frac{531441\,Mt^2\,Rt}{64\,(Rt-Mt)^9}+\frac{177147\,Mt\,(Rt+2\,Mt)\,(2\,Rt+Mt)}{64\,(Rt-Mt)^9}\right)}{531441})\,\hat{}\,(1/3)-(16\left(\frac{\sqrt{3}\,\%i}{2}-\frac{1}{2}\right)$$

$$(Rt-Mt)^6\left(\frac{2187\,Mt\,(2\,Rt+Mt)}{16\,(Rt-Mt)^6}-\frac{729\,(Rt+2\,Mt)^2}{16\,(Rt-Mt)^6}\right))/(6561\,((8\,(Rt-Mt)^6\,sqrt($$

$$\frac{27\left(\frac{19683\,Mt^3\,(2\,Rt+Mt)^3}{16\,(Rt-Mt)^9}-\frac{177147\,Mt^3\,Rt\,(Rt+2\,Mt)\,(2\,Rt+Mt)}{32\,(Rt-Mt)^9}\right)}{4\,(Rt-Mt)^3}-\frac{531441\,Mt^2\,(Rt+2\,Mt)^2\,(2\,Rt+Mt)^2}{256\,(Rt-Mt)^{12}}+$$

$$\frac{531441\,Mt^2\,Rt\,(Rt+2\,Mt)^3}{64\,(Rt-Mt)^{12}}+\frac{14348907\,Mt^4\,Rt^2}{256\,(Rt-Mt)^{12}}))/3^{15/2}-$$

$$\frac{32\,(Rt-Mt)^9\left(-\frac{19683\,(Rt+2\,Mt)^3}{32\,(Rt-Mt)^9}-\frac{531441\,Mt^2\,Rt}{64\,(Rt-Mt)^9}+\frac{177147\,Mt\,(Rt+2\,Mt)\,(2\,Rt+Mt)}{64\,(Rt-Mt)^9}\right)}{531441})\,\hat{}\,(1/3))+\frac{Rt+2\,Mt}{3},t=$$

$$\left(\frac{\sqrt{3}\,\%i}{2}-\frac{1}{2}\right)((8\,(Rt-Mt)^6\,sqrt(\frac{27\left(\frac{19683\,Mt^3\,(2\,Rt+Mt)^3}{16\,(Rt-Mt)^9}-\frac{177147\,Mt^3\,Rt\,(Rt+2\,Mt)\,(2\,Rt+Mt)}{32\,(Rt-Mt)^9}\right)}{4\,(Rt-Mt)^3}-$$

$$\frac{531441\,Mt^2\,(Rt+2\,Mt)^2\,(2\,Rt+Mt)^2}{256\,(Rt-Mt)^{12}}+\frac{531441\,Mt^2\,Rt\,(Rt+2\,Mt)^3}{64\,(Rt-Mt)^{12}}+\frac{14348907\,Mt^4\,Rt^2}{256\,(Rt-Mt)^{12}}))/3^{15/2}-$$

$$\frac{32\,(Rt-Mt)^9\left(-\frac{19683\,(Rt+2\,Mt)^3}{32\,(Rt-Mt)^9}-\frac{531441\,Mt^2\,Rt}{64\,(Rt-Mt)^9}+\frac{177147\,Mt\,(Rt+2\,Mt)\,(2\,Rt+Mt)}{64\,(Rt-Mt)^9}\right)}{531441})\,\hat{}\,(1/3)-(16\left(-\frac{\sqrt{3}\,\%i}{2}-\frac{1}{2}\right)$$

$$(Rt-Mt)^6\left(\frac{2187\,Mt\,(2\,Rt+Mt)}{16\,(Rt-Mt)^6}-\frac{729\,(Rt+2\,Mt)^2}{16\,(Rt-Mt)^6}\right))/(6561\,((8\,(Rt-Mt)^6\,sqrt($$

$$\frac{27\left(\frac{19683\,Mt^3\,(2\,Rt+Mt)^3}{16\,(Rt-Mt)^9}-\frac{177147\,Mt^3\,Rt\,(Rt+2\,Mt)\,(2\,Rt+Mt)}{32\,(Rt-Mt)^9}\right)}{4\,(Rt-Mt)^3}-\frac{531441\,Mt^2\,(Rt+2\,Mt)^2\,(2\,Rt+Mt)^2}{256\,(Rt-Mt)^{12}}+$$

$$\frac{531441\,Mt^2\,Rt\,(Rt+2\,Mt)^3}{64\,(Rt-Mt)^{12}}+\frac{14348907\,Mt^4\,Rt^2}{256\,(Rt-Mt)^{12}}))/3^{15/2}-$$

$$\frac{32\,(Rt-Mt)^9\left(-\frac{19683\,(Rt+2\,Mt)^3}{32\,(Rt-Mt)^9}-\frac{531441\,Mt^2\,Rt}{64\,(Rt-Mt)^9}+\frac{177147\,Mt\,(Rt+2\,Mt)\,(2\,Rt+Mt)}{64\,(Rt-Mt)^9}\right)}{531441})\,\hat{}\,(1/3))+\frac{Rt+2\,Mt}{3},t=(($$

$$27\left(\frac{19683\,Mt^3\,(2\,Rt+Mt)^3}{16\,(Rt-Mt)^9}-\frac{177147\,Mt^3\,Rt\,(Rt+2\,Mt)\,(2\,Rt+Mt)}{32\,(Rt-Mt)^9}\right)$$

But that is just an impression! In fact, Maxima's powerful command "radcan", i.e. "radical canonical", immediately reduces the whole lot to the two simple solutions t=Mt and t=Rt, as we know. Notice that t=Mt is a DOUBLE root, inasmuch as it is the tangent point between the Maximum and the t axis.

```
(%i35) re_discovering_Mt_and_Rt:radcan(Cardan_for_normalized_cubic);
```

(%o35) $[t=Mt, t=Mt, t=Rt]$

APPENDIX 10.B "LIFE_EXPECTANCY_CUBIC_V09_OCTOBER_6_2011.XMCD1"

LIFE EXPECTANCY DROP FROM PALEOLITHIC (-10000 A.D.) TO NEOLITHIC (-3333 A.D.) AND GRECO-ROMAN RECOVERY (0 A.D.)

Author: Claudio Maccone
August 20, 2011, at 18:50

A mathematical "Cubic Law of Recovery": Part 2: Application to Human Lifespan from Paleolithic until Now

$\text{year} := \text{yr} \qquad Mt := -10000 \qquad Rt := 0$

$$a_3 := \frac{27}{4\cdot(Rt - Mt)^3} \qquad a_2 := \frac{-27\cdot(Rt + 2\cdot Mt)}{4\cdot(Rt - Mt)^3} \qquad a_1 := \frac{27\cdot Mt\cdot(2\cdot Rt + Mt)}{4\cdot(Rt - Mt)^3} \qquad a_0 := \frac{-27\cdot Mt^2\cdot Rt}{4\cdot(Rt - Mt)^3}$$

$$\text{Cubic}(t) := a_3\cdot t^3 + a_2\cdot t^2 + a_1\cdot t + a_0$$

$$a_3 = 6.75\times 10^{-12} \qquad a_2 = 1.35\times 10^{-7} \qquad a_1 = 6.75\times 10^{-4} \qquad a_0 = 0$$

$$\text{Cubic}(-10000) = 0 \qquad \text{Cubic}(0) = 0 \qquad \text{Cubic}(2000) = 1.944$$

$$\text{t\_Inflexion} := \frac{Rt + 2Mt}{3} \qquad \text{t\_Inflexion} = -6.667\times 10^3 \qquad \text{Cubic(t\_Inflexion)} = -0.5$$

$$\text{t\_minimum} := \frac{2\cdot Rt + Mt}{3} \qquad \text{t\_minimum} = -3.333\times 10^3 \qquad \text{Cubic(t\_minimum)} = -1$$

$$\text{t\_preMaximum} := \frac{4\cdot Mt - Rt}{3} \qquad \text{t\_preMaximum} = -1.333\times 10^4$$

$\text{Destruction\_Interval} := \frac{2\cdot(\text{Rt} - \text{Mt})}{3}$ $\text{Destruction\_Interval} = 6.667 \times 10^3$

$\text{Recovery\_Interval} := \frac{\text{Rt} - \text{Mt}}{3}$ $\text{Recovery\_Interval} = 3.333 \times 10^3$

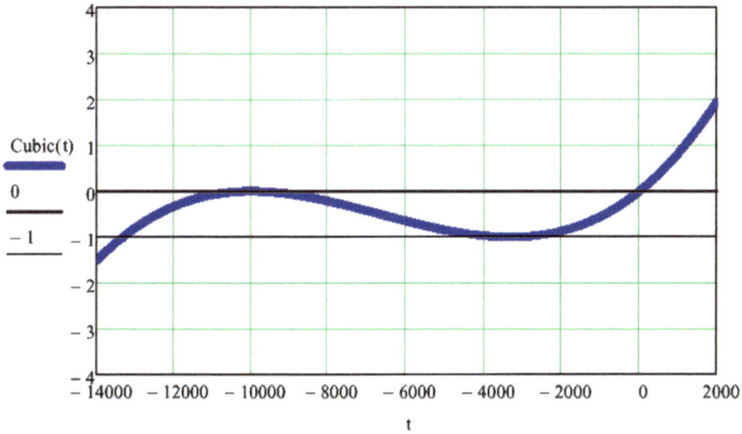

$a := 33\cdot\text{yr} - 20\cdot\text{yr}$ $b := 33\cdot\text{yr}$

$\text{Lifespan}(t) := a\cdot\text{Cubic}(t) + b$ $\text{Lifespan}(2008) = 58.407\cdot\text{yr}$ 2008 Lifespan = 66.12 yr

From the Wikipedia site about "Life Expectancy".

$\text{Lifespan}(-10000) = 33\cdot\text{yr}$ Approximate End of Upper Paleolithic in Eurasia - Mesolithic Begins.

$\text{Lifespan}(-3333) = 20\cdot\text{yr}$ Middle Bronze Age in Eurasia. Lifespan drops to 20 years "because of the introduction of Agriculture".

$\text{Lifespan}(-323) = 30.346\cdot\text{yr}$ Alexander the Great dies aged 33 (human lifespan in his time was nearly the same, i.e. 29.93 years according to our cubic model).

$\text{Lifespan}(0) = 33\cdot\text{yr}$ Jesus Christ is born in 0 A.D. and dies aged 33: exactly the same figure as his expected lifespan accroding to our cubic model.

$\text{Lifespan}(313) = 35.921\cdot\text{yr}$ Roman Emperor Constantine allows Christianity to be accepted as a new religion in the Roman Empire. In about a century Christianity will completely replace Paganism in Roman Empire.

$\text{Lifespan}(800) = 41.188\cdot\text{yr}$ Charlemagne becomes the first "Holy Roman Emperor".

Lifespan(932) = 42.774·yr	Lifespan of 44 years in the worst Dark Ages.
Lifespan(1000) = 43.618·yr	Lifespan of 45 years when the cultural recovery in Western Europe begins.
Lifespan(1250) = 46.882·yr	Emperor Frederick II dies at the age of 55 in 1250.
Lifespan(1324) = 47.898·yr	Marco Polo dies at the age of 70 in 1324.
Lifespan(1492) = 50.29·yr	Columbus discovers America. Lifespan is now about 53.
Lifespan(1558) = 51.263·yr	Emperor Charles V dies at the age of 58 in 1558.
Lifespan(1630) = 52.346·yr	Kepler dies aged 58.
Lifespan(1642) = 52.529·yr	Galileo dies aged 77.
Lifespan(1713) = 53.622·yr	Louis XIV dies aged 76.
Lifespan(1727) = 53.841·yr	Newton dies aged 84.
Lifespan(1750) = 54.201·yr	Bach dies aged 65.
Lifespan(1793) = 54.881·yr	Louis XVI is guillotined aged 38.
Lifespan(1821) = 55.329·yr	Napoleon dies aged 51.

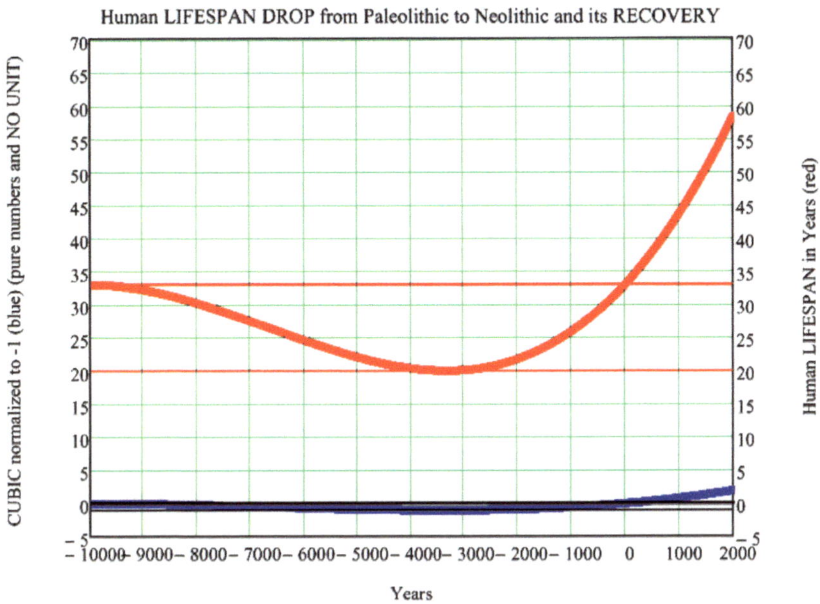

Lifespan(2011) = 58.458·yr

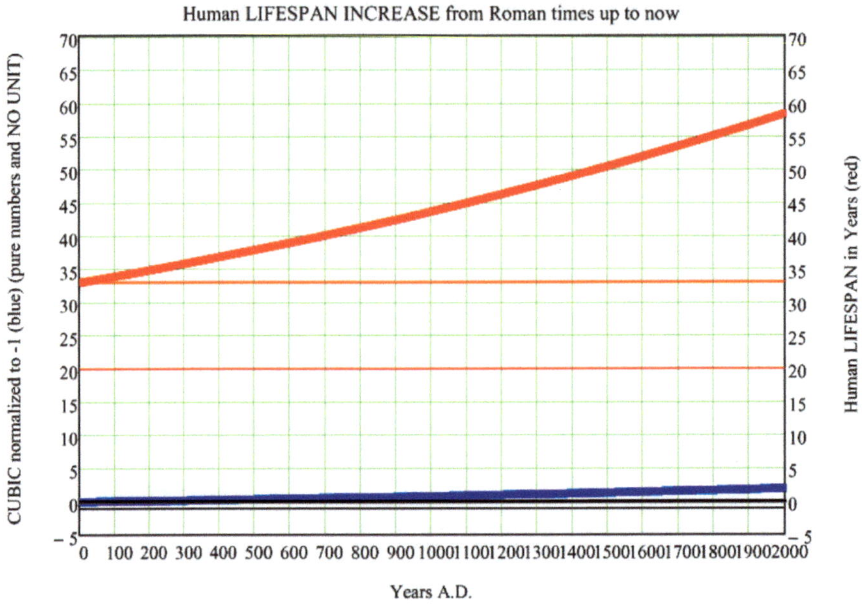
Human LIFESPAN INCREASE from Roman times up to now
CUBIC normalized to -1 (blue) (pure numbers and NO UNIT)
Human LIFESPAN in Years (red)
Years A.D.

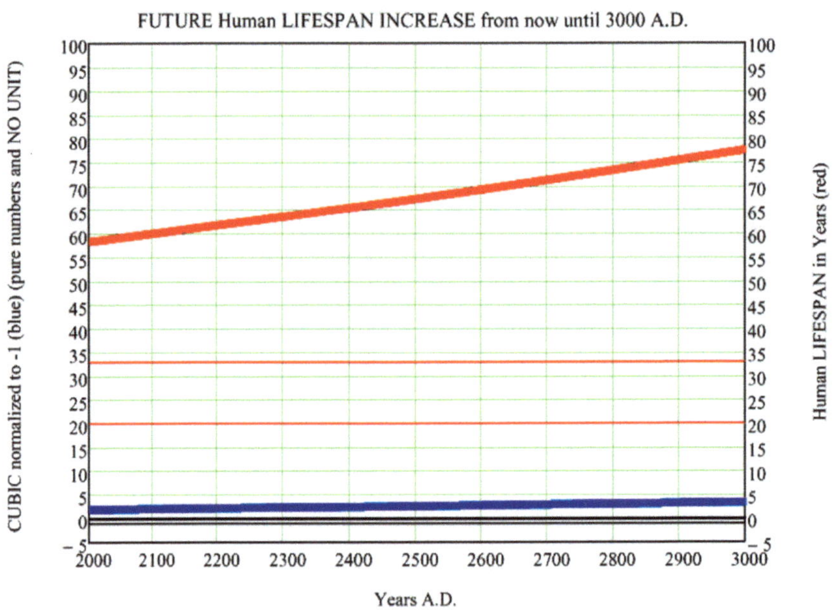
FUTURE Human LIFESPAN INCREASE from now until 3000 A.D.
CUBIC normalized to -1 (blue) (pure numbers and NO UNIT)
Human LIFESPAN in Years (red)
Years A.D.

Lifespan(2008) = 58.407·yr

Lifespan(3000) = 77.489·yr

Lifespan(5000) = 131.719·yr

Lifespan(10000) = 384·yr

Lifespan(50000) = 1.583×10^{4}·yr

Lifespan(100000) = 1.062×10^{5}·yr

Lifespan(1000000) = 8.951×10^{7}·yr

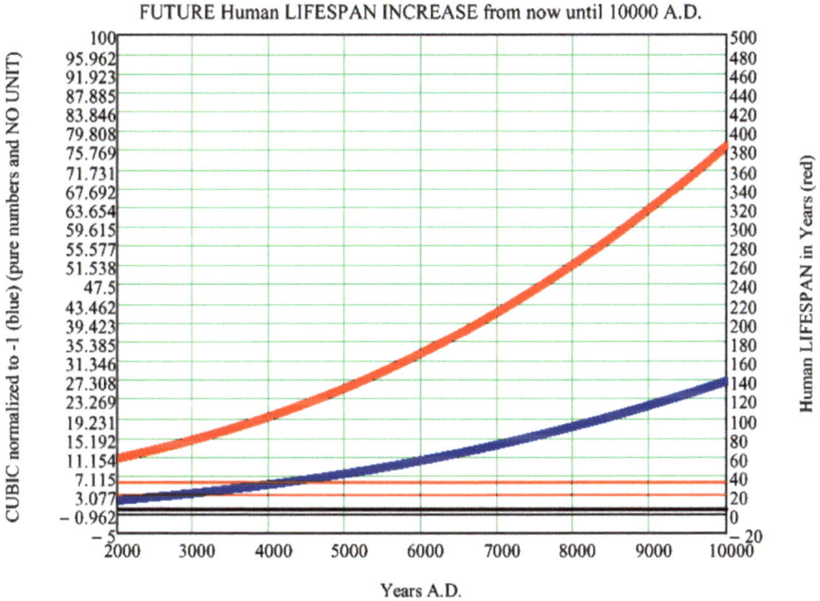

11

Exponential evolution in time as a geometric Brownian motion

11.1 INTRODUCTION

On January 8, 2012, this author came to realize that his statistical Drake equation, described in Chapter 1 and subsequent chapters of this book, is the special static case (i.e., "the picture", so to say) of a more general time-dependent statistical Drake equation (i.e., "the movie", so to say) which we study in this chapter. Also, this result is thus a powerful generalization of all previous results described in the first 10 chapters of this book. This chapter is an introduction to a new exciting mathematical topic, which we call "exponential evolution in time of the statistical Drake equation."

To be precise, the number N in the statistical Drake equation (1.7), yielding the number of extraterrestrial civilizations now existing and communicating in the Galaxy, is replaced in this chapter by a stochastic process $N(t)$, jumping up and down in time like the number e raised to a Brownian motion, but actually in such a way that its mean value keeps increasing exponentially in time as follows

$$\langle N(t) \rangle = N_0 \, e^{\mu t}. \tag{11.1}$$

In (11.1), N_0 and μ are two constants with respect to the time variable t. Their meaning is, respectively:

(1) N_0 is the number of ET communicating civilizations at time $t = 0$ (i.e., "now"), if one decides to regard positive times ($t > 0$) as the future history of the Galaxy and negative times ($t < 0$) as the past history of the Galaxy.
(2) μ is a positive (if the number of ET civilizations increases in time) or negative (if the number of ET civilizations decreases in time) parameter that we call "the drift". Fixing ideas and getting optimistic, we shall suppose $\mu > 0$, and, later in this chapter, we will find that the condition $\mu > \frac{2}{3}\,\sigma^2 > 0$ must hold good.

This evolution in time of $N(t)$ is just what we expect to happen in the Galaxy, where the overall number $N(t)$ of ET civilizations probably does increase in time because of the obvious technological evolution of each civilization. But this $N(t)$ scenario is a stochastic one, rather than a deterministic one, and certainly does not exclude temporary setbacks, like the end of civilizations due to causes as diverse as:

(a) asteroid and comet impacts;
(b) rogue planets or stars, arriving unexpectedly and disrupting the gravitational stability of the planetary system;
(c) supernova explosions that fry entire nearby ET civilizations (just think of AGNs or active nucleus galaxies and ask yourself how many ET civilizations are dying in these galaxies right now?);
(d) ET nuclear wars; and
(e) possibly more causes to bring about civilization end that we do not even know about yet.

Mathematically, we came to define the probability density function (pdf) of this exponentially increasing stochastic process $N(t)$ as the lognormal

$$N(t)\_\text{pdf}(n, N_0, \mu, \sigma, t) = \frac{1}{\sqrt{2\pi}\,\sigma\sqrt{t}\,n} e^{-\frac{\left[\ln(n) - \left(\ln N_0 + \mu t - \frac{\sigma^2 t}{2}\right)\right]^2}{2\sigma^2 t}} \quad \text{for } 0 \le n \le \infty. \tag{11.2}$$

It is easy to prove that this lognormal pdf clearly fulfills the normalization condition

$$\int_0^\infty N(t)\_\ \text{pdf}(n, N_0, \mu, \sigma, t)\, dn = \int_0^\infty \frac{1}{\sqrt{2\pi}\,\sigma\sqrt{t}\,n} e^{-\frac{\left[\ln(n) - \left(\ln N_0 + \mu t - \frac{\sigma^2 t}{2}\right)\right]^2}{2\sigma^2 t}} dn = 1. \tag{11.3}$$

In addition, the mean value of (11.2) does indeed yield the exponential curve (11.1)

$$\int_0^\infty n \cdot N(t)\_\text{pdf}(n, N_0, \mu, \sigma, t)\, dn = \int_0^\infty n \frac{1}{\sqrt{2\pi}\,\sigma\sqrt{t}\,n} e^{-\frac{\left[\ln(n) - \left(\ln N_0 + \mu t - \frac{\sigma^2 t}{2}\right)\right]^2}{2\sigma^2 t}} dn = N_0\, e^{\mu t}. \tag{11.4}$$

The proof of (11.3) and (11.4) is given in Appendix 11.A as the Maxima file "GBM_as_N_of_t_v47". But, at the moment we want the reader to understand why our statistical Drake equation, described in Chapter 1, is just the static version of the stochastic process $N(t)$ whose pdf is given by (11.2).

11.2 OUR STATISTICAL DRAKE EQUATION IS THE STATIC SPECIAL CASE OF $N(t)$

In this section we prove the crucial fact that our statistical Drake equation (1.7) is just "the picture" case of the more general exponentially growing stochastic process $N(t)$ ("the movie") having the lognormal pdf (11.2).

To keep things neat, let us denote both the μ and σ appearing in (11.2) by the subscript "GBM". The latter thus takes the form:

$$N(t)\_\text{pdf}(n, N_0, \mu_{GBM}, \sigma_{GBM}, t) = \frac{1}{\sqrt{2\pi}\,\sigma_{GBM}\sqrt{t}\,n}\, e^{-\frac{\left[\ln(n) - \left(\ln N_0 + \mu_{GBM}\,t - \frac{\sigma_{GBM}^2\,t}{2}\right)\right]^2}{2\sigma_{GBM}^2\,t}}$$

$$\text{for } 0 \le n \le \infty. \quad (11.5)$$

Similarly, let us denote both the μ and σ appearing in the lognormal pdf (1.B.56) of our statistical Drake equation by the subscript "Drake":

$$\text{lognormal\_pdf\_of\_Statistical\_Drake\_Eq}(n, \mu_{Drake}, \sigma_{Drake})$$
$$= \frac{1}{\sqrt{2\pi}\,\sigma_{Drake}\,n}\, e^{-\frac{[\ln(n) - \mu_{Drake}]^2}{2\sigma_{Drake}^2}} \quad \text{for } 0 \le n \le \infty. \quad (11.6)$$

Now, a glance at (11.5) and (11.6) reveals that they can be made to coincide if, and only if, the two simultaneous equations

$$\begin{cases} \sigma_{GBM}\sqrt{t} = \sigma_{Drake} \\ \ln N_0 + \mu_{GBM}\,t - \dfrac{\sigma_{GBM}^2\,t}{2} = \mu_{Drake} \end{cases} \quad (11.7)$$

hold. On the other hand, when we pass (so to say) "from the movie to the picture," the two σ must be the same thing and so must the two μ; that is, one must have:

$$\begin{cases} \sigma_{GBM} = \sigma_{Drake} = \sigma \\ \mu_{GBM} = \mu_{Drake} = \mu. \end{cases} \quad (11.8)$$

Checking the upper equation of (11.8) against the upper equation of (11.7), we are just left with

$$\sqrt{t} = 1 \quad \text{that is} \quad t = 1. \quad (11.9)$$

So, $t = 1$ is the correct numeric value of the time leading "from the movie to the picture." Inserting this into the lower equation of (11.7), and keeping in mind the upper equation of (11.8), the lower equation of (11.7) becomes

$$\ln N_0 + \mu_{GBM} - \frac{\sigma^2}{2} = \mu_{Drake}. \quad (11.10)$$

Since the two μ must also be the same because of the lower equation of (11.8), then (11.10) further reduces to

$$\ln N_0 - \frac{\sigma^2}{2} = 0 \tag{11.11}$$

that is

$$N_0 = e^{\frac{\sigma^2}{2}} \tag{11.12}$$

and the problem of "passing from the movie to the picture" is completely solved.

In conclusion, we have proven the following "movie to picture" theorem.

"Movie to picture" theorem. *The stochastic process $N(t)$ reduces to the random variable N if, and only if, one inserts*

$$\begin{cases} t = 1 \\ \sigma_{GBM} = \sigma_{Drake} = \sigma \\ \mu_{GBM} = \mu_{Drake} = \mu \\ N_0 = e^{\frac{\sigma^2}{2}} \end{cases} \tag{11.13}$$

into the lognormal probability density (11.2) *of the stochastic process $N(t)$.*

11.3 THE $N(t)$ STOCHASTIC PROCESS IS A GEOMETRIC BROWNIAN MOTION

But what is this $N(t)$ stochastic process reducing to the lognormal random variable N in the static case? Well, $N(t)$ is no less than the famous geometric Brownian motion (abbreviated GBM), of paramount importance in the mathematics of finance. In fact, in the so-called Black–Scholes models, $N(t)$ is related to the log return of the stock price. Huge amounts of money all over the world are handled at stock exchanges according to the mathematics of the stochastic process $N(t)$, which is denoted differently as S_t there (where S stands for stock). A concise summary of the GBM is given at *http://en.wikipedia.org/wiki/Geometric_Brownian_motion* on Wikipedia. But we will steer clear of these topics in this book since this book is about SETI rather than stocks.

We just content ourselves with having proven that the GBM used in the mathematics of finance is the same thing as the exponentially increasing process $N(t)$ yielding the number of communicating ET civilizations in the Galaxy!

11.4 PROPERTIES OF THE $N(t)$ GEOMETRIC BROWNIAN MOTION

The Maxima file "GBM_as_N_of_t_v47" that makes up Appendix 11.A contains all the proofs about the $N(t)$ geometric Brownian motion (GBM) stochastic process. We now describe the content of this file.

The definition of GBM as a lognormal pdf in the variable n (= number of ET communicating civilizations in the Galaxy at the time t of Galactic evolution) is given by equations (%i1) through (%o4) and reads:

$$N(t)\_\mathrm{pdf}(n, N_0, \mu, \sigma, t) \equiv \mathrm{GBM}(n, N_0, \mu, \sigma, t)$$

$$= \frac{1}{\sqrt{2\pi}\,\sigma\sqrt{t}\,n} e^{-\frac{\left[\ln(n) - \left(\ln N_0 + \mu t - \frac{\sigma^2 t}{2}\right)\right]^2}{2\sigma^2 t}} \quad \text{for } n \geq 0 \tag{11.14}$$

As demonstrated in Section 11.2, the *static case* of this pdf corresponds to the set of two equations

$$\mathrm{STATIC\_CASE}: \left[t = 1,\ N_0 = e^{\frac{\sigma^2}{2}}\right] \tag{11.15}$$

and this is coded in Maxima equations (%i5) through (%o6).

The normalization condition (11.3) for the (11.14) pdf is proven by equations (%i7) and (%o7).

The important mean value formulas (11.1) and (11.4) for $N(t)$ are proven in Maxima equations (%i8) through (%o10).

Then, the mean value of $N^2(t)$ follows in equations (%i11) through (%o13).

The variance of $N(t)$ is thus given by equations (%i14) and (%o14):

$$\sigma^2_{N(t)} = N_0^2\, e^{2\mu t}\left(e^{\sigma^2 t} - 1\right) \tag{11.16}$$

and the standard deviation (i.e., the square root of this variance) is given by equations (%i15) and (%o16)

$$\sigma_{N(t)} = N_0\, e^{\mu t}\sqrt{e^{\sigma^2 t} - 1}. \tag{11.17}$$

The $N(t)$ peak now follows. This peak moves in time according to the peak abscissa; that is, the mode

$$n_{\mathrm{mode}} \equiv n_{\mathrm{peak}} = N_0\, e^{\mu t}\, e^{-\frac{3\sigma^2 t}{2}}. \tag{11.18}$$

The corresponding peak ordinate is then given by

$$f_{N(t)}(n_{\mathrm{mode}}) = \frac{1}{N_0\sqrt{2\pi}\,\sigma\sqrt{t}} \cdot e^{-\mu t} \cdot e^{\sigma^2 t}. \tag{11.19}$$

These two results about the $N(t)$ peak are demonstrated by Maxima equations (%i17) through (%o20), including the proof that, in the static case, (11.18) and (11.19) reduce to the peak equations given by (1.B.81) and (1.B.82), respectively.

The median of $N(t)$ is computed by Maxima equations (%i21) through (%o27) and turns out to equal

$$\mathrm{median} = m = N_0\, e^{\mu t} e^{-\frac{\sigma^2 t}{2}}. \tag{11.20}$$

In the static case, this reduces to (1.B.89).

The third and fourth moments follow in equations (%i29) through (%o32) of the Maxima code, but we shall not report them here for they are rather unimportant. Actually, it is possible to compute all the moments of the $N(t)$ stochastic process, as shown by equations (%i33) through (%o38) of the Maxima code. Indeed, the k-th moment of $N(t)$ for all integers $k \geq 0$ turns out to be given by:

$$\left\langle N^k(t) \right\rangle = N_0^k \, e^{k\mu t} \, e^{(k^2-k)\frac{\sigma^2 t}{2}} \qquad \text{for } k = 0, 1, 2, 3, 4 \ldots . \tag{11.21}$$

The first four cumulants of $N(t)$ then follow in Maxima equations (%i39) through (%o51), and these lead to the $N(t)$ skewness

$$\frac{K_3}{(K_2)^{\frac{3}{2}}} = \left(e^{\sigma^2 t} + 2\right)\sqrt{e^{\sigma^2 t} - 1} \tag{11.22}$$

and kurtosis

$$\frac{K_4}{(K_2)^2} = e^{4\,\sigma^2 t} + 2\,e^{3\,\sigma^2 t} + 3\,e^{2\,\sigma^2 t} - 6\,. \tag{11.23}$$

As for the two inflexion points of the $N(t)$ pdf (11.14), they are given by the ascending or "adolescence" inflexion point

$$a = N_0\, e^{-\frac{\sigma\sqrt{t}\sqrt{\sigma^2 t + 4}}{2} - 2\sigma^2 t + \mu t} \tag{11.24}$$

and by the descending or "senility" inflexion point, respectively, proven by the equations between (%i54) and (%o58):

$$s = N_0\, e^{\frac{\sigma\sqrt{t}\sqrt{\sigma^2 t + 4}}{2} - 2\sigma^2 t + \mu t} \tag{11.25}$$

It should be noted that, although (11.24) and (11.25) reduce to (6.4) and (6.5), respectively, in the static case (11.15), they should not be regarded as natural generalizations of (6.4) and (6.5) in time. In fact, there is a crucial difference between the b-lognormal pdf (6.3) in time (with $b = 0$) of Chapter 6, and the present lognormal pdf (11.2) in the number n of communicating ET civilizations now existing in the Galaxy. This should always be kept in mind!

Having said this, even the "death time" of the process $N(t)$ can be defined in a way similar to (6.27) (again with $b = 0$), and that leads to the equation

$$d = N_0 \, \frac{\left(\sqrt{\sigma^2 t + 4} + \sigma\sqrt{t}\right)^2 e^{\frac{\sigma\sqrt{t}\sqrt{\sigma^2 t + 4}}{2} - 2\sigma^2 t + \mu t}}{4} \tag{11.26}$$

as proven in equations (%i59) through (%o78) of the Maxima file that makes up Appendix 11.A.

We now wish to discuss a new and unexpected result that we dub the "immortality condition"! The point is as follows.

"Death" (11.26) is clearly to be understood as "death of the number of ET civilizations in the Galaxy" (i.e., the time when no living civilizations survive!). Now, clearly, we do not want this to happen! So, our only hope is to find how to "delay" death (11.26) by pushing it out to infinity! In other words, we now wish to investigate under which conditions the limit of (11.26) for $t \to \infty$ equals infinity! That is, under which conditions do we have the limit

$$\lim_{t\to\infty} d = \lim_{t\to\infty} N_0 \frac{\left(\sqrt{\sigma^2 t + 4} + \sigma\sqrt{t}\right)^2 e^{\frac{\sigma\sqrt{t}\sqrt{\sigma^2 t+4}}{2} - 2\sigma^2 t + \mu t}}{4} = \infty? \tag{11.27}$$

Well, since we have the positive part

$$N_0 \frac{\left(\sqrt{\sigma^2 t + 4} + \sigma\sqrt{t}\right)^2}{4} > 0 \tag{11.28}$$

in (11.26) times the exponential

$$e^{\frac{\sigma\sqrt{t}\sqrt{\sigma^2 t+4}}{2} - 2\sigma^2 t + \mu t} \tag{11.29}$$

the only way to let their products approach infinity is to have the exponential (11.29) approach infinity. So, we must have

$$\lim_{t\to\infty}\left(\frac{\sigma\sqrt{t}\sqrt{\sigma^2 t+4}}{2} - 2\sigma^2 t + \mu\, t\right) = \infty\,. \tag{11.30}$$

Now, a careful look at (11.30) reveals that (11.30) is true if, and only if, one has

$$\begin{aligned}\lim_{t\to\infty}\left(\frac{\sigma\sqrt{t}\sqrt{\sigma^2 t+4}}{2} - 2\sigma^2 t + \mu\, t\right) &= \lim_{t\to\infty}\left(\frac{\sigma\sqrt{t}\sqrt{\sigma^2 t}}{2} - 2\sigma^2 t + \mu\, t\right)\\ &= \lim_{t\to\infty}\left(\frac{\sigma^2 t}{2} - 2\sigma^2 t + \mu\, t\right) = \infty\,.\end{aligned} \tag{11.31}$$

In order for (11.31) to be true, one must have

$$\frac{\sigma^2}{2} - 2\sigma^2 + \mu > 0 \quad \text{that is} \quad \mu > \tfrac{2}{3}\sigma^2\,. \tag{11.32}$$

We have thus reached the important new result that the immortality condition implies $\mu > \frac{2}{3}\sigma^2 > 0$.

In conclusion, Table 11.1 summarizes all the proven properties of the $N(t)$ stochastic process yielding the overall exponential increase in time of the number of communicating ET civilizations in the Galaxy, taking into account both the technological evolution of these civilizations and the sudden "passing away" of some of them due to asteroid impacts, nearby supernova explosions, ET nuclear wars, and so on.

Table 11.1. Summary of the properties of the lognormal distribution that apply to the stochastic process $N(t)$ = exponentially increasing number of ET communicating civilizations in the Galaxy.

Stochastic process	$N(t)$ = *exponentially increasing number of communicating ET civilizations in the Galaxy*
Probability distribution	Lognormal distribution of the geometric Brownian motion (GBM)
Probability density function	$N(t)\_pdf(n, N_0, \mu, \sigma, t) = \frac{1}{\sqrt{2\pi}\,\sigma\sqrt{t}\,n}\, e^{-\frac{\left[\ln(n) - \left(\ln N_0 + \mu t - \frac{\sigma^2 t}{2}\right)\right]^2}{2\sigma^2 t}}$ for $n \geq 0$
Mean value	$\langle N(t) \rangle = N_0\, e^{\mu t}$
Variance	$\sigma^2_{N(t)} = N_0^2\, e^{2\mu t}\left(e^{\sigma^2 t} - 1\right)$
Standard deviation	$\sigma_{N(t)} = N_0\, e^{\mu t}\sqrt{e^{\sigma^2 t} - 1}$
All the moments (i.e., k-th moment)	$\langle N^k(t) \rangle = N_0^k\, e^{k\mu t}\, e^{(k^2 - k)\frac{\sigma^2 t}{2}}$
Mode (= abscissa of the lognormal peak)	$n_{\text{mode}} \equiv n_{\text{peak}} = N_0\, e^{\mu t}\, e^{-\frac{3\sigma^2 t}{2}}$
Value of the mode peak	$f_{N(t)}(n_{\text{mode}}) = \frac{1}{N_0\sqrt{2\pi}\,\sigma\sqrt{t}} \cdot e^{-\mu t} \cdot e^{\sigma^2 t}$
Median (= 50–50 probability value for $N(t)$)	$\text{median} = m = N_0\, e^{\mu t} e^{-\frac{\sigma^2 t}{2}}$
Skewness	$\frac{K_3}{(K_2)^{\frac{3}{2}}} = \left(e^{\sigma^2 t} + 2\right)\sqrt{e^{\sigma^2 t} - 1}$
Kurtosis	$\frac{K_4}{(K_2)^2} = e^{4\sigma^2 t} + 2\, e^{3\sigma^2 t} + 3\, e^{2\sigma^2 t} - 6$
Adolescence point	$a = N_0\, e^{-\frac{\sigma\sqrt{t}\sqrt{\sigma^2 t + 4}}{2} - 2\sigma^2 t + \mu t}$
Senility point	$s = N_0\, e^{\frac{\sigma\sqrt{t}\sqrt{\sigma^2 t + 4}}{2} - 2\sigma^2 t + \mu t}$
Death point	$d = N_0\, \frac{\left(\sqrt{\sigma^2 t + 4} + \sigma\sqrt{t}\right)^2 e^{\frac{\sigma\sqrt{t}\sqrt{\sigma^2 t + 4}}{2} - 2\sigma^2 t + \mu t}}{4}$
Immortality condition	$\mu > \frac{2}{3}\sigma^2 > 0$
Expression of μ in terms of the lower (a_i) and upper (b_i) limits of the Drake uniform input random variable D_i	$\mu = \sum_{i=1}^{7} \langle Y_i \rangle = \sum_{i=1}^{7} \frac{b_i[\ln(b_i) - 1] - a_i[\ln(a_i) - 1]}{b_i - a_i}$
Expression of σ^2 in terms of the lower (a_i) and upper (b_i) limits of the Drake uniform input random variable D_i	$\sigma^2 = \sum_{i=1}^{7} \sigma^2_{Y_i} = \sum_{i=1}^{7}\left(1 - \frac{a_i b_i [\ln(b_i) - \ln(a_i)]^2}{(b_i - a_i)^2}\right)$

11.5 THE NEW STOCHASTIC PROCESS $D(t)$ YIELDING THE CIVILIZATION DISTANCE IN TIME

Readers will remember from Chapter 1 that the most important new result about the statistical Drake equation was discovering the probability distribution of the distance between any two nearby such civilizations, given by the "Maccone" distribution (1.B.113). Formally, the relationship between the two random variables N and ET_distance is given by equation (1.B.104), which we reproduce hereafter for convenience

$$\text{ET\_Distance} = \frac{\sqrt[3]{6\,R_{Galaxy}^2\,h_{Galaxy}}}{\sqrt[3]{N}} = \frac{C}{\sqrt[3]{N}}. \tag{11.33}$$

Well, the next natural question after the introduction of the stochastic process $N(t)$ is then: How does (11.33) generalize in time as well? In other words, replacing the now too long random variable ET_distance by the shorter one D, the question becomes: given the formal relationship between the two stochastic processes $N(t)$ and $D(t)$:

$$D(t) = \frac{C}{\sqrt[3]{N(t)}} \tag{11.34}$$

how can we find the pdf of the new stochastic process $D(t)$?

This question is answered in this section, where we show that the "Maccone stochastic process" (so to say) $D(t)$, yielding the evolution in time of the distance in time between any two nearby communicating civilizations, has the pdf:

$$D(t)\_pdf(r, N_0, \mu, \sigma, t) = \frac{3}{\sqrt{2\pi}\,\sigma\sqrt{t}\,r}\, e^{-\frac{\left[\ln\left(\frac{C^3}{r^3}\right) - \left(\ln N_0 + \mu t - \frac{\sigma^2 t}{2}\right)\right]^2}{2\sigma^2 t}} \quad \text{for } r \geq 0. \tag{11.35}$$

In (11.35) the constant C is given by (1.6) and is related to the Galaxy size by

$$C = \sqrt[3]{6\,R_{Galaxy}^2\,h_{Galaxy}} \approx 28{,}845 \text{ light-years}. \tag{11.36}$$

Intuitively, since the overall average number of ET civilizations $N(t)$ increases in time according to the exponential (11.1), we should expect the average distance between any two nearby communicating civilizations to decrease in time, and this is indeed the case (under the unexpected condition $\mu > \frac{2}{3}\sigma^2$), as we will see in a moment.

The Maxima file "ET_DISTANCE_FOR_N_of_t_v51" that makes up Appendix 11.B contains all the proofs about the new stochastic process $D(t)$. Thus, we see that equations (%i1) through (%o16) of this file lead to the definition (11.35) and to the proof that (11.35) reduces to (1.B.113) in the static case given by (11.15).

Then, all the moments of (11.35) can be computed immediately by Maxima, with the result that the k-th moment, for all integers $k \geq 0$, reads

$$\left\langle D^k(t)\right\rangle = C^k N_0^{-\frac{k}{3}} e^{-\frac{k\mu t}{3}}\, e^{\left(k^2+3k\right)\frac{\sigma^2 t}{18}} \quad \text{for } k = 0, 1, 2, 3, 4 \ldots . \tag{11.37}$$

Proof. See equations (%i17) through (%o24) of the Maxima file in Appendix 11.B.

The particular case $k = 0$ of (11.37) obviously yields the normalization condition of the (11.35) pdf; that is

$$\int_0^\infty D(t)\_pdf(r, N_0, \mu, \sigma, t)\, dr = \int_0^\infty \frac{3}{\sqrt{2\pi}\,\sigma\sqrt{t}\, r}\, e^{-\frac{\left[\ln\left(\frac{C^3}{r^3}\right) - \left(\ln N_0 + \mu t - \frac{\sigma^2 t}{2}\right)\right]^2}{2\sigma^2 t}}\, dr = 1 \tag{11.38}$$

as described by equations (%i25) and (%o25).

More interesting is the time evolution of the distance mean value, given by the particular case $k = 1$ of (11.37); that is

$$\begin{aligned}\langle D(t)\rangle &= \int_0^\infty r\, \frac{3}{\sqrt{2\pi}\,\sigma\sqrt{t}\, r}\, e^{-\frac{\left[\ln\left(\frac{C^3}{r^3}\right) - \left(\ln N_0 + \mu t - \frac{\sigma^2 t}{2}\right)\right]^2}{2\sigma^2 t}} \\ &= C\, N_0^{-\frac{1}{3}} \cdot e^{-\frac{1}{3}\mu t} \cdot e^{\frac{2\sigma^2 t}{9}} = C\, N_0^{-\frac{1}{3}} \cdot e^{-\left(\frac{\mu}{3} - \frac{2\sigma^2}{9}\right) t}.\end{aligned} \tag{11.39}$$

This is proven in equations (%i26) through (%o31) of the Maxima file. Readers should note the very important negative exponential in time

$$e^{-\left(\frac{\mu}{3} - \frac{2\sigma^2}{9}\right) t} \tag{11.40}$$

appearing in the distance mean value (11.39). Intuitively speaking, one should expect, as long as time elapses, the average distance between any two nearby ET civilizations to decrease. In plain words: as long as time elapses, increasingly more new technological civilizations in the Galaxy will reach the maturity level of being able to communicate among themselves (i.e., their physicists will rediscover what humans call the Maxwell equations and thus the existence of radiowaves for communications), and so, statistically speaking, the average distance between communicating civilizations will have to decrease! Now, letting $t \to \infty$ in (11.40) we see that the exponential approaches zero if, and only if

$$\frac{\mu}{3} - \frac{2\sigma^2}{9} > 0 \quad \text{that is} \quad \mu > \frac{2}{3}\sigma^2 = 0.667\,\sigma^2. \tag{11.41}$$

We have thus come to discover a new and unexpected numerical constraint between μ and σ^2, two variables that were expected to be absolutely independent of each other. The constraint is that one must have $\mu > \frac{2}{3}\sigma^2 = 0.667\,\sigma^2$ in order to let our geometric Brownian motion model be realistic for the evolution of life in the Galaxy! We will see in a moment that other properties of the $D(t)$ stochastic process will yield similar numerical constraints between μ and σ^2 as well.

Next, we consider the mean value of $D^2(t)$, which is given by equations (%i32) and (%o32).

The variance of $D(t)$ follows and is proven in equations (%i33) and (%o33), with the result

$$\sigma^2_{D(t)} = C^2 N_0^{-\frac{2}{3}} \cdot e^{-\frac{2\mu t}{3}} \cdot e^{\frac{4\sigma^2 t}{9}} \left(e^{\frac{\sigma^2 t}{9}} - 1\right). \tag{11.42}$$

The $D(t)$ standard deviation is just the square root of (11.42) and thus equations (%i35) and (%o36)

$$\sigma_{D(t)} = C N_0^{-\frac{1}{3}} \cdot e^{-\frac{(3\mu - 2\sigma^2)}{9} t} \cdot \sqrt{e^{\frac{\sigma^2 t}{9}} - 1}. \tag{11.43}$$

The third and fourth moments are found in equations (%i37) through (%o40) of the Maxima file, but we shall not consider them here. Nor shall we consider the first four cumulants of $D(t)$, which are computed in equations (%i41) through (%o50). These lead, however, to the $D(t)$ skewness found in equations (%i47) through (%o55)

$$\frac{K_3}{(K_2)^{\frac{3}{2}}} = \frac{e^{\sigma^2 t} - 3\, e^{\frac{7\sigma^2 t}{9}} + 2\, e^{\frac{2\sigma^2 t}{3}}}{\sqrt{e^{\frac{\sigma^2 t}{9}} - 1}\left(e^{\frac{7\sigma^2 t}{9}} - e^{\frac{2\sigma^2 t}{3}}\right)} \tag{11.44}$$

and kurtosis found in equations (%i56) through (%o57)

$$\frac{K_4}{(K_2)^2} = e^{\frac{4\sigma^2 t}{9}} + 2\, e^{\frac{\sigma^2 t}{3}} + 3\, e^{\frac{2\sigma^2 t}{9}} - 6. \tag{11.45}$$

The time evolution of the distance peak (i.e., the time evolution of the most likely values of the distance) is quite important. In plain words once again, this means that, as long as time elapses and increasingly more new ET civilizations grow up technologically, the average distance between any two nearby such civilizations should decrease. Indeed, this peak distance in time is given by equations (%i58) and (%o62) of the Maxima file, and reads

$$n_{\text{mode}} \equiv n_{\text{peak}} = C\, N_0^{-\frac{1}{3}}\, e^{-\frac{1}{3}\mu t}\, e^{\frac{\sigma^2 t}{18}} = C\, N_0^{-\frac{1}{3}}\, e^{-\left(\frac{\mu}{3} - \frac{\sigma^2}{18}\right) t}. \tag{11.46}$$

Once again, the distance peak abscissa is seen from (11.46) to approach zero as $t \to \infty$ if, and only if, one has

$$\frac{(6\mu - \sigma^2)}{18} > 0 \quad \text{that is} \quad \mu > \tfrac{1}{6}\sigma^2 = 0.167\,\sigma^2 \tag{11.47}$$

which is another constraint between μ and σ^2.

The peak ordinate in time is given by equations (%i63) through (%o65) and reads

$$f_{D(t)}(n_{\text{mode}}) = \frac{3}{\sqrt{2\pi}\,\sigma\sqrt{t}\,C} \cdot e^{\frac{\sigma^2+3\mu}{9}} \cdot e^{-\frac{\left(-\ln(N_0)+\frac{\sigma^2 t}{2}-\mu t+\frac{\sigma^2+3\mu}{3}\right)^2}{2\sigma^2 t}}. \tag{11.48}$$

Finally, the median of $D(t)$ is found in equations (%i66) through (%o72) and reads

$$\text{median} = m = C\,N_0^{-\frac{1}{3}} \cdot e^{-\frac{\mu t}{3}} \cdot e^{\frac{\sigma^2 t}{6}} = C\,N_0^{-\frac{1}{3}} \cdot e^{-\left(\frac{\mu}{3}-\frac{\sigma^2}{6}\right)t}. \tag{11.49}$$

Again, (11.49) yields another constraint between μ and σ^2; that is

$$\frac{(2\mu - \sigma^2)}{6} > 0 \quad \text{that is} \quad \mu > \tfrac{1}{2}\sigma^2 = 0.5\,\sigma^2. \tag{11.50}$$

If we now check all three constraints (11.41), (11.47), and (11.50) against each other, we find that the strongest constraint (i.e., the one that if fulfilled implies all other constraints to be fulfilled as well) is the mean value constraint (11.41), which is

$$\frac{\mu}{3} - \frac{2\sigma^2}{9} > 0 \quad \text{that is} \quad \mu > \tfrac{2}{3}\sigma^2 = 0.667\,\sigma^2 > 0. \tag{11.51}$$

This constraint is just the same as the immortality condition (11.32). We shall thus assume that this constraint between μ and σ is indeed fulfilled in order to let our GBM model of the increasing evolution of the Galaxy be realistic!

Table 11.2 summarizes all the proven properties of the stochastic process $D(t) = \textit{distance}$ decreasing in time between any two nearby ET communicating civilizations in the Galaxy.

11.6 GBM AS THE EXPONENTIALLY INCREASING NUMBER OF HABITABLE PLANETS

Readers will remember that Chapter 3 ("How many planets for man and aliens") utilized the statistical Dole equations (3.2) and (3.3) in just the same mathematical way the Drake equation (1.7) was used in Chapter 1. In other words, both the Drake and Dole statistical equations have the same solution if one lets the number of terms on the right-hand side increase indefinitely, and this solution is the static lognormal pdf (11.6).

Well, the same obviously applies to the extension in time for both the statistical Drake and the statistical Dole equations!

Thus, we immediately conclude that the lognormal pdf (11.2) is also the pdf of the exponentially increasing number of habitable planets for man and aliens in the

Table 11.2. Summary of the properties of the time-evolving "Maccone distribution" of the stochastic process $D(t) = distance$ decreasing in time between any two nearby ET communicating civilizations in the Galaxy.

Stochastic process	$D(t)$ = decreasing-in-time *distance* between any two nearby communicating ET civilizations in the Galaxy as long as increasingly more ET civilizations grow up
Probability distribution	Maccone distribution becoming a Maccone process based on GBM
Probability density function	$D(t)\_pdf(r, N_0, \mu, \sigma, t)$ $= \frac{3}{\sqrt{2\pi}\,\sigma\sqrt{t}\,r} e^{-\frac{\left[\ln\left(\frac{C^3}{r^3}\right) - \left(\ln N_0 + \mu t - \frac{\sigma^2 t}{2}\right)\right]^2}{2\sigma^2\, t}}$ for $r \geq 0$
Mean value	$\langle D(t)\rangle = C\, N_0^{-\frac{1}{3}} \cdot e^{-\frac{1}{3}\mu t} \cdot e^{\frac{2\sigma^2 t}{9}} = C\, N_0^{-\frac{1}{3}} \cdot e^{-\left(\frac{\mu}{3} - \frac{2\sigma^2}{9}\right)\, t}$
Variance	$\sigma^2_{D(t)} = C^2 N_0^{-\frac{2}{3}} \cdot e^{-\frac{2\mu}{3}t} \cdot e^{\frac{4\sigma^2}{9}t}\left(e^{\frac{\sigma^2 t}{9}} - 1\right)$
Standard deviation	$\sigma_{D(t)} = C\, N_0^{-\frac{1}{3}} \cdot e^{-\frac{\mu}{3}t} \cdot e^{\frac{2\sigma^2}{9}t} \cdot \sqrt{e^{\frac{\sigma^2 t}{9}} - 1}$
All the moments (i.e., k-th moment)	$\langle D^k(t)\rangle = C^k N_0^{-\frac{k}{3}} e^{-\frac{k\mu t}{3}}\, e^{\left(k^2+3k\right)\frac{\sigma^2 t}{18}}$
Mode (=abscissa of the distance peak)	$n_{\text{mode}} \equiv n_{\text{peak}} = C\, N_0^{-\frac{1}{3}} e^{-\frac{1}{3}\mu t}\, e^{\frac{\sigma^2 t}{18}} = C\, N_0^{-\frac{1}{3}} e^{-\left(\frac{\mu}{3} - \frac{\sigma^2}{18}\right)t}$
Ordinate of the distance peak	$f_{D(t)}(n_{\text{mode}})$ $= \frac{3}{\sqrt{2\pi}\,\sigma\sqrt{t}\,C} \cdot e^{\frac{\sigma^2+3\mu}{9}} \cdot e^{-\frac{\left(-\ln(N_0) + \frac{\sigma^2 t}{2} - \mu t + \frac{\sigma^2+3\mu}{3}\right)^2}{2\sigma^2 t}}$
Median (= 50–50 probability value for $D(t)$)	$\text{median} = m = C\, N_0^{-\frac{1}{3}} \cdot e^{-\frac{\mu t}{3}} \cdot e^{\frac{\sigma^2 t}{6}}$ $= C\, N_0^{-\frac{1}{3}} \cdot e^{-\left(\frac{\mu}{3} - \frac{\sigma^2}{6}\right)\, t}.$
Skewness	$\frac{K_3}{(K_2)^{\frac{3}{2}}} = \frac{e^{\sigma^2 t} - 3\, e^{\frac{7\sigma^2 t}{9}} + 2\, e^{\frac{2\sigma^2 t}{3}}}{\sqrt{e^{\frac{\sigma^2 t}{9}} - 1}\left(e^{\frac{7\sigma^2 t}{3}} - e^{\frac{2\sigma^2 t}{9}}\right)}$
Kurtosis	$\frac{K_4}{(K_2)^2} = e^{\frac{4\sigma^2 t}{9}} + 2\, e^{\frac{\sigma^2 t}{3}} + 3\, e^{\frac{2\sigma^2 t}{9}} - 6$
Zero distance as $t \to \infty$ condition	$\mu > \frac{2}{3}\sigma^2 > 0$
Expression of μ in terms of the lower (a_i) and upper (b_i) limits of the Drake uniform input random variable D_i	$\mu = \sum_{i=1}^{7} \langle Y_i \rangle = \sum_{i=1}^{7} \frac{b_i[\ln(b_i) - 1] - a_i[\ln(a_i) - 1]}{b_i - a_i}$
Expression of σ^2 in terms of the lower (a_i) and upper (b_i) limits of the Drake uniform input random variable D_i	$\sigma^2 = \sum_{i=1}^{7} \sigma^2_{Y_i} = \sum_{i=1}^{7} \left(1 - \frac{a_i b_i [\ln(b_i) - \ln(a_i)]^2}{(b_i - a_i)^2}\right)$

Galaxy:

$$N_{Hab}(t)\_\mathrm{pdf}\left(n, N_{Hab\_0}, \mu, \sigma, t\right) = \frac{1}{\sqrt{2\pi}\,\sigma\sqrt{t}\,n}\, e^{-\frac{\left[\ln(n) - \left(\ln N_{Hab\_0} + \mu\, t - \frac{\sigma^2 t}{2}\right)\right]^2}{2\,\sigma^2\, t}} \quad \text{for } 0 \le n \le \infty. \tag{11.52}$$

In other words, (11.52) is the pdf of the $N_{Hab}(t)$ stochastic process yielding the number of exponentially increasing habitable planets for man (or for a given type of alien if we change the numerical values of the corresponding terms in the Dole equation) in the Galaxy.

Of course, all other properties proven in the previous sections about ET civilizations are now replaced by similar properties about habitable planets.

Thus, we conclude that the mathematics of GBM will be the key to writing a large computer program about the actual evolution of planets and life in the Galaxy, but that would deserve a much more profound study than we can do right now. The future will tell.

APPENDIX 11.A

GEOMETRIC BROWNIAN MOTION pdf is LOGNORMAL.
Claudio Maccone, February 28th, 2012.

(%i84) kill(all);
(%o0) done

(%i1) assume(t>0, sigma>0);
(%o1) $[t>0,\sigma>0]$

Defining the GEOMETRIC BROWNIAN MOTION (GBM) pdf (i.e. a lognormal pdf in n):

(%i2) GBMexponent(n, N0, mu, sigma, t):=(-(log(n)-(log(N0)+mu*t-sigma^2*t/2))^2/(2*sigma^2*t));

(%o2) $\text{GBMexponent}(n,N0,\mu,\sigma,t):=\dfrac{-\left(\log(n)-\left(\log(N0)+\mu\,t+\dfrac{(-\sigma^2)\,t}{2}\right)\right)^2}{2\,\sigma^2\,t}$

(%i3) GBM(n, N0, mu, sigma, t):=(1/(sqrt(2*%pi)*n*sigma*sqrt(t)))*%e^GBMexponent(n, N0, mu, sigma, t);

(%o3) $\text{GBM}(n,N0,\mu,\sigma,t):=\dfrac{1}{\sqrt{2\,\pi}\,n\,\sigma\,\sqrt{t}}\,\%e^{\text{GBMexponent}(n,N0,\mu,\sigma,t)}$

(%i4) GBM(n, N0, mu, sigma, t);

(%o4) $\dfrac{\%e^{-\dfrac{\left(-\log(N0)+\dfrac{\sigma^2\,t}{2}-\mu\,t+\log(n)\right)^2}{2\,\sigma^2\,t}}}{\sqrt{2}\,\sqrt{\pi}\,n\,\sigma\,\sqrt{t}}$

Defining the STATIC CASE (i.e. when the GBM gets down to the
lognormal Statistical Drake Equation pdf (1.B.56) of Chapter 1).

(%i5) STATIC_CASE:[t=1, N0=%e^(sigma^2/2)];

(%o5) $[t=1,N0=\%e^{\frac{\sigma^2}{2}}]$

Checking that the GBM pdf indeed GETS DOWN to the lognormal pdf
(1.B.56) of the Statistical Drake Equation in the STATIC CASE of N(t).

(%i6) ordinary_lognormal=GBM(n, N0, mu, sigma, t),STATIC_CASE;

(%o6) $\text{ordinary\_lognormal}=\dfrac{\%e^{-\dfrac{(\log(n)-\mu)^2}{2\,\sigma^2}}}{\sqrt{2}\,\sqrt{\pi}\,n\,\sigma}$

NORMALIZATION CONDITION.

(%i7) 'integrate(GBM(n,N0,mu,sigma,t), n, 0, inf)=integrate(GBM(n,N0,mu,sigma,t), n, 0, inf);

(%o7) $$\frac{\int_0^{\infty} \frac{\%e^{-\frac{\left(-\log(N0)-\frac{\sigma^2 t}{2}-\mu t+\log(n)\right)^2}{2\sigma^2 t}}}{n}\,dn}{\sqrt{2}\sqrt{\pi}\,\sigma\sqrt{t}}=1$$

MEAN VALUE

(%i8) 'integrate(n*GBM(n,N0,mu,sigma,t), n, 0, inf)=integrate(n*GBM(n,N0,mu,sigma,t), n, 0, inf);

(%o8) $$\frac{\int_0^{\infty} \%e^{-\frac{\left(-\log(N0)+\frac{\sigma^2 t}{2}-\mu t+\log(n)\right)^2}{2\sigma^2 t}}\,dn}{\sqrt{2}\sqrt{\pi}\,\sigma\sqrt{t}}=\%e^{\mu t}\,N0$$

Very IMPORTANT! The AVERAGE NUMBER of ET Civilizations in the Galaxy INCREASES EXPONENTIALLY IN TIME !

(%i9) def_meanOfNoft:meanOfNoft=rhs(%);

(%o9) meanOfNoft $=\%e^{\mu t}\,N0$

(%i10) mean_value_eq:def_meanOfNoft;

(%o10) meanOfNoft $=\%e^{\mu t}\,N0$

Mean value of the square.

(%i11) 'integrate(n^2*GBM(n,N0,mu,sigma,t), n, 0, inf)=integrate(n^2*GBM(n,N0,mu,sigma,t), n, 0, inf);

(%o11) $$\frac{\int_0^{\infty} n\,\%e^{-\frac{\left(-\log(N0)+\frac{\sigma^2 t}{2}-\mu t+\log(n)\right)^2}{2\sigma^2 t}}\,dn}{\sqrt{2}\sqrt{\pi}\,\sigma\sqrt{t}}=\%e^{(\sigma^2+2\mu)t}\,N0^2$$

(%i12) def_meanOfNoftSquare:meanOfNoft_square=rhs(%);

(%o12) meanOfNoft_square $=\%e^{(\sigma^2+2\mu)t}\,N0^2$

(%i13) second_moment_eq:def_meanOfNoftSquare;

(%o13) meanOfNoft_square $= \%e^{(\sigma^2+2\,\mu)\,t}\,N0^2$

VARIANCE

(%i14) def_variance:variance=factor(expand((rhs(def_meanOfNoftSquare)-rhs(def_meanOfNoft^2))));

(%o14) variance $= \%e^{2\,\mu\,t}\left(\%e^{\sigma^2\,t}-1\right)N0^2$

STANDARD DEDIATION

(%i15) assume(N0>0);
(%o15) [N0>0]

(%i16) def_standard_deviation:standard_deviation=sqrt(rhs(def_variance));

(%o16) standard_deviation $= \%e^{\mu\,t}\sqrt{\%e^{\sigma^2\,t}-1}\;N0$

PEAK, i.e. MODE

(%i17) def_mode:peak_abscissa=rhs(first(solve(diff(GBM(n, N0, mu, sigma, t), n)=0, n)));

(%o17) peak_abscissa $= \%e^{\mu\,t-\frac{3\,\sigma^2\,t}{2}}\;N0$

(%i18) def_peak_ordinate:peak_ordinate=radcan(subst(rhs(def_mode), n, GBM(n, N0, mu, sigma, t)));

(%o18) peak_ordinate $= \dfrac{\%e^{(\sigma^2-\mu)\,t}}{\sqrt{2}\,\sqrt{\pi}\,\sigma\,\sqrt{t}\,N0}$

(%i19) GBM_peak_coordinates:[def_mode, def_peak_ordinate];

(%o19) $\left[\text{peak\_abscissa} = \%e^{\mu\,t-\frac{3\,\sigma^2\,t}{2}}\;N0,\ \text{peak\_ordinate} = \dfrac{\%e^{(\sigma^2-\mu)\,t}}{\sqrt{2}\,\sqrt{\pi}\,\sigma\,\sqrt{t}\,N0}\right]$

(%i20) GBM_peak_coordinates,STATIC_CASE;

(%o20) $\left[\text{peak\_abscissa} = \%e^{\mu-\sigma^2},\ \text{peak\_ordinate} = \dfrac{\%e^{\frac{\sigma^2}{2}-\mu}}{\sqrt{2}\,\sqrt{\pi}\,\sigma}\right]$

MEDIAN

(%i21) assume(median>0);
(%o21) [median>0]

(%i22) def_median:integrate(radcan(expand(GBM(n,N0,mu,sigma,t))), n, 0, median)=1/2;

$$(\%o22)\ \left(NO^{\frac{\sigma^2-2\mu}{2\sigma^2}} \left(\frac{\sqrt{\pi}\,\sigma\,\sqrt{t}\,\%e^{\frac{\mu\log(NO)}{\sigma^2}-\frac{\log(NO)}{2}}}{\sqrt{2}} - \frac{\sqrt{\pi}\,\sigma\,\sqrt{t}\,\%e^{\frac{\mu\log(NO)}{\sigma^2}-\frac{\log(NO)}{2}}\operatorname{erf}\left(\frac{2^{3/2}\log(NO)+(2^{3/2}\mu-\sqrt{2}\,\sigma^2)\,t-2^{3/2}\log(median)}{4\sigma\sqrt{t}}\right)}{\sqrt{2}} \right) \right) / (\sqrt{2}\,\sqrt{\pi}\,\sigma\,\sqrt{t}) = \frac{1}{2}$$

(%i23) median_eq:first(solve(def_median, median));

$$(\%o23)\ median = NO\ \%e^{-\sqrt{2}\,\sigma\,\sqrt{t}\ inverse\_erf\left(1-NO^{\frac{\mu}{\sigma^2}-\frac{1}{2}}\%e^{\frac{\log(NO)}{2}-\frac{\mu\log(NO)}{\sigma^2}}\right)-\frac{\sigma^2 t}{2}+\mu t}$$

Unfortuntely, Maxima "does not see" INSIDE the inverse_erf function that the argument is ZERO, and so it cannot perform the simplification. But we can do so "brutally" as follows.

(%i24) 1-N0^(mu/sigma^2-1/2)*%e^(log(N0)/2-(mu*log(N0))/sigma^2);

$$(\%o24)\ 1-NO^{\frac{\mu}{\sigma^2}-\frac{1}{2}}\%e^{\frac{\log(NO)}{2}-\frac{\mu\log(NO)}{\sigma^2}}$$

(%i25) radcan(%);

(%o25) 0

(%i26) inverse_erf(0);

(%o26) 0

(%i27) median_eq:median=N0*%e^(-(sigma^2*t)/2+mu*t);

$$(\%o27)\ median = \%e^{\mu t-\frac{\sigma^2 t}{2}}\ NO$$

Median in the STATIC_CASE reduces to (1.B.89).

(%i28) median_eq,STATIC_CASE;

$$(\%o28)\ median = \%e^{\mu}$$

THIRD MOMENT.

(%i29) 'integrate(n^3*GBM(n,N0,mu,sigma,t), n, 0, inf)=integrate(n^3*GBM(n,N0,mu,sigma,t), n, 0, inf);

(%o29) $$\frac{\int_0^{\infty} n^2\, \%e^{-\frac{\left(-\log(N0)+\frac{\sigma^2 t}{2}-\mu t+\log(n)\right)^2}{2\sigma^2 t}}\, dn}{\sqrt{2}\sqrt{\pi}\,\sigma\sqrt{t}} = \%e^{(3\sigma^2+3\mu)t}\, N0^3$$

(%i30) third_moment_eq:third_moment=rhs(%);

(%o30) $third\_moment = \%e^{(3\sigma^2+3\mu)t}\, N0^3$

FOURTH MOMENT.

(%i31) 'integrate(n^4*GBM(n,N0,mu,sigma,t), n, 0, inf)=integrate(n^4*GBM(n,N0,mu,sigma,t), n, 0, inf);

(%o31) $$\frac{\int_0^{\infty} n^3\, \%e^{-\frac{\left(-\log(N0)+\frac{\sigma^2 t}{2}-\mu t+\log(n)\right)^2}{2\sigma^2 t}}\, dn}{\sqrt{2}\sqrt{\pi}\,\sigma\sqrt{t}} = \%e^{(6\sigma^2+4\mu)t}\, N0^4$$

(%i32) fourth_moment_eq:fourth_moment=rhs(%);

(%o32) $fourth\_moment = \%e^{(6\sigma^2+4\mu)t}\, N0^4$

ALL MOMENTS (i.e. m-th moment for m=1,2,3,4,5,6...)

(%i33) assume(k>0);
(%o33) [k>0]

(%i34) declare(k, integer);
(%o34) done

(%i35) 'integrate(n^k*GBM(n,N0,mu,sigma,t), n, 0, inf)=integrate(n^k*GBM(n,N0,mu,sigma,t), n, 0, inf);

(%o35) $$\frac{\int_0^{\infty} n^{k-1}\, \%e^{-\frac{\left(-\log(N0)+\frac{\sigma^2 t}{2}-\mu t+\log(n)\right)^2}{2\sigma^2 t}}\, dn}{\sqrt{2}\sqrt{\pi}\,\sigma\sqrt{t}} = \%e^{\frac{2k\log(N0)+((k^2-k)\sigma^2+2k\mu)t}{2}}$$

(%i36) def_k_th_moment:k_th_moment=radcan(rhs(%));

(%o36) $k\_th\_moment = \%e^{\frac{((k^2-k)\sigma^2+2k\mu)t}{2}}\, N0^k$

(%i37) def_k_th_moment,static_case;

(%o37) $k\_th\_moment = \%e^{\frac{((k^2-k)\sigma^2+2k\mu)t}{2}} N0^k$

(%i38) radcan(%);

(%o38) $k\_th\_moment = \%e^{\frac{((k^2-k)\sigma^2+2k\mu)t}{2}} N0^k$

CUMULANTS.

(%i39) mean_value_eq;

(%o39) $meanOfNoft = \%e^{\mu t} N0$

(%i40) def_K1:K[1]=rhs(mean_value_eq);

(%o40) $K_1 = \%e^{\mu t} N0$

(%i41) second_moment_eq;

(%o41) $meanOfNoft\_square = \%e^{(\sigma^2+2\mu)t} N0^2$

(%i42) def_K2:K[2]=rhs(second_moment_eq)-K[1]^2;

(%o42) $K_2 = \%e^{(\sigma^2+2\mu)t} N0^2 - K_1^2$

(%i43) third_moment_eq;

(%o43) $third\_moment = \%e^{(3\sigma^2+3\mu)t} N0^3$

(%i44) def_K3:K[3]=rhs(third_moment_eq)-3*K[1]*K[2]-K[1]^3;

(%o44) $K_3 = \%e^{(3\sigma^2+3\mu)t} N0^3 - 3K_1K_2 - K_1^3$

(%i45) def_K4:K[4]=rhs(fourth_moment_eq)-4*K[1]*K[3]-3*K[2]^2-6*K[2]*K[1]^2-K[1]^4;

(%o45) $K_4 = \%e^{(6\sigma^2+4\mu)t} N0^4 - 4K_1K_3 - 3K_2^2 - 6K_1^2K_2 - K_1^4$

(%i46) def_K2,def_K1;

(%o46) $K_2 = \%e^{(\sigma^2+2\mu)t} N0^2 - \%e^{2\mu t} N0^2$

(%i47) K2_eq:factor(radcan(%));

(%o47) $K_2 = \%e^{2\mu t}\left(\%e^{\sigma^2 t} - 1\right) N0^2$

(%i48) def_K3,def_K2,def_K1;

(%o48) $K_3 = \%e^{(3\,\sigma^2+3\,\mu)\,t}\,N0^3 - \%e^{3\,\mu\,t}\,N0^3 - 3\,\%e^{\mu\,t}\,N0\left(\%e^{(\sigma^2+2\,\mu)\,t}\,N0^2 - \%e^{2\,\mu\,t}\,N0^2\right)$

(%i49) K3_eq:factor(radcan(%));

(%o49) $K_3 = \%e^{3\,\mu\,t}\left(\%e^{\sigma^2\,t}-1\right)^2\left(\%e^{\sigma^2\,t}+2\right)N0^3$

(%i50) def_K4,def_K1,K2_eq,K3_eq;

(%o50) $K_4 = \%e^{(6\,\sigma^2+4\,\mu)\,t}\,N0^4 - 4\,\%e^{4\,\mu\,t}\left(\%e^{\sigma^2\,t}-1\right)^2\left(\%e^{\sigma^2\,t}+2\right)N0^4 - 6\,\%e^{4\,\mu\,t}\left(\%e^{\sigma^2\,t}-1\right)N0^4 - 3\,\%e^{4\,\mu\,t}\left(\%e^{\sigma^2\,t}-1\right)^2 N0^4 - \%e^{4\,\mu\,t}\,N0^4$

(%i51) K4_eq:factor(radcan(%));

(%o51) $K_4 = \%e^{4\,\mu\,t}\left(\%e^{\sigma^2\,t}-1\right)^3\left(\%e^{3\,\sigma^2\,t}+3\,\%e^{2\,\sigma^2\,t}+6\,\%e^{\sigma^2\,t}+6\right)N0^4$

SKEWNESS.

(%i52) Skewness_eq:radcan(K3_eq/(K2_eq)^(3/2));

(%o52) $\frac{K_3}{K_2^{3/2}} = \sqrt{\%e^{\sigma^2\,t}-1}\left(\%e^{\sigma^2\,t}+2\right)$

KURTOSIS.

(%i53) Kurtosis_eq:radcan(K4_eq/(K2_eq)^2);

(%o53) $\frac{K_4}{K_2^2} = \%e^{4\,\sigma^2\,t}+2\,\%e^{3\,\sigma^2\,t}+3\,\%e^{2\,\sigma^2\,t}-6$

INFLEXION POINTS of the GBM lognormal.

(%i54) GBM_lognormal_two_inflexions:inflexions=solve(diff(GBM(n, N0, mu, sigma, t), n, 2)=0, n);

(%o54) $inflexions = [n = \%e^{-\frac{\sigma\,\sqrt{t}\,\sqrt{\sigma^2\,t+4}}{2}-2\,\sigma^2\,t+\mu\,t}\,N0,\ n = \%e^{\frac{\sigma\,\sqrt{t}\,\sqrt{\sigma^2\,t+4}}{2}-2\,\sigma^2\,t+\mu\,t}\,N0]$

So, the TIME-VARYING ADOLESCENCE POINT is given by:

(%i55) def_a_point_in_time_eq:a_point_in_time_eq=rhs(first(rhs(GBM_lognormal_two_inflexions)));

(%o55) $a\_point\_in\_time\_eq = \%e^{-\frac{\sigma\,\sqrt{t}\,\sqrt{\sigma^2\,t+4}}{2}-2\,\sigma^2\,t+\mu\,t}\,N0$

In the STATIC_CASE this reduces to (6.4) with b=0.

(%i56) def_static_a_point:static_a_point=rhs(def_a_point_in_time_eq), STATIC_CASE;

(%o56) static_a_point = $\%e^{-\frac{\sigma\sqrt{\sigma^2+4}}{2}-\frac{3\sigma^2}{2}+\mu}$

Also, the TIME-VARYING SENILITY POINT is given by:

(%i57) def_s_point_in_time_eq:s_point_in_time_eq=rhs(last(rhs(GBM_lognormal_two_inflexions)));

(%o57) s_point_in_time_eq = $\%e^{\frac{\sigma\sqrt{t}\sqrt{\sigma^2 t+4}}{2}-2\sigma^2 t+\mu t}\,N0$

In the STATIC_CASE this reduces to (6.5) with b=0.

(%i58) def_static_s_point:static_s_point=rhs(def_s_point_in_time_eq), STATIC_CASE;

(%o58) static_s_point = $\%e^{\frac{\sigma\sqrt{\sigma^2+4}}{2}-\frac{3\sigma^2}{2}+\mu}$

INTRODUCING DEATH,
defined as the time of interception between the
TANGENT STRAIGHT LINE
TO THE GBM LOGNORMAL AT SENILITY
and the time axis.

(%i59) tangent_eq:(y-y0)=m*(n-n0);

(%o59) $y-y0=m\,(n-n0)$

(%i60) finding_y0:y0=radcan(GBM(rhs(def_s_point_in_time_eq), N0, mu, sigma, t));

(%o60) $y0=\frac{\%e^{\frac{\sigma\sqrt{t}\sqrt{\sigma^2 t+4}+(3\sigma^2-4\mu)\,t-2}{4}}}{\sqrt{2}\sqrt{\pi}\,\sigma\sqrt{t}\,N0}$

(%i61) finding_y0, STATIC_CASE;

(%o61) $y0=\frac{\%e^{\frac{\sigma\sqrt{\sigma^2+4}+3\sigma^2-4\mu-2}{4}-\frac{\sigma^2}{2}}}{\sqrt{2}\sqrt{\pi}\,\sigma}$

(%i62) radcan(%);

(%o62) $y0=\frac{\%e^{\frac{\sigma\sqrt{\sigma^2+4}+\sigma^2-4\mu-2}{4}}}{\sqrt{2}\sqrt{\pi}\,\sigma}$

(%i63) finding_n0:n0=rhs(def_s_point_in_time_eq);

$$\text{(\%o63) } n0=\%e^{\frac{\sigma\sqrt{t}\sqrt{\sigma^2 t+4}}{2}-2\sigma^2 t+\mu t}\,N0$$

(%i64) finding_n0, STATIC_CASE;

$$\text{(\%o64) } n0=\%e^{\frac{\sigma\sqrt{\sigma^2+4}}{2}-\frac{3\sigma^2}{2}+\mu}$$

(%i65) finding_m:m=radcan(subst(rhs(finding_n0), n, diff(GBM(n, N0, mu, sigma, t), n)));

$$\text{(\%o65) } m=-\frac{\left(\sqrt{2}\sqrt{t}\sqrt{\sigma^2 t+4}\,\%e^{\frac{11\sigma^2 t}{4}}-\sqrt{2}\,\sigma\, t\,\%e^{\frac{11\sigma^2 t}{4}}\right)\%e^{-\frac{\sigma\sqrt{t}\sqrt{\sigma^2 t+4}+8\mu t+2}{4}}}{4\sqrt{\pi}\,\sigma^2\, t^{3/2}\, N0^2}$$

(%i66) finding_m, STATIC_CASE;

$$\text{(\%o66) } m=-\frac{\%e^{-\sigma^2}\left(\sqrt{2}\sqrt{\sigma^2+4}\,\%e^{\frac{11\sigma^2}{4}}-\sqrt{2}\,\sigma\,\%e^{\frac{11\sigma^2}{4}}\right)\%e^{-\frac{\sigma\sqrt{\sigma^2+4}+8\mu+2}{4}}}{4\sqrt{\pi}\,\sigma^2}$$

and this is (%o48) of the file "2_b_lognormal_inflexions_and_death_time"

(%i67) radcan(%);

$$\text{(\%o67) } m=-\frac{\left(\sqrt{2}\sqrt{\sigma^2+4}\,\%e^{\frac{7\sigma^2}{4}}-\sqrt{2}\,\sigma\,\%e^{\frac{7\sigma^2}{4}}\right)\%e^{-\frac{\sigma\sqrt{\sigma^2+4}+8\mu+2}{4}}}{4\sqrt{\pi}\,\sigma^2}$$

(%i68) tangent_straight_line_eq:tangent_eq,finding_y0,finding_n0,finding_m;

$$\text{(\%o68) } y-\frac{\%e^{\frac{\sigma\sqrt{t}\sqrt{\sigma^2 t+4}+(3\sigma^2-4\mu)t-2}{4}}}{\sqrt{2}\sqrt{\pi}\,\sigma\sqrt{t}\,N0}=-\frac{\left(\sqrt{2}\sqrt{t}\sqrt{\sigma^2 t+4}\,\%e^{\frac{11\sigma^2 t}{4}}-\sqrt{2}\,\sigma\, t\,\%e^{\frac{11\sigma^2 t}{4}}\right)\%e^{-\frac{\sigma\sqrt{t}\sqrt{\sigma^2 t+4}+8\mu t+2}{4}}\left(n-\%e^{\frac{\sigma\sqrt{t}\sqrt{\sigma^2 t+4}}{2}-2\sigma^2 t+\mu t}\,N0\right)}{4\sqrt{\pi}\,\sigma^2\, t^{3/2}\, N0^2}$$

(%i69) tangent_at_s:radcan(first(solve(tangent_straight_line_eq, y)));

$$\text{(\%o69) } y=(\%e^{-\frac{\sigma\sqrt{t}\sqrt{\sigma^2 t+4}+8\mu t+4}{4}}\,(\sqrt{\%e}\sqrt{t}\left(\sigma\,\%e^{\frac{2\sigma\sqrt{t}\sqrt{\sigma^2 t+4}+(3\sigma^2+4\mu)t}{4}}\,N0+n\,\sigma\,\%e^{\frac{11\sigma^2 t}{4}}\right)+$$
$$\sqrt{\%e}\sqrt{\sigma^2 t+4}\left(\%e^{\frac{2\sigma\sqrt{t}\sqrt{\sigma^2 t+4}+(3\sigma^2+4\mu)t}{4}}\,N0-n\,\%e^{\frac{11\sigma^2 t}{4}}\right)))/(2^{3/2}\sqrt{\pi}\,\sigma^2\, t\, N0^2)$$

(%i70) tangent_at_s,STATIC_CASE;

$$(\%o70)\ y=(\%e^{-\frac{\sigma\sqrt{\sigma^2+4+8\mu+4}}{4}-\sigma^2}\left(\sqrt{\%e}\left(\sigma\,\%e^{\frac{2\sigma\sqrt{\sigma^2+4}+3\sigma^2+4\mu}{4}+\frac{\sigma^2}{2}}+n\,\sigma\,\%e^{\frac{11\sigma^2}{4}}\right)+\sqrt{\%e}\sqrt{\sigma^2+4}\left(\%e^{\frac{2\sigma\sqrt{\sigma^2+4}+3\sigma^2+4\mu}{4}+\frac{\sigma^2}{2}}-n\,\%e^{\frac{11\sigma^2}{4}}\right)\right))/(2^{3/2}\sqrt{\pi}\,\sigma^2)$$

and this is (%050) of the file "2_b_lognormal_inflexions_and_death_time"

(%i71) radcan(%);

$$(\%o71)\ y=(\%e^{-\frac{\sigma\sqrt{\sigma^2+4+8\mu+4}}{4}}\left(\sqrt{\%e}\left(\sigma\,\%e^{\frac{2\sigma\sqrt{\sigma^2+4}+\sigma^2+4\mu}{4}}+n\,\sigma\,\%e^{\frac{7\sigma^2}{4}}\right)+\sqrt{\%e}\sqrt{\sigma^2+4}\left(\%e^{\frac{2\sigma\sqrt{\sigma^2+4}+\sigma^2+4\mu}{4}}-n\,\%e^{\frac{7\sigma^2}{4}}\right)\right))/(2^{3/2}\sqrt{\pi}\,\sigma^2)$$

(%i72) finding_death:solve(subst(0, y, tangent_at_s), n);

$$(\%o72)\ [n=\frac{\%e^{-\frac{11\sigma^2 t}{4}}\left(\sqrt{\sigma^2 t+4}\,\%e^{\frac{\sigma\sqrt{t}\sqrt{\sigma^2 t+4}}{2}+\frac{3\sigma^2 t}{4}+\mu t}+\sigma\sqrt{t}\,\%e^{\frac{\sigma\sqrt{t}\sqrt{\sigma^2 t+4}}{2}+\frac{3\sigma^2 t}{4}+\mu t}\right)N0}{\sqrt{\sigma^2 t+4}-\sigma\sqrt{t}}]$$

NOTE:
we had to shorten the notation to fit it to page width, so ED stands for evolving_death_time:

(%i73) ED_eq:ED=factor(rhs(first(finding_death)));

$$(\%o73)\ ED=\frac{(\sqrt{\sigma^2 t+4}+\sigma\sqrt{t})\,\%e^{\frac{\sigma\sqrt{t}\sqrt{\sigma^2 t+4}}{2}-2\sigma^2 t+\mu t}N0}{\sqrt{\sigma^2 t+4}-\sigma\sqrt{t}}$$

(%i74) def_static_death:static_death=rhs(ED_eq),STATIC_CASE;

$$(\%o74)\ static\_death=\frac{(\sqrt{\sigma^2+4}+\sigma)\,\%e^{\frac{\sigma\sqrt{\sigma^2+4}}{2}-\frac{3\sigma^2}{2}+\mu}}{\sqrt{\sigma^2+4}-\sigma}$$

and this is (%o54) except for multiplying above and below by sqrt(sigma^2+)+sigma.

But let us go back to the TIME EVOLVING DEATH POINT.
just factoring, one gets:

(%i75) simplified_ED:factor(ED_eq);

$$\text{(\%o75) } ED=\frac{\left(\sqrt{\sigma^2\, t+4}+\sigma\,\sqrt{t}\right)\%e^{\frac{\sigma\,\sqrt{t}\,\sqrt{\sigma^2\, t+4}}{2}-2\,\sigma^2\, t+\mu\, t}\, N0}{\sqrt{\sigma^2\, t+4}-\sigma\,\sqrt{t}}$$

On the other hand, the FINAL FORM OF THE TIME-EVOLVING DEATH POINT is:

(%i76) ED_exponent(sigma, mu, t):=((sigma*sqrt(t)*sqrt(sigma^2*t+4))/2-2*sigma^2*t+mu*t);

$$\text{(\%o76) } ED\_exponent(\sigma,\mu,t):=\frac{\sigma\,\sqrt{t}\,\sqrt{\sigma^2\, t+4}}{2}-2\,\sigma^2\, t+\mu\, t$$

(%i77) def_final_ED:ED=((sqrt(sigma^2*t+4)+sigma*sqrt(t))^2*%e^(ED_exponent(sigma, mu, t))*N0)/4;

$$\text{(\%o77) } ED=\frac{\left(\sqrt{\sigma^2\, t+4}+\sigma\,\sqrt{t}\right)^2\%e^{\frac{\sigma\,\sqrt{t}\,\sqrt{\sigma^2\, t+4}}{2}-2\,\sigma^2\, t+\mu\, t}\, N0}{4}$$

In fact, the last two forms are equivalent.

(%i78) radcan(simplified_ED-def_final_ED);

(%o78) 0=0

IMPORTANT ! THE IMMORTALITY CONDITION
(i.e. death point is AT INFINITY for time -> infinity)
is mu > 3 sigma^2 /2 .

(%i79) assume(3*sigma^2-2*mu < 0);

$$\text{(\%o79) } [2\,\mu>3\,\sigma^2]$$

(%i80) limit(def_final_ED, t, inf);

$$\text{(\%o80) } ED=\infty$$

APPENDIX 11.B

THE NEW STOCHASTIC PROCESS D(t),
YIELDING THE EVOLUTION IN TIME
OF THE DISTANCE
BETWEEN ANY TWO NEARBY ET CIVILIZATIONS
IN THE GALAXY.

Author: Claudio Maccone, 27 February 2012.
E-mail: clmaccon@libero.it

```
(%i1) kill(all);
(%o0) done
```

```
(%i1) assume(t>0, sigma>0, N0>0);
(%o1) [t>0,σ>0,N0>0]
```

Defining the GEOMETRIC BROWNIAN MOTION (GBM) pdf (i.e. a lognormal pdf in n):

```
(%i2) GBMexponent(n, N0, mu, sigma, t):=(-(log(n)-(log(N0)+mu*t-sigma^2*t/2))^2/(2*sigma^2*t));
```

(%o2) $\mathrm{GBMexponent}(n,N0,\mu,\sigma,t):=\dfrac{-\left(\log(n)-\left(\log(N0)+\mu\,t+\dfrac{(-\sigma^2)\,t}{2}\right)\right)^2}{2\,\sigma^2\,t}$

```
(%i3) GBM(n, N0, mu, sigma, t):=(1/(sqrt(2*%pi)*n*sigma*sqrt(t)))*%e^GBMexponent(n, N0, mu, sigma, t);
```

(%o3) $\mathrm{GBM}(n,N0,\mu,\sigma,t):=\dfrac{1}{\sqrt{2\,\pi}\,n\,\sigma\,\sqrt{t}}\,\%e^{\mathrm{GBMexponent}(n,N0,\mu,\sigma,t)}$

```
(%i4) GBM(n, N0, mu, sigma, t);
```

(%o4) $\dfrac{\%e^{-\frac{\left(-\log(N0)+\frac{\sigma^2\,t}{2}-\mu\,t+\log(n)\right)^2}{2\,\sigma^2\,t}}}{\sqrt{2}\,\sqrt{\pi}\,n\,\sigma\,\sqrt{t}}$

Defining the STATIC CASE (i.e. when the GBM gets down to the
lognormal Statistical Drake Equation pdf (1.B.56) of Chapter 1).

```
(%i5) STATIC_CASE:[t=1, N0=%e^(sigma^2/2)];
```

(%o5) $[t=1,N0=\%e^{\frac{\sigma^2}{2}}]$

Checking that, in the static case, the GBM pdf gets down to the ordinary lognormal:

(%i6) ordinary_lognormal_of_static_case:ordinary_lognormal=GBM(n, N0, mu, sigma, t),STATIC_CASE;

(%o6) $\text{ordinary\_lognormal} = \frac{\%e^{-\frac{(\log(n)-\mu)^2}{2\sigma^2}}}{\sqrt{2}\sqrt{\pi}\, n\, \sigma}$

Let us nw compute the AVERAGE DISTANCE between
two nearby ET civilizations in the Galaxy.
That is, let us compute the "generalized Distance distribution"
when the GBM replaces the ordinary,
static lognormal distribution of the Static Statistical Drake Equation,
rather than the Exponentially Evolving Statistical Drake Equation.

(%i7) def_GBMmu:GBMmu=log(N0)+mu*t+((-sigma^2)*t)/2;

(%o7) $\text{GBMmu} = \log(N0) - \frac{\sigma^2 t}{2} + \mu t$

(%i8) assume(GBMsigma>0);

(%o8) [GBMsigma>0]

(%i9) def_GBMsigma:GBMsigma=sigma*sqrt(t);

(%o9) $\text{GBMsigma} = \sigma \sqrt{t}$

(%i10) def_substitutions:[def_GBMmu, def_GBMsigma];

(%o10) $[\text{GBMmu} = \log(N0) - \frac{\sigma^2 t}{2} + \mu t, \text{GBMsigma} = \sigma \sqrt{t}]$

(%i11) assume(C>0);

(%o11) [C>0]

In the following we call SD the Static_Distance_pdf,
just to shrink down the too long lines on the right.

(%i12) def_SD:SD=(3/r)*(1/(sqrt(2*%pi)*GBMsigma))*%e^(-(log(C^3/r^3)-GBMmu)^2/(2*GBMsigma^2));

(%o12) $\text{SD} = \frac{3\, \%e^{-\frac{\left(\log\left(\frac{C^3}{r^3}\right) - \text{GBMmu}\right)^2}{2\, \text{GBMsigma}^2}}}{\sqrt{2}\sqrt{\pi}\, \text{GBMsigma}\, r}$

In the following we call ESD the Evolving_Statistical_Distance_pdf,
just to shrink down the too long lines on the right.

(%i13) def_ESD: ESD=rhs(def_SD), def_substitutions;

$$(\%o13)\ ESD=\frac{3\ \%e^{-\frac{\left(-\log(N0)+\log\left(\frac{C^3}{r^3}\right)+\frac{\sigma^2 t}{2}-\mu t\right)^2}{2\sigma^2 t}}}{\sqrt{2}\sqrt{\pi}\, r\, \sigma\sqrt{t}}$$

In the following we call ED the Evolving_Distance_pdf,
just to shrink down the too long lines on the right.

(%i14) EDexponent(n, N0, mu, sigma, t):=(-(-log(N0)+log(C^3/r^3)+(sigma^2*t)/2-mu*t)^2/(2*sigma^2*t));

$$(\%o14)\ EDexponent(n,N0,\mu,\sigma,t):=\frac{-\left(-\log(N0)+\log\left(\frac{C^3}{r^3}\right)+\frac{\sigma^2 t}{2}+(-\mu)\,t\right)^2}{2\sigma^2 t}$$

(%i15) EDpdf(r, N0, mu, sigma, t):=(3*%e^(EDexponent(n, N0, mu, sigma, t))/(sqrt(2*%pi)*r*sigma*sqrt(t)));

$$(\%o15)\ EDpdf(r,N0,\mu,\sigma,t):=\frac{3\ \%e^{EDexponent(n,N0,\mu,\sigma,t)}}{\sqrt{2\pi}\, r\,\sigma\sqrt{t}}$$

(%i16) EDpdf(r, N0, mu, sigma, t);

$$(\%o16)\ \frac{3\ \%e^{-\frac{\left(-\log(N0)+\log\left(\frac{C^3}{r^3}\right)+\frac{\sigma^2 t}{2}-\mu t\right)^2}{2\sigma^2 t}}}{\sqrt{2}\sqrt{\pi}\, r\, \sigma\sqrt{t}}$$

Finding ALL THE DISTANCE MOMENTS, i.e. the k-th moment, with k=0, 1, 2, ...
Defining k :

(%i17) assume(k>0);
(%o17) [k>0]

(%i18) declare(k, integer);
(%o18) done

The following assumption is necessary in order to compute the k-th moment integral:

(%i19) assume(6*log(N0)-18*log(C)+((-2*k-3)*sigma^2+6*mu)*t>0);

$$(\%o19)\ [6\log(N0)-18\log(C)+((-2k-3)\,\sigma^2+6\mu)\,t>0]$$

Computing the k-th moment integral:

(%i20) 'integrate(r^k*EDpdf(r,N0,mu,sigma,t),r,0,inf)=integrate(radcan(r^k*EDpdf(r,N0,mu,sigma,t)),r,0,inf);

(%o20)
$$\frac{3\int_0^{\infty} r^{k-1}\,\%e^{-\frac{\left(-\log(N0)+\log\left(\frac{C^3}{r^3}\right)+\frac{\sigma^2 t}{2}-\mu t\right)^2}{2\sigma^2 t}}\,dr}{\sqrt{2}\sqrt{\pi}\,\sigma\sqrt{t}}=C^{\frac{6\log(N0)+(6\mu-3\sigma^2)t}{2\sigma^2 t}}\,N0^{\frac{\sigma^2-2\mu}{2\sigma^2}}\,\%e^{-\frac{(54\log(C)+((6k+9)\sigma^2-18\mu)t)\log(N0)+((-18k-27)\sigma^2+54\mu)t\log(C)+((-k^2-3k)\sigma^4+6k\mu\sigma^2)t^2}{18\sigma^2 t}}$$

(%i21) defkmom:lhs(%)=radcan(rhs(%));

(%o21)
$$\frac{3\int_0^{\infty} r^{k-1}\,\%e^{-\frac{\left(-\log(N0)+\log\left(\frac{C^3}{r^3}\right)+\frac{\sigma^2 t}{2}-\mu t\right)^2}{2\sigma^2 t}}\,dr}{\sqrt{2}\sqrt{\pi}\,\sigma\sqrt{t}}=\frac{\%e^{\frac{((k^2+3k)\sigma^2-6k\mu)t}{18}}\,C^k}{N0^{k/3}}$$

(%i22) def_evolving_distance_k_th_moment:evolving_distance_k_th_moment=rhs(%);

(%o22)
$$\text{evolving\_distance\_k\_th\_moment}=\frac{\%e^{\frac{((k^2+3k)\sigma^2-6k\mu)t}{18}}\,C^k}{N0^{k/3}}$$

In the following we call defSDk the definition of Static_Distance_k_th_moment, just to shrink down the too long lines on the right.

(%i23) defSDk:static_distance_k_th_moment=rhs(def_evolving_distance_k_th_moment),STATIC_CASE;

(%o23)
$$\text{static\_distance\_k\_th\_moment}=\%e^{\frac{(k^2+3k)\sigma^2-6k\mu}{18}-\frac{k\sigma^2}{6}}\,C^k$$

This is (1.B.115):

(%i24) radcan(defSDk);

(%o24)
$$\text{static\_distance\_k\_th\_moment}=\%e^{\frac{k^2\sigma^2-6k\mu}{18}}\,C^k$$

Let us now derive the special cases k=0,1,2,3,4.

NORMALIZATION CONDITION is the case k = 0.

(%i25) subst(0, k, defkmom);

$$(\%o25)\quad \frac{3\int_{0}^{\infty}\frac{\%e^{-\frac{\left(-\log(N0)+\log\left(\frac{C^{3}}{r^{3}}\right)+\frac{\sigma^{2}t}{2}-\mu t\right)^{2}}{2\sigma^{2}t}}}{r}\,dr}{\sqrt{2}\sqrt{\pi}\,\sigma\sqrt{t}}=1$$

The TIME-EVOLVING DISTANCE MEAN VALUE
is the particular case k=1 of the distance k-th moment (%o21)

(%i26) def_time_evolving_distance_mean_value:subst(1, k, defkmom);

$$(\%o26)\quad \frac{3\int_{0}^{\infty}\%e^{-\frac{\left(-\log(N0)+\log\left(\frac{C^{3}}{r^{3}}\right)+\frac{\sigma^{2}t}{2}-\mu t\right)^{2}}{2\sigma^{2}t}}\,dr}{\sqrt{2}\sqrt{\pi}\,\sigma\sqrt{t}}=\frac{\%e^{\frac{(4\sigma^{2}-6\mu)t}{18}}\,C}{N0^{1/3}}$$

In the following we call m_v_eq the mean value equation,
just to shrink down the too long lines on the right.

(%i27) m_v_eq:time_evolving_distance_mean_value=factor(rhs(def_time_evolving_distance_mean_value));

$$(\%o27)\quad time\_evolving\_distance\_mean\_value=\frac{\%e^{\frac{2\sigma^{2}t}{9}-\frac{\mu t}{3}}\,C}{N0^{1/3}}$$

VERY IMPORTANT: if mu > 2*sigma^2/3, then THE DISTANCE MEAN VALUE
APPROACHES ZERO AS t APPROACHES INFINITY !
In words: as long as time elapses more and more, more and more ET
Civilization are born in the Galaxy. Thus, THEIR DISTANCE FROM ONE
ANOTHER DECREASES. Until it reaches zero for after an infinite amout of time,
because there is an infinite amount of Civilizations in the Galaxy !
But this happens ONLY if mu > 2*sigma^2/3.
Which is thus THE KEY CONSTRAINT AMONG MU AND SIGMA
to let N(t) and D(t) be a realistic model of the evolution of life in the Galaxy!
What a surprise !

(%i28) assume(2*sigma^2-3*mu<0);

(%o28) $[3\mu>2\sigma^{2}]$

(%i29) limit(m_v_eq, t, inf);

(%o29) time_evolving_distance_mean_value=0

In the STATIC_CASE, this becomes:

(%i30) def_time_evolving_distance_mean_value,STATIC_CASE;

$$(\%o30)\quad \frac{3\int_0^{\infty} \%e^{-\frac{\left(\log\left(\frac{C^3}{r^3}\right)-\mu\right)^2}{2\sigma^2}}\,dr}{\sqrt{2}\sqrt{\pi}\,\sigma}=\%e^{\frac{4\sigma^2-6\mu}{18}-\frac{\sigma^2}{6}}\,C$$

This is (1.B.119):

(%i31) def_static_distance_mean_value:static_distance_mean_value=radcan(rhs(%));

$$(\%o31)\quad \text{static\_distance\_mean\_value}=\%e^{\frac{\sigma^2-6\mu}{18}}\,C$$

SECOND MOMENT.
In the following we are going to call TEDMVotS the time_evolving_distance_mean_value_of_the_square.

(%i32) TEDMVotS:subst(2, k,defkmom);

$$(\%o32)\quad \frac{3\int_0^{\infty} r\,\%e^{-\frac{\left(-\log(N0)+\log\left(\frac{C^3}{r^3}\right)+\frac{\sigma^2 t}{2}-\mu t\right)^2}{2\sigma^2 t}}\,dr}{\sqrt{2}\sqrt{\pi}\,\sigma\sqrt{t}}=\frac{\%e^{\frac{(10\sigma^2-12\mu)t}{18}}\,C^2}{N0^{2/3}}$$

VARIANCE of the time-evolving distance,
that hereafter we call TEV for time_evolving_distance_variance.

(%i33) defTEV:time_evolving_distance_variance=factor(radcan(rhs(TEDMVotS)-rhs(m_v_eq)^2));

$$(\%o33)\quad \text{time\_evolving\_distance\_variance}=\frac{\left(\%e^{\frac{\sigma^2 t}{9}}-1\right)\%e^{\frac{4\sigma^2 t}{9}-\frac{2\mu t}{3}}\,C^2}{N0^{2/3}}$$

In the STATIC_CASE, this becomes:

(%i34) def_static_variance:static_variance_of_the_distance=rhs(defTEV),STATIC_CASE;

(%o34) $$\text{static\_variance\_of\_the\_distance} = \%e^{\frac{\sigma^2}{9} - \frac{2\,\mu}{3}} \left(\%e^{\frac{\sigma^2}{9}} - 1\right) C^2$$

and this is (1.B.122), i.e. the static variance of the distance.

STANDARD DEVIATION of the time-evolving distance,
that hereafter we call TEDST, for time_evolving_distance_standard_deviation.

(%i35) def_TEDST:time_evolving_distance_standard_deviation=factor(radcan(sqrt(factor(rhs(defTEV)))));

(%o35) $$\text{time\_evolving\_distance\_standard\_deviation} = \frac{\sqrt{\%e^{\frac{\sigma^2 t}{9}} - 1}\ \%e^{\frac{2\,\sigma^2 t}{9} - \frac{\mu t}{3}}\, C}{NO^{1/3}}$$

In the STATIC_CASE this becomes
(SSDD stands for static_standard_deviation_of_the_distance):

(%i36) def_SSDD:static_standard_deviation_of_the_distance=rhs(def_TEDST),STATIC_CASE;

(%o36) $$\text{static\_standard\_deviation\_of\_the\_distance} = \%e^{\frac{\sigma^2}{18} - \frac{\mu}{3}} \sqrt{\%e^{\frac{\sigma^2}{9}} - 1}\ C$$

and this is (1.B.123), i.e. the static standard deviation of the distance.

THIRD MOMENT
(TEMotD means third_evolving_moment_of_the_distance).

(%i37) TEMotD:subst(3, k, lhs(defkmom))=factor(subst(3, k, rhs(defkmom)));

(%o37) $$\frac{3 \displaystyle\int_0^{\infty} r^2\, \%e^{-\frac{\left(-\log(NO) - \log\left(\frac{C^3}{r^3}\right) + \frac{\sigma^2 t}{2} - \mu t\right)^2}{2\,\sigma^2 t}}\, dr}{\sqrt{2}\sqrt{\pi}\,\sigma\sqrt{t}} = \frac{\%e^{\sigma^2 t - \mu t}\, C^3}{NO}$$

In the STATIC_CASE, this reduces immediately to (1.B.124):

(%i38) TEMotD,STATIC_CASE;

$$\text{(\%o38)}\quad \frac{3\int_0^{\infty} r^2\,\%e^{-\frac{\left(\log\left(\frac{C^3}{r^3}\right)-\mu\right)^2}{2\sigma^2}}\,dr}{\sqrt{2}\sqrt{\pi}\,\sigma}=\%e^{\frac{\sigma^2}{2}-\mu}\,C^3$$

FOURTH MOMENT
(TEFMotD stands for time_evolving_fourth_moment_of_the_distance).

(%i39) TEFMotD:subst(4, k, lhs(defkmom))=factor(subst(4, k, rhs(defkmom)));

$$\text{(\%o39)}\quad \frac{3\int_0^{\infty} r^3\,\%e^{-\frac{\left(-\log(N0)+\log\left(\frac{C^3}{r^3}\right)+\frac{\sigma^2 t}{2}-\mu t\right)^2}{2\sigma^2 t}}\,dr}{\sqrt{2}\sqrt{\pi}\,\sigma\sqrt{t}}=\frac{\%e^{\frac{14\sigma^2 t}{9}-\frac{4\mu t}{3}}\,C^4}{N0^{4/3}}$$

In the STATIC_CASE, this reduces immediately to (1.B.125):

(%i40) TEFMotD, STATIC_CASE;

$$\text{(\%o40)}\quad \frac{3\int_0^{\infty} r^3\,\%e^{-\frac{\left(\log\left(\frac{C^3}{r^3}\right)-\mu\right)^2}{2\sigma^2}}\,dr}{\sqrt{2}\sqrt{\pi}\,\sigma}=\%e^{\frac{8\sigma^2}{9}-\frac{4\mu}{3}}\,C^4$$

First four CUMULANTS.

(%i41) def_K1:K[1]=rhs(m_v_eq);

$$\text{(\%o41)}\quad K_1=\frac{\%e^{\frac{2\sigma^2 t}{9}-\frac{\mu t}{3}}\,C}{N0^{1/3}}$$

(%i42) def_K2:K[2]=rhs(TEDMVotS)-K[1]^2;

$$\text{(\%o42)}\quad K_2=\frac{\%e^{\frac{(10\sigma^2-12\mu)t}{18}}\,C^2}{N0^{2/3}}-K_1^2$$

(%i43) def_K3:K[3]=rhs(TEMotD)-3*K[1]*K[2]-K[1]^3;

(%o43) $$K_3=\frac{\%e^{\sigma^2 t-\mu t}\,C^3}{N0}-3\,K_1\,K_2-K_1^3$$

(%i44) def_K4:K[4]=rhs(TEFMotD)-4*K[1]*K[3]-3*K[2]^2-6*K[2]*K[1]^2-K[1]^4;

(%o44) $$K_4=\frac{\%e^{\frac{14\,\sigma^2 t}{9}-\frac{4\,\mu t}{3}}\,C^4}{N0^{4/3}}-4\,K_1\,K_3-3\,K_2^2-6\,K_1^2\,K_2-K_1^4$$

(%i45) def_K2,def_K1;

(%o45) $$K_2=\frac{\%e^{\frac{(10\,\sigma^2-12\,\mu)\,t}{18}}\,C^2}{N0^{2/3}}-\frac{\%e^{2\left(\frac{2\,\sigma^2 t}{9}-\frac{\mu t}{3}\right)}\,C^2}{N0^{2/3}}$$

(%i46) K2_eq:factor(radcan(%));

(%o46) $$K_2=\frac{\left(\%e^{\frac{\sigma^2 t}{9}}-1\right)\%e^{\frac{4\,\sigma^2 t}{9}-\frac{2\,\mu t}{3}}\,C^2}{N0^{2/3}}$$

(%i47) def_K3,def_K2,def_K1;

(%o47) $$K_3=-\frac{3\,\%e^{\frac{2\,\sigma^2 t}{9}-\frac{\mu t}{3}}\,C\left(\frac{\%e^{\frac{(10\,\sigma^2-12\,\mu)\,t}{18}}\,C^2}{N0^{2/3}}-\frac{\%e^{2\left(\frac{2\,\sigma^2 t}{9}-\frac{\mu t}{3}\right)}\,C^2}{N0^{2/3}}\right)}{N0^{1/3}}+\frac{\%e^{\sigma^2 t-\mu t}\,C^3}{N0}-\frac{\%e^{3\left(\frac{2\,\sigma^2 t}{9}-\frac{\mu t}{3}\right)}\,C^3}{N0}$$

(%i48) K3_eq:factor(radcan(%));

(%o48) $$K_3=\frac{\%e^{-\mu t}\left(\%e^{\sigma^2 t}-3\,\%e^{\frac{7\,\sigma^2 t}{9}}+2\,\%e^{\frac{2\,\sigma^2 t}{3}}\right)C^3}{N0}$$

(%i49) def_K4,def_K1,K2_eq,K3_eq;

(%o49) $$K_4=-\frac{6\left(\%e^{\frac{\sigma^2 t}{9}}-1\right)\%e^{2\left(\frac{2\,\sigma^2 t}{9}-\frac{\mu t}{3}\right)+\frac{4\,\sigma^2 t}{9}-\frac{2\,\mu t}{3}}\,C^4}{N0^{4/3}}+\frac{\%e^{\frac{14\,\sigma^2 t}{9}-\frac{4\,\mu t}{3}}\,C^4}{N0^{4/3}}-$$
$$\frac{3\left(\%e^{\frac{\sigma^2 t}{9}}-1\right)^2\%e^{2\left(\frac{4\,\sigma^2 t}{9}-\frac{2\,\mu t}{3}\right)}\,C^4}{N0^{4/3}}-\frac{\%e^{4\left(\frac{2\,\sigma^2 t}{9}-\frac{\mu t}{3}\right)}\,C^4}{N0^{4/3}}-\frac{4\left(\%e^{\sigma^2 t}-3\,\%e^{\frac{7\,\sigma^2 t}{9}}+2\,\%e^{\frac{2\,\sigma^2 t}{3}}\right)\%e^{\frac{2\,\sigma^2 t}{9}-\frac{4\,\mu t}{3}}\,C^4}{N0^{4/3}}$$

(%i50) K4_eq:factor(radcan(%));

$$\text{(\%o50)}\quad K_4 = \frac{\%e^{-\frac{4\mu t}{3}}\left(\%e^{\frac{14\sigma^2 t}{9}} - 4\,\%e^{\frac{11\sigma^2 t}{9}} - 3\,\%e^{\frac{10\sigma^2 t}{9}} + 12\,\%e^{\sigma^2 t} - 6\,\%e^{\frac{8\sigma^2 t}{9}}\right)C^4}{N0^{4/3}}$$

This completes the calculation of the first four cumulants.

SKEWNESS.

(%i51) Skewness_eq:time_evolving_distance_skewness=factor(radcan(rhs(K3_eq/(K2_eq)^(3/2))));

$$\text{(\%o51)}\quad \mathrm{time\_evolving\_distance\_skewness} = -\frac{\%e^{\sigma^2 t} - 3\,\%e^{\frac{7\sigma^2 t}{9}} + 2\,\%e^{\frac{2\sigma^2 t}{3}}}{\sqrt{\%e^{\frac{\sigma^2 t}{9}} - 1}\left(\%e^{\frac{2\sigma^2 t}{3}} - \%e^{\frac{7\sigma^2 t}{9}}\right)}$$

Notice that the exponentials and radicals cannot be simplified any further:

(%i52) radcan(%);

$$\text{(\%o52)}\quad \mathrm{time\_evolving\_distance\_skewness} = -\frac{\%e^{\sigma^2 t} - 3\,\%e^{\frac{7\sigma^2 t}{9}} + 2\,\%e^{\frac{2\sigma^2 t}{3}}}{\sqrt{\%e^{\frac{\sigma^2 t}{9}} - 1}\left(\%e^{\frac{2\sigma^2 t}{3}} - \%e^{\frac{7\sigma^2 t}{9}}\right)}$$

Let us check that this time-evolving distance skewness reduces to 0
for sigma -> 0:

(%i53) 'limit(lhs(%), sigma, 0)=limit(rhs(%), sigma, 0);

(%o53) time_evolving_distance_skewness=0

In the STATIC_CASE, this time-evolving skewness reduces to:

(%i54) def_static_skewness:static_skewness=rhs(Skewness_eq), STATIC_CASE;

$$\text{(\%o54)}\quad \mathrm{static\_skewness} = -\frac{\%e^{\sigma^2} - 3\,\%e^{\frac{7\sigma^2}{9}} + 2\,\%e^{\frac{2\sigma^2}{3}}}{\sqrt{\%e^{\frac{\sigma^2}{9}} - 1}\left(\%e^{\frac{2\sigma^2}{3}} - \%e^{\frac{7\sigma^2}{9}}\right)}$$

and this is exactly (1.B.132).

(%i55) radcan(%);

(%o55) $$\text{static\_skewness}=\frac{\%e^{\frac{2\,\sigma^2}{9}}+\%e^{\frac{\sigma^2}{9}}-2}{\sqrt{\%e^{\frac{\sigma^2}{9}}-1}}$$

This completes the calculation of the distance skewness.

KURTOSIS.

(%i56) Kurtosis_eq:time_evolving_kurtosis=radcan(rhs(K4_eq/(K2_eq)^2));

(%o56) $$\text{time\_evolving\_kurtosis}=\%e^{\frac{4\,\sigma^2\,t}{9}}+2\,\%e^{\frac{\sigma^2\,t}{3}}+3\,\%e^{\frac{2\,\sigma^2\,t}{9}}-6$$

In the STATIC_CASE this reduces immediately to (1.B.133).

(%i57) def_static_kurtosis:static_kurtosis=rhs(Kurtosis_eq), STATIC_CASE;

(%o57) $$\text{static\_kurtosis}=\%e^{\frac{4\,\sigma^2}{9}}+2\,\%e^{\frac{\sigma^2}{3}}+3\,\%e^{\frac{2\,\sigma^2}{9}}-6$$

Finding the PEAK of the DISTANCE, and its evolution in time.

(%i58) def_r_peak:r_peak=rhs(radcan(last(solve(diff(EDpdf(r,N0,mu,sigma,t), r)=0, r))));

(%o58) $$\text{r\_peak}=\frac{\%e^{\frac{(\sigma^2-6\,\mu)\,t}{18}}\,C}{N0^{1/3}}$$

VERY IMPORTANT: if mu > sigma^2/6, then THE PEAK DISTANCE APPROACHES ZERO AS t APPROACHES INFINITY !
In words: as long as time elapses more and more, more and more ET Civilization are born in the Galaxy. Thus, THEIR DISTANCE FROM ONE ANOTHER DECREASES. Until it reaches zero for after an infinite amout of time, because there is an infinite amount of Civilizations in the Galaxy !
But this happens ONLY if mu > sigma^2/6.
Which is thus THE KEY CONSTRAINT AMONG MU AND SIGMA to let N(t) and D(t) be a realistic model of the evolution of life in the Galaxy!
What a surprise !

(%i59) assume(sigma^2-6*mu<0);

(%o59) $[6\,\mu>\sigma^2]$

(%i60) limit(def_r_peak, t, inf);

(%o60) r_peak = 0

In the STATIC CASE, the DISTANCE PEAK ABSCISSA reduces to (1.B.143):

(%i61) def_r_peak,STATIC_CASE;

(%o61) $r\_peak = \%e^{\frac{\sigma^2 - 6\mu}{18} - \frac{\sigma^2}{6}} C$

(%i62) peak_distance_abscissa:radcan(%);

(%o62) $r\_peak = \%e^{-\frac{\sigma^2 + 3\mu}{9}} C$

Let us now find the DISTANCE PEAK ORDINATE.

(%i63) defDPO:peak_distance_ordinate=subst(rhs(peak_distance_abscissa), r, rhs(def_ESD));

(%o63) $peak\_distance\_ordinate = \frac{3\,\%e^{\frac{\sigma^2+3\mu}{9} - \frac{\left(-\log(N0) + \frac{\sigma^2 t}{2} - \mu t - \frac{\sigma^2+3\mu}{3}\right)^2}{2\sigma^2 t}}}{\sqrt{2}\sqrt{\pi}\,\sigma\sqrt{t}\,C}$

In the STATIC_CASE this reduces to

(%i64) rhs(defDPO),STATIC_CASE;

(%o64) $\frac{3\,\%e^{\frac{\sigma^2+3\mu}{9} - \frac{\left(\frac{\sigma^2+3\mu}{3} - \mu\right)^2}{2\sigma^2}}}{\sqrt{2}\sqrt{\pi}\,\sigma\,C}$

Simplifying this, we find again eq. (1.B.144):

(%i65) radcan(%);

(%o65) $\frac{3\,\%e^{\frac{\sigma^2+6\mu}{18}}}{\sqrt{2}\sqrt{\pi}\,\sigma\,C}$

MEDIAN.

(%i66) assume(time_evolving_median>0);

(%o66) [time_evolving_median>0]

(%i67) def_time_evolving_median:radcan(integrate(EDpdf(r,N0,mu,sigma,t),r,0,time_evolving_median)=1/2);

(%o67) $$\frac{\mathrm{erf}\left(\frac{2^{3/2}\log(N0)-3\,2^{3/2}\log(C)+3\,2^{3/2}\log(time\_evolving\_median)+(2^{3/2}\mu-\sqrt{2}\,\sigma^2)\,t}{4\,\sigma\,\sqrt{t}}\right)+1}{2}=\frac{1}{2}$$

(%i68) time_evolving_median_eq:first(radcan(solve(def_time_evolving_median, time_evolving_median)));

(%o68) $$time\_evolving\_median=\frac{\%e^{\frac{(\sigma^2-2\,\mu)\,t}{6}}\,C}{N0^{1/3}}$$

VERY IMPORTANT: if mu > sigma^2/2, then THE DISTANCE MEDIAN APPROACHES ZERO AS t APPROACHES INFINITY !
In words: as long as time elapses more and more, more and more ET Civilization are born in the Galaxy. Thus, THEIR DISTANCE FROM ONE ANOTHER DECREASES. Until it reaches zero for after an infinite amout of time, because there is an infinite amount of Civilizations in the Galaxy !
But this happens ONLY if mu > sigma^2/2.
Which is thus THE KEY CONSTRAINT AMONG MU AND SIGMA to let N(t) and D(t) be a realistic model of the evolution of life in the Galaxy!
What a surprise !

(%i69) assume(sigma^2-2*mu<0);

(%o69) $[2\,\mu>\sigma^2]$

(%i70) limit(time_evolving_median_eq, t, inf);

(%o70) $time\_evolving\_median=0$

In the STATIC_CASE, this reduces to:

(%i71) def_static_median:static_median=rhs(time_evolving_median_eq), STATIC_CASE;

(%o71) $$static\_median=\%e^{\frac{\sigma^2-2\,\mu}{6}-\frac{\sigma^2}{6}}\,C$$

and this is (1.B.151):

(%i72) radcan(def_static_median);

(%o72) $$static\_median=\%e^{-\frac{\mu}{3}}\,C$$

IMPORTANT CONCLUSION
about the CONSTRAINTS between mu and sigma.

In this file we have derived THREE different constraints
between mu and sigma, coming from, respectively:
1) the MEAN VALUE abscissa: mu > (2/3)*sigma^2 = 0.667*sigma^2,
2) the PEAK abscissa: mu > (1/6)*sigma^2 = 0.167*sigma^2,
3) the MEDIAN abscissa: mu > (1/2)*sigma^2 = 0.5*sigma^2.

Out of these four constraints, the STRONGEST CONSTRAINT,
namely the one that, if fulfilled, is a warranty that all the remaining two
constraints are fulfilled also, is the MEAN VALUE one:

mu > (2/3)*sigma^2 = 0.667*sigma^2.

We shall thus assume that this constraint between mu and sigma is indeed fulfilled
in order to let the Geometric Brownian Motion N(t) and its consequence D(t)
be a realistic model of the the EVOLUTION OF LIFE in the Galaxy !

Part II

Space Missions to Exploit Gravitational Lensing

12

So much gain at 550 AU

12.1 INTRODUCTION

The gravitational focusing effect of the Sun is one of the most amazing discoveries produced by the general theory of relativity. The first paper in this field was published by Albert Einstein in 1936 [1], but his work was virtually forgotten until 1964, when Sydney Liebes of Stanford University [2] gave the mathematical theory of gravitational focusing by a galaxy located between the Earth and a very distant cosmological object, such as a quasar.

In 1978 the first "twin-quasar" image, caused by the gravitational field of an intermediate galaxy, was spotted by the British astronomer Dennis Walsh and his colleagues. Subsequent discoveries of several more examples of gravitational lenses eliminated all doubts about the validity of gravitational focusing as predicted by general relativity.

Von Eshleman of Stanford University then went on to apply the theory to the case of the Sun in 1979 [3]. His paper was the first to suggest the possibility of sending a spacecraft to 550 AU from the Sun to exploit the enormous magnifications provided by the gravitational lens of the Sun, particularly at microwave frequencies, such as the hydrogen line at 1,420 MHz (21 cm wavelength). This is the frequency that all SETI radioastronomers regard as "magic" for interstellar communications, and thus the tremendous potential of the gravitational lens of the Sun for getting in touch with alien civilizations became obvious.

The first experimental SETI radioastronomer in history, Frank Drake (Project Ozma, 1960—see *en.wikipedia.org/wiki/Project_Ozma*), presented a paper on the advantages of using the gravitational lens of the Sun for SETI at the *Second International Bioastronomy Conference* held in Hungary in 1987 [4], as did Nathan "Chip" Cohen of Boston University [5]. Non-technical descriptions of the topic were also given by them in their popular books [6, 7].

However, the possibility of planning and funding a space mission to 550 AU to exploit the gravitational lens of the Sun immediately proved a difficult task. Space scientists and engineers first turned their attention to this goal at the June 18, 1992, *Conference on Space Missions and Astrodynamics* organized in Turin, Italy, led by this author (see Figure 1.1). The relevant proceedings were published in 1994 in the *Journal of the British Interplanetary Society* [8]. Meanwhile, on May 20, 1993 this author also submitted a formal proposal to the European Space Agency (ESA) to fund the space mission design [9]. The optimal direction of space to launch the FOCAL spacecraft was also discussed by Jean Heidmann of Paris Meudon Observatory and the author [10], but it seemed clear that a demanding space mission like this one should not be devoted entirely to SETI. Things like the computation of the parallaxes of many distant stars in the Galaxy, the detection of gravitational waves by virtue of the very long baseline between the spacecraft and the Earth, plus a host of other experiments would complement the SETI utilization of this space mission to 550 AU and beyond. The mission was dubbed "SETISAIL" in earlier papers [11], and "FOCAL" in the proposal submitted to ESA in 1993.

12.2 THE MINIMAL FOCAL DISTANCE OF 550 AU FOR ELECTROMAGNETIC WAVES

The well-known Schwarzschild solution to the Einstein field equations is the mathematical foundation upon which the theory of the gravitational lens of the Sun rests. From it a long string of formulas can be developed. Since those formulas are derived in standard textbooks, we shall simply rewrite without proofs the basic equations needed to explain the advantages provided by the gravitational lens of the Sun, and point the interested reader to [8], [12], and [13] for the relevant mathematical demonstrations.

The geometry of the Sun's gravitational lens is easily described: incoming electromagnetic waves (e.g., arriving from the center of the Galaxy) pass outside the Sun and pass within a certain distance r of its center. Then, the basic result following from the Schwarzschild solution shows that the corresponding deflection angle $\alpha(r)$ at the distance r from the Sun center is given by

$$\alpha(r) = \frac{4GM_{Sun}}{c^2 r}. \tag{12.1}$$

Figure 12.2 depicts the various parameters.

Light rays (i.e., electromagnetic waves) cannot pass through the Sun's interior (whereas gravitational waves and neutrinos can), so the largest deflection angle α occurs for those rays just grazing the Sun's surface (i.e., for $r = r_{Sun}$). This yields the inequality

$$\alpha(r_{Sun}) > \alpha(r) \tag{12.2}$$

with

$$\alpha(r_{Sun}) = \frac{4GM_{Sun}}{c^2 r_{Sun}}. \tag{12.3}$$

1. Stelio Montebugnoli
2. Ed Belbruno
3. Jean Heidmann
4. Jorg Strobl
5. Gregory Matloff
6. Ettore Antona
7. Constance Bangs
8. Renato Pannunzio
9. Sigfrido Leschiutta
10. Giovanni Vulpetti
11. Maria Sarasso
12. Rinaldo Bertone
13. Franco Palutan
14. Vittorio Banfi
15. Mario Pasta
16. Federico Bedarida
17. Luciano Santoro
18. Claudio Maccone

Figure 12.1. First ever conference about the FOCAL space mission to 550 AU, held on June 18, 1992, at the Politecnico di Torino (Turin, Italy).

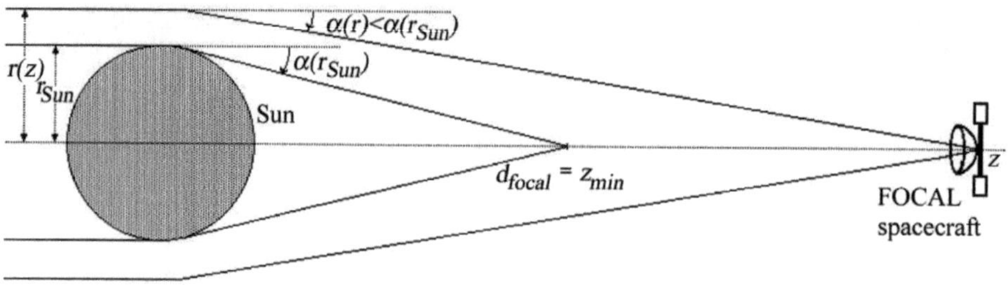

Figure 12.2. Basic geometry of the gravitational lens of the Sun, showing the minimal focal length and the FOCAL spacecraft position.

From the illustration it should be clear that the minimal focal distance d_{focal} is related to the tangent of the maximum deflection angle by the formula

$$\tan(\alpha(r_{Sun})) = \frac{r_{Sun}}{d_{focal}}. \tag{12.4}$$

Moreover, since the angle $\alpha(r_{Sun})$ is very small (its actual value is about 1.75 arcsec), the above expression may be rewritten by replacing the tangent by the small angle itself:

$$\alpha(r_{Sun}) \approx \frac{r_{Sun}}{d_{focal}}. \tag{12.5}$$

Eliminating the angle $\alpha(r_{Sun})$ between equations (12.3) and (12.5), and then solving for the minimal focal distance d_{focal}, one gets

$$d_{focal} \approx \frac{r_{Sun}}{\alpha(r_{Sun})} = \frac{r_{Sun}}{\frac{4GM_{Sun}}{c^2 r_{Sun}}} = \frac{c^2 r_{Sun}^2}{4GM_{Sun}} = \frac{c^2}{4G} \cdot \frac{r_{Sun}^2}{M_{Sun}}. \tag{12.6}$$

This basic result may also be rewritten in terms the Schwarzschild radius

$$r_{Schwarzschild} = \frac{2GM_{Sun}}{c^2} \tag{12.7}$$

yielding

$$d_{focal} \approx \frac{r_{Sun}}{\alpha(r_{Sun})} = \frac{r_{Sun}}{\frac{4GM_{Sun}}{c^2 r_{Sun}}} = \frac{r_{Sun}^2}{2r_{Schwarzschild}}. \tag{12.8}$$

Numerically, one finds

$$d_{focal} \cong 542 \text{ AU} \approx 550 \text{ AU} \approx 3.17 \text{ light-days} \approx \text{half a light-week}. \tag{12.9}$$

This is the fundamental formula yielding the minimal focal distance of the gravitational lens of the Sun (i.e., the minimal distance from the Sun's center that the FOCAL spacecraft must reach in order to get magnified radio pictures of whatever lies on the other side of the Sun with respect to the spacecraft position).

Furthermore, a simple, but very important consequence of the above discussion is that all points on the straight line beyond this minimal focal distance are foci too, because the light rays passing by the Sun farther than the minimum distance have smaller deflection angles and thus come together at an even greater distance from the Sun.

In addition, the very important astronautical consequence of this fact for the FOCAL mission is that it is not necessary to stop the spacecraft at 550 AU. It can go on to almost any distance beyond and focus as well or better. In fact, the farther it goes beyond 550 AU the less distorted the collected radio waves by solar corona fluctuations. The important problem of corona fluctuations and related distortions was studied by Von Eshleman and colleagues at Stanford University (see Section 14.1).

We would like to add here one more result that is very important because it holds well not just for the Sun, but for all stars in general. This we will do without demonstration as it can be found on p. 55 of [12]. Consider a spherical star with radius r_{star} and mass M_{star} that will be called the "focusing star". Suppose also that a light source (i.e., another star or an advanced extraterrestrial civilization) is located at the distance D_{source} from it. We can then ask: How far is the minimal focal distance d_{focal} on the opposite side of the source with respect to the focusing star center? The answer is given by the formula

$$d_{focal} = \frac{r_{star}^2}{\frac{4GM_{star}}{c^2} - \frac{r_{star}^2}{D_{source}}}. \tag{12.10}$$

This is the key to gravitational focusing for a pair of stars and may well be the key for SETI to find extraterrestrial civilizations. It could also be considered for the magnification of a certain source by any star that is perfectly aligned with that source and the Earth: the latter would then be in the same situation as the FOCAL spacecraft except, of course, it would be located much farther out than 550 AU with respect to the focusing, intermediate star. Finally, note that equation (12.10) reduces to equation (12.6) in the limit $D_{source} \to \infty$: that is, (12.6) is the special case of (12.10) for light rays approaching the focusing star from an infinite distance.

12.3 THE (ANTENNA) GAIN OF THE GRAVITATIONAL LENS OF THE SUN

Having thus determined the minimal distance of 550 AU that the FOCAL spacecraft must reach, one now wonders what's the good of going so far out of the Solar System (i.e., how much focusing of light rays is caused by the gravitational field of the Sun). The answer to such a question is provided by the technical notion of "antenna gain", which stems from antenna theory.

A standard formula in antenna theory relates antenna gain, $G_{antenna}$, to the antenna effective area, $A_{effective}$, and to the wavelength λ or the frequency ν by virtue of the equation (see, e.g., [13, p. 6-117 and equation (6-241)] in particular):

$$G_{antenna} = 4\pi \frac{A_{effective}}{\lambda^2}. \tag{12.11}$$

Now, assume the antenna is circular with radius $r_{antenna}$ and assume 50% efficiency as well. Then, the antenna effective area is obviously given by

$$A_{effective} = \frac{A_{physical}}{2} = \frac{\pi r^2_{antenna}}{2}. \tag{12.12}$$

Substituting this back into (12.11) yields antenna gain as a function of the antenna radius and of the observed frequency

$$G_{antenna} = \frac{4\pi A_{effective}}{\lambda^2} = \frac{2\pi A_{physical}}{\lambda^2} = \frac{2\pi^2 r^2_{antenna}}{\lambda^2} = \frac{2\pi^2 r^2_{antenna}}{c^2} \nu^2. \tag{12.13}$$

The important point here is that antenna gain increases with the square of the frequency, thus favoring observations on as high frequencies as possible.

Is anything similar happening for the Sun's gravitational lens as well? Yes is the answer, and the "gain" (one maintains this terminology for convenience) of the gravitational lens of the Sun can be proved to be

$$G_{Sun} = 4\pi^2 \frac{r_{Schwarzschild}}{\lambda} \tag{12.14}$$

or, invoking the expression (12.7) of the Schwarzschild radius

$$G_{Sun} = \frac{8\pi^2 G M_{Sun}}{c^2} \cdot \frac{1}{\lambda} = \frac{8\pi^2 G M_{Sun}}{c^3} \cdot \nu. \tag{12.15}$$

The mathematical proof of equation (12.14) is difficult to achieve.

The author, unsatisfied with the treatment of this key topic given in [1], [3], and [13], turned to three engineers at the engineering school in his home town: Renato Orta, Patrizia Savi, and Riccardo Tascone. To his surprise, in a few weeks they provided a full proof of not just the Sun gain formula (12.14) and (12.15), but also of the focal distance for rays originating from a source at finite distance—see equation (12.10). Their proof is fully described in [12] and is based on the aperture method used to study the propagation of electromagnetic waves, rather than on ray optics.

In the words used by these three authors in the abstract of their paper, they "computed the radiation pattern of the [spacecraft] Antenna + Sun system, which has an extremely high directivity. It has been observed that the focal region of the lens for an incoming plane wave is a half line parallel to the propagation direction starting at a point [550 AU] whose position is related to the blocking effect of the Sun disk [Figure 12.2]. Moreover, a characteristic of this thin lens is that its gain, defined as the magnification factor of the antenna gain, is constant along this half line. In particular, for a wavelength of 21 cm, this lens gain reaches the value of 57.5 dB. Also a measure of the transversal extent of the focal region has been obtained. The performance of this radiation system has been determined by adopting a thin lens model which introduces a phase factor depending on the logarithm of the impact parameter of the incident rays. Then the antenna is considered to be in transmission mode and the radiated field is computed by asymptotic evaluation of the radiation integral in the Fresnel approximation."

12.4 THE COMBINED, TOTAL GAIN ON THE FOCAL SPACECRAFT

One is now able to compute the total gain of the Antenna + Sun system, which is simply obtained by multiplying equations (12.15) and (12.13)

$$G_{total} = G_{Sun} G_{antenna} = \frac{16\pi^2 G M_{Sun} r^2_{antenna}}{c^5} \nu^3. \tag{12.16}$$

Since the total gain increases with the cube of the observed frequency, it favors electromagnetic radiation in the microwave region of the spectrum. Table 12.1 shows numerical data provided by equations (12.15) and (12.14) for five selected frequencies: the hydrogen line at 1,420 MHz and the four frequencies that the Quasat radioastronomy satellite planned to observe, had it been built jointly by ESA and NASA as planned before 1988 [14].

The definition of dB is of course

$$N\,\text{dB} = 10 \log_{10}(N) = 10 \frac{\ln(N)}{\ln(10)}. \tag{12.17}$$

Table 12.1. The gain of the Sun's lens alone, the gain of a spacecraft's (S/C) 12-meter antenna, and the combined gain of the Sun + S/C antenna system at five selected frequencies important in radioastronomy.

Line	*Neutral hydrogen*		*OH radical*		H_2O
Frequency (ν)	1,420 MHz	327 MHz	1.6 GHz	5 GHz	22 GHz
Wavelength (λ)	21 cm	92 cm	18 cm	6 cm	1.35 cm
S/C antenna beamwidth	1.231 deg	5.348 deg	1.092 deg	0.350 deg	0.080 deg
Sun gain	57.4 dB	51.0 dB	57.9 dB	62.9 dB	69.3 dB
12-meter antenna S/C gain	42.0 dB	29.3 dB	43.1 dB	53.0 dB	65.8 dB
Combined sun + S/C gain	99.5 dB	80.3 dB	101.0 dB	115.9 dB	135.1 dB

12.5 THE IMAGE SIZE AT THE SPACECRAFT DISTANCE z

The next important notion to understand is the size of the image of an infinitely distant object created by the Sun's lens at the current spacecraft distance z from the Sun ($z > 550$ AU). We may define such an image size as the distance from the focal axis (i.e., from the spacecraft straight trajectory) at which the gain drops by 6 dB. The formula for this (proven in [8]) is

$$r_{6\,dB} = \frac{\lambda}{\pi^2}\sqrt{\frac{z}{2r_s}} = \frac{c}{2\pi^2\sqrt{GM_{Sun}}}\lambda\sqrt{z} = \frac{c^2}{2\pi^2\sqrt{GM_{Sun}}}\frac{\sqrt{z}}{\nu}. \qquad (12.18)$$

Thus, the image size increases with the spacecraft distance z from the Sun. Table 12.2 provides a quantitative feeling of how the image size changes with spacecraft distance from the Sun.

It is clear that these image size values are very small compared with spacecraft distance from the Earth. This means that if we want to observe a certain point source in the sky, the alignment between this source, the Sun, and the spacecraft position must be extremely precise. In fact, spacecraft tracking must far exceed what we are able to do within the Solar System today. However, this is not true if the source we want to observe is the center of the Galaxy, which is a very broad source: slight changes in the spacecraft trajectory (say, in a spreading spiral shape) would enable us to gradually see much of the Galactic center at the huge resolution provided by the gravitational lens of the Sun.

Table 12.2. Image sizes for a 12-meter antenna, located at distances of 550 AU, 800 AU, and 1,000 AU from the Sun for the five selected frequencies.

Line	*Neutral hydrogen*		*OH radical*		*H_2O*
Frequency (ν)	1,420 MHz	327 MHz	1.6 GHz	5 GHz	22 GHz
Wavelength (λ)	21 cm	92 cm	18 cm	6 cm	1.35 cm
Image size (down 6 dB) at 550 AU	2.498 km	10.847 km	2.217 km	0.709 km	0.161 km
Image size (down 6 dB) at 800 AU	3.033 km	13.169 km	2.691 km	0.861 km	0.196 km
Image size (down 6 dB) at 1,000 AU	3.391 km	14.724 km	3.009 km	0.963 km	0.219 km

12.6 REQUIREMENTS ON THE IMAGE SIZE AND ANTENNA BEAMWIDTH AT THE SPACECRAFT DISTANCE z

There are two "geometrical" requirements that must be fulfilled in order that the combined lens system Sun + FOCAL spacecraft antenna can work optimally:

(1) Size requirement: the full antenna dish of the FOCAL spacecraft must fall well inside the cylindrical region centered along the focal axis and having radius equal to $r_{6\,dB}$ (i.e., the spacecraft feed dish radius must be considerably smaller than $r_{6\,dB}$):

$$r_{antenna} \ll r_{6\,dB} = \frac{c}{2\pi^2\sqrt{GM_{Sun}}}\lambda\sqrt{z} = \frac{c^2}{2\pi^2\sqrt{GM_{Sun}}}\frac{\sqrt{z}}{\nu}. \tag{12.19}$$

(2) Angle requirement: the impact radius circle around the Sun within which electromagnetic waves are focused towards the FOCAL spacecraft must fall well within the antenna beamwidth of the FOCAL spacecraft. Put more technically, the Half-Power Beam Width (= HPBW) (i.e. the angular width of the main lobe of the spacecraft antenna at the half-power level) should be considerably greater than the angle subtended at the spacecraft distance by twice the incident ray impact radius at the Sun

$$HPBW \gg 2\alpha(r) = \frac{8GM_{Sun}}{c^2 r}. \tag{12.20}$$

Tables 12.3 and 12.4 show that both these conditions are fulfilled at the three FOCAL distances from the Sun for the five selected frequencies, respectively.

Table 12.3. Image size vs. antenna radius for a 12-meter antenna located at various distances from the Sun for the five selected frequencies.

Line	*Neutral hydrogen*		*OH radical*		*H_2O*
Frequency (ν)	1,420 MHz	327 MHz	1.6 GHz	5 GHz	22 GHz
Wavelength (λ)	21 cm	92 cm	18 cm	6 cm	1.35 cm
Image size at 550 AU vs. antenna radius	2.498 km $\gg$ 6 m	10.85 km $\gg$ 6 m	2.22 km $\gg$ 6 m	0.71 km $\gg$ 6 m	0.16 km $\gg$ 6 m
Image size at 800 AU vs. antenna radius	3.03 km $\gg$ 6 m	13.17 km $\gg$ 6 m	2.69 km $\gg$ 6 m	0.86 km $\gg$ 6 m	0.20 km $\gg$ 6 m
Image size at 1,000 AU vs. antenna radius	3.39 km $\gg$ 6 m	14.72 km $\gg$ 6 m	3.01 km $\gg$ 6 m	0.96 km $\gg$ 6 m	0.22 km $\gg$ 6 m

Table 12.4. HPBW vs. aspect angle of the Sun for a 12-meter antenna located at various distances from the Sun for the five selected frequencies.

Line	*Neutral hydrogen*		*OH radical*		H_2O
Frequency (ν)	1,420 MHz	327 MHz	1.6 GHz	5 GHz	22 GHz
Wavelength (λ)	21 cm	92 cm	18 cm	6 cm	1.35 cm
HPBW at 550 AU vs. 2α	$1.23154° \gg 1.5 \times 10^{-7\circ}$	$5.34798° \gg 1.5 \times 10^{-7\circ}$	$1.09299° \gg 1.5 \times 10^{-7\circ}$	$0.34976° \gg 1.5 \times 10^{-7\circ}$	$0.07949° \gg 1.5 \times 10^{-7\circ}$
HPBW at 800 AU vs. 2α	$1.23154° \gg 1.5 \times 10^{-7\circ}$	$5.34798° \gg 1.5 \times 10^{-7\circ}$	$1.09299° \gg 1.5 \times 10^{-7\circ}$	$0.34976° \gg 1.5 \times 10^{-7\circ}$	$0.07949° \gg 1.5 \times 10^{-7\circ}$
HPBW at 1,000 AU vs. 2α	$1.23154° \gg 1.5 \times 10^{-7\circ}$	$5.34798° \gg 1.5 \times 10^{-7\circ}$	$1.109299° \gg 1.5 \times 10^{-7\circ}$	$0.34976° \gg 1.5 \times 10^{-7\circ}$	$0.07949° \gg 1.5 \times 10^{-7\circ}$

12.7 ANGULAR RESOLUTION AT THE SPACECRAFT DISTANCE z

The notion of angular resolution of the Sun's lens is relevant to the discussion. Angular resolution is simply defined as the ratio of the image size (at the spacecraft distance z from the Sun) to that distance z. From eq. (12.19),

$$\theta_{resolution}(z) = \frac{r_{6\,dB}}{z} = \frac{c^2}{2\pi^2\sqrt{GM_{Sun}}}\frac{1}{\sqrt{z}\nu}. \tag{12.21}$$

Clearly, angular resolution also depends on the spacecraft distance z from the Sun, and it actually improves (i.e., it gets smaller) as the distance increases beyond 550 AU.

Table 12.5 gives angular resolutions for the same three distances at the same five frequencies. Let us take a moment to ponder over these numbers. The best angular resolutions achieved so far, in visible light, were obtained by the European astrometric satellite Hipparcos, launched in 1989, and decommissioned in 1993. Though the apogee kick motor of Hipparcos didn't fire, forcing technicians to take the software originally written for a circular geostationary orbit and rewrite it for a highly elliptical orbit, the Hipparcos mission has proven a success. The resolutions achieved by Hipparcos are at a level of 2 milliseconds of arc precision. Checking this level against Table 12.5, one can see that the gravitational lens of the Sun plus a (modest) 12-meter antenna would improve angular resolution by about three orders of magnitude (at radio frequencies).

Table 12.5. Angular resolution for three spacecraft distances (550 AU, 800 AU, and 1,000 AU) at the five selected frequencies.

Line	*Neutral hydrogen*		*OH radical*		H_2O
Frequency (ν)	1,420 MHz	327 MHz	1.6 GHz	5 GHz	22 GHz
Wavelength (λ)	21 cm	92 cm	18 cm	6 cm	1.35 cm
Angular resolution at 550 AU S/C distance	6.3458×10^{-6} arcsec	2.7557×10^{-6} arcsec	5.6319×10^{-6} arcsec	1.8022×10^{-6} arcsec	4.0959×10^{-6} arcsec
Angular resolution at 800 AU S/C distance	5.2267×10^{-6} arcsec	2.2697×10^{-6} arcsec	4.6387×10^{-6} arcsec	1.4844×10^{-6} arcsec	3.3736×10^{-6} arcsec
Angular resolution at 1,000 AU S/C distance	4.6749×10^{-6} arcsec	2.0301×10^{-6} arcsec	4.1490×10^{-6} arcsec	1.3277×10^{-6} arcsec	3.0174×10^{-6} arcsec

12.8 SPATIAL RESOLUTION AT THE SPACECRAFT DISTANCE z

Finally, let us turn to the spatial resolution, simply called "resolution" hereafter, of an astronomical object we want examine by help of the gravitational lens of the Sun. It is defined by

$$R_{object} = d_{Sun\text{-}object}\theta_{resolution} = d_{Sun\text{-}object}\frac{c^2}{2\pi^2\sqrt{GM_{Sun}}}\frac{1}{\sqrt{z\nu}}. \tag{12.22}$$

Again, beyond 550 AU the resolution improves (i.e., the angle gets smaller) slowly with increasing spacecraft distance from the Sun. Table 12.6 shows the spatial resolution for a very wide range of distances, from the Oort Cloud to cosmological objects like quasars.

12.9 REFERENCES

[1] A. Einstein, "Lens-like action of a star by the deviation of light in the gravitational field," *Science*, **84** (1936), 506–507.

[2] S. Liebes, Jr., "Gravitational lenses," *Physical Review*, **133** (1964), B835–B844.

[3] V. Eshleman, "Gravitational lens of the Sun: Its potential for observations and communications over interstellar distances," *Science*, **205** (1979), 1133–1135.

[4] F. Drake, "Stars as gravitational lenses," in G. Marx (Ed.), *Proceedings of the Bioastronomy International Conference, Balatonfüred, Hungary, June 22–27, 1987*, pp. 391–394.

[5] N. Cohen, "The pro's and con's of gravitational lenses in CETI," in G. Marx (Ed.), *Proceedings of the Bioastronomy International Conference, Balatonfüred, Hungary, June 22–27, 1987*, p. 395.

Table 12.6. Spatial resolution of astronomical objects at selected distances from the Sun for a spacecraft's 12-meter antenna.

Line	*Neutral hydrogen*		*OH radical*		*H_2O*
Frequency (ν)	1,420 MHz	327 MHz	1.6 GHz	5 GHz	22 GHz
Wavelength (λ)	21 cm	92 cm	18 cm	6 cm	1.35 cm
Resolution at 0.5 lt-yr (Oort Cloud)	145 km	632 km	129 km	41 km	9 km
Resolution at 4.29 lt-yr (α Centauri)	1,248 km	5,422 km	1,108 km	355 km	81 km
Resolution at 10 pc = 32.6 lt-yr	9,576 km	4,158 km	8,499 km	2,719 km	618 km
Resolution at 100 pc = 326 lt-yr	95,75 km	415,8 km	84,98 km	27,19 km	6,180 km
Resolution at 1 kpc = 3,260 lt-yr	957,58 km = 0.006 AU	4,158,330 km = 0.028 AU	849,861 km = 0.005 AU	271,955 km = 0.001 AU	61,808 km = 0.0004 AU
Resolution at 10 kpc = 32,600 lt-yr (Galactic center)	9,575,870 km = 0.06401 AU	41,583,000 km = 0.27797 AU	8,498,610 km = 0.05681 AU	2,719,550 km = 0.01818 AU	618,082 km = 0.00413 AU
Resolution at 50 kpc = 160,000 lt-yr (Magellanic Clouds)	4.78794×10^7 km = 0.32006 AU	2.07917×10^8 km = 1.38984 AU	4.2493×10^7 km = 0.28405 AU	1.3597×10^7 km = 0.0909 AU	3.0903×10^6 km = 0.02066 AU
Resolution at 613 kpc = 1.5 million lt-yr (Andromeda Galaxy M31)	5.82123×10^8 km = 3.89125 AU	2.52788×10^9 km = 16.8978 AU	5.16631×10^8 km = 3.45349 AU	1.65322×10^8 km = 1.10512 AU	3.75732×10^7 km = 0.25116 AU
Resolution at 18,406 pc = 60 million lt-yr ("Jet" Galaxy M87 in Virgo)	1.74636×10^{10} km = 116.738 AU	7.5836×10^{10} km = 506,934 AU	1.5499×10^{10} km = 103.605 AU	4.95968×10^{10} km = 33.1535 AU	1.1272×10^9 km = 7.53488 AU
Resolution at 3.07 million kpc = 10 billion lt-yr (radius of the Universe)	2.91059×10^{12} km = 19,456 AU = 0.30765 lt-yr	1.26393×10^{13} km = 84,489 AU = 1.33598 lt-yr	2.58316×10^{12} km = 17,267 AU = 0.27304 lt-yr	8.2661×10^{11} km = 5525.58 AU = 0.08737 lt-yr	1.8786×10^{11} km = 1255.81 AU = 0.01986 lt-yr

[6] F. Drake and D. Sobel, *Is Anyone Out There?* Delacorte Press, New York, 1992. See pp. 230–234 in particular.
[7] N. Cohen, *Gravity's Lens*, Wiley Science Editions, New York, 1988.
[8] C. Maccone, "Space missions outside the Solar System to exploit the gravitational lens of the Sun," in C. Maccone (Ed.), *Proceedings of the International Conference on Space Missions and Astrodynamics, Turin, Italy, June 18, 1992*, published in *Journal of the British Interplanetary Society*, **47** (1994), 45–52.
[9] C. Maccone, "FOCAL: A new space mission to 550 AU to exploit the gravitational lens of the Sun," a proposal for an M3 Space Mission submitted to the European Space Agency (ESA) on May 20, 1993, on behalf of an international team of scientists and engineers. Later (October 1993) reconsidered by ESA within the Horizon 2000 Plus space mission plan.
[10] J. Heidmann and C. Maccone, "AstroSail and FOCAL: Two extra-Solar System missions to the Sun's gravitational focuses," *Acta Astronautica*, **35** (1994), 409–410.
[11] C. Maccone, "The SETISAIL Project," in G. Seth Shostak (Ed.), *Progress in the Search for Extraterrestrial Life: Proceedings of the 1993 Bioastronomy Symposium, University of California at Santa Cruz, August 16–20, 1993*, published in *Astronomical Society of the Pacific Conference Series*, **74** (1995), 407–417.
[12] R. Orta, P. Savi, and R Tascone, "Analysis of gravitational lens antennas," in C. Maccone (Ed.), *Proceedings of the International Conference on Space Missions and Astrodynamics, Turin, Italy, June 18, 1992*, published in *Journal of the British Interplanetary Society*, **47** (1994), 53–56.
[13] John D. Kraus, *Radio Astronomy*, Second Edition, Cygnus-Quasar Books, Powell, OH, 1966. See pp. 6-115–6-118 in particular.
[14] A. Hawkyard and A. Anselmi, *QUASAT Industrial Phase A Study*, Executive Summary, Aeritalia GSS Report QS-RP-A1-0004, 1988.

Figure 12.3. The principal ideas making up Chapters 12 through 16 of this book were first published in 1997 by the author in a short book entitled *The Sun as a Gravitational Lens: Proposed Space Missions.* Just two jears later, the book was awarded the "1999 Engineering Sciences Book Award" by the International Academy of Astronautics (IAA). The relevant Book Award Diploma is reproduced in the above picture.

13

FOCAL mission to 1,000 AU as an interstellar precursor

13.1 INTRODUCTION

The nearest stellar system, the Alpha Centauri three-star system, is located about 4.40 light-years away. This amounts to 278,261 AU. However, the focus of the gravitational lens of the Sun is found at only 550 AU—or, more generally, at only about 1,000 AU—which is then 278 times closer than our nearest interstellar target. In other words, assuming equal engineering problems, the trip to the Sun's focus takes 278 times less than the time to the nearest stellar target. This makes the Sun focus a reasonable target for our probes to reach within this century. It also plainly appears that, before we send a probe towards any one of the nearest stellar systems, we will need a detailed radio map of that stellar system. In other words, we need a huge radio magnification of all objects located in that neighborhood, and nothing is better than the huge magnification provided by the gravitational lens of the Sun. Thus, sending a preliminary probe to 1,000 AU in the direction opposite to the target stellar system must clearly be done before any interstellar flight to that stellar system is designed, not to mention attempted. In this chapter, a status review is presented about the FOCAL probe to 550 AU or 1,000 AU. The relevant scientific, propulsion, and telecommunication issues are briefly summarized and updated.

13.2 THE AUTHOR'S 2009 BOOK ABOUT THE FOCAL SPACE MISSION

In March 2009 a 400-page, comprehensive book by the author, entitled *Deep Space Flight and Communications: Exploiting the Sun as a Gravitational Lens* [19], was published. This book embodies all the previous material published about the FOCAL space mission and updates it with the view of submitting a formal proposal

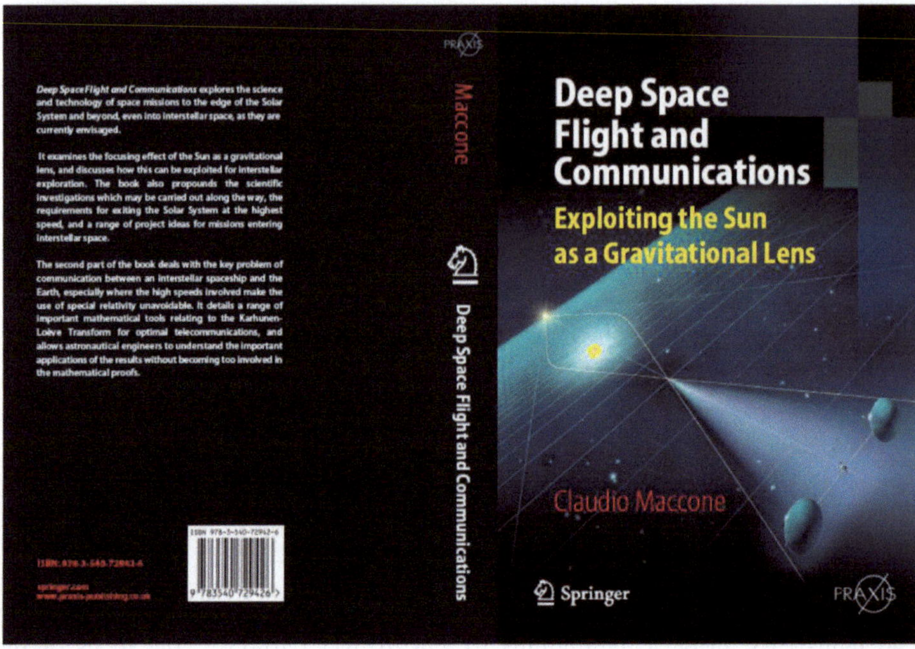

Figure 13.1. Front and back covers of the author's book entitled *Deep Space Flight and Communications: Exploiting the Sun as a Gravitational Lens* published by Springer-Praxis in March 2009 [19].

to NASA about FOCAL. The front and back covers of this book are reproduced in Figure 13.1.

13.3 USING TWO ANTENNAS AND A TETHER TO GET A MUCH LARGER FIELD OF VIEW FOR FOCAL

The goal of this section is to put forward the new notion of a *tethered system* joining together *two antennas* for the FOCAL spacecraft. We are going to show that the length of this tether system does not need to be very long: actually, just a couple of kilometers or so is sufficient to get a radio picture of the big Galactic Black Hole (Sgr A*), and this is a good result because a 2 km tether is certainly technologically feasible. It is important to point out that the tether could possibly be replaced by a truss. This would of course increase the system stability. To build a 2 km long truss in space, however, is a difficult engineering task. We thus prefer to speak about a tethered system rather than a truss system, leaving the actual design to expert engineers.

We start by pointing out the problem caused by plasma fluctuations of the Sun's corona and the effect they have on radio waves passing through the corona itself (as described in Chapters 8 and 9 of the author's 2009 book [19]). Finding a solution to

Figure 13.2. Enlarged part of the front cover of the author's 2009 book [19] showing: (1) the bright radio source at infinity (i.e., the horizon); (2) its radio waves flying by the Sun and made to focus at 550 AU; (3) the FOCAL spacecraft made up of two (say) 12-meter antennas joined to each other by a *tether* and revolving in the orthogonal plane to the spacecraft's velocity vector.

this problem is vital for the success of the FOCAL space mission. We claim the best way to solve the corona problem is through interferometry between the two antennas of the FOCAL spacecraft. Thus, the FOCAL spacecraft, rather than having just one antenna (inflatable and, say, 12 meters in diameter), must have two identical antennas in the new configuration proposed here. This doubles the sensitivity of the system, and introduces the new and fruitful idea of a tether joining each of them to the main cylindrical body of the FOCAL spacecraft (as shown in Figure 13.2).

Thus, the tethered FOCAL system we wish to propose is described as follows:

(1) The whole spacecraft moves away from the Sun along a rectilinear, purely radial trajectory.
(2) When the distance from the Sun is, say, 400 to 500 AU, all "engines" (be they solar sails, nuclear–electric, antimatter) are turned off, so we can assume that, at least beyond 550 AU, the Sun speed of the whole system is uniform.
(3) Uniform speed means no acceleration. So, one can start deploying the tether. The body of the FOCAL spacecraft is supposed to be cylindrical and kept in rotation at a suitable angular speed (i.e., FOCAL is supposed to be spin-stabilized). On two opposite sides of the cylinder, the two packed, inflatable antennas are released from the spacecraft. Moreover, each antenna is tied to the spacecraft by a tether kept taut because of the angular rotation of the whole system.

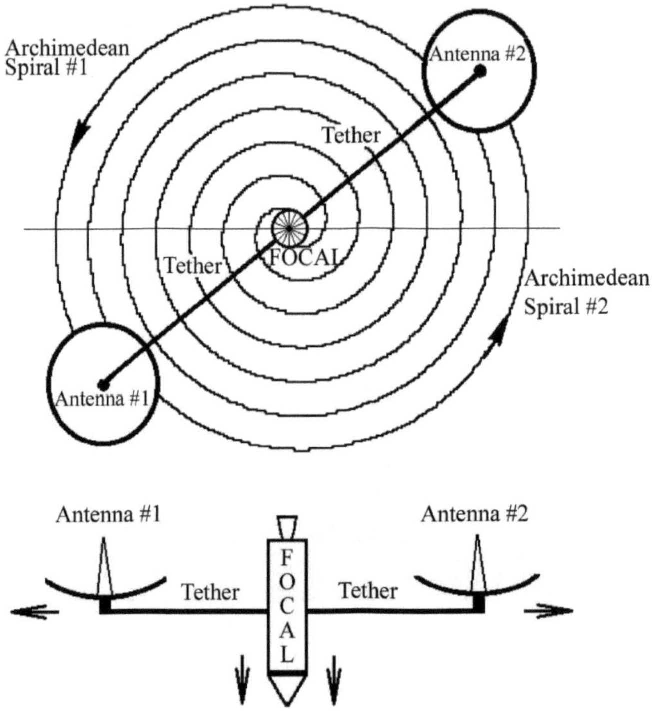

Figure 13.3. Another way of picturing Figure 12.2, this time with the two Archimedean spirals covered by the antennas.

(4) The two antennas are inflated at the very moment they reach the minimal safety distance from the spacecraft.
(5) The two antennas are oriented and pointed toward the Sun. This means that the two antenna axes are parallel or nearly parallel to each other. Then, in practice, a huge isosceles triangle is created in space, having as its basis the distance between the two antennas and as its apex the center of the Sun (at any distance greater than 550 AU).
(6) Slowly, both tethers are deployed such that their lengths are exactly equal on each side of FOCAL. Because of the uniform angular rotation of the whole system, this means that the end-points of the tethers (i.e., the center of each antenna) is made to describe an Archimedean spiral (i.e., a spiral with polar equation $\rho(\theta) = const \cdot \theta$) around the axis of the FOCAL cylindrical spacecraft; see Figure 13.3). What is more, since each antenna is pointing to the Sun, then—on the other side of the Sun, at the distance of the Galactic center (i.e., some 32,000 light-years away)—two huge Archimedean spirals are correspondingly being described around the Galactic center. A huge black hole is suspected to exist here (as depicted in Figure 13.4). This gigantic black hole we call the "Galactic Black Hole" (as described in the next section) (or Sgr A*).

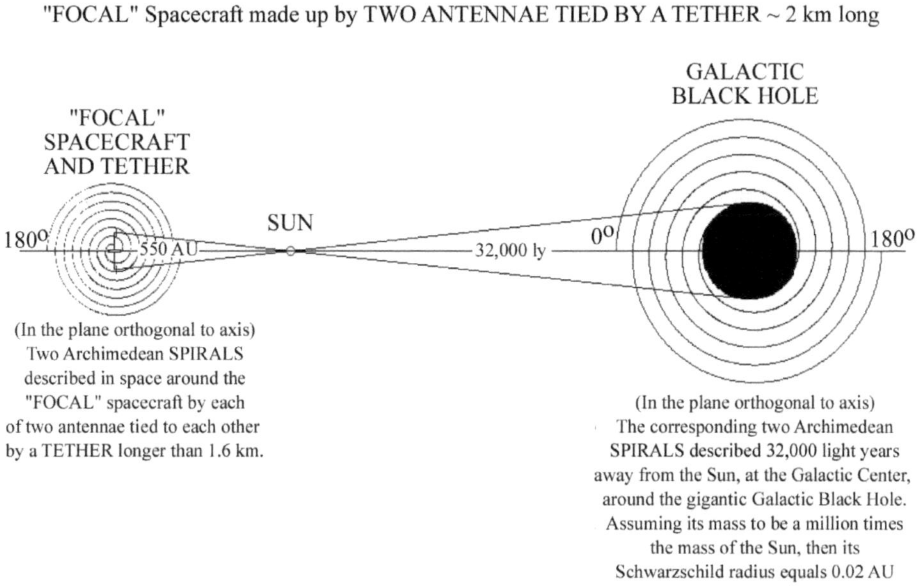

Figure 13.4. Imagine the above two Archimedean spirals in parallel planes that are *orthogonal* to the axis FOCAL, Sun center, and Galactic center. Then, two *similar triangles* relate the FOCAL tether length, the FOCAL spacecraft distance from the Sun, and the size of the Galactic Black Hole as well as its distance from the Sun. They allow us to compute the *minimal tether length*.

13.4 OBSERVING THE GALACTIC BLACK HOLE MAGNIFIED BY VIRTUE OF FOCAL

As we have just described, on the other side of the Sun, at the distance of the Galactic bulge (i.e., some 26,000 to 32,000 light-years away) two huge Archimedean spirals are correspondingly being described around the Galactic center. We provisionally assign to the Galactic Black Hole a mass one million times that of the Sun (as of 2009, its estimated mass is actually 4.31 ± 0.06 Sun masses). Consequently, the Schwarzschild radius of the Galactic Black Hole is a million times larger than the Sun Schwarzschild radius (i.e., it equals $\sim 2.95 \times 10^9$ km $\sim$ 0.01976 AU). The linearity between mass and Schwarzschild radius obviously appears in the definition (11.7).

We are now able to estimate the minimal tether length necessary to include the whole of the Galactic Black Hole within the area encompassed by the two FOCAL Archimedean spirals. Figure 13.4 clearly shows the two similar isosceles triangles: (i) the "small" one, between the tethered FOCAL system and the Sun, and (ii) the "large" one, between the Sun and the Galactic Black Hole. These two triangles immediately yield the proportion:

$$\frac{\text{Minimal Tether Length}}{550\text{ AU}} = \frac{2\, r_{Schwarzschild\_of\_Galactic\_Black\_Hole\_SgrA^*}}{32{,}000\text{ AU}}. \tag{13.1}$$

The Schwarzschild radius of the Galactic Black Hole Sgr A* can simply be given by the Schwarzschild radius formula (11.7)

$$r_{Schwarzschild\_of\_Galactic\_Black\_Hole\_SgrA^*} = \frac{2GM_{Galactic\_Black\_Hole\_SgrA^*}}{c^2}. \tag{13.2}$$

Astronomers have recently estimated the mass of the Galactic Black Hole at some 4 million solar masses (see *http://en.wikipedia.org/wiki/Sagittarius_A**). This source further reveals that, after monitoring stellar orbits around Sagittarius A* for 16 years, Reinhard Genzel, study team leader, announced the following conclusion in 2008: "The stellar orbits in the galactic centre show that the central mass concentration of four million solar masses must be a black hole, beyond any reasonable doubt." Thus, inserting (13.1) into (13.2) and solving for the tether length yields

$$\text{Minimum Tether Length} \approx 5\ \text{km}. \tag{13.3}$$

We thus reach the conclusion that a tether about 10 km long would certainly allow us to see not just the Galactic Black Hole, but also the host of astrophysical phenomena taking place around it, like the "swallowing" of stars by the Galactic Black Hole.

To sum up, the 21st and following centuries are likely to see a host of FOCAL space missions, each devoted to a different stellar target and launched in a different direction out of the Solar System.

13.5 OBSERVING THE THREE ALPHA CENTAURI STARS MAGNIFIED BY VIRTUE OF FOCAL

Alpha Centauri (α Centauri/α Cen) is a triple-star system and the brightest star system in the southern constellation of Centaurus. Alpha Centauri AB (α Cen AB) is a close binary system revolving every 79.91 years. To the unaided eye it appears as a single star, whose total visual magnitude would identify it as the third brightest star in the night sky. As we all know, the triple Alpha Centauri system is the closest star system to the Solar System, the center of gravity of α AB Cen being only 1.34 parsecs or 4.37 light-years away from our Sun. Because of this, the very first truly interstellar space mission will likely be aimed at reaching the Alpha Centauri system, rather than any other nearby star system in the Galaxy.

We learn from *http://en.wikipedia.org/Alpha_centauri* (the Alpha Centauri Wikipedia site) that Alpha Centauri A is the principal member or primary of the binary system, being slightly larger and more luminous than our Sun. It is a Sun-like main sequence star with a similar yellowish-white color, whose stellar classification is spectral type G2 V. From the determined mutual orbital parameters, α Cen A is about 10% more massive than our Sun with a radius about 23% larger. The projected rotational velocity ($v \cdot \sin i$) of this star is $2.7 \pm 0.7\,\text{km} \cdot \text{s}^{-1}$, resulting in an estimated rotational period of 22 days, which gives it a slightly faster rotational period than our Sun's 25 days. Alpha Centauri B is the companion star or secondary, slightly smaller and less luminous than our Sun. This main sequence star is of spectral type K1 V, typified by an orangish-yellow color rather than the whiter primary star. α Cen B is about 90% the mass of the Sun and 14% smaller in radius The projected rotational velocity ($v \cdot \sin i$) is $1.1 \pm 0.8\,\mathrm{km \cdot s^{-1}}$, resulting in an estimated rotational period of 41 days (an earlier estimate gave a similar rotation period of 36.8 days). Although it has a lower luminosity than star A, star B's spectrum emits more energy in the form of X-rays. The light curve of B varies on a short time scale and a flare has been observed.

Finally, Alpha Centauri C, also known as Proxima Centauri, is of spectral class M5Ve or M5VIe, suggesting that it is either a small main sequence star (Type V) or a subdwarf (VI) with emission lines whose B–V color index is +1.81. Its mass is about 0.12 times that of the Sun. Proxima is approximately 12,000 or 13,000 AU distant from α Cen AB and its orbital period around them is of the order of 100,000 to 500,000 years or more (its orbit may even be hyperbolic). Because of this situation, Proxima is indeed the closest star to us at all, its distance being 4.243 ± 0.002 lt-yr (1.3009 ± 0.0005 pc).

We now want to clarify the notion of position angle, usually abbreviated PA and defined as the angular offset in degrees of the secondary star to the primary, relative to the north celestial pole. This is visually described in Figure 13.5, which we have taken from *http://en.wikipedia.org/wiki/File:Positionangle.jpg*

Let us now return to our discussion of the Alpha Centauri system.

Viewed from Earth, the apparent orbit of this binary star system means that the separation and position angle are continuously changing throughout the projected orbit. Observed stellar positions in 2008 were separated by 8.29 arcsec through a PA

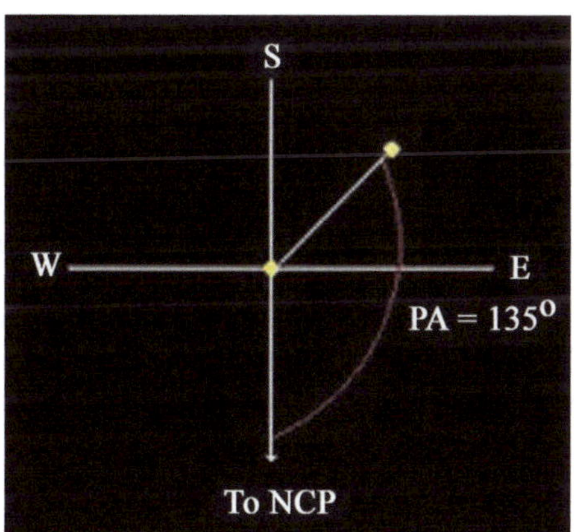

Figure 13.5. How the position angle (PA) is estimated through a telescope eyepiece. The primary star is at the center. If one were observing a hypothetical binary star with a PA of 135°, which means an imaginary line in the eyepiece drawn from the north celestial pole (NCP) to the primary (P), the primary would be offset from the secondary (S) such that the NCP–P–S angle would be 135°. The NCP line is traditionally drawn downward (i.e., with north at the bottom) and PA is measured counterclockwise, from 0 to 359° (from *http://en.wikipedia.org/wiki/Position_angle*).

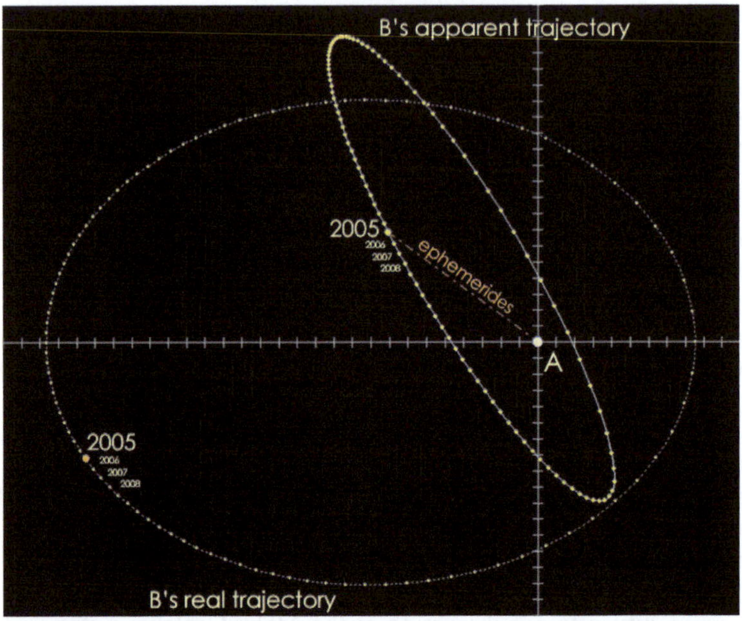

Figure 13.6. Apparent and true orbits of Alpha Centauri B (the secondary) around Alpha Centauri A (the primary). Motion is shown from the A star against the relative orbital motion of B star. The apparent orbit (thin ellipse) is the shape of the orbit as seen by an observer on Earth. The true orbit is the shape of the orbit viewed perpendicular to the plane of the orbital motion (from *http://en.wikipedia.org/wiki/Alpha_Centauri*).

of 237° and reduced to 7.53 arcsec through 241° in 2009. The next closest approach will be in February 2016, at 4.0 arcsec through 300°. The observed maximum separation of these stars is about 22 arcsec, while the minimum distance is a little less than 2 arcsec. The widest separation occurred during February 1976 and the next will be in January 2056 (see Figure 13.6 for B's apparent trajectory with respect to A).

In the true orbit, the closest approach or periastron was in August 1955 and the next will be in May 2035. The farthest orbital separation at apastron last occurred in May 1995 and the next will be in 2075. Thus, the apparent distance between the two stars is presently decreasing.

Turning to the FOCAL space mission once again, the first question we need to answer is: Can we use a tethered system of two antennas to watch the Alpha Centauri system similar to that used to observe the Galactic Black Hole? Unfortunately, the answer is "no", since the tether length would be far too long: of the order of millions of kilometers!

To realize why this is so, just compute the expression

$$550 * \mathrm{AU} * \tan\left(\frac{22 * \mathrm{arcsec}}{2}\right) = 4.388 * 10^6\ \mathrm{km} \tag{13.4}$$

yielding the tether length requested to encompass the view of both A and B at their maximum visual separation of 22 arcsec. Even if we consider the minimal visual separation of 4 arcsec, this is no better:

$$550 * \text{AU} * \tan\left(\frac{4 * \text{arcsec}}{2}\right) = 7.978 * 10^5 \text{ km}. \tag{13.5}$$

So, a tethered system to encompass the whole A–B system is out of the question. But this does not mean that a tethered system can be adopted to visualize each of the two stars separately: it would simply provide too narrow field of view because the whole system it just too close to the Sun. So, we must come up with another idea.

The new idea is to change the FOCAL orbit beyond 550 AU from just an outgoing straight line to a *conical helix of increasing radius.*

This becomes immediately apparent when you consider the Guggenheim Museum in New York City (shown in Figure 13.7) as if it were tilted from right to left.

The profile of this building is a conical helix and, if you look at it from right to left, you have just the conical helix (i.e., the helix is on the surface of a cone with an

Figure 13.7. The Guggenheim Museum in New York City, shown tilted by 90°, as if it were lying horizontally on the ground rather than vertically! Then, viewed from right to left, the profile of this building is a *conical helix of increasing radius.* A FOCAL spacecraft moving right to left along such a trajectory, while the Sun is on the right of this figure at a distance higher than 550 AU, would be able to cover a much larger field of view on the other side. The axis of this conical helix is the straight line between the Sun and the Alpha Cen system of three stars, plus their possible planets, that would just be seen as hugely magnified by the Sun's gravitational lens.

apex at just 550 AU and then an increasingly high radius). This is the modified orbit we propose for FOCAL after 550 AU. But how can such an orbit be achieved in space?

Well, we need a small but continuous thrust like that used in electric propulsion called "field emission electric propulsion" (FEEP) (see *http://en.wikipedia.org/wiki/Field_Emission_Electric_Propulsion* and references therein).

Actually, the acceleration produced by these FEEPs is so small and the times involved in having the FOCAL spacecraft move along its conical helix trajectory so large (decades) that one might well add a tethered system revolving orthogonally to the speed vector, as described for observation of the Galactic Black Hole. The radio image of the Alpha Centauri system would then appear increasingly detailed over the years while FOCAL would travel from 550 to 1,000 AU along its conical helix trajectory.

We stop our description at this point, for the next step would require an accurate engineering design and an excellent astrodynamical calculation of the conical helix. For both space and time reasons we leave its computation to further research. The idea of a conical helix plus a tether we believe to be sound, but again leave its development to further research.

13.6 OBSERVING EXTRASOLAR PLANETS MAGNIFIED BY VIRTUE OF FOCAL

The most important discovery in astronomy since 1995 is probably the discovery of extrasolar planets. As of February 2012, about 2,000 exoplanets have been discovered from both the ground and space (Corot and Kepler space missions). We thus wish to conclude this chapter by providing an example of how the FOCAL space mission would be able to provide greatly magnified images of extrasolar planets. For instance, consider Gliese 581 e (or Gl 581 e), the fourth extrasolar planet found around Gliese 581, an M3V red dwarf star approximately 20.5 light-years distant from Earth in the constellation of Libra. The planet was discovered by a team at the Observatory of Geneva led by Michel Mayor, using the HARPS instrument on the European Southern Observatory 3.6 m (140 in.) telescope in La Silla, Chile. The discovery was announced on April 21, 2009.

Going back to FOCAL, let us consider (11.22) again; that is:

$$R_{object} = d_{Sun\text{-}object}\theta_{resolution} = d_{Sun\text{-}object}\frac{c^2}{2\pi^2\sqrt{GM_{Sun}}}\frac{1}{\sqrt{z\nu}}. \tag{13.6}$$

This is the linear resolution expected of radio pictures of our extrasolar planet provided by FOCAL. In (13.6) we know:

(a) The distance between the Sun and the target star, $d_{Sun\text{-}object}$, given by the *Hipparcos Catalogue*.

(b) The distance z between the Sun and the FOCAL spacecraft after it reached at least 550 AU from the Sun.
(c) The observing frequency ν, which we can choose at will (with many technological constraints) when we design the FOCAL spacecraft dedicated to observe only that particular extrasolar planet.

The key variable is of course the frequency, and (13.6) neatly shows that the higher the frequency, the smaller (i.e., the better) the linear resolution provided by FOCAL. We believe it important to make this crystal clear.

The interested reader may wish to read more by consulting the author's recent book [19], especially Chapter 9 and Section 9.4, where the Sun's coronal effects are also taken into account.

13.7 CONCLUSION

In these few pages we have painted a picture of the FOCAL space mission to 550 AU and beyond to 1,000 AU. However, a number of issues still have to be investigated:

(1) the many scientific aspects related to the mission;
(2) the propulsion tradeoffs necessary to get there in the shortest possible time; and
(3) optimization of the telecommunication link.

Yet, it plainly appears that the Sun's focus at 550 AU is the next most important milestone for humankind to reach in order to be prepared for the following and more difficult task of achieving interstellar flight.

13.8 REFERENCES

[1] A. Einstein, "Lens-like action of a star by the deviation of light in the gravitational field," *Science*, **84** (1936), 506–507.
[2] S. Liebes, Jr., "Gravitational lenses," *Physical Review*, **133** (1964), B835–B844.
[3] V. Eshleman, "Gravitational lens of the Sun: Its potential for observations and communications over interstellar distances," *Science*, **205** (1979), 1133–1135.
[4] F. Drake, "Stars as gravitational lenses," in G. Marx (Ed.), *Proceedings of the Bioastronomy International Conference, Balatonfüred, Hungary, June 22–27, 1987*, pp. 391–394.
[5] N. Cohen, "The pro's and con's of gravitational lenses in CETI," in G. Marx (Ed.), *Proceedings of the Bioastronomy International Conference, Balatonfüred, Hungary, June 22–27, 1987*, p. 395.
[6] F. Drake and D. Sobel, *Is Anyone Out There?* Delacorte Press, New York, 1992. See pp. 230–234 in particular.
[7] N. Cohen, *Gravity's Lens*, Wiley Science Editions, New York, 1988.
[8] C. Maccone, "Space missions outside the Solar System to exploit the gravitational lens of the Sun," in C. Maccone (Ed.), *Proceedings of the International Conference on Space*

Missions and Astrodynamics, Turin, Italy, June 18, 1992, published in the *Journal of the British Interplanetary Society*, **47** (1994), 45–52.

[9] C. Maccone, "FOCAL: A new space mission to 550 AU to exploit the gravitational lens of the Sun," a proposal for an M3 space mission submitted to the European Space Agency (ESA) on May 20, 1993, on behalf of an international team of scientists and engineers. Later (October 1993) reconsidered by ESA within the Horizon 2000 Plus space missions plan.

[10] J. Heidmann and C. Maccone, "AstroSail and FOCAL: Two extra-Solar System missions to the Sun's gravitational focuses," *Acta Astronautica*, **35** (1994), 409–410.

[11] C. Maccone, "The SETISAIL Project," in G. Seth Shostak (Ed.), *Progress in the Search for Extraterrestrial Life: Proceedings of the 1993 Bioastronomy Symposium, University of California at Santa Cruz, August 16–20, 1993*, published in *Astronomical Society of the Pacific Conference Series*, **74** (1995), 407–417.

[12] C. Maccone, *The Sun as a Gravitational Lens: Proposed Space Missions*, Third Edition, IPI Press, Colorado Springs, CO, 2002.

[13] R. Orta, P. Savi, and R Tascone, "Analysis of gravitational lens antennas," in C. Maccone (Ed.), *Proceedings of the International Conference on Space Missions and Astrodynamics, Turin, Italy, June 18, 1992*, published in *Journal of the British Interplanetary Society*, **47** (1994), 53–56.

[14] John D. Kraus, *Radio Astronomy*, Second Edition, Cygnus-Quasar Books, Powell, OH, 1966, pp. 6-115–6-118.

[15] C. Maccone, "Radioactive decay to propel relativistic interstellar probes along a rectilinear hyperbolic motion (Rindler spacetime)," *Acta Astronautica*, **57** (2005), 59–64.

[16] C. Maccone, *Telecommunications, KLT and Relativity*, Volume 1, IPI Press, Colorado Springs, CO, 1994.

[17] C. Maccone, "Relativistic optimized link by KLT," *Journal of the British Interplanetary Society*, **59** (2006), 94–98.

[18] L. Derosa and C. Maccone, "Propulsion tradeoffs for a mission to Alpha Centauri," *Acta Astronautica*, **60** (2007), 711–718.

[19] C. Maccone, *Deep Space Flight and Communications: Exploiting the Sun as a Gravitational Lens*, Springer/Praxis, Heidelberg, Germany/Chichester, U.K., 2009. A 400-page treatise about the FOCAL space mission that embodies and updates all previously published material about FOCAL.

[20] C. Maccone, "Realistic targets at 1,000 AU for interstellar precursor missions," paper #IAC-09.D4.1.4 presented at the *60th International Astronautical Congress (IAC), Daejeon, Republic of Korea, October 12–16, 2009*. Later published in *Acta Astronautica*, **67** (2010), 526–538.

14

Belt of focal spheres between 550 and 17,000 AU

14.1 INTRODUCTION

The gravitational lens of the Sun has a drawback: the solar corona. This is discussed at length in Chapter 8 of [19] and exemplified by Figure 14.1. Electrons in the corona make electromagnetic waves "diverge" and this "pushes the focus out" to distances greater than 550 AU. For instance, at the CMB peak frequency of 160 GHz, the true focus lies at 763 AU. Thus, it would be safer to position the FOCAL spacecraft at 1,000 AU.

We could get rid of all solar corona-related problems, however, if we could reach the sixfold greater distance of 6,077 AU. This is where the focal sphere of Jupiter lies. Jupiter is the second largest mass in the Solar System after the Sun, but in this focal game not only the mass matters: what really matters is the ratio between the radius of the body squared and the mass of the body. In this regard, Jupiter qualifies as the second best choice for a FOCAL space mission, requiring the FOCAL spacecraft to reach 6,077 AU.

Could any of the other planets be used as gravitational lenses? Neptune qualifies in third place with a focal sphere of 13,520 AU and Saturn comes fourth with a focal sphere of 14,420 AU. But the real surprise is the Earth, which qualifies in fifth place with a focal sphere of 15,370 AU. Moreover, the Earth is indeed the best body we could use as a gravitational lens since we know more about its atmosphere than any other planetary atmosphere.

Just to complete the picture, Uranus comes sixth at 16,980 AU and Venus seventh at 17,020 AU. This leaves Mars, Mercury, and the Moon, all with focal spheres at 40,000 AU and beyond.

We have discovered a new *belt* of focal spheres.

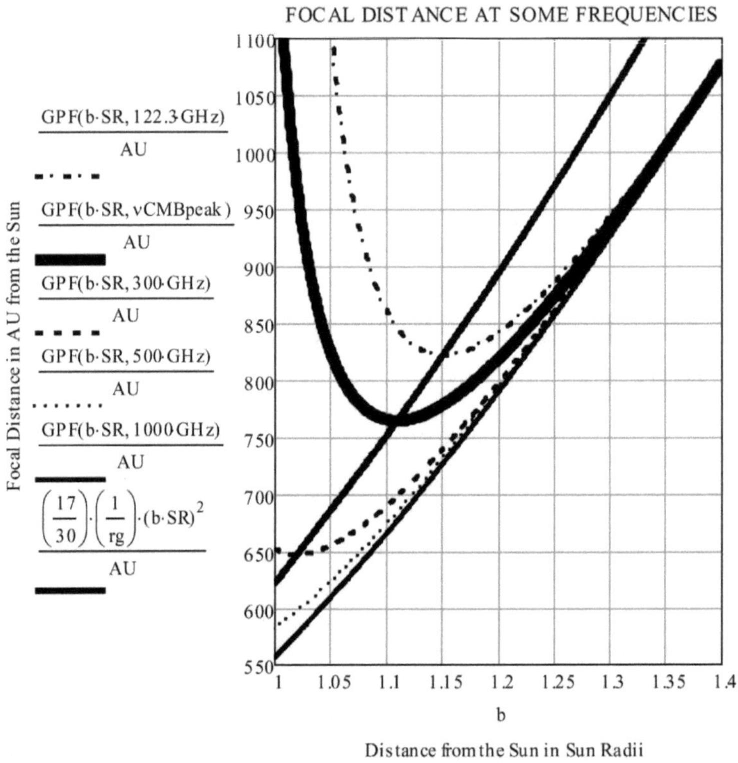

Figure 14.1. How electrons in the Sun's corona "push the true focus out". In this plot we see the (gravity + plasma) Sun's lens focal distance F (in AU) as a function of the impact parameter b (in units of the Sun radius) for all radiowaves focused by the Sun above 120 GHz (plot of equation (8.6-8) in [19]). Impact parameter b is the distance from the Sun center at which radiowaves fly by the Sun and then proceed to focus at distance F from the Sun. Each curve corresponds to radiowaves of a different frequency increasing from 122.3 GHz (top, dash-dot curve) to the CMB peak frequency (thickest solid curve of all) to 1,000 GHz (bottom solid curve). The shifting of the minimum is obvious, and the "locus" of all these minima is just the parabola of equation $F(b) = (17/30) * (1/r_{Schwarzschild}) * (b)^2$ (see eqs. (8.7-9) and (9.4-4) of [19]).

14.2 FROM THE SUN TO THE PLANETS: ALL AS GRAVITATIONAL LENSES

The Sun as a gravitational lens, as we briefly saw at the end of the last section, has the corona problem. This leads us to search for an alternative such as using the planets as gravitational lenses, rather than the Sun, thus avoiding the Sun's corona problem altogether.

The answer to whether this is viable is provided by formula (14.6) yielding the minimal distance d_{Planet} (from the Sun) that the FOCAL spacecraft must reach in order to take advantage of the lensing phenomenon of any given planet.

When rewritten in terms of the planet's radius r_{Planet} and mass M_{planet}, equation (14.6) reads

$$d_{Planet} \approx \frac{c^2}{4G} \cdot \frac{r_{planet}^2}{M_{planet}}. \qquad (14.1)$$

From the point of view of the FOCAL space mission, we can see that what really matters is not the planet radius r_{planet} alone or the planet mass M_{Planet} alone, but rather the ratio of its radius squared to its mass.

In other words, if one inserts the radius squared r_{planet}^2 and mass M_{planet} of each planet into (14.11), one gets for each planet the numerical value of the radius d_{planet} of the relevant planet's focal sphere given by Table 14.1. This table shows the radius of the focal sphere for the naked Sun (i.e., the Sun as if it had *no* corona), the coronal Sun (i.e., the Sun with a conventionally assumed minimal focal distance of 1,000 AU), and all the planets, sorted in the correct order of increasing minimal focal sphere distance from Jupiter onward to Neptune, Saturn, the Earth, Uranus, Venus, Mars, and Mercury in MKS units and light time. The results in Table 14.1 throw up some surprises.

Table 14.1. Radii of the minimal focal spheres of the naked Sun, coronal Sun, Jupiter, Neptune, Saturn, the Earth, Uranus, Venus, Mars, and Mercury in their natural increasing order of distance from the Sun, both in MKS units and in light time (where a light-month is conventionally defined here to be equal to 30 light-days).

Body	*Radius* (km)	*Mass* (kg)	*Focal sphere radius* = $(c^2/(4G))\times$ *(radius²/mass)* (AU)	*Focal sphere radius* = $(c^2/(4G))\times$ *(radius²/mass)*
Naked Sun	6.962×10^5	1.989×10^{30}	548.62 ~ 550	3.169 lt-day
Coronal Sun	?	1.989×10^{30}	1,000 (assumed)	5.778 lt-day
Jupiter	7.159×10^4	1.898×10^{27}	6.077×10^3	1.17 lt-mo
Neptune	2.485×10^4	1.024×10^{26}	1.352×10^4	2.602 lt-mo
Saturn	6.035×10^4	5.684×10^{26}	1.442×10^4	2.777 lt-mo
Earth	6.387×10^3	5.974×10^{24}	1.537×10^4	2.959 lt-mo
Uranus	2.559×10^4	8.683×10^{25}	1.698×10^4	3.269 lt-mo
Venus	6.067×10^3	4.868×10^{24}	1.702×10^4	3.276 lt-mo
Mars	3.404×10^3	6.392×10^{23}	4.081×10^4	7.857 lt-mo
Mercury	2.446×10^3	3.285×10^{23}	4.098×10^4	7.889 lt-mo

After the two focal spheres of the naked Sun at 550 AU and of the coronal Sun conventionally assumed to lie at 1,000 AU (as discussed in the author's book [19]), the Jupiter focal sphere comes next at about 6,100 AU from the Sun (i.e., at about 1.1 light-months). Clearly, this distance would have to be the next target for any FOCAL space mission if we decide that the Sun's corona effects too heavily distort our radio pictures of whatever lies on the other side of the Sun at infinity.

The focal sphere of Neptune comes next at about 13,525 AU from the Sun (i.e., at 2.6 light-months). Again, the fact the Neptune has a smaller focal sphere than all other remaining planets is due to the surprisingly high ratio of the square of its radius to its mass.

Saturn has a focal sphere slightly larger than Neptune's at 14,425 AU (i.e., at 2.777 light-months, even though it looks "bigger" than Neptune in all drawings comparing the sizes of Solar System planets.

But the real surprise is the Earth. In fact, the focal sphere of the Earth turns out to lie at 15,375 AU from the Sun (i.e., about 3 light-months from the Sun). Why is it that Earth beats not only Uranus but all other bodies of the Solar System smaller than itself (like Venus, Mars, and Mercury) as the target for future FOCAL space missions?

Well, the point is that the Earth is the body with the highest density (ratio of mass to volume) in the whole Solar System.

This is far from being insignificant. Having such a high density is the reason the Earth's atmosphere did not "evaporate" into space, thus creating the conditions for life to develop and finally reach the present state of being the only inhabited planet we know.

From the point of view of future FOCAL space missions, there is another basic advantage in reaching the focal sphere of the Earth at 15,500 AU.

We know the Earth's atmosphere better than that of any other planet in the Solar System, of course. So, a FOCAL spacecraft exploiting the Earth as a gravitational lens would operate on much firmer scientific ground than a FOCAL spacecraft exploiting any other planet as a gravitational lens. In fact, to use the same words as *http://en.wikipedia.org/wiki/Planet* "Every planet began its existence in an entirely fluid state; in early formation, the denser, heavier materials sank to the centre, leaving the lighter materials near the surface. Each therefore has a differentiated interior consisting of a dense planetary core surrounded by a mantle which either is or was a fluid. The terrestrial planets are sealed within hard crusts, but in the gas giants the mantle simply dissolves into the upper cloud layers. The terrestrial planets possess cores of magnetic elements such as iron and nickel, and mantles of silicates. Jupiter and Saturn are believed to possess cores of rock and metal surrounded by mantles of metallic hydrogen. Uranus and Neptune, which are smaller, possess rocky cores surrounded by mantles of water, ammonia, methane and other ices. The fluid action within these planets' cores creates a geodynamo that generates a magnetic field."

So, in conclusion, in order to exploit any planet other than the Earth as a gravitational lens, we must estimate the structure of its interior, something that science has done only approximately (as of 2010). The surface and atmosphere of

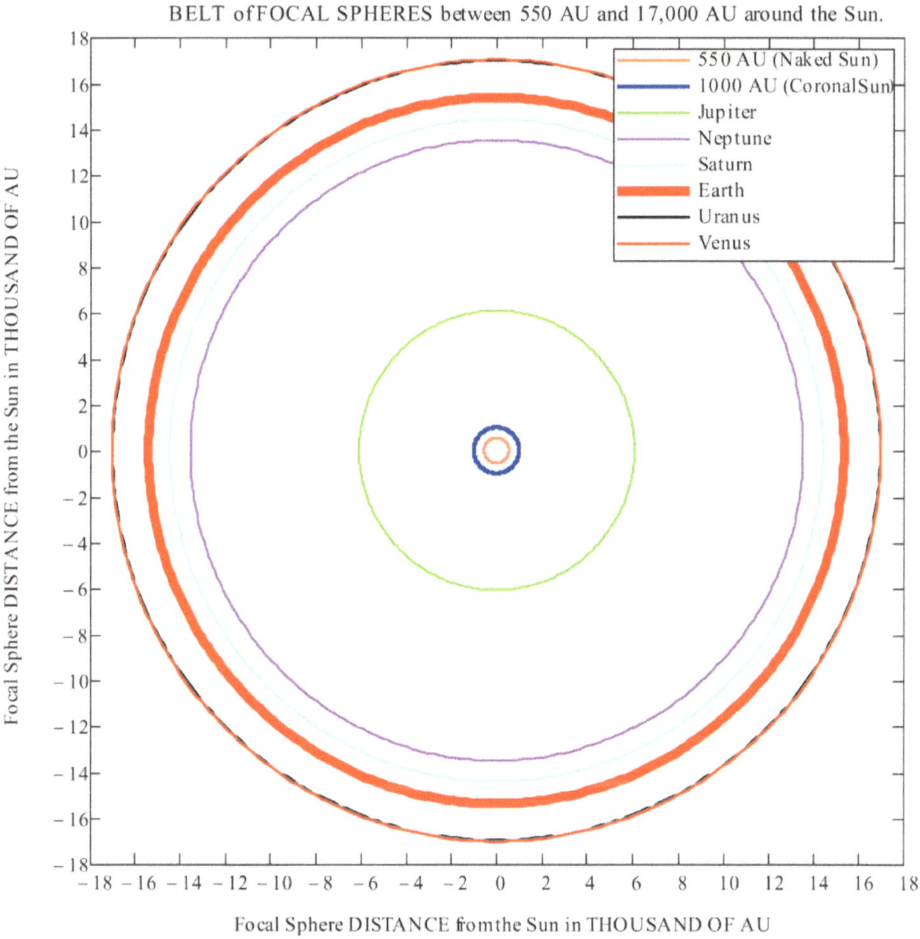

Figure 14.2. The complete *belt* of focal spheres between 550 and 17,000 AU from the Sun, as created by the gravitational lensing effect of the Sun and all planets, shown here to scale. The discovery of this belt of focal spheres, together with the computation of the relevant antenna gains, is the main result put forward in this chapter.

the Earth, on the contrary, are far better known, and that qualifies the focal sphere of the Earth at 15,500 AU as the next realistic target to be reached after, say, 1,000 AU.

Incidentally, 15,500 AU is about a quarter of the width of the Oort Cloud of comets that surround the Sun in all directions (see *http://en.wikipedia.org/wiki/Oort_cloud*). Any FOCAL mission to the Earth's focal sphere would thus automatically imply crossing the inner Oort Cloud, thereby harvesting a multitude of results for cometary science and astrobiology in general.

Figure 14.2 shows the belt of focal spheres around the Sun between 550 and 17,000 AU to scale. The Earth's focal sphere is marked by the thicker solid red circle.

14.3 (ANTENNA) GAIN OF THE GRAVITATIONAL LENSES OF THE SUN AND PLANETS

Having thus determined the minimal distances that the FOCAL spacecraft must reach to exploit both the Sun and all planets as gravitational lenses, we start to question the benefits of going so far out of the Solar System (i.e., what would the degree of light ray focusing be by using such lenses at those minimal distances?). The answer to such a question is provided by the technical notion of antenna gain, simply called "gain" in the remainder of this chapter.

A standard formula in antenna theory relates antenna gain, $G_{antenna}$, to antenna effective area, $A_{effective}$, and to wavelength λ or frequency ν by virtue of the equation (see [14], p. 6-117 and equation (6-241) in particular):

$$d_{focal} \approx \frac{r_{Sun}}{\alpha(r_{Sun})} = \frac{r_{Sun}}{\frac{4GM_{Sun}}{c^2 r_{Sun}}} = \frac{c^2 r_{Sun}^2}{4GM_{Sun}}. \tag{14.2}$$

Now, assume the antenna is circular with radius $r_{antenna}$ and further assume 50% efficiency. Then, the antenna effective area is obviously given by

$$A_{effective} = \frac{A_{physical}}{2} = \frac{\pi\, r_{antenna}^2}{2}. \tag{14.3}$$

Substituting this back into (14.12) yields antenna gain as a function of antenna radius and of observed frequency:

$$G_{antenna} = \frac{4\pi A_{effective}}{\lambda^2} = \frac{2\pi A_{physical}}{\lambda^2} = \frac{2\pi^2 r_{antenna}^2}{\lambda^2} = \frac{2\pi^2 r_{antenna}^2}{c^2} \cdot \nu^2. \tag{14.4}$$

The important point here is that antenna gain increases with the square of frequency, thus favoring observations at frequencies as high as possible. Is anything similar happening for gravitational lenses as well? Yes is the answer, and the gain (one maintains this terminology for convenience) of the gravitational lens of any celestial body can be proved to be

$$G_{body} = 4\pi^2 \frac{r_{Schwaraschild\_of\_the\_body}}{\lambda} \tag{14.5}$$

or, invoking expression (14.7) of the Schwarzschild radius of that body

$$G_{body} = \frac{8\pi^2 G M_{body}}{c^2} \cdot \frac{1}{\lambda} = \frac{8\pi^2 G M_{body}}{c^3} \cdot \nu. \tag{14.6}$$

The mathematical proof of equation (14.15) is difficult to achieve. As already mentioned in Chapter 11 (see p. 268), the author, unsatisfied with the treatment of this key topic in [1], [3], and [13], turned to three engineers at the engineering school in his home town: Renato Orta, Patrizia Savi and Riccardo Tascone. To his surprise, in a few weeks they provided a full proof of not just the celestial body gain formula (14.15), but also of the focal distance for rays originating from a source at a finite distance, equation (14.10). Their proof is fully described in [13] and is based on the

Table 14.2. The *gain* of the Sun's lens alone, the gain of a 12-meter spacecraft (S/C) antenna, and the combined gain of the Sun + S/C antenna system at five selected frequencies important in radioastronomy.

Line	*Neutral hydrogen*		*OH radical*		H_2O
Frequency (ν)	1,420 MHz	327 MHz	1.6 GHz	5 GHz	22 GHz
Wavelength (λ)	21 cm	92 cm	18 cm	6 cm	1.35 cm
S/C antenna beamwidth	1.231 deg	5.348 deg	1.092 deg	0.350 deg	0.080 deg
Sun gain	57.4 dB	51.0 dB	57.9 dB	62.9 dB	69.3 dB
12-meter antenna S/C gain	42.0 dB	29.3 dB	43.1 dB	53.0 dB	65.8 dB
Combined Sun + S/C gain	99.5 dB	80.3 dB	101.0 dB	115.9 dB	135.1 dB

aperture method used to study the propagation of electromagnetic waves, rather than on ray optics (see p. 268 where we quote from the abstract of their paper).

We are now in a position to compute the total gain of the antenna + body system, which is simply obtained by multiplying the two equations yielding spacecraft gain proportional to ν^2 and body gain proportional to ν:

$$G_{total} = G_{body} \cdot G_{antenna} = \frac{16\,\pi^4\, G\, M_{body}\, r^2_{antenna}}{c^5} \cdot \nu^3. \tag{14.7}$$

Since total gain increases with the cube of the observed frequency, it favors electromagnetic radiation in the microwave region of the spectrum. Table 14.2 shows the numerical data provided by the last equation for five selected frequencies: the hydrogen line at 1,420 MHz and the four frequencies that the Quasat radio astronomy satellite planned to observe, had it been built as planned (i.e., jointly by ESA and NASA before 1988), but Quasat was abandoned in 1990 due to lack of funding. The definition of dB is of course:

$$N\ \mathrm{dB} = 10\log_{10} N = \frac{10\ln N}{\ln 10}. \tag{14.8}$$

14.4 KRAUS GAIN VS. DRAKE GAIN: WHICH ONE IS RIGHT?

The definition (14.5) of antenna gain for the gravitational lens of any spherical celestial body is called from now on "Kraus gain":

$$G_{Kraus} = 4\,\pi^2\,\frac{r_{Schwarzschild\_of\_the\_body}}{\lambda}. \tag{14.9}$$

In fact, it appears in the benchmark book *Radio Astronomy* [14] by the late Professor John Daniel Kraus (1910–2004) of Ohio State University (for details about him, see *http://en.wikipedia.org/wiki/John_D._Kraus*). He was the creator of the Big Ear radiotelescope at Columbus, Ohio (see *http://en.wikipedia.org/wiki/Big_Ear*) and an early supporter of SETI. The author had the honor and privilege to meet John Kraus in person at his home on October 16, 1992, just four days after the opening of the NASA SETI program, along with Kraus' pupil Bob Dixon and the then NASA SETI Chief, John Billingham. On that occasion, Kraus presented the author with a free autographed copy of his book, which the author cherishes as the best heritage left by John Kraus to future generations.

But, let us go back to Kraus gain (14.9). Having paid due tribute to the late John Kraus, the author must state that he was never satisfied by the definition of gain in [14] inasmuch as no reasonable proof of (14.15) appears there. It was at this time that the author turned to his colleagues Orta, Tascone, and Savi, who published in 1994 their detailed proof [13], and so the matter seemed to be settled for ever.

But, on February 2, 2010, the author received the following email from Professor Frank Drake:

* *

From: Frank Drake [fdrake@seti.org]
Sent: Tuesday, February 02, 2010 11:31 PM
To: clmaccon@libero.it
Subject: Gain of solar gravitational lens

Hi Claudio:

I have been reading your paper on the radio links enhanced by the solar gravitational lens. Great work! However, I think the sun and other stars produce more gain than stated in equation (14) (I don't know the origin of this equation, so can't analyze it). The gain of an antenna, including the solar lens, is just the ratio of the area of the aperture of the antenna divided by the area of an isotropic radiator of wavelength L (sorry – I don't know how to get greek letters on my computer), which is L squared over 4 pi. The aperture of a gravitational lens is just the area of the Einstein ring. In your Table 14.1, for the 22 Ghz frequency, the gain of the grav. lens is about the same as that of a 12-meter antenna. This implies that the area of the Einstein ring is the same as that of a 12-meter circle. That seems way too small.

To get the area of the Einstein ring one has to find the range of distances from the star from which the radiation diffracted to the focus travels along path lengths which differ by less than a half wavelength. This is certainly many, many wavelengths.

I think equation (14) could be the Voltage gain of the antenna, and not the power gain. It is the power gain which matters, of course. I suspect that the right gain equation has the Schwarzchild radius squared, and the gain equation may be proportional to Sch radius squared/ wavelength squared. This will increase the gain value, say at 22 Ghz, by a factor of a million, which, of course, is a very good thing!

Frank

* *

The author's paper that Drake is referring to is entitled "Interstellar radio links enhanced by exploiting the Sun as a gravitational lens" and was presented by the author at the *60th International Astronautical Congress* held at Daejeon, Republic of Korea, October 12–16, 2009 [20].

Drake's email to the author of February 2, 2010 clearly pointed out that the new Drake gain was to be proportional to the square of the ratio of the Schwarzschild radius to the wavelength, but it did not state any mathematical expression for the missing multiplicative constant in front of such a squared ratio. During his next visit to the SETI Institute of July 16-19, the author personally asked Drake for the expression of this constant, and Drake said "four pi". Thus, in conclusion, from now on we shall call the expression

$$G_{Drake} = 4\pi \left(\frac{r_{Schwarzschild\_of\_the\_body}}{\lambda}\right)^2, \tag{14.10}$$

the Drake gain of the gravitational lens of any spherical celestial body.

In this chapter we shall not argue whether the Drake gain is "right or wrong". We will simply compare the numerical values of the Drake and Kraus gains to see which one is larger, and how much larger, for any given frequency. Were we living in the Middle Ages, we might invoke the Principle of Authority: *Quod Drake dixit, legis habet vigorem* (What Drake said has the force of law).

14.5 DRAKE GAIN VS. KRAUS GAIN FOR THE (NAKED) SUN, JUPITER, AND EARTH

The two definitions of Kraus gain and Drake gain will now be compared numerically to get afeeling of what numbers they really imply in practice. The author wrote a MathCad code for this purpose and the resulting plots are presented in this section. The first step, however, is to convert both (14.19) and (14.20) from wavelengths to frequencies and take into account the definition (14.7) of the Schwarzschild radius. Thus, (14.19) and (14.20) become, respectively:

$$G_{Kraus}(\nu) = \frac{8\pi^2 G\, M_{body}}{c^3} \cdot \nu \tag{14.11}$$

and

$$G_{Drake}(\nu) = \frac{16\,\pi^2 G^2 M_{body}^2}{c^6} \cdot \nu^2. \tag{14.12}$$

In the plane where the horizontal axis is frequency ν, these two curves are a straight line and a parabola, respectively. So, they intersect both at the origin (of no interest to us) and at another point given by equating the two right-hand sides of (14.21) and (14.22) and then solving for ν; that is:

$$\nu_{\text{intercept}} = \frac{\pi\, c^3}{2\, G\, M_{body}}. \tag{14.13}$$

For $\nu < \nu_{\text{intercept}}$ the Kraus gain is higher than the Drake gain, whereas for $\nu > \nu_{\text{intercept}}$ the Drake gain is higher than the Kraus gain. For $\nu = \nu_{\text{intercept}}$ the common value of the two gains is $4\pi^3 = 124.025$, which amounts to 20.935 dB. Thus, we may also conclude that, for gains higher than about 21 dB, the Drake gain is (much) higher than the Kraus gain and this case is, of course, of real interest for all future FOCAL space missions.

Let us next evaluate (14.23) for the three cases of the (naked) Sun, Jupiter, and the Earth. The three intercept frequencies are, respectively:

$$\nu_{\text{intercept\_Sun}} = \frac{\pi c^3}{2 G M_{Sun}} = 318.961 \text{ KHz} \tag{14.14}$$

$$\nu_{\text{intercept\_Jupiter}} = \frac{\pi c^3}{2 G M_{Jupiter}} = 334.116 \text{ MHz} \tag{14.15}$$

$$\nu_{\text{intercept\_Earth}} = \frac{\pi c^3}{2 G M_{Earth}} = 106.182 \text{ GHz.} \tag{14.16}$$

The following three plots show both the Drake and the Kraus gains (in decibels) for all frequencies up to 200 GHz. Figure 14.3 shows the case for the (naked) Sun, whereas Figures 14.4 and 14.5 do the same for Jupiter and the Earth, respectively.

Figures 14.3–14.5 can now be superimposed and the frequency axis can be set to a logarithmic scale to change all previous curves into straight lines. The result is Figure 14.6, enabling direct comparison between all three gravitational lenses for frequencies higher than 1 GHz.

Although Figure 14.6 clearly suggests the approximated numerical values of

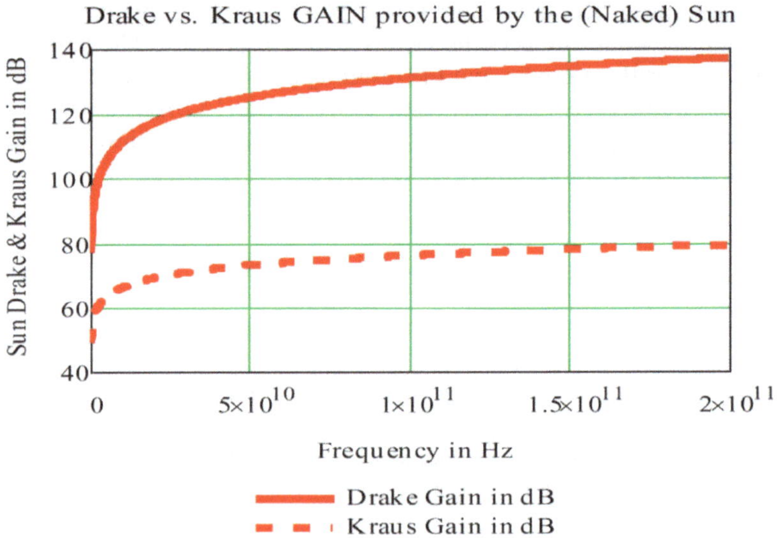

Figure 14.3. The Sun as a gravitational lens: Drake and Kraus gains (in dB) for all frequencies up to 200 GHz.

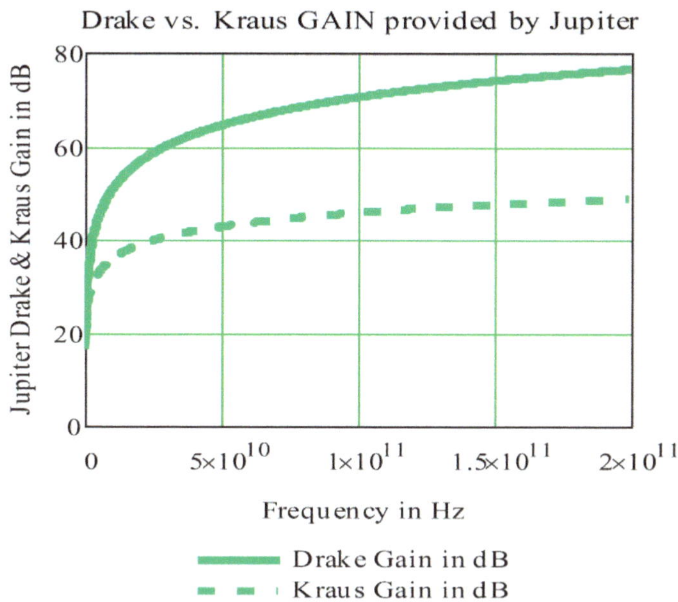

Figure 14.4. Jupiter as a gravitational lens: Drake and Kraus gains (in dB) for all frequencies up to 200 GHz.

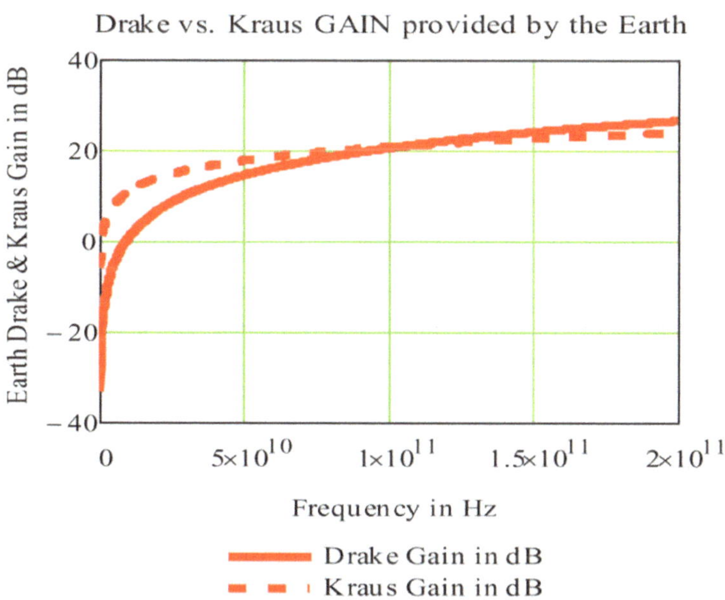

Figure 14.5. The Earth as a gravitational lens: Drake and Kraus gains for all frequencies up to 200 GHz.

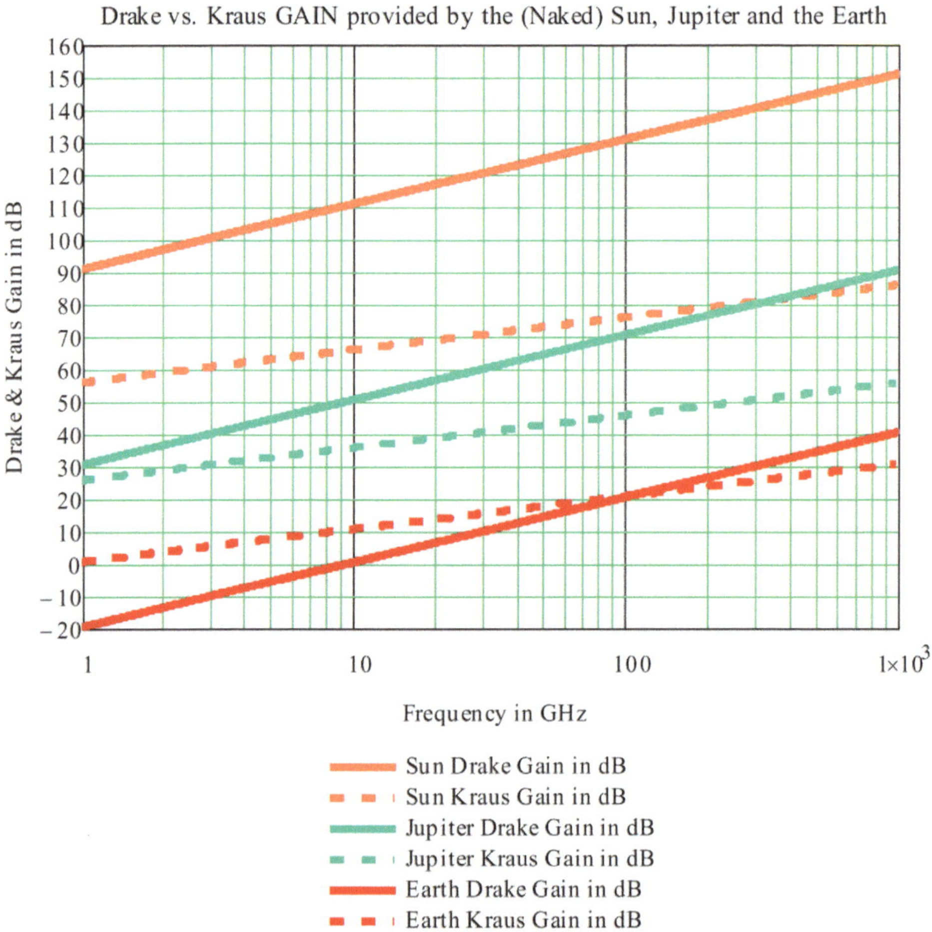

Figure 14.6. Drake and Kraus gains of the (naked) Sun, Jupiter, and the Earth compared.

both the Drake and Kraus gains for all frequencies higher than 1 GHz, Table 14.3 shows the actual numerical values for both the Drake and Kraus gains at five selected frequencies of great importance in astrophysics and cosmology.

All the gain values reported in Table 14.3 clearly show the advantage of the Sun over both Jupiter and the Earth as a gravitational lens:

(1) The minimal focal distance of the Sun (at 550 AU or 1,000 AU) is so much closer than the minimal focal distance of Jupiter at 6,100 AU.
(2) The gain of the Sun (whether Drake of Kraus) is so much higher than the gain of Jupiter (not to mention the miserable gain of the Earth).

Clearly, we must redouble our efforts and try harder with the Sun as the first target for the first FOCAL mission ever. Period.

Table 14.3. Drake and Kraus antenna gains at selected ν for the (naked) Sun, Jupiter, and the Earth as gravitational lenses.

Body gain in decibels	*Hydrogen line* 1.420 GHz	*Water maser line* 22 GHz	*Ka band* 32 GHz	*CMB peak* 160 GHz	*Positronium line* 203 GHz
(Naked) Sun Drake gain	94	118	121	135	137
(Naked) Sun Kraus gain	57	69	71	78	79
Jupiter Drake gain	33	57	60	74	77
Jupiter Kraus gain	27	39	41	48	49
Earth Drake gain	−16	7	10	24	26
Earth Kraus gain	2	14	16	23	24

14.6 USING THE LENSES OF ALL PLANETS TO SWEEP ACROSS THE SKY IN SEARCH OF EXOPLANETS

Let us leave our discussion of FOCAL for a moment and consider instead truly interstellar flight to the nearest star, Alpha Centauri, at a distance of 4.37 light-years (i.e., 276,358 AU).

Several serious scientific studies about a truly interstellar mission have appeared in the literature over the last few decades. They all overlook the bonanza provided by the full belt of focal spheres between 550 and 17,000 AU that the spacecraft must cross in any event on its way to Alpha Centauri.

So, in the last section of this introductory chapter to the belt of focal spheres between 550 and 17,000 AU, we wish to devote some time to considering what can be obtained while crossing our belt of focal spheres.

First of all, while the Sun does not move in the Sun-centered reference frame of the Solar System, all the planets do move. This means that they actually sweep a certain area of the sky, as seen from the spacecraft, such that the spacecraft would enjoys a sort of moving magnifying lens. How many extrasolar planets would fall inside this moving magnifying lens? Well, we don't know precisely, of course, but the over 1,000 exoplanets found to date are a sure sign that many more such exoplanets will be detected by a suitably equipped spacecraft crossing the distances between 550 and 17,000 AU from the Sun thanks to the gravitational lenses of the planets.

Second, apart from finding exoplanets, the moving and magnifying lenses of the planets could also lead to the discovery of many more galactic and extragalactic objects of all kinds.

Finally, direct exploration of the Kuiper Belt and Oort Cloud cannot fail to increase scientific interest in such a crossing of the belt of focal spheres, in ways that are currently premature to describe in detail but will certainly be worked out by the coming generations of space scientists (just as we have done here for the preliminary concepts of lens magnification).

In conclusion, looking back at the work done thus far into the possibility of truly interstellar flight, it seems fair to say that planners of Alpha Centauri missions, in their efforts to reach 277,000 AU, have missed what was at hand at just 17,000 AU. Or, in terms of light time, to reach 4.37 light-years in a single shot, they missed what was just 3 light-months away (like the Earth's focal sphere).

After the discovery of the belt of planetary focal spheres first put forward in this chapter, a revision of theoretical flight plans to Alpha Centauri seems to be desirable.

14.7 CONCLUSIONS

We have discovered the belt of focal spheres created by the gravitational lenses of the planets at distances between 6,100 and 17,000 AU from the Sun in all directions. We have also computed the relevant antenna gains (according to both the equations given by Kraus and later by Drake). Finally, we have pointed out the potential harvest of a multitude of discoveries that moving planet lenses would lead to: from exoplanets to both Galactic and extragalactic objects. Hence, a reconsideration of crossing the distances between 550 and 17,000 AU for all future deep space missions must be in order.

14.8 REFERENCES

[1] A. Einstein, "Lens-like action of a star by the deviation of light in the gravitational field," *Science*, **84** (1936), 506–507.

[2] S. Liebes, Jr., "Gravitational Lenses," *Physical Review*, **133** (1964), B835–B844.

[3] V. Eshleman, "Gravitational lens of the Sun: Its potential for observations and communications over interstellar distances," *Science*, **205** (1979), 1133–1135.

[4] F. Drake, "Stars as gravitational lenses," in G. Marx (Ed.), *Proceedings of the Bioastronomy International Conference, Balatonfüred, Hungary, June 22–27, 1987*, pp. 391–394.

[5] N. Cohen, "The pro's and con's of gravitational lenses in CETI," in G. Marx (Ed.), *Proceedings of the Bioastronomy International Conference, Balatonfüred, Hungary, June 22–27, 1987*, p. 395.

[6] F. Drake and D. Sobel, *Is Anyone Out There?* Delacorte Press, New York, 1992. See pp. 230–234 in particular.

[7] N. Cohen, *Gravity's Lens*, Wiley Science Editions, New York, 1988.

[8] C. Maccone, "Space missions outside the Solar System to exploit the gravitational lens of the Sun," in C. Maccone (Ed.), *Proceedings of the International Conference on Space Missions and Astrodynamics, Turin, Italy, June 18, 1992*, published in *Journal of the British Interplanetary Society*, **47** (1994), 45–52.

[9] C. Maccone, "FOCAL, A new space Mission to 550 AU to exploit the gravitational lens of the Sun," a proposal for an M3 space mission submitted to the European Space Agency (ESA) on May 20, 1993, on behalf of an international team of scientists and

engineers. Later (October 1993) reconsidered by ESA within the Horizon 2000 Plus space missions plan.
[10] J. Heidmann and C. Maccone, "AstroSail and FOCAL: Two extra-Solar System missions to the Sun's gravitational focuses," *Acta Astronautica*, **35** (1994), 409–410.
[11] C. Maccone, "The SETISAIL Project," in G. Seth Shostak (Ed.), *Progress in the Search for Extraterrestrial Life: Proceedings of the 1993 Bioastronomy Symposium, University of California at Santa Cruz, August 16–20, 1993*, published in *Astronomical Society of the Pacific Conference Series*, **74** (1995), 407–417.
[12] C. Maccone, *The Sun as a Gravitational Lens: Proposed Space Missions*, Third Edition, IPI Press, Colorado Springs, CO, 2002.
[13] R. Orta, P. Savi, and R Tascone, "Analysis of gravitational lens antennas," in C. Maccone (Ed.), *Proceedings of the International Conference on Space Missions and Astrodynamics, Turin, Italy, June 18, 1992*, published in *Journal of the British Interplanetary Society*, **47** (1994), 53–56.
[14] John D. Kraus, *Radio Astronomy*, Second Edition, Cygnus-Quasar Books, Powell, OH, 1966, pp. 6-115–6-118.
[15] C. Maccone, "Radioactive decay to propel relativistic interstellar probes along a rectilinear hyperbolic motion (Rindler spacetime)," *Acta Astronautica*, **57** (2005), 59–64.
[16] C. Maccone, *Telecommunications, KLT and Relativity*, Volume 1, IPI Press, Colorado Springs, CO, 1994.
[17] C. Maccone, "Relativistic optimized link by KLT," *Journal of the British Interplanetary Society*, **59** (2006), 94–98.
[18] L. Derosa and C. Maccone, "Propulsion tradeoffs for a mission to Alpha Centauri," *Acta Astronautica*, **60** (2007), 711–718.
[19] C. Maccone, *Deep Space Flight and Communications: Exploiting the Sun as a Gravitational Lens*, Springer/Praxis, Heidelberg, Germany/Chichester, U.K. A 400-page treatise about the FOCAL space mission that embodies and updates all previously published material about FOCAL.
[20] C. Maccone, "Interstellar radio links enhanced by exploiting the Sun as a gravitational lens", paper #IAC-09.D4.1.8 presented at the *49th International Astronautical Congress (IAC), Daejeon, Republic of Korea, October 12–16, 2009*. Later published in *Acta Astronautica*, **68** (2011), 76–84.
[21] C. Maccone, "A new belt beyond Kuiper's: A belt of focal spheres between 550 and 17,000 AU for SETI and science," paper #IAC-10.A4.1.6 presented at the *61st International Astronautical Congress (IAC), Prague, Czech Republic, September 27–October 1, 2010. Later published in Acta Astronautica*, **69** (2011), 939–948.

15

Galactic Internet by star gravitational lensing

15.1 INTRODUCTION

The important question we consider in this chapter is how to *insure* any future interstellar radio links between the Solar System and future interstellar probes.

We show by calculations that only by exploiting the gravitational lens of the Sun as a huge antenna will we be able to achieve such an interstellar link. In other words, we calculate the bit error rate (BER) across interstellar distances both with and without using the gravitational lens effect of the Sun. The conclusion is that only when we exploit the Sun as a gravitational lens will we be able to communicate with our own probes (or with nearby aliens) across the distances of even the nearest stars in the Galaxy,and do so at a reasonable BER.

In addition, we explore the possibility that *radio bridges* between any pair of stars in the Galaxy can be built if the gravitational lenses of both stars are exploited by placing two FOCAL relay satellites on the opposite side of each star, so that a perfect (or nearly perfect) alignment between these four points in space is kept. Only now (2011) are humans realizing that radio bridges between couples of stars in the Galaxy can indeed be constructed.

But, then, some other extraterrestrial civilization in the Galaxy might already have understood this a long time ago. Consequently, a sort of *Galactic Internet* might already be in use in the Galaxy *now*! But humans will be unable to benefit from this Galactic Internet until they reach the minimal focal sphere of their own star, the Sun (i.e., until a human FOCAL probe reaches 550 AU or more).

15.2 THE RADIO LINK

In two recent papers [1, 2], the author proved that radio communications between any pair of stars within our Galaxy are feasible with modest transmitted powers if the gravitational lenses of both stars are exploited. In this introductory section we review the mathematical theory set up in [1] and [2] and get ready to apply it to intergalactic distances rather than just to interstellar distances.

Consider a radio transmitter that radiates a power P_t isotropically and uniformly over a bandwidth B_t. Then, at a distance r it produces a flux density given by

$$\frac{P_t}{B_t\, 4\pi r^2}. \tag{15.1}$$

A receiving antenna of effective aperture A_{er} at a distance r can collect a power given by (15.1) multiplied by both the effective aperture of the receiving antenna and its bandwidth; namely, the received power P_r is given by

$$P_r = \frac{P_t}{B_t\, 4\pi r^2} A_{er}\, B_r. \tag{15.2}$$

It is assumed that the receiving bandwidth B_r is smaller or, at best (in the matched bandwidths case) equal to the transmitting bandwidth B_t (i.e., $B_r \le B_t$).

So far, we have been talking about an isotropic radiator. But let us now assume that the transmitting antenna has directivity D (see, e.g., *http://en.wikipedia.org/wiki/Directivity* or Chapter 1 of [3]), which is an antenna gain:

$$D = \frac{4\pi\, A_{et}}{\lambda^2}. \tag{15.3}$$

The received power P_r is then increased by just such a factor due to the directivity of the transmitting antenna, and so (15.2) must now be replaced by a new equation where the right-hand side is multiplied by such an increased factor; that is

$$P_r = \frac{4\pi\, A_{et}}{\lambda^2} \cdot \frac{P_t}{B_t\, 4\pi r^2} A_{er}\, B_r. \tag{15.4}$$

Rearranging a little, this becomes

$$P_r = \frac{P_t\, A_{et}\, A_{er}}{r^2\, \lambda^2} \cdot \frac{B_r}{B_t}. \tag{15.5}$$

This is the received signal power expression. For the matched bandwidths case (i.e., for $B_r = B_t$), this is called the "Friis transmission formula", since it was first published back in 1946 by the American radio engineer Harald T. Friis (1893–1976) of Bell Labs. In space missions, we know both B_t and B_r exactly of course and so we may construct our spacecraft so that the two bands match exactly (i.e., $B_r = B_t$). Thus, for the case of telecommunications with a spacecraft (but not necessarily for the SETI case) we may well assume the matched bandwidths have (15.5) reducing to

$$P_r = \frac{P_t\, A_{et}\, A_{er}}{r^2\, \lambda^2}. \tag{15.6}$$

Let us now rewrite (15.6) in such a way that we can take into account the gains (i.e., directionalities) of both the transmitting and receiving antennas (i.e., in agreement with (15.5))

$$\begin{cases} G_t = \dfrac{4\pi A_{et}}{\lambda^2} \\ G_r = \dfrac{4\pi A_{er}}{\lambda^2} \end{cases} \quad \text{that is} \quad \begin{cases} A_{et} = \dfrac{G_t \lambda^2}{4\pi} \\ A_{er} = \dfrac{G_r \lambda^2}{4\pi} \end{cases}. \tag{15.7}$$

Inserting the last two expressions into (15.6), we find that (15.6) is turned into

$$P_r = \frac{P_t G_t G_r}{(4\pi)^2 r^2} \cdot \lambda^2. \tag{15.8}$$

This may finally be rewritten in the more traditional form

$$P_r = \frac{P_t G_t G_r}{L(r, \lambda)} \tag{15.9}$$

if one defines

$$L(r, \lambda) = (4\pi)^2 \cdot \frac{r^2}{\lambda^2}, \tag{15.10}$$

which is path loss or path attenuation (i.e., the reduction in power density (attenuation) of electromagnetic waves as they propagate through space). Path loss is a major component in the analysis and design of the link budget of a telecommunication system (see, e.g., *http://www.en.wikipedia.org/wiki/Path_loss*).

Next we define the bit error rate (BER). Then, by virtue of a numerical example, we show that, even at the distance of the nearest star (Alpha Centauri at 4.37 lt-yr) telecommunications would be impossible using the power available to us today for interplanetary space flight. In the next section we shall show that telecommunications would however become feasible if we could take advantage of the magnification provided by the Sun's gravity lens (i.e., if we sent out a FOCAL relay spacecraft to 550 AU; see, e.g., [3] and [4]) for each target star system that we wish to communicate with. This was the key new result presented in [1].

So, let us start by defining the bit error rate or BER. In telecommunication theory an error ratio is the ratio of the number of bits, elements, characters, or blocks incorrectly received to the total number of bits, elements, characters, or blocks sent during a specified time interval. Among these error ratios, the most commonly encountered ratio is the bit error ratio (BER), whish is also called the "bit error rate" (i.e., the number of erroneous bits received divided by the total number of bits transmitted). At *http://www.en.wikipedia.org/wiki/Bit_error_rate* it is shown that the likelihood of bit misinterpretation is

$$p_e = p(0|1)p_1 + p(1|0)p_0 \tag{15.11}$$

(believing that we have received a 0 while it was a 1 or the other way round) is basically given by the complementary error function or $erfc(x)$ as follows

$$BER(d, \nu, P_t) = \frac{1}{2} erfc\left(\sqrt{\frac{E_b(d, \nu, P_t)}{N_0}}\right). \tag{15.12}$$

In this equation one has:

(1) d = distance between the transmitting station on Earth and the receiving antenna in space. For instance, this could be the antenna of a precursor interstellar space probe that was sent out to a distance of a few light-years.
(2) ν = frequency of the electromagnetic waves used in the telecommunication link. The higher this frequency the better it is, since photons are then more energetic ($E = h\nu$).However, the highest ν for spacecraft links today (like the link of the Cassini probe, now at Saturn) are found in the Ka band (i.e., $\nu_{\mathrm{Ka}} \approx 32$ GHz).
(3) P_t is the power in watts transmitted by the Earth antenna, typically a NASA Deep Space Network antenna 70 meters in diameter.
(4) The complementary error function $ercf(x)$ is defined by the integral

$$erfc(x) = \frac{2}{\sqrt{\pi}} \int_x^{\infty} e^{-t^2} dt \tag{15.13}$$

(for more mathematics, see *http://www.en.wikipedia.org/wiki/Complementary_error_function*).
(5) $E_b(d, \nu, P_t)$ is the received energy per bit; that is, the ratio

$$E_b(d, \nu, P_t) = \frac{P_r(d, \nu, P_t)}{\mathrm{Bit\_rate}}. \tag{15.14}$$

(6) Finally, N_0 is given by the Boltzmann constant k multiplied by the noise temperature of space far away from the Sun and from any other star. This "empty space noise temperature" might be assumed to equal, say, 100 K.

This is the analytical structure of the MathCad code that the author wrote to yield the BER. Let us now consider the input values that he used:

(1) Suppose that a human space probe has reached the Alpha Centauri system at the 4.37 light-year distance from the Sun: then, $d = 4.37$ light-years.
(2) Suppose also that the transmitting antenna from the Earth is a typical NASA Deep Space Network (DSN) antenna having a diameter of 70 meters (like those at Goldstone, Madrid, and Canberra), and assume that its efficiency is about 50%.
(3) Suppose that the receiving antenna aboard the spacecraft is 12 meters in diameter (it might be an inflatable space antenna, as we supposed in [3] for the FOCAL spacecraft) and assume 50% efficiency.
(4) Suppose that the link frequency is the Ka band (i.e., 32 GHz), as for the Cassini highest frequency.
(5) Suppose that the bit rate is 32 kbps = 32,000 bit/second. This is the bit rate of ESA's Rosetta interplanetary spacecraft now on its way to a comet.
(6) Finally (this is the most important input assumption), suppose that the transmitting power P_t is moderate: just 40 watts.

Then:

(1) The gain of the transmitting NASA DSN antenna (at this Ka frequency) is about 84 dB.
(2) The gain of the spacecraft antenna is about 69 dB.
(3) Path loss at the distance of Alpha Centauri is 395 dB (a very high indeed path loss with respect to today's interplanetary missions, of course).
(4) The power received by the spacecraft at that distance is 2.90×10^{-23} watt.
(5) The received energy per bit (lowered by the noise temperature of space in between the Sun and Alpha Centauri) is 1.3×10^{-37} joule.
(6) Finally, the BER is 0.49 (i.e., there is a 50% probability of *errors* in the telecommunications between the Earth and the probe at Alpha Centauri) if we use such a small transmitting power!

In other words, if these are the telecommunication links between the Earth and our probe at Alpha Centauri, then this precursor interstellar mission is worthless. See Figures 15.1, 15.2, and 15.3 for the relevant BER plots without and with the Sun as a magnifying gravitational lens

The key point in this example is that, for all calculations, (15.8) and (15.9) were used *without taking the gain of the Sun gravity lens into account*, because this was a *direct* link and *not* a FOCAL mission.

15.3 BIT ERROR RATE AT THE ALPHA CENTAURI DISTANCE ENHANCED BY THE MAGNIFICATION PROVIDED BY THE SUN'S GRAVITY LENS AND THE FOCAL SPACE MISSION

The disappointing BER results of the previous section are totally reversed, however, if we suppose that a FOCAL space mission has been previously sent out to 550 AU in the direction opposite to Alpha Centauri. These topics were fully described by the author in his recent paper [21] and at length in his recent book [19], so we will not repeat them here.

We now have the *magnification* provided by the Sun's gravity lens taking part in the game. Mathematically, this means that we must introduce a third multiplicative gain at the numerator of (15.8). This is the Sun's gravity lens *gain* given by (15.8) of [1]; that is

$$G_{Sun} = 4\pi^2 \frac{r_{Schwarzschild\_of\_Sun}}{\lambda} \tag{15.15}$$

where the Schwarzschild radius of the Sun is given by

$$r_{Schwarzschild\_of\_Sun} = \frac{2\,G\,M_{Sun}}{c^2}. \tag{15.16}$$

This new gain is huge at the Ka band frequency:

$$G_{Sun}(\nu_{Ka}) = 12{,}444{,}837 \approx 70\text{ dB} \tag{15.17}$$

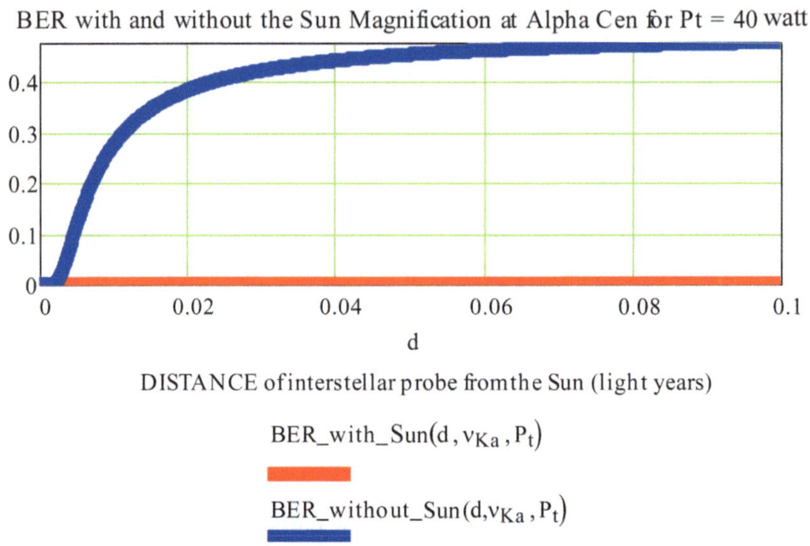

Figure 15.1. The bit error rate (BER) (upper blue curve) tends immediately to the 50% value (BER = 0.5) even at moderate distances from the Sun (0 to 0.1 lt-yr) for a 40-watt transmission from a DSN antenna that is a *direct* transmission (i.e., without using the Sun's magnifying lens). On the contrary (lower red curve) the BER keeps at zero (perfect communications!) if the FOCAL space mission is made, taking advantage of the Sun's magnifying action.

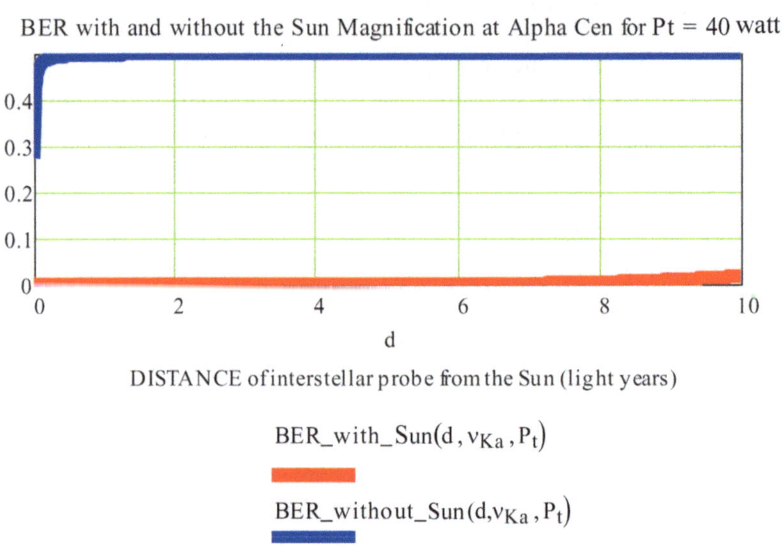

Figure 15.2. Same as Figure 15.1, but for probe distances up to 10 light-years. We see that at about 9 light-years the BER curve no longer remains flat and starts increasing slowly.

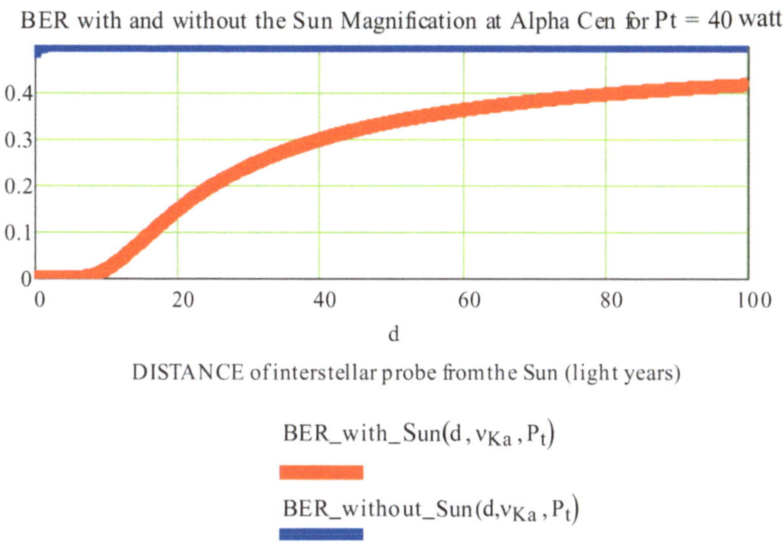

Figure 15.3. Same as Figure 15.2, but for probe distances up to 100 light-years. We see that the Sun–BER increases from 9 light-years on, reaching the dangerous level of 40% (Sun–BER = 0.4) at about 100 light-years. So, at 100 light-years even the Sun's lens cannot cope with this very low transmitted power of 40 watts.

and so the received power (15.9) at Alpha Centauri, with the usual Earth-transmitted power of just 40 watts becomes

$$P_r = 2.9 \cdot 10^{-23} \text{ watts} \tag{15.18}$$

and the relevant BER becomes absolutely acceptable:

$$\text{BER} = 0.000\,000\,526\,387\,845. \tag{15.19}$$

This should convince any doubters that the FOCAL space mission is indispensable to keep the link at interstellar distances equal to or greater than Alpha Centauri.

15.4 THE RADIO BRIDGE BETWEEN THE SUN AND α Cen A USING THEIR GRAVITATIONAL LENSES

In this section we provide another new result: we define the radio bridge between the Sun and α Cen A by using *both* gravitational lenses!

In other words, suppose that in future we will be able to send a probe to α Cen A and suppose that we succeed in placing this probe just on the other side of α Cen A with respect to the Sun and at the minimal focal distance typical of α Cen A. This distance is *not* 550 AU obviously because both the radius and the mass of α Cen A are different (actually slightly higher) than the values of the Sun:

$$\begin{cases} r_{\alpha\_Cen\_A} = 1.227\ r_{Sun} \\ M_{\alpha\_Cen\_A} = 1.100\ M_{Sun}. \end{cases} \tag{15.20}$$

Replacing these values into the minimal focal distance (see eq. (1) of [1]) (obviously rewritten for α Cen A), the relevant minimal focal distance is found

$$d_{focal\_\alpha\_Cen\_A} \approx \frac{c^2 r^2_{\alpha\_Cen\_A}}{4GM_{\alpha\_Cen\_A}} \approx 749\ \text{AU}. \tag{15.21}$$

The Schwarzschild radius for α Cen A is given by

$$r_{Schwarzschild\_\alpha\_Cen\_A} = \frac{2\,G\,M_{\alpha\_Cen\_A}}{c^2} = 3.248\ \text{km}. \tag{15.22}$$

And so the gain, provided by (15.15), turns out to equal

$$G_{\alpha\_Cen\_A}(\nu_{Ka}) = 4\pi^2\ \frac{r_{Schwarzschild\_\alpha\_Cen\_A}}{\lambda_{Ka}} = 13{,}689{,}321 \approx 71\ \text{dB}. \tag{15.23}$$

Incidentally, we chose α Cen A, and not B or C, because it has the highest mass, and so the highest gain, in the whole Alpha Centauri triple system. Future telecommunications between the Sun and the Alpha Centauri system are thus optimized by selecting α Cen A as the star on the other side of which to place a FOCAL spacecraft at the minimal distance of 750 AU. The FOCAL spacecraft would then easily relay its data anywhere within the Alpha Centauri system.

Having found the α Cen A gain (15.22) we are now able to write the new equation corresponding to (15.8) for the Sun–Alpha Centauri bridge. In fact, we must now put at the numerator of (15.8) three gains:

(1) the Sun gain at 32 GHz;
(2) the α Cen A gain at 32 GHz; and
(3) the 12-meter FOCAL antenna gain at 32 GHz raised to the square because there are two such 12-meter antennas: one at 550 AU from the Sun and one at 749 AU from α Cen A, and they must be perfectly aligned with the axis passing through both the Sun and α Cen A.

Thus, the received power given by (15.8) now reads

$$P_r = \frac{P_t\, G_{Sun}\, G_{\alpha\_Cen\_A}\left(G_{12\_meter\_antenna\_at\_Ka}\right)^2}{L(r,\lambda)}, \tag{15.24}$$

where obviously r equals 4.37 light-years and λ corresponds to a 32 GHz frequency.

Let us now go back to the BER and replace (15.8) with (15.24) in the long chain of calculations described in Section 15.1. Since the received power P_r has now changed, clearly both (15.12) and (15.14) yield different numerical results. But now:

(1) The link frequency has been fixed at 32 GHz (Ka band), and so no longer is an independent variable in the game.

(2) Also the distance d has been fixed (it is the distance of α Cen A) and so is no longer an independent variable in the game.
(3) It follows that, in (15.14) and (15.12), the only variable that is free to vary is now the transmitted power, P_t.

Let us rephrase the last sentence in different terms. Practically, we are now studying the BER as a function of transmitted power P_t only and, physically, this means that:

(a) We start by inputting very low transmission powers in watts and discover that the BER is an awful 50% (i.e., telecommunications between the Sun and α Cen are totally disrupted). This is clearly because the energy per bit is so much lower than the empty space noise temperature.
(b) We then increase the transmitted power and at a certain point the BER starts getting smaller than 50%. What is more, it gets smaller and smaller until the transmitted power is so high that the BER drops to zero and telecommunications are just perfect.
(c) This is a surprising revelation for the Sun–Alpha Centauri direct radio bridge exploiting both gravitational lenses, this minimum transmitted power is incredibly small! It actually equals less than 10^{-4} watts (i.e., one tenth of a milliwatt is enough to have perfect communications between the Sun and Alpha Centauri through two 12-meter FOCAL spacecraft antennas) (see Figure 15.4). How is that possible?

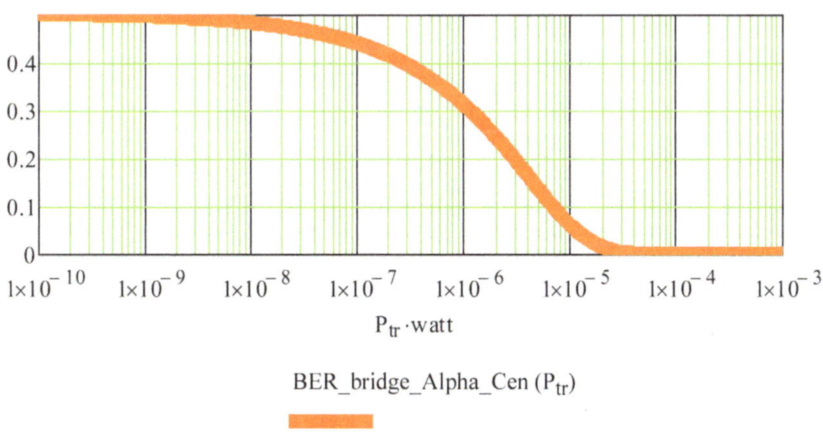

Figure 15.4. Bit error rate (BER) for the double-gravitational-lens radio bridge between the Sun and α Cen A. In other words, there are two gravitational lenses at play here, the Sun's and α Cen A's, and two 12-meter FOCAL spacecraft are supposed to have been put along the two-star axis on opposite sides at or beyond the minimal focal distances of 550 AU and 749 AU, respectively. This radio bridge has an *overall gain so high* that a miserable 10^{-4} watt transmitting power is sufficient to let the BER drop to zero (i.e., for perfect telecommunications). Note also that the scale of the horizontal axis is logarithmic.

(d) Well, it comes down to gravitational lenses enabling humanity to both explore the universe and make links with other stars. Just remember that, in 2009, the discovery of the first extrasolar planet in the Andromeda Galaxy (M31) was announced because of the gravitational lens caused by something in between!

15.5 THE RADIO BRIDGE BETWEEN THE SUN AND BARNARD'S STAR USING THEIR GRAVITATIONAL LENSES

The next closest star to the Sun beyond the triple Alpha Centauri system is Barnard's Star (see *http://www.en.wikipedia.org/wiki/Barnard's_Star*). Let us now repeat for the gravitational lens of Barnard's Star the same calculations that we did in the previous section for α Cen A. Then one has:

$$\begin{cases} d_{Barnard} = 5.98 \text{ light-years} \\ r_{Barnard} = 0.17\ r_{Sun} \\ M_{Barnard} = 0.16\ M_{Sun}. \end{cases} \tag{15.25}$$

Barnard's Star is thus just a small red star that is actually "passing by" the Sun right now; it is not known to have planets around it. As a consequence of the numbers listed in (15.25), one infers that

$$\begin{cases} d_{focal\_Barnard} = 98 \text{ AU} \\ r_{Schwarschild\_Barnard} = 0.47 \text{ km} \\ G_{Barnard}(\nu_{\text{Ka}}) = 1{,}991{,}174. \end{cases} \tag{15.26}$$

The gain, especially, is important to us:

$$G_{Barnard}(\nu_{\text{Ka}}) = 63 \text{ dB}. \tag{15.27}$$

We replace this into Barnard's Star equivalent of (15.24), again supposing that two 12-meter FOCAL spacecraft antennas are placed along the Sun–Barnard straight line at or beyond 550 AU and 100 AU, respectively. The result is a new graph of the BER as a function of transmitted power only (as in Figure 15.5).

15.6 THE RADIO BRIDGE BETWEEN THE SUN AND SIRIUS A USING THEIR GRAVITATIONAL LENSES

The next star we want to consider is Sirius A. This is because Sirius A is a big, massive bluish star and completely different from both α Cen A (which is a Sun-like star) and Barnard's Star (which is a small red star). Taking data from *http://www.en.*

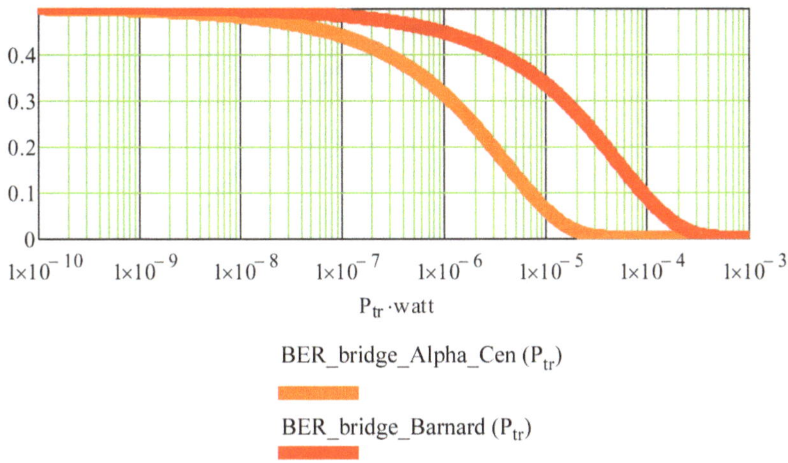

Figure 15.5. Bit error rate (BER) for the double-gravitational-lens radio bridge between the Sun and α Cen A (yellowish curve) plus the same curve for the radio bridge between the Sun and Barnard's Star (reddish curve, just as Barnard's Star is a reddish star): for this to occur, 10^{-3} watts are needed to keep the BER down to zero, because the gain of Barnard's Star is so small when compared with α Cen A's.

wikipedia.org/wiki/Sirius one gets:

$$\begin{cases} d_{Sirius\_A} = 8.6 \text{ light-years} \\ r_{Sirius\_A} = 1.711\ r_{Sun} \\ M_{Sirius\_A} = 2.02\ M_{Sun}. \end{cases} \tag{15.28}$$

From these data one gets:

$$\begin{cases} d_{focal\_Sirius\_A} = 793 \text{ AU} \\ r_{Schwarschold\_Sirius\_A} = 5.96 \text{ km} \\ G_{Sirius\_A}(\nu_{\text{Ka}}) = 251{,}385{,}723. \end{cases} \tag{15.29}$$

The important thing is of course the gain:

$$G_{Sirius\_A}(\nu_{\text{Ka}}) = 74 \text{ dB}. \tag{15.30}$$

Then, one replaces this into the Sirius A equivalent of (15.24), again supposing that two 12-meter FOCAL spacecraft antennas are placed along the Sun–Sirius A straight line at or beyond 550 AU and 793 AU, respectively. The result is a new graph of the BER as a function of transmitted power only (as in Figure 15.6).

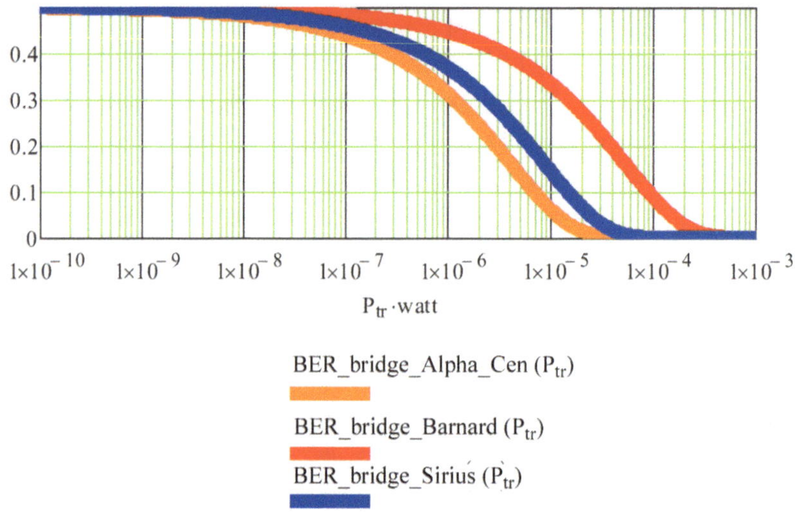

Figure 15.6. Bit error rate (BER) for the double-gravitational-lens radio bridge between the Sun and α Cen A (yellowish curve) plus the same curve for the radio bridge between the Sun and Barnard's Star (reddish curve, just as Barnard's Star is a reddish star) plus the same curve for the radio bridge between the Sun and Sirius A (blue curve, just as Sirius A is a big blue star). From this blue curve we see that only 10^{-4} watts are needed to keep the BER down to zero, because the gain of Sirius A is so big when compared with that of Barnard's Star that it gets closer to α Cen A's gain even if Sirius A is so much farther out than Barnard's Star! In other words, the star's gain and size combined matter even more than its distance!

15.7 THE RADIO BRIDGE BETWEEN THE SUN AND ANOTHER SUN-LIKE STAR LOCATED AT THE GALACTIC BULGE USING THEIR GRAVITATIONAL LENSES

Tempted by the suggestion to increase the distance of the second star continually and see what our calculations yield, we now imagine that the second star is Sun-like (i.e. that it has the same radius and mass exactly as the Sun) but is located inside the Galactic Bulge! This would mean (according to *http://www.en.wikipedia.org/wiki/Milky_Way*) that it is 26,000 light-years away. So, the equivalent of (15.12) and (15.24) now becomes

$$P_r = \frac{P_t \left(G_{Sun\_at\_Ka}\right)^2 \left(G_{12\_meter\_antenna\_at\_Ka}\right)^2}{L(r, \lambda)} \tag{15.31}$$

and the plot of the BER vs. transmitted power is shown in Figure 15.7 as the new pink curve at the far right of the previous three curves of α Cen A (orangish), Barnard's Star (red), and Sirius A (blue). The new pink curve showing the BER of a Sun-like star at the Galactic Bulge is naturally much to the right of the previous three stellar curves inasmuch as the Bulge distance of 26,000 light is so much greater

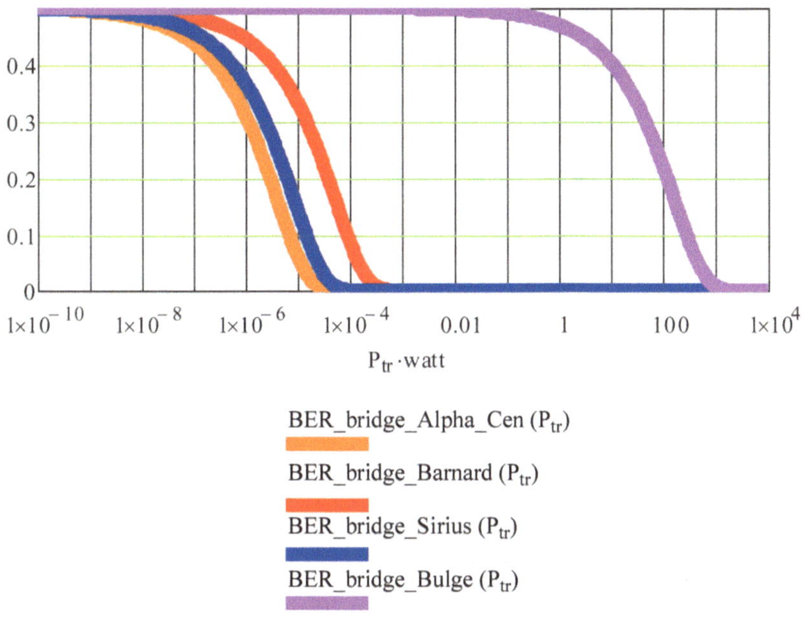

Figure 15.7. Bit error rate (BER) for the double-gravitational-lens radio bridge between the Sun and α Cen A (orangish curve) plus the same curve for the radio bridge between the Sun and Barnard's Star (reddish curve, just as Barnard's Star is a reddish star) plus the same curve of the radio bridge between the Sun and Sirius A (blue curve, just as Sirius A is a big blue star). In addition, far to the right we now have the pink curve showing the BER for the radio bridge between the Sun and another sun (identical in mass and size) located inside the Galactic Bulge at a distance of 26,000 light-years. The radio bridge between these two suns and their two gravitational lenses works perfectly (i.e., BER = 0) if the transmitted power is higher than about 1,000 watts.

than the distances of the three mentioned stars (less than 10 lt-yr away). The horizontal axis scale is much higher now, since the pink BER curve drops to zero only for transmitted power of about 1,000 watts.

15.8 THE RADIO BRIDGE BETWEEN THE SUN AND ANOTHER SUN-LIKE STAR LOCATED INSIDE THE ANDROMEDA GALAXY (M31) USING THEIR GRAVITATIONAL LENSES

We conclude this chapter by calculating the radio bridge between the Sun and another sun in Andromeda! The distance is now 2.5 million light-years, but the bridge would still work if the transmitted power is higher than about 10^7 watts = 10 Megawatts. This is shown by the new cyan curve at the far right in Figure 15.8. Perhaps this idea is not as crazy as it might appear, since recently (June 2009) the first extrasolar planet in the Andromeda Galaxy was announced

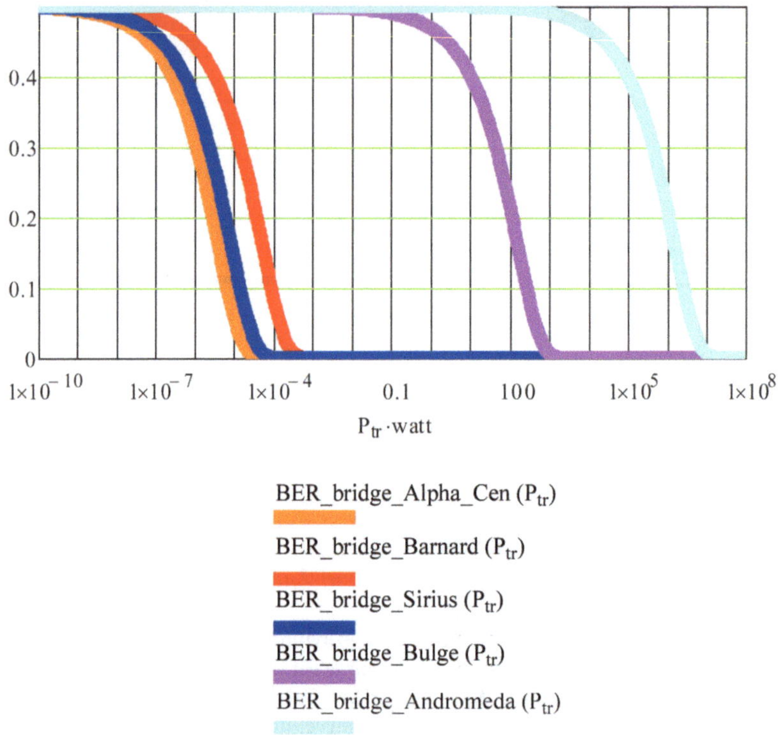

Figure 15.8. The same four bit error rate (BER) curves as shown in Figure 15.7 plus the new cyan curve appearing here at the far right: this is the BER curve of the radio bridge between the Sun and another sun located somewhere in the Andromeda Galaxy (M31). Note that such a radio bridge would work fine (i.e., with BER = 0) if the transmitting power is at least 10^7 watts = 10 Megawatts. This is not as crazy at it might seem if one remembers that recently (June 2009) the discovery of the first extrasolar planet in the Andromeda Galaxy was announced, and the method used for the detection was *gravitational lensing*!

to have been discovered as a result of gravitational lensing (see *http://www.redorbit.com/.../possible_planet_found_outside_our_galaxy/index.html*).

15.9 CONCLUSION

In this chapter we have outlined a FOCAL space mission to 550 AU and beyond to 1,000 AU. A number of issues still remain to be investigated in (1) the science related to the mission, (2) in the propulsion tradeoffs to get there in the shortest possible time, and (3) in optimization of the telecommunications link. Nevertheless, it plainly appears that the Sun focus at 550 AU is the next most important target for humankind to reach in order to be prepared for the following much more difficult task of achieving interstellar flight. In this chapter we have proven, in particular, that the only way to *extend* our telecommunications to spacecraft located in space at least at the distance of Alpha Centauri or greater would be by means of a FOCAL mission. We did this by resorting to the notion of bit error rate, which would need to be zero or nearly zero for *radio bridges* to be built between the Solar System and the destination target (be it a spacecraft or another star system).

15.10 ACKNOWLEDGMENT

The author is indebted to many colleagues for conversations and suggestions but, in particular, he would like to thank Paul Gilster for maintaining his terrific Centauri Dreams website *http://www.centauri-dreams.org*

15.11 REFERENCES

[1] A. Einstein, "Lens-like action of a star by the deviation of light in the gravitational field," *Science*, **84** (1936), 506–507.
[2] S. Liebes, Jr., "Gravitational lenses," *Physical Review*, **133** (1964), B835–B844.
[3] V. Eshleman, "Gravitational lens of the Sun: Its potential for observations and communications over interstellar distances," *Science*, **205** (1979), 1133–1135.
[4] F. Drake, "Stars as gravitational lenses," in G. Marx (Ed.), *Proceedings of the Bioastronomy International Conference, Balatonfüred, Hungary, June 22–27, 1987*, pp. 391–394.
[5] N. Cohen, "The pro's and con's of gravitational lenses in CETI," in G. Marx (Ed.), *Proceedings of the Bioastronomy International Conference, Balatonfüred, Hungary, June 22–27, 1987*, p. 395.
[6] F. Drake and D. Sobel, *Is Anyone Out There?* Delacorte Press, New York, 1992. See pp. 230–234 in particular.
[7] N. Cohen, *Gravity's Lens*, Wiley Science Editions, New York, 1988.
[8] C. Maccone, "Space missions outside the Solar System to exploit the gravitational lens of the Sun," in C. Maccone (Ed.), *Proceedings of the International Conference on Space Missions and Astrodynamics, Turin, Italy, June 18, 1992*, published in *Journal of the British Interplanetary Society*, **47** (1994), 45–52.
[9] C. Maccone, "FOCAL: A new space mission to 550 AU to exploit the gravitational lens of the Sun," a proposal for an M3 space mission submitted to the European Space Agency (ESA) on May 20, 1993, on behalf of an international team of scientists and engineers. Later (October 1993) reconsidered by ESA within the Horizon 2000 Plus space missions plan.
[10] J. Heidmann and C. Maccone, "AstroSail and FOCAL: Two extra-Solar System missions to the Sun's gravitational focuses," *Acta Astronautica*, **35** (1994), 409–410.
[11] C. Maccone, "The SETISAIL Project," in G. Seth Shostak (Ed.), *Progress in the Search for Extraterrestrial Life: Proceedings of the 1993 Bioastronomy Symposium, University of California at Santa Cruz, August 16–20, 1993*, published in *Astronomical Society of the Pacific Conference Series*, **74** (1995), 407–417.
[12] C. Maccone, *The Sun as a Gravitational Lens: Proposed Space Missions*, Third Edition, IPI Press, Colorado Springs, CO, 2002.

[13] R. Orta, P. Savi, and R Tascone, "Analysis of gravitational lens antennas," in C. Maccone (Ed.), *Proceedings of the International Conference on Space Missions and Astrodynamics, Turin, Italy, June 18, 1992*, published in *Journal of the British Interplanetary Society*, **47** (1994), 53–56.

[14] John D. Kraus, *Radio Astronomy*, Second Edition, Cygnus-Quasar Books, Powell, OH, 1966, pp. 6-115–6-118.

[15] C. Maccone, "Radioactive decay to propel relativistic interstellar probes along a rectilinear hyperbolic motion (Rindler spacetime)," *Acta Astronautica*, **57** (2005), 59–64.

[16] C. Maccone, *Telecommunications, KLT and Relativity*, Volume 1, IPI Press, Colorado Springs, CO, 1994.

[17] C. Maccone, "Relativistic optimized link by KLT," *Journal of the British Interplanetary Society*, **59** (2006), 94–98.

[18] L. Derosa and C. Maccone, "Propulsion tradeoffs for a mission to Alpha Centauri," *Acta Astronautica*, **60** (2007), 711–718.

[19] C. Maccone, *Deep Space Flight and Communications: Exploiting the Sun as a Gravitational Lens*, Springer/Praxis, Heidelberg, Germany/Chichester, U.K. A 400-page treatise about the FOCAL space mission that embodies and updates all previously published material about FOCAL.

[20] John D. Kraus, *Radio Astronomy*, Second Edition, Cygnus-Quasar Books, Powell, OH, 1986. See pp. 12-2 and 12-3 in particular.

[21] C. Maccone, "Interstellar radio links enhanced by exploiting the Sun as a gravitational lens," paper #IAC-09.D4.1.8 presented at the *49th International Astronautical Congress (IAC), Daejeon, Republic of Korea, October 12–16, 2009*. Later published in *Acta Astronautica*, **68** (2011), 76–84.

16

Extragalactic Internet by black hole gravitational lensing

16.1 AN INTRODUCTION TO SETI BETWEEN GALAXIES BY EXPLOITING THE SUPERMASSIVE BLACK HOLES LOCATED INSIDE EACH GALAXY AS MAGNIFYING GRAVITATIONAL LENSES

In two recent papers [1, 2] the author proved that radio communications between any pair of stars within our Galaxy are feasible with modest transmitted powers if the gravitational lenses of both stars are exploited. In this chapter we extend those innovative results to the case of radio communications between nearby galaxies. We show that radio communications between galaxies may become feasible if supermassive black holes, usually located at the center of galaxies, are exploited as gravitational lenses. In other words, a massive black hole may be regarded as a huge focusing device for radio waves being transmitted out of that galaxy and/or being received from another galaxy. This happens because a black hole is such a highly massive and compact object that all electromagnetic waves flying by its surface are highly deflected by its gravitational field and made to focus at a comparatively short distance from the black hole itself.

Next we consider the possibility of building radio bridges between our own Galaxy (the Milky Way) and other nearby galaxies. This possibility should be taken seriously because astronomers have known since 1974 that a supermassive black hole called "Sagittarius A*" does exist at the center of our Galaxy. In 2002 its mass was estimated to be of the order of 2.6 million solar masses and in 2008 this estimate was increased to 4.31 million solar masses. Furthermore, in 2004 a team of astronomers reported the discovery of a potential intermediate class black hole called GCIRS 13E orbiting around Sgr A* at about 3 light-years and having an estimated mass of 1,300 solar masses. These two big black holes could be our Galaxy's antennas for communications with alien civilizations in other nearby galaxies.

We mathematically show that the following radio bridges may be created between Sgr A* and the supermassive black holes located at the center of nearby galaxies, respectively:

(1) The Sgr A*–Andromeda (M31) P2 black hole radio bridge at a distance of 2.5 million light-years. The P2 Andromeda black hole is estimated to have a mass of about 40 million solar masses.
(2) The Sgr A*–M32 (a dwarf elliptical galaxy satellite of Andromeda) radio bridge at a distance of 2.65 million light-years. The M32 black hole is estimated to have a mass of about 3 million solar masses.
(3) The Sgr A*–M106 (also called NGC 4258, a spiral galaxy with anomalous arms) radio bridge at about 24 million light years. The M106 black hole is estimated to have a mass of about 40 million solar masses.
(4) The Sgr A*–Sombrero Galaxy (also called M104 or NGC 4594, an unbarred spiral galaxy) radio bridge at a distance of 29.3 million light-years. Its black hole is estimated to have a mass of 1 billion solar masses.
(5) The Sgr A*–M87 radio bridge. M87 is the supergiant elliptical galaxy located at the center of the supercluster of galaxies to which we belong (i.e., the Local Super Cluster), at the edge of which we are located. The distance between M87 and us is 53.5 million light-years in the direction of the constellation of Virgo, which is why M87 and its surrounding clusters of galaxies are sometimes referred to as the Virgo Super Cluster. At the center of M87 is a supermassive black hole estimated to have a mass of 6.4 billion solar masses. M87 is also well known as the "Jet Galaxy" since a jet of energetic plasma originates at the core and extends out at least 5,000 light-years.

The conclusion that we draw from the mathematics describing these radio bridges across huge intergalactic distances is surprising: they all perform better that the simple Sun–α Cen A radio bridge, first studied in detail by the author in [1]. In other words, the power necessary to keep the radio link between Sgr A* and all of the above big black holes in other nearby galaxies is smaller than the power required to keep the radio bridge between the Sun and α Cen A. In yet other words, despite intergalactic distances being huge with respect to ordinary interstellar distances within the Milky Way, black hole masses are so much larger than stellar masses that intergalactic bridges perform better than interstellar bridges. This unexpected and new result might have profound consequences on SETI as currently carried out on Earth. In fact, more advanced civilizations might already have built such intergalactic radio bridges. Thus, our SETI searches should make an allowance for this new situation: the possibility of SETI signals reaching us from other galaxies.

16.2 RADIO BRIDGES BETWEEN THE SUPERMASSIVE BLACK HOLE OF THE MILKY WAY GALAXY (Sgr A*) AND OTHERS

In Chapter 14 we made a detailed study of:

(1) The Sun–α Cen A radio bridge.
(2) The Sun–Barnard Star radio bridge.
(3) The Sun–Sirius A radio bridge.
(4) The radio bridge between the Sun and any Sun-like star located in the Galactic Bulge.
(5) The radio bridge between the Sun and any Sun-like star located inside the Andromeda Galaxy (M31).

The conclusion was that a radio interstellar communications network can indeed be built if the gravitational lenses of all stars involved are exploited.

In this chapter we take a much bolder step still. Given that each big galaxy has a supermassive black hole at its center, we calculate the antenna gain of that supermassive black hole and then build up the huge-distance radio bridge between that black hole and Sagittarius A*, from now on abbreviated Sgr A* (i.e., the supermassive black hole located at the center of our own Milky Way Galaxy).

To understand what Sgr A* is and how it came to be discovered little by little in the last few decades, the author suggests the reader watch the instructive You Tube video entitled *Cosmic Journeys: Supermassive Black Hole in the Milky Way Galaxy* at *http://www.youtube.com/watch?v=KCADH3x56eE.*

After watching that, a reading of the Wikipedia site about Sgr A* would complete one's basic knowledge: *http://en.wikipedia.org/wiki/Sagittarius_A**.

That said, let us now return to our mathematical description of Sgr A* and point out its key properties:

(1) Distance from the Sun in the Milky Way:

$$d_{Sgr\,A^*\_Sun} = 25{,}900\ \text{ly}. \tag{16.1}$$

(2) Mass:

$$M_{Sgr\,A^*} = 4.521 \cdot 10^6 \cdot M_{Sun} = 8.993 \cdot 10^{36}\ \text{kg}. \tag{16.2}$$

(3) Schwarzschild radius (i.e., black hole event horizon):

$$r_{Schwarzschild\_Sgr\,A^*} = \frac{2\,G\,M_{Sgr\,A^*}}{c^2} = 0.089\ \text{AU}. \tag{16.3}$$

(4) Antenna gain of the black hole:

$$G_{Sgr\,A^*}(\nu) = 4\pi^2 \cdot \frac{r_{Schwarzschild\_Sgr\,A^*}}{\lambda(\nu)}. \tag{16.4}$$

(5) Black hole antenna gain at the Ka band frequency:

$$G_{Sgr\,A^*}(\nu_{Ka}) \approx 137.5\ \text{dB}. \tag{16.5}$$

16.3 THE RADIO BRIDGE BETWEEN Sgr A* AND THE M31 ANDROMEDA GALAXY'S P2 BLACK HOLE

The M31 Andromeda Galaxy's nucleus (see *http://en.wikipedia.org/wiki/Andromeda_Galaxy*) is known to have two mass concentrations separated by 1.5 parsecs (4.9 lt-yr). The brighter concentration, designated P1, is offset from the center of the galaxy. The dimmer concentration, P2, falls at the true center of the galaxy and contains a black hole measured at $3\text{–}5 \times 10^7\, M_\odot$ in 1993 and at $1.1\text{–}2.3 \times 10^8\, M_\odot$ in 2005. Figure 16.1 shows the possible double structure (taken from the above Wikipedia site about M31).

So, the data about the bigger (P2) black hole at the exact center of M31 read:

(1) Distance:

$$d_{Sgr\,A^*\_\mathrm{M31P2}} = 2.54 \cdot 10^6 \text{ lt-yr.} \tag{16.6}$$

(2) Mass:

$$M_{\mathrm{M31P2}} = 4 \cdot 10^7 \cdot M_{Sun} = 7.956 \cdot 10^{37} \text{ kg.} \tag{16.7}$$

(3) Schwarzschild radius (i.e., black hole event horizon):

$$r_{Schwarzschild\_\mathrm{M31P2}} = \frac{2\,G\,M_{\mathrm{M31P2}}}{c^2} = 0.78 \text{ AU.} \tag{16.8}$$

(4) Antenna gain of the black hole:

$$G_{\mathrm{M31P2}}(\nu) = 4\pi^2 \cdot \frac{r_{Schwarzschild\_\mathrm{M31P2}}}{\lambda(\nu)}. \tag{16.9}$$

(5) Black hole antenna gain at the Ka band frequency:

$$G_{\mathrm{M31P2}}(\nu_{Ka}) \approx 146.9 \text{ dB.} \tag{16.10}$$

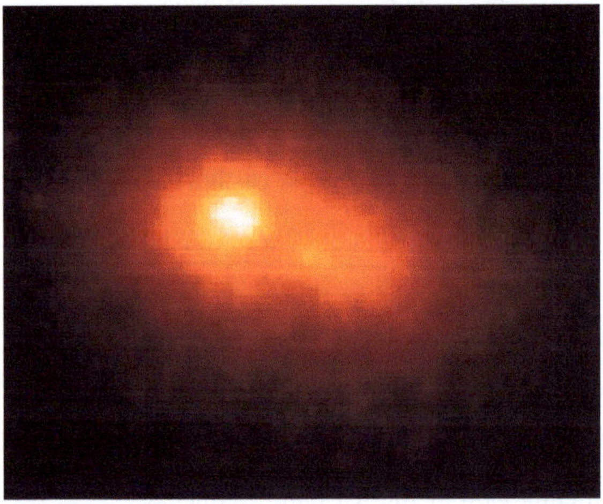

Figure 16.1. HST image of the Andromeda Galaxy core showing a possible double structure (NASA/ESA photo).

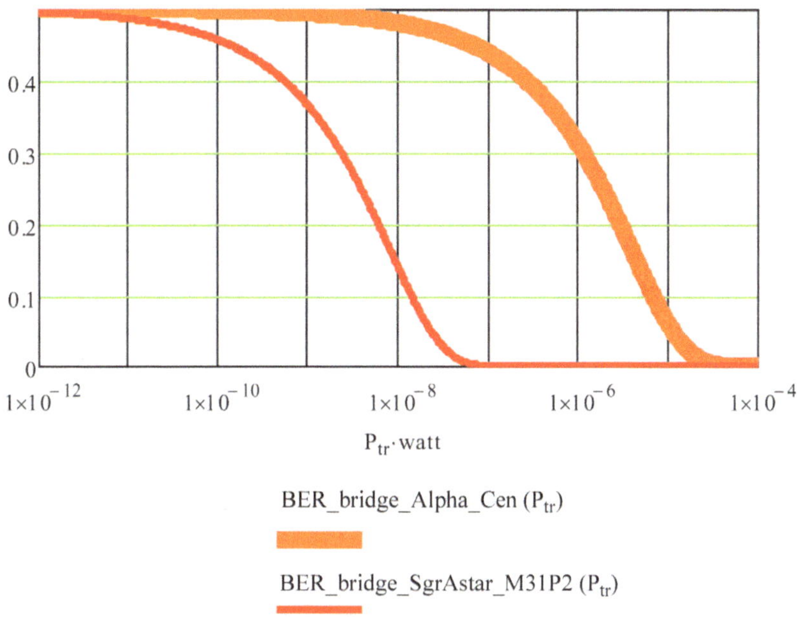

Figure 16.2. Bit error rate (BER) (thin solid curve) for the double-gravitational-lens *intergalactic* radio bridge between the supermassive black hole at the center of our Milky Way Galaxy (Sgr A*) and the corresponding P2 supermassive black hole located at the center of the Andromeda Galaxy (M31P2). We immediately see that this intergalactic radio bridge performs about three orders of magnitude better than the "simple" radio bridge between the Sun and α Cen A (thick solid curve, as in Figure 15.7). This startling new result demonstrates the feasibility of "doing SETI" between any two nearby galaxies, if the two supermassive black holes located at their centers are well exploited for intergalactic communications by those ET civilizations living close to them in their respective galaxies.

Next we write an equation corresponding to (15.24) for received power over the radio bridge between Sgr A* and M31P2; that is:

$$P_r = \frac{P_t \, G_{Sgr\,A^*} \, G_{\mathrm{M31P2}} \left(G_{12\_meter\_antenna\_at\_Ka}\right)^2}{L(r, \lambda)}. \qquad (16.11)$$

After repeating for (16.11) the string of calculations previously done in Section 14.3 to obtain the Sun–α Cen A radio bridge, we finally arrive at the new graph (shown in Figure 16.2). This shows both the same orange curve already found in Figure 14.7 for the Sun–α Cen A radio bridge and a new curve (in red) corresponding to the Sgr A*–M31P2 radio bridge. It is immediately apparent that the Sgr A*–M31P2 radio bridge performs better than the Sun–α Cen A radio bridge, in that 10^{-7} watts or thereabouts are sufficient to keep the link between Sgr A* and M31P2 while 10^{-4} watts are necessary to keep the link in the Sun–α Cen A bridge. This is a startling result when you consider that the Sgr A*–M31P2 distance is 572,082 times larger than the Sun–α Cen A distance. Clearly, this surprising circumstance occurs because the antenna gains of the two supermassive black holes are so much larger than the antenna gains of the Sun and α Cen A that they more than compensate even for the immensely larger distance in between.

16.4 THE RADIO BRIDGE BETWEEN Sgr A* AND THE DWARF ELLIPTICAL GALAXY M32 SMALL SATELLITE OF ANDROMEDA

M32 (see *http://en.wikipedia.org/wiki/Messier_32*) or Messier 32 (also known as NGC 221 and Le Gentil) (see Figure 16.3) is a dwarf elliptical galaxy about 2.49 ± 0.08 Mlt-yr (763 ± 24 kpc) away in the constellation of Andromeda. M32 is a satellite galaxy of the Andromeda Galaxy (M31) and was discovered by Le Gentil in 1749. M32 measures only 6.5 ± 0.2 klt-yr in diameter at its widest point. Like most elliptical galaxies, M32 contains mostly older faint red and yellow stars (like the Sun) with practically no dust or gas and consequently no current star formation. It does, however, show hints of star formation in the relatively recent past, so "earths" could be there, as well as ET civilizations on some of them.

The structure and stellar content of M32 is difficult to explain by traditional galaxy formation models. Recent simulations suggest a new scenario in which the strong tidal field of M31 can transform a spiral galaxy into a compact elliptical. As a small spiral galaxy falls into the central parts of M31, most of the outer layers of the smaller spiral are stripped away. The central bulge of the galaxy is much less affected and retains its morphology. Tidal effects trigger a massive star burst in the core, resulting in the high density of M32 we observe today. There is also evidence that M32 has an outer disk. Most important for us, M32 contains a supermassive black hole, whose mass has been estimated to lie between 1.5 and 5 million solar masses.

Figure 16.3. The dwarf elliptical galaxy M32.

So, let us start with the mathematics of the M32–Milky Way radio bridge by writing down the following equations:

(1) Distance:

$$d_{Sgr\,A^*\_\mathrm{M32}} = 2.49 \cdot 10^6 \text{ lt-yr.} \tag{16.12}$$

(2) Mass:

$$M_{\mathrm{M32}} = 3.25 \cdot 10^6 \cdot M_{Sun} = 6.463 \cdot 10^{36} \text{ kg.} \tag{16.13}$$

(3) Schwarzschild radius (i.e., black hole event horizon):

$$r_{Schwarzschild\_\mathrm{M32}} = \frac{2\,G\,M_{\mathrm{M32}}}{c^2} = 0.06\,\mathrm{AU}. \tag{16.14}$$

(4) Antenna gain of the black hole:

$$G_{\mathrm{M32}}(\nu) = 4\pi^2 \cdot \frac{r_{Schwarzschild\_\mathrm{M32}}}{\lambda(\nu)}. \tag{16.15}$$

(5) Black hole antenna gain at the Ka band frequency:

$$G_{\mathrm{M32}}(\nu_{Ka}) \approx 136.1 \text{ dB}. \tag{16.16}$$

These data yield Figure 16.4, which is the same as Figure 16.2 but for the addition of a new blue curve showing the BER for the Sgr A*–M32 radio bridge. As can be seen, this still performs better than the Sun–α Cen A bridge, though not as well as the Sgr A*–M31P2 radio bridge.

16.5 THE RADIO BRIDGE BETWEEN Sgr A* AND THE M106 GALAXY (NGC 4258) HALF WAY BETWEEN ANDROMEDA AND THE VIRGO SUPERCLUSTER CENTER (M87)

We are now going much farther out than the Andromeda system: we consider the M106 Anomalous Arms Galaxy (or NGC 4258) (see *http://en.wikipedia.org/wiki/Messier_106*) located half way between Andromeda and the center of the local supercluster of galaxies (i.e., the giant elliptical galaxy M87 in Virgo, which we will consider in detail in Section 16.7). As the reader may expect already, we are going to show that both these huge radio bridges are "feasible" for SETI communications.

A composite radio image of M106, taken from the above Wikipedia site, is shown in Figure 16.5.

The astrophysical data for M106 are as follows:

(1) Distance:

$$d_{Sgr\,A^*\_\mathrm{M106}} = 25 \cdot 10^6 \text{ lt-yr.} \tag{16.17}$$

(2) Mass:

$$M_{\mathrm{M106}} = 10^7 \cdot M_{Sun} = 1.988 \cdot 10^{37} \text{ kg.} \tag{16.18}$$

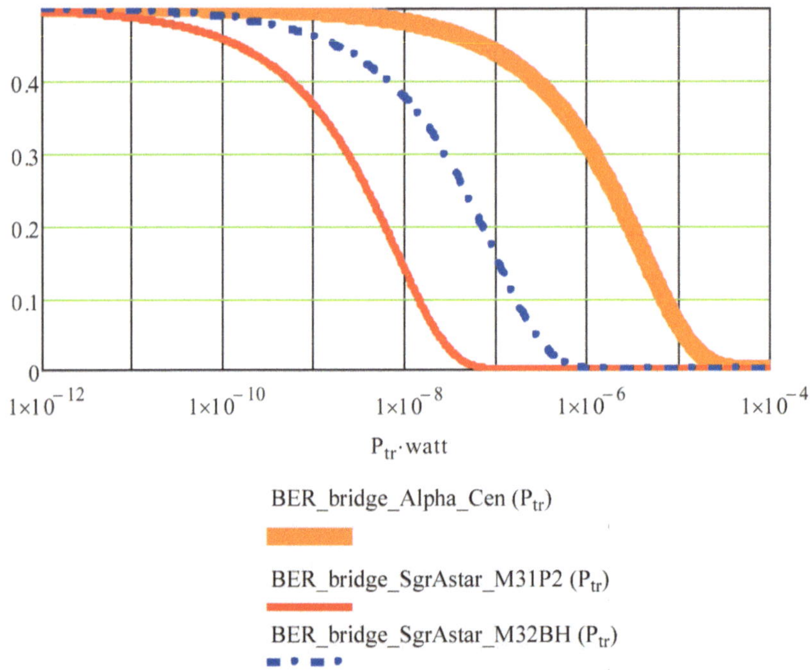

Figure 16.4. Bit error rate (BER) (thin dash–dot curve) for the double-gravitational-lens *intergalactic* radio bridge between the supermassive black hole at the center of our Milky Way Galaxy (Sgr A*) and the corresponding supermassive black hole located at the center of M32, the dwarf elliptical galaxy shown in Figure 16.3 and a satellite of the Andromeda Galaxy M31. We immediately see that this intergalactic radio bridge still performs about 1.5 orders of magnitude better than the "simple" radio bridge between the Sun and α Cen A (thick solid curve, as in Figures 15.7 and 16.2). This new result demonstrates the feasibility of doing SETI between the center of the Milky Way and M32.

Figure 16.5. M106 and its anomalous arms. Composite of IR, X-ray, radio, and visible light view (X-ray, blue; optical, gold; IR, red; radio, purple (image credit NASA/CXC/University of Maryland).

(3) Schwarzschild radius (i.e., black hole event horizon):

$$r_{\mathrm{M106}} = \frac{2\,G\,M_{\mathrm{M106}}}{c^2} = 0.197\,\mathrm{AU}. \tag{16.19}$$

(4) Antenna gain of the black hole:

$$G_{\mathrm{M106}}(\nu) = 4\pi^2 \cdot \frac{r_{Schwarzschild\_\mathrm{M106}}}{\lambda(\nu)}. \tag{16.20}$$

(5) Black hole antenna gain at the Ka band frequency:

$$G_{\mathrm{M106}}(\nu_{Ka}) \approx 140\,\mathrm{dB}. \tag{16.21}$$

These data yield Figure 16.6.

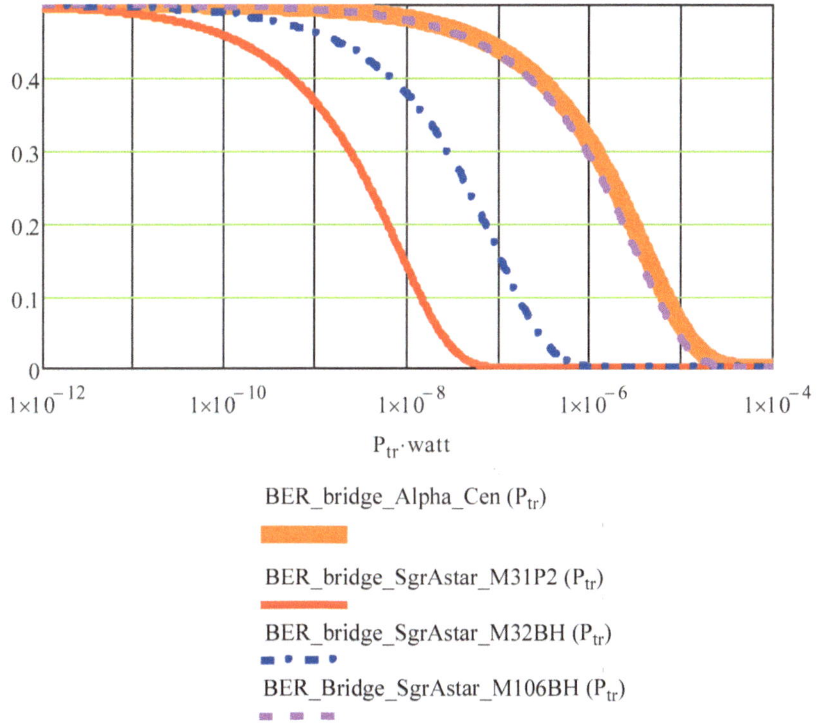

Figure 16.6. Bit error rate (BER) (thin dashed curve) for the double-gravitational-lens *intergalactic* radio bridge between the supermassive black hole at the center of our Milky Way Galaxy (Sgr A*) and the corresponding supermassive black hole located at the center of the M106 Anomalous Arms Galaxy shown in Figure 16.5. We immediately see that this intergalactic radio bridge virtually performs just as well as the "simple" radio bridge between the Sun and α Cen A (thick solid curve, as in Figures 15.7 and 16.2) since the new violet curve almost exactly overlaps the orange curve of the Sun–α Cen A bridge. This result further demonstrates the feasibility of doing SETI between the center of the Milky Way Galaxy and M106.

16.6 THE RADIO BRIDGE BETWEEN Sgr A* AND THE M104 SOMBRERO GALAXY (NGC 4594) AGAIN HALF WAY BETWEEN ANDROMEDA AND THE VIRGO SUPERCLUSTER CENTER (M87)

The M104 Sombrero Galaxy (see Figure 16.7, from *http://en.wikipedia.org/wiki/Sombrero_Galaxy*) is again half way between Andromeda and M87 (i.e., 29.3 million light-years).

In the 1990s a research group led by John Kormendy demonstrated that a supermassive black hole is present within the Sombrero Galaxy. Using spectroscopy data from both the CFHT and the Hubble Space Telescope, the group showed that the speed of revolution of the stars within the center of the galaxy could not be maintained unless a mass one billion times that of the Sun, or $10^9\ M_{\odot}$, is present in the center. This is among the most massive black holes measured in any nearby galaxies.

The following string of data thus emerges:

(1) Distance:

$$d_{Sgr\,A^*\_\mathrm{M104}} = 29.3 \cdot 10^6 \text{ lt-yr}. \tag{16.22}$$

(2) Mass:

$$M_{\mathrm{M104}} = 10^9 \cdot M_{Sun} = 1.98 \cdot 10^{39} \text{ kg}. \tag{16.23}$$

(3) Schwarzschild radius (i.e., black hole event horizon):

$$r_{\mathrm{M104}} = \frac{2\,G\,M_{\mathrm{M104}}}{c^2} = 19.7\,\mathrm{AU}. \tag{16.24}$$

(4) Antenna gain of the black hole:

$$G_{104}(\nu) = 4\pi^2 \cdot \frac{r_{Schwarzschild\_\mathrm{M104}}}{\lambda(\nu)}. \tag{16.25}$$

(5) Black hole antenna gain at the Ka band frequency:

$$G_{\mathrm{M106}}(\nu_{Ka}) \approx 160\,\mathrm{dB}. \tag{16.26}$$

Figure 16.7. The Sombrero Galaxy (M104) as observed by the Spitzer (credit HST/NASA/ESA).

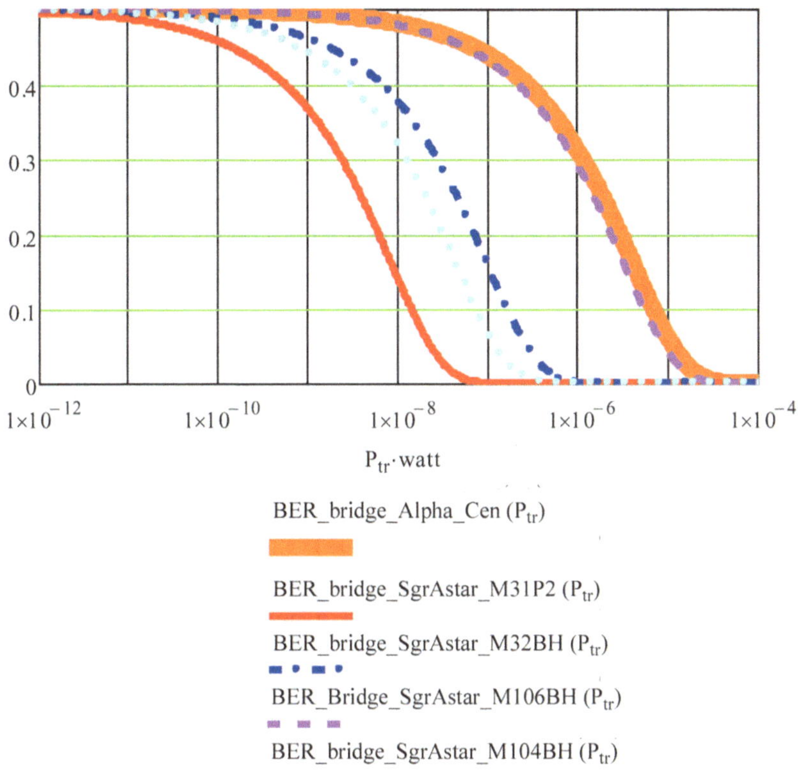

Figure 16.8. Bit error rate (BER) (thin dotted curve) for the double-gravitational-lens *intergalactic* radio bridge between the supermassive black hole at the center of our Milky Way Galaxy (Sgr A*) and the corresponding supermassive black hole located at the center of the M104 Sombrero Galaxy shown in Figure 16.7. This intergalactic radio bridge performs two orders of magnitude better than the "simple" radio bridge between the Sun and α Cen A (thick solid curve, as in Figures 15.7, 16.2, 16.4, and 16.6). This result again demonstrates the feasibility of doing SETI between the center of the Milky Way Galaxy and M104.

These data yield Figure 16.8. This is of course the same as Figure 16.6 except for the new cyan curve for the BER of the Sombrero–Milky Way radio bridge, which again performs about two orders of magnitude better than the Sun–α Cen A radio bridge, though not as well as the M32–Milky Way radio bridge.

16.7 THE RADIO BRIDGE BETWEEN Sgr A* AND THE M87 SUPERGIANT ELLIPTICAL GALAXY (NGC 4486) AT THE VIRGO SUPERCLUSTER CENTER

The farthest we can go is the supergiant elliptical galaxy M87 located at the center of our own Local Supercluster of galaxies (the Milky Way and Andromeda are at its outskirts, or, put another way, we are right at the periphery of the periphery).

We learn from the M87 Wikipedia site (*http://en.wikipedia.org/wiki/Messier_87*) that M87 (Figure 16.9) was discovered in 1781 by French astronomer Charles Messier (1730–1817). The largest and brightest galaxy within the northern Virgo Cluster, it is located about 53.5 million light-years from Earth. Unlike the normal disk-shaped spiral galaxy, M87 has no distinctive dust lanes and has an ellipsoidal shape. At the core is a supermassive black hole, which forms the primary component of an active galactic nucleus that is a strong source of multiwavelength radiation, particularly radio waves. A jet of energetic plasma originates at the core and extends out at least 5,000 light-years, which is why M87 is sometimes called the "Jet Galaxy" (Figure 16.10). The stars in this galaxy form about one sixth of M87's mass. They have a nearly spherical distribution, while the density of stars decreases with increasing distance from the core. The galactic envelope extends out to a radius of about 490,000 light-years, where it has been truncated. Between the stars is a diffuse interstellar medium of gas that has been chemically enriched by elements emitted from evolved stars. Any dust formed within the galaxy is destroyed within 46 million years by X-ray emission from the core, although optical filaments of dust have been observed. Orbiting the galaxy is an abnormally large population of about 12,000 globular clusters, compared with the 150–200 globular clusters orbiting the Milky Way. Since this is the largest giant elliptical galaxy near Earth and is one of the brightest radio sources in the sky, M87 is a popular target for both amateur astronomy observations and professional astronomy study.

The M87 astrophysical data of interest to us are thus:

(1) Distance:

$$d_{Sgr\,A^*\_\mathrm{M87}} = 53.5 \cdot 10^6 \text{ lt-yr}. \tag{16.27}$$

(2) Mass:

$$M_{\mathrm{M87}} = 6.6 \cdot 10^9 \cdot M_{Sun} = 1.98 \cdot 10^{39} \text{ kg}. \tag{16.28}$$

(3) Schwarzschild radius (i.e., black hole event horizon):

$$r_{\mathrm{M87}} = \frac{2\,G\,M_{\mathrm{M87}}}{c^2} = 130\,\mathrm{AU}. \tag{16.29}$$

(4) Antenna gain of the black hole:

$$G_{\mathrm{M87}}(\nu) = 4\pi^2 \cdot \frac{r_{Schwarzschild\_\mathrm{M87}}}{\lambda(\nu)}. \tag{16.30}$$

(5) Black hole antenna gain at the Ka band frequency:

$$G_{\mathrm{M87}}(\nu_{Ka}) \approx 169 \text{ dB}. \tag{16.31}$$

The radio bridge between the huge black hole at the center of M87 and Sgr A* is represented in Figure 16.11 by the new brown curve. We thus see that the M87–Sgr A* radio bridge performs better than almost any other (the exception being the P2 bridge at the center of the Andromeda Galaxy). It performs about 2.5 times better than the Sun–α Cen A bridge.

Figure 16.9. The M87 elliptical supergiant galaxy at the center of the Local Supercluster of galaxies to which we belong.

Figure 16.10. This Hubble Space Telescope photo shows the jet of matter ejected from M87 at nearly light speed and nearly reaching us, as it stretches 5,000 light-years from the galactic core.

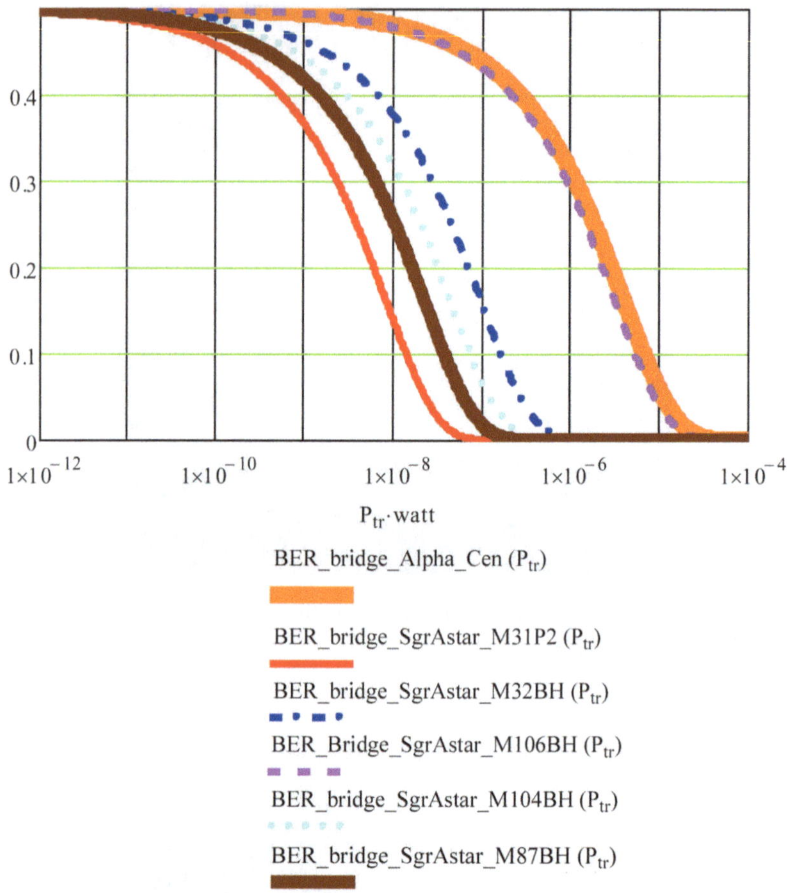

Figure 16.11. Bit error rate (BER) (second thick solid curve from the left) for the double-gravitational-lens *intergalactic* radio bridge between the supermassive black hole at the center of our Milky Way Galaxy (Sgr A*) and the corresponding supermassive black hole located at the center of the M87 Jet Galaxy shown in Figures 16.9 and 16.10. This intergalactic radio bridge performs about 2.5 orders of magnitude better than the "simple" radio bridge between the Sun and α Cen A (orange curve, as in Figures 15.7, 16.2, 16.4, 16.6, and 16.8). This result again demonstrates the feasibility of doing SETI between the center of the Milky Way Galaxy and M87.

16.8 CONCLUSIONS

We have shown that telecommunications between galaxies up to 60 million light-years apart are indeed feasible with modest transmission powers if each galaxy's supermassive black hole is exploited as a gravitational magnifying lens. This is an important advance with respect to "classical" SETI searches, which assume huge transmission powers to be available to ETs. Human SETI searches might therefore be concentrated in those nearby galaxies where a supermassive and central black hole is known to exist.

In addition, this line of thought clearly shows that the central massive black hole of every galaxy is by far the most important "resource" of that galaxy for SETI purposes. In fact, it can be likened to the "central radio station" of that galaxy, to which every civilization in that galaxy could tune in to keep in touch with other aliens living in nearby galaxies.

This idea also explains the Fermi paradox to a certain degree. In fact, one might argue that every important civilization in a galaxy would like to live close to the central supermassive black hole, rather than in the outskirts as we humans do with respect to our Sgr A*.

So, our conclusion is that current human SETI searches should change tack and look for ET signals coming from nearby galaxies known to have a supermassive black hole at their center as well as from Sgr A* and its surroundings.

16.9 REFERENCES

[1] C. Maccone, "Interstellar radio links enhanced by exploiting the Sun as a gravitational lens," *Acta Astronautica*, **68** (2011), 76–84.

[2] C. Maccone, "Interstellar radio links enhanced by exploiting the Sun as a gravitational lens," paper #IAC-09.D4.1.8 presented at the *60th International Astronautical Congress, Daejeon, Republic of Korea, October 12–16, 2009*, and distributed to all participants as a CD-ROM file, but not published in printed form.

[3] C. Maccone, *Deep Space Flight and Communications: Exploiting the Sun as a Gravitational Lens*, Springer/Praxis, Heidelberg, Germany/Chichester, U.K. A 400-page treatise about the FOCAL space mission that embodies and updates all previously published material about FOCAL.

[4] C. Maccone, "SETI among galaxies by virtue of black holes," *Acta Astronautica*, 2012, in press.

Part III

KLT for optimal signal processing

17

A simple introduction to the KLT and BAM-KLT

17.1 INTRODUCTION

This chapter is a simple introduction about using the Karhunen–Loève Transform (KLT) to extract weak signals from noise of any kind. In general, the noise may be colored and over wide bandwidths, and not just white and over narrow bandwidths. We show that the signal extraction can be achieved by the KLT more accurately than by the Fast Fourier Transform (FFT), especially if the signals buried into the noise are very weak, in which case the FFT fails. This superior performance of the KLT happens because the KLT of any stochastic process (both stationary and non-stationary) is defined from the start over a *finite* time span ranging between 0 and a final and *finite* instant T (contrary to the FFT, which is defined over an infinite time span). We then show mathematically that the series of all the eigenvalues of the autocorrelation of the (noise + signal) may be differentiated with respect to T yielding the "Final Variance" of the stochastic process $X(t)$ in terms of a sum of the first-order derivatives of the eigenvalues with respect to T. Finally, we prove that this new result leads to the immediate reconstruction of a signal buried into the thick noise. We have thus put on a strong mathematical foundation a set of very important practical formulae that can be applied to improve SETI, the detection of exoplanets, asteroidal radar, and also other fields of knowledge like economics, genetics, biomedicine, etc. to which the KLT can be equally well applied with success.

17.2 A BIT OF HISTORY

The Karhunen–Loève Transform (KLT) is the most advanced mathematical algorithm available in the year 2008 to achieve both noise filtering and data compression in processing signals of any kind.

It took about two centuries (~1800–2000) for mathematicians to create such a jewel of thought little by little, piece after piece, paper after paper. It is thus difficult to recognize who did what in building up the KLT and at the same time be fair in attributing each individual advance to the appropriate author. In addition, mathematicians, both pure and applied, often speak such a "clumsy" language of their own that even learned scientists sometimes find it hard to understand them. This unfortunate situation hides the esthetic beauty of many mathematical discoveries that were often historically made by their authors more for the joy of opening new lines of thought than for the sake of any immediate application to science and engineering.

In essence, the KLT is a rather new mathematical tool used to improve our understanding of physical phoenomena, far superior to the classical Fourier Transform (FT). The KLT is named for two mathematicians—the Finnish actuary Kari Karhunen (1915–1992) [1] and the French American Michel Loève (1907–1979) [2, 3]—who proved, independently and about the same time (1946), that the series (2) hereafter is *convergent*. Put this way, the KLT looks like a purely mathematical topic, but really this is hardly the case. As early as 1933 the American statistician and economist Harold Hotelling (1895–1973) used the KLT (for discrete time, rather than for continuous time), so that the KLT is sometimes called the "Hotelling Transform". Even much earlier than these three authors the Italian geometer Eugenio Beltrami (1835–1899) discovered as early as 1873 the SVD (Singular Value Decomposition), which is closely related to the KLT in that area of applied mathematics nowadays called Principal Components Analysis (PCA). Unfortunately, a complete historical account about how these contributions developed since 1865—when the English mathematician Arthur Cayley (1821–1895) "invented" matrices—simply does not exist. We only know about "fragments of thought" that impair an overall vision of both the PCA and the KLT.

In Sections 17.3–17.5, we'll derive *heuristically and step-by-step* the many equations that make up for the KLT. We think that this approach is much easier to understand for beginners than what is found in most "pure" mathematical textbooks, and hope that the readers will appreciate our effort to explain the KLT as easily as possible to non-mathematically trained people.

17.3 A HEURISTIC DERIVATION OF THE KL EXPANSION

We start by saying that the KLT was born during the years of World War Two out of the need to merge two different areas of classical mathematics.

(1) The expansion of a deterministic periodic signal $x(t)$ into a basis of orthonormal functions (sines and cosines, in this case), typified by the classical Fourier series—first put forward by the French mathematician Jean Baptiste Joseph Fourier (1768–1830) around 1807,

$$x(t) = \frac{a_0}{2} + \sum_{n=1}^{\infty}[a_n \cos(nt) + b_n \sin(nt)] \qquad (-\pi \le t \le \pi). \tag{17.1}$$

(2) The need to extend this too narrow and deterministic view to probability and statistics. The much larger variety of phenomena called "noise" by physicists and engineers will thus be encompassed by the new transform. This enlarged view means considering a random function $X(t)$ (notice that we denote random quantities by capitals, and that $X(t)$ is also called a "stochastic process of the time"). We now seek to expand this stochastic process onto a set of orthonormal functions $\phi_n(t)$ according to the starting formula

$$X(t) = \sum_{n=1}^{\infty} Z_n \, \phi_n(t) \tag{17.2}$$

which is called the *Karhunen–Loève (KL) expansion of $X(t)$ over the finite time interval* $0 \le t \le T$.

What are then the Z_n and the $\phi_n(t)$ in (17.2)? To find out, let us start by recalling what "orthonormality" means for the Fourier series (17.1). Leonhard Euler (1707–1783) had already laid the first stone towards the Fourier series (17.1) by proving that, if $x(t)$ is assumed to be periodic over the time interval $-\pi \le t \le \pi$, then the coefficients a_n and b_n in (17.1) are obtained from the known function (or "periodic signal") $x(t)$ by virtue of the equations ("Euler formulae"):

$$a_n = \frac{1}{\pi}\int_{-\pi}^{\pi} x(t)\cos(nt)\,dt \qquad b_n = \frac{1}{\pi}\int_{-\pi}^{\pi} x(t)\sin(nt)\,dt. \tag{17.3}$$

If the same result is going to be true for the Karhunen–Loève expansion, the functions of the time, $\phi_n(t)$ in (17.2) must be orthornormal (i.e., both orthogonal and normalized to 1). That is,

$$\int_0^T \phi_m(t)\,\phi_n(t)\,dt = \delta_{mn} \tag{17.4}$$

where the δ_{mn} are the Kronecker symbols, defined by $\delta_{mn} = 0$ for $m \neq n$ and $\delta_{nn} = 1$.

But what then are the Z_n appearing in (17.2)? Well, a random function $X(t)$ can be thought of as something made up of two parts: its behavior in time, represented by the functions $\phi_n(t)$, and its behavior with respect to probability and statistics, which must therefore be represented by the Z_n. In other words, the Z_n must be random variables not changing in time (i.e., "just" random variables and not stochastic processes). By doing so we have actually made one basic, new step ahead: we have found that the KLT separates the probabilistic behavior of the random function $X(t)$ from its behavior in time, a kind of "untypical" separation that is achieved nowhere else in mathematics!

Having discovered that the Z_n are random variables, some trivial consequences follow at once. Let us denote by $E\{\ \}$ the linear operator yielding the average of a random variable or stochastic process. If one takes the average of both sides of the KL expansion (17.2), one then gets (we "freely" interchange here the average operator

$E\{\ \}$ with the infinite summation sign, bypassing the complaints of "subtle" mathematicians!)

$$E\{X(t)\} = \sum_{n=1}^{\infty} E\{Z_n\}\phi_n(t). \tag{17.5}$$

Now, it is not restrictive to suppose that the random function $X(t)$ has a zero mean value in time—namely, that the following equation is identically true for all values of the time t within the interval $0 \leq t \leq T$:

$$E\{X(t)\} \equiv 0. \tag{17.6}$$

In fact, were this not the case, one could replace $X(t)$ by the new random function $X(t) - E\{X(t)\}$ in all the above calculations, thus reverting to the case of a new random function with zero mean value. Thus, in conclusion, the random variables Z_n too must have a zero mean value

$$E\{Z_n\} \equiv 0. \tag{17.7}$$

This equation has a simple consequence: since the variance $\sigma^2_{Z_n}$ of the random variables Z_n is given by

$$\sigma^2_{Z_n} = E\{Z_n^2\} - E^2\{Z_n\} \tag{17.8}$$

by inserting (17.7) into (17.8) we get

$$\sigma^2_{Z_n} = E\{Z_n^2\}. \tag{17.9}$$

At this point, we can make a further step ahead, that has no counterpart in the classical Fourier series: we wish to introduce a new sequence of positive numbers λ_n such that every λ_n is the variance of the corresponding random variable Z_n, that is

$$\sigma^2_{Z_n} = \lambda_n = E\{Z_n^2\} > 0. \tag{17.10}$$

This equation provides the "answer" to the next "natural" question: Do the random variables Z_n fulfill a new type of "orthonormality" somehow similar to what the classical orthonormality (17.4) is for the $\phi_n(t)$? Since we are talking about random variables, the "orthogonality operator" can only be understood in the sense of *statistical independence*. The integral in (17.4) must then be replaced by the average operator $E\{\ \}$ for the random variables Z_n. In conclusion, we found that the random variables Z_n must obey the important equation

$$E\{Z_m Z_n\} = \lambda_n\,\delta_{mn}. \tag{17.11}$$

In this equation, we were forced to introduce the positive λ_n in the right-hand side in order to let (17.11) reduce to (17.10) in the special case $m = n$.

As for the KL equivalent of the Euler formulae (17.3) of the Fourier series, from the KL series (17.2) and the orthonormality (17.4) of the $\phi_n(t)$ one immediately finds that

$$Z_n = \int_0^T X(t)\,\phi_n(t)\,dt. \tag{17.12}$$

In other words: the random variables Z_n are obtained from the given stochastic process $X(t)$ by "projecting" this $X(t)$ over the correspoding eigenvector $\phi_n(t)$. If one likes the language of mathematicians and of quantum physics, then one may say that this projection of $X(t)$ onto $\phi_n(t)$ occurs in the "Hilbert space", which is the infinitely dimensional Euclidean space spanned by the eigenvectors $\phi_n(t)$ so that the square of $\phi_n(t)$ is integrable over the finite time span $0 \le t \le T$.

To sum up, we have actually achieved a remarkable generalization of the Fourier series by defining the Karhunen–Loève expansion (17.2) as the only possible statistical expansion in which all the expansion terms are *uncorrelated* from each other. This word "uncorrelated" comes from the fact that the *autocorrelation* of a random function of the time, $X(t)$, is defined as the mean value of the product of $X(t)$ at two different instants t_1 and t_2:

$$R_{XX}(t_1, t_2) \equiv R_X(t_1, t_2) = E\{X(t_1)X(t_2)\}. \tag{17.13}$$

If we assume, according to (17.6), that the mean value of $X(t)$ vanishes identically in the interval $0 \le t \le T$, the autocorrelation (17.13) reduces to the variance of $X(t)$ when the two instants are the same

$$\sigma^2_{X(t)} = E\{X^2(t)\} = E\{X(t)X(t)\} = R_X(t, t). \tag{17.14}$$

Let us add one final remark about the basic notion of statistical independence of the random viariables Z_n. It can be proven that, while the Z_n in (17.2) always are uncorrelated (by construction), they also are statistically independent if they are Gaussian-distributed random variables. This is fortunately the case for the Brownian motion and for the background noise we face in SETI. So we are not concerned about this subtle mathematical distinction between uncorrelated and statistically independent random variables.

17.4 THE KLT FINDS THE BEST BASIS (EIGEN-BASIS) IN THE HILBERT SPACE SPANNED BY THE EIGENFUNCTIONS OF THE AUTOCORRELATION OF *X(t)*

Up to this point, we have not given any hint about how to find the orthonormal functions of the time, $\phi_n(t)$, and positive numbers λ_n (i.e., the variances of the corresponding uncorrelated random variables Z_n). In this section, we solve this problem by showing that the $\phi_n(t)$ are the eigenfunctions of the autocorrelation $R_X(t_1, t_2) = E\{X(t_1)X(t_2)\}$ and that the λ_n are the corresponding eigenvalues. This is the correct mathematical phrasing of what we are going to prove. However, in order to ease the understanding of the further maths involved hereafter, a "translation" into the language of "common words" is now provided. Consider an object—for instance, a book—and a three-axes rectangular reference frame, oriented in an arbitrary fashion with respect to the book. Then, the classical Newtonian mechanics shows that all the mechanical properties of the book are described by a 3×3 symmetric matrix called the "inertia matrix" (or, more correctly, "inertia tensor") whose elements are, in general, all different from zero. Handling a matrix whose elements are all nonzero is obviously more complicated than handling a matrix where all entries are zeros except for those on the main diagonal (i.e., a *diagonal matrix*). Thus, one may be led to wonder whether a certain transformation of axes exists that changes the inertia matrix of the book into a diagonal matrix. Newtonian mechanics shows then that only *one* privileged orientation of the reference frame with respect to the book exists yielding a *diagonal* inertia matrix: the three axes must then coincide with a set of three axes (parallel to the book edges) called "principal axes" of the book, or "eigenvectors" or "proper vectors" of the inertia matrix of the book. In other words, each body posesses an intrinsic set of three rectangular axes that describes at best its dynamics (i.e., in the most concise form). This was proven again by Euler, and one can always compute the position of the eigenvectors with respect to a generic reference frame by means of a certain mathematical procedure called "finding the eigenvectors of a square matrix".

In a similar fashion, one can describe any stochastic process $X(t)$ by virtue of the statistical quantity called the autocorrelation (or simply the correlation), defined as the mean value of the product of the values of $X(t)$ at two different instants t_1 and t_2, and formally written $E\{X(t_1)X(t_2)\}$. The autocorrelation, obviously symmetric in t_1 and t_2, plays for the stochastic process $X(t)$ just the same role as the inertia matrix for the book example above. Thus, if one first seeks the eigenvectors of the correlation, and then changes the reference frame over to this new set of vectors, one achieves the simplest possible description of the whole (signal + noise) set.

Let us now translate the whole above description into equations. First of all, we must express the autocorrelation $E\{X(t_1)X(t_2)\}$ by virtue of the KL expansion (17.2). This goal is achieved by writing down (17.2) for two different instants, t_1 and t_2, taking the average of their product, and then (freely) interchanging the average and the summations in the right-hand side. The result is

$$E\{X(t_1)X(t_2)\} = \sum_{m=1}^{\infty}\sum_{n=1}^{\infty} \phi_m(t_1)\phi_n(t_2)\, E\{Z_m Z_n\}. \tag{17.15}$$

Taking advantage of the statistical orthogonality of the Z_n, given by (17.11), (17.15) simplifies to

$$E\{X(t_1)X(t_2)\} = \sum_{m=1}^{\infty} \lambda_m \phi_m(t_1)\phi_m(t_2). \tag{17.16}$$

Finally, we now want to let the $\phi_n(t)$ "disappear" from the right-hand side of (17.16) by taking advantage of their orthonormality (17.4). To do so, we multiply both sides of (17.16) by $\phi_n(t_1)$ and then take the integral with respect to t_1 between 0 and T. One then gets:

$$\begin{aligned}\int_0^T E\{X(t_1)X(t_2)\}\phi_n(t_1)\, dt_1 &= \sum_{m=1}^{\infty} \lambda_m \phi_m(t_2) \int_0^T \phi_m(t_1)\phi_n(t_1)\, dt_1 \\ &= \sum_{m=1}^{\infty} \lambda_m \phi_m(t_2)\delta_{mn} = \lambda_n \phi_n(t_2),\end{aligned} \tag{17.17}$$

that is

$$\int_0^T E\{X(t_1)X(t_2)\}\,\phi_n(t_1)\,dt_1 = \lambda_n\,\phi_n(t_2). \tag{17.18}$$

This basic result is an *integral equation*, called by mathematicians "of Fredholm type". Once the correlation $E\{X(t_1)X(t_2)\}$ of $X(t)$ is known, the integral equation (17.18) yields (upon its solution, which may not be easy at all to find analytically!) both the Karhunen–Loève eigenvalues λ_n and the corresponding eigenfunctions $\phi_n(t)$. Readers familiar with quantum mechanics will also recognize in (17.18) a typical "eigenvalue equation" having the kernel $E\{X(t_1)X(t_2)\}$.

Let us finally summarize what we have proven so far in Sections 17.3 and 17.4, and let us use the language of signal processing, which will lead us directly to SETI, the main theme of this chapter.

By adding random noise to a deterministic signal one obtains what is called a "noisy signal" or, in case the signal power is much lower than the noise power, "a signal buried into the noise". The noise + signal is a random function of the time, denoted hereafter by $X(t)$. Karhunen and Loève proved that it is possible to represent $X(t)$ as the infinite series (called the KL expansion) given by (17.2), and this series is convergent. Assuming that the (signal + noise) correlation $E\{X(t_1)X(t_2)\}$ is a known function of t_1 and t_2, then the orthonormal functions $\phi_n(t)$ $(n = 1, 2, \ldots)$ turn out to be just the eigenfunctions of the correlation. These eigenfunctions $\phi_n(t)$ form an orthonormal basis in what physicists and mathematicians call the space of square-integrable functions, also called the Hilbert space. The eigenfunctions $\phi_n(t)$ actually are *the best* possible basis to describe the (signal + noise), much better than any classical Fourier basis made up by sines and cosines only. One can conclude that *the KLT automatically adapts itself to the shape of the (signal + noise), whatever behavior in time it may have, by adopting as a new reference frame in the Hilbert space the basis spanned by the eigenfunctions, $\phi_n(t)$, of the autocorrelation of the (signal + noise), $X(t)$.*

17.5 CONTINUOUS TIME VS. DISCRETE TIME IN THE KLT

The KL expansion in continuous time, t, is what we have described so far. This may be more "palatable" to theoretical physicists and mathematicians inasmuch as it may be related to other branches of physics, or of science in general, in which time obviously must be a continuous variable. For instance, this author spent 15 years of his life (1980–1994) in investigating mathematically the connection between Special Relativity and KLT. The result was the mathematical theory of optimal telecommunications between the Earth and a relativistic spaceship either receding from the Earth or approaching it. Although this may sound like "mathematical science fiction" to some folks (who we would call "short sighted"), the possibility that, in the future, humankind will send out relativistic automatic probes or even manned spaceships, is not unrealistic. Nor is it science fiction to imagine that an *alien* spaceship might approach the Earth slowing down from relativistic speeds to zero speed. So, a mathematical physics book like [4] can make sense. There, the KLT is obtained for any acceleration profile of the relativistic probe or spaceship. The result is that the KL eigenfunctions are Bessel functions of the first kind (suitably modified) and the eigenvalues are determined by the zeros of linear combinations of these Bessel functions and their derivatives, as we shall prove in Chapters 21–27 of this book, and especially in Chapter 22.

Other continuous-time applications of the KLT are to be found in other branches of science, ranging, for instance, from genetics to economics. But, whatever the application may be, if time is a continuous variable, then one must solve the integral equation (17.18), and this may require considerable mathematical skills. In fact, (17.18) is, in general, an integral equation of the Fredholm type, and the usual "iterated nuclei" procedure used to solve Fredholm integral equations may be particularly painful to achieve. The task may be much easier if one is able to reduce the Fredholm integral equation to a Volterra integral equation, in just the way shown in the book [4] for the time-rescaled Brownian motion in relation to Special Relativity.

But let us go back to the time variable t in the KL expansion (17.2). If this variable is *discrete*, rather than continuous, then the picture changes completely. In fact, the integral equation (17.2) now becomes ... a system of simultaneous algebraic equations of the first degree, that can *always* be solved! The difficulty here is that this system of linear equations is *huge*, because the autocorrelation matrix is huge (hundreds or thousands of elements are the rule for autocorrelation matrices in SETI and in other applications, like image processing and the like). Also huge are the eigenvalues of the characteristic equation (i.e., the algebraic equations whose roots are the KL). Can you imagine solving *directly* an algebraic equation of degree 10,000?

So, the KLT is practically impossible to find numerically, unless we resort to *simplifying tricks* of some kind. This is precisely what was done for the SETI-Italia program by this author and his students, strongly supported by Ing. Stelio Montebugnoli and his team [5].

17.6 THE KLT: JUST A LINEAR TRANSFORMATION IN THE HILBERT SPACE

Although we have explained the KL expansion (17.2), we have yet to explain what the *KLT* is! We do so in this section.

The next step towards the KLT proper is the rearrangement of the eigenvalues λ_n in decreasing order of magnitude. Suppose we have done this. Consequently, we also rearrange the eigenfunctions $\phi_n(t)$ so that each eigenfunction keeps corresponding to its own eigenvalue. It can be proved that no mismatch can possibly arise in doing so, inasmuch as each eigenfunction corresponds to one eigenvalue only—namely, it can be proved that there is no degeneracy (contrary to what happens in quantum physics, where, for instance, there is a lot of degeneracy in the eigenfunctions of even the simplest atom of all, the hydrogen atom!). Furthermore, all eigenvalues are positive, and so, once rearranged in decreasing order of magnitude, they form a decreasing sequence where the first eigenvalue is the largest, and is called the "dominant" eigenvalue by mathematicians.

We are now ready to compute the *Direct KLT* of the (signal + noise). Let us use the new set of eigen-axes to describe the (signal + noise). Then, in the new representation, the (signal + noise) is just the Direct KLT of the old (signal + noise). In other words, the KLT is properly called just a *linear* trasformation of axes, and nothing is easier than that! (Incidentally, this accounts for the title of Karhunen's first paper "Über Lineare Methoden in der Wahrscheinlichkeitsrechnung" = "On linear methods in the calculus of probabilities", [1], which obviously refers to the linear character of the transformation of axes in the Hilbert space.)

17.7 A BREAKTHROUGH ABOUT THE KLT: MACCONE'S "FINAL VARIANCE" THEOREM

The importance of the KLT as a mathematical tool superior to the FFT has already been pointed out. However, the implementation of the KLT by a numerical code running on computers has always been a difficult problem. Both François Biraud in France [6] and Bob Dixon in the U.S.A. [16] failed to do so in the 1980s because all computers then available could not make the N^2 calculations required to solve the huge system of simultaneous algebraic equations of the first degree corresponding (in the discrete case) to the integral equation (17.18). At the SETI-Italia facilities at Medicina we faced the same problem, of course. But we did better than our predecessors because we discovered the new theorem about the KLT that we demonstrate in this section and call "the Final Variance theorem". This new theorem seems to be even more important than the rest of research work about the KLT since it solves directly the problem of extracting a weak sinusoidal carrier (a tone) from noise of whatever kind (both colored and white).

The key idea of the Final Variance theorem is to differentiate the first eigenvalue (briefly called the "dominant eigenvalue") of the autocorrelation of the (noise + signal) with respect to the final instant T of the general KLT theory. Remember here that this final instant T simply does not exist in the ordinary Fourier theory, because this T equals infinity according to the Fourier theory. Therefore, the final instant T in itself is possibly the most important "novelty" introduced by the KLT regarding the classical FFT. With respect to T, we may take derivatives (called "final derivatives" in the remainder of this book because they are time derivatives taken with respect to the final instant T) and integrals that have no analogs in the ordinary Fourier theory. The "error" that was made in the past—even by many KLT scholars—was to set $T = 1$, thus obscuring the fundamental novelty represented by the finite, real positive T as a new continuous variable playing in the game! This error made by other scholars clearly appears, for instance, in the Wikipedia site about the "Karhunen–Loève Theorem", *http://en.wikipedia.org/wiki/Karhunen-Loève_theorem*. So, by removing this silly $T = 1$ convention we opened up new prospects for KLT theory, as we now show by proving our "Final Variance theorem".

Consider the eigenfunction expansion of the autocorrelation again—Equation (17.16)—with the traditional dummy index n rewritten instead of m. Upon replacing $t_1 = t_2 = t$, this equation becomes

$$E\{X^2(t)\} = \sum_{n=1}^{\infty} \lambda_n \phi_n^2(t). \tag{17.19}$$

Since the eigenfunctions $\phi_n(t)$ are normalized to 1, we are prompted to integrate both sides of (17.19) with respect to t between 0 and T, so that the integral of the square of the $\phi_n(t)$ becomes just 1:

$$\int_0^T E\{X^2(t)\}dt = \sum_{n=1}^{\infty} \lambda_n \int_0^T \phi_n^2(t)\,dt = \sum_{n=1}^{\infty} \lambda_n. \tag{17.20}$$

On the other hand, since the mean value of $X(t)$ is identically equal to 0, one may now introduce the *variance* $\sigma^2_{X(t)}$ of the stochastic process $X(t)$ defined by

$$\sigma^2_{X(t)} = E\{X^2(t)\} - E^2\{X(t)\} = E\{X^2(t)\}. \tag{17.21}$$

Replacing (17.21) into (17.20), one gets

$$\int_0^T \sigma^2_{X(t)}\,dt = \sum_{n=1}^{\infty} \lambda_n. \tag{17.22}$$

This formula was first given by this author in his 1994 book [4, eq. (1.13), p. 12]. At that time, however, (17.22) was regarded as interesting inasmuch as (upon interchanging the two sides) it proves that the series of all the eigenvalues λ_n is indeed convergent (as one would intuitively expect) and its sum is given by the integral of the variance between 0 and T.

Back in 1994, however, the author did not understand that (17.22) had a more profound meaning: since the final instant T is the upper limit of the time integral on the left-hand side, the right-hand side also must depend on T. In other words, all the eigenvalues λ_n must be some functions of the final instant T:

$$\lambda_n \equiv \lambda_n(T). \tag{17.23}$$

This new remark is vital in order to make further progress. In fact, one is now prompted to let the integral on the left-hand side of (17.22) disappear by differentiating both sides with respect to the final instant T. One thus gets:

$$\sigma^2_{X(T)} = \sum_{n=1}^{\infty} \frac{\partial \lambda_n(T)}{\partial T}. \tag{17.24}$$

This result we call the *Final Variance theorem*. It was discovered by this author in May 2007 and is the key new result put forward in this chapter. It states that for any (either non-stationary or stationary) stochastic process $X(t)$, *the Final Variance* $\sigma^2_{X(T)}$ *is the sum of the series of the first-order partial derivatives of the eigenvalues* $\lambda_n(T)$ *with respect to the final instant* T.

Let us now consider a few particular cases of this theorem that are especially interesting.

(1) In general, only the first N terms of the decreasing sequence of eigenvalues will be retained as "significant" by the user, and all the other terms, from the $(N+1)$th term onward, will be declared to be "just noise". Therefore, the infinite series in (17.24) becomes in practice the finite sum

$$\sigma^2_{X(T)} \approx \sum_{n=1}^{N} \frac{\partial \lambda_n(T)}{\partial T}. \tag{17.25}$$

In numerical simulations, however, one always wants to make computation time as short as possible! Therefore, one might be led to consider the *first* (or *dominant*) eigenvalue only in (17.25); that is

$$\sigma^2_{X(T)} \approx \frac{\partial \lambda_1(T)}{\partial T}. \tag{17.26}$$

This clearly is "the roughest possible" approximation to the full $X(t)$ process since we are actually replacing the full $X(t)$ by its first KLT term $Z_1\phi_1(t)$ only. However, using (17.26) instead of the N-term sum (17.25) is indeed a good shortcut for application of the KLT to the extraction of very weak signals from noise, as we now stress in the very important practical case of stationary processes.

(2) If we restrict our considerations to *stationary* stochastic processes only (i.e., processes for which both the mean value and the variance are constant in time), then (17.25) simplifies even further. In fact, by definition, the stationary processes have the *same final variance at any time (i.e., for stationary processes* σ^2_X *is a constant)*. Then (17.22) immediately shows that, for stationary processes only, all the KLT eigenvalues are *linear* functions of the final instant T:

$$\lambda_n(T) \propto T \quad \textit{for stationary processes only.} \tag{17.27}$$

As a consequence, the first-order partial derivatives of all the λ_n with respect to T for stationary processes are just constants. In yet other words, for stationary processes only, (17.25) becomes

$$\sum_{n=1}^{N} \frac{\partial \lambda_n(T)}{\partial T} \approx \text{a constant with respect to } T. \tag{17.28}$$

In particular, if one sticks again to the first, dominant eigenvalue only (i.e., to the roughest possible approximation), then (17.28) reduces to

$$\frac{\partial \lambda_1(T)}{\partial T} \approx \text{a constant with respect to } T. \tag{17.29}$$

In Section 17.8 we will discuss the deep, practical implications of this result for SETI, extrasolar planet detection, asteroidal radar, and other KLT applications.

(3) Please notice that, for non-stationary processes, the dependence of the eigenvalues on T certainly is non-linear. For instance, for the well-known Brownian motion (i.e., "the easiest of the non-stationary processes"), one has

$$\lambda_n(T) = \frac{4T^2}{\pi^2(2n-1)^2} \quad (n = 1, 2, \ldots) \tag{17.30}$$

and so the dependence on T is quadratic. For the proof, just place the Brownian motion variance $\sigma^2_{B(t)} = t$ into (17.22) and perform the integration, yielding the T^2 directly. Of course, this is in agreement with (17.30), which will be proven in Chapter 21 when we search for the KLT of the standard Brownian motion—see, in particular, (21.21).

(4) Even higher than quadratic is the dependence on T for the eigenvalues of other highly non-stationary processes. For instance, for the zero-mean square of the Brownian motion, the KLT eigenvalues depend cubically on the final instant T, as will be proven in Chapter 24 by Equation (24.60). And so on for more complicated processes, like the time-rescaled squared Brownian motions whose KLT will found in Chapter 24.

17.8 BAM ("BORDERED AUTOCORRELATION METHOD") TO FIND THE NUMERIC KLT OF *STATIONARY* PROCESSES ONLY

The BAM (an acronym for "Bordered Autocorrelation Method") is an alternative numerical technique to evaluate the KLT of stationary processes (only) that may run faster on computers than the traditional full-solving KLT technique described in Section 17.5. The BAM has its mathematical foundation in our Final Variance theorem already proved in Section 17.7. In this section we describe the BAM in detail and provide the results of numerical simulations showing that, by virtue of the BAM, the KLT succeeds in extracting a sinusoidal carrier embedded in a lot of noise when the FFT utterly fails.

Let us start by recalling that the standard, traditional technique to find the KLT of any stochastic process (whether stationary or not) numerically amounts to solving N simultaneous linear algebraic equations whose coefficient matrix is the (huge) autocorrelation matrix. This N^2 amount of calculations is much larger than the $N * \ln(N)$ amount of calculations required by the FFT and that's precisely the reason the FFT has been preferred to the KLT in the last 50 years!

Because of the Final Variance theorem proved in the previous section, however, one is tempted to confine oneself to the study of the dominant eigenvalue, only by virtue of just using (17.29). This means studying (17.29) for different values of the final instant T (i.e., as a function of the final instant T).

Also, we now confine ourselves to a *stationary* $X(t)$ over a *discrete* set of instants $t = 0, \ldots, N$.

In this case, the autocorrelation of $X(t)$ becomes the Toeplitz matrix (for an introduction to the research field of Toeplitz matrices, see the Wikipedia site, *http://en.wikipedia.org/wiki/Toeplitz_matrix*) which we denote by $R_{Toeplitz}$.

$$R_{Toeplitz} = \begin{bmatrix} R_{XX}(0) & R_{XX}(1) & R_{XX}(2) & \cdots & \cdots & R_{XX}(N) \\ R_{XX}(1) & R_{XX}(0) & R_{XX}(1) & \cdots & \cdots & R_{XX}(N-1) \\ R_{XX}(2) & R_{XX}(1) & R_{XX}(0) & \cdots & \cdots & R_{XX}(N-2) \\ \cdots & \cdots & \cdots & R_{XX}(0) & \cdots & \cdots \\ R_{XX}(N) & R_{XX}(N-1) & \cdots & \cdots & R_{XX}(1) & R_{XX}(0) \end{bmatrix}. \tag{17.31}$$

This theorem had already been proven by Bob Dixon and Mike Kline back in 1991 [16], and will not be proven here again. We may choose N at will, but clearly the higher the N, the more accurate the KLT of $X(t)$. On the other hand, the final instant T in the KLT can be chosen at will and now is $T = N$. So, we can regard $T = N$ as a sort of "new time variable" and even take derivatives with respect to it, as we'll do in a moment.

But let us now go back to the Toeplitz autocorrelation (17.31). If we let N vary as a new free variable, that amounts to *bordering it* (i.e., adding one (last) column and one (last) row to the previous correlation). This means solving yet again the system of linear algebraic equations of the KLT for $N + 1$, rather than for N. So, *for each different value of N, we get a new value of the first eigenvalue λ_1 now regarded as a function of N (i.e., $\lambda_1(N)$). Doing this over and over again, for as many values as we wish (or, more correctly, for how many values of N our computer can still handle!) constitutes our BAM, the Bordered Autocorrelation Method.*

But then we know from the Final Variance theorem that $\lambda_1(N)$ is proportional to N. And such a function $\lambda_1(N)$ of course has a derivative, $\partial\lambda_1(N)/\partial N$, that can be computed numerically as a new function of N. And this derivative turns out to be a constant with respect to N. This fact paves the way for a new set of applications of the KLT to all fields of science!

In fact, numeric simulations lead to the results shown in the four plots in Figures 17.1–17.4. The first plot is the ordinary Fourier spectrum of a pure tone at 300 Hz buried in noise with a signal-to-noise ratio of 0.5, abbreviated hereafter as SNR = 0.5. For a definition of the SNR see the Wikipedia site, *http://en.wikipedia.org/wiki/Signal-to-noise_ratio* Please note the following two facts:

(1) This is about as low an SNR can be before the FFT starts failing to denoise a signal, as is well known by electrical and electronic engineers.
(2) This Fourier spectrum is obviously computed by taking the Fourier Transform of the *stationary* autocorrelation of $X(t)$, as is well known from the Wiener–Khinchin theorem (for a concise description of this theorem, see *http://en.wikipedia.org/wiki/Wiener-Khinchin_theorem*).

Notice, however, that this procedure would *not* work for non-stationary $X(t)$ because the Wiener–Khinchin theorem does not apply to non-stationary processes. For

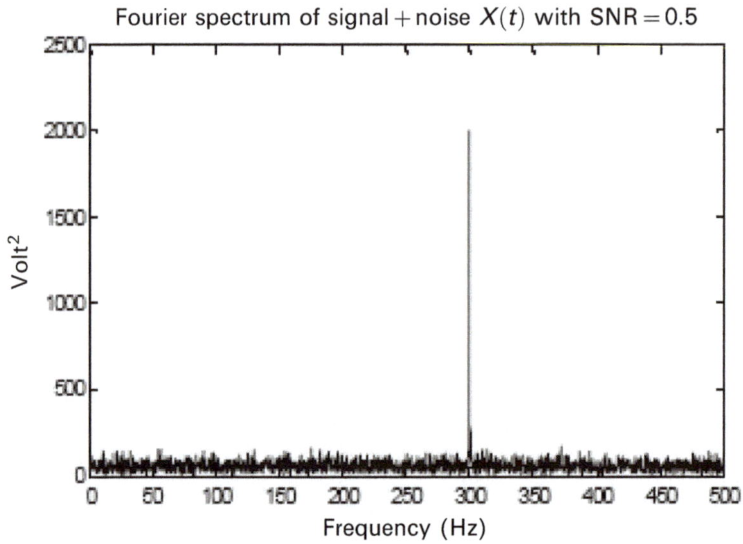

Figure 17.1. Fourier spectrum of a pure tone (i.e., just a sinusoidal carrier) with frequency at 300 Hz buried in stationary noise with a signal-to-noise ratio of 0.5.

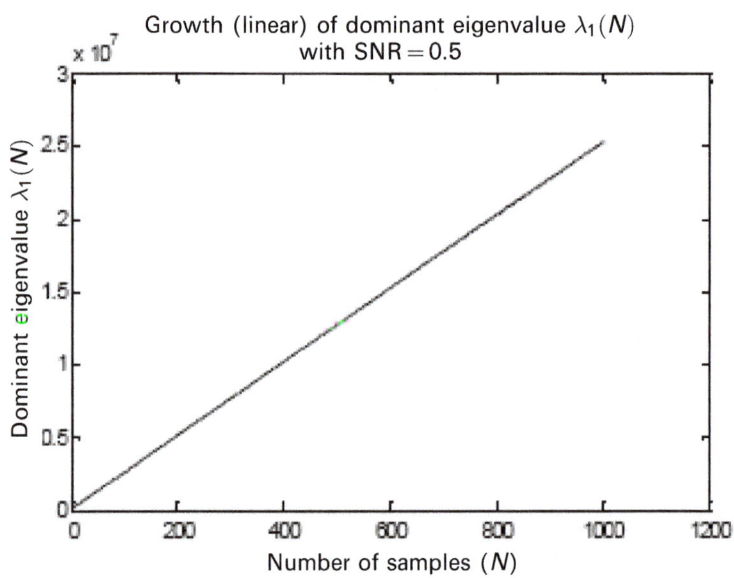

Figure 17.2. The KLT dominant eigenvalue $\lambda_1(N)$ over $N = 1{,}200$ time samples, computed by virtue of the BAM, the Bordered Autocorrelation Method.

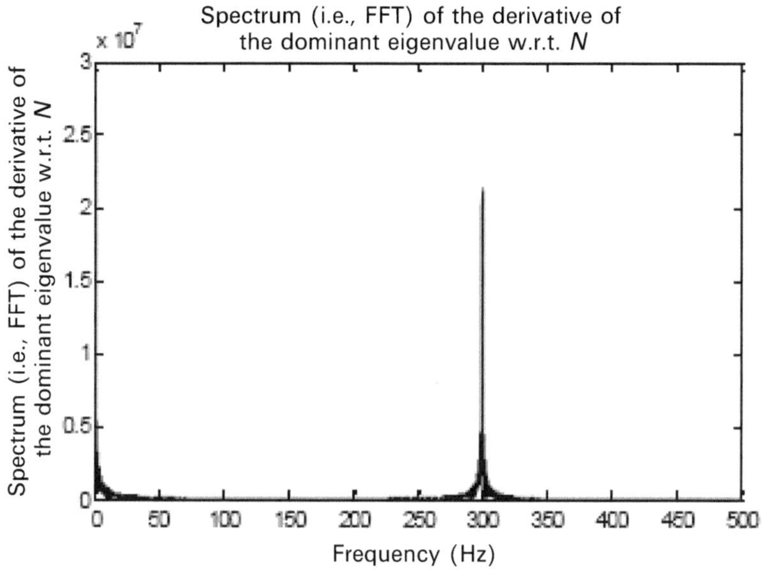

Figure 17.3. The spectrum (i.e., the Fourier Transform) of the *constant* derivative of the KLT dominant eigenvalue $\lambda_1(N)$ with respect to N as given by the BAM. This is clearly a Dirac delta function (i.e., a peak, at 300 Hz), as expected.

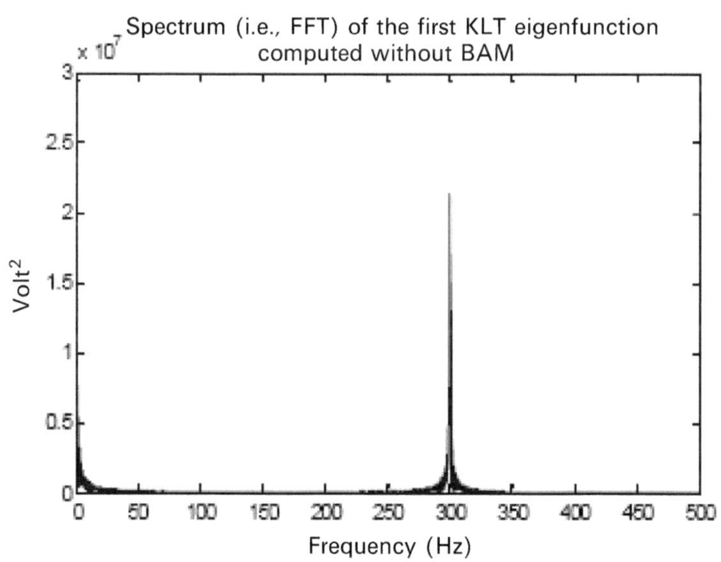

Figure 17.4. The spectrum (i.e., the Fourier Transform) of the first KLT eigenfunction *not* obtained by the BAM, but rather by the very long procedure of solving N linear algebraic equations corresponding, in discrete time, to the integral equation (17.18). Clearly, the result is the same as obtained in Figure 17.3 by the much less time-consuming BAM. So, one can say that adoption of the BAM actually made the KLT "feasible" on small computers by circumventing the difficulty of the N^2 calculations requested by the "straight" KLT theory.

non-stationary processes there are other "tricks" to compute the spectrum from the autocorrelation, like the Wigner–Ville Transform, but we shall not consider them here.

The second plot (Figure 17.2) shows the first (i.e., the dominant) KLT eigenvalue $\lambda_1(N)$ over $N = 1{,}200$ time samples. Clearly, this $\lambda_1(N)$ is proportional to N, as predicted by our Final Variance theorem (17.27).

So, its derivative, $\partial\lambda_1(N)/\partial N$, is a constant with respect to N. But we may then take the Fourier Transform of such a constant and get a Dirac delta function (i.e., a peak just at 300 Hz). In other words, we have KLT-reconstructed the original tone by virtue of the BAM. The third plot (Figure 17.3) shows such a BAM-reconstructed peak.

Finally, this plot is of course identical to the fourth plot (Figure 17.4), showing the ordinary FFT of the first KLT eigenfuction as obtained, not by the BAM, but by solving the full and long system of N algebraic first-degree equations.

Let us now do the same again ... but with an incredibly low SNR of 0.005.

Poor Fourier here is in a mess! Just look at the plot in Figure 17.5! No classical FFT spectrum can be identified at all for such a terribly low SNR!

But for the KLT no problem!

The next plot (Figure 17.6) shows that $\lambda_1(N) \propto N$, as predicted by our Final Variance theorem (17.27).

The third plot (Figure 17.7, *KLT fast* way via the BAM) is the *neat KLT spectrum* of the 300 Hz tone obtained by computing the FFT of the *constant* $\partial\lambda_1(N)/\partial N$.

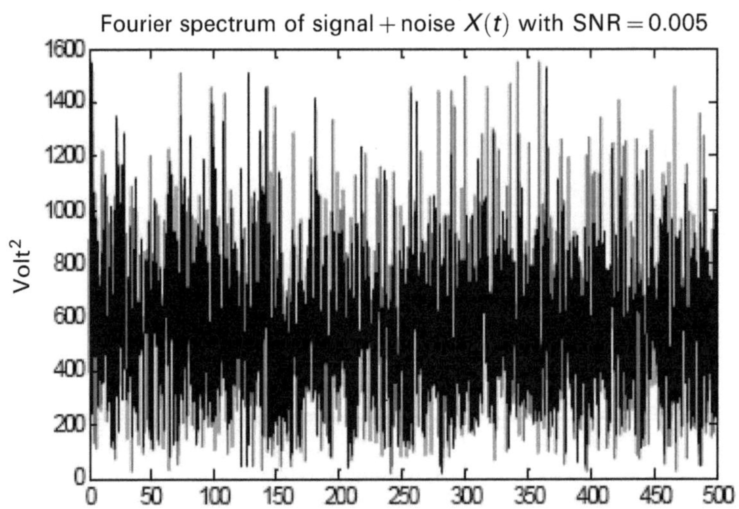

Figure 17.5. Fourier spectrum of a pure tone (i.e., just a sinusoidal carrier) with frequency at 300 Hz buried in stationary noise with the terribly low signal-to-noise ratio of 0.005. This is clearly beyond the reach of the FFT, since we know there should just be one peak only at 300 Hz. Fourier *fails* at such a low SNR.

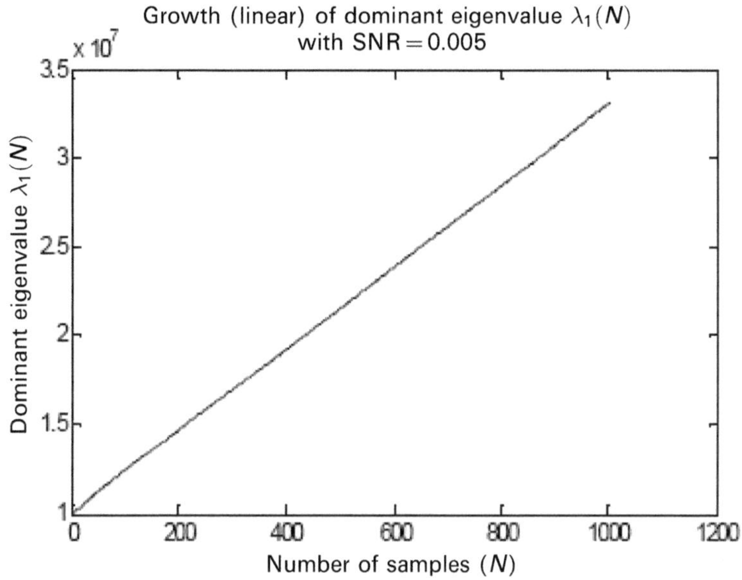

Figure 17.6. The KLT dominant eigenvalue $\lambda_1(N)$ for $N = 1{,}200$ time samples, computed by virtue of the BAM, for the very low SNR = 0.005.

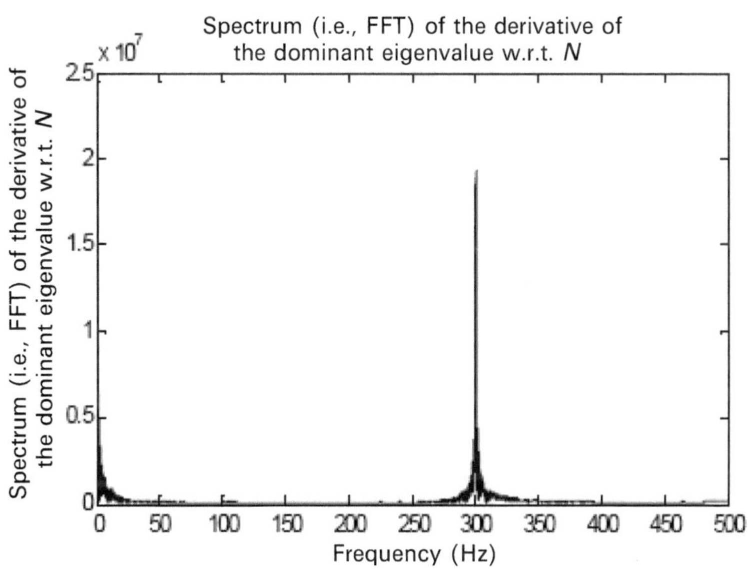

Figure 17.7. The spectrum (i.e., the Fourier Transform) of the *constant* derivative of the KLT dominant eigenvalue $\lambda_1(N)$ with respect to N as given by the BAM. This is a neat Dirac delta function (i.e., it has a peak at 300 Hz, as expected).

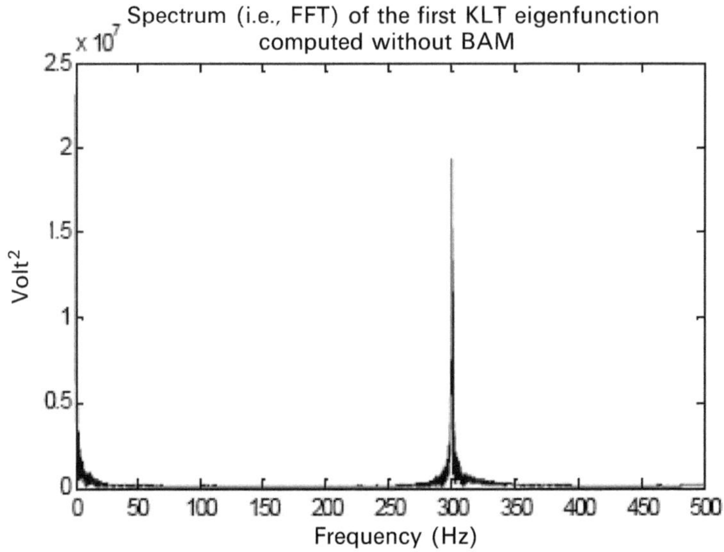

Figure 17.8. The spectrum (i.e., the Fourier Transform) of the first KLT eigenfunction, *not* obtained by the BAM but rather by the very long procedure of solving N linear algebraic equations corresponding, in discrete time, to the integral equation (17.18). Clearly, the result is the same as obtained in Figure 17.7, but this time by the much less time-consuming BAM. So, one can say that the adoption of the BAM actually made the KLT "feasible" on small computers by circumventing the difficulty of N^2 calculations requested by the "straight" KLT theory.

And this is just the same as the last plot (Figure 17.4) of the dominant KLT eigenfunction obtained by the KLT *slow* way of doing N^2 calculations. This proves the superior behavior of the KLT.

17.9 DEVELOPMENTS IN 2007 AND 2008

The numerical simulations described in the previous section were performed at Medicina during the winter 2006–2007 by Francesco Schillirò and Salvatore "Salvo" Pluchino [22]. These simulations suggested in a purely numerical fashion (i.e., without any analytic proof) that the BAM leads to the following result for stationary processes: the ordinary Fourier transform (i.e., "the spectrum" in the common sense, since the processes are supposed to be stationary) of the first-order partial derivative with respect to the final instant T of the dominant eigenvalue, $\frac{\partial \lambda_1(T)}{\partial T}$, is just the frequency of the feeble sinusoidal carrier buried in the mountain of noise. In SETI language, if we are looking for a simple sinusoidal carrier sent by ET and buried in a lot of cosmic noise, then the frequency we are looking for is given by the FFT of $\dfrac{\partial \lambda_1(T)}{\partial T}$.

Why?

No analytic proof of this numerical result was ever found at Medicina. But this author had made the first step towards the then missing analytic proof by proving the Final Variance Theorem in May 2007, and persisted in discussing this "frontier result" with other radioastronomers. One year later, in June 2008, he went to Dwingeloo, the Netherlands, and met with the ASTRON Team working on a possible implementation of SETI on the brand-new LOFAR radiotelescope. Dr. Sarod Yatawatta of ASTRON then made the next step toward the missing analytic proof: he derived an unknown analytic expression for the KLT eigenvalues of the ET sinusoidal carrier [24]. Unfortunately, Dr. Yatawatta made two analytical errors in his derivation (described hereafter), which this author discovered and corrected in September 2008.

In conclusion, the final, correct version of all these equations is explained in the next two sections, and it proves that the Fourier Transform of the first derivative of the KLT eigenvalues with respect to the final instant T is indeed the frequency of the "unknown" ET signal, but only for stationary processes, of course.

For non-stationary processes (i.e., for transient phoenomena as actually happens in practical SETI, since all celestial bodies move, rather than rest), the story is much more complicated, and this author is convinced that a much more refined mathematical investigation has to be made: but this will be our next step, not described in this book yet!

17.10 KLT OF STATIONARY WHITE NOISE

Before we give the analytic proof that the Fourier Transform of $\dfrac{\partial \lambda_1(T)}{\partial T}$ is the frequency of the unknown ET signal, we must understand what the KLT of stationary white noise is.

Stationary white noise is defined as the one "limit" stochastic process that is completely uncorrelated (i.e., the autocorrelation of which is the Dirac delta function). In other words, denoting the stationary white noise by $W(t)$, one has by definition

$$E\{W(t_1)W(t_2)\} = \delta(t_1 - t_2). \tag{17.32}$$

If one now seeks the KLT of stationary white noise, one must of course insert the autocorrelation (17.32) into the KLT integral equation (17.18), getting

$$\lambda_n \phi_n(t_2) = \int_0^T E\{W(t_1)W(t_2)\}\phi_n(t_1)\,dt_1 = \int_0^T \delta(t_1 - t_2)\phi_n(t_1)\,dt_1 = \phi_n(t_2). \tag{17.33}$$

This proves that:

(1) The KLT eigenvalues of stationary white noise are all equal to 1.
(2) *Any* set of orthonormal eigenfunctions $\phi_n(t)$ in the Hilbert space is a suitable basis to represent stationary white noise.

Since *any* set of orthonormal eigenfunctions $\phi_n(t)$ in the Hilbert space is a suitable basis to represent stationary white noise, from now one we shall adopt the easiest possible such basis; that is, the simple Fourier basis made up only by orthonormalized sines over the finite interval $0 \le t \le T$:

$$\phi_n(t) = \sqrt{\frac{2}{T}} \sin\left(\frac{2\pi n}{T} t\right) \equiv W_n(t). \tag{17.34}$$

Of course, this set of basis functions fulfills the orthonormality condition

$$\int_0^T W_m(t) W_n(t)\, dt = \int_0^T \sqrt{\frac{2}{T}} \sin\left(\frac{2\pi m}{T} t\right) \cdot \sqrt{\frac{2}{T}} \sin\left(\frac{2\pi n}{T} t\right) dt = \delta_{mn}. \tag{17.35}$$

This property will be used in the next section, where we give the proof that the Fourier Transform of $\frac{\partial \lambda_n(T)}{\partial T}$ is indeed (twice) the frequency of the unknown ET sinusoidal carrier buried in white, cosmic noise. We conclude this section by pointing out the first analytical error made by Dr. Yatawatta in his personal communication to this author [24]: he forgot to put the square root in (17.34). This of course means that his further results were flawed, even more so since he made a second analytical error later, which we shall not describe. But the key ideas behind his proof were perfectly correct, and we shall describe them in the next section.

17.11 KLT OF AN ET SINUSOIDAL CARRIER BURIED IN WHITE, COSMIC NOISE

Consider a new stochastic process $S(t)$ made up by the sum of stationary white noise $W(t)$ plus an alien ET sinusoidal carrier of amplitude a and frequency $\nu = \frac{\omega}{2\pi}$; that is,

$$S(t) = W(t) + a \sin(\omega t). \tag{17.36}$$

What is the KLT of such a (signal + noise) process? This is the central problem of SETI, of course.

To find the answer, first build up the autocorrelation of this process:

$$\begin{aligned} E\{S(t_1)S(t_2)\} &= E\{W(t_1)W(t_2)\} + a^2 \sin(\omega t_1) \sin(\omega t_2) \\ &\quad + aE\{W(t_1) \sin(\omega t_2)\} + aE\{W(t_2) \sin(\omega t_1)\}. \end{aligned} \tag{17.37}$$

The last two terms in (17.37) represent the two cross-correlations between the white noise and the sinusoidal signal. It is reasonable to assume that the white noise and the signal are uncorrelated, and so we shall simply replace these two cross-correlations by zero. The autocorrelation (17.37) of the (signal + noise) stochastic process $S(t)$ thus becomes

$$E\{S(t_1)S(t_2)\} = E\{W(t_1)W(t_2)\} + a^2 \sin(\omega t_1)\sin(\omega t_2). \tag{17.38}$$

In order to proceed, we now make use of the eigenfunction expansion of the autocorrelation (17.16), which, replaced into (17.38), changes it into

$$\sum_{m=1}^{\infty} \lambda_{S_m} S_m(t_1) S_m(t_2) = \sum_{m=1}^{\infty} \lambda_{W_m} W_m(t_1) W_m(t_2) + a^2 \sin(\omega t_1)\sin(\omega t_2). \tag{17.39}$$

In the last equation, the $S_m(t)$ clearly are the (unknown) eigenfunctions of the (signal + noise) process $S(t)$, and the λ_{S_m} are (unknown) corresponding eigenvalues. In the right-hand side, the λ_{W_m} are the eigenvalues of the stationary white noise, which we know to be equal to 1, but, for the sake of clarity, let us keep the symbol λ_{W_m} rather than replacing it by 1.

To proceed further, we now must get rid of both t_1 and t_2 in (17.39), and there is only one way to do so: use the orthonormality of the eigenfuctions appearing in (17.39). We shall do so in a moment. Before, however, let us make the following practical consideration: since the signal is much waker than the noise (by assumption) (i.e., the signal-to-noise ratio is much smaller than 1, or SNR $\ll$ 1), then, numerically speaking, the (signal + noise) eigenfunctions $S_m(t)$ must not differ very much from the pure white noise eigenfunctions $W_m(t)$. And, similarly, the (signal + noise) eigenvalues λ_{S_m} must not differ very much from the corresponding pure white noise eigenvalues λ_{W_m}. In other words, the hypothesis that SNR $\ll$ 1 amounts to the two approximate equations

$$\left.\begin{aligned} S_m(t) &\approx W_m(t) \\ \lambda_{S_m} &\approx \lambda_{W_m} = 1. \end{aligned}\right\} \tag{17.40}$$

Of course, only the first of these two equations will play a role in the two integrations that we are now going to perform: once with respect to t_1 and once with respect to t_2, and both over the interval $0 \le t \le T$. As a consequence, the new orthonormality condition (nearly) holds:

$$\int_0^T S_m(t_1) W_n(t_1)\, dt_1 \approx \delta_{mn} \tag{17.41}$$

and, similarly,

$$\int_0^T S_k(t_2) W_n(t_2)\, dt_2 \approx \delta_{kn} \tag{17.42}$$

So, let us now multiply both sides of (17.39) by $W_n(t_1)$ and integrate with respect to t_1 between 0 and T. Because of (17.41) and (17.35) one has:

$$\sum_{n=1}^{\infty} \lambda_{S_n} S_n(t_2) \approx \sum_{n=1}^{\infty} \lambda_{W_n} W_n(t_2) + a^2 \sin(\omega t_2) \int_0^T W_n(t_1)\sin(\omega t_1)\, dt_1 \tag{17.43}$$

The good point is that the integral appearing in the right-hand side of this equation can be found. In fact, replacing $W_n(t_1)$ by virtue of (17.34) and integrating, one gets

$$\sum_{k=1}^{\infty} \lambda_{S_k} S_k(t_2) \approx \sum_{k=1}^{\infty} \lambda_{W_k} W_k(t_2) + a^2 \sin(\omega t_2) \cdot \frac{2\sqrt{2}\pi n \sqrt{T} \sin(\omega T)}{\omega^2 T^2 - 4\pi^2 n^2} \tag{17.44}$$

We next multiply this equation by $W_n(t_2)$ and integrate with respect to t_2 between 0 and T. Because of (17.42) and (17.35), (17.44) becomes:

$$\lambda_{S_n} \approx \lambda_{W_n} + a^2 \frac{2\sqrt{2}\pi n \sqrt{T} \sin(\omega T)}{\omega^2 T^2 - 4\pi^2 n^2} \int_0^T W_n(t_2) \sin(\omega t_2)\, dt_2. \tag{17.45}$$

Again, the integral in the last equation can be computed—it is actually the same integral as in (17.43)—and so the conclusion is

$$\lambda_{S_n} \approx \lambda_{W_n} + a^2 \frac{8\pi^2 n^2 T \sin^2(\omega T)}{(\omega^2 T^2 - 4\pi^2 n^2)^2}. \tag{17.46}$$

This is Yatawatta's main result (corrected by Maccone). Let us now point out clearly that the eigenvalues on the left are a function of the final instant T; that is,

$$\lambda_{S_n}(T) \approx \lambda_{W_n} + a^2 \frac{8\pi^2 n^2 T \sin^2(\omega T)}{(\omega^2 T^2 - 4\pi^2 n^2)^2}. \tag{17.47}$$

This equation clearly shows that

(1) For $T \to 0$, the fraction in the right-hand side approaches zero, and so the eigenvalues of the signal + noise approach the pure white noise eigenvalues (as is intuitively obvious).
(2) For $n \to \infty$, again the fraction in the right-hand side approaches zero, and so the eigenvalues of the signal + noise approach the pure white noise eigenvalues (as again is intuitively obvious). This result may justify numerically the practical approximation made by the Medicina engineers when they confined their simulations to the first eigenvalue only (roughest approximation). In other words, the dominant eigenvalue of the signal + noise is given by

$$\lambda_{S_1}(T) \approx \lambda_{W_1} + a^2 \frac{8\pi^2 T \sin^2(\omega T)}{(\omega^2 T^2 - 4\pi^2)^2} = 1 + a^2 \frac{8\pi^2 T \sin^2(\omega T)}{(\omega^2 T^2 - 4\pi^2)^2}. \tag{17.48}$$

This completes our analysis of the KLT of a sinusoidal carrier buried in white, cosmic noise.

17.12 ANALYTIC PROOF OF THE BAM–KLT

We are now ready for the analytic proof of the BAM-KLT method.

Let us first re-write (17.47) in a form in which the pure white noise eigenvalues are replaced by 1:

$$\lambda_{S_n}(T) \approx 1 + a^2 \frac{8\pi^2 n^2 T \sin^2(\omega T)}{(\omega^2 T^2 - 4\pi^2 n^2)^2}. \tag{17.49}$$

We then notice that the final instant T appears three times in the right-hand side of the last equation:

(1) once in the numerator outside the sine;
(2) once in the numerator inside the sine;
(3) once in the denominator.

Therefore, the partial derivative of (17.49) with respect to T will be made up by the sum of three terms:

(1) One term with the derivative of the T in the numerator (i.e., 1 times the sine square). This brings a term in the cosine of TWICE the sine argument, since one obviously has
$$\sin^2(\omega T) = \tfrac{1}{2} - \tfrac{1}{2}\cos(2\omega T). \tag{17.50}$$
(2) One term with the derivative of the T inside the sine. This brings a term in the sine of TWICE the sine argument, because one has
$$2\sin(\omega T)\cos(\omega T) = \sin(2\omega T). \tag{17.51}$$
(3) One term with the derivative of the T in the denominator. This does not bring any term in either the sine or the cosine, but just a rational function of T that we shall give in a moment. In fact, we now prefer to skip the lengthy and tedious steps leading to the derivative of (17.49) with respect to T and just give the final result.

In conclusion, the derivative of (17.49) with respect to T is given by the following sum of three terms:

$$\frac{\partial \lambda_{S_n}(T)}{\partial T} \approx \mathrm{Coeff}_1(T)\cdot\sin(2\omega T) + \mathrm{Coeff}_2(T)\cdot\cos(2\omega T) + \mathrm{Coeff}_3(T) \tag{17.52}$$

where the three coefficients turn out to be (after lengthy calculations)

$$\left.\begin{aligned} \mathrm{Coeff}_1(T) &= a^2\frac{8\pi^2 n^2\omega T}{(\omega^2T^2 - 4\pi^2n^2)^2}, \\ \mathrm{Coeff}_2(T) &= a^2\frac{4\pi^2n^2(3\omega^2T^2 + 4\pi^2n^2)}{(\omega^2T^2 - 4\pi^2n^2)^3}, \\ \mathrm{Coeff}_3(T) &= -a^2\frac{4\pi^2n^2(3\omega^2T^2 + 4\pi^2n^2)}{(\omega^2T^2 - 4\pi^2n^2)^3}. \end{aligned}\right\} \tag{17.53}$$

But the right-hand side of (17.52) is no more than ... the simple Fourier series expansion of $\frac{\partial \lambda_{S_n}(T)}{\partial T}$. Moreover, (17.52) shows that $\frac{\partial \lambda_{S_n}(T)}{\partial T}$ is a *periodic* function of T with frequency $2\omega T$. We conclude that: The Fourier transform of $\frac{\partial \lambda_{S_n}(T)}{\partial T}$ equals *twice* the frequency of the buried alien sinusoidal carrier. In other words, the frequency of the alien signal is a HALF of the frequency found by taking the Fourier transform of $\frac{\partial \lambda_{S_n}(T)}{\partial T}$.

And the BAM–KLT method is thus proved analytically.

17.13 KLT SIGNAL-TO-NOISE RATIO (SNR) AS A FUNCTION OF THE FINAL T, EIGENVALUE INDEX n, AND ALIEN FREQUENCY ν

We now derive a consequence from the eigenvalue relationship (17.47) dealing with the signal-to-noise ratio (abbreviated SNR) in the KLT theory. We shall call it the "KLT–SNR Theorem". The proof is as follows.

Consider Equation (17.10), showing that the eigenvalues λ_n of any KL expansion are actually the variances of the zero-mean corresponding uncorrelated (i.e., orthogonal, in the probabilistic sense) random variables Z_n. If we apply this to the KLT of stationary unitary white noise, described in Section 17.10, the conclusion is that the λ_{W_m} are the mean values of the square of the corresponding orthogonal (i.e., uncorrelated random variables $Z^2_{W_n}$)

$$\lambda_{W_m} = E\{Z^2_{W_n}\}. \tag{17.54}$$

Now, the definition of the signal-to-noise ratio (which we prefer to denote SNR, rather than S/R) of a sinusoidal signal with amplitude a buried in the noise with amplitude Z_{W_n} is just:

$$\text{SNR} = \frac{\text{power of the signal}}{\text{power of the noise}} = \frac{a^2}{E\{Z^2_{W_n}\}} = \frac{a^2}{\lambda_{W_m}}. \tag{17.55}$$

This definition can now be inserted into (17.47) divided by λ_{W_m}; that is,

$$\frac{\lambda_{S_n}(T)}{\lambda_{W_n}} \approx 1 + \frac{a^2}{\lambda_{W_n}} \cdot \frac{8\pi^2 n^2 T \sin^2(\omega T)}{(\omega^2 T^2 - 4\pi^2 n^2)^2}, \tag{17.56}$$

with the result that (17.56) is changed into

$$\frac{\lambda_{S_n}(T)}{\lambda_{W_n}} \approx 1 + \text{SNR} \cdot \frac{8\pi^2 n^2 T \sin^2(\omega T)}{(\omega^2 T^2 - 4\pi^2 n^2)^2}. \tag{17.57}$$

Solving this for SNR yields

$$\text{SNR}(T, n, \omega) \approx \left(\frac{\lambda_{S_n}(T)}{\lambda_{W_n}} - 1\right) \cdot \frac{(\omega^2 T^2 - 4\pi^2 n^2)^2}{8\pi^2 n^2 T \sin^2(\omega T)}. \tag{17.58}$$

For SETI applications, it may be preferable to re-express the last formula directly in terms of the "alien" frequency $\nu = \frac{\omega}{2\pi}$, instead of ω. Equation (17.58) is thus changed into

$$\text{SNR}(T, n, \nu) \approx \left(\frac{\lambda_{S_n}(T)}{\lambda_{W_n}} - 1\right) \cdot \frac{2\pi^2(\nu^2 T^2 - n^2)^2}{n^2 T \sin^2(2\pi\nu T)}. \tag{17.59}$$

This is our KLT–SNR Theorem. Since the quantity

$$\left(\frac{\lambda_{S_n}(T)}{\lambda_{W_n}} - 1\right) > 0 \tag{17.60}$$

has a positive numeric value just slighty above zero, from (17.59) we conclude that

$$\begin{cases} \mathrm{SNR}(T, n, \nu) = O(T^3) & \text{as } T \to \infty \\ \mathrm{SNR}(T, n, \nu) = O(n^2) & \text{as } n \to \infty \\ \mathrm{SNR}(T, n, \nu) = O(\nu^4) & \text{as } \nu \to \infty. \end{cases} \tag{17.61}$$

These equations yield the "pace of increase" of the KLT–SNR, and should be of importance in writing down the numeric codes for the actual implementation of the KLT.

17.14 GLOBAL NAVIGATION SATELLITE SYSTEMS (GNSS) IMPROVED BY THE KLT

GNSS is an acronym for Global Navigation Satellite Systems referring to the main four satellite navigation systems (i.e., GPS, GLONASS, Compass, and Galileo) already constructed or under construction in the U.S.A., Russia, China, and Europe, respectively. Several regional navigation satellite systems are also part of GNSS (see, e.g., the Wikipedia site: *http://en.wikipedia.org/wiki/Satellite_navigation*). All the GNSS face the difficult mathematical problem of extracting weak signals from background noise. If background noise was white over the allotted bandwidths, then the usual FFT would be the ideal mathematical solution to this problem. But, in reality, background noise is far from being white because of the huge amount of radio frequency interference (RFI) produced as a result of the growth of civilization, which then expands into space but fails to fade away sufficiently.

In June 2011 there was an important development for the application of the KLT to GNSS: a young telecommunication expert at the European Space Agency, Dr. Arkadiusz Szumski, published a review article [27] about this author's previous 2008 paper on the KLT and BAM-KLT [26]. Szumski also wrote *ex-novo* suitable computer codes, carried out several numerical simulations, and proved beyond any engineer's doubt that the KLT is enormously superior to the traditional FFT, especially for the GNSS. Space companies are thus likely to take the KLT seriously from now on. This author regrets that his lifetime of mathematical discoveries in this field will hardly bring him any economic benefit, but maybe some accolades.

Anyway, let us now briefly report on some key aspects of Szumski's paper [27].

Szumski considers wideband signal detection by the KLT. He describes a case where SNR $= -12$ dB (i.e., SNR $= -0.063$ where signal power is 6.3% of noise power). This is far too much for FFT. We do not reproduce the plots obtained by Szumski, but just confine ourselves to repeating his conclusive words: "To summarize the simulation results, one can observe that the KLT technique indeed is able to detect wideband signals even in the presence of strong noise, whereas the STFT

(Short Time Fourier Transform) and Wigner–Ville methods clearly fail." Szumski then continues: "We also analyzed a chirp signal with a wide-frequency boundary. This type of signal was chosen to enable us to evaluate the performance of KLT in detecting a dynamic, non-stationary signal. Figure 8 (of Szumski's paper [27]) shows the KLT spectrogram and the power spectrum. For a signal with SNR = -12 dB the spectrogram is a little frayed, but still readable. The KLT is able to detect the chirp signal in the noise."

In the section entitled "BAM-KLT: A step closer to fast KLT" of his paper [27], Szumski wrote: "Again, the biggest drawback of the KLT is its complexity and its resulting high computational burden. As with the Fourier transform, however, which became popular when its fast implementation (the FFT) became available, the KLT has the potential to experience a similar boost if a fast KLT implementation is discovered. Maccone has already presented an innovative way of using the KLT, which paves the way towards a faster algorithm." The article then continues by enlisting the key new formula of the BAM-KLT (i.e., the "Final Variance Theorem" (16.24)). Also reported by Szumski is its first-term-only (i.e. "dominant eigenvalue only") particular case (16.26).

In January 2012 this author personally thanked Szumski for having written the simulation codes based on the paper [27]. Hopefully, this will attract increasingly more attention to the KLT and BAM-KLT techniques by those who had previously not yet realized their terrific potential of replacing the FFT by the KLT and BAM-KLT for GPS, GLONASS, Galileo, and Compass. Szumski, however, did not consider the relativistic KLT described by this author in this book and in his 2009 book *Deep Space Flight and Communications*: ESA engineers pull back from showing interest in "*Star Trek* telecommunications", as that still smacks of sci-fi too much!

17.15 HOW TO EAVESDROP ON ALIEN CHAT

Following the Paris *First IAA Workshop on Searching for Life Signatures* (held at UNESCO, Paris, September 22–26, 2008, and organized by this author), the British popular science magazine *New Scientist* published the following article on October 30, 2008, that well summarizes the key features of the present scientific discussion.

How to eavesdrop on alien chat

30 October 2008

From *New Scientist* Print Edition.

Jessica Griggs

ET, phone ... each other? If aliens really are conversing, we are not picking up what they are saying. Now one researcher claims to have a way of tuning in to alien cellphone chatter.

On Earth, the signal used to send information via cellphones has evolved from a single carrier wave to a "spread spectrum" method of transmission. It's more efficient, because chunks of information are essentially carried on multiple low-powered carrier waves, and more secure because the waves continually change frequency so the signal is harder to intercept.

It follows that an advanced alien civilisation would have made this change too, but the search for extraterrestrial life (SETI) is not listening for such signals, says Claudio Maccone, co-chair of the SETI Permanent Study Group based in Paris, France.

An algorithm known as the Fast Fourier Transform (FFT) is the method of choice for extracting an alien signal from cosmic background noise. However, the technique cannot extract a spread spectrum signal. Maccone argues that SETI should use an algorithm known as the Karhunen–Loève Transform (KLT), which could find a buried conversation with a signal-to-noise ratio 1000 times lower than the FFT.

A few people have been "preaching the KLT" since the early 1980s but until now it has been impractical as it involves computing millions of simultaneous equations, something even today's supercomputers would struggle with. At a recent meeting in Paris called Searching for Life Signatures, Maccone presented a mathematical method to get around this burden and suggested that the KLT should be programmed into computers at the new Low Frequency Array telescope in the Netherlands and the Square Kilometre Array telescope, due for completion in 2012.

Seth Shostak at the SETI Institute in California agrees that the KLT might be the way to go but thinks we shouldn't abandon existing efforts yet. "It is likely that for their own conversation they use a spread-spectrum method but it is not terribly crazy to assume that to get our attention they might use a 'ping' signal that has a lot of energy in a narrow band—the kind of thing the FFT could find."

"It is likely that aliens use the same spread-spectrum method of transmission as us on their cellphones."

From issue 2680 of *New Scientist* magazine,
30 October 2008, p. 14.

17.16 CONCLUSIONS

Let us summarize the main results of this chapter.

When the stochastic process $X(t)$ is stationary (i.e., it has both mean value and variance constant in time), then there are two alternative ways to compute the first KLT dominant eigenfunction (i.e., the roughest approximation to the full KLT expansion, which may be "enough" for practical applications!):

(1) (long way)—either you compute the first eigenvalue from the autocorrelation and then solve the huge (N^2) system of linear equations to get the first eigenfunction;
(2) (short way = BAM)—or you compute the derivative of the first eigenvalue with respect to $T = N$ and then Fourier-transform it to get the first eigenfunction.

In practical, numerical simulations of the KLT it may be much less time-consuming to choose option (2) rather than option (1).

In either case, the KLT of a given stationary process can retrieve a sinusoidal carrier out of the noise for values of the signal-to-noise ratio (SNR) that are three orders of magnitude lower than those that the FFT can still filter out. In other words, while the FFT (at best) can filter out signals buried in noise with an SNR of about 1 or so, the KLT can, say, filter out signals that have an SNR of, say, 0.001 or so.

This is the superior achievement of the KLT over the FFT.

The BAM (Bordered Autocorrelation Method) is an alternative numerical technique to evaluate the KLT of stationary processes (only) that may run faster on computers than the traditional full-solving KLT technique. In this chapter we have provided the results of numerical simulations that show, by virtue of the BAM, how the KLT succeeds in extracting a sinusoidal carrier embedded in a lot of noise when the FFT utterly fails.

17.17 ACKNOWLEDGMENTS

The author is indebted to many radioastronomers and scientists who helped him over the years to work out what is now the BAM–KLT method. Principal among them are Ing. Stelio Montebugnoli and his SETI-Italia Team, Dr. Mike Garrett and his ASTRON Team (in particular Dr. Sarod Yatawatta), Dr. Jill Tarter and the SETI Institute Team (in particular Drs. Seth Shostak and Doug Vakoch). Also, the Paris *SETI Conference of September 22–26, 2008*, organized by this author at UNESCO, was possible only through the full support of the Secretary General of the IAA, Dr. Jean-Michel Contant, and of the newly-born French SETI community. Finally, a number of other young and not-so-young folks continued to support this author in his efforts for SETI over the years, and their help is hereby gratefully acknowledged.

17.18 REFERENCES

[1] K. Karhunen, "Über lineare Methoden in der Wahrscheinlichkeitsrechnung," *Ann. Acad. Sci. Fennicae, Series A 1, Math. Phys.*, **37** (1946), 3–79.

[2] M. Loève, "Fonctions Aléatoires de Second Ordre," *Rev. Sci.*, **84**(4) (1946), 195–206.

[3] M. Loève, *Probability Theory: Foundations, Random Sequencies*, Van Nostrand, Princeton, NJ, 1955.

[4] C. Maccone, *Telecommunications, KLT and Relativity*, Volume 1, IPI Press, Colorado Springs, CO, 1994, ISBN # 1-880930-04-8. This book embodies the results of some 30 research papers published by the author about the KLT in the 15-year span 1980–1994 in peer-reviewed journals.

[5] S. Montebugnoli, C. Bortolotti, D. Caliendo, A. Cattani, N. D'Amico, A. Maccaferri, C. Maccone, J. Monari, A. Orlati, P. P. Pari *et al.*, "SETI-Italia 2003 Status Report and First Results of a KL Transform Algorithm for ETI Signal Detection," paper IAC-03-IAA.9.1.02 presented at the *2003 International Astronautical Congress* held in Bremen, Germany, September 29–October 3, 2003.

[6] F. Biraud, "SETI at the Nançay Radio-telescope," *Acta Astronautica*, **10** (1983), 759–760.

[7] C. Maccone, "Advantages of the Karhunen–Loève Transform over Fast Fourier Transform for Planetary Radar and Space Debris Detection," *Acta Astronautica*, **60** (2007), 775–779.

17.19 ANNOTATED BIBLIOGRAPHY

In addition to the above references, we would like to offer an "enlightened" list of a few key references about the KLT, subdivided according to the field of application.

Early papers by the author about the KLT in mathematics, physics, and the theory of relativistic interstellar flight, subdivided by journals

Il Nuovo Cimento

[8] C. Maccone, "Special Relativity and the Karhunen–Loève Expansion of Brownian Motion," *Nuovo Cimento, Series B*, **100** (1987), 329–342.

Bollettino dell'Unione Matematica Italiana

[9] C. Maccone, "Eigenfunctions and Energy for Time-Rescaled Gaussian Processes," *Bollettino dell'Unione Matematica Italiana, Series 6*, **3-A** (1984), 213–219;

[10] C. Maccone, "The Time-Rescaled Brownian Motion $B(t^{2H})$," *Bollettino dell'Unione Matematica Italiana, Series 6*, **4-C** (1985), 363–378; C. Maccone, "The Karhunen–Loève Expansion of the Zero-Mean Square Process of a Time-Rescaled Gaussian Process," *Bollettino dell'Unione Matematica Italiana, Series 7*, **2-A** (1988), 221–229.

Journal of the British Interplanetary Society

[12] C. Maccone, "Relativistic Interstellar Flight and Genetics," *Journal of the British Interplanetary Society*, **43** (1990), 569–572.

Acta Astronautica

[13] C. Maccone, "Relativistic Interstellar Flight and Gaussian Noise," *Acta Astronautica*, **17**(9) (1988), 1019–1027.

[14] C. Maccone, "Relativistic Interstellar Flight and Instantaneous Noise Energy," *Acta Astronautica*, **21**(3) (1990), 155–159.

KLT for data compression

[15] C. Maccone, "The Data Compression Problem for the 'Gaia' Astrometric Satellite of ESA," *Acta Astronautica*, **44**(7–12) (1999), 375–384.

Some important papers about the KLT for SETI

[16] R. S. Dixon, and M. Klein, "On the detection of unknown signals," *Proceedings of the Third Decennial US–USSR Conference on SETI* held at the University of California at Santa Cruz, August 5–9, 1991. Later published in the *Astronomical Society of the Pacific (ASP) Conference Series* (Seth Shostak, Ed.), **47** (1993), 128–140.

[17] C. Maccone, *Karhunen–Loève versus Fourier Transform for SETI*, Lecture Notes in Physics, Springer-Verlag, Vol. 390 (1990), pp. 247-253. These are the *Proceedings* (J. Heidmann and M. Klein, Eds.) *of the Third Bioastronomy Conference* held in Val Cenis, Savoie, France, June 18–23, 1990.

After these seminal works were published, the importance of the KLT for SETI was finally acknowledged by SETI Institute experts in

[18] R. Eckers, K. Cullers, J. Billingham, and L. Scheffer (Eds.), *SETI 2020*, SETI Institute, Mountain View, CA, 2002, p. 234, note 13. The authors say: "Currently (2002) only the Karhunen Loeve (KL) transform [Mac94] shows potential for recognizing the difference between incidental radiation technology and white noise. The KL transform is too computationally intensive for the present generation of systems. The capability for using the KL transform should be added to future systems when computational requirements become affordable."

The paper [Mac94] referred to in the *SETI 2020* statement mentioned above is

[19] C. Maccone, "The Karhunen–Loève Transform: A Better Tool than the Fourier Transform for SETI and Relativity," *Journal of the British Interplanetary Society*, **47** (1994), 1.

An early paper about the KLT for SETI-Italia

[20] S. Montebugnoli, and C. Maccone, "SETI-Italia Status Report 2001", a paper presented at the 2001 *IAF Conference* held in Toulouse, France, October 1–5, 2001.

An early paper about the possibility of a "fast" KLT

[21] A. K. Jain, "A Fast Karhunen–Loève Transform for a Class of Random Processes," *IEEE Trans. Commun.*, **COM-24** (1976), 1023–1029.

Recent papers about the KLT and BAM–KLT

[22] F. Schillirò, S. Pluchino, C. Maccone, and S. Montebugnoli, *La KL Transform: considerazioni generali sulle metodologie di analisi ed impiego nel campo della Radio-astronomia*, Istituto Nazionale di Astrofisica (INAF)/Istituto di Radioastronomia (IRA), Technical Report, January 2007 [in Italian].
[23] C. Maccone, "Innovative SETI by the KLT," *Proceedings of the Bursts, Pulses and Flickering Conference* held at Kerastari, Greece, June 13–18, 2007. Available at POS (Proceedings of Science) website *http://pos.sissa.it//archive/conferences/056/034/Dynamic2007_034.pdf*
[24] S. Yatawatta, pers. commun., June 17, 2008.

A recent paper about the KLT for relativistic interstellar flight

[25] C. Maccone, "Relativistic Optimized Link by KLT," *Journal of the British Interplanetary Society*, **59** (2006), 94–98.
[26] C. Maccone, "The KLT (Karhunen Loève Transform) to extend SETI searches to broad-band and extremely feeble signals," *Acta Astronautica*, **67** (2010), 1427–1439. This is the

key review paper about both the classical KLT and the BAM-KLT that was used by Szumski [27] to apply them to GNSS.

[27] Arkadiusz Szumski, "Finding the interference: The Karhunen–Loève Transform as an instrument to detect weak RF signals," *InsideGNSS* (Working Papers section), May/June 2011, 56–63.

APPENDIX 17.A

BAM-KLT to extract a SINUSOIDAL carrier out of white noise,
and relevant SNR equation.

Author: Claudio Maccone (e-mail: clmaccon@libero.it).

This commented Maxima file was completed on August 8th, 2011.

This file embodies all calculations provided by the author
at the pages 168 thru 175 of his book entitled "Deep Space Flight
and Communications", published by Praxis-Springer in 2009.
In the following, this book will simply be referred to as "DSF".

```
(%i1) kill(all);
(%o0) done
```

Declare POSITIVE all physical variables to follow.
Declare m and n to be POSITIVE INTEGERS, i.e. they are the indexes
of the KLT eigenvalues and eigenfunctions, i.e. n, m = 1, 2, 3...

```
(%i1) assume(m>0, n>0,T>0,omega>0,a>0);
(%o1) [m>0,n>0,T>0,ω>0,a>0]
```

```
(%i2) declare([m,n],integer);
(%o2) done
```

Define the simplest possible set of BASIS FUNCTIONS
in the Hilbert space, and check both their ORTHOGONALITY
(for m different from n) and their NORMALIZATION CONDITION
(for m equal to n).

```
(%i3) basis_functions:phi[n](t):=sqrt(2/T)*sin((2*%pi*n*t)/T);
```

(%o3) $\phi_n(t):=\sqrt{\frac{2}{T}}\,\sin\left(\frac{2\pi n t}{T}\right)$

```
(%i4) orthogonality:'integrate(phi[n](t)*phi[m](t), t, 0, T)=integrate(phi[n](t)*phi[m](t),
t, 0, T);
```

(%o4) $\frac{2\int_0^T \sin\left(\frac{2\pi m t}{T}\right)\sin\left(\frac{2\pi n t}{T}\right)dt}{T}=0$

(%i5) normalization_integral:'integrate(phi[n](t)*phi[n](t), t, 0, T)=integrate(phi[n](t)*phi[n](t), t, 0, T);

(%o5) $$\frac{2\int_0^T \sin\left(\frac{2\pi n t}{T}\right)^2 dt}{T}=1$$

Next comes the computation of tha autocorrelation of both the pure white noise W(t) and of S(t) = W(t) + a sin(omega t), corresponding to equations (10.37) (10.38) and (10.39) on the pages 170 and 171 of DSF.

Then, the most important term in the autocorrelation of S(t) turns out to be the "autocorrelation of the sinusoidal carrier", given by the extra_term_1 hereafter:

(%i6) extra_term_1:a^2*sin(omega*t[1])*sin(omega*t[2]);

(%o6) $a^2 \sin(t_1\,\omega)\sin(t_2\,\omega)$

We now form eq. (10.43) of DSF (page 171) multipling the extra_term_1 times the sinusoidal basis function phi sub n(t). That is:

(%i7) extra_term_2:extra_term_1*phi[n](t[1]);

(%o7) $$\frac{\sqrt{2}\,a^2 \sin(t_1\,\omega)\sin(t_2\,\omega)\sin\left(\frac{2\pi t_1 n}{T}\right)}{\sqrt{T}}$$

Next we integrate le last function with respect to t1 between 0 and T, so as to let t1 disappear. The integration is easily performed by Maxima:

(%i8) extra_term_3:'integrate(extra_term_2, t[1], 0, T)=integrate(extra_term_2, t[1], 0, T);

(%o8) $$\frac{\sqrt{2}\,a^2 \sin(t_2\,\omega)\int_0^T \sin(t_1\,\omega)\sin\left(\frac{2\pi t_1 n}{T}\right)dt_1}{\sqrt{T}}=\frac{2^{3/2}\pi a^2 n \sin(t_2\,\omega)\sqrt{T}\sin(\omega T)}{\omega^2 T^2-4\pi^2 n^2}$$

Similarly, in order to let t2 disappear also, we multiply the last result by phi sub n (t2) ...

(%i9) extra_term_4:rhs(extra_term_3)*phi[n](t[2]);

(%o9) $$\frac{4\pi a^2 n \sin(t_2\,\omega)\sin\left(\frac{2\pi t_2 n}{T}\right)\sin(\omega T)}{\omega^2 T^2-4\pi^2 n^2}$$

... and then integrate with respect to t2 between 0 and T. One thus gets:

(%i10) extra_term_5:'integrate(extra_term_4, t[2], 0, T)=integrate(extra_term_4, t[2], 0, T);

(%o10) $$\frac{4\,\pi\,a^2\,n\int_0^T \sin(t_2\,\omega)\,\sin\left(\frac{2\,\pi\,t_2\,n}{T}\right)dt_2\,\sin(\omega\,T)}{\omega^2\,T^2-4\,\pi^2\,n^2}=\frac{8\,\pi^2\,a^2\,n^2\,T\,\sin(\omega\,T)^2}{(\omega^2\,T^2-4\,\pi^2\,n^2)^2}$$

Let us call FINAL_TERM the last result, corresponding to the last term in eqs (10.46) and (10.47) in DSF, page 172.

(%i11) result:FINAL_TERM=rhs(extra_term_5);

(%o11) $$\mathrm{FINAL\_TERM}=\frac{8\,\pi^2\,a^2\,n^2\,T\,\sin(\omega\,T)^2}{(\omega^2\,T^2-4\,\pi^2\,n^2)^2}$$

In loose terms, the FINAL_TERM is "the n-th eigenvalue due to the tiny sinusoid embedded in the thick white noise". So, let us check that:
1) It tends to zero for T tending to infinity, i.e. when the the final instant T becomes larger and larger (as in the Fourier Transform, where T=infinity) then the n-th eigenvalue due to the tiny sinusoid is dwarfed by the white noise and actually diasppears:

(%i12) 'limit(rhs(result), T, inf)=limit(rhs(result), T, inf);

(%o12) $$8\,\pi^2\,a^2\,n^2\left(\lim_{T\to\infty}\frac{T\,\sin(\omega\,T)^2}{(\omega^2\,T^2-4\,\pi^2\,n^2)^2}\right)=0$$

2) The same tends to zero again if one considers eigenvalues of order n higher and higher. This is obvious since this is just the CONVERGENCE of the KLT expansion, i.e. smaller and smaller eigenvalues for higher and higher n:

(%i13) 'limit(rhs(result), n, inf)=limit(rhs(result), n, inf);

(%o13) $$8\,\pi^2\,a^2\,T\,\sin(\omega\,T)^2\left(\lim_{n\to\infty}\frac{n^2}{(\omega^2\,T^2-4\,\pi^2\,n^2)^2}\right)=0$$

The FINAL_TERM has a sine square. In view of reducing the FINAL_TERM to a Fourier series expansion, where sines and cosines are raised to the power ONE only, we now convert this sine square to a cosine of twice value:

(%i14) factor(result);

(%o14) $FINAL\_TERM = \frac{8\,\pi^2\,a^2\,n^2\,T\,\sin(\omega\,T)^2}{(\omega\,T-2\,\pi\,n)^2\,(\omega\,T+2\,\pi\,n)^2}$

(%i15) factor(trigreduce(result));

(%o15) $FINAL\_TERM = -\frac{4\,\pi^2\,a^2\,n^2\,T\,(\cos(2\,\omega\,T)-1)}{(\omega\,T-2\,\pi\,n)^2\,(\omega\,T+2\,\pi\,n)^2}$

In conclusion the n-th eigenvalue is now A FUNCTION OF THE FINAL INSTANT, T:

(%i16) nth_eigenvalue:rhs(result);

(%o16) $\frac{8\,\pi^2\,a^2\,n^2\,T\,\sin(\omega\,T)^2}{(\omega^2\,T^2-4\,\pi^2\,n^2)^2}$

(%i17) depends(%lambda, T);

(%o17) $[\lambda(T)]$

(%i18) def_nth_eigenvalue:%lambda[n]=nth_eigenvalue;

(%o18) $\lambda_n = \frac{8\,\pi^2\,a^2\,n^2\,T\,\sin(\omega\,T)^2}{(\omega^2\,T^2-4\,\pi^2\,n^2)^2}$

But then one may take the DERIVATIVE of this n-th eigenvalue WITH RESPECT TO THE FINAL INSTANT T:

(%i19) nth_eigenvalue_derivative:distrib(rat(trigreduce(ratsimp(diff(def_nth_eigenvalue, T))), sin(2*omega*T)));

(%o19) $$\frac{d}{d\,T}\,\lambda_n = \frac{(8\,\pi^2\,a^2\,n^2\,\omega^3\,T^3-32\,\pi^4\,a^2\,n^4\,\omega\,T)\,\sin(2\,\omega\,T)}{\omega^6\,T^6-12\,\pi^2\,n^2\,\omega^4\,T^4+48\,\pi^4\,n^4\,\omega^2\,T^2-64\,\pi^6\,n^6} + \frac{(12\,\pi^2\,a^2\,n^2\,\omega^2\,T^2+16\,\pi^4\,a^2\,n^4)\,\cos(2\,\omega\,T)}{\omega^6\,T^6-12\,\pi^2\,n^2\,\omega^4\,T^4+48\,\pi^4\,n^4\,\omega^2\,T^2-64\,\pi^6\,n^6} - \frac{12\,\pi^2\,a^2\,n^2\,\omega^2\,T^2}{\omega^6\,T^6-12\,\pi^2\,n^2\,\omega^4\,T^4+48\,\pi^4\,n^4\,\omega^2\,T^2-64\,\pi^6\,n^6} - \frac{16\,\pi^4\,a^2\,n^4}{\omega^6\,T^6-12\,\pi^2\,n^2\,\omega^4\,T^4+48\,\pi^4\,n^4\,\omega^2\,T^2-64\,\pi^6\,n^6}$$

The last expression is obviously made by THREE PARTS:
1) The coefficient of the SINE,
2) The coefficient of the COSINE,
3) The KNOWN TERM.
Maxima can easily separate them by virtue of the "coeff" command,
but a little more work is requested to cast the KNOWN_TERM into
its simplest form:

```
(%i20) sine_coefficient:sine_coeff=factor(coeff(rhs(nth_eigenvalue_derivative),sin(2*omega*T)));
```

(%o20) $$\text{sine\_coeff}=\frac{8\,\pi^2\,a^2\,n^2\,\omega\,T}{(\omega\,T-2\,\pi\,n)^2\,(\omega\,T+2\,\pi\,n)^2}$$

```
(%i21) cosine_coefficient:cosine_coeff=factor(coeff(rhs(nth_eigenvalue_derivative),cos(2*omega*T)));
```

(%o21) $$\text{cosine\_coeff}=\frac{4\,\pi^2\,a^2\,n^2\,(3\,\omega^2\,T^2+4\,\pi^2\,n^2)}{(\omega\,T-2\,\pi\,n)^3\,(\omega\,T+2\,\pi\,n)^3}$$

```
(%i22) known_term_1:'known_term_1=subst(0, sin(2*omega*T), rhs(nth_eigenvalue_derivative));
```

(%o22) $$\text{known\_term\_1}=\frac{(12\,\pi^2\,a^2\,n^2\,\omega^2\,T^2+16\,\pi^4\,a^2\,n^4)\cos(2\,\omega\,T)}{\omega^6\,T^6-12\,\pi^2\,n^2\,\omega^4\,T^4+48\,\pi^4\,n^4\,\omega^2\,T^2-64\,\pi^6\,n^6}-\frac{12\,\pi^2\,a^2\,n^2\,\omega^2\,T^2}{\omega^6\,T^6-12\,\pi^2\,n^2\,\omega^4\,T^4+48\,\pi^4\,n^4\,\omega^2\,T^2-64\,\pi^6\,n^6}-\frac{16\,\pi^4\,a^2\,n^4}{\omega^6\,T^6-12\,\pi^2\,n^2\,\omega^4\,T^4+48\,\pi^4\,n^4\,\omega^2\,T^2-64\,\pi^6\,n^6}$$

```
(%i23) known_term_2:'known_term_2=subst(0, cos(2*omega*T), rhs(known_term_1));
```

(%o23) $$\text{known\_term\_2}=-\frac{12\,\pi^2\,a^2\,n^2\,\omega^2\,T^2}{\omega^6\,T^6-12\,\pi^2\,n^2\,\omega^4\,T^4+48\,\pi^4\,n^4\,\omega^2\,T^2-64\,\pi^6\,n^6}-\frac{16\,\pi^4\,a^2\,n^4}{\omega^6\,T^6-12\,\pi^2\,n^2\,\omega^4\,T^4+48\,\pi^4\,n^4\,\omega^2\,T^2-64\,\pi^6\,n^6}$$

```
(%i24) known_term:'known_term=factor(ratsimp(rhs(known_term_2)));
```

(%o24) $$\text{known\_term}=-\frac{4\,\pi^2\,a^2\,n^2\,(3\,\omega^2\,T^2+4\,\pi^2\,n^2)}{(\omega\,T-2\,\pi\,n)^3\,(\omega\,T+2\,\pi\,n)^3}$$

This completes our proof that the BAM-KLT is indeed based upon the
DERIVATIVE OF THE N-TH EIGENVALUE WRT THE FINAL INSTANT, T.

Finally, we want to derive the expression of the SIGNAL-TO-NOISE RATIO
(SNR) for the BAM-KLT.
We start by rewriting the relatioship between the three eigenvalues
(of the (sinusoid+ white noise), of the pure sinusoid, and of the pure white noise:

(%i25) towards_KLT_SNR_theorem_1:%lambda[Sn]=
%lambda[Wn]+(8*%pi^2*a^2*n^2*T*sin(omega*T)^2)/(omega^2*T^2-4*%pi^2*n^2)^2;

(%o25) $$\lambda_{Sn}=\frac{8\,\pi^2\,a^2\,n^2\,T\,\sin(\omega\,T)^2}{(\omega^2\,T^2-4\,\pi^2\,n^2)^2}+\lambda_{Wn}$$

Dividing by the pure white noise eigenvalue...

(%i26) towards_KLT_SNR_theorem_2:distrib(towards_KLT_SNR_theorem_1/%lambda[Wn]);

(%o26) $$\frac{\lambda_{Sn}}{\lambda_{Wn}}=\frac{8\,\pi^2\,a^2\,n^2\,T\,\sin(\omega\,T)^2}{\lambda_{Wn}\,(\omega^2\,T^2-4\,\pi^2\,n^2)^2}+1$$

allows us to introduce the SNR DEFINITION...

(%i27) def_SNR:SNR=a^2/(%lambda[Wn]);

(%o27) $$SNR=\frac{a^2}{\lambda_{Wn}}$$

(%i28) a_square_vs_SNR:first(solve(def_SNR, a^2));

(%o28) $$a^2=\lambda_{Wn}\,SNR$$

...into the above relationship among the three eigenvalues:

(%i29) towards_KLT_SNR_theorem_3:towards_KLT_SNR_theorem_2, a_square_vs_SNR;

(%o29) $$\frac{\lambda_{Sn}}{\lambda_{Wn}}=\frac{8\,\pi^2\,n^2\,SNR\,T\,\sin(\omega\,T)^2}{(\omega^2\,T^2-4\,\pi^2\,n^2)^2}+1$$

(%i30) towards_KLT_SNR_theorem_4:towards_KLT_SNR_theorem_3-1;

(%o30) $$\frac{\lambda_{Sn}}{\lambda_{Wn}}-1=\frac{8\,\pi^2\,n^2\,SNR\,T\,\sin(\omega\,T)^2}{(\omega^2\,T^2-4\,\pi^2\,n^2)^2}$$

From this, eq. (10.58) on page 174 of DSF follows at once.
Shifting now to frequencies...

(%i31) def_nu_alien:omega=(2*%pi*nu);

(%o31) $$\omega=2\,\pi\,\nu$$

... the relationship among the three eigenvalues reads:

(%i32) towards_KLT_SNR_theorem_4:towards_KLT_SNR_theorem_3, def_nu_alien;

(%o32)
$$\frac{\lambda_{Sn}}{\lambda_{Wn}}=\frac{8\,\pi^2\,n^2\,SNR\,T\,\sin(2\,\pi\,v\,T)^2}{\left(4\,\pi^2\,v^2\,T^2-4\,\pi^2\,n^2\right)^2}+1$$

(%i33) towards_KLT_SNR_theorem_5:towards_KLT_SNR_theorem_4-1;

(%o33)
$$\frac{\lambda_{Sn}}{\lambda_{Wn}}-1=\frac{8\,\pi^2\,n^2\,SNR\,T\,\sin(2\,\pi\,v\,T)^2}{\left(4\,\pi^2\,v^2\,T^2-4\,\pi^2\,n^2\right)^2}$$

and again eq. (10.59) of DSF follows from this immediately.

18

KLT of radio signals from relativistic spaceships in uniform and decelerated motion

18.1 INTRODUCTION

It is well known that in special relativity two time variables exist: the coordinate time t, which is the time measured in the fixed reference frame, and the proper time τ, which is the time shown by a clock rigidly connected to the moving body. They are related by

$$\tau(t) = \int_0^t \sqrt{1 - \frac{v^2(s)}{c^2}}\, ds \tag{18.1}$$

where $v(t)$ is the body velocity and c is the speed of light (see [1, p. 44]).

The remainder of this book, starting with the present chapter, is devoted to the relativistic interpretation of Brownian motion whose time variable is the proper time, $B(\tau)$, rather than the coordinate time, $B(t)$ and to find the KLT of $B(\tau)$. The bulk of these results was given by the author in a purely mathematical form, with no reference to relativity, in [2]. The KLT is also explained in detail in Chapter 16 and Chapters 21–25. However, to enable the reader to read Chapters 18–14 independently of Chapter 16 and Chapters 21–25, a summary of that work is now given in a form suitable for the physical developments that will follow in Chapters 12–14.

Consider standard Brownian motion (Wiener–Lévy process) $B(t)$, with mean zero, variance t, and initial condition $B(0) = 0$, as described in Chapter 21.

A white noise integral is the process $X(t)$ defined by

$$X(t) = \int_0^t f(s)\, dB(s) \tag{18.2}$$

where $f(t)$ is assumed to be continuous and non-negative. Evidently, $X(0) = 0$, and it can be proved (see (21.35) or, equivalently, [3, pp. 84–87]) that

$$X(t) = B\left(\int_0^t f^2(s)\, ds\right). \tag{18.3}$$

Thus, $X(t)$ is a time-rescaled Gaussian process, with mean zero and

$$E\{X(t_1)X(t_2)\} = \int_0^{t_1 \wedge t_2} f^2(s)\, ds \tag{18.4}$$

as autocorrelation (covariance); $t_1 \wedge t_2$ denotes the minimum (smallest) t_1 and t_2.

Now the KLT theorem (see [4, pp. 262–271]) states that

$$X(t) = \sum_{n=1}^{\infty} Z_n \phi_n(t) \quad (0 \le t \le T) \tag{18.5}$$

where (1) the functions $\phi_n(t)$ are the autocorrelation eigenfunctions to be found from

$$\int_0^T E\{X(t_1)X(t_2)\}\phi_n(t_2)\, dt_2 = \lambda_n \phi_n(t_1) \tag{18.6}$$

where the constants λ_n are the corresponding eigenvalues; and (2) the Z_n are orthogonal random variables, with mean zero and variance λ_n; that is:

$$E\{Z_m Z_n\} = \lambda_n \delta_{mn}. \tag{18.7}$$

This theorem is valid for any continuous-parameter second-order process with mean zero and known autocorrelation. The series (18.5) converges in mean square, and uniformly in t. Finally, if $X(t)$ is Gaussian—as in Equations (18.2) and (18.3)—the random variables Z_n are also Gaussian, and since they are orthogonal they are independent.

After these preliminaries, we can state the main result of [2] (Maccone First KLT Theorem, fully proven in Chapter 22).

The white noise integral (18.2), or the equivalent time-rescaled Gaussian process (18.3), has the KLT expansion:

$$X(t) = \sum_{n=1}^{\infty} Z_n N_n \sqrt{f(t) \int_0^t f(s)\, ds} \cdot J_{\nu(t)}\left(\gamma_n \frac{\int_0^t f(s)\, ds}{\int_0^T f(s)\, ds}\right). \tag{18.8}$$

Here

(1) the order of the Bessel functions $\nu(t)$ is not a constant, but the time function

$$\nu(t) = \sqrt{-\frac{\chi^3(t)}{f^2(t)} \cdot \frac{d}{dt}\left[\frac{\chi'(t)}{f^2(t)}\right]} \tag{18.9}$$

with

$$\chi(t) = \sqrt{f(t) \int_0^t f(s)\, ds}. \tag{18.10}$$

(2) The constants γ_n are the (increasing) positive zeros of

$$\chi'(T) \cdot J_{\nu(T)}(\gamma_n) + \chi(T) \cdot \left[\frac{f(T) \cdot \gamma_n}{\int_0^T f(s)\, ds} J'_{\nu(T)}(\gamma_n) + \frac{\partial J_{\nu(T)}(\gamma_n)}{\partial \nu} \nu'(T) \right] = 0. \tag{18.11}$$

In general, (18.11) can only be solved numerically.

(3) The normalization constants N_n follow from the normalization condition

$$N_n^2 \left[\int_0^T f(s)\, ds \right]^2 \cdot \int_0^1 x [J_{\nu((x))}(\gamma_n x)]^2\, dx = 1 \tag{18.12}$$

where the new Bessel functions order $\nu((x))$ is (18.9) changed by aid of the transformation

$$\int_0^t f(s)\, ds = x \int_0^T f(s)\, ds.$$

(4) The eigenvalues are determined by

$$\lambda_n = \left[\int_0^T f(s)\, ds \right]^2 \frac{1}{(\gamma_n)^2}. \tag{18.13}$$

(5) The Gaussian random variables Z_n are independent and orthogonal, and have zero mean and variance λ_n.

The proof of this theorem may be sketched as follows: first, the Volterra-type integral equation (18.6) is transformed into a differential equation with two boundary conditions; and, second, the latter is reduced to the standard Bessel differential equation by means of two changes of variables. The full proof is given in Chapter 22.

Let us now go back to relativity. Since from (18.3) it plainly appears that the rescaled time of the new Brownian motion is given by

$$\int_0^t f^2(s)\, ds \tag{18.14}$$

we merely have to equate (18.1) and (18.14) to get the relationship among the arbitrary time-rescaling function $f(t)$ and the arbitrary body velocity $v(t)$:

$$\int_0^t f^2(s)\, ds = \int_0^t \sqrt{1 - \frac{v^2(s)}{c^2}}\, ds. \tag{18.15}$$

By differentiating and taking the positive square root, it follows that:

$$f(t) = \left[1 - \frac{v^2(t)}{c^2}\right]^{\frac{1}{4}}. \tag{18.16}$$

This formula is the starting point to study the KLT expansion (18.8) for a relativistic body, like a relativistic spacecraft or spaceship moving in a radial direction away or towards the Earth.

Inversion of (18.16) leads at once to:

$$v(t) = c\sqrt{1 - f^4(t)}. \tag{18.17}$$

Now, the reality of the motion requires the radicand to be non-negative, whence, taking the positive sign in front of all square roots, we find

$$f(t) \le 1. \tag{18.18}$$

This is the fundamental upper bound imposed on the "arbitrary" function $f(t)$ by special relativity. In other words, as the speed of light can in no case be exceeded, so $f(t)$ must not exceed 1.

As already pointed out, the lower bound on $f(t)$, required by the presence of the radicals in (18.8) and (18.10), is zero. Therefore

$$0 \le f(t) \le 1 \quad (0 \le t \le T) \tag{18.19}$$

is the physical range of the (otherwise arbitrary) function $f(t)$.

We also need to point out the Newtonian limit of the results. By this we mean the limit as $c \to \infty$. Then, as we see from (18.16),

$$\lim_{c\to\infty} f(t) = 1 \tag{18.20}$$

and the time-rescaled process under consideration reduces to standard Brownian motion, $B(t)$. This agrees, of course, with (18.1), stating that the proper time τ becomes the same as the coordinate time t in the Newtonian limit $c \to \infty$.

Finally, we want to hint at how the shape of the eigenfunctions $\phi_n(t)$ may be determined even without knowing their analytical expression. This possibility is a consequence of the Sonine–Pólya theorem, which is explored in Section 22.5, for the non-relativistic case. The reader is referred there for the details, and here we merely confine ourselves to the relativistic version of the results. From (18.16) and (22.61) one finds:

$$\frac{d\ln f(t)}{dt} = \frac{1}{4}\frac{d}{dt}\left[\ln\left(1 - \frac{v^2(t)}{c^2}\right)\right] = -\frac{1}{2c^2}\cdot\frac{v(t)\dfrac{dv(t)}{dt}}{1 - \dfrac{v^2(t)}{c^2}}$$

$$= (\text{negative})\cdot v(t)\frac{dv(t)}{dt}. \tag{18.21}$$

Thus, not only the velocity $v(t)$, but also its derivative (i.e., acceleration taken with respect to the coordinate time, t) determines the shape (i.e., the stability) of the $\phi_n(t)$. The resulting Table 18.1 follows from this and Table 22.1.

Table 18.1. Stability criterion for the relativistic eigenfunctions $\phi_n(t)$.

Sign of the velocity $v(t)$	Sign of the coordinate acceleration $dv(t)/dt$	Shape of the KL eigenfunctions $\phi_n(t)$	Description when T is finite	Description when T is infinite
Positive	Negative	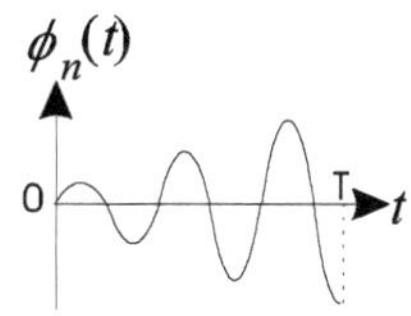	Divergent	Asymptotic unstable
Negative	Positive	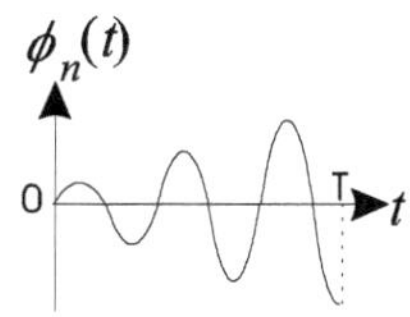	Divergent	Asymptotic unstable
Positive	Positive	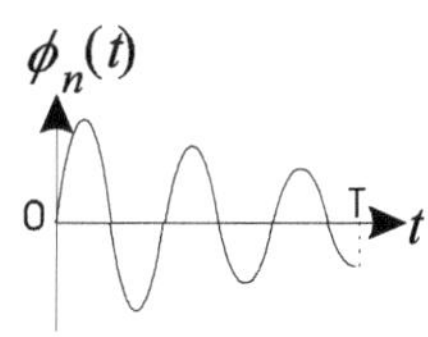	Convergent	Asymptotic stable
Negative	Negative	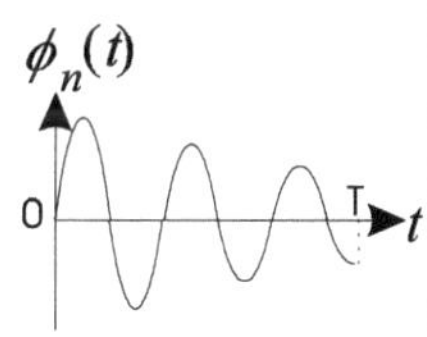	Convergent	Asymptotic stable

18.2 UNIFORM MOTION

The simplest possible case of (18.16) is when the velocity $v(t)$ is a constant (i.e., the body's motion is uniform). Then $f(t)$ is a constant K as well

$$f(t) = \left[1 - \frac{v^2(t)}{c^2}\right]^{\frac{1}{4}} = K. \tag{18.22}$$

Let us now recall the property of the Brownian motion called self-similarity to the order 1/2 and expressed by the formula $B(ct) = \sqrt{c}B(t)$ where c is any real positive constant—see (21.6) for the relevant proof. From this and from (18.3), one gets at once

$$X(t) = B\left(\int_0^t K^2\, ds\right) = B(K^2 t) = KB(t). \qquad (18.23)$$

Thus, the *uniform* proper-time Brownian motion $B(\tau) = X(t)$ equals the *uniform* coordinate-time Brownian motion $B(t)$ multiplied by the *constant* K, which is

$$B(\tau) = \left[1 - \frac{v^2}{c^2}\right]^{\frac{1}{4}} B(t). \qquad (18.24)$$

The KL expansion of $B(\tau)$ is, of course, the same as that of $B(t)$ apart from the multiplicative factor K. And the relevant eigenfunctions are just sines.

To provide an example of how the machinery outlined in Section 18.1 actually works, we shall now prove this result, also proved in Section 21.3 (or in [4, p. 280]). From (18.10):

$$\chi(t) = \sqrt{K \int_0^t K\, ds} = K\sqrt{t} \qquad (18.25)$$

and

$$\chi'(t) = \frac{K}{2\sqrt{t}}. \qquad (18.26)$$

The order $\nu(t)$ of the Bessel functions is then found from (18.9):

$$\nu(t) = \sqrt{-\frac{\chi^3(t)}{f^2(t)}\frac{d}{dt}\left[\frac{\chi'(t)}{f^2(t)}\right]} = \sqrt{-\frac{K^3 t^{\frac{3}{2}}}{K^2}\frac{d}{dt}\left[\frac{1}{2K\sqrt{t}}\right]}$$
$$= \sqrt{-Kt^{\frac{3}{2}}\left(-\frac{t^{-\frac{3}{2}}}{4K}\right)} = \sqrt{\frac{1}{4}} = \frac{1}{2} \qquad (18.27)$$

where both the time t and the constant K have vanished from the result. Simplifications of this kind (further examples will be given in Sections 18.3 and 12.4) are vital to make the mathematical investigations feasible. Since $\nu = \frac{1}{2}$, the relevant Bessel function is [6, p. 54]

$$J_{\frac{1}{2}}(x) = \sqrt{\frac{2}{\pi x}} \sin x. \qquad (18.28)$$

Thus, from (18.5), (18.27), and (18.28), the KL expansion follows:

$$X(t) = \sqrt{K \int_0^t K\, ds} \sum_{n=1}^{\infty} Z_n N_n J_{\frac{1}{2}}\left(\gamma_n \frac{\int_0^t K\, ds}{\int_0^T K\, ds}\right)$$
$$= K \sum_{n=1}^{\infty} Z_n N_n \sqrt{\frac{2T}{\pi\gamma_n}} \sin\left(\gamma_n \frac{t}{T}\right). \qquad (18.29)$$

In this expression the normalization constants N_n are yet to be found. To this end, we must know the γ_n given by (18.11). That is,

$$\frac{K}{2\sqrt{T}}J_{\frac{1}{2}}(\gamma_n) + K\sqrt{T}\left[\frac{K\gamma_n}{K\int_0^T ds}J'_{\frac{1}{2}}(\gamma_n)\right] = 0 \tag{18.30}$$

or, simplifying,

$$\tfrac{1}{2}J_{\frac{1}{2}}(\gamma_n) + \gamma_n J'_{\frac{1}{2}}(\gamma_n) = 0. \tag{18.31}$$

But this is a special case of the more general Bessel functions formula (see [5, p. 11, entry (54)]:

$$\nu J_\nu(z) + zJ'_\nu(z) = zJ_{\nu-1}(z) \tag{18.32}$$

so that (18.31) actually amounts to

$$J_{-\frac{1}{2}}(\gamma_n) = 0 \tag{18.33}$$

since $\gamma_n \neq 0$. One now has (see [6, p. 55, entry (6)])

$$J_{-\frac{1}{2}}(x) = \sqrt{\frac{2}{\pi x}}\cos x \tag{18.34}$$

so that (18.33) finally becomes the boundary condition:

$$\cos\gamma_n = 0. \tag{18.35}$$

In this case we find the exact γ_n expression to be

$$\gamma_n = n\pi - \frac{\pi}{2} \quad (n = 1, 2, \ldots). \tag{18.36}$$

Reverting now to the normalization constants N_n, (18.12) yields

$$\begin{aligned} 1 &= N_n^2\left[\int_0^T K\,ds\right]^2\int_0^1 x\left[J_{\frac{1}{2}}(\gamma_n x)\right]^2 dx \\ &= N_n^2K^2T^2\frac{2}{\pi\gamma_n}\int_0^1 \sin^2(\gamma_n x)\,dx \\ &= N_n^2K^2T^2\frac{1}{\pi\gamma_n^2}[\gamma_n - \sin\gamma_n\cos\gamma_n] = \frac{N_n^2K^2T^2}{\pi\gamma_n} \end{aligned} \tag{18.37}$$

from which

$$N_n = \frac{\sqrt{\gamma_n}\sqrt{\pi}}{KT}. \tag{18.38}$$

As for the eigenvalues λ_n, from (18.13) they are given by

$$\lambda_n = \frac{K^2T^2}{\gamma_n^2} \tag{18.39}$$

and these are also the variances of the independent Gaussian random variables Z_n. It is interesting to point out that the property

$$\sigma_{cZ}^2 = c^2 \sigma_Z^2 \tag{18.40}$$

and (18.39) yield the following proportionality among the proper-time random variables Z_n and the coordinate-time random variables Z_n^0—corresponding to the case $\nu(t) \equiv \frac{1}{2}$, or, from (18.16), $f(t) \equiv 1$:

$$Z_n = K Z_n^0. \tag{18.41}$$

Thus, the KL expansion of the proper-time Brownian motion is

$$B(\tau) = \sum_{n=1}^{\infty} Z_n \sqrt{\frac{2}{T}} \sin\left(\gamma_n \frac{t}{T}\right) = K \sum_{n=1}^{\infty} Z_n^0 \sqrt{\frac{2}{T}} \sin\left(\gamma_n \frac{t}{T}\right) = KB(t) \tag{18.42}$$

and (18.24) is found once again. In other words, passing from one inertial reference frame to another, the random variables Z_n just change their variance according to (18.41), whereas the time eigenfunctions remain the same. In Section 18.5 total energy will also be discussed.

18.3 DECELERATED MOTION

This and the remaining sections are devoted to the case when the proper time is proportional to a real positive power of the coordinate time, namely

$$\tau = Ct^{2H} \quad (t \geq 0) \tag{18.43}$$

C being a constant that will be determined immediately, and H being a real variable whose range has yet to be found. The factor 2 in the exponent is introduced for convenience. By checking (18.43) against (18.1), differentiating, and taking the square root, one gets

$$f(t) = \sqrt{2HC}\, t^{H-\frac{1}{2}}. \tag{18.44}$$

Inserting this into (18.17), the resulting velocity radical reads

$$v(t) = c\sqrt{1 - (2HC)^2 t^{2(2H-1)}}. \tag{18.45}$$

In order to have a real velocity, the inequality

$$(2HC)^2 t^{2(2H-1)} \leq 1 \tag{18.46}$$

must be valid. Moreover, the initial instant is conventionally zero, and the final instant is T, so that the range of H is necessarily greater than one-half. By setting $t = T$, the constant C is determined so as $v(T) = 0$, and one gets

$$C = \frac{1}{2HT^{2H-1}}. \tag{18.47}$$

One can now understand the physical meaning of the motion we are studying. Initially ($t = 0$) the spaceship is traveling at the speed of light. Then it starts decelerating until it stops at the final instant $t = T$. Actually, if we let H vary, we have a family of curves in the $t, v(t, H)$ plane. But we have to be careful: the tangent to all such curves at $t = 0$ must be *horizontal* in order to preserve the physical reality when the spaceship starts decelerating from c to lower speeds (i.e., there cannot be any sudden "speed jump"). Thus, differentiating (18.45)—with C given by (18.47)—with respect to t and then setting $t = 0$, one discovers that the condition on H given $H > \frac{1}{2}$ must physically be replaced by the stronger condition:

$$4H - 3 > 0 \quad \text{hence } H > \tfrac{3}{4} = 0.75. \tag{18.48}$$

An important special case of $v(t, H)$ occurs when $H = 1$: in fact, $v(t)$ is then the upper-right quarter of an ellipse. One also easily infers that, for $1 < H < \infty$, all $v(t)$ curves lie above this arc of ellipse. In the (physically meaningless) limit case $H \to \infty$ the $v(t, H)$ "curve" would be the upper-right quarter of a rectangle. Figure 18.1 shows this set of $v(t, H)$ curves representing the decelerated motion for different values of H.

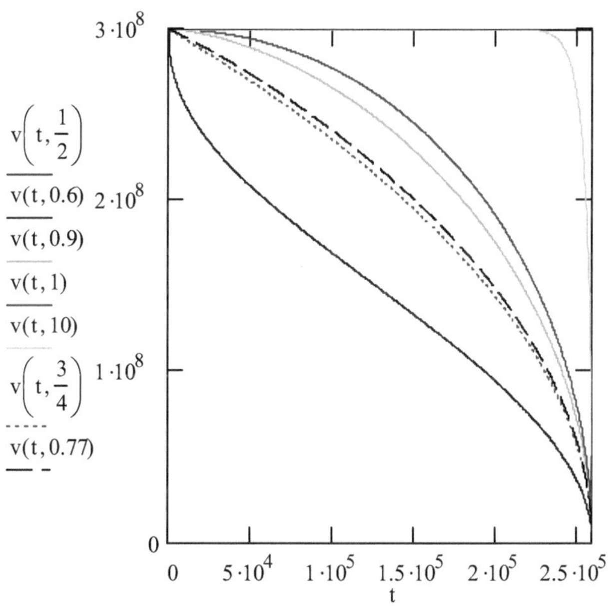

Figure 18.1. Decelerated motion of a relativistic spaceship approaching the Earth at the speed of light c down to speed zero in the finite time interval $0 \le t \le T$. We dubbed this spaceship the *Independence Day* (alien) spaceship. For instance, let $T = 3$ days of coordinate time (i.e., time elapsed on Earth). At the initial instant $t = 0$ (when the deceleration starts) all the curves $v(t, H)$ must have their tangents horizontal (to avoid bumps aboard the spaceship) and that yields the physical constraint: $H > \frac{3}{4} = 0.75$. The above plots show just this fact in a neat, graphic fashion: (1) all solid curves have $H > \frac{3}{4}$ and horizontal tangent at $t = 0$, so they are acceptable; (2) the dividing line is the dash-dotted curve corresponding to $H = \frac{3}{4}$, and one can see that it does not have a horizontal tangent at $t = 0$; (3) all the lower curves (dotted) are not allowed since they don't have a horizontal tangent at $t = 0$.

In conclusion, the function $f(t)$ is defined by the real positive power

$$f(t) = \frac{t^{H-\frac{1}{2}}}{T^{H-\frac{1}{2}}} \quad (0 \le t \le T). \tag{18.49}$$

From (18.17) and (18.49) we see that the velocity $v(t)$ is given by

$$v(t) = c\sqrt{1 - \left(\frac{t^{2H-1}}{T^{2H-1}}\right)^2} \quad (0 \le t \le T) \tag{18.50}$$

One can now understand the physical meaning of the motion we are studying. Initially ($t = 0$) the particle is traveling at the speed of light, then it starts decelerating until it stops at the final instant $t = T$. Actually, (18.50) represents a family of curves on the $(t, \nu(t))$ plane if we let H vary according to (18.48). The particular case $H = \frac{1}{2}$ represents standard Brownian motion. Another important special case of (18.50) occurs when $H = 1$: in fact $v(t)$ is then an ellipse. One also easily infers that, for $\frac{1}{2} < H < 1$ the curve lies below the arc of ellipse, whereas for $1 < H < \infty$ the curve lies above it. In the (physically meaningless) limit case $H \to \infty$ the curve would be half a rectangle.

Let us now turn to the KL expansion of the decelerated Brownian motion

$$X(t) = B\left(\frac{t^{2H}}{2HT^{2H-1}}\right) = \frac{1}{\sqrt{2H}T^{H-\frac{1}{2}}} B(t^{2H}). \tag{18.51}$$

Integrating (18.49), we get

$$\int_0^t f(s)\, ds = \frac{t^{H+\frac{1}{2}}}{(H+\frac{1}{2})T^{H-\frac{1}{2}}}. \tag{18.52}$$

Then, by virtue of (18.49) and (18.52), the function $\chi(t)$ defined by (18.10) reads:

$$\chi(t) = \frac{t^H}{\sqrt{H+\frac{1}{2}}\,T^{H-\frac{1}{2}}} \tag{18.53}$$

thus

$$\chi'(t) = \frac{Ht^{H-1}}{\sqrt{H+\frac{1}{2}}\,T^{H-\frac{1}{2}}}. \tag{18.54}$$

Moreover, from (18.49) and (18.54), one finds the expressions

$$\frac{\chi'(t)}{f^2(t)} = \frac{T^{H-\frac{1}{2}}Ht^{-H}}{\sqrt{H+\frac{1}{2}}} \tag{18.55}$$

and

$$\frac{d}{dt}\left[\frac{\chi'(t)}{f^2(t)}\right] = \frac{T^{H-\frac{1}{2}}H(-H)t^{-H-1}}{\sqrt{H+\frac{1}{2}}} \tag{18.56}$$

and, from (18.49) and (18.53),

$$\frac{\chi^3(t)}{f^2(t)} = \frac{t^{H+1}}{(H+\frac{1}{2})\sqrt{H+\frac{1}{2}}\,T^{H-\frac{1}{2}}}. \tag{18.57}$$

The Bessel functions order can now be found from (18.9), (18.56), and (18.57):

$$\nu = \frac{2H}{2H+1}. \tag{18.58}$$

Note that both the time t and the constant T disappear identically, and the order of the Bessel functions is a constant, rather than a function of the time t. Moreover, by letting $H = \frac{3}{4}$ and $H \to \infty$, respectively, we see that the range of ν is rather limited: $\frac{3}{5} \le \nu \le 1$.

Our next task is to find the meaning of the constants γ_n. Upon substituting (18.52), (18.53), and (18.54) into (18.11), along with $\nu'(t) = 0$ one gets, after simplifying any multiplicative factors,

$$\frac{2H}{2H+1} J_\nu(\gamma_n) + \gamma_n J'_\nu(\gamma_n) = 0. \tag{18.59}$$

By virtue of (18.58), (18.59) is equivalent to

$$\nu J_\nu(\gamma_n) + \gamma_n J'_\nu(\gamma_n) = 0. \tag{18.60}$$

Once again the Bessel functions property (18.32) may be applied, and

$$\gamma_n J_{\nu-1}(\gamma_n) = 0. \tag{18.61}$$

Since $\gamma_n \neq 0$,

$$J_{\nu-1}(\gamma_n) = 0. \tag{18.62}$$

Thus, the γ_n are the real positive zeros, arranged in ascending order of magnitude, of the Bessel function of order $\nu - 1$. No formula yielding these zeros explicitly is known. Yet it is possible to find an approximated expression for them by means of the asymptotic formula for $J_\nu(x)$ (see [8, p. 134]).

$$\lim_{x\to\infty} J_\nu(x) = \lim_{x\to\infty} \sqrt{\frac{2}{\pi x}} \cos\left(x - \frac{\nu\pi}{2} - \frac{\pi}{4}\right). \tag{18.63}$$

In fact, from (18.58) one first gets

$$\nu - 1 = -\frac{1}{2H+1}. \tag{18.64}$$

Second, (18.62) and (18.64), checked against (18.63), yield

$$0 = J_{\nu-1}(\gamma_n) \approx \sqrt{\frac{2}{\pi\gamma_n}} \cos\left(\gamma_n + \frac{\pi}{2(2H+1)} - \frac{\pi}{4}\right) \tag{18.65}$$

hence

$$\gamma_n + \frac{\pi}{2(2H+1)} - \frac{\pi}{4} \approx n\pi - \frac{\pi}{2} \quad (n = 1, 2, \ldots) \tag{18.66}$$

and finally

$$\gamma_n \approx n\pi - \frac{\pi}{4} - \frac{\pi}{2(2H+1)} \quad (n = 1, 2, \ldots) \tag{18.67}$$

The first 32 approximated γ_n, obtained by means of (18.67), appear in Table 18.2, for various values of $H \geq \frac{1}{2}$. In the Brownian case $H = \frac{1}{2}$ (18.67) is an exact formula, in that it coincides with (18.36). We are reminded that these γ_n give the pace of convergence of the KL expansion, inasmuch as the standard deviations of the Gaussian random variables Z_n depend inversely on the γ_n by virtue of (18.13).

Eventually, the normalization constants N_n follow from (18.12) and (18.52):

$$1 = N_n^2 \frac{T^2}{(H+\frac{1}{2})^2} \int_0^1 x J_\nu^2(\gamma_n x)\, dx. \tag{18.68}$$

This integral is calculated within the framework of the Dini series (see [5, p. 71]) and the result is

$$\int_0^1 x J_\nu^2(\gamma_n x)\, dx = \frac{1}{2\gamma_n^2} [\gamma_n^2 J_\nu'^2(\gamma_n) + (\gamma_n^2 - \nu^2) J_\nu^2(\gamma_n)]. \tag{18.69}$$

This formula, however, may be greatly simplified upon eliminating $\gamma_n J_\nu'(\gamma_n)$ taken from (18.60). In fact, one finds

$$\gamma_n^2 J_\nu'^2(\gamma_n) = \nu^2 J_\nu^2(\gamma_n) \tag{18.70}$$

and (18.68), by virtue of (18.69) and (18.70), becomes

$$1 = N_n^2 \frac{T^2}{(H+\frac{1}{2})^2} \cdot \frac{J_\nu^2(\gamma_n)}{2}. \tag{18.71}$$

Thus

$$N_n = \frac{(H+\frac{1}{2})\sqrt{2}}{T|J_\nu(\gamma_n)|}. \tag{18.72}$$

This is the exact expression of the normalization constants. An approximated expression can be found upon inserting both (18.67) and (18.58) into the approximated

Table 18.2. Approximate values of the constants γ_n.

	$H = 0.5$ Brownian	$H = 0.6$	$H = 0.7$	$H = 0.8$	$H = 0.9$	$H = 1.0$	$H = \infty$
$n = 1$	1.571	1.642	1.702	1.752	1.795	1.833	2.356
$n = 2$	4.712	4.784	4.843	4.894	4.937	4.974	5.498
$n = 3$	7.854	7.925	7.985	8.035	8.078	8.116	8.639
$n = 4$	11.00	11.07	11.13	11.18	11.22	11.26	11.78
$n = 5$	14.14	14.21	14.27	14.32	14.37	14.40	14.92
$n = 6$	17.28	17.36	17.41	17.46	17.50	17.54	18.06
$n = 7$	20.42	20.50	20.55	20.60	20.64	20.68	21.20
$n = 8$	23.56	23.63	23.69	23.74	23.79	23.82	24.35
$n = 9$	26.70	26.77	26.83	26.88	26.93	26.96	27.49
$n = 10$	27.84	27.92	27.98	30.03	30.07	30.11	30.63
$n = 11$	32.99	33.06	33.12	33.17	33.21	33.25	33.77
$n = 12$	36.13	36.20	36.26	36.31	36.35	36.39	36.91
$n = 13$	37.27	37.34	37.40	37.45	37.49	37.53	40.05
$n = 14$	42.41	42.48	42.54	42.59	42.64	42.67	43.20
$n = 15$	45.55	45.62	45.68	45.73	45.78	45.81	46.34
$n = 16$	48.69	48.77	48.83	48.88	48.92	48.96	47.48
$n = 17$	51.84	51.91	51.97	52.02	52.06	52.10	52.62
$n = 18$	54.98	55.05	55.11	55.16	55.20	55.24	55.76
$n = 19$	58.12	58.19	58.25	58.30	58.34	58.38	58.90
$n = 20$	61.26	61.33	61.39	61.44	61.48	61.52	62.05
$n = 21$	64.40	64.47	64.53	64.58	64.63	64.66	65.19
$n = 22$	67.54	67.62	67.67	67.72	67.77	67.81	68.33
$n = 23$	70.69	70.76	70.82	70.87	70.91	70.95	71.47
$n = 24$	73.83	73.90	73.96	74.01	74.05	74.09	74.61
$n = 25$	76.97	77.04	77.10	77.15	77.19	77.23	77.75
$n = 26$	80.11	80.18	80.24	80.29	80.33	80.37	80.90
$n = 27$	83.25	83.32	83.38	83.43	83.48	83.51	84.04
$n = 28$	86.39	86.46	86.52	86.57	86.62	86.66	87.18
$n = 29$	87.53	87.61	87.67	87.72	87.76	87.80	90.32
$n = 30$	92.68	92.75	92.81	92.86	92.90	92.94	93.46
$n = 31$	95.82	95.90	95.95	96.00	96.04	96.08	96.60
$n = 32$	98.96	97.0	97.0	97.1	97.1	97.2	97.75

(18.63) for $J_\nu(\gamma_n)$:

$$|J_\nu(\gamma_n)| \approx \left|\sqrt{\frac{2}{\pi\gamma_n}}\cos\left(n\pi - \frac{\pi}{4} - \frac{\pi}{2(2H+1)} - \frac{\pi 2H}{2(2H+1)} - \frac{\pi}{4}\right)\right|$$

$$\approx \sqrt{\frac{2}{\pi\gamma_n}}|\cos(n\pi - \pi)| \approx \sqrt{\frac{2}{\pi\gamma_n}}. \tag{18.73}$$

By substituting this into (18.72) and using (18.67) for the γ_n, it follows that

$$N_n \approx \frac{\pi}{T}(H + \tfrac{1}{2})\sqrt{n - \frac{1}{4} - \frac{1}{2(2H+1)}}. \tag{18.74}$$

These are the approximated normalization constants.

A similar procedure applies to the eigenvalues λ_n. In fact, from (18.13) and (18.52) we get the exact formula

$$\lambda_n = \frac{T^2}{(H + \frac{1}{2})^2} \cdot \frac{1}{(\gamma_n)^2} \tag{18.75}$$

whereas from (18.75) and (18.67) we get the approximated formula

$$\lambda_n \approx \frac{T^2}{(H + \frac{1}{2})^2} \cdot \frac{1}{\pi^2\left(n - \frac{1}{4} - \frac{1}{2(2H+1)}\right)^2}. \tag{18.76}$$

These are the variances of the independent Gaussian random variables Z_n.

Let us now summarize all the results found in the present section by writing two KL expansions: the exact one

$$X(t) = \frac{\sqrt{2H+1}\,t^H}{T^{H+\frac{1}{2}}} \sum_{n=1}^{\infty} Z_n \frac{1}{|J_\nu(\gamma_n)|} J_\nu\left(\gamma_n \frac{t^{H+\frac{1}{2}}}{T^{H+\frac{1}{2}}}\right) \tag{18.77}$$

and the approximated one

$$X(t) \approx \frac{\sqrt{2H+1}\,t^{\frac{H}{2}-\frac{1}{4}}}{T^{\frac{H}{2}+\frac{1}{4}}} \sum_{n=1}^{\infty} Z_n \cos\left(\gamma_n \frac{t^{H+\frac{1}{2}}}{T^{H+\frac{1}{2}}} - \frac{2H\pi}{2(2H+1)} - \frac{\pi}{4}\right). \tag{18.78}$$

18.4 CHECKING THE KLT OF DECELERATED MOTION BY MATLAB SIMULATIONS

Just look at Figure 18.2.

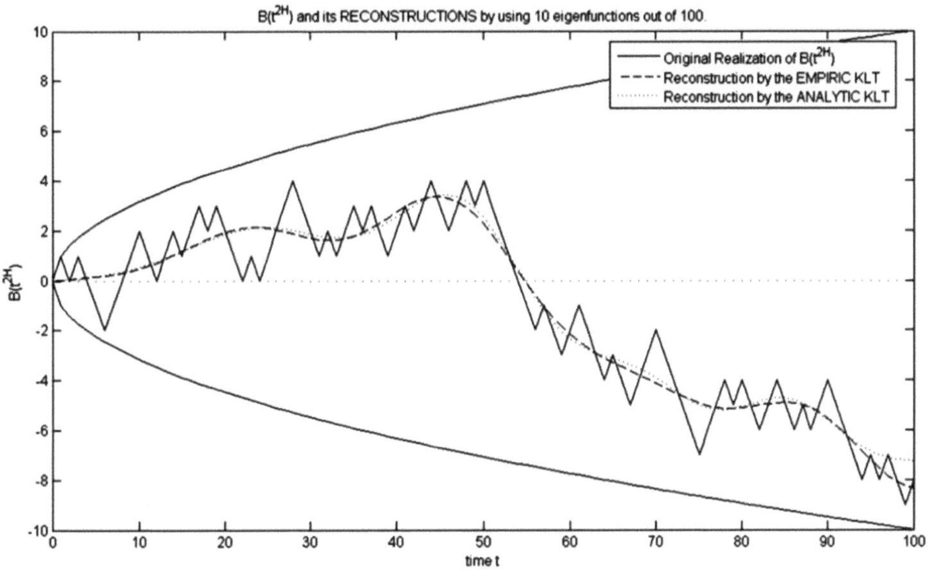

Figure 18.2. The time-rescaled Brownian motion $X(t)$ of (18.78) vs. time t simulated as a random walk over 100 time instants. This $X(t)$ represents the "noisy signal" received on Earth (whence the use of the coordinate time t = Earth time) from a relativistic spaceship approaching the Earth in a decelerated motion, as in the movie *Independence Day*. Next to the "bumpy curve" of $X(t)$, two more "smooth curves" are shown that *interpolate at best* the bumpy $X(t)$. These two curves are the KLT reconstruction of $X(t)$ *by using the first ten eigenfunctions only*. It is important to note that the two smooth curves are *different* in this case because the KLT expansion (18.78) is *approximated*. Actually, it is an approximated KLT expansion because the asymptotic expansion of the Bessel functions (18.63) was used. So, the two curves are different from each other, but both still interpolate $X(t)$ at best. Note that, were we taking into account the full set of 100 KLT eigenfuctions—rather than just 10—then the *empirical* reconstruction would overlap $X(t)$ *exactly*, but the *analytic* reconstruction would not because of the use of the asymptotic expansion (18.63) of the Bessel functions.

18.5 TOTAL ENERGY OF THE NOISY SIGNAL FROM RELATIVISTIC SPACESHIPS IN DECELERATED AND UNIFORM MOTION

A thorough study of the total energy of the noisy signals emitted by relativistic spaceships in decelerated motion (and of the uniform motion, in particular) is allowed by the results obtained in Sections 18.2, 18.3, and 21.10 in Chapter 21.

Our first goal will be to get the characteristic function (i.e., the Fourier transform) of the random variable "total energy", defined by (21.47). In fact, inserting the eigenvalues (18.75) into (21.51), it follows that

$$\Phi_{\varepsilon}(\zeta) = \left[\prod_{n=1}^{\infty}\left(1 - \frac{2iT^2\zeta}{(H+\frac{1}{2})^2\gamma_n^2}\right)\right]^{-\frac{1}{2}}. \tag{18.79}$$

On the other hand,

$$J_\nu(z) = \frac{\left(\frac{z}{2}\right)^\nu}{\Gamma(\nu+1)} \cdot \prod_{n=1}^{\infty} \left(1 - \frac{z^2}{j_{\nu,n}^2}\right) \tag{18.80}$$

is the infinite product expansion for $J_\nu(z)$ [6, p. 498], and the constants $j_{\nu,n}$ evidently are the real positive zeros of $J_\nu(z)$, arranged in ascending order of magnitude. Then, keeping in mind (18.62), we can let the two infinite products (18.79) and (18.80) coincide by setting

$$\gamma_n = j_{\nu-1,n} \tag{18.81}$$

and

$$z^2 = \frac{2iT^2\zeta}{(H+\frac{1}{2})^2} \quad \text{or} \quad z = T\frac{\sqrt{2i\zeta}}{H+\frac{1}{2}}. \tag{18.82}$$

Solving for $\Phi_\varepsilon(\zeta)$, one gets

$$\Phi_\varepsilon(\zeta) = \frac{1}{\sqrt{\Gamma(\nu)\left[\frac{T\sqrt{2i\zeta}}{2H+1}\right]^{1-\nu} J_{\nu-1}\left(\frac{T\sqrt{2i\zeta}}{H+\frac{1}{2}}\right)}} \tag{18.83}$$

which is the exact expression for the characteristic function of the total energy distribution, ε. An approximated expression can also be derived using the asymptotic expression for the Bessel function (18.63); one then gets

$$\Phi_\varepsilon(\zeta) \approx \frac{1}{\sqrt{\frac{\Gamma(\nu)}{\sqrt{\pi}}\left[\frac{T\sqrt{2i\zeta}}{2H+1}\right]^{\frac{1}{2}-\nu} \cos\left(\frac{T\sqrt{2i\zeta}}{H+\frac{1}{2}} - \frac{\nu\pi}{2} + \frac{\pi}{4}\right)}}. \tag{18.84}$$

In the standard Brownian case $H = \frac{1}{2}$ (hence $\nu = \frac{1}{2}$ and one can apply the formula $\Gamma(\frac{1}{2}) = \sqrt{\pi}$), both (18.83) and (18.84) become

$$\Phi_\varepsilon(\zeta) = \frac{1}{\sqrt{\cos(T\sqrt{2i\zeta})}}. \tag{18.85}$$

This result is due to Cameron and Martin, who published it in 1944 [9].

Our next goal is the computation of all the total energy cumulants, given by (21.56). To this end, consider the series

$$\sum_{n=1}^{\infty} \frac{1}{(\gamma_n)^{2k}} \equiv S_{2k,\nu-1} \equiv \sigma_{\nu-1}^{(k)} \quad (k = 1, 2, \ldots) \tag{18.86}$$

where the notation $S_{2k,\nu-1}$ is used on [5, p. 61], while the notation $\sigma_{\nu-1}^{(k)}$ is used on [6, p. 502]. Then

$$\sum_{k=1}^{\infty} S_{2k,\nu-1} x^{2k-1} = \frac{J_\nu(x)}{2J_{\nu-1}(x)} \tag{18.87}$$

is the power series in x, with coefficients $S_{2k,\nu-1}$, whose proof is given on [5, p. 61].

From the formula that yields any coefficient of a power series, it follows that the coefficients $S_{2k,\nu-1}$ of the power series in x on the left side of (18.87) are given by

$$S_{2k,\nu-1} = \frac{1}{(2k-1)!} \lim_{x \to 0^+} \left[\frac{d^{2k-1}}{dx^{2k-1}} \left(\frac{J_\nu(x)}{2J_{\nu-1}(x)} \right) \right] \tag{18.88}$$

and the sum of the series (18.86) is obtained. Finally, by virtue of (21.56) and (18.88) we conclude that all the cumulants of the total energy are

$$\begin{aligned} K_n &= \frac{2^{n-1} T^{2n}}{(H+\frac{1}{2})^{2n}} \cdot \frac{(n-1)!}{(2n-1)!} \lim_{x \to 0^+} \left[\frac{d^{2n-1}}{dx^{2n-1}} \left(\frac{J_\nu(x)}{2J_{\nu-1}(x)} \right) \right] \\ &= \frac{2^{n-1} T^{2n}}{(H+\frac{1}{2})^{2n}} \cdot (n-1)! \cdot \sigma_{\nu-1}^{(n)} \quad (n = 1, 2, \ldots) \end{aligned} \tag{18.89}$$

where the quantities $\sigma_{\nu-1}^{(1)}$, $\sigma_{\nu-1}^{(2)}$, $\sigma_{\nu-1}^{(3)}$, $\sigma_{\nu-1}^{(4)}$, $\sigma_{\nu-1}^{(5)}$, and $\sigma_{\nu-1}^{(6)}$ appear on [6, p. 502]—ν is to be replaced by H via (18.58).

Having found all the cumulants, we can now derive the expressions of the most interesting statistical parameters of the total energy ε.

(1) Mean value of the total energy:

$$K_1 = E\{\varepsilon\} = \frac{T^2}{2H(2H+1)}. \tag{18.90}$$

(2) Variance of the total energy:

$$K_2 = \sigma_\varepsilon^2 = \frac{T^4}{2H^2(2H+1)(4H+1)}. \tag{18.91}$$

(3) Third total energy cumulant:

$$K_3 = \frac{T^6}{H^3(2H+1)(3H+1)(4H+1)}. \tag{18.92}$$

(4) Fourth total energy cumulant:

$$K_4 = \frac{3(11H+3)T^8}{H^4(2H+1)(3H+1)(4H+1)^2(8H+3)}. \tag{18.93}$$

(5) Skewness of the total energy distribution:

$$\frac{K_3}{(K_3)^{\frac{3}{2}}} = \frac{2^{\frac{3}{2}}\sqrt{2H+1}\sqrt{4H+1}}{3H+1}. \tag{18.94}$$

(6) Kurtosis (or excess) of the total energy distribution:

$$\frac{K_4}{(K_2)^2} = \frac{12(2H+1)(11H+3)}{(3H+1)(8H+3)}. \tag{18.95}$$

Since $H \geq \frac{1}{2}$ we infer from (18.94) that the skewness ranges from $\frac{8}{5}\sqrt{3} = 2.7712813$ (for $H = \frac{1}{2}$) to $\frac{8}{3} = 2.6666667$ for $H \to \infty$. In addition, from (18.95) we find that the kurtosis ranges from $\frac{408}{35} = 11.657143$ for $H = \frac{1}{2}$ to 11 for $H \to \infty$. Therefore, we may conclude that the total energy peak is narrow for any $H \geq \frac{1}{2}$.

The ordinary Brownian motion case of all the previous results is noteworthy, and, relativistically speaking, corresponds to the uniform motion of the moving reference frame with zero velocity (i.e., no motion at all). In fact, by substituting $H = \frac{1}{2}$, $\nu = \frac{1}{2}$ and both (18.28) and (18.34) into (18.89), we find all the Brownian motion total energy cumulants

$$K_n = 2^{n-2} T^{2n} (n-1)! \frac{1}{(2n-1)!} \lim_{x \to 0^+} \left[\frac{d^{2n-1} \tan x}{dx^{2n-1}} \right]. \tag{18.96}$$

Evidently, the last two terms are the $(2n-1)$th coefficient in the MacLaurin expansion of $\tan x$, that reads [5, p. 51]

$$\tan x = \sum_{n=1}^{\infty} \frac{1}{(2n)!} 2^{2n} (2^{2n} - 1)(-1)^{n+1} B_{2n} x^{2n-1}, \tag{18.97}$$

where the B_{2n} are the Bernoulli numbers, a table of which is found, for instance, on [7, p. 810]. Thus, by inserting the coefficients of (18.97) into (18.96), we get all the cumulants of the total energy of standard Brownian motion:

$$K_n = T^{2n} \frac{(n-1)!}{(2n)!} 2^{3n-2} (2^{2n} - 1)(-1)^{n+1} B_{2n}. \tag{18.98}$$

In particular, we have:

(1) mean value of the total energy

$$K_1 = E\{\varepsilon\} = \frac{T^2}{2}; \tag{18.99}$$

(2) variance of the total energy

$$K_2 = \sigma_\varepsilon^2 = \frac{T^4}{3}; \tag{18.100}$$

(3) skewness of the total energy distribution

$$\text{skewness} = \frac{8}{5}\sqrt{3} = 2.7712812921102; \tag{18.101}$$

(4) kurtosis (or excess) of the total energy distribution

$$\text{kurtosis} = \frac{408}{35} = 11.657. \tag{18.102}$$

18.6 *INDEPENDENCE DAY* MOVIE: EXPLOITING THE KLT TO DETECT AN ALIEN SPACESHIP APPROACHING THE EARTH IN DECELERATED MOTION

Everybody remembers the 1996 movie *Independence Day* (see *http://en.wikipedia.org/wiki/Independence_Day_%28film%29*): huge alien spaceships first appear close to Moon and move slowly to prepare for the final attack! It is to be believed, however, that if they move slowly when they are at the Moon distance, they must have moved much, much faster when they were in the open interstellar space in order to cover the vast interstellar distances (please note that here we stick to special relativity only, and do not wish to consider "exotic" mathematical tricks like wormholes, stemming out of general relativity).

In other words, the alien spaceships must have *decelerated* in some way from (say) the speed of light c to zero speed with respect to the Earth. Well, in this section we are going to study the *decelerated* signals emitted by the aliens while they approach the Earth, and work out some equations about the energy of such signals that might help us to dectect an alien invasion much in advance thanks to the KLT developed in this chapter (in the movie *Independence Day*, on the contrary, aliens are already at the Moon distance when humans detect them!).

To adjust our theory to the problem, first consider a trivial Newtonian problem: How long would it take to decelerate from speed c to 0 at the uniform deceleration of just $1g = 9.8\,\mathrm{m/s^{-2}}$? The trivial calculation yieds about 1 year (in Earth time) and the distance at which the deceleration must start is 30,000 AU, or about half a light year (Oort cloud distance) . Should aliens and/or their gadgets withstand decelerations of $2g$, the overall deceleration time would take about half a year, and it should start at the closer distance of $7{,}600\,\mathrm{AU} = 0.12\,\mathrm{lt\text{-}yr}$ from Earth.

Let us now go back to the relativistic decelerated speed $v(t)$ given by (18.50) and consider the radial distance $r(t)$ covered by the spacecraft during the deceleration phase:

$$\frac{dr(t)}{dt} = v(t) = c\sqrt{1 - \left(\frac{t^{2H-1}}{T^{2H-1}}\right)^2}; \tag{18.103}$$

that is

$$R_H(t) = \int_0^{T_H} r(t)\,dt = \int_0^{T_H} c\sqrt{1 - \left(\frac{t^{2H-1}}{T^{2H-1}}\right)^2}\,dt. \tag{18.104}$$

Unfortunately, this integral cannot be computed in a closed form, and we are thus prevented from fully extending our investigation to any value of H larger than $\frac{3}{4}$. We shall thus confine ourselves to the two values $H = \frac{3}{4}$ and $H = 1$, for which one finds

$$R_{\frac{3}{4}}(T) = \int_0^{T_H} r(t)\,dt = \int_0^{T_H} c\sqrt{1 - \frac{t}{T}}\,dt = \tfrac{2}{3}cT_{\frac{3}{4}} = 0.66cT_{\frac{3}{4}} \tag{18.105}$$

and

$$R_1(T) = \int_0^{T_H} r(t)\, dt = \int_0^{T_H} c\sqrt{1-\left(\frac{t}{T}\right)^2}\, dt = \frac{\pi}{4} c T_1 = 0.78 c T_1, \quad (18.106)$$

respectively.

Next we are going to focus only on (18.105) because this is the case where the deceleration of the alien spacecraft is "smoothest" (i.e., less sudden).

The total mean energy emitted by the alien spacecraft in the form of electromagnetic waves (= signals + noise) during the time $T_{\frac{3}{4}}$ is given by (18.90) with $H = \frac{3}{4}$; that is

$$K_1 = E\{\varepsilon\} = \frac{T_{\frac{3}{4}}^2}{2H(2H+1)} = \frac{4}{15} T_{\frac{3}{4}}^2 = 0.266 T_{\frac{3}{4}}^2. \quad (18.107)$$

The variance of the total energy is given by (18.91) again with $H = \frac{3}{4}$

$$K_2 = \sigma_\varepsilon^2 = \frac{T_{\frac{3}{4}}^4}{2H^2(2H+1)(4H+1)} = \frac{4}{45} T_{\frac{3}{4}}^4 = 0.088 T_{\frac{3}{4}}^4. \quad (18.108)$$

Thus, the total mean energy of the electromagnetic waves emitted by the approaching alien spacecraft lies within the range

$$E\{\varepsilon\} \pm \sigma_\varepsilon = \frac{4}{15} T_{\frac{3}{4}}^2 \pm \frac{2}{3\sqrt{5}} T_{\frac{3}{4}}^2 = (0.266 \pm 0.298) T_{\frac{3}{4}}^2. \quad (18.109)$$

This is the "energy bandwidth" upon which any detector of electromagnetic radiation emitted by the alien spacecraft must be built.

The topics discussed in this section were first presented by the author in October 1994 at the *International Astronautical Congress*, held in Jerusalem [10].

18.7 REFERENCES

[1] P. G. Bergmann, *Introduction to the Theory of Relativity*, Prentice-Hall, Englewood Cliffs, NJ, 1942.

[2] C. Maccone, "Eigenfunctions and Energy for Time-Rescaled Gaussian Processes," *Bollettino dell'Unione Matematica Italiana*, Series 6, **3-A** (1984), 213–217.

[3] L. Arnold, *Stochastic Differential Equations*, Wiley, New York, 1973.

[4] R. Ash, *Information Theory*, Interscience, New York, 1965.

[5] A. Erdélyi, W. Magnus, F. Oberhettinger, and F. G. Tricomi, *Higher Transcendental Functions*, Vol. 2, McGraw-Hill, New York, 1953.

[6] G. N. Watson, *A Treatise on the Theory of Bessel Functions*, Cambridge University Press, 1966.

[7] M. Abramowitz, and I. Stegun, *Handbook of Mathematical Functions*, Dover, New York, 1970.

[8] N. N. Lebedev, *Special Functions and Their Applications*, Dover, New York, 1972.

[9] R. H. Cameron, and W. R. Martin, "The Wiener Measure of Hilbert Neighborhoods in the Space of Real Continuous Functions," *Journal of Mathematics and Physics of the M.I.T.*, **23** (1944), 195–201.

[10] C. Maccone, "Mathematical Algorithm to Detect Alien Spacecraft Approaching the Earth," paper IAA-94-IAA.4.1.656 presented at the *International Astronautical Congress held in Jerusalem, Israel, October 11, 1994.*

19

KLT of radio signals from relativistic spaceships in hyperbolic motion

19.1 INTRODUCTION

A spaceship, traveling at a constant acceleration g in its own reference frame, exemplifies the relativistic interstellar flight. If a Gaussian noise (Brownian motion) is emitted in units of the spaceship's proper time, it undergoes a time rescaling when measured in units of the coordinate time. This noise is studied in this chapter in terms of its KL expansion. All topics discussed in this chapter were first published by the author between 1988 and 1990 [1, 2].

19.2 HYPERBOLIC MOTION

A classical topic in special relativity is the so-called *hyperbolic motion*, first considered by Minkowski in 1908 [3], and discussed in most textbooks (see [4, p. 41].[1] Spaceflight did not exist in the time of Minkowski, so he believed that his formulas about the hyperbolic motion could only be applied to the physics of elementary particles then known to exist, such as electrons. Here, however, we shall give the topic of hyperbolic motion a space-travel cut, in view of the applications to telecommunications that will be made in the rest of this book.

Imagine a spacecraft traveling faster and faster with respect to its own reference frame, so that the crew experience a constant acceleration that, for their maximum comfort, we assume numerically equal to $g = 9.8\ \mathrm{m/s^2}$. The longitudinal force (see [5, p. 205]) is

$$f_{\parallel} = \left[1 - \frac{v^2(t)}{c^2}\right]^{-\frac{3}{2}} m\frac{dv(t)}{dt} \tag{19.1}$$

[1] The adjective "hyperbolic" refers to the fact that the $x(t)$ curve in the (x, t) plane is a hyperbola—given by Equation (13.24)—and that hyperbolic functions are used in the analysis.

and so we must find the unknown $v(t)$ in the differential equation

$$\left[1 - \frac{v^2(t)}{c^2}\right]^{-\frac{3}{2}} m \frac{dv(t)}{dt} = mg. \tag{19.2}$$

Separating the variables, and setting $v(t) = c \sin \Omega(t)$, one easily finds

$$\Omega(t) = \arctan\left(\frac{g}{c}t\right), \tag{19.3}$$

whence

$$v(t) = c \sin\left[\arctan\left(\frac{g}{c}t\right)\right] \tag{19.4}$$

but

$$\sin[\arctan x] = \sqrt{\frac{x^2}{1+x^2}} \tag{19.5}$$

so that the velocity $v(t)$ in (11.16) is given by

$$v(t) = \frac{gt}{\sqrt{1 + \left(\frac{g}{c}t\right)^2}}. \tag{19.6}$$

Note that as $t \to \infty$, (19.6) gives $v(t) \to c$, as one would expect. The function $f(t)$ for the hyperbolic motion is then found from (11.16) and (19.6)

$$f(t) = \frac{1}{\left[1 + \left(\frac{g}{c}t\right)^2\right]^{\frac{1}{4}}}. \tag{19.7}$$

Unfortunately, it is quite difficult to handle this function. For instance, its integral

$$\int \frac{dx}{[1+x^2]^{\frac{1}{4}}} \tag{19.8}$$

can be shown to be expressed by hypergeometric functions inasmuch as it is a binomial integral, but not of an elementary type. Thus, we will not attempt to study (19.7) directly, but shall consider its asymptotic expansion in Section 19.4.

A few more results, however, can still be derived from (19.7). In fact, one has (see [6, p. 86])

$$\begin{aligned}\tau(t) = \int_0^t f^2(s)\, ds = \int_0^t \frac{ds}{\sqrt{1+\left(\frac{g}{c}s\right)^2}} &= \frac{c}{g} \operatorname{arcsinh}\left(\frac{g}{c}t\right) \\ &= \frac{c}{g} \ln\left[\frac{g}{c}t + \sqrt{1+\left(\frac{g}{c}t\right)^2}\right]. \end{aligned} \tag{19.9}$$

Thus, the time-rescaled Brownian motion corresponding to the hyperbolic motion of special relativity is

$$X(t) = B(\tau) = B\left(\frac{c}{g}\operatorname{arcsinh}\left(\frac{g}{c}t\right)\right)$$
$$= B\left(\frac{c}{g}\ln\left[\frac{g}{c}t + \sqrt{1 + \left(\frac{g}{c}t\right)^2}\right]\right). \quad (19.10)$$

We shall simply refer to it as the hyperbolic motion.

19.3 TOTAL ENERGY OF SIGNALS FROM RELATIVISTIC SPACESHIPS IN HYPERBOLIC MOTION

In this section we shall show that it is possible (by virtue of the formulas derived in Chapter 21) to compute both the mean total energy and total energy variance of the signals emitted by relativistic spaceships in hyperbolic motion.

Let us start with the mean total energy (21.60). This, by substituting (19.9), takes the form of the definite integral

$$E\{\varepsilon\} = \int_0^T dt \int_0^t f^2(s)\, ds = \frac{c}{g}\int_0^T \operatorname{arcsinh}\left(\frac{g}{c}t\right) dt$$
$$= \frac{c^2}{g^2}\left[x\operatorname{arcsinh}(x) - \sqrt{1+x^2}\right]_0^{\frac{g}{c}T} \quad (19.11)$$

where we make use of the substitution $(gt)/c = x$ and of [6, p. 88, entry 4.6.43]. Thus, the mean total energy of the hyperbolic motion is

$$E\{\varepsilon\} = \frac{c^2}{g^2}\left[\frac{gT}{c}\operatorname{arcsinh}\left(\frac{gT}{c}\right) - \sqrt{1 + \left(\frac{gT}{c}\right)^2} + 1\right]. \quad (19.12)$$

It is also possible to derive a closed-form expression for the total energy variance starting from (21.62) and (19.9), but the calculations are more involved. To this end, let us first note that

$$\int \operatorname{arcsinh}^2(s)\, ds = s\operatorname{arcsinh}^2(s) - 2\sqrt{1+s^2}\operatorname{arcsinh}(s) + 2s + C \quad (19.13)$$

This result can be used to prove the more complicated expression

$$\int x\operatorname{arcsinh}^2(x)\, dx = \tfrac{1}{2}x^2\operatorname{arcsinh}^2(x) - x\sqrt{1+x^2}\operatorname{arcsinh}(x)$$
$$+ \frac{x^2}{2} + \int \sqrt{1+x^2}\operatorname{arcsinh}(x)\, dx + C. \quad (19.14)$$

This leads us to compute a further integral

$$\int \sqrt{1+x^2}\,\operatorname{arcsinh}(x)\,dx = \tfrac{1}{4}[(2\sqrt{1+x^2}\,\operatorname{arcsinh}(x) - x)x + \operatorname{arcsinh}^2(x)] + C. \tag{19.15}$$

These preliminary results enable us to tackle σ_ε defined in (21.62) using (19.9)

$$\sigma_\varepsilon^2 = 4\int_0^T dt \int_0^t du \left[\int_0^u f^2(s)\,ds\right]^2 = 4\left(\frac{c}{g}\right)^2 \int_0^T dt \int_0^t du\, \operatorname{arcsinh}^2\left(\frac{g}{c}u\right). \tag{19.16}$$

Now (19.13) and the substitution $((g/c)u = s)$ change this into

$$\sigma_\varepsilon^2 = 4\left(\frac{c}{g}\right)^3 \int_0^T dt \left[s\,\operatorname{arcsinh}^2(s) - 2\sqrt{1+s^2}\,\operatorname{arcsinh}(s) + 2s\right]_0^{\left(\frac{g}{c}t\right)}$$

$$= 4\left(\frac{c}{g}\right)^3 \left[\int_0^T \left(\frac{g}{c}t\right)\operatorname{arcsinh}^2\left(\frac{g}{c}t\right) dt - 2\int_0^T \sqrt{1+\left(\frac{g}{c}t\right)^2}\operatorname{arcsinh}\left(\frac{g}{c}t\right) dt + 2\int_0^T \frac{g}{c}t\,dt\right].$$

The further substitution $(g/c)t = x$ and (19.14) yield

$$\sigma_\varepsilon^2 = 4\left(\frac{c}{g}\right)^4 \left[\frac{1}{2}\left(\frac{g}{c}T\right)^2 \operatorname{arcsinh}^2\left(\frac{g}{c}T\right) - \left(\frac{g}{c}T\right)\sqrt{1+\left(\frac{g}{c}T\right)^2}\operatorname{arcsinh}\left(\frac{g}{c}T\right) + \frac{3}{2}\left(\frac{g}{c}T\right)^2\right.$$
$$\left. - \int_0^{\left(\frac{g}{c}T\right)} \sqrt{1+x^2}\,\operatorname{arcsinh}(x)\,dx\right]$$

hence the integral (19.15) finally yields

$$\sigma_\varepsilon^2 = \left(\frac{c}{g}\right)^4 \left[2\left(\frac{g}{c}T\right)^2 \operatorname{arcsinh}^2\left(\frac{g}{c}T\right) - 4\left(\frac{g}{c}T\right)\sqrt{1+\left(\frac{g}{c}T\right)^2}\operatorname{arcsinh}\left(\frac{g}{c}T\right)\right.$$
$$\left. +6\left(\frac{g}{c}T\right)^2 - 2\left(\frac{g}{c}T\right)\sqrt{1+\left(\frac{g}{c}T\right)^2}\operatorname{arcsinh}\left(\frac{g}{c}T\right) + \left(\frac{g}{c}T\right)^2 - \operatorname{arcsinh}\left(\frac{g}{c}T\right)\right].$$

Rearranging, the total energy variance for the hyperbolic motion is obtained

$$\sigma_\varepsilon^2 = \left(\frac{c}{g}\right)^4 \left\{\left[2\left(\frac{g}{c}T\right)^2 - 1\right] \operatorname{arcsinh}^2\left(\frac{g}{c}T\right)\right.$$
$$\left. -6\left(\frac{g}{c}T\right)\sqrt{1+\left(\frac{g}{c}T\right)^2}\operatorname{arcsinh}\left(\frac{g}{c}T\right) + 7\left(\frac{g}{c}T\right)^2\right\}. \tag{19.17}$$

19.4 KLT FOR SIGNALS EMITTED IN ASYMPTOTIC HYPERBOLIC MOTION

The obvious asymptotic formula

$$\lim_{x\to\infty} \sqrt{1+x^2} = \lim_{x\to\infty} x$$

and its consequence

$$\lim_{x\to\infty} \ln[x + \sqrt{1 + x^2}] = \lim_{x\to\infty} \ln[2x] \tag{19.18}$$

form the starting point to investigate the asymptotic hyperbolic motion. In fact, from (19.10), we see that, when $t \to \infty$, $X(t)$ approaches

$$B\left(\frac{c}{g}\ln\left(2\frac{g}{c}t\right)\right). \tag{19.19}$$

This we shall call the asymptotic hyperbolic motion and shall study it thoroughly.

By comparing (19.19) against (21.40), we immediately find

$$\int_0^t f^2(s)\,ds = \frac{c}{g}\ln\left(2\frac{g}{c}t\right). \tag{19.20}$$

Then, differentiating and taking the square root, we are led to

$$f(t) = \sqrt{\frac{c}{g}}\frac{1}{\sqrt{t}}. \tag{19.21}$$

This is the $f(t)$ function for the hyperbolic motion.

Integrating (19.21), one then gets

$$\int_0^t f(s)\,ds = 2\sqrt{\frac{c}{g}}\sqrt{t}. \tag{19.22}$$

By virtue of (19.21) and (19.22) the $\chi(t)$ function defined by (11.10) reads

$$\chi(t) = \sqrt{f(t)\int_0^t f(s)\,ds} = \sqrt{2\frac{c}{g}} \tag{19.23}$$

a constant. This circumstance is vital in order to develop the asymptotic hyperbolic case, inasmuch as it simplifies things greatly. In fact, from

$$\chi'(t) = 0 \tag{19.24}$$

and from (11.9), it can be seen at once that $\nu(t)$ vanishes identically

$$\nu(t) = 0 \tag{19.25}$$

(i.e., the order of the Bessel functions is zero). Thus, the KL expansion is given by functions of the form

$$J_0\left(\gamma_n \frac{\int_0^t f(s)\,ds}{\int_0^T f(s)\,ds}\right) = J_0\left(\gamma_n \frac{\sqrt{t}}{\sqrt{T}}\right). \tag{19.26}$$

Our next task is to find the meaning of the constants γ_n, formally given as the real positive zeros of (11.11). Letting $\chi'(t) = 0$ and $\nu'(t) = 0$, and getting rid of all multiplicative factors, one easily sees that (11.11) simplifies to

$$J_0'(\gamma_n) = 0. \tag{19.27}$$

Thus, the γ_n are the positive zeros, arranged in ascending order of magnitude, of the derivative of $J_0(x)$. In other words, they are the abscissas of the maxima and minima of $J_0(x)$, which are known to follow each other alternately. However, a different interpretation of the γ_n follows from the Bessel function property (see [7, p. 12, entry (55) (set $\nu = 0$)]

$$J'_\nu(x) = \frac{\nu}{x} J_\nu(x) - J_{\nu+1}(x). \tag{19.28}$$

In fact, (19.27) now becomes equivalent to

$$J_1(\gamma_n) = 0 \tag{19.29}$$

and one may also say that the γ_n are the real positive zeros of $J_1(x)$. The first 40 among them are listed in [8, p. 748], and one finds, for instance,

$$\gamma_1 = 3.8317060 \quad \gamma_2 = 7.0155867 \quad \gamma_{40} = 126.4461387. \tag{19.30}$$

No explicit formula yielding these zeros exactly is known. However, it is possible to get an approximated expression by setting $\nu = 1$ into the asymptotic formula for $J_\nu(x)$ (see [9, p. 134])

$$\lim_{x\to\infty} J_\nu(x) = \lim_{x\to\infty} \sqrt{\frac{2}{\pi x}} \cos\left(x - \frac{\nu\pi}{2} - \frac{\pi}{4}\right) \tag{19.31}$$

from which

$$\cos\left(\gamma_n - \frac{3\pi}{4}\right) \approx 0 \tag{19.32}$$

or

$$\gamma_n - \frac{3\pi}{4} \approx n\pi - \frac{\pi}{2} \quad (n = 1, 2, \ldots). \tag{19.33}$$

Thus,

$$\gamma_n \approx n\pi + \frac{\pi}{4}. \tag{19.34}$$

We may see how good this approximation is by setting $n = 1, 2, \ldots, 40$

$$\gamma_1 \approx 3.9269908 \quad \gamma_2 \approx 7.0685835 \quad \gamma_{40} \approx 126.4491 \tag{19.35}$$

and checking these results against (19.30). Of course, the agreement improves with increasing n. As for the eigenvalues λ_n, they are related to the γ_n by (11.13)

$$\lambda_n = \frac{4cT}{g} \frac{1}{(\gamma_n)^2} \tag{19.36}$$

and are also variances of the independent Gaussian random variables Z_n.

Finally, we turn to the normalization constants N_n that are obtained from (11.12) after inserting (19.22) and (19.25). The resulting condition for N_n is

$$1 = N_n^2 \frac{4cT}{g} \int_0^1 x[J_0(\gamma_n x)]^2\, dx. \tag{19.37}$$

This integral of (19.37) is calculated within the framework of the Dini expansion in series of Bessel functions (see [7, p. 71]), and one finds

$$\begin{aligned}1 &= N_n^2 \frac{4cT}{g}\left\{\frac{1}{2}\left[J_0'^2(\gamma_n) + \left(1 - \frac{0}{\gamma_n^2}\right)J_0^2(\gamma_n)\right]\right\} \\ &= N_n^2 \frac{2cT}{g}[J_0'^2(\gamma_n) + J_0^2(\gamma_n)] = N_n^2 \frac{2cT}{g} J_0^2(\gamma_n)\end{aligned} \tag{19.38}$$

where (19.27) was used in the last step. Solving with respect to N_n requires the introduction of the modulus of $J_0(\gamma_n)$, and one has

$$N_n = \frac{\sqrt{g}}{\sqrt{2cT}|J_0(\gamma_n)|}. \tag{19.39}$$

This is the exact expression of the normalization constants.

For an approximated expression for N_n, we substitute the Bessel function in its asymptotic form (19.31) with γ_n given in (19.34):

$$|J_0(\gamma_n)| \approx \left|\sqrt{\frac{2}{\pi\gamma_n}}\cos\left(\gamma_n - \frac{\pi}{4}\right)\right| = \sqrt{\frac{2}{\pi\gamma_n}}|\cos(n\pi)| = \sqrt{\frac{2}{\pi\gamma_n}}. \tag{19.40}$$

Then, from (19.39) and (19.40) we get the approximated N_n:

$$N_n \approx \frac{\pi}{2}\sqrt{\frac{g}{cT}}\sqrt{n + \tfrac{1}{4}}. \tag{19.41}$$

All the results obtained in this section may now be summarized by writing the exact KL expansion

$$B\left(\frac{c}{g}\ln\left(2\frac{g}{c}t\right)\right) = \sum_{n=1}^{\infty} Z_n \frac{\sqrt{c}}{\sqrt{g}} \cdot \frac{1}{\sqrt{2}\sqrt{T}|J_0(\gamma_n)|} J_0\left(\gamma_n \frac{\sqrt{t}}{\sqrt{T}}\right) \tag{19.42}$$

and the approximated expansion—found by virtue of (19.31) and (19.41)

$$B\left(\frac{c}{g}\ln\left(2\frac{g}{c}t\right)\right) = \sum_{n=1}^{\infty} Z_n \frac{\sqrt{c}}{\sqrt{g}} \cdot \frac{1}{\sqrt{2}T^{\frac{1}{4}}t^{\frac{1}{4}}} \cos\left(\gamma_n \frac{\sqrt{t}}{\sqrt{T}} - \frac{\pi}{4}\right). \tag{19.43}$$

The physical range of validity of (19.42) and (19.43) is provided by the relativistic condition (11.7). Since, from (19.21)

$$f^4(t) = \frac{c^2}{g^2} \cdot \frac{1}{t^2}, \tag{19.44}$$

(11.7) yields the velocity of the asymptotic hyperbolic motion

$$v(t) = c\sqrt{1 - \frac{c^2}{g^2} \cdot \frac{1}{t^2}}. \tag{19.45}$$

In order to have a non-negative radicand, the inequality

$$\frac{c^2}{g^2} \cdot \frac{1}{t^2} \leq 1 \tag{19.46}$$

must hold, meaning

$$t \geq \frac{c}{g} = 3.0612245 \cdot 10^7 \text{ s} \approx 0.96996974 \text{ years} \approx 1 \text{ year}. \tag{19.47}$$

Thus, the asymptotic approximation to the hyperbolic motion holds only after about 1 year of travel. Since any trip to even the nearest stars will certainly last longer than that, this approximation may be reagarded as physically acceptable.

19.5 CHECKING THE KLT OF ASYMPTOTIC HYPERBOLIC MOTION BY MATLAB SIMULATIONS

Just look at Figure 19.1.

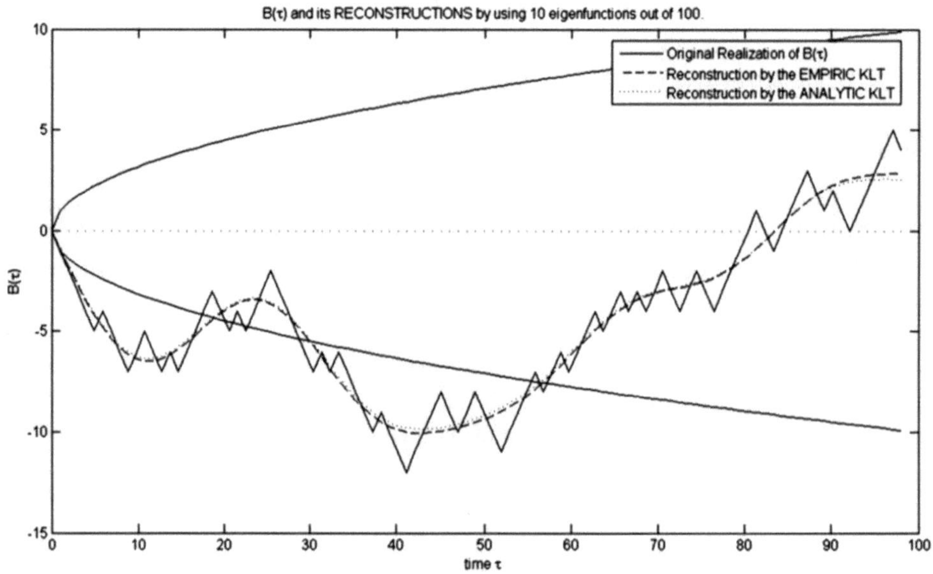

Figure 19.1. The time-rescaled Brownian motion $X(t)$ of (19.43) vs. time t simulated as a random walk over 100 time instants. This $X(t)$ represents the "noisy signal" received on Earth (whence the use of the coordinate time $t =$ Earth time) from a relativistic spaceship moving away from the Earth in an asymptotic hyperbolic motion, as in the science fiction novel *Tau Zero*. Next to the "bumpy curve" of $X(t)$, two more "smooth curves" are shown that *interpolate at best* the bumpy $X(t)$. These two curves are the KLT reconstruction of $X(t)$ *by using the first ten eigenfunctions only*. It is important to note that the two smooth curves are *different* in this case because the KLT expansion (19.43) is *approximated.* Actually, it is an approximated KLT expansion because the asymptotic expansion of the Bessel functions (19.31) was used. So, the two curves are different from each other, but both still interpolate $X(t)$ at best. Note that, were we taking into account the full set of 100 KLT eigenfuctions—rather than just 10—then the *empirical* reconstruction would overlap $X(t)$ *exactly*, but the *analytic* reconstruction would not because of the use of the asymptotic expansion (19.31) of the Bessel functions.

19.6 SIGNAL TOTAL ENERGY AS A STOCHASTIC PROCESS OF T

Formulas (21.60) and (19.20) enable us to obtain the total energy mean value

$$E\{\varepsilon_{Asy}\} = \int_0^T dt \int_0^t f^2(s)\,ds = \frac{c}{g}\int_0^T \ln\left(\frac{2g}{c}t\right)dt. \tag{19.48}$$

The substitution $x = (2g/c)t$ then results in

$$E\{\varepsilon_{Asy}\} = \frac{1}{2}\left(\frac{c}{g}\right)^2 [x(\ln x - 1)]_0^{(\frac{2g}{c})T} = \frac{cT}{g}\left[\ln\left(\frac{2g}{c}T\right) - 1\right].$$

Thus, the asymptotic mean total energy reads

$$E\{\varepsilon_{Asy}\} = \frac{cT}{g}\left[\ln\left(\frac{2g}{c}T\right) - 1\right]. \tag{19.49}$$

Note that the same asymptotic result is obtained from the exact expression (19.12) upon substituting arcsinh by log, and disregarding all the $+1$ that disappear for large T.

Next let us turn to the asymptotic total energy variance by resorting to

$$\int \ln^2 x\,dx = x\ln^2 x - 2x\ln x + 2x + C \tag{19.50}$$

$$\int x\ln^2 x\,dx = \frac{x^2}{2}\ln^2 x - \frac{x^2}{2}\ln x + \frac{x^2}{4} + C \tag{19.51}$$

$$\int x\ln x\,dx = \frac{x^2}{2}\ln x - \frac{x^2}{4} + C. \tag{19.52}$$

In fact, inserting (19.20) into the expression for σ_ε^2 in (21.62), one finds

$$\begin{aligned}\sigma^2_{\varepsilon_{Asy}} &= 4\int_0^T dt\int_0^t du\left[\int_0^u f^2(s)\,ds\right]^2 \\ &= 4\left(\frac{c}{g}\right)^2\int_0^T dt\int_0^t \ln^2\left(\frac{2g}{c}u\right)du\end{aligned} \tag{19.53}$$

whence the substitution $[(2g)/c]u = x$ and the integral in (19.50) yield

$$\sigma^2_{\varepsilon_{Asy}} = 2\left(\frac{c}{g}\right)^3\int_0^T dt\left[x\ln^2 x - 2x\ln x + 2x\right]_0^{\frac{2g}{c}t}.$$

The further substitution $[(2g/c)]t = x$ now leads to the couple of integrals (19.51) and (19.52)

$$\sigma^2_{\varepsilon_{Asy}} = \left(\frac{c}{g}\right)^4 \left[\int_0^{\frac{2g}{c}T} x \ln^2 x \, dx - 2\int_0^{\frac{2g}{c}T} x \ln x \, dx + 2\int_0^{\frac{2g}{c}T} x \, dx\right]$$

$$= \left(\frac{c}{g}\right)^4 \left[\frac{x^2}{2}\ln^2 x - \frac{x^2}{2}\ln x - x^2 \ln x + \frac{x^2}{4} + \frac{x^2}{2} + x^2\right]_0^{\frac{2g}{c}T}$$

$$= \left(\frac{c}{g}\right)^4 \frac{1}{4}\left[x^2(2\ln^2 x - 6\ln x + 7)\right]_0^{\frac{2g}{c}T}$$

$$= \left(\frac{cT}{g}\right)^2 \left[2\ln^2\left(\frac{2g}{c}T\right) - 6\ln\left(\frac{2g}{c}T\right) + 7\right].$$

Thus, the asymptotic total energy variance reads

$$\sigma^2_{\varepsilon_{Asy}} = \left(\frac{cT}{g}\right)^2 \left[2\ln^2\left(\frac{2g}{c}T\right) - 6\ln\left(\frac{2g}{c}T\right) + 7\right]. \tag{19.54}$$

Note that just as (19.49) is the asymptotic version of (19.12), so (19.54) is the asymptotic form of (19.17), and could have been found by substituting arcsinh by log, and forgetting all the additive +1 that are dwarfed for large T.

The square root of (19.74) is the asymptotic total energy standard deviation

$$\sigma_{\varepsilon_{Asy}} = \pm\frac{cT}{g}\sqrt{2\ln^2\left(\frac{2g}{c}T\right) - 6\ln\left(\frac{2g}{c}T\right) + 7}. \tag{19.55}$$

Setting

$$\frac{gT}{c} = x \tag{19.56}$$

we see that the radicand of (19.55) is the quadratic in $\xi \equiv \ln x$

$$2\xi^2 - 6\xi + 7 > 0. \tag{19.57}$$

This is positive for any ξ because $\Delta = -20 < 0$.

Let us regard the noise asymptotic total energy as a stochastic process of T. The process behavior in time is characterized by its mean value curve (19.49) and by the upper and lower (mean value ± standard deviation) curves given by

$$E\{\varepsilon_{Asy}\} \pm \sigma_{\varepsilon_{Asy}}. \tag{19.58}$$

The first column of Table 19.1 shows the numerical values of the independent variable x defined by (19.56) ranging from 0 to 20. In units of time, T ranges from 0 to 20 years since

$$\frac{c}{g} \approx 3.0612245 \cdot 10^7 \text{ s} \approx 0.96699947 \text{ years} \approx 1 \text{ year}. \tag{19.59a}$$

Table 19.1. Noise asymptotic total energy.

$$x = \frac{gT}{c}$$

$$M = x(\ln(2x) - 1)$$

$$V = x^2(2\ln^2(2x) - 6\ln(2x) + 7)$$

x	M	$M - \sqrt{V}$	$M + \sqrt{V}$
0	0	0	0
1	−0.30685	−2.25673	1.643024
2	0.772588	−2.40600	3.951178
3	2.375278	−2.52698	7.277545
4	4.317766	−2.80572	11.44125
5	6.512925	−3.21883	16.24468
6	8.909439	−3.73346	21.55234
7	11.47340	−4.32692	27.27327
8	14.18070	−4.98419	33.34561
9	17.01334	−5.69497	39.72166
10	19.95732	−6.45182	46.36646
11	23.00146	−7.24919	53.25213
12	26.13664	−8.08277	60.35607
13	29.35525	−8.94912	67.65963
14	32.65086	−9.84541	75.14714
15	36.01796	−18.7693	82.80522
16	39.45177	−11.7188	90.62235
17	42.94812	−12.6922	98.58846
18	46.50334	−13.6880	106.6947
19	50.11413	−14.7049	114.9332
20	53.77758	−15.7418	123.2970

The second column gives the numerical values of the asymptotic mean value (19.49) of the noise total energy apart from a factor $(c/g)^2$. The third and fourth columns, respectively, show the values of the lower (minus sign) and upper (plus sign) curves (19.58), again apart from a factor $(c/g)^2$.

One may check the above asymptotic total energy results against the corresponding exact results derived at the end of Section 19.3. Table 19.2 shows the same items as Table 19.1, but is calculated by using the exact total energy variance (19.17). We see that the agreement is not as good for very small values of T, while it increases for increasing T, and the dispersion of the total energy around its mean value increases roughly by the same amount as the total energy itself.

The conclusion to this section is that the KL eigenfunction expansion has been derived for the noise emitted by a spaceship traveling at a constantly accelerated relativistic motion. Though the mathematical difficulties forced us to confine ourselves to the asymptotic theory for values of time larger than 1 year, the study of the noise total energy (where both asymptotic and exact results can be obtained) shows that the errors of the asymptotic version are not very large.

19.7 INSTANTANEOUS NOISE ENERGY FOR ASYMPTOTIC HYPERBOLIC MOTION: PREPARATORY CALCULATIONS

In Chapter 24, as well as in [2], the process $Y(t)$ defined by

$$Y(t) = X^2(t) - E\{X^2(t)\} \tag{19.59b}$$

was considered. According to (24.35), the KL eigenfunction expansion of that process reads

$$Y(t) = \sum_{n=1}^{\infty} \tilde{Z}_n \tilde{N}_n \sqrt{\tilde{f}(t) \int_0^t \tilde{f}(s)\, ds} \cdot J_{\tilde{\nu}(t)} \left(\tilde{\gamma}_n \frac{\int_0^t \tilde{f}(s)\, ds}{\int_0^T \tilde{f}(s)\, ds} \right), \tag{19.60}$$

where the function $\tilde{f}(t)$ is defined in terms of $f(t)$ by (24.24). That is,

$$\tilde{f}(t) \equiv 2f(t) \sqrt{\int_0^t f^2(z)\, dz}. \tag{19.61}$$

This section is devoted to finding the KL expansion of the zero-mean square process $Y(t)$, in the asymptotic hyperbolic case, and its physical meaning for relativistic interstellar flight will be examined in the coming section. In this section we just pave the mathematical way to the coming section by performing the necessary calculations.

Table 19.2. Noise exact total energy.

$$x = \frac{gT}{c}$$

$$M = x\ln(x + \sqrt{1+x^2}) - \sqrt{1+x^2} + 1$$

$$V = (2x^2 - 1)[\ln^2(x + \sqrt{1+x^2}) - 6x\sqrt{1+x^2}\ln(x + \sqrt{1+x^2}) + 7x^2]$$

0	0	0	0
1	0.467160	−0.07884	1.013160
2	1.651202	−0.31139	3.613797
3	3.293061	−0.67013	7.256257
4	5.255744	−1.12713	11.63862
5	7.463172	−1.66295	16.58929
6	9.867916	−2.26419	22.00002
7	12.43777	−2.92124	27.79679
8	15.14952	−3.62692	33.92596
9	17.98561	−4.37568	40.34690
10	20.93235	−5.16310	47.02780
11	23.97876	−5.98558	53.94311
12	27.11583	−6.84016	61.07182
13	30.33603	−7.72432	68.39640
14	33.633301	−8.63592	75.90195
15	37.00130	−9.57310	83.57571
16	40.43615	−18.5342	91.40657
17	43.93342	−11.5179	99.38482
18	47.48945	−12.5229	107.5018
19	51.10098	−13.5481	115.7500

According to (19.61), we must first obtain the function $\tilde{f}(t)$, which follows at once from (19.20) and (19.21)

$$\tilde{f}(t) = \frac{2c}{g}\sqrt{\frac{\ln\left(\frac{g}{c}t\right)}{t}}. \tag{19.62}$$

We now proceed to construct the complicated expression (4.26), or, alternatively, (3.50) with $f(t)$ substituted by $\tilde{f}(t)$, to find the time-dependent order $\tilde{\nu}(t)$. But a glance at (24.26) and (19.62) shows that considerable analytical difficulties are involved. For instance, evaluation of the integral appearing in (22.50) with $f(t)$ substituted by $\tilde{f}(t)$, namely

$$\int_0^t \tilde{f}(s)\,ds = \int_0^t \frac{2c}{g}\sqrt{\frac{\ln\left(2\frac{g}{c}s\right)}{s}}\,ds \tag{19.63}$$

does not seem to be feasible in terms of elementary transcendental functions.

Nevertheless, these difficulties may be overcome by keeping in mind that we are seeking the asymptotic version of (19.62) for large values of time. Therefore, one is led to consider the limit

$$\lim_{t\to\infty}\tilde{f}(t) = \lim_{t\to\infty}\frac{2c}{g}\sqrt{\frac{\ln\left(2\frac{g}{c}t\right)}{t}} = \frac{2c}{g}\sqrt{\lim_{t\to\infty}\frac{\ln\left(2\frac{g}{c}t\right)}{t}} = \frac{\infty}{\infty} \tag{19.64}$$

where the indefinite form forces us to resort to L'Hospital's rule, and yields

$$\lim_{t\to\infty}\tilde{f}(t) = \frac{2c}{g}\sqrt{\lim_{t\to\infty}\frac{\left(2\frac{g}{c}\right)(1)}{\left(2\frac{g}{c}t\right)(1)}} = \frac{2c}{g}\lim_{t\to\infty}\frac{1}{\sqrt{t}} = \lim_{t\to\infty}\frac{2c}{g}\frac{1}{\sqrt{t}}.$$

Concluding the calculation at the last limit, and checking this against the initial limit, we thus obtain the following "ultimate" asymptotic version of (19.62), which from now on we shall regard as the asyptotic replacement to (19.62) for large values of time t

$$\tilde{f}(t) = \frac{2c}{g}\frac{1}{\sqrt{t}}. \tag{19.65}$$

This formula is simple enough for us to perform the remaining calculations involved with the KL expansion (19.60).

Beginning with the computation of the Bessel function order (22.50) with $f(t)$ substituted by $\tilde{f}(t)$, (19.65) yields at once

$$\ln\tilde{f}(t) = \ln\left(\frac{2c}{g}\right) - \tfrac{1}{2}\ln t \tag{19.66}$$

whence

$$\frac{d\ln\tilde{f}(t)}{dt} = -\frac{1}{2}\frac{1}{t} \tag{19.67}$$

$$\frac{d^2\ln\tilde{f}(t)}{dt^2} = +\frac{1}{2}\frac{1}{t^2} \tag{19.68}$$

and

$$\int_0^t \tilde{f}(s)\,ds = \frac{2c}{g}\int_0^t \frac{1}{\sqrt{s}}\,ds = \frac{4c}{g}\sqrt{t}. \tag{19.69}$$

Therefore, (22.50), with $f(t)$ substituted by $\tilde{f}(t)$, yields

$$\begin{aligned}\tilde{\nu}(t) &= \sqrt{\frac{1}{4}+\left[\frac{\frac{4c}{g}\sqrt{t}}{\frac{2c}{g}\frac{1}{\sqrt{t}}}\right]^2\left\{\frac{3}{4}\left(-\frac{1}{2}\frac{1}{t}\right)^2-\frac{1}{2}\left(\frac{1}{2}\frac{1}{t^2}\right)\right\}}\\ &= \sqrt{\frac{1}{4}+[2t]^2\left\{\frac{3}{16}\frac{1}{t^2}-\frac{1}{4}\frac{1}{t^2}\right\}} = \sqrt{\frac{1}{4}+4t^2\cdot\frac{1}{t^2}\left\{\frac{3}{16}-\frac{1}{4}\right\}}.\end{aligned} \tag{19.70}$$

The time variable is thus seen to disappear from the last formula, leaving

$$\tilde{\nu}(t) = \sqrt{\frac{1}{4}+4\left\{\frac{3-4}{16}\right\}} = \sqrt{\frac{1}{4}+\frac{-1}{4}} = \sqrt{0} = 0.$$

That is, the order of the Bessel function vanishes identically

$$\tilde{\nu}(t) = 0 \tag{19.71}$$

and this circumstance helps to simplify further calculations considerably. Intuitively speaking, (19.71) is quite a reasonable result. In fact, on the one hand, the corresponding Bessel function order in the KL expansion of the $X(t)$ process vanished too

$$\nu(t) = 0, \tag{19.72}$$

which is Equation (19.25), or eq. (68) in [1]. On the other hand, (19.71) truly mirrors the asymptotic character of the KL expansion under consideration, since the Bessel function of order zero is the only Bessel function of the first kind to have its initial value equal to one rather than zero, pointing out the non-validity of this theory for values of time near to the origin.

Let us now proceed to finding the function $\tilde{\chi}(t)$ defined by (24.25). By virtue of (19.65) and (19.69), it follows that

$$\tilde{\chi}(t) = \sqrt{\frac{2c}{g}\frac{1}{\sqrt{t}}\cdot\frac{4c}{g}\sqrt{t}} = 2\sqrt{2}\frac{c}{g}. \tag{19.73}$$

Once again, the time variable cancels out from the last formula, yielding a constant rather than a time function. An immediate consequence of (19.73) is, of course,

$$\tilde{\chi}'(t) = 0 \tag{19.74}$$

which helps to simplify further calculations also.

Reverting now to the KL expansion of (19.60), we see that the Bessel function must have the form

$$J_0\left(\tilde{\gamma}_n \frac{\int_0^t \tilde{f}(s)\, ds}{\int_0^T \tilde{f}(s)\, ds}\right) = J_0\left(\tilde{\gamma}_n \frac{\sqrt{t}}{\sqrt{T}}\right). \tag{19.75}$$

Our next task is to find the meaning of the constants $\tilde{\gamma}_n$, given by (24.27). As $\tilde{\chi}'(t) = 0$ and $\tilde{\nu}'(t) = 0$, and getting rid of all multiplicative factors, one easily sees that (24.27) yields

$$J_0'(\tilde{\gamma}_n) = 0. \tag{19.76}$$

Thus, the $\tilde{\gamma}_n$ are the positive zeros, arranged in ascending order of magnitude, of the derivative of $J_0(x)$. In other words, they are the abscissas of the maxima and minima of $J_0(x)$, that are known to follow each other alternately. However, a different interpretation of the $\tilde{\gamma}_n$ follows from (see [7, p. 12, entry 55 ($\nu = 0$ must be set)]

$$J_\nu'(x) = \frac{\nu}{x} J_\nu(x) - J_{\nu+1}(x). \tag{19.77}$$

Using (19.77), (19.76) now becomes equivalent to

$$J_1(\tilde{\gamma}_n) = 0 \tag{19.78}$$

and one may also say that the $\tilde{\gamma}_n$ are the real positive zeros of $J_1(x)$ The first 40 among them are listed in [8, p. 748], and one finds, for instance:

$$\tilde{\gamma}_1 = 3.8317060 \quad \tilde{\gamma}_2 = 7.0155867 \quad \tilde{\gamma}_{40} = 126.4461387. \tag{19.79}$$

No explicit formula yielding these zeros exactly is known. However, it is possible to get an approximated expression for them on setting $\nu = 1$ into the asymptotic formula for $J_\nu(x)$ (see [9, p. 134])

$$\lim_{x\to\infty} J_\nu(x) = \lim_{x\to\infty} \sqrt{\frac{2}{\pi x}} \cos\left(x - \frac{\nu\pi}{2} - \frac{\pi}{4}\right) \tag{19.80}$$

getting

$$\cos\left(\tilde{\gamma}_n - \frac{3\pi}{4}\right) \approx 0 \tag{19.81}$$

whence

$$\tilde{\gamma}_n - \frac{3\pi}{4} \approx n\pi - \frac{\pi}{2} \quad (n = 1, 2, \ldots) \tag{19.82}$$

and finally

$$\tilde{\gamma}_n \approx n\pi + \frac{\pi}{4}. \tag{19.83}$$

We can see how good this approximation is by setting $n = 1, 2, \ldots, 40$,

$$\tilde{\gamma}_1 \approx 3.9269908 \quad \tilde{\gamma}_2 \approx 7.0685835 \quad \tilde{\gamma}_{40} \approx 126.4491 \tag{19.84}$$

and checking these results against (19.79): the agreement improves with increasing n.

As for the eigenvalues $\tilde{\lambda}_n$, they are related to the $\tilde{\gamma}_n$ by (24.29), and, by virtue of (19.69), take the form

$$\tilde{\lambda}_n = \frac{16c^2T}{g^2}\frac{1}{(\tilde{\gamma}_n)^2}. \tag{19.85}$$

Finally, we turn to the normalization constants $\tilde{N}_n$ that can be discovered from (24.28) by inserting (19.69) and (19.71). Therefore

$$1 = \tilde{N}_n^2 \frac{16c^2T}{g^2}\int_0^1 x[J_0(\tilde{\gamma}_n x)]^2\,dx. \tag{19.86}$$

This integral is evaluated within the framework of the Dini expansion in the series of Bessel functions [5, p. 71], and one finds

$$\begin{aligned} 1 &= \tilde{N}_n^2 \frac{16c^2T}{g^2}\left\{\frac{1}{2}\left[J_0'^2(\tilde{\gamma}_n) + \left(1 - \frac{0}{\tilde{\gamma}_n^2}\right)J_0^2(\tilde{\gamma}_n)\right]\right\} \\ &= \tilde{N}_n^2 \frac{8c^2T}{g^2}[J_0'^2(\tilde{\gamma}_n) + J_0^2(\tilde{\gamma}_n)] = \tilde{N}_n^2 \frac{8c^2T}{g^2} J_0^2(\tilde{\gamma}_n) \end{aligned} \tag{19.87}$$

where (19.76) was used in the last step. Solving for $\tilde{N}_n$ requires the introduction of the modulus of $J_0(\tilde{\gamma}_n)$, and one has

$$\tilde{N}_n = \frac{g}{2\sqrt{2}c\sqrt{T}|J_0(\tilde{\gamma}_n)|}. \tag{19.88}$$

This is the exact expression of the normalization constants. We can, however, derive an approximated expression for them on substituting the Bessel function by virtue of (19.80) and (19.83)

$$|J_0(\tilde{\gamma}_n)| \approx \left|\sqrt{\frac{2}{\pi\tilde{\gamma}_n}}\cos\left(\tilde{\gamma}_n - \frac{\pi}{4}\right)\right| = \sqrt{\frac{2}{\pi\tilde{\gamma}_n}}|\cos(n\pi)| = \sqrt{\frac{2}{\pi\tilde{\gamma}_n}}. \tag{19.89}$$

Thus, from (19.88), by virtue of (19.89) and (19.83), we get the approximated $\tilde{N}_n$:

$$\tilde{N}_n \approx \frac{\pi}{4}\frac{g}{c\sqrt{T}}\sqrt{n + \tfrac{1}{4}} \tag{19.90}$$

which completes our set of preliminary calculations.

19.8 KL EXPANSION FOR THE INSTANTANEOUS ENERGY OF THE NOISE EMITTED BY A RELATIVISTIC SPACESHIP

When dealing with a noise represented by a stochastic process $X(t)$, an important distinction is between its instantaneous energy, given by the square process

$$X^2(t) \tag{19.91}$$

and the total energy, given by the stochastic integral of the instantanous energy (19.91) over the finite time span, $0 \le t \le T$ during which the noise is observed:

$$I = \int_0^T X^2(s)\, ds. \tag{19.92}$$

This section is devoted to finding the KL expansion of the process (19.91), whereas both mean value and variance of the random variable (19.92) have already been obtained in Section 19.3, as well as in section 5 of [1]. A related paper, [10], may also be consulted.

Let us then consider the mean value of (19.91), given by (21.59); that is,

$$E\{X^2(t)\} = \int_0^t f^2(s)\, ds, \tag{19.93}$$

where E denotes mean value operator, or ensemble average. By virtue of (19.20), (19.93) is changed into

$$E\{X^2(t)\} = \frac{c}{g}\ln\left(2\frac{g}{c}t\right). \tag{19.94}$$

Thus, the zero-mean square process $Y(t)$, defined by (19.59), takes the form

$$Y(t) = X^2(t) - \frac{c}{g}\ln\left(2\frac{g}{c}t\right) \tag{19.95}$$

whence

$$X^2(t) = \frac{c}{g}\ln\left(2\frac{g}{c}t\right) + Y(t). \tag{19.96}$$

Let us now consider the KL expansion of the $Y(t)$ process. By sustituting into (19.60) the normalization constants (19.90), the $\tilde{\chi}(t)$ function (19.73), and the Bessel function (19.75), we come up with

$$Y(t) = \sum_{n=1}^{\infty} \tilde{Z}_n \frac{g}{2\sqrt{2}c\sqrt{T}|J_0(\tilde{\gamma}_n)|} 2\sqrt{2}\frac{c}{g} J_0\left(\tilde{\gamma}_n \frac{\sqrt{t}}{\sqrt{T}}\right)$$

from which both c and g disappear, yielding

$$Y(t) = \sum_{n=1}^{\infty} \tilde{Z}_n \frac{1}{\sqrt{T}|J_0(\tilde{\gamma}_n)|} J_0\left(\tilde{\gamma}_n \frac{\sqrt{t}}{\sqrt{T}}\right). \tag{19.97}$$

Thus, by virtue of (19.96) and (19.97), we conclude that the exact KL expansion of the instantaneous energy $X^2(t)$ reads

$$X^2(t) = \frac{c}{g}\ln\left(2\frac{g}{c}t\right) + \sum_{n=1}^{\infty} \tilde{Z}_n \frac{1}{\sqrt{T}|J_0(\tilde{\gamma}_n)|} \frac{2\sqrt{2}c}{g} J_0\left(\tilde{\gamma}_n \frac{\sqrt{t}}{\sqrt{T}}\right). \tag{19.98}$$

From this exact expansion we may also derive an approximated one by resorting to the usual asymptotic formula (19.80) for both the Bessel functions appearing in (19.98). The result is

$$X^2(t) \approx \frac{c}{g}\ln\left(2\frac{g}{c}t\right) + \sum_{n=1}^{\infty} \tilde{Z}_n \frac{1}{T^{\frac{1}{4}}t^{\frac{1}{4}}} \frac{2\sqrt{2}c}{g} \cos\left(\tilde{\gamma}_n \frac{\sqrt{t}}{\sqrt{T}} - \frac{\pi}{4}\right), \tag{19.99}$$

which, after substituting the $\tilde{\gamma}_n$ by the approximated version (19.83), takes the final form

$$X^2(t) \approx \frac{c}{g}\ln\left(2\frac{g}{c}t\right) + \sum_{n=1}^{\infty} \tilde{Z}_n \frac{1}{T^{\frac{1}{4}}t^{\frac{1}{4}}} \frac{2\sqrt{2}c}{g} \cos\left(\left(n\pi + \frac{\pi}{4}\right)\frac{\sqrt{t}}{\sqrt{T}} - \frac{\pi}{4}\right). \tag{19.100}$$

This is the approximated (i.e., asymptotic) KL expansion of the noise instantaneous energy for large values of time. The computational advantage of (19.100) over (19.98) is that the Bessel functions have been substituted by a cosine.

19.9 CONCLUSION

A surprising property of both the instantaneous energy KL expansions (19.98) and (19.100) is revealed by checking them, respectively, against the corresponding KL expansions (19.42) and (19.43) of the noise process $X(t)$. In fact, on the one hand, one should note that the $\tilde{\gamma}_n$ (19.78) of the $Y(t)$ process are just the same as the $\tilde{\gamma}_n$ of the $X(t)$ process, given by (19.29), inasmuch as both are the real positive zeros of $J_1(t)$. Moreover, a glance shows that (19.98) has just the same eigenfunctions as (19.42), and (19.100) as (19.43). Therefore, we reach the unexpected conclusion that, when dealing with the noise emitted by a relativistic spaceship in asymptotic hyperbolic motion, the best orthonormal basis in the Hilbert space (i.e., the basis spanned by the eigenfunctions) is the same for both the noise and its own zero-mean instantaneous energy. Alternatively, if we prefer to give up the zero-mean restriction, we may say that the noise and its own instantaneous energy share parallel optimal reference frames, or bases, in the Hilbert space. This unusual feature should bear consequences in the design of a correct signal analysis procedure to filter out the noise received on Earth from a relativistically moving spaceship in asymptotic hyperbolic motion.

19.10 REFERENCES

[1] C. Maccone, "Relativistic Interstellar Flight and Gaussian Noise," *Acta Astronautica*, **17** (1988), 1019–1027.

[2] C. Maccone, "Relativistic Interstellar Flight and Instantaneous Noise Energy," *Acta Astronautica*, **21** (1990), 155–159.
[3] H. Minkowski, *Raum und Zeit*, English translation in *The Principle of Relativity*, Dover, New York, 1952, in particular pp. 85–86.
[4] W. Rindler, *Special Relativity*, Oliver & Boyd, Edinburgh, 1960.
[5] H. Goldstein, *Classical Mechanics*, Addison-Wesley, Reading, MA, 1950.
[6] M. Abramowitz, and I. Stegun, *Handbook of Mathematical Functions*, Dover, New York, 1970.
[7] A. Erdélyi, W. Magnus, F. Oberhettinger, and F. G. Tricomi, *Higher Transcendental Functions*, Vol. 2, McGraw-Hill, New York, 1953.
[8] G. N. Watson, *A Treatise on the Theory of Bessel Functions*, Cambridge University Press, 1966.
[9] N. N. Lebedev, *Special Functions and Their Applications*, Dover, New York, 1972.
[10] G. M. Anderson, and D. T. Greenwood, "Relativistic Flight with a Constant Thrust Rocket," *Astronautica Acta*, **16** (1970), 153–158.

20

KLT of radio signals from relativistic spaceships in arbitrary motion

20.1 INTRODUCTION

In three papers [1–3] the present author applied the concept of time-rescaled Brownian motion to aspects of relativistic interstellar flight ranging from communication theory to genetics. The content of the present chapter was fully published in paper form in [11]. In particular, the Gaussian noise (Brownian motion) $X(t)$, emitted by a relativistic spaceship traveling at a *constant* acceleration g in its own reference frame, was shown to be—see Equation (12.10) or [1, eq. (53)]

$$X(t) = B\left(\frac{c}{g}\operatorname{arcsinh}\left(\frac{g}{c}t\right)\right) = B\left(\frac{c}{g}\ln\left[\frac{g}{c}t + \sqrt{1+\left(\frac{g}{c}t\right)^2}\right]\right), \tag{20.1}$$

where c is the speed of light, $B(t)$ denotes standard Brownian motion with mean zero, variance t, and initial condition $B(0) = 0$, and time ranges within the finite interval $0 \le t \le T$.

An approximated (i.e., asymptotic) version of (20.1) for large values of time follows by ignoring the additive $+1$ under the root sign that is dwarfed by the other terms, and reads

$$B\left(\frac{c}{g}\ln\left(2\frac{g}{c}t\right)\right). \tag{20.2}$$

For this stochastic process, it was proved in Equation (10.42) that the KL eigenfunction expansion is given by

$$B\left(\frac{c}{g}\ln\left(2\frac{g}{c}t\right)\right) = \sum_{n=1}^{\infty} Z_n \frac{1}{\sqrt{T}|J_0(\gamma_n)|} J_0\left(\gamma_n \frac{\sqrt{t}}{\sqrt{T}}\right), \tag{20.3}$$

where

(1) The constants γ_n ($n = 1, 2, \ldots$), appearing in the argument of the Bessel function of order 0, $J_0(\ldots)$, are the (infinite) real positive zeros of the Bessel function of order 1; that is

$$J_1(\gamma_n) = 0. \tag{20.4}$$

(2) The eigenvalues λ_n are expressed in terms of the constants γ_n by the formula

$$\lambda_n = \frac{4cT}{g}\frac{1}{(\gamma_n)^2}.$$

(3) The Z_n are a set of orthogonal Gaussian random variables with mean zero and variance λ_n; that is

$$E\{Z_m Z_n\} = \lambda_n \delta_{mn}. \tag{20.5}$$

By resorting to the asymptotic expansion of the Bessel function of the first kind for large values of its argument—see Chapter 12, Equation (12.80)—it is also possible to derive an approximated KL eigenfunction expansion reading

$$B\left(\frac{c}{g}\ln\left(2\frac{g}{c}t\right)\right) \approx \sum_{n=1}^{\infty} Z_n \frac{1}{T^{\frac{1}{4}}t^{\frac{1}{4}}}\cos\left(\gamma_n\frac{\sqrt{t}}{\sqrt{T}} - \frac{\pi}{4}\right). \tag{20.6}$$

Here again, the meaning of the quantities appearing in (20.6) is the same as for those appearing in (20.3), with the only exception that the γ_n are now replaced by the approximated formula

$$\gamma_n \approx n\pi + \frac{\pi}{4}. \tag{20.7}$$

All the above results apply if, and only if, the spaceship acceleration in its own reference frame (proper acceleration) equals a *constant*, here denoted by g. Indeed, customary terminology such as "relativistic rocket", "relativistic interstellar flight", and the like, is almost always understood in this restricted constant g sense. However, for the sake of completeness it is desirable to extend the above results of Equations (20.1)–(20.7) to the general case where the constant g is replaced by an arbitrary time function

$$gA(t). \tag{20.8}$$

Here the time function $A(t)$ has the dimensions of a purely numerical factor, and represents the non-constant spaceship acceleration measured in units of g with respect to the spaceship reference frame. Clearly, the constant g case can now be regarded as the particular case for which the $A(t)$ function is given by

$$A(t) = 1. \tag{20.9}$$

The present chapter is devoted to generalization of the mentioned results (20.1)–(20.7) to the non-constant g case, and to exploring how many of these equations can be cast in a closed analytical form without resorting to numerical techniques for solving them.

20.2 ARBITRARY SPACESHIP ACCELERATION

20.2.1 Relativistic interstellar flight with an arbitrary spaceship acceleration profile

Analogous to the constant g case in (10.2), the differential equation of motion for a relativistic spaceship moving with arbitrary acceleration with respect to its own reference frame reads

$$\left[1 - \frac{v^2(t)}{c^2}\right]^{-\frac{3}{2}} \frac{dv(t)}{dt} = gA(t) \tag{20.10}$$

with the initial condition

$$v(0) = 0. \tag{20.11}$$

The left-hand side of (20.10) is the so-called "longitudinal force" of special relativity seen before as (12.1), and the right-hand side is just (20.8). The variables in (20.10) may be separated to achieve the analytical expression of the unknown velocity $v(t)$ by setting

$$v(t) = c \sin \Omega(t). \tag{20.12}$$

In fact, integrating both sides, one gets

$$c \tan \Omega(t) = g \int_0^t A(s)\, ds \tag{20.13}$$

whence an inversion yields

$$\Omega(t) = \arctan\left(\frac{g}{c} \int_0^t A(s)\, ds\right). \tag{20.14}$$

Replacing this into (20.12), we get for the velocity

$$v(t) = c \sin\left(\arctan\left(\frac{g}{c} \int_0^t A(s)\, ds\right)\right). \tag{20.15}$$

But

$$\sin(\arctan(x)) = \frac{x}{\sqrt{1+x^2}} \tag{20.16}$$

so that, in conclusion, the spaceship velocity is given by

$$v(t) = \frac{g \displaystyle\int_0^t A(s)\, ds}{\sqrt{1 + \left(\dfrac{g}{c} \displaystyle\int_0^t A(s)\, ds\right)^2}}. \tag{20.17}$$

In the constant g case, this reduces to (12.6), namely

$$v(t) = \frac{gt}{\sqrt{1 + \left(\dfrac{g}{c} t\right)^2}}. \tag{20.18}$$

The distance $x(t)$ covered by the spaceship up to time t and measured by Earth standards is defined by

$$x(t) = \int_0^t v(s)\,ds \tag{20.19}$$

with the consequential initial condition

$$x(0) = 0. \tag{20.20}$$

To perform the integration (20.19) for the function (20.17) in a closed form, we note that one has

$$\frac{d}{dt}\sqrt{1+\left(\frac{g}{c}\int_0^t A(s)\,ds\right)^2} = \frac{\frac{g}{c}\int_0^t A(s)\,ds}{\sqrt{1+\left(\frac{g}{c}\int_0^t A(s)\,ds\right)^2}} \cdot \frac{g}{c}A(t). \tag{20.21}$$

This enables us to rewrite (20.17) in the form

$$v(t) = \frac{c^2}{g}\frac{1}{A(t)}\frac{d}{dt}\sqrt{1+\left(\frac{g}{c}\int_0^t A(s)\,ds\right)^2} \tag{20.22}$$

whence integration (by parts on the right-hand side) yields the required distance:

$$x(t) = \frac{c^2}{g}\left[\frac{\sqrt{1+\left(\frac{g}{c}\int_0^t A(s)\,ds\right)^2}-1}{A(t)} + \int_0^t \frac{\frac{dA(s)}{ds}}{A^2(s)}\sqrt{1+\left(\frac{g}{c}\int_0^s A(w)\,dw\right)^2}\,ds\right] \tag{20.23}$$

In the constant g case, this reduces to (see, for example, [5, p. 199, eq. (13)])

$$x(t) = \frac{c^2}{g}\left[\sqrt{1+\left(\frac{g}{c}t\right)^2}-1\right]. \tag{20.24}$$

Finally, the proper time τ (i.e., the time measured aboard the spaceship) is given by

$$\tau(t) = \int_0^t \sqrt{1-\frac{v^2(s)}{c^2}}\,ds. \tag{20.25}$$

On replacing (20.17), some rearranging yields

$$\tau(t) = \int_0^t \frac{1}{\sqrt{1+\left(\frac{g}{c}\int_0^s A(w)\,dw\right)^2}}\,ds. \tag{20.26}$$

In the constant g case, both integrals in (20.26) can be performed analytically, whence Equation (12.9) is found again

$$\tau(t) = \int_0^t \frac{ds}{\sqrt{1+\left(\frac{g}{c}s\right)^2}} = \frac{c}{g}\operatorname{arcsinh}\left(\frac{g}{c}t\right) = \frac{c}{g}\ln\left[\frac{g}{c}t + \sqrt{1+\left(\frac{g}{c}t\right)^2}\right]. \quad (20.27)$$

20.2.2 KL expansion of the Gaussian noise emitted by a spaceship having an arbitrary acceleration profile

We now turn to the problem of finding the KL expansion for the Brownian motion (Gaussian noise) emitted when the spaceship moves according to the arbitrary acceleration law $gA(t)$. From a purely mathematical point of view, the formulas solving this problem were obtained by the author in [6]. However, that paper only dealt with analytical developments and not with their application to special relativity, not to mention starflight. It thus appears appropriate to recast the content of [6] here so as to make it fit the new developments required by introduction of the generic acceleration function $A(t)$.

The important point expressed in Section 21.8 is that the time-dilation effect of special relativity induces a time-rescaling in the Brownian motion argument that may be expressed by the formula

$$B(\tau) = B\left(\int_0^t f^2(s)\, ds\right), \quad (20.28)$$

where the function $f(t)$ is called the "time-rescaling" function (see [6, p. 213, eq. (1.1)]. From this and the definition (20.25) of proper time, one gets

$$\int_0^t f^2(s)\, ds = \int_0^t \sqrt{1 - \frac{v^2(s)}{c^2}}\, ds \quad (20.29)$$

whence, differentiating and taking the square root, one finds

$$f(t) = \left[1 - \frac{v^2(t)}{c^2}\right]^{\frac{1}{4}}. \quad (20.30)$$

This is the relationship between the Brownian motion time-rescaling function, $f(t)$, and the spaceship velocity, $v(t)$. Inversion of (20.30) immediately leads to

$$v(t) = c\sqrt{1 - f^4(t)}. \quad (20.31)$$

We next develop the relationship between the time-rescaling function, $f(t)$, and the acceleration, $gA(t)$. Inserting (20.17) into (20.30), with a few reductions, yields

$$f(t) = \frac{1}{\left[1+\left(\frac{g}{c}\int_0^t A(s)\, ds\right)^2\right]^{\frac{1}{4}}} = \left[1+\left(\frac{g}{c}\int_0^t A(s)\, ds\right)^2\right]^{-\frac{1}{4}}. \quad (20.32)$$

This formula will be extensively used in the rest of the present chapter in either of the above forms. To invert it, one first solves for the integral

$$\int_0^t A(s)\, ds = \frac{c}{g}\sqrt{\frac{1}{f^4(t)} - 1} \tag{20.33}$$

and then differentiates with respect to t, getting

$$A(t) = -\frac{2c}{g} \frac{\dfrac{df(t)}{dt}}{f^3(t)\sqrt{1 - f^4(t)}}. \tag{20.34}$$

We are now ready to state the main theorem of the present section.

Theorem 1 *The KL expansion for the Gaussian noise emitted by a spaceship moving at an arbitrary proper acceleration $gA(t)$ is given by*

$$X(t) \equiv B(\tau) = \left[1 + \left(\frac{g}{c}\int_0^t A(s)\, ds\right)^2\right]^{-\frac{1}{8}} \cdot \sqrt{\int_0^t \frac{ds}{\left[1 + \left(\frac{g}{c}\int_0^s A(w)\, dw\right)^2\right]^{\frac{1}{4}}}} \cdot \sum_{n=1}^{\infty} Z_n N_n J_{\nu(t)}\left(\gamma_n \frac{\displaystyle\int_0^t \frac{ds}{\left[1 + \left(\frac{g}{c}\int_0^s A(w)\, dw\right)^2\right]^{\frac{1}{4}}}}{\displaystyle\int_0^T \frac{ds}{\left[1 + \left(\frac{g}{c}\int_0^s A(w)\, dw\right)^2\right]^{\frac{1}{4}}}}\right). \tag{20.35}$$

Here

(1) The Bessel function of the first kind appearing in (20.35) has the time-dependent order

$$\nu(t) = \sqrt{-\chi^3(t)\sqrt{1 + \left(\frac{g}{c}\int_0^t A(s)\, ds\right)^2} \cdot \frac{d}{dt}\left[\chi'(t)\sqrt{1 + \left(\frac{g}{c}\int_0^t A(s)\, ds\right)^2}\right]} \tag{20.36}$$

where the auxiliary function $\chi(t)$ has been defined

$$\chi(t) = \left[1 + \left(\frac{g}{c}\int_0^t A(s)\, ds\right)^2\right]^{-\frac{1}{8}} \cdot \sqrt{\int_0^t \frac{ds}{\left[1 + \left(\frac{g}{c}\int_0^s A(w)\, dw\right)^2\right]^{\frac{1}{4}}}}. \tag{20.37}$$

(2) The real positive zeros, arranged in ascending order of magnitude, of the equation

$$\chi'(T) \cdot J_{\nu(T)}(\gamma_n) + \chi(T) \cdot \left\{ \frac{\left[1+\left(\frac{g}{c}\int_0^t A(s)\,ds\right)^2\right]^{-\frac{1}{4}} \cdot \gamma_n}{\int_0^T \left[1+\left(\frac{g}{c}\int_0^s A(w)\,dw\right)^2\right]^{-\frac{1}{4}} ds} J'_{\nu(T)}(\gamma_n) + \frac{J_{\nu(T)}(\gamma_n)}{\nu}\nu'(T) \right\} = 0 \quad (20.38)$$

define the constants γ_n.

(3) The normalization condition fulfilled by the eigenfunctions reads

$$1 = N_n^2 \left[\int_0^T \left[1+\left(\frac{g}{c}\int_0^s A(w)\,dw\right)^2\right]^{-\frac{1}{4}} ds\right]^2 \cdot \int_0^1 x J^2_{\nu((x))}(\gamma_n x)\,dx \quad (20.39)$$

and defines the normalization constants N_n. In (20.39) the new transformed order $\nu((x))$ is obtained from the order $\nu(t)$ of (20.36) via the transformation

$$\int_0^t \left[1+\left(\frac{g}{c}\int_0^s A(w)\,dw\right)^2\right]^{-\frac{1}{4}} ds = x \int_0^T \left[1+\left(\frac{g}{c}\int_0^s A(w)\,dw\right)^2\right]^{-\frac{1}{4}} ds. \quad (20.40)$$

(4) The eigenvalues are given by

$$\lambda_n = \left[\int_0^T \left[1+\left(\frac{g}{c}\int_0^s A(w)\,dw\right)^2\right]^{-\frac{1}{4}} ds\right]^2 \frac{1}{(\gamma_n)^2}. \quad (20.41)$$

(5) Finally, the Gaussian random variables Z_n are orthogonal with mean zero and variance λ_n, as in (20.5).

The proof of Theorem 1 will not be repeated here, for it is just the same as the proof we give in Appendix G (see [6, p. 214, theorem 1.1]):[1] one just needs replace $f(t)$ by the arbitrary function $A(t)$ by virtue of their relationship given by (20.32).

20.2.3 Total noise energy

One of the main advantages of the KL expansion over other expansion types is that it allows analytical computation of the stochastic integral

$$\varepsilon = \int_0^T X^2(s)\,ds. \quad (20.42)$$

Physically, this integral represents the total noise energy over the finite time span $0 \le t \le T$ during which the noise is observed on Earth. Mathematically, the expression in (20.42) is a random variable whose cumulants K_n, for $n = 1, 2, \ldots$, are

[1] First published in 1984.

obtained in Chapter 21, Equation (21.56)—see also [1, eqs. (21)–(30)]—and read, when rewritten in terms of the $A(t)$ function

$$K_n = 2^{n-1}(n-1)!\left[\int_0^T \left[1+\left(\frac{g}{c}\int_0^s A(w)\,dw\right)^2\right]^{-\frac{1}{4}} ds\right]^{2n} \sum_{m=1}^{\infty} \frac{1}{(\gamma_m)^{2n}}. \tag{20.43}$$

The mean value of the total noise energy may be directly expressed in terms of the $A(t)$ function by the procedure proven in [1, p. 1021]. The result is eq. (34) of [1], which, translated and generalized by virtue of (20.32), now takes the form

$$E\{\varepsilon\} = \int_0^T dt \int_0^t \frac{1}{\sqrt{1+\left(\frac{g}{c}\int_0^s A(w)\,dw\right)^2}}\,ds. \tag{20.44}$$

In the constant g case, this reduces to Equation (12.12) (or [1, eq. (55)]), namely

$$E\{\varepsilon\} = \frac{c^2}{g^2}\left[\frac{gT}{c}\operatorname{arcsinh}\left(\frac{gT}{c}\right) - \sqrt{1+\left(\frac{gT}{c}\right)^2}+1\right]. \tag{20.45}$$

Total noise energy variance may also be expressed in a similar fashion, as described in Chapter 21, Equation (21.62) (or [1, eqs. (56)–(60)]). When generalized and recast in terms of $A(t)$ by virtue of (20.32), this variance reads

$$\sigma_\varepsilon^2 = 4\int_0^T dt \int_0^t du \left[\int_0^u \frac{1}{\sqrt{1+\left(\frac{g}{c}\int_0^s A(w)\,dw\right)^2}}\,ds\right]^2 \tag{20.46}$$

and reduces to (12.17) (or [1, eq. (60)]) in the constant g case; that is

$$\sigma_\varepsilon^2 = \left(\frac{c}{g}\right)^4\left\{\left[2\left(\frac{g}{c}T\right)^2-1\right]\operatorname{arcsinh}^2\left(\frac{g}{c}T\right) - 6\left(\frac{g}{c}T\right)\sqrt{1+\left(\frac{g}{c}T\right)^2}\operatorname{arcsinh}\left(\frac{g}{c}T\right)+7\left(\frac{g}{c}T\right)^2\right\}. \tag{20.47}$$

20.2.4 KL expansion of noise instantaneous energy

The instantaneous energy of noise is the square process

$$B^2(\tau) = X^2(t) \tag{20.48}$$

and we are now going to determine its KL expansion in terms of the arbitrary acceleration $A(t)$. In a strictly mathematical sense, the problem of finding the KL expansion of (20.48) was solved by the author in [7], with no reference to special relativity or starflight. The application of these results to the constant g case of starflight was found in [2], which will now be generalized to encompass the general $A(t)$ acceleration case.

Consider the process $Y(t)$ defined by

$$Y(t) = X^2(t) - E\{X^2(t)\}. \tag{20.49}$$

It is natural to call this *the zero-mean square process of time-rescaled Brownian motion* $X(t)$, for (20.49) is just the square of $X(t)$ centered around the latter's mean value.

Define now the new function

$$\tilde{f}(t) = 2f(t)\sqrt{\int_0^t f^2(s)\,ds}. \tag{20.50}$$

Recast in terms of $A(t)$ by virtue of (20.32), this becomes

$$\tilde{f}(t) = 2\left[1 + \left(\frac{g}{c}\int_0^t A(s)\,ds\right)^2\right]^{-\frac{1}{4}} \cdot \sqrt{\int_0^t \left[1 + \left(\frac{g}{c}\int_0^s A(w)\,dw\right)^2\right]^{-\frac{1}{2}} ds}. \tag{20.51}$$

Then, Equation (20.35) (or [7, eq. (3.24)][2]) is the proof of the following:

Theorem 2 *The KL expansion of the zero-mean instantaneous energy of the Gaussian noise emitted by a spaceship moving at proper arbitrary acceleration* $A(t)$ *is given by*

$$X^2(t) - E\{X^2(t)\} = Y(t) = \sqrt{\tilde{f}(t)\int_0^t \tilde{f}(s)\,ds} \cdot \sum_{n=1}^{\infty} \tilde{Z}_n \tilde{N}_n J_{\tilde{\nu}(t)}\left(\tilde{\gamma}_n \frac{\int_0^t \tilde{f}(s)\,ds}{\int_0^T \tilde{f}(s)\,ds}\right). \tag{20.52}$$

Here

(1) The Bessel function of the first kind appearing in (20.52) has the time-dependent order

$$\tilde{\nu}(t) = \sqrt{-\frac{\tilde{\chi}^3(t)}{\tilde{f}^2(t)} \cdot \frac{d}{dt}\left[\frac{\tilde{\chi}'(t)}{\tilde{f}^2(t)}\right]} \tag{20.53}$$

where the auxiliary $\tilde{\chi}(t)$ function is given by

$$\tilde{\chi}(t) = \sqrt{\tilde{f}(t)\int_0^t \tilde{f}(s)\,ds}. \tag{20.54}$$

Alternatively, it is possible to express the order by virtue of the single formula

$$\tilde{\nu}(t) = \sqrt{\frac{1}{4} + \left[\frac{\int_0^t \tilde{f}(s)\,ds}{\tilde{f}(t)}\right]^2 \cdot \left\{\frac{3}{4}\left[\frac{d\ln\tilde{f}(t)}{dt}\right]^2 - \frac{1}{2}\frac{d^2\ln\tilde{f}(t)}{dt^2}\right\}}. \tag{20.55}$$

proved by the author in the appendix to [8] and in (22.50).

[2] First published in 1988.

(2) The real positive zeros, arranged in ascending order of magnitude, of the equation

$$\tilde{\chi}'(T)\cdot J_{\tilde{\nu}(T)}(\tilde{\gamma}_n)+\tilde{\chi}(T)\cdot\left[\frac{\tilde{f}(T)\cdot\tilde{\gamma}_n}{\int_0^T \tilde{f}(s)\,ds}J'_{\tilde{\nu}(T)}(\tilde{\gamma}_n)+\frac{\partial J_{\tilde{\nu}(T)}(\tilde{\gamma}_n)}{\partial\tilde{\nu}}\tilde{\nu}'(T)\right]=0 \quad (20.56)$$

define the constants $\tilde{\gamma}_n$.

(3) The normalization condition fulfilled by the eigenfunctions reads

$$1=\tilde{N}_n^2\left[\int_0^T \tilde{f}(s)\,ds\right]^2\int_0^1 xJ^2_{\tilde{\nu}((x))}(\tilde{\gamma}_n x)\,dx \quad (20.57)$$

and defines the normalization constants $\tilde{N}_n$. In (20.57) the new transformed order $\tilde{\nu}((x))$ is obtained from the order $\tilde{\nu}(t)$ of either (20.53) or (20.55) via the transformation

$$\int_0^t \tilde{f}(s)\,ds = x\int_0^T \tilde{f}(s)\,ds. \quad (20.58)$$

(4) The eigenvalues are given by

$$\tilde{\lambda}_n=\left[\int_0^T \tilde{f}(s)\,ds\right]^2\frac{1}{(\tilde{\gamma}_n)^2}. \quad (20.59)$$

(5) Finally, the gamma-type probability density

$$f_{\tilde{Z}_n}(z)=\frac{1}{\sqrt{\pi}(2\tilde{\lambda}_n)^{\frac{1}{4}}}e^{-\frac{z+\sqrt{\frac{\tilde{\lambda}_n}{2}}}{\sqrt{2\tilde{\lambda}_n}}}\cdot\left[z+\sqrt{\frac{\tilde{\lambda}_n}{2}}\right]^{-\frac{1}{2}}U\left(z+\sqrt{\frac{\tilde{\lambda}_n}{2}}\right), \quad (20.60)$$

where $U(x)$ is the unit step function, is the probabilistic law according to which the random variables $\tilde{Z}_n$ are distributed.

20.3 ASYMPTOTIC ARBITRARY SPACESHIP ACCELERATION

20.3.1 Asymptotic motion with arbitrary acceleration

The apparently complicated mathematical nature of most results derived so far, in particular the two KL expansions of Theorems 1 and 2, may seem to give little hope for their closed-form application to interesting cases of starflight. However, this is not the case. In fact, if we confine ourselves to the *asymptotic* for $t \to \infty$ motion at an arbitrary acceleration, all the forgoing results get easier. They will be further simplified in the power-like acceleration case that we are going to present in Section 20.4. So, let us now examine under what circumstances the asymptotic approximation is physically acceptable.

The primary idea behind performing the asymptotic approximation is to replace the exact radical

$$\sqrt{1+\left(\frac{g}{c}\int_0^t A(s)\,ds\right)^2} \tag{20.61}$$

by what is left of it when the +1 term is dwarfed by the other term; that is

$$\sqrt{1+\left(\frac{g}{c}\int_0^t A(s)\,ds\right)^2} \approx \sqrt{\left(\frac{g}{c}\int_0^t A(s)\,ds\right)^2} = \frac{g}{c}\left|\int_0^t A(s)\,ds\right|. \tag{20.62}$$

In other words, we suppose that

$$\left(\frac{g}{c}\int_0^t A(s)\,ds\right)^2 \gg 1 \tag{20.63}$$

whence we find the *asymptotic condition*

$$\left|\int_0^t A(s)\,ds\right| > \frac{c}{g} \approx 3.061 \cdot 10^7 \text{ seconds} \approx 0.9699 \text{ years} \approx 1 \text{ year}. \tag{20.64}$$

In the constant g case, (20.64) reduces to

$$t > \frac{c}{g} \approx 1 \text{ year} \tag{20.65}$$

which is (12.47) (or [1, eq. (90)]) and tells us that the asymptotic approximation holds only after about one year of travel. Since the left-hand side of (20.64) should not numerically differ too much from that of (20.65) for whatever "reasonable" $A(t)$ function we may adopt, and since any trip to even the nearest stars will certainly take longer than one year, we regard the asymptoticity condition (20.64) as physically acceptable.

In the remainder of this chapter we shall denote all asymptotic formulas derived from the corresponding exact formulas by the "*Asy*" suffix.

The first formula that we must asymptotically simplify with the aid of (20.62) is (20.28), for this is the argument for the time-rescaled Brownian motion $B(\tau)$ and leads to the asymptotic time-rescaling function "*Asy*". Thus, (20.28), (20.32), and (20.62) yield

$$B(\tau_{Asy}) \equiv B\left(\int_0^t \frac{1}{\frac{g}{c}\left|\int_0^s A(w)\,dw\right|}\,ds\right) = B\left(\int_0^t f^2_{Asy}(s)\,ds\right) \tag{20.66}$$

whence we infer

$$f^2_{Asy}(t) = \frac{1}{\frac{g}{c}\left|\int_0^t A(s)\,ds\right|} \tag{20.67}$$

and, finally, taking the square root, the asymptotic time-rescaling function

$$f_{Asy}(t) = \frac{1}{\sqrt{\frac{g}{c}}\sqrt{\left|\int_0^t A(s)\,ds\right|}} \tag{20.68}$$

is obtained.

In the constant g case, this reduces to (12.21) (or [1, eq. (64)]), namely

$$f_{Asy}(t) = \sqrt{\frac{c}{g}}\frac{1}{\sqrt{t}}. \tag{20.69}$$

The spaceship asymptotic velocity is obtained from (20.68) and (20.31), and reads

$$v_{Asy}(t) = c\sqrt{1 - f_{Asy}^4(t)} = c\sqrt{1 - \frac{1}{\left(\frac{g}{c}\int_0^t A(s)\,ds\right)^2}}. \tag{20.70}$$

In the constant g case, this reduces to (12.45) (or [1, eq. (88)]); that is

$$v_{Asy}(t) = c\sqrt{1 - \frac{c^2}{g^2}\cdot\frac{1}{t^2}}. \tag{20.71}$$

It should be noted that the reality of the radicand in (20.70) again yields the same asymptoticity condition (20.64) as did (20.63).

20.3.2 Asymptotic KL expansion for noise

The asymptotic version of Theorem 1 is obtained on rewriting formulas (20.35)–(20.41) by the aid of the asymptotic radical (20.62). One then gets

Theorem 3 *The asymptotic KL expansion for the Gaussian noise emitted by a spaceship moving at a proper acceleration $gA(t)$ is given by*

$$X_{Asy}(t) \equiv B_{Asy}(\tau) = \sqrt{\frac{c}{g}}\left|\left|\int_0^t A(s)\,ds\right|\right|^{-\frac{1}{4}} \cdot \sqrt{\int_0^t \frac{ds}{\sqrt{\left|\int_0^s A(w)\,dw\right|}}} \cdot \sum_{n=1}^{\infty} Z_{n_{Asy}} N_{n_{Asy}} J_{\nu(t)_{Asy}}\left(\gamma_{n_{Asy}} \frac{\int_0^t \frac{ds}{\sqrt{\left|\int_0^s A(w)\,dw\right|}}}{\int_0^T \frac{ds}{\sqrt{\left|\int_0^s A(w)\,dw\right|}}}\right). \tag{20.72}$$

Here

(1) The Bessel function of the first kind appearing in (20.72) has the time-dependent order

$$\nu_{Asy}(t) = \frac{g}{c}\sqrt{-\chi^3_{Asy}(t)\left|\int_0^t A(s)\,ds\right|}\cdot\sqrt{\frac{d}{dt}\left[\chi'_{Asy}(t)\left|\int_0^t A(s)\,ds\right|\right]}. \quad (20.73)$$

where the auxiliary function $\chi_{Asy}(t)$ has been defined

$$\chi_{Asy}(t) = \sqrt{\frac{c}{g}}\left[\left|\int_0^t A(s)\,ds\right|\right]^{-\frac{1}{4}}\cdot\sqrt{\int_0^t \frac{ds}{\sqrt{\left|\int_0^s A(w)\,dw\right|}}}. \quad (20.74)$$

(2) The real positive zeros, arranged in ascending order of magnitude, of the equation

$$\chi'_{Asy}(T)\cdot J_{\nu_{Asy}(T)}(\gamma_{n_{Asy}}) + \chi_{Asy}(T)\cdot$$
$$\cdot\left\{\frac{\left[\left|\int_0^t A(s)\,ds\right|\right]^{-\frac{1}{2}}\gamma_{n_{Asy}}}{\int_0^T\left[\left|\int_0^s A(w)\,dw\right|\right]^{-\frac{1}{2}}ds}J'_{\nu_{Asy}(T)}(\gamma_{n_{Asy}}) + \frac{\partial J_{\nu_{Asy}(T)}(\gamma_{n_{Asy}})}{\partial\nu_{Asy}}\nu'_{Asy}(T)\right\} = 0. \quad (20.75)$$

define the constants $\gamma_{n_{Asy}}$.

(3) The normalization condition fulfilled by the eigenfunctions reads

$$1 = N^2_{n_{Asy}}\frac{c}{g}\left[\int_0^T\left[\left|\int_0^s A(w)\,dw\right|\right]^{-\frac{1}{2}}ds\right]^2\cdot\int_0^1 xJ^2_{\nu_{Asy}((x))}(\gamma_{n_{Asy}}x)\,dx \quad (20.76)$$

and defines the normalization constants $N_{n_{Asy}}$. In (20.76) the new transformed order $\nu_{Asy}((x))$ is obtained from the order $\nu_{Asy}(t)$ of (20.73) via the transformation

$$\int_0^t\left[\left|\int_0^s A(w)\,dw\right|\right]^{-\frac{1}{2}}ds = x\int_0^T\left[\left|\int_0^s A(w)\,dw\right|\right]^{-\frac{1}{2}}ds. \quad (20.77)$$

(4) The eigenvalues are given by

$$\lambda_{n_{Asy}} = \frac{c}{g}\left[\int_0^T\left[\left|\int_0^s A(w)\,dw\right|\right]^{-\frac{1}{2}}ds\right]^2\frac{1}{(\gamma_{n_{Asy}})^2}. \quad (20.78)$$

(5) Finally, the Gaussian random variables $Z_{n_{Asy}}$ are orthogonal with mean zero and variance $\lambda_{n_{Asy}}$.

20.3.3 Asymptotic total noise energy

In the asymptotic version, characterized by (20.62), the asymptotic total noise energy cumulants (20.43) take the form

$$K_{n_{Asy}} = 2^{n-1}(n-1)!\left(\frac{c}{g}\right)^{n} \cdot \left[\int_0^T \left[\left|\int_0^s A(w)\,dw\right|\right]^{-\frac{1}{2}} ds\right]^{2n} \sum_{m=1}^{\infty} \frac{1}{(\gamma_{m_{Asy}})^{2n}}. \tag{20.79}$$

The asymptotic total noise energy derived from (20.44) and (20.62) reads

$$E\{\varepsilon_{Asy}\} = \frac{c}{g}\int_0^T dt \int_0^t \frac{1}{\left|\int_0^s A(w)\,dw\right|} ds. \tag{20.80}$$

Finally, asymptotic total noise energy variance follows from (20.46) and (20.62), yielding

$$\sigma^2_{\varepsilon_{Asy}} = 4\left(\frac{c}{g}\right)^2 \int_0^T dt \int_0^t du \left[\int_0^u \frac{1}{\left|\int_0^s A(w)\,dw\right|} ds\right]^2. \tag{20.81}$$

20.3.4 Asymptotic KL expansion for noise instantaneous energy

Using (20.62) once again, the content of Theorem 2 is turned into the asymptotic KL expansion for noise instantaneous energy. Let us start by defining the new function, corresponding to (20.50) in the asymptotic limit,

$$\tilde{f}_{Asy}(t) = 2f_{Asy}(t)\sqrt{\int_0^t f^2_{Asy}(s)\,ds}. \tag{20.82}$$

Expressed in terms of $A(t)$ by virtue of (20.68), this becomes

$$\tilde{f}_{Asy}(t) = 2\frac{c}{g}\left[\left|\int_0^t A(s)\,ds\right|\right]^{-\frac{1}{2}} \cdot \sqrt{\int_0^t \frac{1}{\left|\int_0^s A(w)\,dw\right|} ds}. \tag{20.83}$$

Theorem 2 then immediately yields the following:

Theorem 4 *The asymptotic KL expansion of the zero-mean instantaneous energy of the Gaussian noise emitted by a spaceship moving at a proper acceleration $gA(t)$ is given by*

$$X^2_{Asy}(t) - E\{X^2_{Asy}(t)\} = Y_{Asy}(t)$$

$$= \sqrt{\tilde{f}_{Asy}(t)\int_0^t \tilde{f}_{Asy}(s)\,ds} \cdot \sum_{n=1}^{\infty} \tilde{Z}_{n_{Asy}} \tilde{N}_{n_{Asy}} J_{\tilde{\nu}_{Asy}(t)}\left(\tilde{\gamma}_{n_{Asy}} \frac{\int_0^t \tilde{f}_{Asy}(s)\,ds}{\int_0^T \tilde{f}_{Asy}(s)\,ds}\right). \tag{20.84}$$

Here

(1) The Bessel function of the first kind appearing in (20.84) has the time-dependent order

$$\tilde{\nu}_{Asy}(t) = \sqrt{-\frac{\tilde{\chi}^3_{Asy}(t)}{\tilde{f}^2_{Asy}(t)} \cdot \frac{d}{dt}\left[\frac{\tilde{\chi}'_{Asy}(t)}{\tilde{f}^2_{Asy}(t)}\right]} \tag{20.85}$$

where the auxiliary function is defined by

$$\tilde{\chi}_{Asy}(t) = \sqrt{\tilde{f}_{Asy}(t) \int_0^t \tilde{f}_{Asy}(s)\, ds}. \tag{20.86}$$

Alternatively, it is possible to express the order by virtue of the single formula

$$\tilde{\nu}_{Asy}(t) = \sqrt{\frac{1}{4} + \left[\frac{\int_0^t \tilde{f}_{Asy}(s)\, ds}{\tilde{f}_{Asy}(t)}\right]^2 \cdot \left\{\frac{3}{4}\left[\frac{d \ln \tilde{f}_{Asy}(t)}{dt}\right]^2 - \frac{1}{2}\frac{d^2 \ln \tilde{f}_{Asy}(t)}{dt^2}\right\}} \tag{20.87}$$

proved in Chapter 22, Equation (22.50).

(2) The real positive zeros, arranged in ascending order of magnitude, of the equation

$$\tilde{\chi}'_{Asy}(T) \cdot J_{\tilde{\nu}_{Asy}(T)}(\tilde{\gamma}_{n_{Asy}}) + \tilde{\chi}_{Asy}(T) \cdot \left[\frac{\tilde{f}_{Asy}(T) \cdot \tilde{\gamma}_{n_{Asy}}}{\int_0^T \tilde{f}_{Asy}(s)\, ds} J'_{\tilde{\nu}_{Asy}(T)}(\tilde{\gamma}_{n_{Asy}}) + \frac{J_{\tilde{\nu}_{Asy}(T)}(\tilde{\gamma}_{n_{Asy}})}{\tilde{\nu}_{Asy}} \tilde{\nu}'_{Asy}(T)\right] = 0 \tag{20.88}$$

define the constants $\tilde{\gamma}_{n_{Asy}}$.

(3) The normalization condition fulfilled by the eigenfunctions reads

$$1 = \tilde{N}^2_{n_{Asy}} \left[\int_0^T \tilde{f}_{Asy}(s)\, ds\right]^2 \cdot \int_0^1 x J^2_{\tilde{\nu}_{Asy}((x))}(\tilde{\gamma}_{n_{Asy}} x)\, dx \tag{20.89}$$

and defines the normalization constants $\tilde{N}_{n_{Asy}}$. In (20.89) the new transformed order $\tilde{\nu}_{Asy}((x))$ is obtained from the order $\tilde{\nu}_{Asy}(x)$ of either (20.85) or (20.87) via the transformation

$$\int_0^1 \tilde{f}_{Asy}(s)\, ds = x \int_0^T \tilde{f}_{Asy}(s)\, ds. \tag{20.90}$$

(4) The eigenvalues are given by

$$\tilde{\lambda}_{n_{Asy}} = \left[\int_0^T \tilde{f}_{Asy}(s)\, ds\right]^2 \frac{1}{(\tilde{\gamma}_{n_{Asy}})^2}. \tag{20.91}$$

(5) Finally, the gamma-type probability density

$$f_{\tilde{Z}_{n_{Asy}}}(z) = \frac{1}{\sqrt{\pi}(2\tilde{\lambda}_{n_{Asy}})^{\frac{1}{4}}} e^{-\frac{z+\sqrt{\frac{\tilde{\lambda}_{n_{Asy}}}{2}}}{\sqrt{2\tilde{\lambda}_{n_{Asy}}}}} \cdot \left[z + \sqrt{\frac{\tilde{\lambda}_{n_{Asy}}}{2}}\right]^{-\frac{1}{2}} U\left(z + \sqrt{\frac{\tilde{\lambda}_{n_{Asy}}}{2}}\right), \tag{20.92}$$

where $U(x)$ is the unit step function, is the probability law obeyed by the random variables $\tilde{Z}_{n_{Asy}}$.

20.4 POWER-LIKE ASYMPTOTIC SPACESHIP ACCELERATION

20.4.1 Asymptotic motion with power-like acceleration

There exists a particular case of the spaceship acceleration $A(t)$ for which all the asymptotic formulas previously worked out, including the KL expansions of Theorems 3 and 4, can be analytically developed in a closed form. This is the case where the acceleration changes in time like some real power α of time; that is

$$A(t) = t^{\alpha}. \tag{20.93}$$

The exponent α is subjected to the limitations

$$-1 < \alpha < 1 \tag{20.94}$$

that we shall determine in the present section.

In what follows, we shall denote all the formulas derived from the power-like hypothesis (20.93) by a "$|P$" suffix. Moreover, since we are going to apply the power-like hypothesis (20.93) to asymptotic formulas only, in practice the new suffix will be "$Asy|P$".

Note also that the asymptotic constant g case, already studied in detail in [1, 2], corresponds to the particular case

$$\alpha = 0 \tag{20.95}$$

of the asymptotic power-like acceleration theory developed in the present and following sections. All the results of [1, 2] are thus considerably generalized in the present chapter.

Turning next to the general formulas for the asymptotic power-like acceleration case, consider first the asymptotic time-rescaling function $f_{Asy|P}(t)$ defined by (20.68). By virtue of (20.93) one gets

$$f_{Asy|P}(t) = \frac{1}{\sqrt{\frac{g}{c}}\sqrt{\left|\int_0^t s^{\alpha}\, ds\right|}} = \frac{1}{\sqrt{\frac{g}{c}}\frac{\sqrt{t^{1+\alpha}}}{\sqrt{1+\alpha}}}. \tag{20.96}$$

That is,

$$f_{Asy|P}(t) = \frac{\sqrt{c}\sqrt{1+\alpha}}{\sqrt{g}} t^{-\frac{1+\alpha}{2}}. \tag{20.97}$$

Clearly, the integration in (20.96) converges, and $\sqrt{1+\alpha}$ is not complex if, and only if

$$1 + \alpha > 0. \tag{20.98}$$

On the other hand, the presence in Theorem 4 of many integrals of the type

$$\int_0^t f_{Asy|P}(s)\, ds = \frac{\sqrt{c}\sqrt{1+\alpha}}{\sqrt{g}} \int_0^t s^{-\frac{1+\alpha}{2}} ds = \frac{2\sqrt{c}\sqrt{1+\alpha}}{\sqrt{g}(1-\alpha)} t^{\frac{1-\alpha}{2}} \tag{20.99}$$

clearly requires

$$1 - \alpha > 0. \tag{20.100}$$

Thus, the limitations (20.98) and (20.100) yield (20.94).

The asymptotic spaceship velocity follows from (20.70) and (20.97), and reads

$$v_{Asy|P}(t) = c\sqrt{1 - \frac{c^2(1+\alpha)^2}{g^2} \cdot \frac{1}{t^{2(1+\alpha)}}}. \tag{20.101}$$

Thus, to avoid complex variables, the inequality

$$\frac{c^2(1+\alpha)^2}{g^2} \cdot \frac{1}{t^{2(1+\alpha)}} \leq 1 \tag{20.102}$$

must apply, whence we infer the *power-like asymptotic condition*, generalizing (20.65)

$$t \geq \left[\frac{c(1+\alpha)}{g}\right]^{\frac{1}{1+\alpha}} \approx [1 \text{ year} \cdot (1+\alpha)]^{\frac{1}{1+\alpha}}. \tag{20.103}$$

In other words, in order that the asymptotic power-like acceleration case be physically acceptable, the time elapsed since departure must be larger than the $1/(1+\alpha)$ power of $(1+\alpha)$ years.

20.4.2 Power-like asymptotic KL expansion for noise

The power-like acceleration (20.93) has the remarkable advantage over other acceleration types that its KL expansions stated in Theorems 3 and 4 can be fully obtained in the closed form. As a matter of fact, most of the purely mathematical features of the content of the present section appeared in [9], with no reference to either physics or engineering applications. In [8, pp. 333–339], the material of the present section first appeared cast in the language of special relativity, but it was oriented to particle physics rather than to starflight: the terminology "decelerated motion" appearing there refers to deceleration with respect to the Earth, not with respect to a moving point. We believe that recasting and completing the mentioned material of [8, 9] into the framework of the *proper* acceleration $A(t)$, as required by starflight, will be of help to scientists and engineers wishing to refer to a single book chapter rather than to a widely scattered literature.

Theorem 5 *The asymptotic KL expansion for the Gaussian noise emitted by a spaceship moving at a proper power-like acceleration is given by*

$$X_{Asy|P}(t) \equiv B_{Asy|P}(\tau)$$
$$= \frac{\sqrt{1-\alpha}\, t^{-\frac{\alpha}{2}}}{T^{\frac{1-\alpha}{2}}} \cdot \sum_{n=1}^{\infty} Z_{n_{Asy|P}} \frac{1}{|J_{\nu_{Asy|P}}(\gamma_{n_{Asy|P}})|} J_{\nu_{Asy|P}}\left(\gamma_{n_{Asy|P}} \frac{t^{\frac{1-\alpha}{2}}}{T^{\frac{1-\alpha}{2}}}\right). \quad (20.104)$$

Here

(1) The Bessel function of the first kind appearing in (20.104) has the constant order

$$\nu_{Asy|P} = \frac{|\alpha|}{1-\alpha}. \quad (20.105)$$

The auxiliary function defined by (20.74) now reads

$$\chi_{Asy|P}(t) = \sqrt{\frac{c}{g}} \sqrt{2} \frac{\sqrt{1+\alpha}}{\sqrt{1-\alpha}} t^{-\frac{\alpha}{2}}. \quad (20.106)$$

(2) The constants $\gamma_{n_{Asy|P}}$ are the real positive zeros, arranged in ascending order of magnitude, of the Bessel function of the first kind of order $\nu + 1$. That is,

$$J_{\nu_{Asy|P}+1}(\gamma_{n_{Asy|P}}) = 0. \quad (20.107)$$

(3) The normalization constants for the eigenfunctions are given by

$$N_{n_{Asy|P}} = \sqrt{\frac{g}{c}} \frac{1-\alpha}{\sqrt{2}\sqrt{1+\alpha}\, T^{\frac{1-\alpha}{2}} |J_{\nu_{Asy|P}}(\gamma_{n_{Asy|P}})|}. \quad (20.108)$$

(4) The eigenvalues are given by

$$\lambda_{n_{Asy|P}} = \frac{c}{g} \frac{4T^{1-\alpha}(1+\alpha)}{(1-\alpha)^2} \cdot \frac{1}{\gamma^2_{n_{Asy|P}}}. \quad (20.109)$$

(5) Finally, the Gaussian random variables $Z_{n_{Asy|P}}$ are orthogonal with mean zero and variance $\lambda_{n_{Asy|P}}$, as in (20.5).

We shall just highlight the proof of the foregoing theorem, omitting all lengthy calculations. The $\chi_{Asy|P}(t)$ function of (20.106) immediately follows from (20.74) and (20.93). Then, a vital simplification occurs: the time disappears identically from (20.73), yielding the constant order (20.105). This, in turn, simplifies the writing of (20.75), which, after several reductions, takes the final form

$$-\frac{|\alpha|}{1-\alpha} J_{\nu_{Asy|P}}(\gamma_{n_{Asy|P}}) + \gamma_{n_{Asy|P}} J'_{\nu_{Asy|P}}(\gamma_{n_{Asy|P}}) = 0. \quad (20.110)$$

Let us now recall the Bessel function property (for a proof, see [10, p. 12, entry (55)])

$$zJ'_\nu(z) - \nu J_\nu(z) = -zJ_{\nu+1}(z). \quad (20.111)$$

By virtue of (20.105), the left-hand sides of (20.110) and (20.111) coincide, yielding

$$-\gamma_{n_{Asy|P}} J_{\nu_{Asy|P}+1}(\gamma_{n_{Asy|P}}) = 0 \tag{20.112}$$

whence, as $\gamma_{n_{Asy|P}} \neq 0$, (20.107) is obtained.

Let us next turn to the normalization constants (20.108). In this regard, we note that a formula arising from the Dini expansion in a series of Bessel functions (see [10, p. 71, entry (49)]), that is

$$\int_0^1 xJ_\nu^2(\gamma_n x)\,dx = \frac{1}{2\gamma_n^2}\{\gamma_n^2 J_\nu'^2(\gamma_n) + (\gamma_n^2 - \nu^2)J_\nu^2(\gamma_n)\} \tag{20.113}$$

may be simplified, by virtue of (20.107) and (20.112), to yield

$$\int_0^1 xJ_{\nu_{Asy|P}}^2(\gamma_{n_{Asy|P}} x)\,dx = \frac{J_{\nu_{Asy|P}}^2(\gamma_{n_{Asy|P}})}{2}. \tag{20.114}$$

This and (20.76), where the order transformation (20.77) is not required because the order (20.105) is a constant, lead then to (20.108).

Finally, the eigenvalues (20.109) are obtained at once from (20.78) and (20.93). In the constant g case, use of (20.95) shows that formulas (20.104)–(20.109) reduce to the corresponding formulas appearing in section 4 of [1].

20.4.3 Approximated power-like asymptotic KL expansion for noise

Theorem 5 is perhaps still too complicated for engineering purposes inasmuch as the Bessel functions may not by easy to evaluate numerically. It seems thus useful to derive the *approximated* version of Theorem 5 that follows on replacing the Bessel function of the first kind by means of its asymptotic expansion for large values of its argument

$$\lim_{x\to\infty} J_\nu(x) = \lim_{x\to\infty}\sqrt{\frac{2}{\pi x}}\cos\left(x - \frac{\nu\pi}{2} - \frac{\pi}{4}\right). \tag{20.115}$$

In the following we shall denote all the formulas derived for this approximated case by a further "$|A$" suffix. Thus, as we are considering the asymptotic power-like approximated case, in practice the whole suffix will be "$Asy|P|A$". When applied to (20.107), (20.115) yields

$$\cos\left(\gamma_{n_{Asy|P|A}} - \frac{(\nu_{Asy|P}+1)\pi}{2} - \frac{\pi}{4}\right) \approx 0 \tag{20.116}$$

whence

$$\gamma_{n_{Asy|P|A}} - \frac{(\nu_{Asy|P}+1)\pi}{2} - \frac{\pi}{4} \approx n\pi - \frac{\pi}{2} \quad (n = 1, 2, \ldots). \tag{20.117}$$

Thus, invoking (20.105), we get the following approximated expression for the constants $\gamma_{n_{Asy|P|A}}$

$$\gamma_{n_{Asy|P|A}} \approx n\pi + \frac{\pi}{4} + \frac{\nu_{Asy|P}\pi}{2} = \pi\left(n + \frac{1}{4} + \frac{|\alpha|}{2(1-\alpha)}\right). \tag{20.118}$$

We now set out to determine the approximated normalization constants $N_{n_{Asy|P|A}}$. First, the Bessel function appearing in (20.108) has to be replaced by its asymptotic version (20.115), yielding

$$|J_{\nu_{Asy|P}}(\gamma_{n_{Asy|P|A}})| \approx \left|\sqrt{\frac{2}{\pi\gamma_{n_{Asy|P|A}}}}\right| \cdot \left|\cos\left(\gamma_{n_{Asy|P|A}} - \frac{\nu_{Asy|P}\pi}{2} - \frac{\pi}{4}\right)\right|$$
$$= \sqrt{\frac{2}{\pi\gamma_{n_{Asy|P|A}}}}|\cos(n\pi)| = \sqrt{\frac{2}{\pi\gamma_{n_{Asy|P|A}}}}. \tag{20.119}$$

Then, (20.108) and (20.119) yield the approximated normalization constants $N_{n_{Asy|P|A}}$

$$N_{n_{Asy|P|A}} \approx \frac{\pi}{2}\sqrt{\frac{g}{cT^{1-\alpha}}\frac{1-\alpha}{\sqrt{1+\alpha}}} \cdot \sqrt{n + \frac{1}{4} + \frac{|\alpha|}{2(1-\alpha)}}. \tag{20.120}$$

As for the eigenvalues, one merely has to insert (20.118) into (20.109) to get

$$\lambda_{n_{Asy|P|A}} \approx \frac{4cT^{1-\alpha}(1+\alpha)}{g(1-\alpha)^2} \cdot \frac{1}{\pi^2\left(n + \frac{1}{4} + \frac{|\alpha|}{2(1-\alpha)}\right)^2}. \tag{20.121}$$

These also are the variances of Gaussian zero-mean orthogonal Gaussian random variables $Z_{n_{Asy|P|A}}$.

In conclusion, we have proven the following:

Theorem 6 *The approximated asymptotic KL expansion for the Gaussian noise emitted by a spaceship moving at a proper power-like acceleration is given by*

$$X_{Asy|P|A}(t) \equiv B_{Asy|P|A}(\tau)$$
$$\approx \frac{\sqrt{1-\alpha}}{T^{\frac{1-\alpha}{4}}t^{\frac{1+\alpha}{4}}} \cdot \sum_{n=1}^{\infty} Z_{n_{Asy|P|A}} \cos\left(\gamma_{n_{Asy|P|A}} \frac{t^{\frac{1-\alpha}{2}}}{T^{\frac{1-\alpha}{2}}} - \frac{\pi}{4} - \frac{|\alpha|}{1-\alpha} \cdot \frac{\pi}{2}\right). \tag{20.122}$$

In the constant g case, this theorem reduces to (12.43) (or [1, eq. (86)]).

20.4.4 Power-like asymptotic total noise energy

The results of Section 20.3.3 on asymptotic total noise energy will now be applied to the power-like acceleration (20.93). However, the required convergence of all integrals that we have to use will force us to restrict the range of validity of the exponent

$$-1 < \alpha < 0. \tag{20.123}$$

Note that the upper inequality implies that the constant g case, characterized by (20.95), cannot be derived as a particular case of the theory developed within the present section. For instance, consider the integral

$$\int_0^t f^2_{Asy|P}(s)\, ds = \frac{c(1+\alpha)}{g}\left[\frac{s^{-\alpha}}{-\alpha}\right]_0^t = \frac{c(1+\alpha)}{g}\frac{t^{-\alpha}}{-\alpha}. \tag{20.124}$$

Clearly, the last step is possible only if α is negative, whence the upper limitation in (20.123).

The integral in (20.124) is required for computation of the mean total noise energy (20.80). In fact, invoking (20.93), one obtains

$$E\{\varepsilon_{Asy|P}\} = \int_0^T dt \int_0^t f^2_{Asy|P}(s)\, ds = \frac{c}{g}\frac{1+\alpha}{(-\alpha)(1-\alpha)} T^{1-\alpha}. \tag{20.125}$$

The power-like asymptotic total noise energy variance, derived from (20.81) by virtue of (20.93) and (20.124), reads

$$\sigma^2_{\varepsilon_{Asy|P}} = 4\int_0^T dt \int_0^t du \left[\int_0^u f^2_{Asy|P}(s)\, ds\right]^2 = 2\frac{c^2}{g^2}\frac{(1+\alpha)^2}{\alpha^2(\alpha-1)^2(\alpha-2)(2\alpha-3)} T^{2(2-\alpha)}. \tag{20.126}$$

whence, by taking the square root, we get the power-like asymptotic total noise energy standard deviation

$$\sigma_{\varepsilon_{Asy|P}} = \pm\sqrt{2}\frac{c}{g}\cdot\frac{1+\alpha}{\alpha(\alpha-1)\sqrt{(\alpha-2)(2\alpha-3)}} T^{2-\alpha}. \tag{20.127}$$

Finally, the power-like asymptotic total noise energy cumulants (20.79) take the form

$$K_{n_{Asy|P}} = 2^{3n-1}(n-1)!\left(\frac{c}{g}\right)^n \frac{(1+\alpha)^n}{(1-\alpha)^{2n}} T^{n(1-\alpha)} \cdot \sum_{m=1}^{\infty} \frac{1}{(\gamma_{m_{Asy|P}})^{2n}}. \tag{20.128}$$

This formula holds for the full α range (20.94), rather than for the limited range (20.123), for (20.124) was not required for its derivation.

20.4.5 Power-like asymptotic KL expansion for noise instantaneous energy

Just as Theorem 5 of Section 20.4.2 was the power-like subcase of Theorem 3 of Section 20.3.2, so it is natural to seek a Theorem 7 to be the power-like subcase of Theorem 4 of Section 20.3.4. In the present section we shall prove that this guess is true. However, as in the previous section, we shall have to restrict the range of validity of the exponent α to

$$-1 < \alpha < 0. \tag{20.129}$$

Once again the upper inequality implies that the constant g case, characterized by (20.95), cannot be derived as a particular case of the theory now to be established.

To prove the upper inequality in (20.129), rewrite the definition (20.82) in terms of the new $\tilde{f}_{Asy|P}(t)$ function

$$\tilde{f}_{Asy|P}(t) = 2f_{Asy|P}(t)\sqrt{\int_0^t f^2_{Asy|P}(s)\,ds}. \tag{20.130}$$

Invoking (20.97), it is observed that the convergence of the integral just requires the upper inequality in (20.129):

$$\int_0^t f^2_{Asy|P}(s)\,ds = \frac{c(1+\alpha)}{g}\left[\frac{s^{-\alpha}}{-\alpha}\right]_0^t = \frac{c(1+\alpha)}{g}\frac{t^{-\alpha}}{-\alpha}. \tag{20.131}$$

Then, inserting (20.97) and (20.131) into (20.130), we get the explicit expression

$$\tilde{f}_{Asy|P}(t) = \frac{c}{g}\frac{2(1+\alpha)}{\sqrt{-\alpha}}t^{-\frac{2\alpha+1}{2}}. \tag{20.132}$$

Finally, the integral of (20.132) reads

$$\int_0^t \tilde{f}_{Asy|P}(s)\,ds = \frac{c}{g}\cdot\frac{4(1+\alpha)}{\sqrt{-\alpha}(1-2\alpha)}t^{\frac{1-2\alpha}{2}}. \tag{20.133}$$

We are now ready to state the following theorem:

Theorem 7 *The power-like asymptotic KL expansion of the zero-mean instantaneous energy of the Gaussian noise emitted by a spaceship moving at a proper acceleration $gA(t)$ is given by*

$$\begin{aligned} Y_{Asy|P}(t) &\equiv X^2_{Asy|P}(\tau) - E\{X^2_{Asy|P}(\tau)\} \\ &= \sqrt{1-2\alpha}\frac{t^{-\alpha}}{T^{\frac{1}{2}-\alpha}}\cdot\sum_{n=1}^{\infty}\tilde{Z}_{n_{Asy|P}}\tilde{N}_{n_{Asy|P}}J_{\tilde{\nu}_{Asy|P}}\left(\tilde{\gamma}_{n_{Asy|P}}\frac{t^{\frac{1}{2}-\alpha}}{T^{\frac{1}{2}-\alpha}}\right). \end{aligned} \tag{20.134}$$

Here

(1) The Bessel function of the first kind appearing in (20.134) has the constant order

$$\tilde{\nu}_{Asy|P} = \frac{|2\alpha|}{|2\alpha-1|}. \tag{20.135}$$

The auxiliary function $\tilde{\chi}_{Asy|P}(t)$ defined by (20.86), by the aid of (20.132) and (20.133), takes the form

$$\tilde{\chi}_{Asy|P}(t) = \frac{c}{g}\cdot\frac{2\sqrt{2}(1+\alpha)}{\sqrt{-\alpha}\sqrt{1-2\alpha}}t^{-\alpha}. \tag{20.136}$$

(2) The constants $\tilde{\gamma}_{n_{Asy|P}}$ are the real positive zeros, arranged in ascending order of magnitude, of the Bessel function of the first kind of order $(\tilde{\nu}_{n_{Asy|P}} - 1)$; that is

$$J_{\tilde{\nu}_{n_{Asy|P}}-1}(\tilde{\gamma}_{n_{Asy|P}}) = 0. \tag{20.137}$$

(3) The normalization constants for the eigenfunctions are given by

$$\tilde{N}_{n_{Asy|P}} = \frac{g}{c}\frac{\sqrt{2}\sqrt{-\alpha}(1-2\alpha)T^{\alpha-\frac{1}{2}}}{4(1+\alpha)|J_{\tilde{\nu}_{n_{Asy|P}}}(\tilde{\gamma}_{n_{Asy|P}})|}. \tag{20.138}$$

(4) The eigenvalues are given by

$$\tilde{\lambda}_{n_{Asy|P}} = \frac{c^2}{g^2}\frac{16(1+\alpha)^2 T^{1-2\alpha}}{(-\alpha)(1-2\alpha)^2}\cdot\frac{1}{(\tilde{\gamma}_{n_{Asy|P}})^2}. \tag{20.139}$$

(5) Finally, the probability density followed by the orthogonal random variables $\tilde{Z}_{n_{Asy|P}}$ is the gamma-type:

$$f_{\tilde{Z}_{n_{Asy|P}}}(z) = \frac{1}{\sqrt{\pi}(2\tilde{\lambda}_{n_{Asy|P}})^{\frac{1}{4}}}\, e^{-\frac{z+\sqrt{\frac{\tilde{\lambda}_{n_{Asy|P}}}{2}}}{\sqrt{2\tilde{\lambda}_{n_{Asy|P}}}}}\cdot\left[z+\sqrt{\frac{\tilde{\lambda}_{n_{Asy|P}}}{2}}\right]^{-\frac{1}{2}} U\left(z+\sqrt{\frac{\tilde{\lambda}_{n_{Asy|P}}}{2}}\right) \tag{20.140}$$

where $U(x)$ is the unit step function.

We now sketch the proof of Theorem 7. The $\tilde{\chi}_{Asy|P}(t)$ function of (20.136) immediately follows from the definition (20.86) by invoking (20.132) and (20.133). Then, just as happened for Theorem 5, time disappears *identically* from (20.85) or (20.87) yielding the constant order (20.135). Whether this simplification (which is essential in order to simplify all subsequent analytical developments) occurs just "by chance" or has any deeper meaning is unknown to this author at the present time. The reader, however, may wish to speculate on the following unpublished result: the analytical technique used in [6, 7] to obtain the KL expansions of $B(\tau)$ and $B^2(\tau)$, respectively, *cannot* be extended to $B^3(\tau)$ and $B^4(\tau)$. Thus, it would appear that the mentioned simplifications, occurring in the $B(\tau)$ and $B^2(\tau)$ cases, are indeed a "lucky circumstance".

Continuing the proof of our Theorem 7, we see that the constancy of $\tilde{\nu}_{Asy|P}$ simplifies the writing of (20.88), and, after several reductions, it takes the form

$$-\frac{2\alpha}{2\alpha-1}J_{\tilde{\nu}_{Asy|P}}(\tilde{\gamma}_{n_{Asy|P}}) + \tilde{\gamma}_{n_{Asy|P}} J'_{\tilde{\nu}_{Asy|P}}(\tilde{\gamma}_{n_{Asy|P}}) = 0. \tag{20.141}$$

Then the Bessel function property (see [10, p. 11, entry (19.54)])[3]

$$zJ'_{\tilde{\nu}}(z) + \tilde{\nu}J_{\tilde{\nu}}(z) = zJ_{\tilde{\nu}-1}(z) \tag{20.142}$$

and (20.141) yield

$$-\tilde{\gamma}_{n_{Asy|P}} J_{\tilde{\nu}_{Asy|P}}(\tilde{\gamma}_{n_{Asy|P}}) = 0 \tag{20.143}$$

from which

$$\tilde{\gamma}_{n_{Asy|P}} \neq 0 \tag{20.144}$$

and (20.137) is obtained.

[3] Note that this is not the same as (20.111).

The normalization constants (20.138) come next. Just as for the proof of (20.108), the Dini expansion formula for the Bessel functions (20.113) may be invoked to perform the integration in (20.89), with the result

$$\int_0^1 x J^2_{\tilde{\nu}_{Asy|P}}(\tilde{\gamma}_{n_{Asy|P}} x)\, dx = \frac{J^2_{\tilde{\nu}_{Asy|P}}(\tilde{\gamma}_{n_{Asy|P}} x)}{2}. \tag{20.145}$$

Equation (20.89) then leads to (20.138) by invoking (20.133) and (20.145). The eigenvalues (20.139) are immediately obtained from (20.91) and (20.133). Finally, the probability density (20.140) is the same as (20.60), apart from the replacement of $\tilde{\lambda}_{n_{Asy}}$ by $\tilde{\lambda}_{n_{Asy|P}}$.

20.4.6 Approximated power-like asymptotic KL expansion for noise instantaneous energy

Just as Theorem 6 is the asymptotic approximation of Theorem 5, so in the present section we are going to derive a Theorem 8 that is the asymptotic approximation to Theorem 7. Again, the leading idea is to replace all Bessel functions by their simpler asymptotic form (20.115), which has the advantage of expressing them in terms of a cosine.

As in Section 20.4.3, all formulas derived for the asymptotic approximation will be denoted by a further "$|A$" suffix. Thus, the whole suffix will now be "$Asy|P|A$".

When applied to (20.137), (20.115) yields

$$\cos\left(\tilde{\gamma}_{n_{Asy|P|A}} - \frac{(\tilde{\nu}_{Asy|P} - 1)\pi}{2} - \frac{\pi}{4}\right) \approx 0 \tag{20.146}$$

whence

$$\tilde{\gamma}_{n_{Asy|P|A}} - \frac{(\tilde{\nu}_{Asy|P} - 1)\pi}{2} - \frac{\pi}{4} \approx n\pi - \frac{\pi}{2} \quad (n = 1, 2, \ldots) \tag{20.147}$$

Invoking (20.135), we now get the following approximated expression for the $\tilde{\gamma}_{n_{Asy|P|A}}$

$$\tilde{\gamma}_{n_{Asy|P|A}} \approx (n-1)\pi + \frac{\pi}{4} + \frac{\tilde{\nu}_{Asy|P}\pi}{2} = \pi\left(n - \frac{3}{4} + \frac{|\alpha|}{|2\alpha - 1|}\right). \tag{20.148}$$

The approximated normalization constants $\tilde{N}_{n_{Asy|P|A}}$ come next. However, the Bessel function appearing in (20.138) must first be replaced by its asymptotic version (20.115), yielding

$$\begin{aligned} |J_{\tilde{\nu}_{Asy|P}}(\tilde{\gamma}_{n_{Asy|P|A}})| &\approx \left|\sqrt{\frac{2}{\pi\tilde{\gamma}_{n_{Asy|P|A}}}}\right| \left|\cos\left(\tilde{\gamma}_{n_{Asy|P}} - \frac{\tilde{\nu}_{Asy|P}\pi}{2} - \frac{\pi}{4}\right)\right| \\ &= \sqrt{\frac{2}{\pi\tilde{\gamma}_{n_{Asy|P|A}}}} |\cos((n-1)\pi)| = \sqrt{\frac{2}{\pi\tilde{\gamma}_{n_{Asy|P|A}}}} \end{aligned} \tag{20.149}$$

Then, (20.138) and (20.149) yield the approximated normalization constants $\tilde{N}_{n_{Asy|P|A}}$

$$\tilde{N}_{n_{Asy|P|A}} \approx \frac{g}{c} \frac{\sqrt{-\alpha}(1-2\alpha)T^{\alpha-\frac{1}{2}}}{4(1+\alpha)} \pi \cdot \sqrt{n - \frac{3}{4} + \frac{|\alpha|}{|2\alpha - 1|}}. \tag{20.150}$$

As for the eigenvalues, one merely has to substitute (20.148) into (20.139) to get

$$\tilde{\lambda}_{n_{Asy|P|A}} \approx \frac{c^2}{g^2} \frac{16(1+\alpha)^2 T^{1-2\alpha}}{\pi^2(-\alpha)(1-2\alpha)^2} \cdot \frac{1}{\left(n - \frac{3}{4} + \frac{|\alpha|}{|2\alpha - 1|}\right)^2}. \tag{20.151}$$

These are also the variances of Gaussian zero-mean orthogonal random variables $\tilde{Z}_{n_{Asy|P|A}}$ whose probability density is the same as (20.140).

In conclusion, we have proved the following:

Theorem 8 *The approximated asymptotic KL expansion of the instantaneous energy of the Gaussian noise emitted by a spaceship moving at a power-like acceleration is given by*

$$\begin{aligned} Y^2_{Asy|P|A}(t) &\equiv X^2_{Asy|P|A}(\tau) - E\{X^2_{Asy|P|A}(\tau)\} \\ &\approx \sqrt{1-2\alpha}\frac{t^{-\frac{\alpha}{2}-\frac{1}{4}}}{T^{\frac{1}{4}-\frac{\alpha}{2}}} \cdot \sum_{n=1}^{\infty} Z_{n_{Asy|P|A}} \cos\left(\gamma_{Asy|P|A} \frac{t^{\frac{1}{2}-\alpha}}{T^{\frac{1}{2}-\alpha}} - \frac{\pi}{4} - \frac{|2\alpha|}{|2\alpha-1|} \cdot \frac{\pi}{2}\right). \end{aligned} \tag{20.152}$$

20.5 CONCLUSION

Eight theorems have been proved concerning application of the KL eigenfunction expansion to radio communications between the Earth and a relativistic spaceship. First, we obtained the KL expansion for the noise emitted by the spaceship moving at an *arbitrary* acceleration with respect to its own reference frame. The results of the previous Chapter 12, where this acceleration was supposed to remain constant ($= g$), have thus been fully generalized. Yet the KL expansion for the general case is not much more complicated than for constant g: in all cases, in fact, the eigenfunctions are Bessel functions of the first kind, and the eigenvalues are the zeros of such Bessel functions. The noise total energy cumulants, mean value, and variance were obtained as byproducts of this KL expansion.

A second KL expansion, already studied in Chapter 12 for constant g, was also fully generalized to arbitrary acceleration: this is the instantaneous energy (i.e., the zero-mean square noise) eigenfunction expansion. Again, the eigenfunctions are Bessel functions of the first kind, and the eigenvalues are their zeros.

All these results undergo remarkable simplifications when one considers the *asymptotic* acceleration of the spaceship about one year after departure from Earth. In particular, if the accelaration behaves like a power of time, the formulas are simple enough to allow Bessel functions to be replaced by suitable sinusoids. The way is thus paved for computer numeric simulations, good for engineering design.

The KL expansion is *optimal* among all possible transforms inasmuch as the eigenfunction set is, by definition, the best orthonormal basis in the Hilbert space. Finding this basis only by means of numerical techniques may be time-consuming, even on fast computers, for one has to determine both the eigenvalues and eigenvectors of the large symmetric matrix representing noise autocorrelation. However, we have solved this problem by providing explicit—if only asymptotically approximated—eigenfunctions and eigenvalues directly expressed in terms of the acceleration of the emitting source. The first step has thus been taken in designing the optimal signal analysis procedure to filter out the noise received on Earth from a relativistically moving spaceship.

20.6 REFERENCES

[1] C. Maccone, "Relativistic Interstellar Flight and Gaussian Noise," *Acta Astronautica*, **17**(9) (1988), 1019–1027.

[2] C. Maccone, "Relativistic Interstellar Flight and Instantaneous Noise Energy," *Acta Astronautica*, **21**(3) (1990), 155–159.

[3] C. Maccone, "Relativistic Interstellar Flight and Genetics," *Journal of the British Interplanetary Society*, **43** (1990), 569–572.

[4] E. Mallove, and G. Matloff, *The Starflight Handbook: A Pioneer's Guide to Interstellar Travel*, Wiley, New York, 1989.

[5] D. Goldsmith, *The Quest for Extraterrestrial Life: A Book of Readings*, University Science Books, Mill Valley, CA, 1980.

[6] C. Maccone, "Eigenfunctions and Energy for Time-Rescaled Gaussian Processes," *Bollettino dell'Unione Matematica Italiana*, Series 6, **3-A** (1984), 213–219.

[7] C. Maccone, "The Karhunen–Loève Expansion of the Zero-Mean Square Process of a Time-Rescaled Gaussian Process," *Bollettino dell'Unione Matematica Italiana*, Series 7, **2-A** (1988), 221–229.

[8] C. Maccone, "Special Relativity and the Karhunen–Loève Expansion of Brownian Motion," *Il Nuovo Cimento*, Series B, **100** (1987), 329–341.

[9] C. Maccone, "The Time-Rescaled Brownian Motion $B(t^{2H})$," *Bollettino dell'Unione Matematica Italiana*, Series 6, **4-C** (1985), 363–378.

[10] A. Erdélyi, W. Magnus, F. Oberhettinger, and F. G. Tricomi, *Higher Transcendental Functions*, Vol. 2, McGraw Hill, New York, 1953.

[11] C. Maccone, "Relativistic Interstellar Flight Communication Theorems," *Acta Astronautica*, **26**(2) (1992), 105–117.

21

Brownian motion and its time rescaling

21.1 INTRODUCTION

Let us now change the topic of the discussion, and consider *Brownian motion*. Since Brownian motion has been investigated by physicists and mathematicians for about a century, a number of aspects of both theoretical and practical interest have been brought to light, and a large book would thus be required to cover them. In this book we shall confine ourselves to a very particular feature that we will call "time rescaling".

By time rescaling we mean making a change in the Brownian motion time variable in such a way that the new time variable does not elapse uniformly. The new resulting Brownian motion is mathematically called *time-inhomogeneous Brownian motion*, and will be called time-rescaled Brownian motion in the present book. Thus, our time rescaling is a suitable nonlinear transformation, which may be adapted to represent a host of scientific phenomena in the fields of mathematics and physics as well as in those of economics and biology.

For instance, it is evident that the time-rescaling transformation is of interest in the theory of relativity, inasmuch as two time variables, "coordinate time" and "proper time", exist in relativity, and Brownian motion may then be a function of either of them. In addition to restricting our presentation to time-rescaled Brownian motion, we want to restrict it further to those aspects of Brownian motion that are related to the KLT.

Now, the basic result proved in this book (Maccone First KLT Theorem, proved in Chapter 22) shows that the KL eigenfunctions for time-rescaled Brownian motion are Bessel functions of the first kind having nonconstant order. This appears to be an original contribution, not only with regard to time-rescaled Brownian motion, but also to the theory of Bessel functions of the first kind. In fact, several properties of Bessel functions that are well-known to hold good for constant order are naturally extended here to a general, time-dependent order.

21.2 BROWNIAN MOTION ESSENTIALS

Brownian motion—or, better, *standard Brownian motion*—is the easiest and most important non-stationary Gaussian process. We shall denote it by $B(t)$, and define it as the stochastic process with probability density function (of the first order) equal to the well-known Gaussian

$$f_{B(t)}(x) = \frac{1}{\sqrt{2\pi}\sqrt{t}} e^{-\frac{x^2}{2t}}. \tag{21.1}$$

From this definition, it immediately follows that the mean value of Brownian motion is identically zero; that is, it equals zero for all values of time

$$E\{B(t)\} = \int_{-\infty}^{\infty} x f_{B(t)}(x)\, dx = 0 \tag{21.2}$$

and taking $t \to 0$ in (21.1), it follows that $B(t)$ fulfills the initial condition

$$B(0) = 0 \tag{21.3}$$

In other words, at $t = 0$, $B(t)$ becomes deterministic with precise value $B(0) = 0$. Another way of stating this is to say that as $t \to 0$ the Gaussian density (21.1) approaches the Dirac delta function $\delta(x)$.

Sometimes, Brownian motion is called the *Wiener process*, or the *Wiener–Lévy process*, according to the aspects of the topic that the authors desire to stress. We are not going to prove here the mathematical properties of Brownian motion. The interested reader may find, for instance, a nice presentation of them in [2, pp. 292–293]. Instead, we just highlight the main results, as well as those special features that will be used in the remainder of the present book.

The relevant variance and standard deviation are immediately seen to be given, respectively, by

$$\sigma^2_{B(t)} = t \tag{21.4}$$

$$\sigma_{B(t)} = \pm\sqrt{t}. \tag{21.5}$$

If we plot $B(t)$ against t, the resulting graph is a continuous curve, randomly moving above and below the time axis, in such a way that the standard deviation curve (21.5) is a parabola having its vertex at the origin and its axis coinciding with the time axis. Figure 21.1 illustrates this.

An interesting property of Brownian motion which we use in this book is its self-similarity to the order 1/2, expressed by the formula

$$B(ct) = \sqrt{c}B(t) \quad (c > 0) \tag{21.6}$$

To understand why this is true, just replace $t \to ct$ into the Gaussian density (21.3), and then rearrange as follows:

$$f_{B(ct)}(x) = \frac{1}{\sqrt{c}} \frac{1}{\sqrt{2\pi}\sqrt{t}} e^{-\frac{1}{2t}\left(\frac{x}{\sqrt{c}}\right)^2} = f_{\sqrt{c}B(t)}(x). \tag{21.7}$$

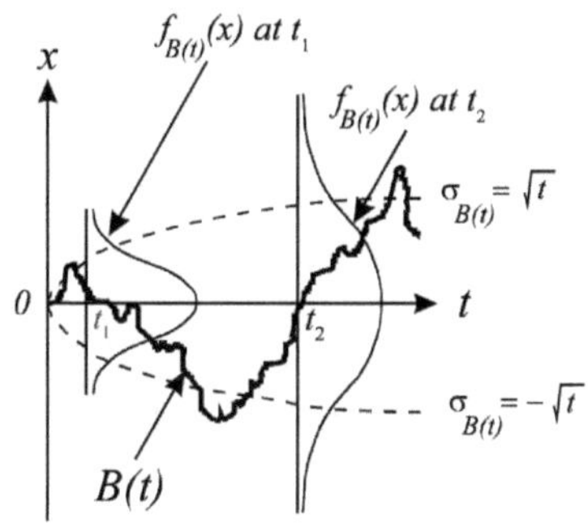

Figure 21.1. Illustration of Brownian motion $B(t)$ vs. t. The dotted lines are the upper and lower branches of the standard deviation parabola, having the equation $\pm\sqrt{t}$.

To complete this overview of Brownian motion properties, we state without proof (e.g., [2, p. 293]) that the autocorrelation of Brownian motion is given by

$$E\{B(t_1)B(t_2)\} = \begin{cases} t_1 & \text{for } t_1 < t_2 \\ t_2 & \text{for } t_1 > t_2 \end{cases}. \tag{21.8}$$

This circumstance may be re-phrased by introducing a new symbol, called the minimum (= smallest of) and denoted $\wedge$, so that (21.8) takes the form

$$E\{B(t_1)B(t_2)\} = t_1 \wedge t_2. \tag{21.9}$$

Let us now express the minimum in terms of the unit step function, defined by

$$U(t) = \begin{cases} 0 & \text{for } t < 0 \\ 1 & \text{for } t > 0 \end{cases}. \tag{21.10}$$

The unit step function is clearly discontinuous at the origin, and its value there is not defined by (21.10).[1] Moreover, the derivative of the unit step function (in the sense of the theory of distributions) is the Dirac delta function familiar to physicists and engineers alike (see [3, pp. 255–282])

$$\frac{dU(t)}{dt} = \delta(t). \tag{21.11}$$

Let us now go back to the minimum $t_1 \wedge t_2$. An easy but essential result about the minimum $t_1 \wedge t_2$ is that it can be re-written in terms of the unit step function as

[1] Actually, within the mathematically rigorous context of the *theory of distributions* it can be shown that the value of (21.10) at the origin may be any value between zero and one, for this point is a "topological" one. However, we shall not elaborate further topics because it is irrelevant to the current purpose.

follows:

$$t_1 \wedge t_2 = t_1 U(t_2 - t_1) + t_2 U(t_1 - t_2). \tag{21.12}$$

A generalization of this is

$$F(t_1 \wedge t_2) = F(t_1)\, U(t_2 - t_1) + F(t_2)\, U(t_1 - t_2) \tag{21.13}$$

where $F(\ldots)$ is *any* function of the minimum $t_1 \wedge t_2$. The full power of (21.13) in helping one to get rid of a number of apparent difficulties related to minimum will show up in Chapter 24 (particularly its equations).

21.3 KLT OF BROWNIAN MOTION

We are now ready to compute the KL expansion by solving the integral equation (10.18) for standard Brownian motion. This exercise yields a fundamental insight into the mathematical methods that will later be developed to find new results.

In practice, the integral equation (10.18) must be solved with the autocorrelation (21.9) of Brownian motion. However, dealing with the minimum $t_1 \wedge t_2$ as if it was a continuous function of both its arguments may not be easy, so that we replace it by the equivalent expression (21.12) in terms of unit step functions. The integral equation (10.18) then becomes

$$\begin{aligned} \lambda_n \phi_n(t_1) &= \int_0^T [t_1 U(t_2 - t_1) + t_2 U(t_1 - t_2)] \phi_n(t_2)\, dt_2 \\ &= t_1 \int_{t_1}^T \phi_n(t_2)\, dt_2 + \int_0^{t_1} t_2 \phi_n(t_2)\, dt_2. \end{aligned} \tag{21.14}$$

If $t_1 = 0$, the right-hand side of the last expression vanishes, and we get the initial condition for the eigenfunctions

$$\phi_n(0) = 0. \tag{21.15}$$

Differentiating both sides of (21.14) with respect to t_1 reduces the integral equation to a differential equation. This procedure actually amounts to realizing that the integral equation (10.18) is not a *Fredholm-type* for the particular autocorrelation (21.9) of Brownian motion, but is actually a *Volterra-type*. We thus get

$$\lambda_n \frac{d\phi_n(t_1)}{dt_1} = \int_{t_1}^T \phi_n(t_2)\, dt_2 - t_1 \phi_n(t_1) + t_1 \phi_n(t_1). \tag{21.16}$$

The last two terms cancel each other (and similar cancellations will prove to be vital in future calculations), so that, by setting $t_1 = T$, (21.16) yields the end point condition for the eigenfunctions

$$\left.\frac{d\phi_n(t)}{dt}\right|_{t=T} \equiv \phi'_n(T) = 0. \tag{21.17}$$

The presence of a remaining integral in (21.16) suggests performing a further differentiation with respect to t_1 changing (21.16) into the differential equation

$$\frac{d^2\phi_n(t)}{dt^2} + \frac{1}{\lambda_n}\phi_n(t) = 0. \tag{21.18}$$

This is the harmonic oscillator differential equation, whose general integral is a linear combination of a sine and a cosine

$$\phi_n(t) = A\sin\left(\frac{t}{\sqrt{\lambda_n}}\right) + B\cos\left(\frac{t}{\sqrt{\lambda_n}}\right) \tag{21.19}$$

where A and B are integration constants.

Determining the integration constants and λ_n from the two boundary conditions (21.15) and (21.17) and the normalization condition (10.4), it follows that

$$\phi_n(t) = \sqrt{\frac{2}{T}}\sin\left(\frac{\pi(2n-1)}{2T}t\right) \tag{21.20}$$

$$\lambda_n = \frac{4T^2}{\pi^2(2n-1)^2} \quad (n = 1, 2, \ldots). \tag{21.21}$$

These are just the KL expansion coefficients for standard Brownian motion. We may thus formally write the KL eigenfunction expansion of standard Brownian motion by substituting (21.20) into (10.2)

$$B(t) = \sqrt{\frac{2}{T}}\sum_{n=1}^{\infty} Z_n \sin\left(\frac{\pi(2n-1)}{2T}t\right). \tag{21.22}$$

Our purpose is to generalize this basic result (21.22) to forms of Brownian motion for which time does not elapse uniformly. This leads to many applications of interest in physics and in relativistic spaceflight.

21.4 WHITE NOISE AS THE DERIVATIVE OF BROWNIAN MOTION WITH RESPECT TO TIME

White noise $W(t)$ is an important notion with which physicists and engineers alike are acquainted. It is usually introduced as the one stochastic process whose power spectrum is a constant—in practice over a finite range of frequencies. That is, for which

$$P(\omega) \equiv \int_{-\infty}^{\infty} e^{i\omega t}W^2(t)\,dt$$

is a constant.

A stochastic process is said to be *stationary* if its autocorrelation is a function of the time difference $(t_2 - t_1)$ *alone*, rather than being a function of t_1 and t_2 independently. Then, the *Wiener–Khinchin theorem* guarantees that the power spectrum and the autocorrelation of the stationary processes are the Fourier transforms of each other. Thus

$$E\{W(t_1)W(t_2)\} = \int_{-\infty}^{\infty} P(\omega)\, e^{-i\omega(t_2-t_1)} d\omega$$
$$= const \int_{-\infty}^{\infty} e^{-i\omega(t_2-t_1)} d\omega = const\, \delta(t_2 - t_1)$$

and, apart from the constant, the autocorrelation of white noise is the delta function.

In this section we will show that the derivative of Brownian motion $B(t)$ is white noise in the sense that if we define $\dot{B}(t)$ such that

$$E\{\dot{B}(t_1)\dot{B}(t_2)\} \equiv \frac{\partial^2 E\{B(t_1)B(t_2)\}}{\partial t_1\, \partial t_2} \tag{21.23}$$

then

$$E\{\dot{B}(t_1)\dot{B}(t_2)\} = \delta(t_1 - t_2) \tag{21.24}$$

just like *white noise.*

To do this we first recall from (21.9) that

$$E\{B(t_1)B(t_2)\} = t_1 \wedge t_2.$$

We now form the left member as in (21.23) and use the expression from (21.12) to get

$$E\{\dot{B}(t_1)\dot{B}(t_2)\} \equiv \frac{\partial^2 E\{B(t_1)B(t_2)\}}{\partial t_1\, \partial t_2} = \frac{\partial^2 (t_1 \wedge t_2)}{\partial t_1\, \partial t_2}$$
$$= \frac{\partial^2}{\partial t_1\, \partial t_2} [t_1 U(t_2 - t_1) + t_2 U(t_1 - t_2)]$$

using the fact that $dU(x)/dx = \delta(x)$

$$= \frac{\partial}{\partial t_2} [U(t_2 - t_1) - t_1 \delta(t_2 - t_1) + t_2 \delta(t_1 - t_2)]$$

using the fact that $x\delta(x) = 0$

$$= \frac{\partial}{\partial t_2} [U(t_2 - t_1) + (t_2 - t_1)\, \delta(t_2 - t_1)] = \frac{\partial}{\partial t_2} U(t_2 - t_1)$$
$$= \delta(t_2 - t_1) \tag{21.25}$$

as required. Thus (21.24) is proved. The interpretation of white noise as the derivative of Brownian motion will be used in Section 21.6 to prove basic results about the time rescaling of Brownian motion.

21.5 INTRODUCTION TO TIME RESCALING

The rest of this chapter explores what happens to Brownian motion if one lets the time elapse according to an arbitrary law, rather than uniformly. We will call this *time-rescaling* the Brownian motion. Others call it time-inhomogeneous Brownian motion, among other things.

The results of the present chapter are not original. They have been known for years, but have seldom been exploited. We exploited them to the advantage of physics in Chapters 11 through 13 by establishing their relationship to the theory of relativity. Further scientific applications of these results may occur in mathematical genetics, and we dealt with some of them in Chapter 14.

The idea underpinning our work is that the Gaussian character of Brownian motion is not altered if time is arbitrarily rescaled. That is, all the Gaussian properties remain the same for the rescaled process.

21.6 THE WHITE NOISE INTEGRAL AND ITS AUTOCORRELATION

The term *white noise integral* refers to any stochastic process having the form

$$X(t) \equiv \int_0^t f(s)\, dB(s) \tag{21.26}$$

where $B(t)$ is standard Brownian motion (with zero mean and variance t) and for the time being the function $f(t)$ may be any arbitrary function continuous over the real positive axis. Since the integral in (21.26) is over a stochastic process $B(t)$, and hence difficult to define, it makes more sense to use the alternate definition

$$X(t) \equiv \int_0^t f(s)\, \dot{B}(s)\, ds \tag{21.27}$$

and $\dot{B}(t)$ is the white noise by virtue of Section 21.5.

A few words about notation: $f(t)$ shall henceforth be the (almost) arbitrary scaling function of time, while $f_X(x)$ with a subscript will always refer to the probability density of the stochastic process $X(t)$—specifically the time-rescaled Gaussian probability density in this book. While in Chapter 10 $X(t)$ referred to an arbitrary stochastic process, in this section and in the rest of the Appendix it will refer to the white noise integral (21.26).

The stochastic process $X(t)$ defined by the white noise integral (21.26) is a Gaussian process. This should be obvious from the fact that the linear integral operator in (21.26) acts only on time and not on the statistical nature of the process $dB(s) = \dot{B}(s)\, ds$, which is Gaussian. However, a more sophisticated argument to realize that this is indeed the case proceeds as follows:

(1) When one has two (independent) Gaussian random variables X_1 and X_2 (with respective means m_1 and m_2 and variances σ_1^2 and σ_2^2), the random variable *sum*

$X_1 + X_2$ is again Gaussian (with mean $m_1 + m_2$ and variance $\sigma_1^2 + \sigma_2^2$)—for a proof of this well-known fact, see, for instance, [2, p. 250].

(2) Clearly, the same argument can be extended to the sum of any finite number of Gaussian random variables, which is thus one more Gaussian random variable.

(3) Finally, the integral is the limit of a sum, so that a white noise integral like (21.1) also is a Gaussian random variable (i.e., the stochastic process $X(t)$ is Gaussian).

Now from (21.26) the initial condition for the $X(t)$ process is immediately seen to be $X(0) = 0$. By taking the mean value of both sides of (21.27) and interchanging the mean value operator with the integral sign, it is also evident from (21.27) that

$$E\{X(t)\} = E\left\{\int_0^t f(s)\dot{B}(s)\,ds\right\} = \int_0^t f(s)\,E\{\dot{B}(s)\}\,ds$$

$$= \int_0^t f(s)\frac{d}{ds}[E\{B(s)\}]\,ds = 0 \qquad (21.28)$$

since Brownian motion satisfies the standard condition $E\{B(t)\} = 0$ for all t.

Let us now compute the autocorrelation of the process (21.27):

$$E\{X(t_1)X(t_2)\} = E\left\{\int_0^{t_1} f(s)\,\dot{B}(s)\,ds\int_0^{t_2} f(t)\,\dot{B}(t)\right\}$$

$$= \int_0^{t_1} ds\,f(s)\int_0^{t_2} f(t)\,E\{\dot{B}(s)\dot{B}(t)\}\,dt$$

by virtue of (21.24)

$$= \int_0^{t_1} ds\,f(s)\int_0^{t_2} f(t)\,\delta(t-s)\,dt.$$

Now the inner integral differs from zero only if the singularity of the delta function lies between zero and t_2—that is, only if one has $0 \le s \le t_2$. By resorting to the unit step function, we can rewrite this in the form

$$E\{X(t_1)X(t_2)\} = \int_0^{t_1} ds\,f(s)[f(s)U(t_2 - s)] = \int_0^{t_1\wedge t_2} f^2(s)\,ds \qquad (21.29)$$

where the notion of minimum (= smallest of) t_1 and t_2 was used in the last step. In conclusion, autocorrelation of the white-noise integral (21.26) or (21.27) reads

$$E\{X(t_1)X(t_2)\} = \int_0^{t_1\wedge t_2} f^2(s)\,ds. \qquad (21.30)$$

The variance of $X(t)$ can now be found at once by noticing that

$$\sigma_{X(t)}^2 = E\{X^2(t)\} - E^2\{X(t)\} = E\{X(t)X(t)\}. \qquad (21.31)$$

Then, one merely has to set $t_1 = t_2 = t$ into the autocorrelation (21.30), and use the obvious formula $t \wedge t = t$ to get

$$\sigma^2_{X(t)} = \int_0^t f^2(s)\, ds. \tag{21.32}$$

The square root of this is the standard deviation

$$\sigma_{X(t)} = \pm\sqrt{\int_0^t f^2(s)\, ds}. \tag{21.33}$$

Equation (21.33) defines two curves on the $(t, X(t))$-plane, named standard deviation curves, that lie above and below the (zero) mean value axis, and are quite helpful for qualitative understanding of the behavior of the $X(t)$ process in time.

21.7 TIME RESCALING AND GAUSSIAN PROPERTIES OF $X(t)$

In this section we are going to prove formally that the Gaussian process $X(t)$ defined by (21.27) is just a time-rescaled version of Brownian motion. To this end, consider the following two processes:

(1) The process $X(t)$ defined by (21.27).
(2) The Brownian motion $B(t)$, where the ordinary time variable t is replaced by a rescaled time variable expressed by

$$\int_0^t f^2(s)\, ds. \tag{21.34}$$

That is, we want to consider time-rescaled Brownian motion

$$B\left(\int_0^t f^2(s)\, ds\right). \tag{21.35}$$

Evidently, both processes (21.1) and (21.9) fulfill the same initial condition

$$X(0) = 0, \quad B(0) = 0 \tag{21.36}$$

and also have the same mean value (zero):

$$E\{X(t)\} = 0, \quad E\left\{B\left(\int_0^t f^2(s)\, ds\right)\right\} = 0. \tag{21.37}$$

The crucial point lies in processes (21.27) and (21.35) also having the same autocorrelation (i.e., the relationship holds)

$$E\{X(t_1)X(t_2)\} = E\left\{B\left(\int_0^{t_1} f^2(s)\, ds\right)B\left(\int_0^{t_2} f^2(s)\, ds\right)\right\}. \tag{21.38}$$

To prove this, one first notices that the right-hand side of (21.38)—that is, the autocorrelation of the time-rescaled Brownian motion $B(\ldots)$—by virtue of (21.9) is the minimum

$$E\left\{B\left(\int_0^{t_1} f^2(s)\,ds\right)B\left(\int_0^{t^2} f^2(s)\,ds\right)\right\} = \left(\int_0^{t_1} f^2(s)\,ds\right) \wedge \left(\int_0^{t_2} f^2(s)\,ds\right). \quad (21.39)$$

Second, that this minimum just equals the integral

$$\int_0^{t_1 \wedge t_2} f^2(s)\,ds \quad \text{or} \quad E\{X(t_1)X(t_2)\}$$

which, in turn, is the autocorrelation (21.30), completing the proof of (21.38).

To sum up, we have shown that the Gaussian processes (21.27) and (21.35) have the same initial condition, the same mean value, and the same autocorrelation. Because of its Gaussian nature, time-rescaled Brownian motion has all its higher moments fully determined by its first two moments (see, e.g., [2, section 9.3]). Moreover, the same fact must be true for the process $X(t)$ as well, and, since *all* the moments of the two processes are identical, we conclude that the two processes coincide completely

$$X(t) \equiv B\left(\int_0^t f^2(s)\,ds\right). \quad (21.40)$$

This basic result reveals the nature of time-inhomogeneous, or time-rescaled, Brownian motion: all the action of the $f(t)$ function consists in altering the time behavior of the original Brownian motion, $B(t)$, but not in changing its Gaussian character. In fact, the probability density of $X(t)$ is

$$f_{X(t)}(x) = \frac{1}{\sqrt{2\pi}\sigma_{X(t)}} e^{-\frac{x^2}{2\sigma^2_{X(t)}}} \quad (21.41)$$

expressing a Gaussian whose variance changes in time according to (21.32). When rewritten explicitely in terms of the $f(t)$ function, (21.41) reads

$$f_{X(t)}(x) = \frac{1}{\sqrt{2\pi}\sqrt{\int_0^t f^2(s)\,ds}} e^{-\frac{x^2}{2\int_0^t f^2(s)\,ds}}. \quad (21.42)$$

Also, one may notice that, in agreement with the initial condition $X(0) = 0$, both (21.41) and (21.42) tend to the delta function, $\delta(x)$, for $t \to 0$. Finally, it follows immediately that the Gaussian (21.42) reduces to the familiar Gaussian density of Brownian motion in the case $f(t) = 1$:

$$f_{X(t)}(x) = \frac{1}{\sqrt{2\pi}\sqrt{t}} e^{-\frac{x^2}{2t}}. \quad (21.43)$$

As for the higher moments of the Gaussian process $X(t)$, they can be found at once from the corresponding higher moments of the Gaussian density (21.43) simply by replacing t by virtue of the variance (21.32). The proofs of the Gaussian-higher-moments formulas can be found, for instance, in [2, section 5.4]. Thus, the higher moments of $X(t)$ read, for even and odd values of n, respectively

$$E\{X^n(t)\} = \begin{cases} 1 \cdot 3 \cdot \ldots \cdot (n-1) \cdot \left[\int_0^t f^2(s)\, ds \right]^{n/2}, & \text{even } n \\ 0, & \text{odd } n \end{cases}. \tag{21.44}$$

21.8 ORTHOGONAL INCREMENTS FOR NONOVERLAPPING TIME INTERVALS

Consider four consecutive instants t_1, t_2, t_3, t_4 on the positive time axis, starting from left to right. The two time intervals, ranging from t_1 to t_2, and from t_3 to t_4, respectively, can be arranged in any one of three ways: completely disjoint intervals, contiguous intervals (i.e., with $t_2 = t_3$), and overlapping intervals.

Let us then consider the corresponding two increments taken by $X(t)$ over the two time intervals

$$[X(t_2) - X(t_1)] \tag{21.45}$$

and

$$[X(t_4) - X(t_3)]. \tag{21.46}$$

By virtue of (21.30), we may write the mean value of their product as

$$E\{[X(t_2) - X(t_1)][X(t_4) - X(t_3)]\}$$
$$= \int_0^{t_2 \wedge t_4} f^2(s)\, ds - \int_0^{t_2 \wedge t_3} f^2(s)\, ds - \int_0^{t_1 \wedge t_4} f^2(s)\, ds + \int_0^{t_1 \wedge t_3} f^2(s)\, ds.$$

Owing to the relative positions of the four points t_1, t_2, t_3, t_4, the four integrals in the last formula may or may not cancel against each other. Therefore, in the three cases stated above, one gets, respectively,

$$E\{[X(t_2) - X(t_1)][X(t_4) - X(t_3)]\} = \begin{cases} 0, & \text{for completely disjoint intervals} \\ 0, & \text{for continuous intervals} \\ \int_{t_3}^{t_4} f^2(s)\, ds, & \text{for overlapping intervals.} \end{cases}$$

We summarize these results by saying that the increments of the process $X(t)$ are orthogonal for non-overlapping time intervals only.

21.9 AN APPLICATION OF THE KLT: FINDING THE TOTAL ENERGY OF $X(t)$

One of the most important applications of the KL expansion (10.2) is the calculation of the stochastic integral

$$\varepsilon \equiv \int_0^T X^2(t)\, dt \tag{21.47}$$

representing the total energy of the time-rescaled Brownian motion $X(t)$. In fact, ε is a random variable that, by virtue of the orthonormality property of the eigenfunctions $\phi_n(t)$

$$\int_0^T \phi_m(t)\phi_n(t)\,dt = \delta_{mn} \tag{21.48}$$

is immediately seen to be given by the series

$$\varepsilon = \int_0^T \sum_{m=1}^{\infty}\sum_{n=1}^{\infty} Z_m Z_n \phi_m(t)\phi_n(t)\,dt = \sum_{n=1}^{\infty} Z_n^2. \tag{21.49}$$

This series expansion for the total energy ε can be further investigated by means of Fourier transforms, which are called *characteristic functions* in probability theory. Let us set $i = \sqrt{-1}$ and define the characteristic function $\Phi_{X(t)}(\zeta)$ of the stochastic process $X(t)$ with two alternative notations (see [4, p. 153]):

(1) the traditional Fourier transform integral, here applied to the probability density $f_{X(t)}(x)$ of the stochastic process $X(t)$; and
(2) the probabilistic notation exploiting the notion of mean value

$$\Phi_{X(t)}(\zeta) \equiv \int_{-\infty}^{\infty} e^{i\zeta x} f_{X(t)}(x)\,dx = E\{e^{i\zeta x}\}.$$

The use of characteristic functions (i.e., Fourier transforms) in probability theory simplifies things greatly, and was initiated by the French mathematician Paul Lévy in the 1920s. For instance, consider the simple integral

$$\int_0^{\infty} e^{i\zeta x}\, e^{-\alpha x}\,dx = \frac{1}{\alpha - i\zeta} \qquad \alpha > 0.$$

From this it follows that the exponential probability density

$$f(x) = \alpha\, e^{-\alpha x} U(x)$$

has the characteristic function

$$\Phi(\zeta) = \frac{\alpha}{\alpha - i\zeta}.$$

By differentiating the above integral n times with respect to ζ, one gets

$$\int_0^{\infty} e^{i\zeta x}\, e^{-\alpha x} x^n\,dx = \frac{n!}{(\alpha - i\zeta)^{n+1}}.$$

This prompts us to take a "bold" step: replacement of the positive integer n by the *real* variable ν (it actually is better to replace n by $\nu - 1$). To prove that this is correct would take be a step too far, so we will skip the proof here; however, the reader should be aware that the extension of the factorial from discrete to continuum values is given by definition of Euler's gamma function $\Gamma(\nu)$ (first conceived by Euler around 1744)

$$n! \to (\nu - 1)! = \Gamma(\nu) \equiv \int_0^\infty e^{-t} t^{\nu-1} dt.$$

With this extension to the continuum, the integral above becomes

$$\int_0^\infty e^{i\zeta x} e^{-\alpha x} x^{\nu-1} \, dx = \frac{\Gamma(\nu)}{(\alpha - i\zeta)^\nu}.$$

The so-called "gamma-type" probability density is defined by

$$f_{X,\mathrm{gamma}}(x) = \frac{\alpha^\nu}{\Gamma(\nu)} e^{-\alpha x} x^{\nu-1} U(x).$$

The characteristic function of the gamma density is immediately seen from the above integral to be

$$\Phi_{\mathrm{gamma}}(\zeta) = \int_0^\infty e^{i\zeta x} \frac{\alpha^\nu}{\Gamma(\nu)} e^{-\alpha x} x^{\nu-1} U(x) \, dx = \frac{\alpha^\nu}{(\alpha - i\zeta)^\nu} = \frac{1}{\left(1 - i\dfrac{\zeta}{\alpha}\right)^\nu}.$$

Let us now go back to the random variables Z_n^2. From the KL expansion of $X(t)$ we already know that the Z_n are Gaussian with mean zero and variance equal to the eigenvalues λ_n. On the other hand, a famous theorem in the theory of probability—for the proof, see, for instance, [4, p. 130]—states that if X is a Gaussian random variable with mean value zero and standard deviation σ, the square of this random variable obeys a distribution of the gamma type given by

$$f_{X^2}(x) = \frac{1}{\sqrt{2\pi}\sigma} e^{-\frac{x}{2\sigma^2}} \frac{1}{\sqrt{x}} U(x).$$

Checking this against the general gamma-type density previously studied, it is seen that the density of the square random variables Z_n^2 is a particular gamma-type density (sometimes also called a χ^2-type density) having

$$\begin{cases} \sigma = \sqrt{\lambda_n} \\ \nu = \dfrac{1}{2} \\ \alpha = \dfrac{1}{2\sigma^2} \end{cases}$$

and, in conclusion, we see that the characteristic function of each Z_n^2 is given by

$$\Phi_{Z_n^2}(\zeta) = \frac{1}{\left(1 - i\dfrac{\zeta}{\alpha}\right)^\nu} = \frac{1}{(1 - 2i\lambda_n\zeta)^{1/2}} = (1 - 2i\lambda_n\zeta)^{-1/2}. \tag{21.50}$$

Another useful property of the Fourier transforms that we need is the convolution theorem, which states that the Fourier transform of a convolution is the product of Fourier transforms. (This theorem was first proved by the French mathematician Duhamel around 1833.) In probability theory, convolution of the two densities of the independent random variables X and Y is very important, because it is just the density of the random variable sum, $X + Y$; that is,

$$f_{X+Y}(z) = \int_{-\infty}^{\infty} f_X(z-t) f_Y(t)\, dt.$$

This result, plus the convolution theorem, clearly imply that for the corresponding characteristic functions of the *independent* random variables X and Y one has

$$\Phi_{X+Y}(\zeta) = \Phi_X(\zeta)\, \Phi_Y(\zeta).$$

Let us apply this to the random variables Z_n that we know already to be independent because they are orthogonal and Gaussian. From the series (21.49) it just follows that the energy characteristic function is found as the infinite product

$$\Phi_\varepsilon(\zeta) = \prod_{n=1}^{\infty} \Phi_{Z_n^2}(\zeta) = \prod_{n=1}^{\infty} (1 - 2i\lambda_n \zeta)^{-1/2}. \tag{21.51}$$

In order to deal with series rather than with products, it is convenient to introduce the so-called second characteristic function $\Psi_\varepsilon(\zeta)$, simply defined as the natural logarithm of the (first) characteristic function

$$\Psi_\varepsilon(\zeta) = \ln \Phi_\varepsilon(\zeta). \tag{21.52}$$

Applying this definition to the infinite product (21.51), the latter is changed into the infinite series

$$\Psi_\varepsilon(\zeta) = -\frac{1}{2} \sum_{n=1}^{\infty} \ln(1 - 2i\lambda_n \zeta). \tag{21.53}$$

Differentiating n times, we get

$$\Psi_\varepsilon^{(n)}(\zeta) = \frac{1}{2} (2i)^n\, (n-1)! \sum_{n=1}^{\infty} \frac{(\lambda_m)^n}{(1 - 2i\lambda_m \zeta)^n}. \tag{21.54}$$

We will now introduce the statistical quantity called the *cumulants*, denoted K_n, of a random variable X. Once again, this may be a notion that not all readers may be familiar with, because it is mainly used by applied statisticians. Therefore, we do not regard it a waste of time to briefly describe how cumulants get the intuitive justification that later leads to their formal mathematical definition in (21.55).

Consider the familiar Maclaurin expansion of the exponential

$$e^x = \sum_{k=0}^{\infty} \frac{x^k}{k!}$$

and use this to expand the complex exponential appearing in the definition of the characteristic function of a generic random variable X:

$$\Phi_X(\zeta) = \int_{-\infty}^{\infty} e^{i\zeta x} f_X(x)\,dx$$
$$= \int_{-\infty}^{\infty} \sum_{k=1}^{\infty} \frac{(i\zeta x)^k}{k!} f_X(x)\,dx = \sum_{k=1}^{\infty} \frac{(i\zeta)^k}{k!} \int_{-\infty}^{\infty} x^k f_X(x)\,dx = \sum_{k=1}^{\infty} \frac{i^k m_k}{k!} \zeta^k$$

where we used the definition of the absolute kth moment; that is,

$$m_k = \int_{-\infty}^{\infty} x^k f_X(x)\,dx.$$

The above Maclaurin expansion for the characteristic function is evidently related to the absolute kth moment by the formula

$$\Phi_X^{(k)}(0) = i^k m_k$$

meaning that, if the characteristic function of a certain probability density is known, all the absolute moments can be computed by differentiating the characteristic function and then setting the independent variable to zero

$$m_k = \frac{\Phi_X^{(k)}(0)}{i^k}.$$

Precisely the same definition as this, applied to the second characteristic function rather than to the first, yields all the cumulants K_n—whence the moments may be found (see [5, p. 27]). That is,

$$K_n = \frac{\Psi_\varepsilon^{(n)}(0)}{i^n}. \tag{21.55}$$

Besides this formal definition, however, there is much more that one might say about cumulants. For instance, it is easy to prove that the mean value of the random variable ε is given by

$$E\{\varepsilon\} = K_1 = \Psi_\varepsilon'(0)$$

and the variance is given by

$$\sigma_\varepsilon^2 = K_2 = -\Psi_\varepsilon''(0).$$

Reverting now to our problem of determining the total energy distribution ε of $X(t)$, by setting $\zeta = 0$ into (21.54) and making use of (22.48) (to be proven in Chapter 22) we get

$$K_n = 2^{n-1}(n-1)! \left[\int_0^T f(s)\,ds \right]^{2n} \sum_{m=1}^{\infty} \frac{1}{(\gamma_m)^{2n}} \tag{21.56}$$

yielding all the cumulants of the energy distribution of $X(t)$.

The mean energy of ε is a special case $n = 1$ of (21.56)

$$E\{\varepsilon\} = K_1 = \left[\int_0^T f(s)\,ds \right]^2 \sum_{m=1}^{\infty} \frac{1}{(\gamma_m)^2}. \tag{21.57}$$

However, one can also write the following expression for $E\{\varepsilon\}$:

$$E\{\varepsilon\} = E\left\{\int_0^T X^2(t)\,dt\right\} = \int_0^T E\{X^2(t)\}dt. \tag{21.58}$$

On the other hand, (21.30) with $t_1 = t_2 = t$ leads at once to

$$E\{X^2(t)\} = \int_0^t f^2(s)\,ds. \tag{21.59}$$

Thus, (21.58) yields for the mean energy of $X(t)$

$$E\{\varepsilon\} = \int_0^T dt \int_0^t f^2(s)\,ds. \tag{21.60}$$

Of course, (21.57) and (21.60) are equivalent. Therefore, one finds

$$\sum_{m=1}^{\infty} \frac{1}{(\gamma_m)^2} = \frac{\int_0^T dt \int_0^t f^2(s)\,ds}{\left[\int_0^T f(s)\,ds\right]^2}. \tag{21.61}$$

We conclude that, although no explicit expression for the γ_n is known, it is possible to sum a series like (21.61) involving them. The explanation of this apparent paradox must lie deep in the theory of Bessel functions, and is unknown to mathematicians at this time.

Having found an explicit formula for the mean energy in terms of $f(t)$ such as (21.60), it is natural to seek a similar expression yielding energy variance. We shall now prove that it reads

$$\sigma_\varepsilon^2 = 4\int_0^T dt \int_0^t dv \left[\int_0^v f^2(s)\,ds\right]^2. \tag{21.62}$$

Though this result is simple, its proof is not so. The starting point is, of course, the definition of variance as

$$\sigma_\varepsilon^2 = E\{\varepsilon^2\} - E^2\{\varepsilon\}. \tag{21.63}$$

The second term is known by (21.60). The first term is calculated to be

$$E\{\varepsilon^2\} = E\left\{\int_0^T X^2(t)\,dt \int_0^T X^2(s)\,ds\right\} = \int_0^T dt \int_0^T ds\, E\{X^2(t)X^2(s)\} \tag{21.64}$$

In order to rewrite the integrand, note that $X(t)$ and $X(s)$ are Gaussian with zero mean. Therefore, one is allowed to apply the following property

$$\begin{aligned} E\{X^2(t)X^2(s)\} &= E\{X^2(t)\}E\{X^2(s)\} + 2E^2\{X(t)X(s)\} \\ &= \int_0^t f^2(x)\,dx \int_0^s f^2(y)\,dy + 2\left[\int_0^{t\wedge s} f^2(z)\,dz\right]^2 \end{aligned} \tag{21.65}$$

where (21.30) and (21.32) were used in the last step. Thus, (21.64) yields

$$\begin{aligned} E\{\varepsilon^2\} &= \int_0^T dt \int_0^T ds \left\{ \int_0^t f^2(x)\,dx \int_0^s f^2(y)\,dy + 2\left[\int_0^{t\wedge s} f^2(z)\,dz\right]^2 \right\} \\ &= \int_0^T dt \int_0^t f^2(x)\,dx \int_0^T ds \int_0^s f^2(y)\,dy + 2\int_0^T dt \int_0^T ds \left[\int_0^{t\wedge s} f^2(z)\,dz\right]^2 \\ &= E\{\varepsilon\}E\{\varepsilon\} + 2\int_0^T dt \int_0^T ds \left[\int_0^{t\wedge s} f^2(z)\,dz\right]^2 \end{aligned} \tag{21.66}$$

where (21.60) was used twice to rewrite the first term. A remarkable simplification now occurs. On inserting (21.66) into (21.63): the two terms $E^2\{\varepsilon\} - E^2\{\varepsilon\}$ cancel, and the energy variance becomes

$$\sigma_\varepsilon^2 = 2\int_0^T dt \int_0^T ds \left[\int_0^{t\wedge s} f^2(x)\,dx\right]^2. \tag{21.67}$$

The next difficulty lies in handling the minimum, $t \wedge s$, which is greatly simplified by the use of the unit step function:

$$U(t) = \begin{cases} 1 & \text{for } t > 0 \\ 0 & \text{for } t < 0 \end{cases}. \tag{21.68}$$

In fact, the minimum may be rewritten as a sum of two terms, resulting in

$$\begin{aligned} \sigma_\varepsilon^2 &= 2\int_0^T dt \int_0^T ds \left\{ \left[\int_0^t f^2(z)\,dz\right]^2 U(s-t) + \left[\int_0^s f^2(z)\,dz\right]^2 U(t-s) \right\} \\ &= 2\int_0^T dt \left[\int_0^t f^2(z)\,dz\right]^2 \int_0^T ds U(s-t) + 2\int_0^T ds \left[\int_0^s f^2(z)\,dz\right]^2 \int_0^T dt U(t-s). \end{aligned} \tag{21.69}$$

The two terms differ only in the variables t and s, so

$$\sigma_\varepsilon^2 = 4\int_0^T dt \left[\int_0^t f^2(z)\,dz\right]^2 \int_t^T ds = 4\int_0^T (T-t) \left[\int_0^t f^2(z)\,dz\right]^2 dt.$$

Finally, we use integration by parts to get:

$$\begin{aligned} \sigma_\varepsilon^2 &= 4\left[(T-t)\int_0^t dv \left[\int_0^v f^2(z)\,dz\right]^2\right]_{t=0}^{t=T} + 4\int_0^T dt \int_0^t dv \left[\int_0^v f^2(z)\,dz\right]^2 \\ &= 4\int_0^T dt \int_0^t dv \left[\int_0^v f^2(z)\,dz\right]^2 \end{aligned}$$

which is recognized to be the same as (21.62), the desired result.

21.10 REFERENCES

[1] R. Ash, *Information Theory*, Interscience, Wiley, New York, 1965.
[2] A. Papoulis, *Probability, Random Variables, and Stochastic Processes*, McGraw-Hill, New York, 1965.
[3] A. Papoulis, *The Fourier Integral and Its Applications*, McGraw-Hill, New York, 1962.
[4] A. Papoulis, *Signal Analysis*, McGraw-Hill, New York, 1977.
[5] E. Lukacs, *Characteristic Functions*, Griffin, London, 1970.

22

Maccone first KLT theorem: KLT of all time-rescaled Brownian motions

22.1 INTRODUCTION

In Section 21.4 the problem of finding the KL expansion of standard Brownian motion was solved completely. That was possible because the differential equation for the eigenfunctions was just a simple harmonic oscillator equation, whose solution is trivial.

In the present chapter we face the problem of finding the KL expansion for the time-rescaled Brownian motion defined in Chapter 21. This problem is rather involved from the analytical point of view, but we are going to show that it can be solved completely by resorting to an unusual type of Bessel function of the first kind whose order is not constant in time.[1]

Some mathematicians might thus be tempted to explore these new special functions more in depth. Physicists might do so for their applications, some of which will be brought to light in the forthcoming chapters.

22.2 SELF-ADJOINT FORM OF A SECOND-ORDER DIFFERENTIAL EQUATION

When we present the KL eigenfunction expansion of the $X(t)$ process in Section 22.3, the calculations will be lengthy. Therefore, it appears convenient to isolate within the present section the content and proof of a lemma that will later be used in Section 22.3. This lemma deals with the general self-adjoint form of a linear differential equation of the second order, and it will help putting a certain differential equation in Section 22.3 into its own self-adjoint form.

[1] The original results appearing in the present chapter were first published by the author in 1984 in [1].

Consider the most generic linear homogeneous ordinary differential equation of the second order with non-constant coefficients

$$A(t)\, y''(t) + B(t)y'(t) + C(t)y(t) = 0. \tag{22.1}$$

Since the $A(t)$ coefficient cannot vanish identically, we can divide the entire equation by it, illustrating that it actually has two independent coefficients rather than three:

$$y''(t) + \frac{B(t)}{A(t)}y'(t) + \frac{C(t)}{A(t)}y(t) = 0. \tag{22.2}$$

We are now going to prove that any differential equation of the form (22.2) may be put into its own self-adjoint form that reads

$$\frac{d}{dt}[P(t)y'(t)] + R(t)y(t) = 0 \tag{22.3}$$

which is the well-known *Sturm–Liouville* form of the equation. In other words, we must prove that the new coefficients $P(t)$ and $R(t)$ may uniquely be expressed in terms of the old coefficients $A(t)$, $B(t)$, and $C(t)$.

To this end, the derivative in (22.3) is expanded to give

$$P(t)y''(t) + P'(t)y'(t) + R(t)y(t) = 0 \tag{22.4}$$

and the equation divided by $P(t)$ (which cannot vanish identically) to give

$$y''(t) + \frac{P'(t)}{P(t)}y'(t) + \frac{R(t)}{P(t)}y(t) = 0. \tag{22.5}$$

By comparing (22.2) and (22.5), the following pair of simultaneous equations is found:

$$\begin{cases} \dfrac{B(t)}{A(t)} = \dfrac{P'(t)}{P(t)} = \dfrac{d}{dt}[\ln P(t)] & (22.6) \\[2ex] \dfrac{C(t)}{A(t)} = \dfrac{R(t)}{P(t)}. & (22.7) \end{cases}$$

Now (22.6) may be integrated at once, yielding the solution

$$\begin{cases} P(t) = e^{\int \frac{B(t)}{A(t)}\, dt} & (22.8) \\[2ex] R(t) = P(t)\dfrac{C(t)}{A(t)} = \dfrac{C(t)}{A(t)} e^{\int \frac{B(t)}{A(t)}\, dt} & (22.9) \end{cases}$$

These are the required formulas yielding $P(t)$ and $R(t)$ in terms of $A(t)$, $B(t)$, and $C(t)$, and the theorem is thus proved.

Let us now put this theorem to work by reducing the following differential equation to its own self-adjoint form:

$$\frac{\chi(t)}{f^2(t)}y_n''(t) + \left\{\frac{\chi'(t)}{f^2(t)} + \frac{d}{dt}\left[\frac{\chi(t)}{f^2(t)}\right]\right\}y_n'(t) + \left\{\frac{d}{dt}\left[\frac{\chi'(t)}{f^2(t)}\right] + \frac{\chi(t)}{\lambda_n}\right\}y_n(t) = 0. \tag{22.10}$$

Evidently, (22.10) coincides with (22.1) if we set

$$A(t) = \frac{\chi(t)}{f^2(t)} \tag{22.11}$$

$$B(t) = \frac{\chi'(t)}{f^2(t)} + \frac{d}{dt}\left[\frac{\chi(t)}{f^2(t)}\right] \tag{22.12}$$

$$C(t) = \frac{d}{dt}\left[\frac{\chi'(t)}{f^2(t)}\right] + \frac{\chi(t)}{\lambda_n} \tag{22.13}$$

and if the unknown functions $y_n(t)$ with $y(t)$ can be identified.

We now want to find the new pair of coefficients $P(t)$ and $Q(t)$ defined by (22.8) and (22.9), respectively. Dividing (22.12) by (22.11), we get

$$\frac{B(t)}{A(t)} = \frac{\dfrac{\chi'(t)}{f^2(t)} + \dfrac{d}{dt}\left[\dfrac{\chi(t)}{f^2(t)}\right]}{\dfrac{\chi(t)}{f^2(t)}}. \tag{22.14a}$$

The key to further steps lies in noticing that both terms are logarithmic derivatives. Hence

$$\begin{aligned} \frac{B(t)}{A(t)} &= \frac{\chi'(t)}{\chi(t)} + \frac{\dfrac{d}{dt}\left[\dfrac{\chi(t)}{f^2(t)}\right]}{\dfrac{\chi(t)}{f^2(t)}} \\ &= \frac{d}{dt}[\ln\chi(t)] + \frac{d}{dt}\left[\ln\left[\frac{\chi(t)}{f^2(t)}\right]\right] \\ &= \frac{d}{dt}\left[\ln\chi(t) + \ln\left[\frac{\chi(t)}{f^2(t)}\right]\right] \\ &= \frac{d}{dt}\left[\ln\left[\chi(t)\frac{\chi(t)}{f^2(t)}\right]\right] = \frac{d}{dt}\left[\ln\left[\frac{\chi^2(t)}{f^2(t)}\right]\right]. \end{aligned} \tag{22.14b}$$

By inserting (22.14b) into (22.8), we can now find the coefficient $P(t)$:

$$P(t) = e^{\int \frac{B(t)}{A(t)}\,dt} = e^{\int \frac{d}{dt}\left[\ln\left[\frac{\chi^2(t)}{f^2(t)}\right]\right]dt} = e^{\ln\left[\frac{\chi^2(t)}{f^2(t)}\right]} = \frac{\chi^2(t)}{f^2(t)}. \tag{22.15}$$

The coefficient $R(t)$ is also found by virtue of (22.9):

$$R(t) = \frac{C(t)}{A(t)}P(t) = \chi(t)\left\{\frac{d}{dt}\left[\frac{\chi'(t)}{f^2(t)}\right] + \frac{\chi(t)}{\lambda_n}\right\}. \tag{22.16}$$

In conclusion, the self-adjoint version of the differential equation (22.10) reads

$$\frac{d}{dt}\left[\frac{\chi^2(t)}{f^2(t)}y_n'(t)\right] + \chi(t)\left\{\frac{d}{dt}\left[\frac{\chi(t)}{f^2(t)}\right] + \frac{\chi(t)}{\lambda_n}\right\}y_n(t) = 0. \tag{22.17}$$

22.3 EXACT SOLUTION OF THE INTEGRAL EQUATION FOR KLT EIGENFUNCTIONS OF ALL BROWNIAN MOTIONS OF WHICH THE TIME IS NOT ELAPSING UNIFORMLY

In the present section we completely solve the problem of finding the KLT of any time-rescaled Brownian motion (i.e., of any Gaussian process, in practice).

Our proof develops in three subsequent steps that may be summarized as follows:

(1) The KLT integral equation (10.18), with the time-rescaled Brownian motion autocorrelation (21.30), turns out to be a Volterra-type (and not Fredholm-type) integral equation. In other words, it can be transformed into a linear differential equation of the second order in the unknown KLT eigenfunctions jointly with two boundary conditions on the eigenfunctions that we will soon discover: an initial condition and a final condition.

(2) Back in 1984 this author discovered (and published) the fact that this differential equation can actually be reduced to the standard Bessel differential equation by virtue of two analytical transformations: (i) suitably changing the unknown function and (ii) suitably changing the independent (time) variable. The latter time transformation is actually a general (non-linear) time-rescaling paving the way to the applications of our results in the theory of relativistic telecommunications.

(3) The general form of the KLT eigenfunctions for all the time-rescaled Brownian motions (i.e., Gaussian processes) is thus the product of a time-rescaled Bessel function of the first kind $J_{\nu(t)}(t)$ (where the order ν may itself depend on time, in some cases) multiplied by another time function that represents one more rescaling in time. As for the KLT eigenvalues, we will show that they are essentially the zeros of certain linear combinations of the $J_{\nu(t)}(t)$ and their derivatives.

Let us start from the KLT integral equation (10.18) with the autocorrelation (21.4)

$$\int_0^T \left[\int_0^{t_1\wedge t_2} f^2(s)\,ds\right]\phi_n(t_2)\,dt_2 = \lambda_n\phi_n(t_1). \tag{22.18}$$

A first consequence of (22.18) is easily found by setting $t_1 = 0$. In fact, because of the minimum $t_1 \wedge t_2 = 0 \wedge t_2 = 0$, the entire left-hand side of (22.18) vanishes, and one is left with

$$\phi_n(0) = 0. \tag{22.19}$$

This is the initial condition fulfilled by the eigenfunctions.

Let us now proceed toward the full solution to (22.18). A typical feature of the Volterra-type integral equations is that any such equation may be changed into a differential equation with two boundary conditions. For instance, in the case of (22.18), we can temporarily set

$$m(t_1) \equiv t_1 \wedge t_2 \tag{22.20}$$

to denote the minimum $t_1 \wedge t_2$ as a function of t_1. Then we can use the Leibniz theorem for the differentiantion of an integral to get

$$\frac{\partial}{\partial t_1}\int_0^{m(t_1)} f^2(s)\,ds = \int_0^{m(t_1)} \frac{\partial f^2(s)}{\partial t_1}\,ds + f^2(t_1)\frac{dm(t_1)}{dt_1} - f^2(0)\frac{d0}{dt_1}$$
$$= f^2(t_1)\frac{dm(t_1)}{dt_1} = f^2(t_1)\,U(t_2 - t_1) \tag{22.21}$$

where use of (21.36) was made in the last step. It follows that if both sides of (22.18) are differentiated with respect to t_1, (22.20) yields

$$\int_0^T \phi_n(t_2) f^2(t_1)\,U(t_2 - t_1)\,dt_2 = \lambda_n \phi_n'(t_1). \tag{22.22}$$

That is,

$$f^2(t_1)\int_{t_1}^T \phi_n(t_2)\,dt_2 = \lambda_n \phi_n'(t_1). \tag{22.23}$$

A glance at (22.23) shows that setting $t_1 = T$ results in

$$\phi_n'(T) = 0. \tag{22.24}$$

This is the final condition fulfilled by the eigenfunctions.

Let us now rewrite (22.23) in the form

$$\int_T^{t_1} \phi_n(t_2)\,dt_2 = -\lambda_n \frac{\phi_n'(t_1)}{f^2(t_1)}.$$

In order to let the integral disappear, we must differentiate both sides of this with respect to t_1. That finally yields, with some rearranging,

$$\frac{d}{dt}\left[\frac{1}{f^2(t)}\phi_n'(t)\right] + \frac{1}{\lambda_n}\phi_n(t) = 0. \tag{22.25}$$

This is the differential equation fulfilled by the eigenfunctions. It is already cast into its own self-adjoint form.

Having thus changed the integral equation (22.18) into the differential equation (22.25), we must now solve the latter subject to the boundary conditions (22.19) and (22.24). Now a great result comes. We discovered that it is possible to reduce (22.25) to the standard Bessel differential equation (see, e.g., [2, p. 4])

$$\frac{d}{dx}\left[x\frac{dy(x)}{dx}\right] + \left[x - \frac{\nu^2}{x}\right]y(x) = 0 \tag{22.26}$$

on replacing the eigenfunctions by means of a product of two unknown functions like

$$\phi_n(t) = \chi(t) y_n(t). \tag{22.27}$$

Then differentiating (22.27), and inserting it into (22.25), the latter is turned into

$$\frac{d}{dt}\left[\frac{\chi'(t)}{f^2(t)}y_n(t) + \frac{\chi(t)}{f^2(t)}y_n'(t)\right] + \frac{\chi(t)}{\lambda_n}y_n(t) = 0. \tag{22.28}$$

Performing differentiations with some rearranging of the terms, (22.28) becomes

$$\frac{\chi(t)}{f^2(t)}y_n''(t)+\left\{\frac{\chi'(t)}{f^2(t)}+\frac{d}{dt}\left[\frac{\chi(t)}{f^2(t)}\right]\right\}y_n'(t)+\left\{\frac{d}{dt}\left[\frac{\chi'(t)}{f^2(t)}\right]+\frac{\chi(t)}{\lambda_n}\right\}y_n(t)=0.$$

This differential equation is just (22.10), and, as we already know from Section 22.2, it may be cast into its own self-adjoint form given by (22.17)

$$\frac{d}{dt}\left[\frac{\chi^2(t)}{f^2(t)}y_n'(t)\right]+\chi(t)\left\{\frac{d}{dt}\left[\frac{\chi'(t)}{f^2(t)}\right]+\frac{\chi(t)}{\lambda_n}\right\}y_n(t)=0. \tag{22.29}$$

By so doing, we have performed one of the two allowed substitutions in any differential equation like (22.28), i.e., changing the unknown function. The only remaining change allowed by the theory is a change of the independent variable (i.e., time). Changing time means time rescaling. Thus, let us make this time rescaling $x=\psi(t)$ on the Bessel differential equation (22.26), turning it into

$$\frac{d}{dt}\left[\frac{\psi(t)}{\psi'(t)}y'(t)\right]+[\psi^2(t)-\nu^2]\frac{\psi'(t)}{\psi(t)}y(t)=0. \tag{22.30}$$

Now, the differential equations (22.29) and (22.30) must coincide in order to yield the solution $y_n(t)=$ Bessel function of $\psi_n(t)$ By equating the coefficients of each term, we get a pair of simultaneous equations:

$$\begin{cases}\dfrac{\chi^2(t)}{f^2(t)}=\dfrac{\psi_n(t)}{\psi_n'(t)} & (22.31)\\[2ex] \chi(t)\left\{\dfrac{d}{dt}\left[\dfrac{\chi'(t)}{f^2(t)}\right]+\dfrac{\chi(t)}{\lambda_n}\right\}=[\psi_n^2(t)-\nu^2]\dfrac{\psi_n'(t)}{\psi_n(t)}. & (22.32)\end{cases}$$

Our next task is the full solution of this pair of simultaneous equations, in terms of the only known function $f(t)$ Let us start by inspecting (22.31). Its left-hand side does not depend on the subscript (i.e., variable) n. We thus infer that the same thing must happen to the right-hand side of (22.31). Hence, we must have the following functional dependence:

$$\psi_n(t)=l_n\,\theta(t). \tag{22.33}$$

In other words, a new constant l_n has been introduced as well as a new time-dependent function $\theta(t)$ to perform the separation of the variables t and n. Let us now rewrite the simultaneous equations (22.31) and (22.32) by aid of (22.33) to get the new pair of simultaneous equations

$$\begin{cases}\dfrac{\chi^2(t)}{f^2(t)}=\dfrac{\theta(t)}{\theta'(t)} & (22.34)\\[2ex] \chi(t)\left\{\dfrac{d}{dt}\left[\dfrac{\chi'(t)}{f^2(t)}\right]+\dfrac{\chi(t)}{\lambda_n}\right\}=l_n^2\theta(t)\,\theta'(t)-\nu^2\dfrac{\theta'(t)}{\theta(t)}. & (22.35)\end{cases}$$

Once again, only two terms depend on n in (22.35), causing (22.35) to split into three more simultaneous equations. Adding these to the previous equation (22.34), we now get a set of four simpler simultaneous equations

$$\begin{cases} \dfrac{\chi^2(t)}{f^2(t)} = \dfrac{\theta(t)}{\theta'(t)} & (22.36) \\ \chi(t)\dfrac{d}{dt}\left[\dfrac{\chi'(t)}{f^2(t)}\right] = -\nu^2\dfrac{\theta'(t)}{\theta(t)} & (22.37) \\ \chi^2(t) = \theta(t)\theta'(t) & (22.38) \\ \dfrac{1}{\lambda_n} = l_n^2. & (22.39) \end{cases}$$

To solve this set, we eliminate $\chi^2(t)$ between (22.36) and (22.38), finding an easy differential equation in the unknown function $\theta(t)$

$$\theta(t)\,\theta'(t) = f^2(t)\frac{\theta(t)}{\theta'(t)}.$$

This differential equation may be solved at once by separation of variables, yielding the solution

$$\theta(t) = \int_0^t f(s)\,ds \tag{22.40}$$

where the plus sign must be taken in front of all square roots because the rescaled time is always positive. Equations (22.33), (22.39), and (22.40) then yield the $\psi_n(t)$ function

$$\psi_n(t) = \frac{1}{\sqrt{\lambda_n}}\int_0^t f(s)\,ds \tag{22.41}$$

while Equations (22.38) and (22.40) yield the expression

$$\chi(t) = \sqrt{f(t)\int_0^t f(s)\,ds} \tag{22.42}$$

for the $\chi(t)$ function. Finally, Equations (22.37) and (22.36) yield, for the order ν of the Bessel functions, the new time function

$$\nu(t) = \sqrt{-\frac{\chi^3(t)}{f^2(t)}\frac{d}{dt}\left[\frac{\chi'(t)}{f^2(t)}\right]}. \tag{22.43}$$

Only after several attempts was this author able to further transform this expression into an easier one. That will be presented in Section 22.4.

So far, we have been discussing Bessel functions without actually determining the precise kind that applies to the particular problem we are facing. We now do so by noting that in the standard Brownian motion case, described by the functions

$$\begin{cases} f(t) \equiv 1 \\ \chi(t) = \sqrt{t} \\ \nu(t) = +\dfrac{1}{2} \end{cases}$$

one gets just the classical sine eigenfunctions (21.20) if, and only if, Bessel functions of the first kind are chosen. Therefore, only Bessel functions of the first kind fit our problem, and only they will be retained among all the possible kinds of Bessel functions, making the solution to the differential equation (22.26) read

$$y_n(t) = J_{\nu(t)}(\psi_n(t)) = J_{\nu(t)}\left(\frac{1}{\sqrt{\lambda_n}}\int_0^t f(s)\,ds\right).$$

Next, we turn to the orthogonality property for the eigenfunctions $\phi_n(t)$

$$\begin{aligned} 0 &= \int_0^T \phi_m(t)\phi_n(t)\,dt = \int_0^T N_m\chi(t)y_m(t)\cdot N_n\chi(t)y_n(t)\,dt \\ &= N_m N_n \int_0^T dt\, f(t)\int_0^T ds\, f(s) J_{\nu(t)}\left(\frac{1}{\sqrt{\lambda_m}}\int_0^t f(s)\,ds\right) J_{\nu(t)}\left(\frac{1}{\sqrt{\lambda_n}}\int_0^t f(s)\,ds\right) \end{aligned} \tag{22.44}$$

holding for $m \neq n$. Note that the new, unknown normalization constants N_m had to be introduced here. We now wish to prove that (22.44) is just a disguised form of the orthogonality condition (holding for $m \neq n$)

$$\int_0^1 x\, J_\nu(g_m x)\, J_\nu(g_n x)\,dx = 0 \quad \text{for } m \neq n$$

which is known to be fulfilled by Bessel functions of the first kind $J_\nu(x)$. This is called the *Dini orthogonality condition* because it can be used to expand an arbitrary function over the set of Bessel functions $J_\nu(x)$, and this series expansion is called the "Dini series" because it was proved for the first time by the Italian mathematician Ulisse Dini of Pisa in 1877. For a thorough description of these topics in the theory of Bessel functions, the reader may wish to consult [2, p. 70, entry (48)].

We now substitute a new variable x for t in (22.44) by virtue of the equation

$$x\int_0^T f(s)\,ds = \int_0^t f(s)\,ds. \tag{22.45}$$

It is immediately seen that for $t \to 0$ then $x \to 0$, and for $t \to T$ then $x \to 1$. Differentianting (22.45), we get

$$dx \int_0^T f(s)\,ds = f(t)\,dt.$$

Hence

$$dt = \frac{\displaystyle\int_0^T f(s)\,ds}{f(t)}\,dx.$$

When this is replaced into (22.44), a "miracle" occurs (i.e., the time-rescaling function $f(t)$ is canceled out from the integrand). This is a hint that we are on the right track to reach the Dini orthogonality condition. In fact, by virtue of the new definition

$$\gamma_n = \frac{1}{\sqrt{\lambda_n}}\int_0^T f(s)\,ds \tag{22.46}$$

the orthogonality expressed by (22.44) becomes

$$\int_0^1 x\, J_{\nu((x))}(\gamma_m x)\, J_{\nu((x))}(\gamma_n x)\, dx = 0.$$

This is, indeed, the desired Dini orthogonality condition rewritten in the notation required by our needs.

Also, from the integral equation (22.18) it appears that the eigenfunctions $\phi_n(t)$ may be multiplied by the arbitrary constant called the normalization constant and denoted by N_n. The numerical values of these normalization constants may be established in a fully arbitrary fashion. However, it is customary to determine them so that the eigenfunctions have "unit length" in the Hilbert space; that is,

$$1 = \int_0^T \phi_n^2(t)\, dt = N_n^2 \left[\int_0^T f(s)\, ds\right]^2 \int_0^1 x\,[J_{\nu((x))}(\gamma_n x)]^2\, dx \tag{22.47}$$

where the new Bessel function of order $\nu((x))$ is given by (22.43) with the substitution (22.45). The normalization condition (22.47) is the case $m = n$ of the orthogonality condition (22.44), and the right-hand side of (22.47) is a definite integral that must be computed, either analytically or numerically, in order to determine the normalization constants N_n.

To complete the solution to the integral equation (22.18), we still need to find the eigenvalues λ_n. From (22.46) it follows that:

$$\lambda_n = \left[\int_0^T f(s)\, ds\right]^2 \frac{1}{(\gamma_n)^2}. \tag{22.48}$$

This formula establishes a one-to-one relationship between the eigenvalues λ_n and the unknown constants γ_n. Finding the λ_n thus means finding the γ_n, but how can that be done? A clue comes from consideration of the standard Brownian motion case developed in Section 21.3. There Equation (21.22) virtually determines the λ_n by virtue of the final condition (21.17), which is the same as (22.24). Since the initial condition (22.19) is identically fulfilled by any Bessel function of the first kind, we are forced to resort to the final condition (22.24). This final condition plus the expression (22.27) for the eigenfunctions $\phi_n(t)$ yield, after one differentiation

$$\chi'(T) J_{\nu(T)}(\gamma_n) + \chi(T)\left[\frac{f(T)\gamma_n}{\displaystyle\int_0^T f(s)\, ds} J'_{\nu(T)}(\gamma_n) + \frac{\partial J_{\nu(T)}(\gamma_n)}{\partial \nu}\nu'(T)\right] = 0. \tag{22.49}$$

This is a linear combination of the Bessel functions and their partial derivatives, whose zeros are the required γ_n. Finding an analytical expression for the zeros of (22.49) is, in general, impossible, and one must thus resort to a numerical solution of (22.49). However, there may exist some particular cases of the $f(t)$ function for which the zeros of (22.49) can be found, albeit in an approximated form. In Chapters 23–25 we will give examples of how that can be done.

22.4 A SIMPLER FORMULA FOR BESSEL FUNCTION ORDER

As mentioned before, the expression (22.43) for the order of Bessel functions

$$\nu(t) = \sqrt{-\frac{\chi^3(t)}{f^2(t)}\frac{d}{dt}\left[\frac{\chi'(t)}{f^2(t)}\right]}$$

is cumbersome because of the expression of (22.42)

$$\chi(t) = \sqrt{f(t)\int_0^t f(s)\,ds}$$

in terms of the only known function $f(t)$.

It would thus be desirable to have a single formula yielding the order $\nu(t)$ directly in terms of the $f(t)$ function, without invoking the $\chi(t)$ function. The present section is devoted to proving that such a formula reads

$$\nu(t) = \sqrt{\frac{1}{4} + \left[\frac{\int_0^t f(s)\,ds}{f(t)}\right]^2 \left\{\frac{3}{4}\left[\frac{d\ln f(t)}{dt}\right]^2 - \frac{1}{2}\frac{d^2\ln f(t)}{dt^2}\right\}}. \tag{22.50}$$

Since the proof is rather lengthy, we set, for convenience

$$f \equiv f(t) \tag{22.51}$$

$$f' \equiv \frac{df(t)}{dt} \tag{22.52}$$

$$f'' \equiv \frac{d^2f(t)}{dt^2} \tag{22.53}$$

$$\int \equiv \int_0^t f(s)\,ds. \tag{22.54}$$

Then, from (22.42) it follows that:

$$\ln\chi(t) = \frac{1}{2}\ln f + \frac{1}{2}\ln\int \tag{22.55}$$

$$\frac{d\ln\chi(t)}{dt} = \frac{1}{2}\frac{f'}{f} + \frac{1}{2}\frac{f}{\int} = \frac{f'\int + f^2}{2f\int} \tag{22.56}$$

$$\frac{d^2\ln\chi(t)}{dt^2} = \frac{f''f\int^2 - f'^2\int^2 + f'f^2\int - f^4}{2f^2\int^2} \tag{22.57}$$

The order $\nu(t)$ of (22.43), by virtue of (22.56) and (22.57), now reads

$$\begin{aligned}
\nu^2(t) &= -\frac{\chi^3(t)}{f^2(t)}\frac{d}{dt}\left[\frac{\chi'(t)}{f^2(t)}\right] \\
&= -\frac{\chi^3(t)}{f^2(t)}\frac{d}{dt}\left[\frac{\chi(t)}{f^2(t)}\frac{d\ln\chi(t)}{dt}\right] \\
&= -\frac{\chi^3(t)}{f^2(t)}\left\{\frac{d}{dt}\left[\frac{\chi(t)}{f^2(t)}\right]\frac{d\ln\chi(t)}{dt} + \frac{\chi(t)}{f^2(t)}\frac{d^2\ln\chi(t)}{dt^2}\right]\right\} \\
&= -\frac{1}{2}\frac{d}{dt}\left[\left(\frac{\chi(t)}{f^2(t)}\right)^2\right]\frac{f'\int + f^2}{2} - \frac{f''f\int^2 - f'^2\int^2 + f'f^2\int - f^4}{2f^4} \\
&= -\frac{1}{2}\frac{d}{dt}\left[\frac{\int\int}{f^3}\right]\frac{f'\int + f^2}{2} - \frac{f''f\int^2 - f'^2\int^2 + f'f^2\int - f^4}{2f^4} \\
&= -\frac{1}{4f^4}\left[-f^4 - 5f'^2\int^2 + 2f''f\int^2\right] \\
&= \frac{1}{4} + \frac{3}{4}\left[\frac{f'\int}{f^2}\right]^2 + \frac{1}{2}\left[\frac{f'}{f}\right]^2\left[\frac{\int}{f}\right]^2 - \frac{1}{2}\frac{f''}{f}\left[\frac{\int}{f}\right]^2 \\
&= \frac{1}{4} + \frac{3}{4}\left[\frac{f'\int}{f^2}\right]^2 - \frac{1}{2}\left[\frac{\int}{f}\right]^2\left[\frac{f''f - f'^2}{f^2}\right] \\
&= \frac{1}{4} + \frac{3}{4}\left[\frac{\int}{f}\right]^2\left[\frac{f'}{f}\right]^2 - \frac{1}{2}\left[\frac{\int}{f}\right]^2\frac{d^2\ln f(t)}{dt^2} \\
&= \frac{1}{4} + \left[\frac{\int}{f}\right]^2\left\{\frac{3}{4}\left[\frac{d\ln f(t)}{dt}\right]^2 - \frac{1}{2}\frac{d^2\ln f(t)}{dt^2}\right\}
\end{aligned} \tag{22.58}$$

and (22.50) is proved.

22.5 STABILITY CRITERION FOR EIGENFUNCTIONS

An amazing feature of eigenfunctions $\phi_n(t)$ is that their behavior in time may be predicted even without knowing their actual analytical expression, as we prove in the present section. The starting point is the *Sonine–Pólya theorem* (see [2, p. 205]), which states that, if in the differential equation

$$\frac{d}{dt}[K(x)y'(x)] + \phi(x)y(x) = 0 \tag{22.59}$$

Table 22.1. Stability criterion for eigenfunctions $\phi_n(t)$.

Sign of the logarithmic derivative of $f(t)$	Shape of the KL eigenfunctions $\phi_n(t)$	Description when T is finite	Description when T is infinite
Positive	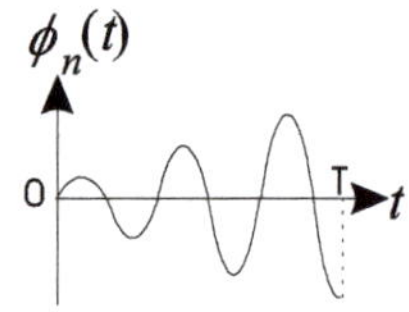	Divergent	Asymptotic unstable
Zero	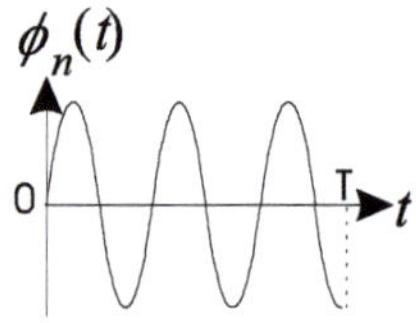	Sine/cosine type	Simply stable
Negative	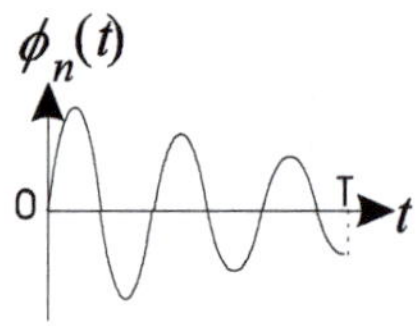	Convergent	Asymptotic stable

$K(x)$ and $\phi(x)$ are positive and continuously differentiable, and if $K(x)\phi(x)$ is monotonic, then the successive (relative) maxima of $|y(x)|$ form an increasing or decreasing sequence according as $K(x)\phi(x)$ is decreasing or increasing.

We are going to apply this theorem to the differential equation (22.25), which reads

$$\frac{d}{dt}\left[\frac{1}{f^2(t)}\phi_n'(t)\right] + \frac{1}{\lambda_n}\phi_n(t) = 0.$$

By checking this against (22.59), one evidently finds the correspondences

$$\begin{cases} x = t \\ y(x) = \phi_n(t) \\ K(x) = \dfrac{1}{f^2(t)} > 0 \\ \phi(x) = \dfrac{1}{\lambda_n} = \left[\dfrac{\gamma_n}{\displaystyle\int_0^T f(s)\,ds}\right]^2 = \text{constant} > 0 \end{cases} \tag{22.60}$$

where (22.48) was used in the last formula of (22.60). We thus see that the requirements of the Sonine–Pólya theorem are fulfilled only if $f^2(t)$ is assumed to be monotonic, which will indeed be the case in all physical applications we are going to consider in the book.

Now, it is important to know whether $K(x)\phi(x)$ is decreasing or increasing—namely, whether its derivative is negative or positive. With easy steps from (22.60), it follows that

$$\begin{aligned}\frac{d}{dt}\left[\frac{1}{f^2(t)}\frac{1}{\lambda_n}\right] &= \frac{1}{\lambda_n}\frac{-2f(t)f'(t)}{f^4(t)} \\ &= -\frac{2}{\lambda_n f^2(t)}\frac{f'(t)}{f(t)} = \begin{pmatrix}\text{negative}\\ \text{quantity}\end{pmatrix}\frac{d\ln f(t)}{dt}.\end{aligned} \tag{22.61}$$

Combining this with the Sonine–Pólya theorem statement, we get the scheme shown in Table 22.1. In Chapter 11 we extended this stability criterion to the theory of relativity: see Table 11.1.

22.6 REFERENCES

[1] C. Maccone, "Eigenfunctions and Energy for Time-Rescaled Gaussian Processes," *Bollettino dell'Unione Matematica Italiana*, Series 6, **3-A** (1984), 212–219.

[2] A. Erdélyi, W. Magnus, F. Oberhettinger, and F. G. Tricomi, *Higher Transcendental Functions*, Vol. 2, McGraw-Hill, New York, 1952.

23

KLT of the $B(t^{2H})$ time-rescaled Brownian motion

23.1 INTRODUCTION

The topics considered in the present chapter are twofold: on the one hand, they can be regarded as a particular application of the results obtained in Chapters 21–22 to a case that allows analytic calculations to be easily carried through to completion; on the other hand, new light is shed on the theory of certain H-self-similar stochastic processes, in the wake of the celebrated results obtained by Benoit B. Mandelbrot in his theory of fractals.

23.2 TIME-RESCALED BROWNIAN MOTION $B(t^{2H})$

Consider the process $B_{PH}(t)$—where subscript PH refers to the "power H"—defined by the white noise integral

$$B_{PH}(t) \equiv \int_0^t \frac{s^{H-\frac{1}{2}}}{\Gamma\left(H+\frac{1}{2}\right)} dB(s). \tag{23.1}$$

By setting

$$f(t) = \frac{t^{H-\frac{1}{2}}}{\Gamma\left(H+\frac{1}{2}\right)} \tag{23.2}$$

the definite integral of the square of this function from 0 to T is given by

$$\int_0^t f^2(s)\, ds = \int_0^t \frac{s^{2H-1}}{\Gamma^2\left(H+\frac{1}{2}\right)} ds = \frac{t^{2H}}{2H\,\Gamma^2\left(H+\frac{1}{2}\right)}.$$

Equation (21.40) shows that (23.1) is the same as time-rescaled Brownian motion

$$B_{PH}(t) \equiv B\left(\frac{t^{2H}}{2H\,\Gamma^2\left(H+\frac{1}{2}\right)}\right). \tag{23.3}$$

This formula, by virtue of the self-similarity to the order 1/2 expressed by (21.6), namely

$$B(ct) = \sqrt{c}B(t), \quad c \geq 0 \tag{23.4}$$

becomes

$$B_{PH}(t) \equiv \frac{1}{\sqrt{2H}\,\Gamma\left(H+\frac{1}{2}\right)} B(t^{2H}), \quad H \geq 0. \tag{23.5}$$

We thus see that $B_{PH}(t)$ is essentially a new Brownian motion whose time variable does not elapse uniformly. Rather, it is accelerated t^{2H}-like for $H > 1/2$, and decelerated t^{2H}-like for $H < 1/2$.

The application of (23.4) to (23.5) immediately yields the important self-similarity property

$$B_{PH}(ct) = c^H B_{PH}(t), \quad c \geq 0 \tag{23.6}$$

which is called the *self-similarity to the order H*, or *H-self-similarity of* $B_{PH}(ct)$—again, the subscript *PH* ("power *H*") in our notation reminds us of this.

Let us now consider two more processes:

$$B_{LH}(t) \equiv \int_0^t \frac{(t-s)^{H-\frac{1}{2}}}{\Gamma\left(H+\frac{1}{2}\right)} dB(s) \tag{23.7}$$

and

$$B_{WH}(t) \equiv \int_{-\infty}^t \frac{(t-s)^{H-\frac{1}{2}}}{\Gamma\left(H+\frac{1}{2}\right)} dB(s), \tag{23.8}$$

both of which are H-self-similar (as can easily be proved):

$$B_{LH}(ct) = c^H B_{LH}(t), \tag{23.9}$$

$$B_{WH}(ct) = c^H B_{WH}(t). \tag{23.10}$$

The process $B_{LH}(t)$ is the Riemann–Liouville fractional integral to the order $H - 1/2$ of Brownian motion: it is an integral for $H > 1/2$ and a derivative for $H < 1/2$ (see 1, p. 115], [2, Vol. 2, p. 181], and [3, the whole book][1]). The definition (23.7) was given in 1953 by the French mathematician Paul Lévy [4, p. 357], who confined himself to finding the process variance without further investigations. The process

[1] However, only fractional integrals of functions—and not of stochastic processes—are considered in these works.

$B_{WH}(t)$ is the Weyl fractional integral to the order $H - 1/2$ of Brownian motion, and was first considered in 1940 by Kolmogorov [5]. In 1965 Mandelbrot used its self-similarity to account for a hydrological law discovered in 1949 by Hurst [6]. Ever since, Mandelbrot and co-workers were mainly responsible for developing the computer applications of the process $B_{WH}(t)$ (notably in [7]). An excellent list of references to this and related topics, updated to 1982, can be found in Mandelbrot's book about fractals [8]. It should also be noted that a detailed analytical study of both processes (23.7) and (23.8) offers considerable difficulties. It is possible to show that (23.1) may be regarded as the first-order approximation to both (23.7) and (23.8). As a result of this we believe our results are interesting, particularly with regard to the energy distribution of the process $B_{PH}(t)$, the study of which can be carried on with a considerable amount of details (Section 23.4) by exploiting the KLT of $B_{PH}(t)$ that we derive in the next section.

23.3 KL EXPANSION OF $B_{PH}(t)$

We are now going to derive the KL expansion for $B_{PH}(t)$ as in (23.1). Evidently, we are dealing with the special case of $f(t)$ given by

$$f(t) = \frac{t^{H-\frac{1}{2}}}{\Gamma\left(H + \frac{1}{2}\right)} \tag{23.11}$$

Standard Brownian motion corresponds to the special case $H = 1/2$ of all results to follow, because at the denominator of (23.11) one then uses the formula $\Gamma(1) = 1$.

Let us first consider autocorrelation and variance. From (23.11) and (21.30) integration easily yields the required autocorrelation of $B_{PH}(t)$

$$E\{B_{PH}(t_1)B_{PH}(t_2)\} = \int_0^{t_1 \wedge t_2} f^2(s)\, ds = \frac{(t_1 \wedge t_2)^{2H}}{2H\,\Gamma^2\left(H + \frac{1}{2}\right)}. \tag{23.12}$$

Substituting $t_1 = t_2 = t$ into (23.12) immediately gives the variance

$$\sigma^2_{B_{PH}(t)} = \frac{t^{2H}}{2H\,\Gamma^2\left(H + \frac{1}{2}\right)}. \tag{23.13}$$

Unfortunately, a much longer proof is required to derive the following explicit formula for the KL expansion of $B_{PH}(t)$:

$$B_{PH}(t) \equiv X(t) = (2H+1) \cdot \frac{t^H}{T^{H+\frac{1}{2}}} \cdot \sum_{n=1}^{\infty} Z_n \frac{1}{|J_\nu(\gamma_n)|} J_\nu\left(\gamma_n \frac{t^{H+\frac{1}{2}}}{T^{H+\frac{1}{2}}}\right). \tag{23.14}$$

Here

(1) The order ν of the Bessel functions $J_\nu(\ldots)$ is given by

$$\nu = \frac{2H}{2H+1}. \tag{23.15}$$

Hence it does not depend on time, and this circumstance is vital to all forthcoming developments because it simplifies things greatly.

(2) The exact constants γ_n are the real positive zeros of the Bessel function of order $\nu - 1$ arranged in ascending order of magnitude

$$J_{\nu-1}(\gamma_n) = 0. \tag{23.16}$$

(3) The exact normalization constants N_n are given by

$$N_n = \frac{\left(H+\frac{1}{2}\right)\Gamma\left(H+\frac{1}{2}\right)\sqrt{2}}{T^{H+\frac{1}{2}}|J_\nu(\gamma_n)|}. \tag{23.17}$$

(4) The exact eigenvalues λ_n depend on the γ_n according to

$$\lambda_n = \frac{T^{2H+1}}{\left(H+\frac{1}{2}\right)^2\Gamma^2\left(H+\frac{1}{2}\right)}\frac{1}{\gamma_n^2}. \tag{23.18}$$

(5) The Z_n are Gaussian random variables with mean zero and variance λ_n.

To start the proof, let us consider the function $\chi(t)$, defined by (22.42), which by virtue of (23.11) takes the form

$$\chi(t) = \frac{t^H}{\sqrt{H+\frac{1}{2}\,\Gamma\left(H+\frac{1}{2}\right)}}. \tag{23.19}$$

Then (22.43), or, equivalently, (22.50), yields the order of the Bessel functions, which after some calculations is seen to be given exactly by (23.15). Since the latter is a constant, we have

$$\nu'(t) = 0, \tag{23.20}$$

so that the term with $\nu'(T)$ in (22.49) vanishes

$$\chi'(T)\cdot J_{\nu(T)}(\gamma_n) + \chi(T)\frac{f(T)\cdot\gamma_n}{\int_0^T f(s)\,ds}J'_{\nu(T)}(\gamma_n) = 0. \tag{23.21}$$

Inserting (23.15) and (23.19) into (23.21), after a few steps one finds

$$\frac{2H}{2H+1}J_\nu(\gamma_n) + \gamma_n J'_\nu(\gamma_n) = 0 \tag{23.22}$$

which can be shown to be (see [2, p. 11, entry (54)]) the same as

$$\gamma_n J_{\nu-1}(\gamma_n) = 0. \tag{23.23}$$

The γ_n cannot vanish, so (23.16) is proved. Finally, (23.17) for the normalization constants N_n and (23.18) for the eigenvalues λ_n follow from (22.47) and (22.48), respectively, on replacing the time-rescaling function (23.11) and integrating, as we shall prove in a moment.

But, let us go back to (23.23), which states that the γ_n are the real positive zeros, arranged in ascending order of magnitude, of the Bessel function of order $\nu - 1$. No formula explicitly yielding these zeros exactly is known. Yet it is possible to find an approximated expression for them by aid of the asymptotic formula for $J_\nu(x)$

$$\lim_{x\to\infty} J_\nu(x) = \lim_{x\to\infty} \sqrt{\frac{2}{\pi x}} \cos\left(x - \frac{\nu\pi}{2} - \frac{\pi}{4}\right). \tag{23.24}$$

In fact, from (23.15) one first gets

$$\nu - 1 = -\frac{1}{2H+1}. \tag{23.25}$$

Second, (23.24) and (23.25), checked against (23.23), yield

$$0 = J_{\nu-1}(\gamma_n) \approx \sqrt{\frac{2}{\pi\gamma_n}} \cos\left(\gamma_n + \frac{\pi}{2(2H+1)} - \frac{\pi}{4}\right). \tag{23.26}$$

Hence

$$\gamma_n + \frac{\pi}{2(2H+1)} - \frac{\pi}{4} \approx n\pi - \frac{\pi}{2} \quad (n = 1, 2, \ldots) \tag{23.27}$$

and finally

$$\gamma_n \approx n\pi - \frac{\pi}{4} - \frac{\pi}{2(2H+1)} \quad (n = 1, 2, \ldots). \tag{23.28}$$

The first 32 approximated γ_n, obtained by means of (23.28), appear in Table 23.1, for various values of $H \geq 1/2$. In the Brownian case $H = 1/2$, (23.28) is an exact formula. We recall that these γ_n give the pace of convergence of the KL expansion, inasmuch as the standard deviations of the Gaussian random variables Z_n depend inversely on the squares of the γ_n by virtue of (23.18).

Next we want to prove (23.17) for the normalization constants N_n. From (22.47) and (23.11) it follows that:

$$1 = N_n^2 \cdot \frac{T^{2H+1}}{\left(H + \frac{1}{2}\right)^2 \Gamma^2\left(H + \frac{1}{2}\right)} \int_0^1 x J_\nu^2(\gamma_n x)\, dx. \tag{23.29}$$

This integral is calculated within the framework of the Dini series, and the result is

$$\int_0^1 x J_\nu^2(\gamma_n x)\, dx = \frac{1}{2\gamma_n^2}\left[\gamma_n^2 J_\nu'^2(\gamma_n) + (\gamma_n^2 - \nu^2) J_\nu^2(\gamma_n)\right]. \tag{23.30}$$

Table 23.1. Approximate values of the constants γ_n.

	$H = 0.5$ Brownian	$H = 0.6$	$H = 0.7$	$H = 0.8$	$H = 0.9$	$H = 1.0$	$H = \infty$
$n = 1$	1.571	1.642	1.702	1.752	1.795	1.833	2.356
$n = 2$	4.712	4.784	4.843	4.894	4.937	4.974	5.498
$n = 3$	7.854	7.925	7.985	8.035	8.078	8.116	8.639
$n = 4$	11.00	11.07	11.13	11.18	11.22	11.26	11.78
$n = 5$	14.14	14.21	14.27	14.32	14.37	14.40	14.92
$n = 6$	17.28	17.36	17.41	17.46	17.50	17.54	18.06
$n = 7$	20.42	20.50	20.55	20.60	20.64	20.68	21.20
$n = 8$	23.56	23.63	23.69	23.74	23.79	23.82	24.35
$n = 9$	26.70	26.77	26.83	26.88	26.93	26.96	27.49
$n = 10$	27.84	27.92	27.98	30.03	30.07	30.11	30.63
$n = 11$	32.99	33.06	33.12	33.17	33.21	33.25	33.77
$n = 12$	36.13	36.20	36.26	36.31	36.35	36.39	36.91
$n = 13$	37.27	37.34	37.40	37.45	37.49	37.53	40.05
$n = 14$	42.41	42.48	42.54	42.59	42.64	42.67	43.20
$n = 15$	45.55	45.62	45.68	45.73	45.78	45.81	46.34
$n = 16$	48.69	48.77	48.83	48.88	48.92	48.96	47.48
$n = 17$	51.84	51.91	51.97	52.02	52.06	52.10	52.62
$n = 18$	54.98	55.05	55.11	55.16	55.20	55.24	55.76
$n = 19$	58.12	58.19	58.25	58.30	58.34	58.38	58.90
$n = 20$	61.26	61.33	61.39	61.44	61.48	61.52	62.05
$n = 21$	64.40	64.47	64.53	64.58	64.63	64.66	65.19
$n = 22$	67.54	67.62	67.67	67.72	67.77	67.81	68.33
$n = 23$	70.69	70.76	70.82	70.87	70.91	70.95	71.47
$n = 24$	73.83	73.90	73.96	74.01	74.05	74.09	74.61
$n = 25$	76.97	77.04	77.10	77.15	77.19	77.23	77.75
$n = 26$	80.11	80.18	80.24	80.29	80.33	80.37	80.90
$n = 27$	83.25	83.32	83.38	83.43	83.48	83.51	84.04
$n = 28$	86.39	86.46	86.52	86.57	86.62	86.66	87.18
$n = 29$	87.53	87.61	87.67	87.72	87.76	87.80	90.32
$n = 30$	92.68	92.75	92.81	92.86	92.90	92.94	93.46
$n = 31$	95.82	95.90	95.95	96.00	96.04	96.08	96.60
$n = 32$	98.96	97.0	97.0	97.1	97.1	97.2	97.75

This formula, however, may be greatly simplified by eliminating $\gamma_n J'_\nu(\gamma_n)$ from (23.22). In fact, one finds

$$\gamma_n^2 J'^2_\nu(\gamma_n) = \nu^2 J^2_\nu(\gamma_n) \tag{23.31}$$

and (23.29), by virtue of (23.31), becomes

$$1 = N_n^2 \frac{T^{2H+1}}{\left(H+\frac{1}{2}\right)^2 \Gamma^2\left(H+\frac{1}{2}\right)} \cdot \frac{J_\nu^2(\gamma_n)}{2}. \tag{23.32}$$

Thus

$$N_n = \frac{\left(H+\frac{1}{2}\right)\Gamma\left(H+\frac{1}{2}\right)\sqrt{2}}{T^{H+\frac{1}{2}}|J_\nu(\gamma_n)|}. \tag{23.33}$$

This is the exact expression of the normalization constants. An approximated expression can be found on inserting both (23.28) for the approximated γ_n and (23.15) for the exact ν into the asymptotic expansion (23.24) for $J_\nu(\gamma_n)$

$$\begin{aligned} |J_\nu(\gamma_n)| &\approx \left|\sqrt{\frac{2}{\pi\gamma_n}} \cos\left(n\pi - \frac{\pi}{4} - \frac{\pi}{2(2H+1)} - \frac{\pi 2H}{2(2H+1)} - \frac{\pi}{4}\right)\right| \\ &= \sqrt{\frac{2}{\pi\gamma_n}} |\cos(n\pi - \pi)| = \sqrt{\frac{2}{\pi\gamma_n}}. \end{aligned}$$

By substituting this into the exact expression (23.33) for N_n and using the approximated expression (23.28) for the γ_n, it follows that:

$$N_n \approx \frac{\pi}{T^{H+\frac{1}{2}}} \left(H+\frac{1}{2}\right) \Gamma\left(H+\frac{1}{2}\right) \sqrt{n - \frac{1}{4} - \frac{1}{2(2H+1)}}. \tag{23.34}$$

These are the approximated normalization constants.

A similar procedure applies to eigenvalues λ_n. In fact, from (22.48) and (23.11) we get the exact formula

$$\lambda_n = \frac{T^{2H+1}}{\left(H+\frac{1}{2}\right)^2 \Gamma^2\left(H+\frac{1}{2}\right)} \cdot \frac{1}{(\gamma_n)^2}. \tag{23.35}$$

Finally from (23.35) and (23.28) we get the approximated formula

$$\lambda_n \approx \frac{T^{2H+1}}{\left(H+\frac{1}{2}\right)^2 \Gamma^2\left(H+\frac{1}{2}\right)} \cdot \frac{1}{\pi^2\left(n - \frac{1}{4} - \frac{1}{2(2H+1)}\right)^2}. \tag{23.36}$$

These are the variances of the independent Gaussian random variables Z_n.

We may now summarize all the results found in the present section by writing two KL expansions: the exact one

$$X(t) = \frac{(2H+1)\,t^H}{T^{H+\frac{1}{2}}} \sum_{n=1}^{\infty} Z_n \frac{1}{|J_\nu(\gamma_n)|} J_\nu\left(\gamma_n \frac{t^{H+\frac{1}{2}}}{T^{H+\frac{1}{2}}}\right) \tag{23.37}$$

and the approximated one

$$X(t) = \frac{\sqrt{2}\left(H+\frac{1}{2}\right)t^{\frac{H}{2}-\frac{1}{4}}}{T^{\frac{H}{2}+\frac{1}{4}}} \sum_{n=1}^{\infty} Z_n \sin\left(\gamma_n \frac{t^{H+\frac{1}{2}}}{T^{H+\frac{1}{2}}} - \frac{2\pi H}{2(2H+1)} - \frac{\pi}{4}\right). \tag{23.38}$$

23.4 TOTAL ENERGY OF $B_{PH}(t)$

In 1944 Cameron and Martin [9] proved that the random variable

$$I = \int_0^T B^2(t)\,dt, \tag{23.39}$$

which is the total energy of standard Brownian motion, has the characteristic function (i.e., Fourier transform)

$$\Phi_I(\zeta) = E\{e^{i\zeta I}\} = \frac{1}{\sqrt{\cos(T\sqrt{2i\zeta})}.} \tag{23.40}$$

We are now going to prove that the above result is generalized by our following result: the total energy of $B_{PH}(t)$ given, in analogy to (23.39), by the random variable

$$I_H = \int_0^T B_{PH}^2(t)\,dt \tag{23.41}$$

has the characteristic function

$$\Phi_{I_H}(\zeta) = \frac{1}{\sqrt{\Gamma(\nu)\left[\dfrac{T^{H+\frac{1}{2}}\sqrt{2i\zeta}}{2\Gamma\left(H+\frac{3}{2}\right)}\right]^{1-\nu} \cdot J_{\nu-1}\left(\dfrac{T^{H+\frac{1}{2}}\sqrt{2i\zeta}}{\Gamma\left(H+\frac{3}{2}\right)}\right)}}. \tag{23.42}$$

To prove (23.42), consider (21.51), where the λ_n are now given by (23.18) and the γ_n by (23.23). That is,

$$\Phi_{I_H}(\zeta) = \prod_{n=1}^{\infty}(1-2i\lambda_n\zeta)^{-\frac{1}{2}} = \left[\prod_{n=1}^{\infty}\left(1 - \frac{T^{2H+1}2i\zeta}{\gamma_n^2\Gamma^2\left(H+\frac{3}{2}\right)}\right)\right]^{-\frac{1}{2}}.$$

Let us check it against the infinite product expressing $J_\nu(z)$

$$J_\nu(z) = \frac{\left(\frac{z}{2}\right)^\nu}{\Gamma(\nu+1)} \prod_{n=1}^{\infty} \left(1 - \frac{z^2}{j_{\nu,n}^2}\right)$$

whose proof appears on p. 498 of [10]. On the one hand, the constants $j_{\nu,n}$ in the last equation are evidently the real positive zeros of $J_\nu(z)$, arranged in ascending order of magnitude, while, on the other hand, the γ_n are the zeros of $J_{\nu-1}(x)$ by (23.23). Thus, we can let the last two infinite products coincide by setting

$$\begin{cases} \gamma_n = j_{\nu-1,n} \\ z^2 = \dfrac{T^{2H+1} 2i\zeta}{\Gamma^2\left(H + \frac{3}{2}\right)}. \end{cases} \tag{23.43}$$

Equation (23.42) for the characteristic function of the total energy of $B_{PH}(t)$ is thus proved. It can be inverted numerically by a computer so as to yield the total energy distribution of $B_{PH}(t)$ to any degree of accuracy.

As for the the cumulants (and hence also the moments) of the total energy distribution of $B_{PH}(t)$, they are given by the expression for K_n in (21.56) with $f(t)$ given by (23.11)

$$K_n = 2^{n-1}(n-1)! \frac{T^{n(2H+1)}}{\left[\Gamma\left(H+\frac{3}{2}\right)\right]^{2n}} \sum_{m=1}^{\infty} \frac{1}{(\gamma_m)^{2n}}. \tag{23.44}$$

Moreover, although an explicit expression for the γ_n (other than the approximated (23.28)) is unknown, it is possible to sum the series appearing in (23.44). In fact, we are now going to prove that all the cumulants of the total energy distribution of $B_{PH}(t)$ are given by

$$\begin{aligned} K_n &= \frac{2^{n-1} T^{n(2H+1)}}{\left[\Gamma\left(H+\frac{3}{2}\right)\right]^{2n}} \cdot \frac{(n-1)!}{(2n-1)!} \lim_{x\to 0^+} \left[\frac{d^{2n-1}}{dx^{2n-1}} \left(\frac{J_\nu(x)}{2J_{\nu-1}(x)}\right)\right] \\ &= \frac{2^{n-1} T^{n(2H+1)}}{\left[\Gamma\left(H+\frac{3}{2}\right)\right]^{2n}} \cdot (n-1)! \cdot \sigma_{\nu-1}^{(n)} \end{aligned} \tag{23.45}$$

where the quantities $\sigma_\nu^{(1)}$, $\sigma_\nu^{(2)}$, $\sigma_\nu^{(3)}$, $\sigma_\nu^{(4)}$, $\sigma_\nu^{(5)}$, and $\sigma_\nu^{(8)}$ appear on p. 502 of [10]. ν is to be replaced by H via (23.15). In fact, consider the series

$$\sum_{n=1}^{\infty} \frac{1}{(\gamma_n)^{2k}} \equiv S_{2k,\nu-1} \equiv \sigma_{\nu-1}^{(k)} \tag{23.46}$$

where the notation $S_{2k,\nu-1}$ is used on p. 61 of [11], while the notation $\sigma_{\nu-1}^{(k)}$ is used on p. 502 of [10]. Then

$$\sum_{k=1}^{\infty} S_{2k,\nu-1} x^{2k-1} = \frac{J_\nu(x)}{2J_{\nu-1}(x)} \tag{23.47}$$

is the power series in x, with coefficients $S_{2k,\nu-1}$, whose proof is given on p. 61 of [11]. Therefore, the coefficients are

$$S_{2k,\nu-1} = \frac{1}{(2k-1)!} \lim_{x \to 0^+} \left[\frac{d^{2k-1}}{dx^{2k-1}} \left(\frac{J_\nu(x)}{2J_{\nu-1}(x)} \right) \right] \tag{23.48}$$

and the sum of the series (23.46) is obtained.

Having found all the cumulants, we can derive expressions of the most interesting statistical parameters of total energy.

(1) Mean value of total energy

$$K_1 = E\{\varepsilon\} = \frac{T^{2H+1}}{2H(2H+1)}. \tag{23.49}$$

(2) Variance of total energy

$$K_2 = \sigma_\varepsilon^2 = \frac{T^{4H+2}}{2H^2(2H+1)(4H+1)}. \tag{23.50}$$

(3) Third total energy cumulant

$$K_3 = \frac{T^{3(2H+1)}}{H^3(2H+1)(3H+1)(4H+1)}. \tag{23.51}$$

(4) Fourth total energy cumulant

$$K_4 = \frac{3(11H+3)T^{4(2H+1)}}{H^4(2H+1)(3H+1)(4H+1)^2(8H+3)}. \tag{23.52}$$

(5) Skewness of total energy distribution

$$\frac{K_3}{(K_2)^{\frac{3}{2}}} = \frac{2^{\frac{3}{2}}\sqrt{2H+1}\sqrt{4H+1}}{3H+1}. \tag{23.53}$$

(6) Kurtosis (or excess) of total energy distribution

$$\frac{K_4}{(K_2)^2} = \frac{12(2H+1)(11H+3)}{(3H+1)(8H+3)}. \tag{23.54}$$

Since $H \geq 0$ we infer from (23.53) that skewness is at least 2.828, and from (23.54) that kurtosis ranges from 12 for $H = 0$ to 11 for $H \to \infty$. Therefore, we may conclude that the total energy peak is narrow for any $H \geq 0$.

The standard Brownian motion case of all the previous results is noteworthy. In fact, by replacing $H = 1/2$, $\nu = 1/2$, and the expression of both $J_{\frac{1}{2}}(x)$ and $J_{-\frac{1}{2}}(x)$ appearing on pp. 54 and 55 of [10], one then gets

$$K_n = 2^{n-2} T^{2n} (n-1)! \frac{1}{(2n-1)!} \lim_{x \to 0^+} \left[\frac{d^{2n-1} \tan x}{dx^{2n-1}} \right]. \tag{23.55}$$

Evidently, the last two terms are the $(2n-1)$th coefficient in the MacLaurin expansion of $\tan x$, which reads [11, Vol. 1, p. 51]

$$\tan x = \sum_{n=1}^{\infty} \frac{1}{(2n)!} 2^{2n} (2^{2n} - 1)(-1)^{n+1} B_{2n} x^{2n-1} \tag{23.56}$$

where the B_{2n} are Bernoulli numbers, a table of which can be found, for instance, in [12, p. 810]. Thus, by inserting the coefficients of (23.56) into (23.55), we get all the cumulants of the total energy of standard Brownian motion

$$K_n = T^{2n} \frac{(n-1)!}{(2n)!} 2^{3n-2} (2^{2n} - 1)(-1)^{n+1} B_{2n}. \tag{23.57}$$

In particular, we have

(1) Mean value of total energy

$$K_1 = E\{\varepsilon\} = \frac{T^2}{2}. \tag{23.58}$$

(2) Variance of total energy

$$K_2 = \sigma_\varepsilon^2 = \frac{T^4}{3}. \tag{23.59}$$

(3) Skewness of total energy distribution

$$\text{skewness} = \frac{8}{5}\sqrt{3}. \tag{23.60}$$

(4) Kurtosis (or excess) of total energy distribution

$$\text{kurtosis} = \frac{408}{35}. \tag{23.61}$$

23.5 REFERENCES

[1] I. M. Gel'fand, and G. E. Shilov, *Generalized Functions*, Vol. 1, Academic Press, New York, 1964.

[2] A. Erdélyi, W. Magnus, F. Oberhettinger, and F. G. Tricomi, *Tables of Integral Transforms*, Vol. 2, McGraw-Hill, New York, 1954.

[3] K. B. Oldham, and J. Spanier, *The Fractional Calculus*, Academic Press, New York, 1974.

[4] P. Lévy, "Random Functions: General Theory with Special Reference to Laplacian Random Functions," *University of California Publications in Statistics*, **1** (1953), 331–390.

[5] A. N. Kolmogorov, "Wienersche Spiralen und einige andere interessante Kurven im Hilbertschen Raum," *Comptes Rendus (Doklady) Académie des Sciences de l'URSS (N. S.)*, **26** (1940), 115–118.
[6] H. E. Hurst, "Long-term Storage Capacity of Reservoirs," *Transactions of the American Society of Civil Engineers*, **116** (1951), 770–808.
[7] B. B. Mandelbrot, and J. W. Van Ness, "Fractional Brownian Motions, Fractional Noises and Applications," *SIAM Review*, **10** (1968), 422–437.
[8] B. B. Mandelbrot, *The Fractal Geometry of Nature*, Freeman, San Francisco, 1982.
[9] R. H. Cameron, and W. R. Martin, "The Wiener Measure of Hilbert Neighborhoods in the Space of Real Continuous Functions," *Journal of Mathematics and Physics of the M.I.T.*, **23** (1944), 195–209.
[10] G. N. Watson, *A treatise on the Theory of Bessel Functions*, Cambridge University Press, 1966.
[11] A. Erdélyi, W. Magnus, F. Oberhettinger, and F. G. Tricomi, *Higher Transcendental Functions*, Vol. 2, McGraw-Hill, New York, 1953.
[12] M. Abramowitz, and I. Stegun, *Handbook of Mathematical Functions*, Dover, New York, 1970.
[13] A. Papoulis, *Probability, Random Variables and Stochastic Processes*, McGraw-Hill, New York, 1965.

24

Maccone second KLT theorem: KLT of all time-rescaled square Brownian motions

24.1 INTRODUCTION

A surprising feature of the KL expansion obtained in Chapter 22 is that the same analytical solution valid for the $X(t)$ process can be carried over to the $X^2(t)$ process. In other words, to keep within the easy framework of standard Brownian motion $B(t)$, if we know the KL expansion of $B(t)$, then we may also find the KL expansion of $B^2(t)$. The latter will actually be computed at the end of the present chapter, but, as mentioned above, the general proof is valid for any time-rescaled Brownian motion $X^2(t)$. The results proved in this Appendix were discovered by the author in 1988 and published in [1].

24.2 AUTOCORRELATION OF ANY ZERO-MEAN SQUARE PROCESS

The present chapter is devoted to the study of the process

$$Y(t) \equiv X^2(t) - E\{X^2(t)\}. \tag{24.1}$$

Since its mean value is obviously zero, we may call it the zero-mean square process of the time-rescaled Gaussian process $X(t)$. In this section we want to derive the autocorrelation of $Y(t)$. To this end, let us introduce the *function* (not process)

$$m(t) \equiv E\{X^2(t)\}. \tag{24.2}$$

With this notation, the autocorrelation of $Y(t)$ is by definition

$$\begin{aligned} E\{Y(t)Y(s)\} &= E\{[X^2(t) - m(t)][X^2(s) - m(s)]\} \\ &= E\{X^2(t)X^2(s)\} - m(s)E\{X^2(t)\} - m(t)E\{X^2(s)\} + m(s)m(t). \end{aligned} \tag{24.3}$$

The first term in this expression is given by a classical result in the theory of Gaussian processes (see [2, p. 374])

$$E\{X^2(t)X^2(s)\} = E\{X^2(t)\}E\{X^2(s)\} + 2E^2\{X(t)X(s)\}. \tag{24.4}$$

Using (24.2) and (24.4), we now reduce (24.3) as follows

$$\begin{aligned} E\{Y(t)Y(s)\} &= m(t)m(s) + 2E^2\{X(t)X(s)\} - m(t)m(s) - m(t)m(s) + m(t)m(s) \\ &= 2E^2\{X(t)X(s)\}. \end{aligned}$$

That is, the following easy result holds: the autocorrelation of the zero-mean square process is twice the square of the autocorrelation of the process:

$$E\{Y(t)Y(s)\} = 2E^2\{X(t)X(s)\}. \tag{24.5}$$

Even better, we can make use of Equation (21.29) to rewrite the autocorrelation:

$$E\{Y(t)Y(s)\} = 2\left[\int_0^{t\wedge s} f^2(z)\,dz\right]^2. \tag{24.6}$$

But, according to (21.13), any function $F(t \wedge s)$ of the minimum $t \wedge s$ may, by use of the unit step function

$$U(t) = \begin{cases} = 1 & \text{for } t > 0 \\ = 0 & \text{for } t < 0 \end{cases} \tag{24.7}$$

be written in the form

$$F(t \wedge s) = F(t)U(s - t) + F(s)U(t - s). \tag{24.8}$$

Therefore, the autocorrelation (24.6) is finally rewritten

$$E\{Y(t)Y(s)\} = 2\left[\int_0^{t} f^2(z)\,dz\right]^2 U(s-t) + 2\left[\int_0^{s} f^2(z)\,dz\right]^2 U(t-s) \tag{24.9}$$

and will be used in this form in Section 24.3.

From Equation (24.6) the variance of the $Y(t)$ process is immediately found by setting $s = t$. Since $t \wedge t = t$ and $E\{Y(t)\} = 0$, it follows that:

$$\sigma^2_{Y(t)} = E\{Y^2(t)\} = 2\left[\int_0^{t} f^2(z)\,dz\right]^2. \tag{24.10}$$

The standard deviation is just the square root of the above

$$\sigma_{Y(t)} = \pm\sqrt{2}\int_0^t f^2(z)\,dz. \tag{24.11}$$

24.3 KLT OF ANY ZERO-MEAN TIME-RESCALED SQUARE PROCESS

Knowledge of the autocorrelation (24.9) enables us to consider

$$\int_0^T E\{Y(t)Y(s)\}\tilde{\phi}_n(s)\,ds = \tilde{\lambda}_n\tilde{\phi}_n(t). \tag{24.12}$$

This is the integral equation whose kernel is (24.9) and whose solutions, $\tilde{\phi}_n(t)$ and $\tilde{\lambda}_n$, are the eigenfunctions and eigenvalues of the KL expansion, respectively. Though (24.12) looks like a Fredholm-type equation, it is actually a Volterra-type equation, and, as such, it can be reduced to a differential equation by differentiating twice. Moreover, (24.12) and its derivative to the first order also yield the two boundary conditions.

Let us start by inserting the right-hand side of (24.9) into (24.12)

$$2\int_0^T ds\left[\int_0^t f^2(z)\,dz\right]^2 U(s-t)\tilde{\phi}_n(s) + 2\int_0^T ds\left[\int_0^s f^2(z)\,dz\right]^2 U(t-s)\tilde{\phi}_n(s) = \tilde{\lambda}_n\tilde{\phi}_n(t). \tag{24.13}$$

Using the properties of the unit step function, the above can be written

$$2\left[\int_0^t f^2(z)\,dz\right]^2 \int_t^T \tilde{\phi}_n(s)\,ds + 2\int_0^t \left[\int_0^s f^2(z)\,dz\right]^2 \tilde{\phi}_n(s)\,ds = \tilde{\lambda}_n\tilde{\phi}_n(t). \tag{24.14}$$

The left-hand side of (24.14) vanishes for $t = 0$ making

$$\tilde{\phi}_n(0) = 0 \tag{24.15}$$

as the initial condition fulfilled by the eigenfunctions $\tilde{\phi}_n(t)$.

According to the general procedure for solving Volterra-type integral equations, we can now differentiate both sides of (24.14) with respect to t to get

$$4\int_0^t f^2(z)\,dz\cdot f^2(t)\cdot\int_t^T \tilde{\phi}_n(s)\,ds - 2\left[\int_0^t f^2(z)\,dz\right]^2\tilde{\phi}_n(t) + 2\left[\int_0^t f^2(z)\,dz\right]^2\tilde{\phi}_n(t) = \tilde{\lambda}_n\tilde{\phi}'_n(t). \tag{24.16}$$

Luckily enough, the last two terms on the left-hand side of (24.16) cancel leaving the simpler equation

$$-4\int_0^t f^2(z)\,dz\cdot f^2(t)\cdot\int_T^t \tilde{\phi}_n(s)\,ds = \tilde{\lambda}_n\tilde{\phi}'_n(t). \tag{24.17}$$

This equation may be written more conveniently by dividing both sides by the terms involving $f(t)$ and $\tilde{\lambda}_n$:

$$-\frac{1}{\tilde{\lambda}_n}\int_T^t \tilde{\phi}_n(s)\,ds = \frac{\tilde{\phi}'_n(t)}{4f^2(t)\displaystyle\int_0^t f^2(z)\,dz}. \tag{24.18}$$

On setting $t = T$ the left-hand side of (24.18) vanishes, yielding the final condition fulfilled by the eigenfunctions

$$\tilde{\phi}'_n(T) = 0. \tag{24.19}$$

Moreover, we can make the integral on the left-hand side of (24.18) disappear by once more differentiating both sides with respect to t. The differential equation for the eigenfunctions $\tilde{\phi}_n(t)$ is thus obtained

$$\frac{d}{dt}\left[\frac{\tilde{\phi}'_n(t)}{4f^2(t)\int_0^t f^2(z)\,dz}\right] + \frac{1}{\tilde{\lambda}_n}\tilde{\phi}_n(t) = 0. \tag{24.20}$$

We must now solve the differential equation (24.20) jointly with the two boundary condition (24.15) and (24.19). Before proceding, however, the following important remark will save a lot of work.

Recall from Section 22.3 (see also [3]) that the full solution to the differential equation (22.25), that is,

$$\frac{d}{dt}\left[\frac{\phi'_n(t)}{f^2(t)}\right] + \frac{1}{\lambda_n}\phi_n(t) = 0, \tag{24.21}$$

could be found. In other words, we were able to solve (24.21), jointly with the two boundary conditions (22.19) and (22.24); that is, respectively

$$\phi_n(0) = 0 \tag{24.22}$$

and

$$\phi'_n(T) = 0. \tag{24.23}$$

Now, the important remark we are referring to is that (24.15) is identical to (24.22), and (24.19) is identical to (24.23). Moreover, (24.21) corresponds to (24.20) if $f(t)$ is formally replaced by

$$f^2(t) \to 4f^2(t)\int_0^t f^2(z)\,dz.$$

In other words, the whole mathematical solution of (24.20) coincides with the solution of (24.21) if the replacement

$$\tilde{f}(t) \equiv 2f(t)\sqrt{\int_0^t f^2(z)\,dz} \tag{24.24}$$

is performed. *This result is fundamental.* In fact, we can now use all the apparatus created in Chapter 22 for the KL expansion of the $X(t)$ process to find the KL expansion of the $Y(t)$ process.

Start by finding the new $\tilde{\chi}(t)$ function, corresponding to the $\chi(t)$ function defined by (22.42). To this end, we merely have to substitute (24.24) into (22.42), and get

$$\tilde{\chi}(t) = 2\sqrt{f(t)\sqrt{\int_0^t f^2(z)\,dz}\int_0^t f(s)\sqrt{\int_0^s f^2(z)\,dz\,ds}}. \tag{24.25}$$

Next comes the new order $\tilde{\nu}(t)$ of the Bessel functions of the first kind, which from (22.43) and (24.24) turns out to be equal to

$$\tilde{\nu}(t) = \sqrt{\frac{\tilde{\chi}^3(t)}{-4f^2(t)\int_0^t f^2(z)\,dz}\frac{d}{dt}\left[\frac{\tilde{\chi}'(t)}{4f^2(t)\int_0^t f^2(z)\,dz}\right]}. \tag{24.26}$$

Consider now the new constants $\tilde{\gamma}_n$. From (22.49) and (24.24) one can conclude that they are the real positive zeros, arranged in ascending order of magnitude, of the equation

$$\tilde{\chi}'(T)J_{\tilde{\nu}(T)}(\tilde{\gamma}_n) + \tilde{\chi}(T)\frac{f(T)\sqrt{\int_0^T f^2(z)\,dz}\cdot\tilde{\gamma}_n}{\int_0^T f(s)\sqrt{\int_0^s f^2(z)\,dz}\,ds}J'_{\tilde{\nu}(T)}(\tilde{\gamma}_n) + \tilde{\chi}(T)\frac{\partial J_{\tilde{\nu}(T)}(\tilde{\gamma}_n)}{\partial\tilde{\nu}}\tilde{\nu}'(T) = 0. \tag{24.27}$$

Clearly, this equation is very difficult to solve analytically, even in elementary cases where the function $f(t)$ is particularly simple. Thus, in practice the $\tilde{\gamma}_n$ will have to be found numerically.

As for the normalization constants $\tilde{N}_n$, these also must be computed numerically from the normalization condition that follows from (22.47) and (24.24), namely

$$1 = \tilde{N}_n^2\left[2\int_0^T f(s)\sqrt{\int_0^s f^2(z)\,dz}\,ds\right]^2\int_0^1 xJ^2_{\tilde{\nu}((x))}(\tilde{\gamma}_n x)\,dx. \tag{24.28}$$

The eigenvalues $\tilde{\lambda}_n$ are related to the constants $\tilde{\gamma}_n$ (known from (24.27)) by a formula that follows from (22.48) and (24.24):

$$\tilde{\lambda}_n = 4\left[\int_0^T f(s)\sqrt{\int_0^s f^2(z)\,dz}\,ds\right]^2\frac{1}{(\tilde{\gamma}_n)^2}. \tag{24.29}$$

Finally, we need to find the probability distribution of the random variables $\tilde{Z}_n$, which are obviously *not* Gaussian. To this end, (24.2) and (21.32) yield

$$m(t) = E\{X^2(t)\} = \int_0^t f^2(z)\,dz = \sigma^2_{X(t)}. \tag{24.30}$$

This time function (not process) is the variance of $X(t)$ because $X(t)$ has zero mean. But $X(t)$ is Gaussian. Therefore, $X^2(t)$ is chi-square distributed, having the probability density

$$f_{X^2(t)}(x) = \frac{1}{\sqrt{2\pi m(t)}}e^{-\frac{x}{2m(t)}}x^{-\frac{1}{2}}U(x). \tag{24.31}$$

Now from the definition (24.1), we get

$$f_{Y(t)}(y) = f_{X^2(t)-m(t)}(y) = f_{X^2(t)}(y + m(t)). \tag{24.32}$$

By virtue of (24.31) and (24.32), we have now proved that the process $Y(t)$ has the (gamma-type) probability density

$$f_{Y(t)}(y) = \frac{1}{\sqrt{2\pi m(t)}} e^{-\left[\frac{y+m(t)}{2m(t)}\right]} [y + m(t)]^{-1/2} U(y + m(t)). \tag{24.33}$$

The random variables $\tilde{Z}_n$ must also have a gamma-type probability distribution like the process $Y(t)$, for it is well known that the convolution of two gamma densities is again a gamma density. Thus

$$f_{\tilde{Z}_n}(z) = \frac{1}{\sqrt{\pi}(2\tilde{\lambda}_n)^{1/4}} e^{-\frac{z+\sqrt{\frac{\tilde{\lambda}_n}{2}}}{\sqrt{2\tilde{\lambda}_n}}} \left[z + \sqrt{\frac{\tilde{\lambda}_n}{2}}\right]^{-1/2} U\left(z + \sqrt{\frac{\tilde{\lambda}_n}{2}}\right) \tag{24.34}$$

is the probability density function (gamma-type) of the random variables $\tilde{Z}_n$, and is found from (24.33) by formally replacing the variance $2m^2(t)$ by $\tilde{\lambda}_n$.

In conclusion, we have derived the KL expansion of the zero-mean square process of $X(t)$:

$$\begin{aligned} Y(t) &\equiv X^2(t) - E\{X^2(t)\} \\ &= \sum_{n=1}^{\infty} \tilde{Z}_n \tilde{N}_n \cdot \tilde{\chi}(t) J_{\tilde{\nu}(t)} \left(\tilde{\gamma}_n \frac{\int_0^t f(s) \sqrt{\int_0^s f^2(z)\, dz}\, ds}{\int_0^T f(s) \sqrt{\int_0^s f^2(z)\, dz}\, ds} \right). \end{aligned} \tag{24.35}$$

24.4 KLT OF SQUARE BROWNIAN MOTION

Standard Brownian motion is a particular case of the foregoing theory characterized by

$$f(t) \equiv 1. \tag{24.36}$$

Therefore, the KL expansion of square standard Brownian motion can be found by merely substituting (24.36) into all the formulas developed in the present chapter.

We start by forming the $\tilde{f}(t)$ function defined by (24.24),

$$\tilde{f}(t) = 2\sqrt{t}. \tag{24.37}$$

Using (24.25), it further follows that:

$$\tilde{\chi}(t) = \sqrt{2\sqrt{t}\int_0^t 2\sqrt{z}\, dz} - \sqrt{\frac{8}{3}}\, t \tag{24.38}$$

along with its derivative

$$\tilde{\chi}'(t) = \sqrt{\frac{8}{3}}. \tag{24.39}$$

Then, (24.26) yields the order of the Bessel functions

$$\tilde{\nu}(t) = \sqrt{-\frac{\left(\sqrt{\frac{8}{3}}t\right)^3}{4t} \cdot \frac{d}{dt}\left[\frac{\sqrt{\frac{8}{3}}}{4t}\right]} = \sqrt{\frac{4}{9}} = \frac{2}{3}, \tag{24.40}$$

which is a *constant*, resulting in the conclusion that (24.35), (24.37), and (24.40) yield the orthogonal eigenfunctions

$$\tilde{\phi}_n(t) = \tilde{N}_n \cdot \sqrt{\frac{8}{3}}t \cdot J_{\frac{2}{3}}\left(\tilde{\gamma}_n \frac{t^{\frac{3}{2}}}{T^{\frac{3}{2}}}\right). \tag{24.41}$$

Let us now discover the meaning of the set of constants $\tilde{\gamma}_n$. From

$$\tilde{\nu}'(t) = 0 \tag{24.42}$$

as well as from (24.36), (24.38), and (24.39), we see that (24.27) is changed into

$$\sqrt{\frac{8}{3}} J_{\frac{2}{3}}(\tilde{\gamma}_n) + \sqrt{\frac{8}{3}} T \frac{\sqrt{T}\tilde{\gamma}_n}{\frac{2}{3} T^{\frac{3}{2}}} J'_{\frac{2}{3}}(\tilde{\gamma}_n) = 0. \tag{24.43}$$

Rearranging, this reduces to

$$\frac{2}{3} J_{\frac{2}{3}}(\tilde{\gamma}_n) + \tilde{\gamma}_n J'_{\frac{2}{3}}(\tilde{\gamma}_n) = 0, \tag{24.44}$$

that is, by virtue of (24.40),

$$\tilde{\nu} J_{\tilde{\nu}}(\tilde{\gamma}_n) + \tilde{\gamma}_n J'_{\tilde{\nu}}(\tilde{\gamma}_n) = 0. \tag{24.45}$$

This, however, is just the left-hand side of an important formula in the theory of Bessel functions (see [4, p. 11, entry (54)]) stating that

$$\tilde{\nu} J_{\tilde{\nu}}(\tilde{\gamma}_n) + \tilde{\gamma}_n J'_{\tilde{\nu}}(\tilde{\gamma}_n) = \tilde{\gamma}_n J_{\tilde{\nu}-1}(\tilde{\gamma}_n). \tag{24.46}$$

Therefore, (24.45) amounts to

$$\tilde{\gamma}_n J_{\tilde{\nu}-1}(\tilde{\gamma}_n) = 0 \tag{24.47}$$

and, since the constants $\tilde{\gamma}_n$ may not vanish,

$$J_{\tilde{\nu}-1}(\tilde{\gamma}_n) = 0. \tag{24.48}$$

We have thus found the meaning of the set of constants $\tilde{\gamma}_n$: they are the (infinite) real positive zeros of the Bessel function of the first kind and of order $-1/3$

$$J_{-\frac{1}{3}}(\tilde{\gamma}_n) = 0. \tag{24.49}$$

No explicit formula yielding these zeros is known. However, it is possible to get a good numerical approximation for them by substituting $\nu = 1/3$ and $x = \tilde{\gamma}_n$ into the asymptotic formula for $J_\nu(x)$ (see [5, p. 134]), which reads

$$\lim_{x\to\infty} J_\nu(x) = \lim_{x\to\infty}\sqrt{\frac{2}{\pi x}}\cos\left(x - \frac{\nu\pi}{2} - \frac{\pi}{4}\right). \tag{24.50}$$

This results in

$$J_{-\frac{1}{3}}(\tilde{\gamma}_n) \to \sqrt{\frac{2}{\pi\tilde{\gamma}_n}}\cos\left(\tilde{\gamma}_n - \frac{\pi}{2}\left[-\frac{1}{3}\right] - \frac{\pi}{4}\right). \tag{24.51}$$

Since the zeros of the cosine are $(n\pi - \pi/2)$, it follows from (24.49) that:

$$\tilde{\gamma}_n - \frac{\pi}{2}\left[-\frac{1}{3}\right] - \frac{\pi}{4} \approx n\pi - \frac{\pi}{2} \tag{24.52}$$

or

$$\tilde{\gamma}_n \approx \pi\left(n - \frac{5}{12}\right). \tag{24.53}$$

Next, we want to determine the normalization constants defined by (24.28). That is,

$$1 = \tilde{N}_n^2\left[\int_0^T \tilde{f}(s)\,ds\right]^2 \int_0^1 xJ_{\frac{2}{3}}^2(\tilde{\gamma}_n x)\,dx. \tag{24.54}$$

Let us now replace (24.36) and a definite integral calculated within the framework of the Dini expansion in a series of Bessel functions that appear on p. 71 of [4]; the last expression is then turned into

$$1 = \tilde{N}_n^2\frac{16}{9}T^3\left\{\frac{1}{2}\left[J_{\frac{2}{3}}^{\prime 2}(\tilde{\gamma}_n) + \left(1 - \frac{\left(\frac{2}{3}\right)^2}{\tilde{\gamma}_n^2}\right)J_{\frac{2}{3}}^2(\tilde{\gamma}_n)\right]\right\}.$$

Now, by virtue of (24.44), we may let the derivative of the Bessel function disappear, and one gets

$$1 - \tilde{N}_n^2\frac{8}{9}T^3 \cdot J_{\frac{2}{3}}^2(\tilde{\gamma}_n).$$

Solving (24.54) for the normalization constants requires introduction of the modulus, and so

$$\tilde{N}_n = \frac{3\sqrt{2}}{4T^{\frac{3}{2}}}\frac{1}{\left|J_{\frac{2}{3}}(\tilde{\gamma}_n)\right|} \tag{24.55}$$

is the exact expression of the normalization constants.

A good approximation for the normalization constants (24.55) can be obtained by using the asymptotic expression (24.51) for the Bessel function appearing in the denominator of (24.55), namely

$$|J_{\frac{2}{3}}(\tilde{\gamma}_n)| \approx \left|\sqrt{\frac{2}{\pi\tilde{\gamma}_n}}\cos\left(\tilde{\gamma}_n - \frac{\pi}{2}\left[\frac{2}{3}\right] - \frac{\pi}{4}\right)\right|. \tag{24.56}$$

Inserting the approximated formula (24.53) for the $\tilde{\gamma}_n$ into the cosine argument yields

$$|J_{\frac{2}{3}}(\tilde{\gamma}_n)| \approx \sqrt{\frac{2}{\pi\tilde{\gamma}_n}}\left|\cos\left(\pi\left[n - \frac{5}{12}\right] - \frac{\pi}{3} - \frac{\pi}{4}\right)\right|$$

$$\approx \sqrt{\frac{2}{\pi\tilde{\gamma}_n}}|\cos(\pi[n-1])| \approx \sqrt{\frac{2}{\pi\tilde{\gamma}_n}}|(-1)^{n+1}| \approx \sqrt{\frac{2}{\pi\tilde{\gamma}_n}}.$$

Thus

$$|J_{\frac{2}{3}}(\tilde{\gamma}_n)| \approx \sqrt{\frac{2}{\pi\tilde{\gamma}_n}}. \tag{24.57}$$

Substituting this expression in (24.55) makes

$$\tilde{N}_n \approx \frac{3\sqrt{2}}{4T^{\frac{3}{2}}}\frac{\sqrt{\pi}\sqrt{\tilde{\gamma}_n}}{\sqrt{2}} \approx \frac{3\pi}{4T^{\frac{3}{2}}}\sqrt{n - \frac{5}{12}}. \tag{24.58}$$

Similarly, (24.29) and (24.29) plus (24.53), respectively, yield the exact and approximated expressions for the eigenvalues

$$\tilde{\lambda}_n = \frac{16}{9}T^3\frac{1}{(\tilde{\gamma}_n)^2}, \tag{24.59}$$

and

$$\tilde{\lambda}_n = \frac{16}{9}T^3\frac{1}{\pi^2\left(n - \frac{5}{12}\right)^2}. \tag{24.60}$$

In conclusion, by virtue of (24.55), (24.38), and (24.40), we have proven that the exact expression of the KL eigenfunctions reads

$$\tilde{\phi}_n(t) = \tilde{N}_n\tilde{\chi}(t)J_{\frac{2}{3}}\left(\tilde{\gamma}_n\frac{t^{\frac{3}{2}}}{T^{\frac{3}{2}}}\right) = \frac{\sqrt{3}t}{T^{\frac{3}{2}}|J_{\frac{2}{3}}(\tilde{\gamma}_n)|}J_{\frac{2}{3}}\left(\tilde{\gamma}_n\frac{t^{\frac{3}{2}}}{T^{\frac{3}{2}}}\right). \tag{24.61}$$

The approximated form of (24.61) may be obtained by using the approximate expression for N_n in (24.58), the approximate expression (24.57) for the Bessel function in the denominator, and the asymptotic expression for the Bessel function in (24.51). The result is

$$\tilde{\phi}_n(t) \approx \frac{\sqrt{3}t}{T^{\frac{3}{2}}\sqrt{\frac{2}{\pi\tilde{\gamma}_n}}}\sqrt{\frac{2}{\pi\left(\tilde{\gamma}_n\frac{t^{\frac{3}{2}}}{T^{\frac{3}{2}}}\right)}}\cos\left(\tilde{\gamma}_n\frac{t^{\frac{3}{2}}}{T^{\frac{3}{2}}} - \frac{\pi\tilde{\nu}}{2} - \frac{\pi}{4}\right). \tag{24.62}$$

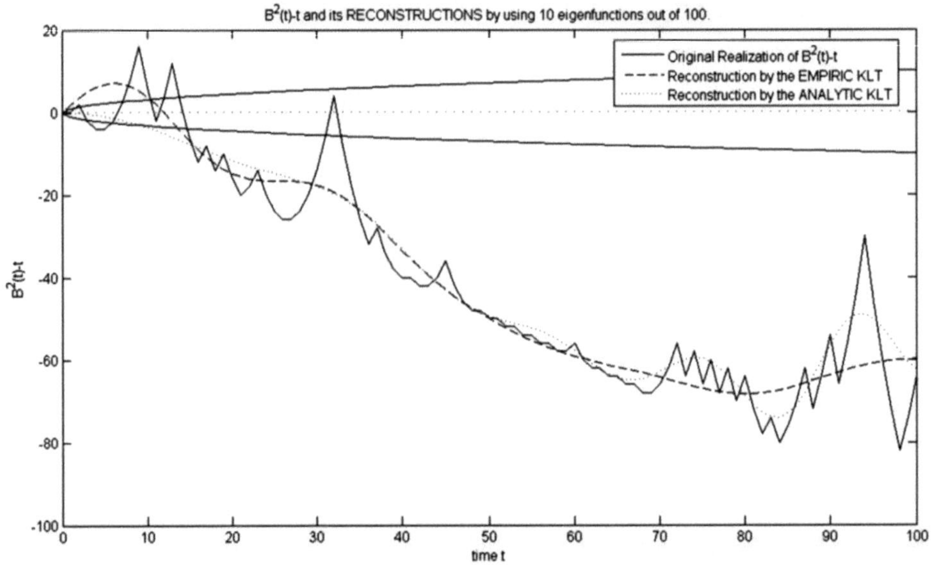

Figure 24.1. The *zero-mean* square Brownian motion $B^2(t) - t = X(t)$ vs. time t simulated as a random walk over 100 time instants. Next to the "bumpy curve" of $X(t)$, two more "smooth curves" are shown that *interpolate at best* the bumpy $X(t)$. These two curves are the KLT reconstruction of $B^2(t) - t$ *by using the first ten eigenfunctions only*. It is important to note that the two smooth curves are *different* in this case because the KLT expansion (24.66) is *approximated.* Actually, it is an approximated KLT expansion because the asymptotic expansion of the Bessel functions (24.50) was used. So, the two curves are different from each other, but both still interpolate $X(t)$ at best. Note that, were we taking into account the full set of 100 KLT eigenfuctions—rather than just 10—then the *empiric* reconstruction would overlap $X(t)$ *exactly*, but the *analytic* reconstruction would not because of the use of the asymptotic expansion (24.50) of the Bessel functions.

Substituting $\tilde{\nu} = 1/3$ from (24.40) and $\tilde{\gamma}_n = \pi(n - 5/12)$ from (24.53), the approximate expression for the KL eigenfunctions is

$$\tilde{\phi}_n(t) \approx \sqrt{3}\frac{t^{\frac{1}{4}}}{T^{\frac{3}{4}}} \cos\left(\pi\left[\left(n - \frac{5}{12}\right)\frac{t^{\frac{3}{2}}}{T^{\frac{3}{2}}} - \frac{7}{12}\right]\right). \tag{24.63}$$

In summary, we may now write the full formula (24.35) for the exact KL eigenfunction expansion for the square Brownian motion. In fact, on inserting $t_1 = t_2 = t$ into (21.8), we see that

$$E\{B^2(t)\} = t, \tag{24.64}$$

and (24.38), (24.55), and (24.64) yield the exact KL expansion for $B^2(t)$

$$B^2(t) - t = \sum_{n=1}^{\infty} \tilde{Z}_n \frac{\sqrt{3}t}{T^{\frac{3}{2}}|J_{\frac{2}{3}}(\tilde{\gamma}_n)|} J_{\frac{2}{3}}\left(\tilde{\gamma}_n \frac{t^{\frac{3}{2}}}{T^{\frac{3}{2}}}\right) \tag{24.65}$$

The corresponding approximate expression is

$$B^2(t) - t \approx \sum_{n=1}^{\infty} \tilde{Z}_n \sqrt{3} \frac{t^{\frac{1}{4}}}{T^{\frac{3}{4}}} \cos\left(\pi\left[\left(n - \frac{5}{2}\right)\frac{t^{\frac{3}{2}}}{T^{\frac{3}{2}}} - \frac{7}{12}\right]\right). \tag{24.66}$$

Formulas (24.65) and (24.66) were published by the author in the last section of [1], but unfortunately they contained a slight computational error. The formulas appearing above are correct.

24.5 CHECKING THE KLT OF THE SQUARE BROWNIAN MOTION BY MATLAB SIMULATIONS

Just look at Figure 24.1.

24.6 REFERENCES

[1] C. Maccone, "The Karhunen–Loève Expansion of the Square of a Time-Rescaled Gaussian Process," *Bollettino dell'Unione Matematica Italiana*, Series 7, **2-A** (1988), 221–229.

[2] A. Papoulis, *Signal Analysis*, McGraw-Hill, New York, 1977.

[3] C. Maccone, "Eigenfunctions and Energy for Time-Rescaled Gaussian Processes," *Bollettino dell'Unione Matematica Italiana*, Series 6, **3-A** (1984), 213–219.

[4] A. Erdélyi, W. Magnus, F. Oberhettinger, and F. G. Tricomi, *Higher Transcendental Functions*, Vol. 2, McGraw-Hill, New York, 1953.

[5] N. N. Lebedev, *Special Functions and Their Applications*, Dover, New York, 1972.

25

KLT of the $B^2(t^{2H})$ time-rescaled square Brownian motion

25.1 INTRODUCTION

Just as Chapter 23 showed the application of the general results obtained in Chapter 22 to the particular time-rescaled stochastic process $B(t^{2H})$, the present appendix investigates the application of the general results obtained in Chapter 24 about the square time-rescaled process $B^2(\ldots)$ to the particular process $B^2(t^{2H})$.

Before doing so, however, we regard it useful to review briefly the main results so far obtained. Consider then the general time-rescaled Brownian motion given by (21.40); that is,

$$X(t) = B\left(\int_0^t f^2(s)\, ds\right). \tag{25.1}$$

In Appendix G, as well as in [1], it was proved that the KL eigenfunction expansion of (25.1) reads

$$X(t) = \sum_{n=1}^{\infty} Z_n N_n \sqrt{f(t) \int_0^t f(s)\, ds} \cdot J_{\nu(t)}\left(\gamma_n \frac{\int_0^t f(s)\, ds}{\int_0^T f(s)\, ds}\right) \tag{25.2}$$

and converges in mean square, and uniformly in t, for $0 \le t \le T$.

Here

(1) The time-dependent order $\nu(t)$ of the Bessel function of the first kind $J_\nu(\ldots)$ is given by

$$\nu(t) = \sqrt{-\frac{\chi^3(t)}{f^2(t)} \frac{d}{dt}\left[\frac{\chi'(t)}{f^2(t)}\right]}, \tag{25.3}$$

where the auxiliary function $\chi(t)$ is defined by

$$\chi(t) = \sqrt{f(t) \int_0^t f(s)\, ds}. \tag{25.4}$$

Moreover, in Section 22.4 as well as in [2], the following straightforward expression for $\nu(t)$, in terms of the time-rescaling function $f(t)$, was proved:

$$\nu(t) = \sqrt{\frac{1}{4} + \left[\frac{\int_0^t f(s)\, ds}{f(t)}\right]^2 \left\{\frac{3}{4}\left[\frac{d \ln f(t)}{dt}\right]^2 - \frac{1}{2}\frac{d^2 \ln f(t)}{dt^2}\right\}}. \tag{25.5}$$

(2) The constants γ_n appearing in the argument of the Bessel function in (25.2) are the real positive zeros, arranged in ascending order of magnitude, of the equation

$$\chi'(T) \cdot J_{\nu(T)}(\gamma_n) + \chi(T) \left[\frac{f(T) \cdot \gamma_n}{\int_0^T f(s)\, ds} J'_{\nu(T)}(\gamma_n) + \frac{\partial J_{\nu(T)}(\gamma_n)}{\partial \nu} \nu'(T)\right] = 0. \tag{25.6}$$

In general, (25.6) cannot be solved for the γ_n analytically, and one has to do so numerically. However, some particular case of the time-rescaling function $f(t)$ may exist for which (25.6) can be solved analytically. The present chapter is devoted to one such important case.

(3) The normalization constants N_n follow from the normalization condition

$$N_n^2 \left[\int_0^T f(s)\, ds\right]^2 \int_0^1 x[J_{\nu((x))}(\gamma_n x)]^2\, dx = 1, \tag{25.7}$$

where the new transformed order $\nu((x))$ is obtained from the order $\nu(t)$ of either (25.3) or (25.5) via the transformation

$$\int_0^t f(s)\, ds = x \int_0^T f(s)\, ds. \tag{25.8}$$

(4) The eigenvalues λ_n are determined by the γ_n (known from (25.6)) according to

$$\lambda_n = \left[\int_0^T f(s)\, ds\right]^2 \frac{1}{(\gamma_n)^2}. \tag{25.9}$$

(5) Finally, the Z_n are independent and orthogonal Gaussian random variables having mean zero and variance equal to the eigenvalues λ_n; that is,

$$E\{Z_m Z_n\} = \lambda_n \delta_{mn}. \tag{25.10}$$

Let us next turn to the further process $Y(t)$ defined by (24.1); that is,

$$Y(t) = X^2(t) - E\{X^2(t)\}. \tag{25.11}$$

It is natural to call this the zero-mean square process of the time rescaled Brownian motion given by (25.1), for (25.11) is just the square of (25.1) centered around the latter's mean value.

In Chapter 24, as well as in [3], the author proved that the KL eigenfunction expansion of the process (25.11) reads:

$$Y(t) = \sum_{n=1}^{\infty} \tilde{Z}_n \tilde{N}_n \sqrt{\tilde{f}(t) \int_0^t \tilde{f}(s)\, ds} \cdot J_{\tilde{\nu}(t)}\left(\tilde{\gamma}_n \frac{\int_0^t \tilde{f}(s)\, ds}{\int_0^T \tilde{f}(s)\, ds}\right). \tag{25.12}$$

By checking (25.12) against (25.2), one sees that the KL expansion of the zero-mean square process $Y(t)$ is formally identical to the KL expansion of the original $X(t)$ process, with only two exceptions.

(1) The time-rescaling function $f(t)$ of the $X(t)$ is now replaced by the new function (no longer called "time rescaling") (for the proof, see (24.20) through (24.24))

$$\tilde{f}(t) = 2f(t)\sqrt{\int_0^t f^2(s)\, ds}. \tag{25.13}$$

(2) The gamma-type probability density

$$f_{\tilde{Z}_n}(z) = \frac{1}{\sqrt{\pi}(2\tilde{\lambda}_n)^{\frac{1}{4}}} e^{-\frac{z+\sqrt{\frac{\tilde{\lambda}_n}{2}}}{\sqrt{2\tilde{\lambda}_n}}} \left[z + \sqrt{\frac{\tilde{\lambda}_n}{2}}\right]^{-\frac{1}{2}} U\left(z + \sqrt{\frac{\tilde{\lambda}_n}{2}}\right), \tag{25.14}$$

where $U(x)$ is the unit step function, is followed by the random variables Z_n (for the proof, see (24.34)).

Consequently, we infer that

(1) the Bessel function of the first kind appearing in (25.12) has the time-dependent order

$$\tilde{\nu}(t) = \sqrt{-\frac{\tilde{\chi}^3(t)}{\tilde{f}^2(t)} \frac{d}{dt}\left[\frac{\tilde{\chi}'(t)}{\tilde{f}^2(t)}\right]}, \tag{25.15}$$

where the auxiliary function $\tilde{\chi}(t)$ has been defined by

$$\tilde{\chi}(t) = \sqrt{\tilde{f}(t) \int_0^t \tilde{f}(s)\, ds} \tag{25.16}$$

in analogy to (25.3) and (25.4), respectively. Alternatively, it is possible to express the order by virtue of a single formula, corresponding to (25.5)

$$\tilde{\nu}(t) = \sqrt{\frac{1}{4} + \left[\frac{\int_0^t \tilde{f}(s)\, ds}{\tilde{f}(t)}\right]^2 \left\{\frac{3}{4}\left[\frac{d \ln \tilde{f}(t)}{dt}\right]^2 - \frac{1}{2}\frac{d^2 \ln \tilde{f}(t)}{dt^2}\right\}}. \tag{25.17}$$

(2) The real positive zeros, $\tilde{\gamma}_{n'}$, arranged in ascending order of magnitude, of the equation corresponding to (25.6), are defined by

$$\tilde{\chi}'(T) \cdot J_{\tilde{\nu}(T)}(\tilde{\gamma}_n) + \tilde{\chi}(T) \left[\frac{\tilde{f}(T) \cdot \tilde{\gamma}_n}{\int_0^T \tilde{f}(s)\, ds} J'_{\tilde{\nu}(T)}(\tilde{\gamma}_n) + \frac{\partial J_{\tilde{\nu}(T)}(\tilde{\gamma}_n)}{\partial \tilde{\nu}_t} \tilde{\nu}'(T) \right] = 0. \quad (25.18)$$

(3) The normalization condition, analogous to (25.7), reads

$$1 = \tilde{N}_n^2 \left[\int_0^T \tilde{f}(s)\, ds \right]^2 \int_0^1 x[J_{\tilde{\nu}((x))}(\tilde{\gamma}_n x)]^2\, dx \quad (25.19)$$

and defines the normalization constants $\tilde{N}_n$. In (25.19) the new transformed order $\tilde{\nu}((x))$ is obtained from the order $\tilde{\nu}(t)$ of either (25.15) or (25.17) via the transformation

$$\int_0^t \tilde{f}(s)\, ds = x \int_0^T \tilde{f}(s)\, ds. \quad (25.20)$$

(4) Finally, the eigenvalues are given by an expression analogous to (25.9); that is,

$$\tilde{\lambda}_n = \left[\int_0^T \tilde{f}(s)\, ds \right]^2 \frac{1}{(\tilde{\gamma}_n)^2}. \quad (25.21)$$

25.2 PREPARATORY CALCULATIONS ABOUT $B^2(t^{2\alpha+1})$

In the present section we pave the way to mathematically finding the KL eigenfunction expansion of the square process $B^2(t^{2\alpha+1})$.

Let the time-rescaling function $f(t)$ be a real-exponent power of time, multiplied by a generic real constant C

$$f(t) = Ct^{\alpha}. \quad (25.22)$$

The range of the exponent α is determined by the condition that the following pair of definite integrals, appearing in (25.2) and (25.1), respectively, must converge:

$$\int_0^t f(s)\, ds = \frac{C}{\alpha + 1} t^{\alpha+1}, \quad (25.23)$$

$$\int_0^t f^2(s)\, ds = \frac{C^2}{2\alpha + 1} t^{2\alpha+1}. \quad (25.24)$$

Evidently, the stricter condition on α is due to the convergence of (25.24); that is,

$$\alpha > -\frac{1}{2}. \quad (25.25)$$

Let us now go back to the problem of finding the KL eigenfunction expansion of the square process $B^2(t^{2\alpha+1})$. To this end, we must first form the new function defined by (25.13), which, by virtue of (25.1) and (25.3), turns out to be

$$\tilde{f}(t) = Kt^{2\alpha+\frac{1}{2}}, \tag{25.26}$$

where the new constant K is introduced to simplify things a little

$$K = \frac{2C^2}{\sqrt{2\alpha+1}}. \tag{25.27}$$

Our first task is to find the corresponding order of the Bessel function, alternatively defined by (25.15) plus (25.16), or by (25.17). Choosing the latter, we are led to compute the logarithm of (25.26)

$$\ln \tilde{f}(t) = \ln K + \left(2\alpha + \frac{1}{2}\right) \ln t \tag{25.28}$$

the logarithmic derivatives

$$\frac{d \ln \tilde{f}(t)}{dt} = \frac{\left(2\alpha + \frac{1}{2}\right)}{t}, \tag{25.29}$$

and

$$\frac{d^2 \ln \tilde{f}(t)}{dt^2} = -\frac{\left(2\alpha + \frac{1}{2}\right)}{t^2}, \tag{25.30}$$

and the integral

$$\int_0^t \tilde{f}(s)\, ds = \frac{K}{2\alpha + \frac{3}{2}} t^{2\alpha+\frac{3}{2}}. \tag{25.31}$$

Note that the condition (25.25) on α still holds, since the converge of (25.31) only requires that

$$\alpha > -\frac{3}{4}. \tag{25.32}$$

On substituting these results into the square of (25.17), the latter is turned into

$$\begin{aligned} \tilde{\nu}^2(t) &= \frac{1}{4} + \left[\frac{\frac{K}{2\alpha+\frac{3}{2}} t^{2\alpha+\frac{3}{2}}}{Kt^{2\alpha+\frac{1}{2}}}\right]^2 \cdot \left\{\frac{3}{4}\frac{\left(2\alpha+\frac{1}{2}\right)^2}{t^2} + \frac{1}{2}\frac{\left(2\alpha+\frac{1}{2}\right)}{t^2}\right\} \\ &= \frac{1}{4} + \frac{t^2}{\left(2\alpha+\frac{3}{2}\right)^2} \cdot \frac{\left(2\alpha+\frac{1}{2}\right)}{t^2}\left\{\frac{3}{4}\left(2\alpha+\frac{1}{2}\right) + \frac{1}{2}\right\}. \end{aligned} \tag{25.33}$$

The time t now cancels out in the expression, leaving

$$\tilde{\nu}^2(t) = \frac{1}{4} + \frac{\left(2\alpha + \frac{1}{2}\right)}{\left(2\alpha + \frac{3}{2}\right)^2} \frac{1}{4} \left\{ 3\left(2\alpha + \frac{1}{2}\right) + 2 \right\}$$

$$= \frac{\left(2\alpha + \frac{3}{2}\right)^2 + \left(2\alpha + \frac{1}{2}\right)\left\{6\alpha + \frac{3}{2} + 2\right\}}{4\left(2\alpha + \frac{3}{2}\right)^2} = \frac{(2\alpha + 1)^2}{\left(2\alpha + \frac{3}{2}\right)^2}.$$

That is,

$$\tilde{\nu} = \frac{2\alpha + 1}{2\alpha + \frac{3}{2}}, \tag{25.34}$$

with the immediate consequence

$$\tilde{\nu}'(t) = 0. \tag{25.35}$$

This circumstance helps simplify upcoming calculations considerably.

Let us now turn to finding the roots of (25.18) that define the constants $\tilde{\gamma}_n$. By virtue of (25.35), (25.18) takes the simpler form

$$\tilde{\chi}'(T) \cdot J_{\tilde{\nu}(T)}(\tilde{\gamma}_n) + \tilde{\chi}(T) \frac{\tilde{f}(T) \cdot \tilde{\gamma}_n}{\int_0^T \tilde{f}(s)\, ds} J'_{\tilde{\nu}(T)}(\tilde{\gamma}_n) = 0. \tag{25.36}$$

We next compute the function defined by (25.16)

$$\tilde{\chi}(t) = \sqrt{\tilde{f}(t) \int_0^t \tilde{f}(s)\, ds} = \frac{K}{\sqrt{2\alpha + \frac{3}{2}}} t^{2\alpha+1} \tag{25.37}$$

and its derivative

$$\tilde{\chi}'(t) = \frac{K(2\alpha + 1)}{\sqrt{2\alpha + \frac{3}{2}}} t^{2\alpha}. \tag{25.38}$$

Then, (25.38) and (25.37), after rearranging, change (25.15) into

$$\frac{2\alpha + 1}{2\alpha + \frac{3}{2}} J_{\tilde{\nu}}(\tilde{\gamma}_n) + \tilde{\gamma}_n J'_{\tilde{\nu}}(\tilde{\gamma}_n) = 0. \tag{25.39}$$

By virtue of (25.34), the above may be rewritten as

$$\tilde{\nu} J_{\tilde{\nu}}(\tilde{\gamma}_n) + \tilde{\gamma}_n J'_{\tilde{\nu}}(\tilde{\gamma}_n) = 0. \tag{25.40}$$

This form of the equation corresponds to the Bessel functions property appearing in [4, p. 12, entry (55)]. Since zero may not be a root of (25.18), this amounts to

$$J_{\tilde{\nu}-1}(\tilde{\gamma}_n) = 0. \tag{25.41}$$

The meaning of the set of constants $\tilde{\gamma}_n$ has now been found: these are the (infinite) real positive zeros, arranged in ascending order of magnitude, of the Bessel function of the first kind and order given by

$$\tilde{\nu} - 1 = -\frac{1}{4\alpha + 3}. \tag{25.42}$$

No analytic formula explicitly yielding these zeros is known. However, a good approximated expression for them may be found by resorting to the asymptotic (for $x \to \infty$) expansion for the Bessel function of the first kind (see [4, p. 134]), which reads

$$J_\nu(x) \to \sqrt{\frac{2}{\pi x}} \cos\left(x - \frac{\pi\nu}{2} + \frac{\pi}{4}\right). \tag{25.43}$$

In fact, by replacing (25.42), (25.43) takes the form

$$J_{-\frac{1}{4\alpha+3}}(\tilde{\gamma}_n) \to \sqrt{\frac{2}{\pi\tilde{\gamma}_n}} \cos\left(\tilde{\gamma}_n - \frac{\pi}{2}\left[-\frac{1}{4\alpha+3}\right] - \frac{\pi}{4}\right). \tag{25.44}$$

It then follows from (25.41) that:

$$\tilde{\gamma}_n - \frac{\pi}{2}\left[-\frac{1}{4\alpha+3}\right] - \frac{\pi}{4} \approx n\pi - \frac{\pi}{2} \tag{25.45}$$

and finally

$$\tilde{\gamma}_n \approx \pi\left[n - \frac{1}{4} - \frac{1}{2(4\alpha+3)}\right] \quad \text{valid for } n = 1, 2, \ldots. \tag{25.46}$$

Next, we want to determine the normalization constants defined by (25.19); that is,

$$1 = \tilde{N}_n^2 \left[\int_0^T \tilde{f}(s)\, ds\right]^2 \int_0^1 x[J_{\tilde{\nu}((x))}(\tilde{\gamma}_n x)]^2\, dx. \tag{25.47}$$

Let us now replace (25.26) and the integral that is calculated within the framework of the Dini expansion in a series of Bessel functions (see [4, p. 71]); (25.47) thus becomes

$$1 = \tilde{N}_n^2 \left[\frac{KT^{2\alpha+\frac{3}{2}}}{2\alpha + \frac{3}{2}}\right]^2 \left\{\frac{1}{2}\left[J_{\tilde{\nu}}'^2(\tilde{\gamma}_n) + \left(1 - \frac{\tilde{\nu}^2}{\tilde{\gamma}^2}\right) J_{\tilde{\nu}}^2(\tilde{\gamma}_n)\right]\right\}.$$

Now, by virtue of (25.40), the derivative of the Bessel function disappears, and one gets

$$1 = \tilde{N}_n^2 \left[\frac{KT^{2\alpha+\frac{3}{2}}}{2\alpha + \frac{3}{2}}\right]^2 \left\{\frac{1}{2} J_{\tilde{\nu}}^2(\tilde{\gamma}_n)\right\}$$

or

$$\tilde{N}_n = \frac{\sqrt{2}\left(2\alpha + \frac{3}{2}\right)}{KT^{2\alpha+\frac{3}{2}}} \frac{1}{|J_{\tilde{\nu}}(\tilde{\gamma}_n)|} \tag{25.48}$$

as the exact expression of the normalization constants.

A well-approximated expression of the normalization constants (25.48) may be obtained by first working out the asymptotic expression—via (25.43)—for the Bessel function appearing in the denominator of (25.48). To this end, (25.43) and (25.34) yield

$$|J_{\tilde{\nu}}(\tilde{\gamma}_n)| \approx \left| \sqrt{\frac{2}{\pi\tilde{\gamma}_n}} \cos\left(\tilde{\gamma}_n - \frac{\pi}{2}\left[\frac{2\alpha+1}{2\alpha+\frac{3}{2}}\right] - \frac{\pi}{4}\right) \right| \tag{25.49}$$

and the approximated formula (25.46) produces

$$\begin{aligned} |J_{\tilde{\nu}}(\tilde{\gamma}_n)| &\approx \sqrt{\frac{2}{\pi\tilde{\gamma}_n}} \left| \cos\left(\pi\left[n - \frac{1}{4} - \frac{1}{2(4\alpha+3)}\right] - \frac{\pi}{2}\left[\frac{2\alpha+1}{2\alpha+\frac{3}{2}}\right] - \frac{\pi}{4}\right) \right| \\ &= \sqrt{\frac{2}{\pi\tilde{\gamma}_n}} \left| \cos\left(\pi\left[n - \frac{1}{2} - \frac{1+2(2\alpha+1)}{2(4\alpha+3)}\right]\right) \right| \\ &= \sqrt{\frac{2}{\pi\tilde{\gamma}_n}} |\cos(\pi[n-1])| \\ &= \sqrt{\frac{2}{\pi\tilde{\gamma}_n}} |(-1)^{n+1}| = \sqrt{\frac{2}{\pi\tilde{\gamma}_n}}. \end{aligned} \tag{25.50}$$

By substituting (25.50) and (25.46) into (25.48), the desired approximated expression for the normalization constants becomes

$$\begin{aligned} \tilde{N}_n &\approx \frac{\sqrt{2}\left(2\alpha + \frac{3}{2}\right)}{KT^{2\alpha+\frac{3}{2}}} \cdot \frac{\sqrt{\pi}\sqrt{\tilde{\gamma}_n}}{\sqrt{2}} \\ &= \frac{\pi\left(2\alpha + \frac{3}{2}\right)}{KT^{2\alpha+\frac{3}{2}}} \sqrt{n - \frac{1}{4} - \frac{1}{2(4\alpha+3)}}. \end{aligned} \tag{25.51}$$

Similarly, (25.21) yields the exact eigenvalues

$$\tilde{\lambda}_n = \frac{K^2 T^{4\alpha+3}}{\left(2\alpha + \frac{3}{2}\right)^2} \cdot \frac{1}{(\tilde{\gamma}_n)^2}, \tag{25.52}$$

while (25.46) gives the approximate expression

$$\tilde{\lambda}_n \approx \frac{K^2 T^{4\alpha+3}}{\left(2\alpha+\frac{3}{2}\right)^2} \cdot \frac{1}{\pi^2\left(n-\frac{1}{4}-\frac{1}{2(4\alpha+3)}\right)^2}. \tag{25.53}$$

In conclusion, using (25.48), (25.37) and (25.34), we have proven that the exact expression of the KL eigenfunctions reads

$$\tilde{\phi}_n(t) = \tilde{N}_n \cdot \tilde{\chi}(t) \cdot J_{\tilde{\nu}}\left(\tilde{\gamma}_n \frac{t^{2\alpha+\frac{3}{2}}}{T^{2\alpha+\frac{3}{2}}}\right) = \frac{\sqrt{2}\sqrt{2\alpha+\frac{3}{2}}\, t^{2\alpha+1}}{T^{2\alpha+\frac{3}{2}}|J_{\tilde{\nu}}(\tilde{\gamma}_n)|} J_{\tilde{\nu}}\left(\tilde{\gamma}_n \frac{t^{2\alpha+\frac{3}{2}}}{T^{2\alpha+\frac{3}{2}}}\right). \tag{25.54}$$

The approximated counterparts to (25.54) may be obtained by resorting to the exact expression (25.37), to the approximated expression (25.51) for the normalization constants, to the approximated (25.50) for the Bessel function at the denominator, and finally to (25.43) for the asymptotic expansion of the Bessel function of the first kind. The result is

$$\begin{aligned} \tilde{\phi}_n(t) &\approx \frac{\sqrt{4\alpha+3}\, t^{2\alpha+1}}{T^{2\alpha+\frac{3}{2}}\sqrt{\frac{2}{\pi\tilde{\gamma}_n}}} \cdot \sqrt{\frac{2}{\pi\left(\tilde{\gamma}_n \frac{t^{2\alpha+\frac{3}{2}}}{T^{2\alpha+\frac{3}{2}}}\right)}} \cdot \cos\left(\tilde{\gamma}_n \frac{t^{2\alpha+\frac{3}{2}}}{T^{2\alpha+\frac{3}{2}}} - \frac{\pi\tilde{\nu}}{2} - \frac{\pi}{4}\right) \\ &\approx \sqrt{4\alpha+3}\frac{t^{\alpha+\frac{1}{4}}}{T^{\alpha+\frac{3}{4}}} \cos\left(\tilde{\gamma}_n \frac{t^{2\alpha+\frac{3}{2}}}{T^{2\alpha+\frac{3}{2}}} - \frac{\pi(2\alpha+1)}{4\alpha+3} - \frac{\pi}{4}\right), \end{aligned} \tag{25.55}$$

where (25.34) was used to replace the order in the last expression. Finally, using (25.46) we get

$$\tilde{\phi}_n(t) \approx \sqrt{4\alpha+3}\frac{t^{\alpha+\frac{1}{4}}}{T^{\alpha+\frac{3}{4}}} \cdot \cos\left(\pi\left[\left(n-\frac{1}{4}-\frac{1}{2(4\alpha+3)}\right)\frac{t^{2\alpha+\frac{3}{2}}}{T^{2\alpha+\frac{3}{2}}} - \frac{12\alpha+7}{4(4\alpha+3)}\right]\right) \tag{25.56}$$

as the approximate expression for the KL orthonormalized eigenfunctions.

25.3 KL EXPANSION OF THE SQUARE PROCESS $B^2(t^{2H})$

In the present section, we derive the full expression for the KL expansion of the $B^2(t^{2H})$ process by resorting to the results obtained in Section 25.2. The notation in this section is thus consistent with that adopted in Appendix H for the $B(t^{2H})$ process, and this will allow useful comparisons to be made, particularly with regard to the self-similarity of both processes.

We start by setting

$$2\alpha+1 = 2H \tag{25.57}$$

from which

$$\alpha = H - \frac{1}{2}. \tag{25.58}$$

The range (25.25) is then replaced by the new range

$$H > 0. \tag{25.59}$$

The advantage of this notation lies in that the Gaussian process

$$X(t) = B\left(\int_0^t f^2(s)\,ds\right) = B\left(\frac{C^2}{2H}t^{2H}\right) \tag{25.60}$$

now plainly reveals its H-self-similarity. By this, we mean the $\frac{1}{2}$-self-similarity of the Brownian motion expressed by the formula (21.6); that is,

$$B(c \cdot t) = \sqrt{c} \cdot B(t) \tag{25.61}$$

(valid for any real positive constant c) is carried over to the process $X(t)$ by way of the generalization

$$X(c \cdot t) = c^H \cdot X(t). \tag{25.62}$$

The proof of this fact immediately follows from (25.60) and (25.61)

$$X(c \cdot t) = B\left(\frac{C^2}{2H} \cdot c^{2H} \cdot t^{2H}\right) = c^H \cdot B\left(\frac{C^2}{2H}t^{2H}\right) = c^H \cdot X(t). \tag{25.63}$$

Since the publication of Benoit Mandelbrot's book about fractals [5], the importance of H-self-similarity (25.63) is apparent. An investigation about a possible relationship between the processes studied in the present chapter and Mandelbrot's fractional Brownian motions deserves much deeper investigation. We confine ourselves to pointing out the further $2H$-self-similarity fulfilled by the $Y(t)$ process of (25.11)

$$Y(c \cdot t) = c^{2H} \cdot Y(t). \tag{25.64}$$

The relevant proof follows at once from (25.11) and (25.62):

$$\begin{aligned} Y(c \cdot t) &= X^2(c \cdot t) - E\{X^2(c \cdot t)\} = c^{2H}X^2(t) - E\{c^{2H}X^2(t)\} \\ &= c^{2H}[X^2(t) - E\{X^2(t)\}] = c^{2H}Y(t). \end{aligned} \tag{25.65}$$

Let us now turn to the KL expansion of the $Y(t)$ process. The autocorrelation of the $X(t)$ process is given by (21.30), that is

$$E\{X(t_1)X(t_2)\} = \int_0^{t_1 \wedge t_2} f^2(s)\,ds. \tag{25.66}$$

Upon setting $t_1 = t_2 = t$ in (25.66), the obvious formula $t \wedge t = t$, (25.22) and (25.24) yield

$$E\{X^2(t)\} = \int_0^t f^2(s)\,ds = \frac{C^2}{2H}t^{2H}. \tag{25.67}$$

Therefore, the definition (25.11) of $Y(t)$, by virtue of (25.60) and (25.67), becomes

$$Y(t) = X^2(t) - E\{X^2(t)\} = \frac{C^2}{2H}[B^2(t^{2H}) - t^{2H}]. \tag{25.68}$$

Since the KL expansion (25.12) has the form

$$Y(t) = \sum_{n=1}^{\infty} \tilde{Z}_n \tilde{\phi}_n(t) \tag{25.69}$$

from (25.68) and (25.69) the following KL expansion for the process $B^2(t^{2H})$ is inferred:

$$B^2(t^{2H}) - t^{2H} = \frac{2H}{C^2} \sum_{n=1}^{\infty} \tilde{Z}_n \tilde{\phi}_n(t). \tag{25.70}$$

Inserting now the exact orthonormalized eigenfunctions (25.54), rewritten by aid of the substitution (25.22), we get the explicit KL expansion

$$B^2(t^{2H}) - t^{2H} = \frac{2H}{C^2} \sum_{n=1}^{\infty} \tilde{Z}_n \frac{\sqrt{4H+1}\, t^{2H}}{T^{2H+\frac{1}{2}} |J_{\tilde{\nu}}(\tilde{\gamma}_n)|} J_{\tilde{\nu}}\left(\tilde{\gamma}_n \frac{t^{2H+\frac{1}{2}}}{T^{2H+\frac{1}{2}}}\right). \tag{25.71}$$

Here

(1) The order of the Bessel function of the first kind is constant in time, and reads

$$\tilde{\nu} = \frac{2H}{2H + \dfrac{1}{2}}. \tag{25.72}$$

(2) The set of constants $\tilde{\gamma}_n$ are defined as the real positive zeros of the Bessel function of the first kind and of order given by (25.34) minus 1; that is,

$$J_{\tilde{\nu}-1}(\tilde{\gamma}_n) = 0. \tag{25.73}$$

The approximate expression is derived by substituting the approximated orthonormalized eigenfunctions (25.56) into (25.70):

$$B^2(t^{2H}) - t^{2H} \approx \frac{2H}{C^2} \sum_{n=1}^{\infty} \tilde{Z}_n \sqrt{4H+1} \frac{t^{H-\frac{1}{4}}}{T^{H+\frac{1}{4}}} \cdot \cos\left(\pi\left[\left(n - \frac{1}{4} - \frac{1}{2(4H+1)}\right) \frac{t^{2H+\frac{1}{2}}}{T^{2H+\frac{1}{2}}} - \frac{12H+1}{4(4H+1)}\right]\right). \tag{25.74}$$

25.4 CHECKING THE KLT OF $B^2(t^{2H})$ BY MATLAB SIMULATIONS

Just look at Figure 25.1.

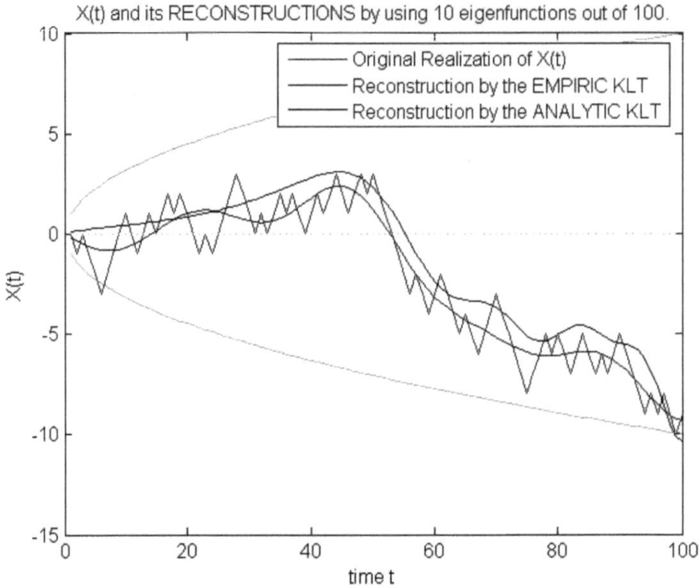

Figure 25.1. The *zero-mean* time-rescaled square Brownian motion $B^2(t^{2H}) - t^{2H} = X(t)$ vs. time t simulated as a random walk over 100 time instants. Next to the "bumpy curve" of $X(t)$, two more "smooth curves" are shown that *interpolate at best* the bumpy $X(t)$. These two curves are the KLT reconstruction of $X(t)$) *by using the first ten eigenfunctions only*. It is important to note that the two smooth curves are *different* in this case because the KLT expansion (25.74) is *approximated.* Actually, it is an approximated KLT expansion because the asymptotic expansion of the Bessel functions (25.43) was used. So, the two curves are different from each other, but both still interpolate $X(t)$ at best. Note that were we taking into account the full set of 100 KLT eigenfuctions, rather than just 10, then the *empiric* reconstruction would overlap $X(t)$ *exactly*, but the *analytic* reconstruction would not because of the use of the asymptotic expansion (25.43) of the Bessel functions.

25.5 REFERENCES

[1] C. Maccone, "Eigenfunctions and Energy for Time-Rescaled Gaussian Processes," *Bollettino dell'Unione Matematica Italiana*, Series 6, **3-A** (1984), 213–219.

[2] C. Maccone, "Special Relativity and the Karhunen–Loève Expansion of Brownian Motion," *Il Nuovo Cimento*, **100-B** (1987), 329–341.

[3] C. Maccone, "The Karhunen–Loève Expansion of the Square of a Time-Rescaled Gaussian Process," *Bollettino dell'Unione Matematica Italiana*, Series 7, **2-A** (1988), 221–229.

[4] A. Erdelyi, W. Magnus, F. Oberhettinger, and F. G. Tricomi, *Higher Transcendental Functions*, Vol. 2, McGraw-Hill, New York, 1953.

[5] B. Mandelbrot, *The Fractal Geometry of Nature*, Freeman, San Francisco, 1982.

26

Maccone third KLT theorem: Asymptotic KLT of GBM

26.1 INTRODUCTION

Geometric Brownian motion (GBM) (also known as exponential Brownian motion) is a continuous time stochastic process in which the logarithm of the randomly varying quantity follows a Brownian motion. It is used in mathematical finance to model stock prices in the Black–Scholes model (see *http://en.wikipedia.org/wiki/Geometric_Brownian_motion*).

In Chapter 12, we showed that GBM is the natural extension in time of our statistical Drake equation, described in Chapter 1. As such, GBM may be used to investigate a host of natural phenomena related to SETI: from the statistical Drake equation to the evolution and increase of the number of planets for man and aliens as described in Chapter 3.

The KLT of GBM is unknown to the best of this author's knowledge. However, in September 2007 this author was able to discover an asymptotic (as $t \to \infty$) KLT of GMB based on his own theory of time-rescaled Gaussian processes $X(t)$ as developed in Chapter 22. Our asymptotic KLT of GMB is published in this chapter for the first time ever.

26.2 TIME-RESCALING FUNCTION FOR THE EXPONENTIAL AUTOCORRELATION $e^{t_1 \wedge t_2}$

Consider the exponential autocorrelation $e^{t_1 \wedge t_2}$, with

$$e^{t_i \wedge t_2} \quad \text{with} \quad (t_1 \wedge t_2) \equiv \min(t_1, t_2) = \begin{cases} t_1 & \text{if } t_1 < t_2 \\ t_2 & \text{if } t_2 < t_1. \end{cases} \qquad (26.1)$$

Then, the KLT integral equation (16.18) for the autocorrelation (26.1) reads

$$\int_0^T e^{t_1 \wedge t_2} \varphi_n(t_2) dt_2 = \lambda_n \varphi_n(t_1). \tag{26.2}$$

In order to expand this integral, we must resort to expressing $e^{t_1 \wedge t_2}$ in terms of the unit step function $U(t)$; that is

$$e^{t_1 \wedge t_2} = e^{t_1} U(t_2 - t_1) + e^{t_2} U(t_1 - t_2). \tag{26.3}$$

Inserting (26.3) into (26.2) yields:

$$\begin{aligned} \lambda_n \varphi_n(t_1) &= \int_0^T e^{t_1 \wedge t_2} \varphi_n(t_2) dt_2 = \int_0^T [e^{t_1} U(t_2 - t_1) + e^{t_2} U(t_1 - t_2)] \varphi_n(t_2) dt_2 \\ &= e^{t_1} \int_0^T U(t_2 - t_1) \varphi_n(t_2) dt_2 + \int_0^T e^{t_2} U(t_1 - t_2) \varphi_n(t_2) dt_2 \\ &= e^{t_1} \int_{t_1}^T \varphi_n(t_2) dt_2 + \int_0^{t_1} e^{t_2} \varphi_n(t_2) dt_2. \end{aligned} \tag{26.4}$$

At this point we may differentiate both sides of (26.4) with respect to t_1 with the result that all integrals but one cancel against each other:

$$\lambda_n \frac{d\varphi_n(t_1)}{dt_1} = e^{t_1} \int_{t_1}^T \varphi_n(t_2) dt_2 - e^{t_1} \varphi_n(t_1) + e^{t_1} \varphi_n(t_1) = e^{t_1} \int_{t_1}^T \varphi_n(t_2) dt_2. \tag{26.5}$$

Note that, upon setting $t_1 = T$ in the last equation, the integral on the right-hand side vanishes, and one is left with the final condition (i.e., the boundary condition at the final instant $t = T$) on the derivative of the eigenfunctions:

$$\phi_n'(T) = 0. \tag{26.6}$$

Next, we may divide (26.5) by e^{t_1} and then differentiate both sides once more with respect to t_1. After rearranging, and replacing t_1 by the more traditional t, the result is

$$\frac{d}{dt}\left[\frac{1}{e^t} \frac{d\varphi_n(t)}{dt}\right] + \frac{1}{\lambda_n} \varphi_n(t) = 0. \tag{26.7}$$

This is the differential equation yielding the KLT eigenfunctions $\varphi_n(t)$ for time-rescaled Brownian motion having the autocorrelation (26.1). Checking it against the general differential equation (22.25) for all time-rescaled eigenfunctions, we immediately infer that the time-rescaling function $f(t)$ for the case under consideration is:

$$f^2(t) = e^t \quad \text{whence} \quad f(t) = e^{\frac{t}{2}}. \tag{26.8}$$

This is the key result we needed in order to find the KLT of geometric Brownian motion.

26.3 APPROXIMATED (ASYMPTOTIC FOR $t \to \infty$) KLT EIGENVALUES AND EIGENFUNCTIONS

Having found the time-rescaling function (26.8), in order to find the relevant KLT we just need to follow up the procedure described in Section 18.1. This is fully described in the Maxima file that makes up Appendix 26.A, which we now explain in detail for the reader's benefit. The two integrals

$$\int_0^t f^2(s)\,ds = \int_0^t e^s\,ds = e^t - 1 \tag{26.9}$$

and

$$\int_0^t f(s)\,ds = \int_0^t e^{\frac{s}{2}}\,ds = 2(e^{\frac{t}{2}} - 1) \tag{26.10}$$

are immediately found by Maxima as (%o2) and (%o3), respectively.

Then comes the Bessel function order $\nu(t)$. It is given by (18.9) or by (22.50), whichever one prefers: we rewrite them here for convenience. Equation (18.19) is the same as (22.43) and that reads

$$\nu(t) = \sqrt{-\frac{\chi^3(t)}{f^2(t)}\frac{d}{dt}\left[\frac{\chi'(t)}{f^2(t)}\right]}. \tag{26.11}$$

Equation (22.50) reads

$$\nu(t) = \sqrt{\frac{1}{4} + \left[\frac{\int_0^t f(s)\,ds}{f(t)}\right]^2 \left\{\frac{3}{4}\left[\frac{d\ln f(t)}{dt}\right]^2 - \frac{1}{2}\frac{d^2\ln f(t)}{dt^2}\right\}}. \tag{26.12}$$

Before finding these expressions for $\nu(t)$, however, one needs to find the $\chi(t)$ function defined by (18.10), which Maxima yields as:

$$\chi(t) = \sqrt{2}\,e^{\frac{t}{4}}\sqrt{e^{\frac{t}{2}} - 1}. \tag{26.13}$$

This is equation (%o18) of the Maxima file. The derivative of (26.13) is given by equation (%o19), and the derivative of (26.13) under the square root of Bessel function order $\nu(t)$ in (18.9) is given by equation (%o20), while (%o21) yields the two derivatives of (26.13) needed for the Bessel function order $\nu(t)$ given by (18.9). After that, the Bessel function order $\nu(t)$ is found by Maxima in either of the following two equivalent forms, corresponding to (26.11) and (26.12), respectively:

$$\nu(t) = \frac{e^{-\frac{t}{2}}\sqrt{4e^t - 6e^{\frac{t}{2}} + 3}}{2} \quad \text{or} \quad \nu(t) = \sqrt{\frac{3e^{-t}(e^{\frac{t}{2}} - 1)^2}{4} + \frac{1}{4}} \tag{26.14}$$

also corresponding to (%o4) and (%o5), respectively. Now, by carefully looking at the second equation of (26.14), one comes to realize that this $\nu(t)$ function starts at $t = 1/2$ and then is nearly like a horizontal straight line with a constant ordinate equal to 1 (see the relevant plot in Figure 26.1).

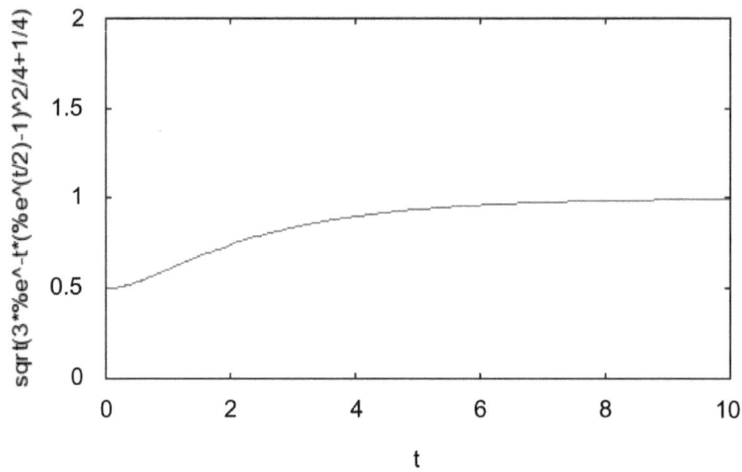

Figure 26.1. The KLT Bessel function order $\nu(t)$ is practically the same as the horizontal straight line of ordinate 1 that starts a little way from the origin and then on to infinity. In other words, we may *replace* the Bessel function order $\nu(t)$ by the number 1 in all the following equations to get the KLT asymptotic expansion of geometric Brownian motion for $t \to \infty$. This discovery simplifies things greatly for all subsequent KLT calculations for such time rescaling.

Our next (and rather difficult) task is to find how we can express the constants γ_n (even if approximated for $t \to \infty$) as zeros of the complicated equation (18.11), which we rewrite here for convenience

$$\chi'(T) J_{\nu(T)}(\gamma_n) + \chi(T) \left[\frac{f(T)\gamma_n}{\int_0^T f(s)\, ds} J'_{\nu(T)}(\gamma_n) + \frac{\partial J_{\nu(T)}(\gamma_n)}{\partial \nu} \nu'(T) \right] = 0. \quad (26.15)$$

Let us first compute the various functions of the final instant T appearing in (26.15):

(1) The derivative $\nu'(T)$ is clearly equal to zero (since, asymptotically, one has $\nu(t) = 1$). Thus, the last term in brackets in (26.15) vanishes, and one is left with

$$\chi'(T) J_{\nu(T)}(\gamma_n) + \chi(T) \frac{f(T)\gamma_n}{\int_0^T f(s)\, ds} J'_{\nu(T)}(\gamma_n) = 0. \quad (26.16)$$

(2) The coefficient of the first term in (26.16) is computed by equations (%i24) and (%o24).

(3) The remaining coefficient of the second term in (26.16) is computed by equations (%i25) through (%o27).

By inserting all these functions of the final instant T into (26.16) , the latter takes the form given by (%o28), which we reproduce here as an image

$$\frac{\mathrm{J1\ prime}\ \gamma_n\ \%\mathrm{e}^{\frac{3\mathrm{T}}{4}}}{\sqrt{2}\sqrt{\%\mathrm{e}^{\mathrm{T}/2}-1}} + \frac{\mathrm{J1}\ \%\mathrm{e}^{\mathrm{T}/4}(2\ \%\mathrm{e}^{\mathrm{T}/2}-1)}{2^{3/2}\sqrt{\%\mathrm{e}^{\mathrm{T}/2}-1}} = 0. \quad (26.17)$$

When simplified, (26.17) takes the form given by (%o29); that is

$$\gamma_{\rm n} = -\frac{\%{\rm e}^{-\frac{T}{2}}(2\,{\rm J1}\,\%{\rm e}^{T/2} - {\rm J1})}{2\,{\rm J1\_prime}} \tag{26.18}$$

which, after rearranging, becomes (%o31)

$$2\,{\rm J1} + 2\,{\rm J1\_prime}\,\gamma_{\rm n} = {\rm J1}\,\%{\rm e}^{-\frac{T}{2}}. \tag{26.19}$$

This equation may indeed be further simplified if one lets $T \to \infty$. This is acceptable inasmuch as we are seeking the asymptotic KLT of GBM for $t \to \infty$. Then, in this asymptotic approximation of the KLT for large values of time, (26.19) further reduces to (%o33); that is

$$ {\rm J1} + {\rm J1\_prime}\,\gamma_{\rm n} = 0 \tag{26.20}$$

When rewritten extensively in terms of Bessel functions (i.e., looking at (26.16)) this equation really amounts to

$$J_1(\gamma_n) + \gamma_n J_1'(\gamma_n) = 0. \tag{26.21}$$

Can this be simplified any further? Yes, since it is actually just a special case of the typical Bessel function property

$$J_\nu(z) + zJ_\nu'(z) = zJ_{\nu-1}(z), \tag{26.22}$$

proven in [1, p. 55, entry (6)]. Thus, inserting $\nu = 1$ and $z = \gamma_n$ into (26.22), (26.21) is changed into

$$0 = J_1(\gamma_n) + \gamma_n J_1'(\gamma_n) = \gamma_n J_0(\gamma_n). \tag{26.23}$$

Since the γ_n may not equal zero, (26.23) finally boils down to

$$J_0(\gamma_n) = 0 \tag{26.24}$$

proving the key result that the constants γ_n are just the zeros of the Bessel function $J_0(t)$. Unfortunately, the analytic Bessel function theory, as given in [1], shows that there is no analytic expression for the zeros of any Bessel function $J_\nu(t)$. But, worry not! We are actually seeking the asymptotic KLT of GBM for $t \to \infty$. Thus, we may resort to the asymptotic expansion formula for all Bessel functions of the first kind, (18.63), which we rewrite here for convenience

$$\lim_{x\to\infty} J_\nu(x) = \lim_{x\to\infty} \sqrt{\frac{2}{\pi x}} \cos\left(x - \frac{\nu\pi}{2} - \frac{\pi}{4}\right). \tag{26.25}$$

Inserting $\nu = 0$ and $x = \gamma_n$ into (26.25), (26.24) becomes

$$\cos\left(\gamma_n - \frac{\pi}{4}\right) = 0 \tag{26.26}$$

with $n = 1, 2, 3, \ldots$ and not $n = 0$ since the counting of eigenvalues λ_n starts at 1. This means that (26.26) yields

$$\gamma_n - \frac{\pi}{4} = n\pi - \frac{\pi}{2} \tag{26.27}$$

and finally

$$\gamma_n = \pi(n - \tfrac{1}{4}) \quad \text{for} \quad n = 1, 2, 3 \ldots \tag{26.28}$$

These are the long-sought approximated constants γ_n for values of t away from the origin (i.e., for $t \to \infty$).

Let us now turn to the eigenvalues λ_n, given by (22.48), which we rewrite here for convenience

$$\lambda_n = \left[\int_0^T f(s)\, ds\right]^2 \frac{1}{(\gamma_n)^2}. \tag{26.29}$$

Inserting (26.10) and (26.28) into (26.29), the latter becomes the expression of GBM eigenvalues λ_n for values of t and T away from the origin (i.e., for $t \to \infty$ and $T \to \infty$):

$$\lambda_n = \frac{4(e^{\frac{T}{2}} - 1)^2}{(\gamma_n)^2} = \frac{4(e^{\frac{T}{2}} - 1)^2}{\pi^2 (n - \frac{1}{4})^2}. \tag{26.30}$$

Of course, (26.30) shows that $\lambda_n \to 0$ for $n \to \infty$, which it has to be in order to have a convergent KL expansion (16.2) for GBM. All these results about GBM eigenvalues λ_n are proven in equations (%i37) through (%o40) of the Maxima file. But, the eigenvalues λ_n are also important for the BAM-KLT, as described in Sections 16.7 through 16.15. On the one hand, while it is true that the BAM-KLT only holds rigorously for stationary processes and that the GBM is certainly not a stationary process, if the "drift" constant μ has a numeric value around zero, then (12.1) is reasonably close to a horizontal straight line, and so BAM-KLT conditions may apply in the first approximation. Therefore, equations (%i41) through (%o50) of the Maxima file yield the first four derivatives with respect to T of both the eigenvalues (26.30) and of the dominant (=first) eigenvalue λ_1 only, showing that they all increase exponentially in time like e^T. More would be needed to explore the BAM-KLT of GBM, but we prefer to stop at this point.

Our next topic of investigation is the normalization constants N_n given by (18.12) (i.e., (22.47)), which we rewrite here for convenience

$$1 = \int_0^T \varphi_n^2(t)\, dt = N_n^2 \left[\int_0^T f(s)\, ds\right]^2 \int_0^1 x[J_{\nu((x))}(\gamma_n x)]^2 dx. \tag{26.31}$$

The integral involving the Bessel function $J_1(\gamma_n x)$ is calculated within the framework of the Dini series in the theory of Bessel functions, and has already been used in equation (23.30), which we rewrite here for convenience

$$\int_0^1 x J_\nu^2(\gamma_n x)\, dx = \frac{1}{2\gamma_n^2}[\gamma_n^2 J_\nu'^2(\gamma_n) + (\gamma_n^2 - \nu^2) J_\nu^2(\gamma_n)]. \tag{26.32}$$

In the $\nu = 1$ case, (26.32) becomes

$$\int_0^1 x J_1^2(\gamma_n x)\, dx = \frac{1}{2\gamma_n^2}[\gamma_n^2 J_1'^2(\gamma_n) + (\gamma_n^2 - 1) J_1^2(\gamma_n)]. \tag{26.33}$$

This formula may however be simplified by removing the expression of $\gamma_n^2 J_1'^2(\gamma_n)$ obtained by virtue of (26.21) from it. In fact, (26.21) yields

$$[\gamma_n J_1'(\gamma_n)]^2 = [-J_1(\gamma_n)]^2 = J_1^2(\gamma_n). \tag{26.34}$$

Now, inserting (26.34) into the right-hand side of (26.33), the latter is turned into

$$\int_0^1 x J_1^2(\gamma_n x)\, dx = \tfrac{1}{2} J_1^2(\gamma_n). \tag{26.35}$$

Inserting this into (26.31) and inserting (26.10) as well, (26.31) becomes

$$1 = 2N_n^2 (e^{\frac{T}{2}} - 1)^2 J_1^2(\gamma_n). \tag{26.36}$$

This is equation (%o52) of the Maxima file, which we rewrite here for convenience as an image

$$2\, N_n^2\, J1^2 (\%e^{T/2} - 1)^2 = 1 \tag{26.37}$$

Of course, we must now solve this for N_n discarding the negative root and introducing the absolute value for $|J_1(\gamma_n)|$. We then get the expression

$$N_n = \frac{1}{\sqrt{2}(e^{\frac{T}{2}} - 1)|J_1(\gamma_n)|} \tag{26.38}$$

for N_n. This corresponds to equation (%o54) of the Maxima file, which we rewrite here as an image for convenience

$$\mathrm{N_n} = \frac{1}{\sqrt{2}\, \mathrm{J1}\, (\%\mathrm{e}^{\mathrm{T}/2} - 1)} \tag{26.39}$$

Is it still possible to find a better expression for $|J_1(\gamma_n)|$? The answer is yes if we once again use the asymptotic expansion (26.25) for $J_1(\gamma_n)$. Thus, inserting $\nu = 1$ and $x = \gamma_n$ into (26.25) and invoking (26.28) as well, (26.25) yields

$$\begin{aligned} |J_1(\gamma_n)| &= \sqrt{\frac{2}{\pi\gamma_n}} \left|\cos\left(\gamma_n - \frac{\pi}{2} - \frac{\pi}{4}\right)\right| = \sqrt{\frac{2}{\pi\gamma_n}} \left|\cos\left(n\pi - \frac{\pi}{4} - \frac{\pi}{2} - \frac{\pi}{4}\right)\right| \\ &= \sqrt{\frac{2}{\pi\gamma_n}} |\cos(n\pi - \pi)| = \sqrt{\frac{2}{\pi\gamma_n}} |(-1)^{n-1}| = \sqrt{\frac{2}{\pi\gamma_n}}. \end{aligned} \tag{26.40}$$

Inserting the last expression (26.40) into (26.38), the final expression of the normalization constants N_n is thus found

$$N_n = \frac{\sqrt{\pi}\,\sqrt{\gamma_n}}{2(e^{\frac{T}{2}} - 1)} = \frac{\pi}{2(e^{\frac{T}{2}} - 1)} \sqrt{n - \tfrac{1}{4}} \tag{26.41}$$

where (26.28) was also used. This is equation (%o57) of the Maxima file.

We are now ready to write the full KL (asymptotic) expansion of GBM. This is given by (18.8), which we rewrite here for convenience

$$X(t) = \sum_{n=1}^{\infty} Z_n N_n \sqrt{f(t) \int_0^t f(s)\, ds} \cdot J_{\nu(t)}\left(\gamma_n \frac{\int_0^t f(s)\, ds}{\int_0^T f(s)\, ds}\right). \tag{26.42}$$

In other words, the asymptotic eigenfunctions $\phi_n(t)$ are given by

$$\phi_n(t) = N_n \sqrt{f(t) \int_0^t f(s)\, ds} \cdot J_{\nu(t)}\left(\gamma_n \frac{\int_0^t f(s)\, ds}{\int_0^T f(s)\, ds}\right). \tag{26.43}$$

Equations (%i58) through (%o62) of the Maxima file compute all terms in (26.43) and come up with the expression (%o62), which we rewrite here for convenience as an image

$$\phi_n = \frac{\sqrt{\pi}\,\sqrt{\gamma_n}\ \%e^{t/4}\,\sqrt{\%e^{t/2}-1}\ J1\left(\dfrac{\gamma_n(\%e^{t/2}-1)}{\%e^{T/2}-1}\right)}{\sqrt{2}\,(\%e^{T/2}-1)} \tag{26.44}$$

In this expression one Bessel function still exists, namely

$$J1\left(\frac{\gamma_n(\%e^{t/2}-1)}{\%e^{T/2}-1}\right) \tag{26.45}$$

Can we make it easier? Yes, if we resort once again to the asymptotic expansion (26.25). Then, inserting $\nu = 1$ and the argument appearing inside (26.45) into (26.25), the asymptotic expression of (26.45) turns out to equal equation (%o63) of the Maxima file; that is

$$\frac{\sqrt{2}\sqrt{\dfrac{\%e^{T/2}-1}{\gamma_n(\%e^{t/2}-1)}}\cos\left(\dfrac{\gamma_n(\%e^{t/2}-1)}{\%e^{T/2}-1} - \dfrac{3\pi}{4}\right)}{\sqrt{\pi}}. \tag{26.46}$$

When inserted into (26.44), (26.46) leads to remarkable simplifications: in fact, the two γ_n outside the cosine cancel each other, and the radicals get simplified too, so that one is left with equation (%o66) of the Maxima file; that is

$$\Phi_n = \frac{\%e^{t/4}\cos\left(\dfrac{\gamma_n(\%e^{t/2}-1)}{\%e^{T/2}-1} - \dfrac{3\pi}{4}\right)}{\sqrt{\%e^{T/2}-1}}. \tag{26.47}$$

This is the long-sought expression for asymptotic (for $t \to \infty$) KLT eigenfunctions.

The explicit expression (26.28) of the asymptotic γ_n may also be invoked, with the result that (26.47) is changed into equation (%o67) of the Maxima file; that is

$$\Phi_{\mathrm{n}} = \frac{\%\mathrm{e}^{\mathrm{t}/4} \cos\left(\frac{\pi(\mathrm{n}-\frac{1}{4})(\%\mathrm{e}^{\mathrm{t}/2}-1)}{\%\mathrm{e}^{\mathrm{T}/2}-1} - \frac{3\pi}{4}\right)}{\sqrt{\%\mathrm{e}^{\mathrm{T}/2}-1}}. \tag{26.48}$$

This completes our investigation of KLT asymptotic (for $t \to \infty$) eigenvalues and eigenfunctions for the time-rescaling function given by (26.8).

26.4 KLT OF GEOMETRIC BROWNIAN MOTION $N(t)$

Whenever one looks for the KL expansion of a stochastic process $X(t)$, this process $X(t)$ must have zero mean, as was clearly assumed in elementary KLT theory given in Sections (16.1) through (16.6).

Does the geometric Brownian motion $N(t)$ described in Chapter 11 have a zero mean? Of course not, because its mean is the exponential curve in time having equation $N_0\, e^{\mu t}$. So, the process $X(t)$ that we have studied in this chapter equals $N(t) - N_0\, e^{\mu t}$; namely, one has

$$X(t) = N(t) - N_0\, e^{\mu t}. \tag{26.49}$$

On the other hand, the eigenfunctions (26.47) are given by

$$\phi_n(t) = \frac{e^{\frac{t}{4}}}{\sqrt{e^{\frac{T}{2}} - 1}} \cos\left(\gamma_n \frac{e^{\frac{t}{2}} - 1}{e^{\frac{T}{2}} - 1} - \frac{3\pi}{4}\right). \tag{26.50}$$

Merging (26.49) and (26.50), we may thus state the following result: The asymptotic ($t \to \infty$) KL expansion of the stochastic GBM process studied in Chapter 11 is given by

$$N(t) = N_0\, e^{\mu t} + \sum_{n=1}^{\infty} Z_n \phi_n(t) = N_0\, e^{\mu t} + \sum_{n=1}^{\infty} Z_n \frac{e^{\frac{t}{4}}}{\sqrt{e^{\frac{T}{2}} - 1}} \cos\left(\gamma_n \frac{e^{\frac{t}{2}} - 1}{e^{\frac{T}{2}} - 1} - \frac{3\pi}{4}\right). \tag{26.51}$$

In (26.51), the constants γ_n are given by

$$\gamma_n = \pi(n - \tfrac{1}{4}) \quad \text{for} \quad n = 1, 2, 3 \ldots \tag{26.52}$$

and the eigenvalues λ_n are given by

$$\lambda_n = \frac{4(e^{\frac{T}{2}} - 1)^2}{(\gamma_n)^2} = \frac{4(e^{\frac{T}{2}} - 1)^2}{\pi^2(n - \frac{1}{4})^2}. \tag{26.53}$$

This completes our asymptotic KL expansion of geometric Brownian motion for large values of time.

26.5 REFERENCE

[1] G. N. Watson, *A Treatise on the Theory of Bessel Functions*, Cambridge University Press, 1966.

APPENDIX 26.A

GEOMETRIC BROWNIAN MOTION
asymptotic KLT for t -> infinity.

Namely, asymptotic KL expansion of the "exponential Brownian motion" (as we prefer to call it) e^B(t) for t tending to infinity.

Discovered by Claudio Maccone on September 21th, 2007, but never published until 2012, here for the first time.

This Maxima commented code was completed on January 1st, 2012.

```
(%i1) kill(all);
(%o0) done
```

Defining the TIME-RESCALING FUNCTION f(t) for Geometric Brownian Motion (please see Chapter 26 for the relevant justifiction).

```
(%i1) def_f:f(t):=%e^(t/2);
```

(%o1) $f(t):=\%e^{t/2}$

Computing the integral of the square of this time-rescaling function f(t) (in the language of special relativity, this is the proper time tau of the new time-rescaled Brownian motion B(tau), as given by eq. (18.3), but forget about special relativity here! In fact, here we are just interested to TIME-RESCALING as a "mathematical trick" to find the KLT of Geometric Brownian Motion for t tending to infinity !).

```
(%i2) integral_of_f_of_t_square:'integrate('f(s)^2, s, 0, t)=integrate(f(s)^2, s, 0, t);
```

(%o2) $\int_0^t f(s)^2 ds = \%e^t - 1$

Computing the integral of the (NOT squared!) time-rescaling function.

```
(%i3) int_f:'integrate('f(s), s, 0, t)=factor(integrate(f(s), s, 0, t));
```

(%o3) $\int_0^t f(s) ds = 2\,(\%e^{t/2} - 1)$

FIRST DISCOVERY leading to the KL expansion of e^B(t):

e^B(t) may be rewritten as the TIME-RESCALED Brownian motion B(e^t-1), that is

$$e^{B(t)} - e^{t/2} = B(e^{t} - 1) .$$

Please see Chapter 26 for the relevant justification.

==

SECOND DISCOVERY:
For the time-rescaled Brownian motion with f(t)=e^(t/2), the ORDER nu(t) of the Bessel functions of the first kind J sub nu(t) IS A FUNTION OF t STARTING AT t=1/2 and then BEING NEARLY LIKE THE t=1 HORIZONTAL STRAIGHT LINE, i.e the asymptote t=1.

Namely, the function nu(t) for this f(t)=e^(t/2) is the function, written here below in its two different but equivalent forms, following from (18.10) and (22.50), respectively:

```
(%i4) first_form_of_Bessel_Function_order_nu:%nu(t)=(%e^(-t/2)*sqrt(4*%e^t-6*%e^(t/2)+3))/2;
```

(%o4) $$\nu(t)=\frac{\%e^{-\frac{t}{2}}\sqrt{4\,\%e^{t}-6\,\%e^{t/2}+3}}{2}$$

```
(%i5) second_form_of_Bessel_Function_order_nu:%nu(t)=sqrt((3*%e^(-t)*(%e^(t/2)-1)^2)/4+1/4);
```

(%o5) $$\nu(t)=\sqrt{\frac{3\,\%e^{-t}\,(\%e^{t/2}-1)^{2}}{4}+\frac{1}{4}}$$

These are of course the same thing:

```
(%i6) radcan(first_form_of_Bessel_Function_order_nu-second_form_of_Bessel_Function_order_nu);
(%o6) 0=0
```

And so we decide to assume as "definition" the second one, following from (22.50):

```
(%i7) def_of_nu:second_form_of_Bessel_Function_order_nu;
```

(%o7) $$\nu(t)=\sqrt{\frac{3\,\%e^{-t}\,(\%e^{t/2}-1)^{2}}{4}+\frac{1}{4}}$$

By carefully looking at the second version of this time function nu(t), one comes to realize that this nu(t) IS A FUNCTION OF t STARTING AT t=1/2 and then BEING NEARLY LIKE THE HORIZONTAL STRAIGHT LINE OF ORDINATE 1, i.e the nu=1 horizontal asymptote. In other words still, this function nu(t) (in its second version, here), when plotted for t ranging between 0 and infinity, looks like:

```
(%i8) wxplot2d([sqrt((3*%e^(-t)*(%e^(t/2)-1)^2)/4+1/4)], [t,0,10], [y,0,2])$
```

(%t8)

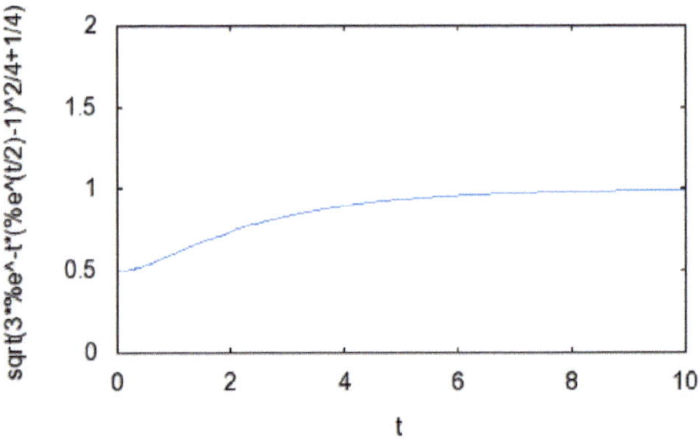


```
(%i9) limit(def_of_nu, t, 0);
```

(%o9) $\lim_{t \to 0} \nu(t) = \frac{1}{2}$

```
(%i10) limit(def_of_nu, t, inf);
```

(%o10) $\lim_{t \to \infty} \nu(t) = 1$

The numerical values of this nu(t) also reassure us of this asymptotic behaviour approaching 1, even for "moderate" initial values of t, between t=0 and, say, t=6 or t=8. Thus, ONE MIGHT APPROXIMANTELY (FOR t -> infinity) REPLACE nu(t) JUST BY ONE=1 !

```
(%i11) ev(subst(4,t,rhs(def_of_nu)), numer);
(%o11) 0.90040757677378

(%i12) ev(subst(6,t,rhs(def_of_nu)), numer);
(%o12) 0.96290106531289

(%i13) ev(subst(8,t,rhs(def_of_nu)), numer);
(%o13) 0.9862951579714
```

(%i14) ev(subst(10,t,rhs(def_of_nu)), numer);

(%o14) 0.99495081760291

(%i15) ev(subst(50,t,rhs(def_of_nu)), numer);

(%o15) 0.99999999998958

Let us now go ahead by evaluating the new function chi(t) given by eq. (18.10), its derivative, and all other functions making up the definition (18.10), that is:

(%i16) def_of_chi:%chi(t):=sqrt(f(t)*'integrate(f(s), s, 0, t));

(%o16) $\chi(t):=\sqrt{f(t)\int_0^t f(s)\,ds}$

(%i17) depends(%chi,t);

(%o17) $[\chi(t)]$

(%i18) function_chi:%chi=factor(sqrt(f(t)*integrate(f(s), s, 0, t)));

(%o18) $\chi=\sqrt{2}\,\%e^{t/4}\sqrt{\%e^{t/2}-1}$

(%i19) derivative_of_chi:factor(radcan(diff(function_chi,t)));

(%o19) $\frac{d}{d\,t}\chi=\frac{\%e^{t/4}\,(2\,\%e^{t/2}-1)}{2^{3/2}\sqrt{\%e^{t/2}-1}}$

(%i20) to_be_differentiated_in_eq_18_9:derivative_of_chi/f(t)^2;

(%o20) $\left(\frac{d}{d\,t}\chi\right)\%e^{-t}=\frac{\%e^{-\frac{3\,t}{4}}\,(2\,\%e^{t/2}-1)}{2^{3/2}\sqrt{\%e^{t/2}-1}}$

(%i21) derivative_under_root:factor(diff(to_be_differentiated_in_eq_18_9,t));

(%o21) $\left(\frac{d^2}{d\,t^2}\chi-\frac{d}{d\,t}\chi\right)\%e^{-t}=-\frac{\%e^{-\frac{3\,t}{4}}\,(4\,\%e^{t}-6\,\%e^{t/2}+3)}{2^{7/2}\,(\%e^{t/2}-1)^{3/2}}$

(%i22) from_def_18_10:%nu(t)=radcan(sqrt(-((rhs(function_chi)^3/(f(t)^2)))*rhs(derivative_under_root)));

(%o22) $\nu(t)=\frac{\%e^{-\frac{t}{2}}\sqrt{4\,\%e^{t}-6\,\%e^{t/2}+3}}{2}$

This is the function nu(t) as given by the definition (18.10).
Alternatively, the SAME function is of course found by resorting to (22.50):

(%i23) from_definition_22_50:%nu(t)=sqrt(1/4+((((rhs(int_f))/(rhs(def_f)))^2)*((3/4)*(diff(log(rhs(def_f)),t)^2) -(1/2)*diff(log(rhs(def_f)),t,2))));

(%o23) $$\nu(t)=\sqrt{\frac{3\,\%e^{-t}\,(\%e^{t/2}-1)^2}{4}+\frac{1}{4}}$$

In conclusion, we are now fully convinced that the Bessl function order nu(t) may be practically replaced by just nu(t) ~ 1 for all t >> 0.

Then, our next goal is to use the KLT exact theory given by the equations (18.8) thru (18.13) replacing nu = 1.

We thus start with (18.11).

Before doing so, however, we must compute all functions appearing in (18.11) as they are at the FINAL INSTANT t=T. These functions of T are, respecively:

(%i24) chi_dash_of_T:rhs(subst(T, t, derivative_of_chi));

(%o24) $$\frac{\%e^{T/4}\,(2\,\%e^{T/2}-1)}{2^{3/2}\sqrt{\%e^{T/2}-1}}$$

(%i25) chi_of_T:subst(T, t, rhs(function_chi));

(%o25) $$\sqrt{2}\,\%e^{T/4}\sqrt{\%e^{T/2}-1}$$

(%i26) f_of_T:rhs(subst(T, t, def_f));

(%o26) $$\%e^{T/2}$$

(%i27) intfofT:rhs(subst(T, t, int_f));

(%o27) $$2\,(\%e^{T/2}-1)$$

Thus, inserting all these functions into (18.11), it is turned into the following equation (notice that the time derivative of nu(t) is ZERO because nu(t)=1= a constant, so the corresponding term in (18.11) is ZERO, and it disappears from (18.11)).

(%i28) eq_18_11_for_f:chi_dash_of_T*J1+chi_of_T*(((f_of_T*%gamma[n]*J1_prime)/(intfofT)))=0;

(%o28) $$\frac{J1\_prime\,\gamma_n\,\%e^{\frac{3T}{4}}}{\sqrt{2}\sqrt{\%e^{T/2}-1}}+\frac{J1\,\%e^{T/4}\,(2\,\%e^{T/2}-1)}{2^{3/2}\sqrt{\%e^{T/2}-1}}=0$$

To simplify this equation, let us first solve it with respect to gamma sub n.

```
(%i29) eq_18_11_bis:first(solve(eq_18_11_for_f, %gamma[n]));
```

(%o29) $\gamma_n = -\dfrac{\%e^{-\frac{T}{2}}\,(2\,J1\,\%e^{T/2} - J1)}{2\,J1\_prime}$

Then isolate the only term with the "final time" T.

```
(%i30) eq_18_11_ter:expand(eq_18_11_bis*denom(rhs(eq_18_11_bis)));
```

(%o30) $2\,J1\_prime\,\gamma_n = J1\,\%e^{-\frac{T}{2}} - 2\,J1$

```
(%i31) eq_18_11_quater:eq_18_11_ter+2*J1;
```

(%o31) $2\,J1 + 2\,J1\_prime\,\gamma_n = J1\,\%e^{-\frac{T}{2}}$

Now remember that we are seeking an ASYMPTOTIC KLT for t->infinity. This already played a crucial role in assuming that the the ORDER nu of the Bessel functions J sub nu (t) equals just 1. On the other hand, to suppose that t-> infinity really implies supposing the same for the final instant T, i.e. we are supposing that T->infinity also. Thus, applying this remark to the last equation, we see that THE EXPONENTIAL IN T TENDS TO ZERO, and so we are just left with the following simplified form for equation (18.11).

```
(%i32) asymptotic_eq_18_11:limit(eq_18_11_quater, T, inf);
```

(%o32) $2\,J1 + 2\,J1\_prime\,\gamma_n = 0$

Getting rid of the common factor 2, this becomes:

```
(%i33) asymptotic_eq_18_11_bis:factor(asymptotic_eq_18_11)/2;
```

(%o33) $J1 + J1\_prime\,\gamma_n = 0$

Now one more important and highly simplifying remark.
In the theory of Bessel fuctions, one can prove the formula

nu J sub nu (z) + z J prime sub nu (z) = z J sub (nu-1) (z)

(for the proof, see G. N. Watson, A Treatise on the Theory of Bessel functions, Cambridge University Press, 1966, page 55, entry (6)).
Since in our case nu=1, this boils down to:

J1 (z) + z J1 prime (z) = z J0 (z).

Now, replacing z in this equation by the gamma sub n of our last equation, we immediately conclude that THE LAST EQUATION 18_11 AMOUNTS TO

J0 (gamma sub n) = 0.

This result is fundamental, since it allows us to FIND the asymptotic gamma sub n.
In fact one more formula in te theory of Bessel functions is the ASYMPOTIC EXPANSION OF J sub n (t) FOR t->infinity. It reads:

limit J sub nu (x) = limit sqrt(2/(%pi*x)) * cos(x - nu*%pi/2 -%pi/4).
x->infinity x->infinity

Applying this asympotitc expansion to J0(x), we find that the condition
J0 (gamma sub n) = 0, asymptotically for t->infinity is the same as the condition

cos(gamma sub n - %pi/4) = 0, with n=1, 2, 3, ...

THIS ALLOWS US TO FIND THE gamma sub n AS

gamma sub n - %pi/4 = n %pi - %pi/2

AND FINALLY:

```
(%i34) assume(n>0);
(%o34) [n>0]
```

```
(%i35) declare(n, integer);
(%o35) done
```

```
(%i36) def_gamma_sub_n: %gamma[n]=%pi*(n-1/4);
```

(%o36) $\gamma_n = \pi \left(n - \frac{1}{4} \right)$

THESE ARE THE LONG-SOUGHT gamma sub n, and they yield
the ASYMPTOTIC KLT EIGENVALUES lambda sub n.
In fact, according to (18.13), the KLT EIGENVALUES lambda sub n
are given by:

```
(%i37) eq_18_13_for_the_KLT_eigenvalues:%lambda[n]=(intfofT)^2/(%gamma[n])^2;
```

(%o37) $\lambda_n = \frac{4\,(\%e^{T/2}-1)^2}{\gamma_n^2}$

that is, inserting here the expression just found for the gamma sub n

```
(%i38) def_ASYMPTOTIC_KLT_EIGENVALUES:eq_18_13_for_the_KLT_eigenvalues,def_gamma_sub_n;
```

(%o38) $\lambda_n = \frac{4\,(\%e^{T/2}-1)^2}{\pi^2\left(n-\frac{1}{4}\right)^2}$

Clearly, the KLT eigenvalues approach zero as n becomes arbitrary large:

```
(%i39) limit(def_ASYMPTOTIC_KLT_EIGENVALUES, n, inf);
```

(%o39) $\lim_{n\to\infty} \lambda_n = 0$

Notice also that, when the final instant is supposed to be zero, i.e. when
there is NO KLT AT ALL (!) the found eigenvalues all (i.e. independently
of n) downgrade to zero:

```
(%i40) limit(def_ASYMPTOTIC_KLT_EIGENVALUES, T, 0);
```

(%o40) $\lambda_n = 0$

This is an important CHECK that what we have done so far is CORRECT!

But the really important thing for the BAM-KLT is how the eigenvalues
depend on the FINAL INSTANT T. Let us expand both squares in (%o37):

```
(%i41) new_eigenvalues:factor(ratsimp(expand(def_ASYMPTOTIC_KLT_EIGENVALUES)));
```

(%o41) $\lambda_n = \frac{64\,(\%e^{T}-2\,\%e^{T/2}+1)}{\pi^2\,(4\,n-1)^2}$

This clearly shows an EXPONENTIAL INCREASE IN THE TIME T for all eigenvalues. And since the derivative of %e^T with respect to T is just the same, i.e. %e^T, THE SAME EXPONENTIAL INCREASE IN TIME T IS TRUE FOR ALL TIME DERIVATIVES of the lambda sub n also:

```
(%i42) 'diff(lhs(new_eigenvalues), T)=diff(rhs(new_eigenvalues), T);
```

$$(\%o42)\quad \frac{d}{d\,T}\lambda_n = \frac{64\,(\%e^T - \%e^{T/2})}{\pi^2\,(4\,n-1)^2}$$

```
(%i43) 'diff(lhs(new_eigenvalues), T, 2)=diff(rhs(new_eigenvalues), T, 2);
```

$$(\%o43)\quad \frac{d^2}{d\,T^2}\lambda_n = \frac{64\left(\%e^T - \frac{\%e^{T/2}}{2}\right)}{\pi^2\,(4\,n-1)^2}$$

```
(%i44) 'diff(lhs(new_eigenvalues), T, 3)=diff(rhs(new_eigenvalues), T, 3);
```

$$(\%o44)\quad \frac{d^3}{d\,T^3}\lambda_n = \frac{64\left(\%e^T - \frac{\%e^{T/2}}{4}\right)}{\pi^2\,(4\,n-1)^2}$$

```
(%i45) 'diff(lhs(new_eigenvalues), T, 4)=diff(rhs(new_eigenvalues), T, 4);
```

$$(\%o45)\quad \frac{d^4}{d\,T^4}\lambda_n = \frac{64\left(\%e^T - \frac{\%e^{T/2}}{8}\right)}{\pi^2\,(4\,n-1)^2}$$

Let us now concentrate on the DOMINANT EIGENVALUE lambda sub 1:

```
(%i46) dominant_eigenvalue:subst(1, n, new_eigenvalues);
```

$$(\%o46)\quad \lambda_1 = \frac{64\,(\%e^T - 2\,\%e^{T/2} + 1)}{9\,\pi^2}$$

Its first four derivatives with respect to T read:

```
(%i47) 'diff(lhs(dominant_eigenvalue), T)=diff(rhs(dominant_eigenvalue), T);
```

$$(\%o47)\quad \frac{d}{d\,T}\lambda_1 = \frac{64\,(\%e^T - \%e^{T/2})}{9\,\pi^2}$$

(%i48) 'diff(lhs(dominant_eigenvalue), T, 2)=diff(rhs(dominant_eigenvalue), T, 2);

(%o48) $$\frac{d^2}{d\,T^2}\lambda_1 = \frac{64\left(\%e^{T} - \frac{\%e^{T/2}}{2}\right)}{9\,\pi^2}$$

(%i49) 'diff(lhs(dominant_eigenvalue), T, 3)=diff(rhs(dominant_eigenvalue), T, 3);

(%o49) $$\frac{d^3}{d\,T^3}\lambda_1 = \frac{64\left(\%e^{T} - \frac{\%e^{T/2}}{4}\right)}{9\,\pi^2}$$

(%i50) 'diff(lhs(dominant_eigenvalue), T, 4)=diff(rhs(dominant_eigenvalue), T, 4);

(%o50) $$\frac{d^4}{d\,T^4}\lambda_1 = \frac{64\left(\%e^{T} - \frac{\%e^{T/2}}{8}\right)}{9\,\pi^2}$$

Once again, we see that the dominant eigenvalue and all its derivatives wrt to T basically increase like %e^T.
It would be nice to explore the consequences of this fact for the BAM-KLT of Geometric Brownian Motion, but we don't have the time to do so right now.

Let next turn to the NORMALIZATION CONSTANTS N sub n of our asymptotic KLT of Geometric Brownian Motion for large values of T.
They are given by equation (18.12). In this equation we already know the integral of f(t) between 0 and T raised to the square, that is:

(%i51) (intfofT)^2;

(%o51) $4\left(\%e^{T/2} - 1\right)^2$

The difficulty, however, is to compute the integral of the Bessel function J1 appearing in (18.18). Now the reader should please read carefully the mathematical steps given in equations (18.68), (18.69), (18.70) and (18.71).
Although these four equations refer to the case where the time-rescaling function f(t) is given by (18.49), rather than by (%o1) as here now, the steps leading the the calculation of the integral of the Bessel function J1 are the same.
In conclusion, that integral is given by J1 SQUARED OVER 2, and so (18.18) in our present case yields:

(%i52) eq_18_12:N[n]^2*((intfofT)^2)*J1^2/2=1;

(%o52) $2\,N_n^2\,J1^2\left(\%e^{T/2} - 1\right)^2 = 1$

The next step is of course to solve the last equation with respect to N sub n, so that, by obviously discarding the negative root, one is left with:

(%i53) solve(eq_18_12, N[n]);

(%o53) $[N_n = -\frac{1}{\sqrt{2\ J1}\sqrt{\%e^T - 2\ \%e^{T/2} + 1}}, N_n = \frac{1}{\sqrt{2\ J1}\sqrt{\%e^T - 2\ \%e^{T/2} + 1}}]$

(%i54) def_N_sub_n:factor(radcan(last(%)));

(%o54) $N_n = \frac{1}{\sqrt{2\ J1}\ (\%e^{T/2} - 1)}$

The last equation corresponds to (18.72) for the case of f(t) given by (18.49).
Now the idea is that, using the asymptotic expansion (18.63) for all Bessel functions, the J1 appearing in (%o54) is really given by the following (%o56). The proof of this, however, is the same as the proof given by (18.73), that the reader should carefully read.

(%i55) asymptotic_J1:J1=rootscontract(sqrt(2/(%pi*%gamma[n])));

(%o55) $J1 = \sqrt{\frac{2}{\pi\ \gamma_n}}$

The, upon inserting (%o56) into (%o54), the latter becomes:

(%i56) N_sub_n_bis:def_N_sub_n,asymptotic_J1;

(%o56) $N_n = \frac{\sqrt{\pi}\ \sqrt{\gamma_n}}{2\ (\%e^{T/2} - 1)}$

and finally, inserting the gamma sub n of (%o36) into this, the NORMALIZATION CONSTANTS N sub n turn out to be given by:

(%i57) N_sub_n_ter:N_sub_n_bis,def_gamma_sub_n;

(%o57) $N_n = \frac{\pi\ \sqrt{n - \frac{1}{4}}}{2\ (\%e^{T/2} - 1)}$

Let us now write the ASYMPTOTIC KLT EIGENFUNCTIONS phi sub n (t).
According to (18.8), these phi sub n (t) are the product of three terms:
1) the Normalization constants N sub n
2) the fuction of the time chi(t)
3) the Bessel function J sub nu(t) (gamma sub n * integral of f between 0 and t OVER integral of f between 0 and T), hereafter abbreviated J1_with_argument.
Thus, the eigenfunctions phi sub n (t) read:

(%i58) def_eigenfunctions:phi [n]=rhs(N_sub_n_bis)*rhs(function_chi)*J1_with_argument;

(%o58) $$\phi_n = \frac{\sqrt{\pi}\ \mathrm{J1\_with\_argument}\ \sqrt{\gamma_n}\ \%e^{t/4}\ \sqrt{\%e^{t/2}-1}}{\sqrt{2\ (\%e^{T/2}-1)}}$$

Let us now find the actual argument of J1_with_argument. Since one has:

(%i59) int_f;

(%o59) $$\int_0^t f(s)\,ds = 2\ (\%e^{t/2}-1)$$

(%i60) intfofT;

(%o60) $$2\ (\%e^{T/2}-1)$$

then the function J1_with_argument is given by:

(%i61) def_J1_with_argument:J1_with_argument=J1(%gamma[n]*rhs(int_f)/intfofT);

(%o61) $$\mathrm{J1\_with\_argument} = \mathrm{J1}\left(\frac{\gamma_n\ (\%e^{t/2}-1)}{\%e^{T/2}-1}\right)$$

Thus, the Asymptotic KLT Eigenfuntions of the Geometric Brownian motion read (with the gamma sub n given by (%036)):

(%i62) def_eigenfunctions_bis:def_eigenfunctions,def_J1_with_argument;

(%o62) $$\phi_n = \frac{\sqrt{\pi}\ \sqrt{\gamma_n}\ \%e^{t/4}\ \sqrt{\%e^{t/2}-1}\ \mathrm{J1}\left(\frac{\gamma_n\ (\%e^{t/2}-1)}{\%e^{T/2}-1}\right)}{\sqrt{2\ (\%e^{T/2}-1)}}$$

But, in reality, we can still simplify this expression by taking once again advantage of the asymptotic expansion (18.63) for the Bessel function J1. Thus, setting nu=1 into (18.63) we see that J1_with_argument of (%061) becomes:

(%i63) aJ1a:sqrt(2/(%pi*(%gamma[n]*rhs(int_f)/intfofT)))*cos(%gamma[n]*rhs(int_f)/intfofT-%pi/2-%pi/4);

(%o63) $$\frac{\sqrt{2}\ \sqrt{\frac{\%e^{T/2}-1}{\gamma_n\ (\%e^{t/2}-1)}}\ \cos\left(\frac{\gamma_n\ (\%e^{t/2}-1)}{\%e^{T/2}-1}-\frac{3\ \pi}{4}\right)}{\sqrt{\pi}}$$

(%i64) def_asy_eigf_bis:asymptotic_phi[n]=rhs(N_sub_n_bis)*rhs(function_chi)*aJ1a;

(%o64) $$\text{asymptotic\_phi}_n = \frac{\sqrt{\gamma_n}\ \%e^{t/4}\ \sqrt{\%e^{t/2}-1}\ \sqrt{\frac{\%e^{T/2}-1}{\gamma_n\ (\%e^{t/2}-1)}}\ \cos\left(\frac{\gamma_n\ (\%e^{t/2}-1)}{\%e^{T/2}-1}-\frac{3\ \pi}{4}\right)}{\%e^{T/2}-1}$$

But radicals in front of the cosine may be simplified as follows:

(%i65) radcan(def_asy_eigf_bis);

(%o65) $$\text{asymptotic\_phi}_n = \frac{\%e^{t/4}\ \cos\left(\frac{3\ \pi\ \%e^{T/2}-4\ \gamma_n\ \%e^{t/2}+4\ \gamma_n-3\ \pi}{4\ \%e^{T/2}-4}\right)}{\sqrt{\%e^{T/2}-1}}$$

In conclusion, we may prefer to keep the expression (%o63) for the cosine

(%i66) def_asy_eigf_ter:Phi[n]=(%e^(t/4)/sqrt(%e^(T/2)-1))*cos(%gamma[n]*rhs(int_f)/intfofT-%pi/2-%pi/4);

(%o66) $$\Phi_n = \frac{\%e^{t/4}\ \cos\left(\frac{\gamma_n\ (\%e^{t/2}-1)}{\%e^{T/2}-1}-\frac{3\ \pi}{4}\right)}{\sqrt{\%e^{T/2}-1}}$$

(%i67) def_asy_eigf_with_explicit_n:def_asy_eigf_ter,def_gamma_sub_n;

(%o67) $$\Phi_n = \frac{\%e^{t/4}\ \cos\left(\frac{\pi\left(n-\frac{1}{4}\right)(\%e^{t/2}-1)}{\%e^{T/2}-1}-\frac{3\ \pi}{4}\right)}{\sqrt{\%e^{T/2}-1}}$$

These are our simplified eigenfunctions of the asymptotic KLT of Geometric Brownian Motion for t->infinity.

27

A Matlab code for KLT simulations

27.1 INTRODUCTION

After so much mathematics, it is natural to think of some computer code capable of simulating the KLTs derived analytically in this book.

The well-known Matlab environment is well-suited for such simulations inasmuch as it can handle both the eigenvalues and the eigenvectors of symmetric matrices that are at the heart of the KLT.

But this author is hardly an expert in Matlab programming! So he turned to one of his pupils, Dr. Nicolò Antonietti, and together they wrote the set of Matlab routines described in this appendix. This does not mean that such routines are "optimized", nor even that they are error-free! Thus, readers of this book might wish to improve on our work, and we would be most grateful if they could let us have their new Matlab codes by sending them to the author's e-mail address: *clmaccon@libero.it*. Thanks!

27.2 THE MAIN FILE "STANDARD_BROWNIAN_MOTION_MAIN.M"

The main file of our Matlab 7.1 set of routines is called "Standard_Brownian_Motion_MAIN.m" and is listed hereafter. It has plenty of comments, so our reader should not have difficulties in following what is going on.

```
% Matlab 7.1 Code for the SIMULATION OF THE KLT.
% Authors: Drs. Nicolò Antonietti & Claudio Maccone.
% This version was completed in January 2008.

% Clear the memory & worksheet.
clear all, close all, clc
```

```
% Define the initial value of the loop variable "runtime" as the
% initial cpu time. At the end of this simulation, the runtime will
% be equal to the final cpu time minus the initial cpu time.
run_time = cputime;

% Do you prefer to run a new simulation (creating a new and different
% realization of the stochastic process X(t)) or do you wish to load an
% existing data file (produced by a previous simulation) ?
[t, n, Input_Process_data, flag] = input_data_toggle;

% Decide how many eigenfuctions in the KL expansion you wish to take into
% account for the reconstruction of the process X(t) in the time interval
% between 0 and T. Clearly, the number of eigenfunctions taken into account
% is at most equal to the number of instants considered in the simulation.
% In practice, however, you may wish to use FEWER eigenfunctions, or even
% just VERY FEW eigenfuctions. The reconstruction of X(t) will thus be
% rougher and rougher, but the computation burden will still be affordable
% by your machine. THIS IS THE TRADE-OFF that the KLT offers to you as
% A LOSSY COMPRESSION ALGORITHM.
How_many_eigenfunctions = input('How many eigenfunctions ? \n');

% Computation of the ANALYTIC autocorrelation matrix of the Brownian
% motion, defined as min(t1, t2). This autocorrelation matrix is fed into
% the code only if you previously selected to run an entirely new
% simulation. If you previously selected to load a pre-existing data file
% (as it happens in all EXPERIMENTAL applications of the KLT), then the
% data file of the values of X(t), for t ranging between 0 and T, is fed
% into the code.
% The autocorrelation of the Brownian motion of size n is defined as
% min(t1, t2) by the function (i.e. by the subroutine)
% Brownian_Autocorrelation(n), hereby called by the Main code.
Autocorrelation_matrix = Brownian_Autocorrelation(t);

% The next step is the most important step in this Main code.
% By virtue of the "eigs" subroutine of Matlab, we avoid getting entangled
% in the computation of the eigenvalues Lambda and of the eigenfunctions Phi
% of the KLT. Quite simply, we feed in the Autocorrelation matrix (whether
% it was ANALYTIC or NUMERIC = EXPERIMENTAL) and "eigs" returns both Lambda
% and Phi! Clearly, in non-Matlab simulations, this "eigs" routine must be
% very carefully written!
[Phi,Lambda] = eigs(Autocorrelation_matrix,How_many_eigenfunctions);

% We now compute the EMPIRIC KLT (as opposed to the ANALYTIC KLT derived in
% the book analytically) for the simulation of X(t) under consideration.
%
% This EMPIRIC KLT we obtain in the following LOOP by:
%
% 1) PROJECTING the vector of the Input_Process_data (i.e. the vector
% representing the stochastic process X(t) to be KL-expanded) ONTO THE
% RELEVANT i-th EIGENVECTOR Phi(i). THIS PROJECTION IS THE RANDOM VARIABLE
% Z(i) of the KL expansion (as it follows by INVERTING the KL expansion,
% just as one does for the Fourier series).
%
% 2) DEFINING the ith term of the KL expansion as the product of Z(i) times
% Phi(i).
%

for i = 1:How_many_eigenfunctions,
  Z(i) = Input_Process_data.' * Phi(:,i);
  KL_EXPANSIONs_ith_term(:,i) = Z(i) * Phi(:,i);
end
```

```
% We now create the DATA VECTOR of the EMPIRIC RECONSTRUCTION of X(t)
% achieved by the KLT numerically. This is simply the sum of all the
% KL_EXPANSIONs_ith_term obtained in the previous step of this Main code.
% EMPIRIC_Data_vector = sum(KL_EXPANSIONs_ith_term,2);
EMPIRIC_Data_vector = sum(KL_EXPANSIONs_ith_term,2);

% Next we create the DATA VECTOR of the ANALYTIC RECONSTRUCTION of X(t) as
% given by the formulae mathematically demonstrated in the book.
% This requires a separate routine (named hereafter ANALYTIC_KLT) to be
% called up by this Main code. The text of this routine clearly changed
% according to which formula in the book we refer to.
if flag == 1
  [ANALYTIC_Data_vector] = ANALYTIC_KLT(Input_Process_data, n,
How_many_eigenfunctions, t, Lambda);
elseif flag == 2
  [ANALYTIC_Data_vector] = ANALYTIC_KLT_decelerated(Input_Process_data, n,
How_many_eigenfunctions, t, Lambda);
elseif flag == 3
  [ANALYTIC_Data_vector] = ANALYTIC_KLT_square_brow_motion(Input_Process_data, n,
How_many_eigenfunctions, t, Lambda);
elseif flag == 4
  [ANALYTIC_Data_vector] =
ANALYTIC_KLT_square_brow_dec_motion(Input_Process_data, n,
How_many_eigenfunctions, t, Lambda);
elseif flag == 5
  [ANALYTIC_Data_vector, t] = ANALYTIC_KLT_uniform_rel(Input_Process_data, n,
How_many_eigenfunctions, t, Lambda);
end

% Plot the EIGENVALUES of the EMPIRIC RECONSTRUCTION of X(t).
h2 = figure;

% Plots of:
% 1) The ORIGINAL REALIZATION of X(t).
% 2) The EMPIRIC RECONSTRUCTION OF X(t) by the KLT.
% 3) The ANALYTIC RECONSTRUCTION OF X(t) by the KLT.
graphic(Input_Process_data, EMPIRIC_Data_vector, ANALYTIC_Data_vector, flag,
How_many_eigenfunctions, t, n)

% Save the Input_Process_data as the Matlab file i01mat.
save i01 Input_Process_data

% How long it took to do all these calculations.
run_time = cputime - run_time
```

27.3 THE FILE "INPUT_DATA_TOGGLE.M"

The file "input_data_toggle.m" allows the user to choose which stochastic process to select for the KLT computation. Here is this file's listing.

```
% This subroutine allows you either to:
% 1) Create a brand-new REALIZATION of the input stochastic process X(t) or
% 3) Load an existing matlab file (.mat) where the input variables are
% saved
% 2) Load an existing file with all the numeric data of the input
% stochastic process. Clearly, this arises when you do EXPERIMENTAL work, %
such as getting the input of a radiotelescope, etc.
```

```
function [time_range,final_instant_T,process_values_vector, flag] = input_data_toggle

while true,
  flag = input('What process is going to be analized? \n 1. A standard brownian motion
from a still source \n 2. A standard brownian motion from a source in a decelerated
motion \n 3. A square standard brownian motion from a still source \n 4. A square
standard brownian motion from a source in decelerated motion \n 5. A uniform
relativistic motion \n');
  if flag==1 | flag==2 | flag==3 | flag==4 | flag==5 |flag==6
    break
  end
end

while true,
  case_number = input(' Enter 1 to create a NEW REALIZATION of the Brownian
motion X(t).\n Enter 2 to load an existing Brownian motion matlab file (.mat). \n Enter
3 to load an existing Brownian motion data file. \n');
  if case_number==1 | case_number==2 | case_number==3
    break
  end
end
  if case_number == 1
    % Creating the NEW REALIZATION of the stochastic process X(t)
    final_instant_T = input('Please, type the final time unit. (Suggested: no more than
1000) \n');
    time_range = (1:1:final_instant_T)';
    process_values_vector = process_path(final_instant_T);
    % Plot the Original Realization of X(t) to be later expanded and
    % reconstructed by virtue of the KLT.
    t = [0; time_range];
    random_walk = [0, process_values_vector];
    h0 = figure;
    parabola = sqrt(t);
    plot(t, random_walk,'-k', t,parabola,'-k', t,-parabola,'-k', t,0,'-k'), title(['REALIZATION of
B(t) over ', num2str(final_instant_T), ' time instants.']), xlabel('time t'), ylabel('X(t)')
    process_values_vector = process_values_vector';

    % Loading an existing matlab file (.mat) with its input stochastic
    % process.
  elseif case_number == 2
    nome = input('Please TYPE the full path and file name.\n','s');
    load(nome);
    process_values_vector = Input_Process_data;
    final_instant_T = length(Input_Process_data);
    time_range = (1:1:final_instant_T)';

    % Plot the Original Realization of X(t) to be later expanded and
    % reconstructed by virtue of the KLT.
    t = [0; time_range];
    random_walk = [0, process_values_vector'];
    h0 = figure;
    parabola = sqrt(t);
    plot(t, random_walk,'-k', t,parabola,'-k', t,-parabola,'-k', t,0,'-k'), title(['REALIZATION of
B(t) over ', num2str(final_instant_T), ' time instants.']), xlabel('time t'), ylabel('X(t)')

    % Loading an existing EXPERIMENTALLY OBTAINED input stochastic
    % process.
  elseif case_number == 3
    nome = input('Please TYPE the full path and file name.\n','s');
    fid = fopen(nome);
    A = fscanf(fid, '%4d %4d', [2, inf]);
```

```
    A = A';
    time_range = A(:,1);
    final_instant_T = length(time_range);
    process_values_vector = A(:,2);
    fclose(fid);

    % Plot the Original Realization of X(t) to be later expanded and
    % reconstructed by virtue of the KLT.
    t = [0; time_range];
    random_walk = [0, process_values_vector'];
    h0 = figure;
    parabola = sqrt(t);
    plot(t, random_walk,'-k', t,parabola,'-k', t,-parabola,'-k', t,0,'-k'), title(['REALIZATION of
B(t) over ', num2str(final_instant_T), ' time instants.']), xlabel('time t'), ylabel('X(t)')
  end

  if flag == 1
    process_values_vector = process_values_vector;

  elseif flag == 2
    process_values_vector = process_values_vector;

  elseif flag == 3
    process_values_vector = process_values_vector.^2 - time_range;

  elseif flag == 4
    process_values_vector = process_values_vector.^2;

  elseif flag == 5
    process_values_vector = process_values_vector;

  end
```

27.4 THE FILE "BROWNIAN_AUTOCORRELATION.M"

The file "Brownian_Autocorrelation.m" simply translates the Brownian motion autocorrelation formula (i.e., the minimum—smallest—of t_1 and t_2) into a Matlab file ready for further applications. Please notice the "lucky circumstance" that the Brownian motion autocorrelation is known in its analytical form, rather than in some purely numerical form. It is by virtue of this analytical form, coded in the routine below, that all the KLT simulations described in this appendix can be performed.

```
% This subroutine computes the AUTOCORRELATION of the Brownian motion B(t)
% by translating its analytical definition min(t1, t2) into a numeric
% matrix. The entries of such a matrix are each the MINIMUM between the
% relevant row and column numbers.

function C = corr_brow(t)

n = length(t)
C = zeros(n);

for row = 1:n,
  for column = 1:n,
    C(row, column) = min(t(row),t(column));
  end
end
```

27.5 THE FILE "PROCESS_PATH.M"

The subroutine "process_path.m" creates the Brownian motion RANDOM WALK. In other words, this subroutine adds +1 to or subtracts −1 from the value of the $B(t)$ stochastic process at every new instant t. Its Matlab listing is the following.

```
% Subroutine creating the RANDOM WALK path of the Brownian motion B(t).
function X = process_path(T)

% Set to zero all the initial values of the T-element vector B that will
% contain the (random) values of the X(t) process when the new realization
% of X(t) will have been computed.
X = zeros(1,T);

% Create a vector with random entries and as many elements as are the time
% instants between 1 and the final instant T.
random_vector = rand(1,T);

for i = 1:T
  while random_vector(i) == 0.5
    random_vector(i) = rand(1);
  end

  if random_vector(i) < 0.5
    Increment(i) = - 1;
  elseif random_vector(i) > 0.5
    Increment(i) = + 1;
  end

  if i == 1
    X(i) = 0 + Increment(i);
  else
    X(i) = X(i-1) + Increment(i);
  end
end
```

27.6 THE FILE "GRAPHIC.M"

The subroutine "graphic.m" provides all the graphic instructions enabling Matlab to DRAW the plot of the selected stochastic process as a function of time. There are basically THREE "curves" for each realization of the Brownian motion—or time-rescaled Brownian motion—drawn by our Matlab code:

(1) The original "peaky" Brownian motion "curve", which is the actual Brownian motion realization.
(2) Its EMPIRIC RECONSTRUCTION, performed by the KLT, BY USING ONLY A FEW (OR EVEN ALL) THE KLT EIGENFUNCTIONS.
(3) Its ANALYTIC RECONSTRUCTION, performed by the KLT, BY USING ONLY A FEW (OR EVEN ALL) THE KLT EIGENFUNCTIONS.

In this way, we may clearly see the extent to which the two KLT RECONSTRUCTIONS actually FIT the original Brownian motion process.

This is one of the neater results provided by our simulations code, inasmuch as it "proves" in a neat, graphical way, "how good" the KLT is according to the number of eigenfunctions that we wish to take into account.

```
function graphic(Input_Process_data, EMPIRIC_Data_vector, ANALYTIC_Data_vector, flag,
How_many_eigenfunctions, t, N)

if flag == 1
t = [0; t];
Input_Process_data = [0; Input_Process_data];
EMPIRIC_Data_vector = [0; EMPIRIC_Data_vector];
ANALYTIC_Data_vector = [0; ANALYTIC_Data_vector];

plot(t, Input_Process_data,'-k', t ,EMPIRIC_Data_vector,'-k', t, ANALYTIC_Data_vector,':k'),

title(['B(t) and its RECONSTRUCTIONS by using ', num2str(How_many_eigenfunctions), '
eigenfunctions out of ', num2str(N), '.']),
xlabel('time t'), ylabel('B(t)'), legend('Original Realization of B(t)', 'Reconstruction by the
EMPIRIC KLT', 'Reconstruction by the ANALYTIC KLT')
hold on
parabola = sqrt(t);
plot(t, parabola,'-k', t, -parabola,'-k', t, 0, '-k')
hold off

elseif flag == 2
t = [0; t];
Input_Process_data = [0; Input_Process_data];
EMPIRIC_Data_vector = [0; EMPIRIC_Data_vector];
ANALYTIC_Data_vector = [0; ANALYTIC_Data_vector];

plot(t, Input_Process_data,'-k', t ,EMPIRIC_Data_vector,'-k', t, ANALYTIC_Data_vector,':k'),

title(['B(t^{2H}) and its RECONSTRUCTIONS by using ',
num2str(How_many_eigenfunctions), ' eigenfunctions out of ', num2str(N), '.']),
xlabel('time t'), ylabel('B(t^{2H})'), legend('Original Realization of B(t^{2H})',
'Reconstruction by the EMPIRIC KLT', 'Reconstruction by the ANALYTIC KLT')
hold on
parabola = sqrt(t);
plot(t, parabola, '-k', t, -parabola, '-k', t, 0, '-k')
hold off

elseif flag == 3
t = [0; t];
Input_Process_data = [0; Input_Process_data];
EMPIRIC_Data_vector = [0; EMPIRIC_Data_vector];
ANALYTIC_Data_vector = [0; ANALYTIC_Data_vector];

plot(t, Input_Process_data,'-k', t ,EMPIRIC_Data_vector,'-k', t, ANALYTIC_Data_vector,':k'),

title(['B^2(t)-t and its RECONSTRUCTIONS by using ',
num2str(How_many_eigenfunctions), ' eigenfunctions out of ', num2str(N), '.']),
xlabel('time t'), ylabel('B^2(t)-t'), legend('Original Realization of B^2(t)-t', 'Reconstruction
by the EMPIRIC KLT', 'Reconstruction by the ANALYTIC KLT')
hold on
parabola = sqrt(t);
plot(t, parabola, '-k', t, -parabola, '-k', t, 0, '-k')
hold off
```

```
elseif flag == 4
t = [0; t];
Input_Process_data = [0; Input_Process_data];
EMPIRIC_Data_vector = [0; EMPIRIC_Data_vector];
ANALYTIC_Data_vector = [0; ANALYTIC_Data_vector];

plot(t, Input_Process_data,'-k', t ,EMPIRIC_Data_vector,'-k', t, ANALYTIC_Data_vector,':k'),

title(['B^2(t^{2H}) and its RECONSTRUCTIONS by using ',
num2str(How_many_eigenfunctions), ' eigenfunctions out of ', num2str(N), '.']),
xlabel('time t'), ylabel('B^2(t^{2H})'), legend('Original Realization of B^2(t^{2H})',
'Reconstruction by the EMPIRIC KLT', 'Reconstruction by the ANALYTIC KLT')
hold on
parabola = sqrt(t);
plot(t, parabola, '-k', t, -parabola, '-k', t, 0, '-k')
hold off

elseif flag == 5
t = [0; t];
Input_Process_data = [0; Input_Process_data];
EMPIRIC_Data_vector = [0; EMPIRIC_Data_vector];
ANALYTIC_Data_vector = [0; ANALYTIC_Data_vector];

plot(t, Input_Process_data,'-k', t ,EMPIRIC_Data_vector,'-k', t, ANALYTIC_Data_vector,':k'),

title(['B(\tau) and its RECONSTRUCTIONS by using ',
num2str(How_many_eigenfunctions), ' eigenfunctions out of ', num2str(N), '.']),
xlabel('time \tau'), ylabel('B(\tau)'), legend('Original Realization of B(\tau)',
'Reconstruction by the EMPIRIC KLT', 'Reconstruction by the ANALYTIC KLT')
hold on
parabola = sqrt(t);
plot(t, parabola, '-k', t, -parabola, '-k', t, 0, '-k')
hold off

end
```

27.7 THE FILE "ANALYTIC_KLT.M"

The file "ANALYTIC_KLT.m" finds the KLT of the ordinary Brownian motion $B(t)$ as described and proven in Section 21.3. We recall here that this is an EXACT ANALYTICAL solution of the KLT integral equation (10.18). Therefore, the reconstruction of $B(t)$ by the EMPIRIC KLT and the reconstruction of $B(t)$ by the ANALYTIC KLT are exactly the same thing; that is, the two reconstructed curves just OVERLAP EXACTLY in the $B(t)$ plot. This happens independently of the number of eigenfunctions that we decide to take into account for the reconstruction of $B(t)$.

This subroutine's listing is as follows.

```
% This subroutine computes the ANALYTIC RECONSTRUCTION of X(t) according to
% the analytic (either exact or approximated formulae given in the book.
function [ANALYTIC_Data_vector] = ANALYTIC_KLT(Input_Process_data, N,
How_many_eigenfunctions, t, Lambda)

ANALYTIC_Data_vector = zeros (N,1);
```

```
% Final instant T, i.e. the largest value in the "t" set.
T = N;

% This is the KEY SUBROUTINE YIELDING THE ANALYTIC RECONSTRUCTION of X(t).
for n = 1:How_many_eigenfunctions,
  arg = t.*pi * (2 * n -1) / (2*(T+1));
  NN = sqrt(2 /(T+1));
  lambda(n) = 4 * (T+1)^2 / (((pi)^2) * (2*n - 1)^2);
  phi = NN*sin(arg);
  zed = sum(Input_Process_data.*phi);
  ANALYTIC_Data_vector = ANALYTIC_Data_vector + zed * phi;
end

x = [1:1:How_many_eigenfunctions];

% Plot the EIGENVALUES of the EMPIRIC AND ANALYTIC RECONSTRUCTIONS of X(t).
figure;
plot(x, diag(Lambda(1:n,1:n)), '-k', x, lambda, ':k'), title('EIGENVALUES of the EMPIRIC
and ANALYTIC Reconstruction of B(t).'), ylabel('Eigenvalues');
legend('EMPIRIC eigenvalues','ANALYTIC eigenvalues')
```

Now we would like to show how all this performs graphically.

Figure 27.1 shows a simple realization of the ordinary Brownian motion $B(t)$ over 500 time instants, starting from the origin of the axes according to the initial condition $B(0) = 0$—see Equation (21.3). The solid horizontal parabola with axis coinciding with the time axis is of course the STANDARD DEVIATION parabola, having the two equations $\pm\sqrt{t}$ according to Equation (21.5). In plain words, the ordinary Brownian motion $B(t)$ "oscillates at random" above and below its mean value (i.e., the time axis, since $E\{B(t)\} \equiv 0$) and its average distance from the mean value equals approximately the standard deviation $\pm\sqrt{t}$. But this does not mean at all that $B(t)$ will always stay above or below the time axis: it actually shifts periodically, as this simulation clearly shows over just 500 time instants.

Figure 27.2 shows how well the KLT reconstructs the given stochastic process according to the number of eigenfunctions taken into account for the reconstruction. It should be clear that if the user employs ALL of the KLT eigenfunctions for the reconstruction, the EMPIRICAL reconstruction will overlap exactly the original realization (100% reconstruction). In contrast, the ANALYTICAL reconstruction will overlap the original realization exactly ONLY if the solution to the KLT integral equation (10.18) is EXACT. This is precisely the case for the ordinary Brownian motion, as described in Section 21.3.

27.8 THE FILE "ANALYTIC_KLT_SQUARE_BROW_MOTION.M"

The subroutine "ANALYTIC_KLT_square_brow_motion" finds the KLT of the square of the ordinary Brownian motion—that is, $B^2(t)$—as described and proven in Section 24.4.

However, we must point out a "surprise" here: the mean value of the $B^2(t)$ stochastic process is NOT zero, but, rather, it is t. Thus, we may NOT compute the KLT of the $B^2(t)$ process alone, but, rather, we must compute the KLT of the new

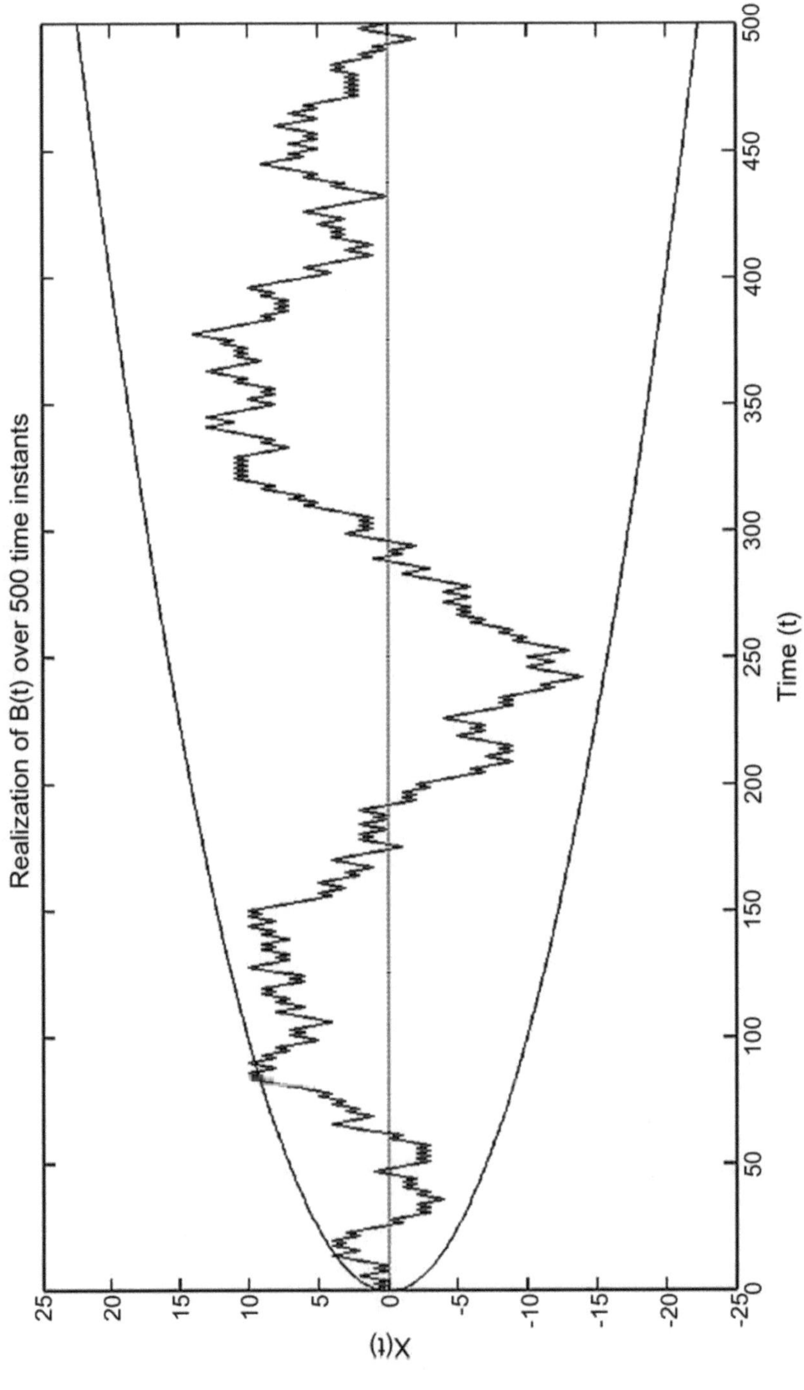

Figure 27.1. A simple realization of the ordinary Brownian motion over 500 time instants.

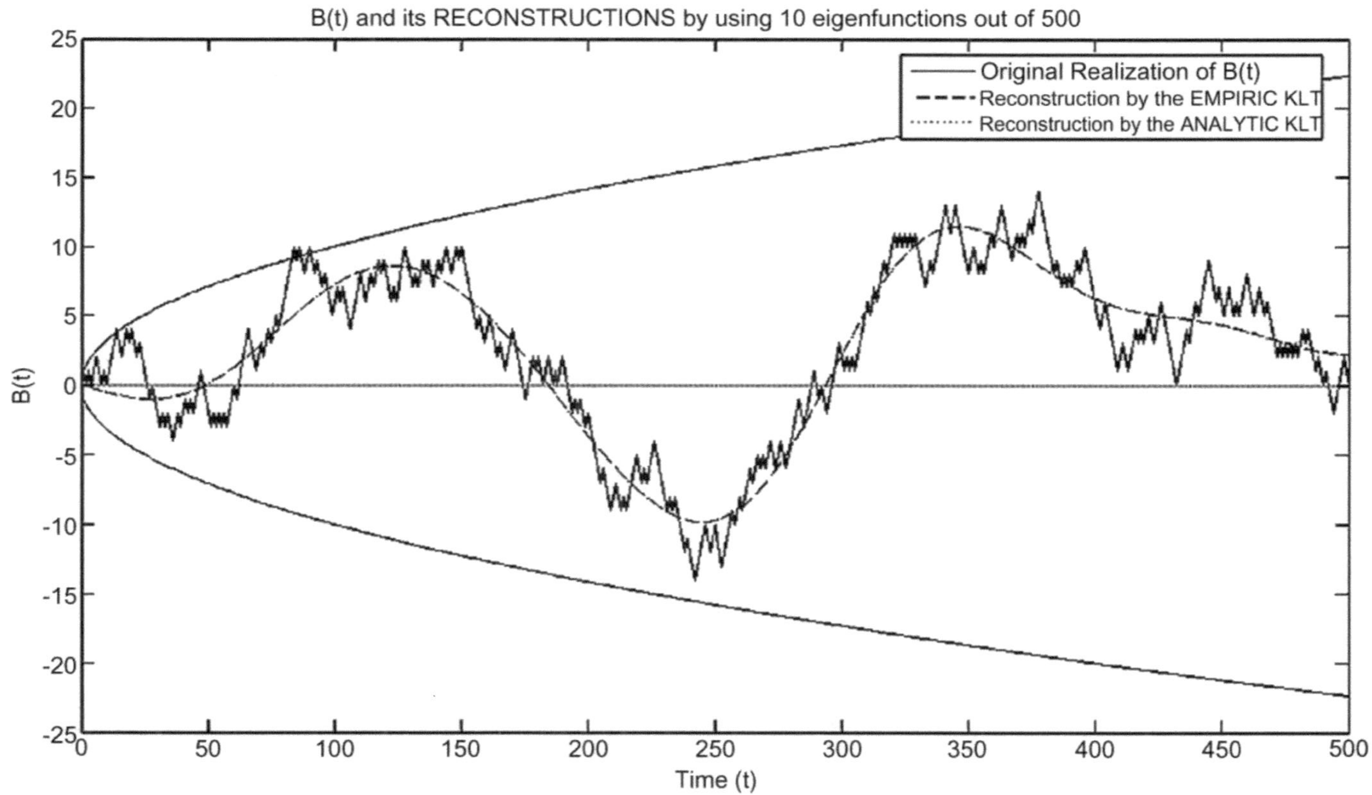

Figure 27.2. The same ordinary Brownian motion realization over 500 time instants, as shown in the previous Figure 27.1, plus its two KLT *reconstructions* made by taking into account just the first 10 eigenfunctions of 500. These two curves actually *coincide* with the single, "smooth curve" *interpolating* the "peaky" Brownian motion realization. They coincide because, as shown in Section 21.3, the analytical KLT for the ordinary Brownian motion is an *exact* solution, rather than a numerically approximate solution. So, the *empiric and analytical reconstructions* cannot fail to coincide!

process $X(t) = B^2(t) - t$, the mean value of which is indeed zero, so that the assumption $E\{X(t)\} = 0$, upon which all the KLT theorems of Chapter 10 are based, is indeed fulfilled. Having said this, the listing is as follows.

```
% This subroutine computes the ANALYTIC RECONSTRUCTION of X(t) according to
% the analytic (either exact or approximated formulae given in the book.
function [ANALYTIC_Data_vector] = ANALYTIC_KLT(Input_Process_data, N,
How_many_eigenfunctions, t, Lambda)

ANALYTIC_Data_vector = zeros (N,1);

% Final instant T, i.e. the largest value in the "t" set.
T = N;

%Input_Process_data = Input_Process_data - mean(Input_Process_data);

% This is the KEY SUBROUTINE YIELDING THE ANALYTIC RECONSTRUCTION of
X^2(t). for n = 1:How_many_eigenfunctions,
  gamma(n) = pi * (n-5/12);
  arg = gamma(n) * t.^(3/2)/(T+1)^(3/2);
  NN = sqrt(3)*t /(T+1)^(3/2) / abs(besselj(2/3, gamma(n)));
  lambda(n) = 16/9*(T+1)^3 / ((gamma(n))^2);
  phi = NN.*besselj(2/3,arg);
  zeta = sum(Input_Process_data.*phi);
  ANALYTIC_Data_vector = ANALYTIC_Data_vector + zeta * phi ;
end

x = [1:1:How_many_eigenfunctions];

% Plot the EIGENVALUES of the EMPIRIC RECONSTRUCTION of X(t).
figure;
plot(x, diag(Lambda(1:n,1:n)), '-k', x, lambda, ':k'), title('EIGENVALUES of the EMPIRIC
and ANALYTIC Reconstruction of B^2(t)-t.'), ylabel('Eigenvalues');
legend('EMPIRIC eigenvalues','ANALYTIC eigenvalues')
```

Let us now see the graphs produced by this routine.

Figure 27.3 shows a new, simple realization of the ordinary Brownian motion over 500 time instants (different from the realization shown in Figure 27.1).

But the graphs shown in Figure 27.4 are "unexpected"! In fact, the vertical axis now plots $X(t) = B^2(t) - t$ for the same realization of $B(t)$ shown in Figure 27.3, and one sees that the numerical values on the vertical axis are of course much higher than those in Figure 27.3. For instance, the parabola neatly drawn in Figure 27.3 is now so much "squashed" in Figure 27.4 that it looks nearly like two "parallel" straight lines above and below the time axis!

There is one more important feature of Figure 27.4 that we must point out. The two reconstructed lines (drawn by taking into account just the first 10 eigenfunctions out of 500) DO NOT COINCIDE EXACTLY because the analytical solution given by Equation (24.53) is an APPROXIMATE analytical solution, and NOT an exact one. In fact, Equation (24.50) is the ASYMPTOTIC (for $t \to \infty$) expansion for the Bessel functions $J_\nu(t)$ and so it cannot give accurate values of $J_\nu(t)$ near the origin $t \approx 0$. This explains why, for $t \approx 0$, the two reconstructions shown in Figure 27.4 are actually different from each other.

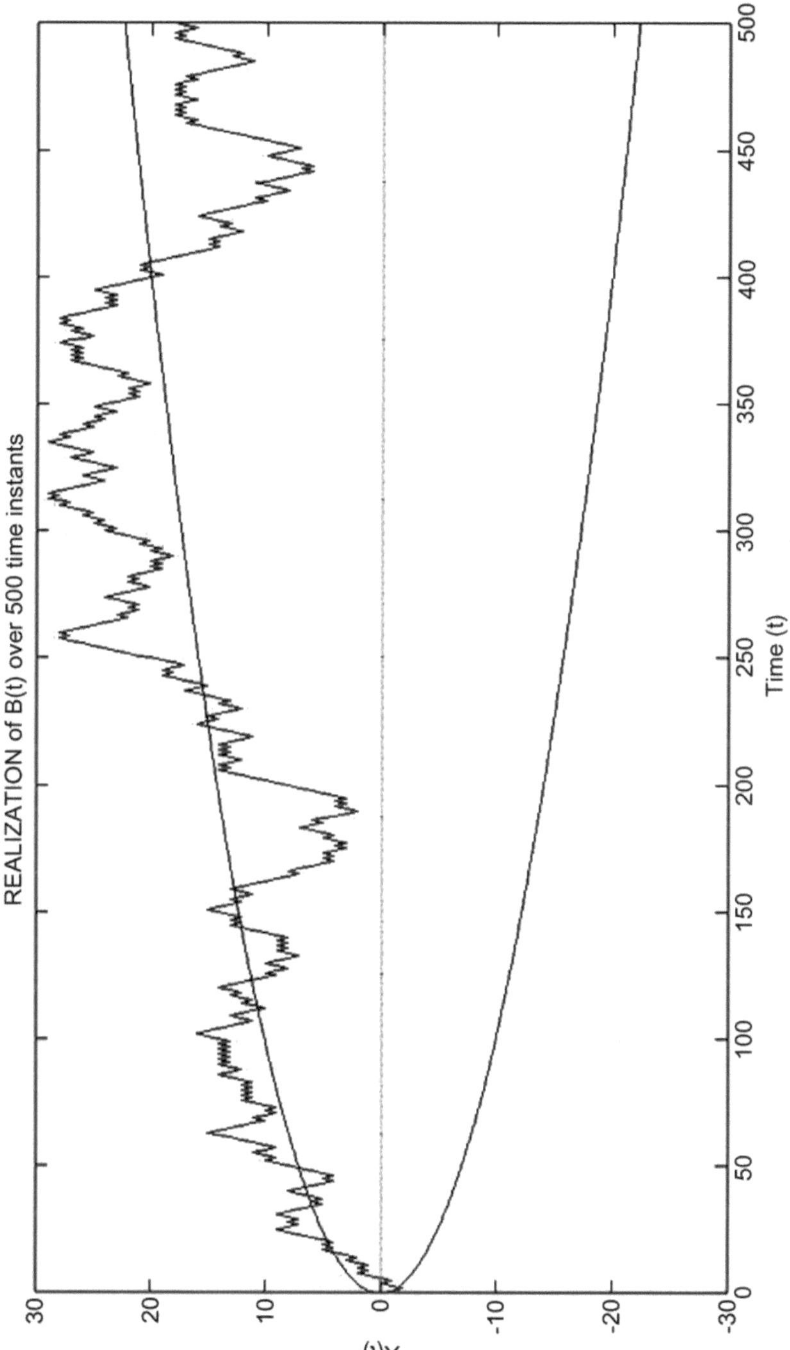

Figure 27.3. A new, simple realization of the ordinary Brownian motion $B(t)$ over 500 time instants.

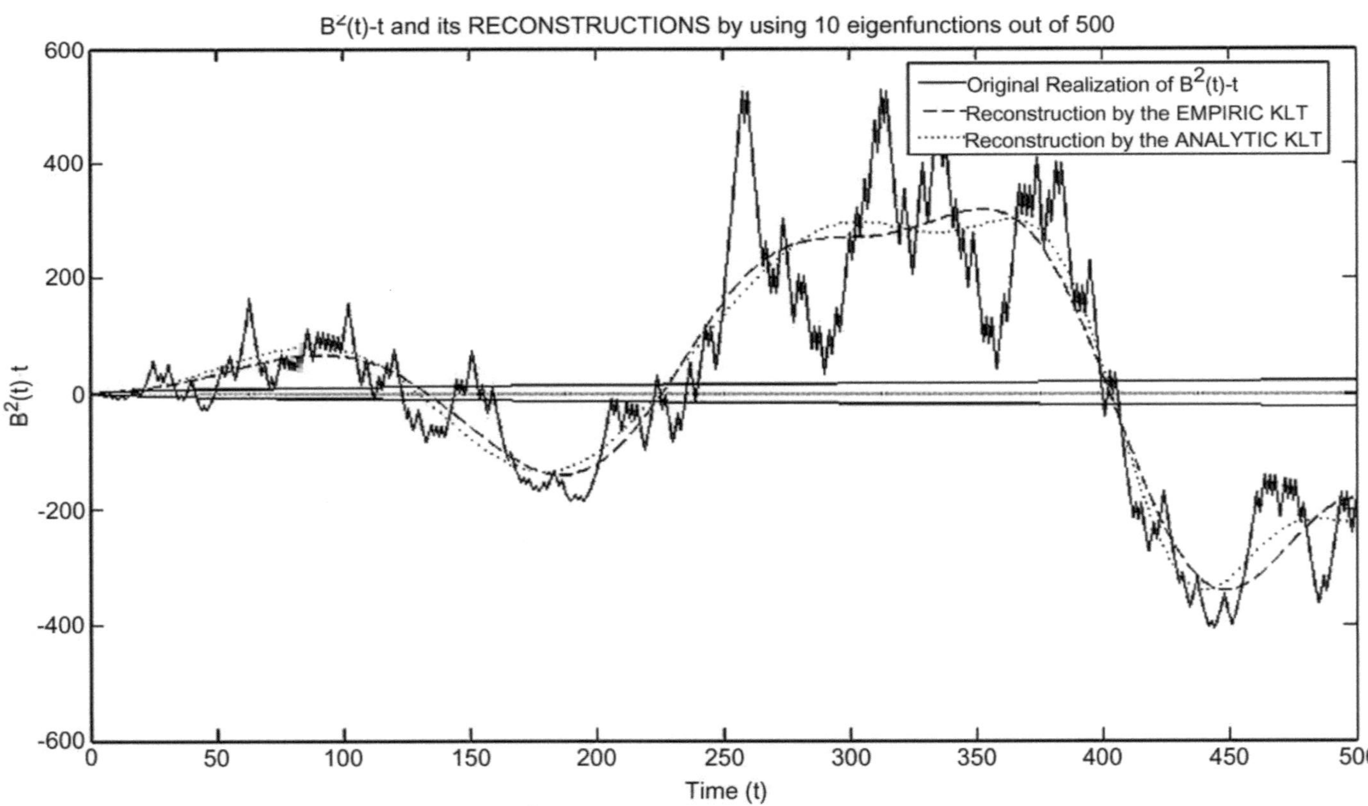

Figure 27.4. Showing the stochastic process $X(t) = B^2(t) - t$ for the same realization of $B(t)$ shown in Figure 27.3. Notice that the standard deviation parabola of Figure 27.3 is now very much "squashed" around the time axis because of the changes in the vertical scale of the diagram. Also, the two KLT-reconstructed curves (using only the first 10 eigenfunctions out of 500), do not overlap exactly for small values of time t because the analytical reconstruction is just an approximate (i.e., inexact) formula. In fact, the asymptotic expansion formula for the Bessel functions (24.50) was used in the mathematical derivation. But for high values of time, the two reconstructions of course overlap.

27.9 THE FILE "ANALYTIC_KLT_UNIFORM_REL.M"

Let us now turn to the RELATIVISTIC KLT, as described and proven in Chapters 11, 12, and 13. The simplest possible case of relativistic motion is of course UNIFORM motion (i.e., the motion of a spaceship at a constant speed which is also a significant fraction of the speed of light). The KLT for signals received back on Earth from a spaceship moving (away from or towards the Earth) with a UNIFORM motion was obtained in Section 11.2. The subroutine "ANALYTIC_KLT_uniform_ rel.m" translates the results of Section 11.2 into a Matlab file. Its listing is as follows.

```
% This subroutine computes the ANALYTIC RECONSTRUCTION of X(t) according to
% the analytic (either exact or approximated formulae given in the book.
function [ANALYTIC_Data_vector, tau] = ANALYTIC_KLT_uniform_rel(Input_Process_data,
N, How_many_eigenfunctions, t, Lambda)

ANALYTIC_Data_vector = zeros (N,1);

% Final instant T, i.e. the largest value in the "t" set.
T = N;

ratio = input('What ratio of the speed of light is the uniform velocity? (For instance
0.2) \n');
K = (1-ratio^2)^(1/4);

% This is the KEY SUBROUTINE YIELDING THE ANALYTIC RECONSTRUCTION of X(t).
for n = 1:How_many_eigenfunctions,
  gamma(n) = n*pi - pi/2;
  arg = gamma(n) * t / (T+1);
  NN = K * sqrt(2 /(T+1));
  lambda(n) = K^2 * T^2 / gamma(n)^2;
  phi = NN*sin(arg);
  zed = sum(Input_Process_data.*phi);
  ANALYTIC_Data_vector = ANALYTIC_Data_vector + zed * phi;
end

tau = sqrt(1-ratio^2)*t;
x = [1:1:How_many_eigenfunctions];

% Plot the EIGENVALUES of the EMPIRIC RECONSTRUCTION of X(t).
figure;
plot(x, diag(Lambda(1:n,1:n)), '-k', x, lambda, ':k'), title('EIGENVALUES of the EMPIRIC
and ANALYTIC Reconstruction of B(\tau).'), ylabel('Eigenvalues');
legend('EMPIRIC eigenvalues','ANALYTIC eigenvalues')
```

Again, we now want to see the graphs produced by this subroutine.

Figure 27.5 shows the new realization of the Brownian motion (different of course from those in Figures 27.1 and 27.3). Figure 27.6 shows the PROPER time τ on the horizontal axis (i.e., the time measured aboard the spaceships). The proper time τ is related to the coordinate time t by virtue of Equation (11.1). Since we have assumed in Figure 27.5 that t ranges from 0 to 500, then we must compute the integral (11.1) for $t = 500$. Also, we must select at which fraction of the speed of light our spaceship is advancing in empty space. To fix ides, let us assume that it advances at

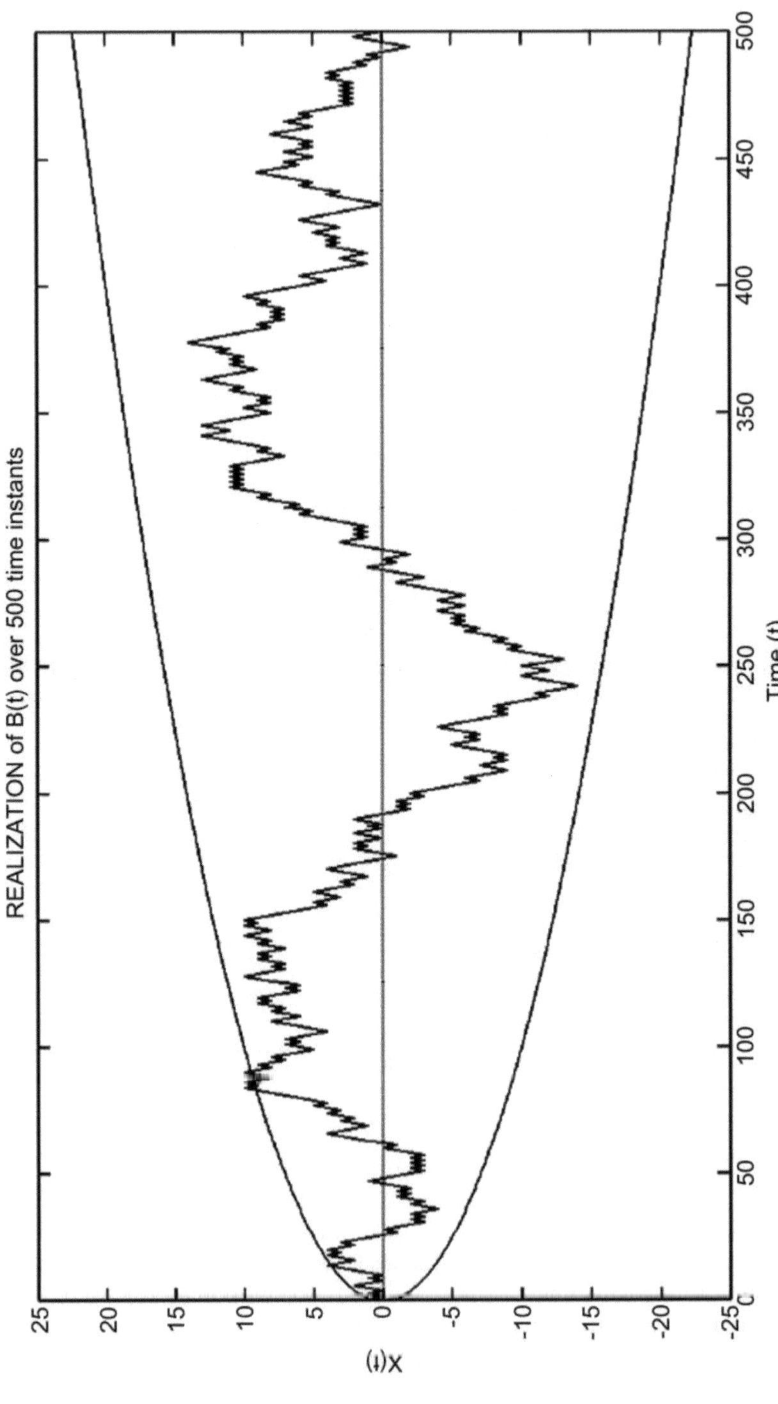

Figure 27.5. A new, simple realization of the ordinary Brownian motion $B(t)$ over 500 time instants.

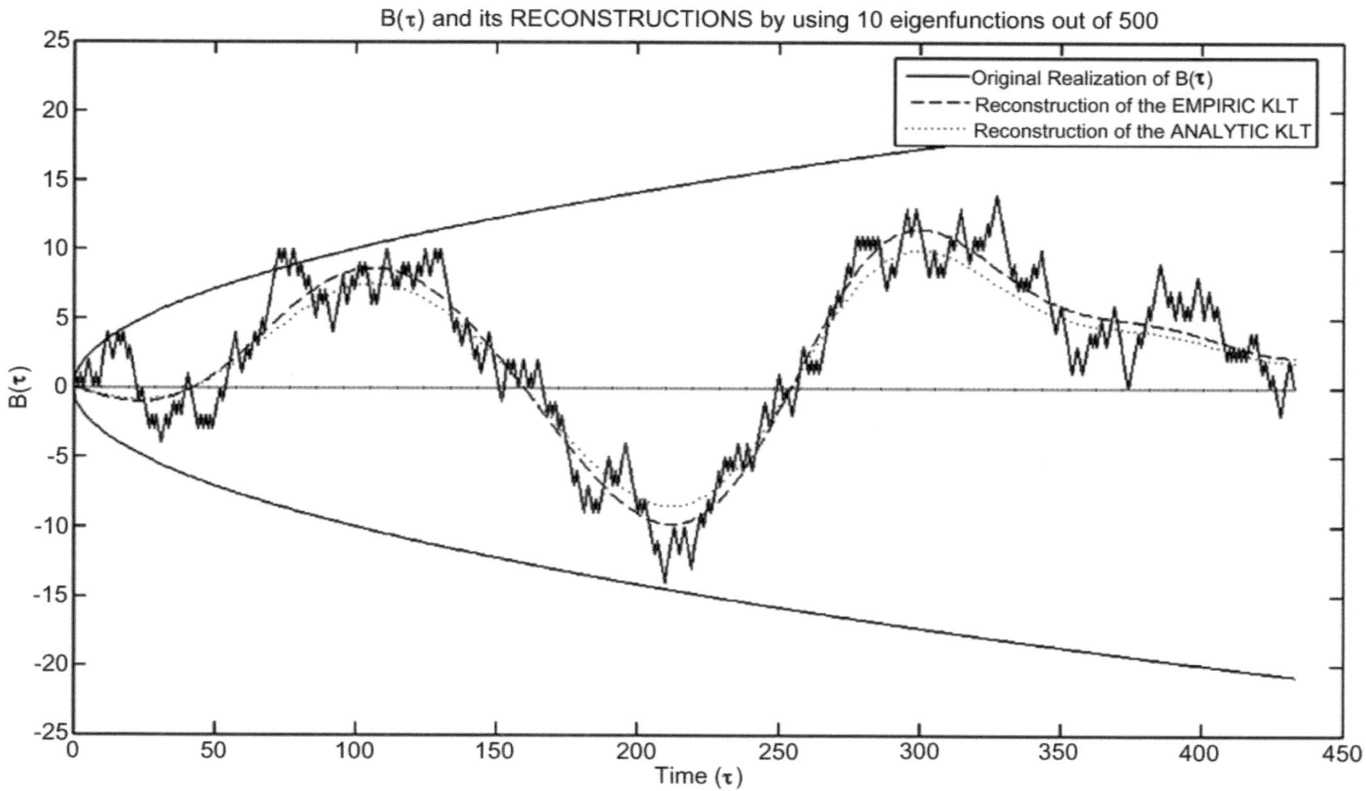

Figure 27.6. The time dilation effect of special relativity forces the proper time τ to range only from 0 to 433—because of Equation (11.1). Having said that, this figure shows the original input Brownian motion as in Figure 27.5 plus the two reconstructed KLT curves by taking into account only the first 10 eigenfunctions.

50% of the speed of light. Equation (11.1) then yields $\tau = 433.013 \approx 433$. Thus, the τ axis in Figure 27.6 ranges only from 0 to 433.

27.10 CONCLUSIONS

We have provided the readers with a Matlab code showing the KLT of the Brownian motion in a graphical fashion. This code might be extended and improved in a number of ways, of course. Any volunteers?

28

KLT applications and a Fortran code

Stephane Dumas[1]

KLT is a modern data analysis tools—a black box that is widely used but poorly understood. It is also known as Principal Component Analysis (PCA), Singular Value Decomposition (SVD), Singular Spectral Analysis (SSA), and Pisarenko's Method. Applications of the Karhunen–Loève Transform (KLT) are numerous in several fields of research, as this short list illustrates:

- In biology, it is used to classify bacteria by means of surface-enhanced Raman spectroscopy.
- In genetics, it is used to analyze and classify genes.
- In geology, it is used to study earthquakes. Geophysicists have used SSA to analyze a wide variety of time series such as solar oscillations, precipitation, stream flow and sea surface temperature, the chemical constituents of ice cores, global temperature, magnetosphere dynamics, and suspended sediment concentration in an estuary.
- In mathematics, it is used to study chaos theory.
- In climatology, it is used to study the fluctuation of temperature over area and time.
- In astrophysics, KLT methods are used in the search for acoustic oscillations of the Sun. They are also used to classify variable stars (3,200 stars according to 51 features).

[1] Stephane Dumas is a Canadian physicist working for the Canadian government in the field of modeling and simulation. He has worked with simulated integrated optical devices for a few years. His expertise covers such subjects as physics, mathematics, astronomy, SETI, cryptography, the theory of information/communication/data compression. He is currently the President of the Quebec Center of the Royal Astronomical Society of Canada (RASC) and is a member of the Society of Amateur Radio Astronomers (SARA) and of the Planetary Society. He is also the SETI League Coordinator for east Canada. At the moment he is doing a master's degree in physics/astrobiology (see *http://www3.sympatico.ca/stephane_dumas/CETI/biographies.html*).

KLT's capacity to predict and fill gaps in missing data will not be discussed in this chapter.

28.1 THE EIGENPROBLEM

The core of KLT is its capacity to solve the eigenproblem. An $N \times N$ matrix $\mathbf{A}$ is said to have an eigenvector $\vec{x}$ and a corresponding eigenvalue λ if

$$\mathbf{A}\vec{x} = \lambda\vec{x} \tag{28.1}$$

which can be solved if and only if

$$\det(\mathbf{A} - \lambda\mathbf{I}) = \begin{vmatrix} A_{11} - \lambda_1 & A_{12} & \dots & A_{1n} \\ A_{21} & A_{22} - \lambda_2 & \dots & A_{2n} \\ \vdots & \vdots & \ddots & \vdots \\ A_{n1} & A_{n2} & \dots & A_{nn} - \lambda_n \end{vmatrix} = 0 \tag{28.2}$$

Suppose that we have a series of $\vec{x}$ composed of p random variables and that we are interested in analyzing the structure of the covariance or correlation between p variables. If p is small or the structure is simple, the solution may be easy. However, this is not often the case and we end up with a large p and a complex structure. Working in the eigenspace corresponding to this dataset gives us the capability to manipulate a smaller dataset. Figure 28.1 illustrates this for $p = 2$.

Matrix $\mathbf{A}$ is the covariance, or correlation, matrix of the dataset described by the vectors $\vec{x}_i$.

28.1.1 Correlation vs. covariance

Variance is a measure of how far a set of numbers is spread out. The covariance of two datasets, X and Y, is given in equation (28.3)

$$\sigma_{XY}^2 = \frac{\sum_{i=1}^{N}(X_i - \overline{X})(Y_i - \overline{Y})}{N} \tag{28.3}$$

where X_i and Y_i are the i-th element of their respective vector, and $\overline{X}$ and $\overline{Y}$ are the expected value of each vector (or mean value).

Covariance is sensitive to the scale of measurement adopted and is not a good measure of coherence—being multiplied by any scale factor introduced in one variable or another. For example, if X and Y are measured in inches, covariance will be 144 times what it would be if they were measured in feet. Furthermore, covariance provides little information on data if the variance of some variables is too large compared with others.

To obtain a measure of coherence that does not have this defect, the correlation coefficient is used (equation 28.4):

$$\rho_{X,Y} = \frac{E[(X - \overline{X})(Y - \overline{Y})]}{\sigma_X \sigma_Y} \tag{28.4}$$

Correlation is linked to convolution. The correlation of two functions, denoted

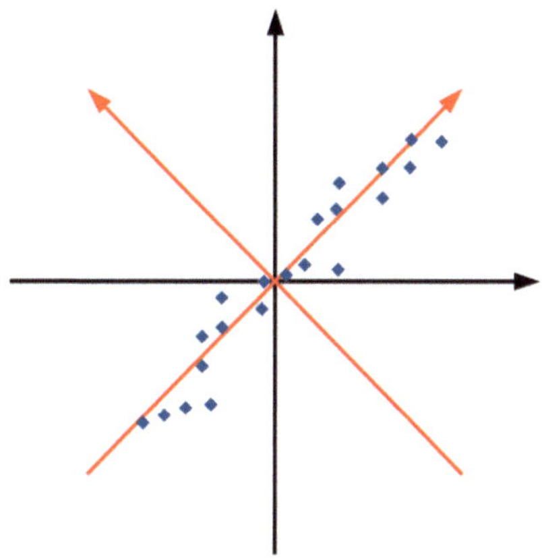

Figure 28.1. Illustration of the eigenspace with $p = 2$. In the original space (black), the data require p coordinates to be located. While in the eigenspace (red), the data require only one coordinate.

corr(g, h), is defined by equation (28.5) which is also the definition of convolution:

$$\text{corr}(g, h) \equiv \int_{-\infty}^{\infty} g(\tau + t)h(\tau)dt = G(f)H^*(f) \tag{28.5}$$

where $G(f)$ is the Fourier transform of g, and H^* is the complex conjugate of the Fourier transform of h. The correlation of a function with itself is called its autocorrelation:

$$\text{corr}(g, g) = |G(f)|^2 \tag{28.6}$$

also known as the Wiener–Khinchin theorem, which states that the power spectral density of a wide-sense stationary random process is the Fourier transform of the corresponding autocorrelation function.

It is possible to compute the entire autocorrelation matrix using the FFT, which makes its calculations faster than when the regular method is employed.

Even if correlation is a better indicator of coherence, covariance is sometimes used to solve the eigenproblem. The use of covariance or correlation in KLT is dependent on the kind of analysis and dataset. Autocorrelation is better suited to analyze time series while covariance is used to study the temperature fluctuation of a region over a given period of time.

28.1.2 Solving the problem numerically

Solving the eigenproblem involves N linear algebraic equations of N unknowns. The complexity of the problem grows very fast with the number of λ taken into account. Several numerical methods exist to find the solution.

28.1.3 The Jacobi method

One of the most popular methods to solve the system of linear equations is the Jacobi method [1]. It consists of a sequence of orthogonal similarity transformations. It is named for Carl Gustav Jacob Jacobi, who first proposed the method in 1846, but only became widely used in the 1950s with the advent of computers. Each transformation (e.g., Jacobi rotation) is just a plane rotation designed to annihilate one of the off-diagonal matrix elements. With each rotation, the matrix becomes increasingly more diagonal.

The algorithm is relatively easy to implement on a computer but its performance decreases dramatically with increasing values of N (i.e., the order of the matrix to be processed).

28.1.4 The QR algorithm

The QR transformation was developed in the late 1950s by John G. F. Francis (England) and by Vera N. Kublanovskaya (U.S.S.R.). The basic idea behind the QR algorithm is that any real matrix can be decomposed in the form:

$$\mathbf{A} = \mathbf{Q} \cdot \mathbf{R} \tag{28.7}$$

where $\mathbf{Q}$ is orthogonal and $\mathbf{R}$ is upper triangular. The algorithm is based on the tridiagonal decomposition of matrix $\mathbf{A}$. Matrix $\mathbf{A}$ must be symmetric. This method is harder to implement but faster than the Jacobi method.

28.1.5 The Lanczos algorithm

The Lanczos algorithm, created by Cornelius Lanczos (1950), is a technique that can be used to solve certain large sparse symmetric eigenproblems. The technique is very fast and appropriate to large matrices. It involves partial tridiagonalization of a given matrix $\mathbf{A}$. However, unlike some techniques (e.g., Jacobi, QR, Householder, and LU decomposition), no intermediate full submatrices are generated.

Algorithm 28.1. The Lanczos recursion

$v_0 = 0$;
$\beta_1 = 1$;
for $i = 1 \rightarrow m$ **do**
  $\beta_{i+1} v_{i+1} = \mathbf{A} v_i - \alpha_i v_i - \beta_i v_{i-1}$
  $\alpha_i = v_i^T \mathbf{A} v_i$
  $\beta_{i+1} = v_{i+1}^T \mathbf{A} v_i$
end for

Extremal eigenvectors (i.e., Φ_1 and Φ_n) tend to emerge long before the end of the tridiagonalization. This means the algorithm does not have to be executed on the full range of λ.

The basic Lanczos algorithm involves a series of recursions to tridiagonalize a matrix, given a real symmetric $N \times N$ matrix $\mathbf{A}$ and a unit vector v_i, typically generated randomly. The Lanczos matrices $\mathbf{T}_j$ (for $j = 1, \ldots, m$) are defined by the following recursion.

Table 28.1. Comparison between Jacobi, QR algorithm, and Lanczos. Calculations involve a time series vector of N points, which requires a matrix $\mathbf{R}$ of $N \times N$. The Jacobi and QR algorithm are taken from Numerical Recipes. In this example, the Lanczos algorithm is executed using only two cores of the CPU.

N	*Jacobi*	*QR*	*Modified Lanczos*
500	0 min 12.52 s	0 min 1.00 s	0 min 0.09 s
1,000	2 min 58.04 s	0 min 26.25 s	0 min 0.31 s
2,000	32 min 37.68 s	4 min 24.23 s	0 min 1.51 s
4,000	n/a	48 min 0.78 s	0 min 4.56 s
8,000	n/a	n/a	0 min 12.20 s
16,000	n/a	n/a	0 min 34.58 s
32,000	n/a	n/a	2 min 55.42 s
64,000	n/a	n/a	9 min 47.65 s

For each j, the corresponding Lanczos matrix T_j is defined as the real symmetric and tridiagonal matrix with diagonal entries α_i and subdiagonal entries β_{i+1}. The matrices T_j will ultimately provide the eigenvectors of $\mathbf{A}$.

A complete description of the algorithm can be found in [2] and [3].

The algorithm can be written in the form of a computer program and be combined with high-performance calculation mathematical libraries (e.g., BLAS[2], LAPACK[3], and Numerical Recipes[4]) to increase its performance.

Table 28.1 shows a series of comparisons between Jacobi, QR, and Lanczos algorithms. The calculations were performed on the same matrix $\mathbf{A}$. The time for each calculation is a mean value over several executions.

Since the Lanczos algorithm does not directly modify the original matrix (as opposed to other algorithms), it is possible to process extremely large matrices, with a few modifications of the basic algorithm. These modifications were implemented in the code written by the author. Using this modified code, the eigenproblem related to a Brownian motion was tested up to 1,000,000 points. Figure 28.2 shows the results (involving an autocorrelation matrix of 10^{12} elements). The calculations took 5 h 9 min on a regular workstation (e.g., a quad-core CPU and 8 Gb of RAM). Such a large matrix is impossible to hold completely in the memory of a workstation since it requires 7.62×10^6 Mb of memory. This is one of the strengths of the modified Lanczos algorithm, which does not require the full matrix to be in memory.

[2] *http://www.netlib.org/blas/*
[3] *http://www.netlib.org/lapack/*
[4] *http://www.nr.com/*

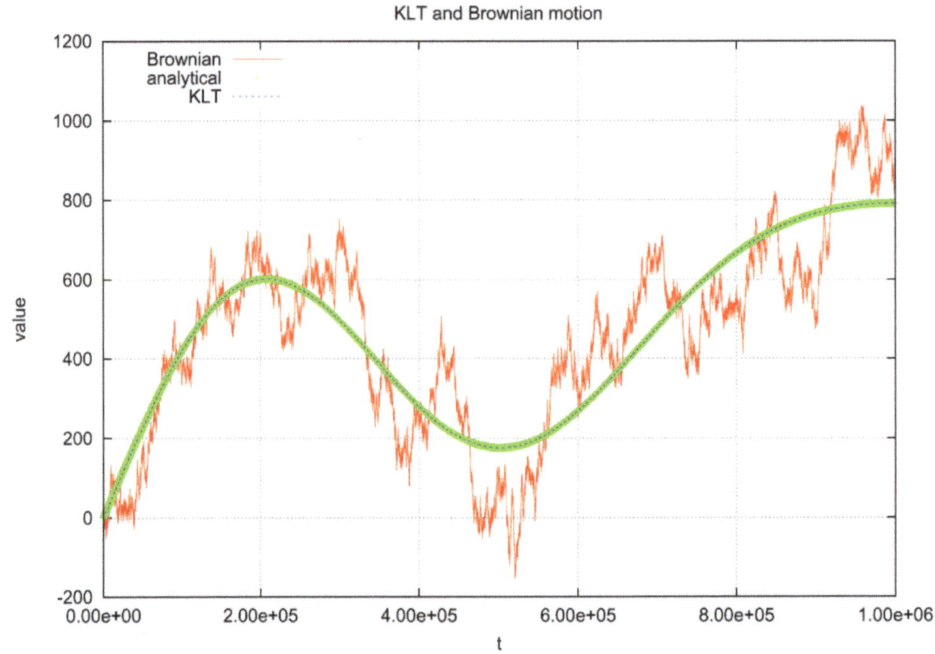

Figure 28.2. KLT applied to a Brownian motion dataset of 1 million points. The KLT curve is made of 10 eigenvectors.

28.2 DATA COMPRESSION

A major feature of solving the eigenproblem is the capacity to reduce the number of dimensions to represent the original dataset.

The number of dimensions required to represent the original information in the eigenspace depends only on the magnitude of their respective eigenvalues. By plotting them, from the largest to the smallest, one can select the largest q needed to describe the original information. Figure 28.3 shows a typical representation of the first few largest eigenvalues. The first few eigenvalues (λ_1) are very large and the amplitude drops very fast. Only a few λ could be used to represent a large dataset. The eigenvectors associated with small λ bring little information to the final solution.

Figure 28.4 shows an example of data compression. The whole signal can be represented by just three eigenvalues. The original data can be approximated by summing the first three eigenvectors into a single vector. Each eigenvector can be built by using its corresponding eigenvalues.

Figure 28.5 illustrates a more complex scenario when the original dataset is a Brownian motion. As data are more complex than a single sine curve, more eigenvectors are required to approximate them. Compression of an even more complex signal can be performed, as the next section shows, by applying KLT to infrared spectroscopy.

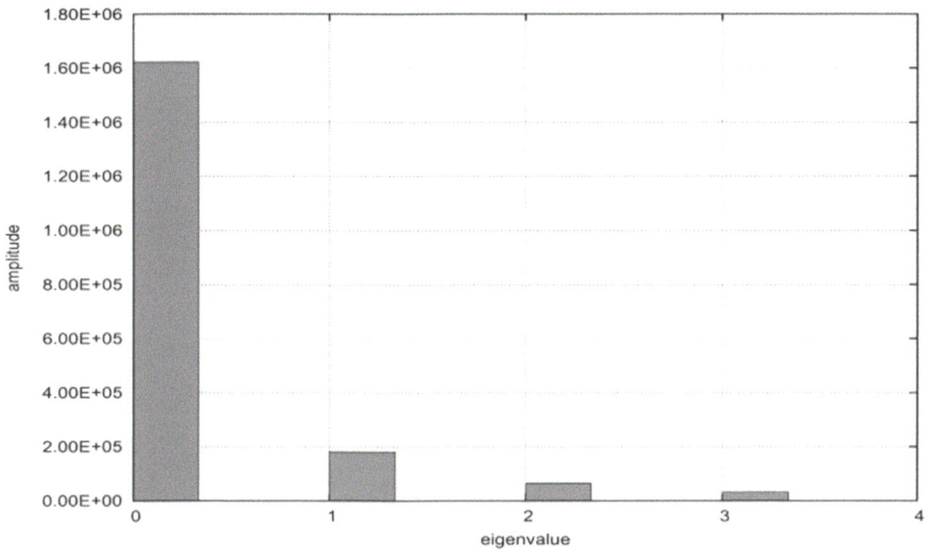

Figure 28.3. Typical eigenvalues plotted by their amplitude.

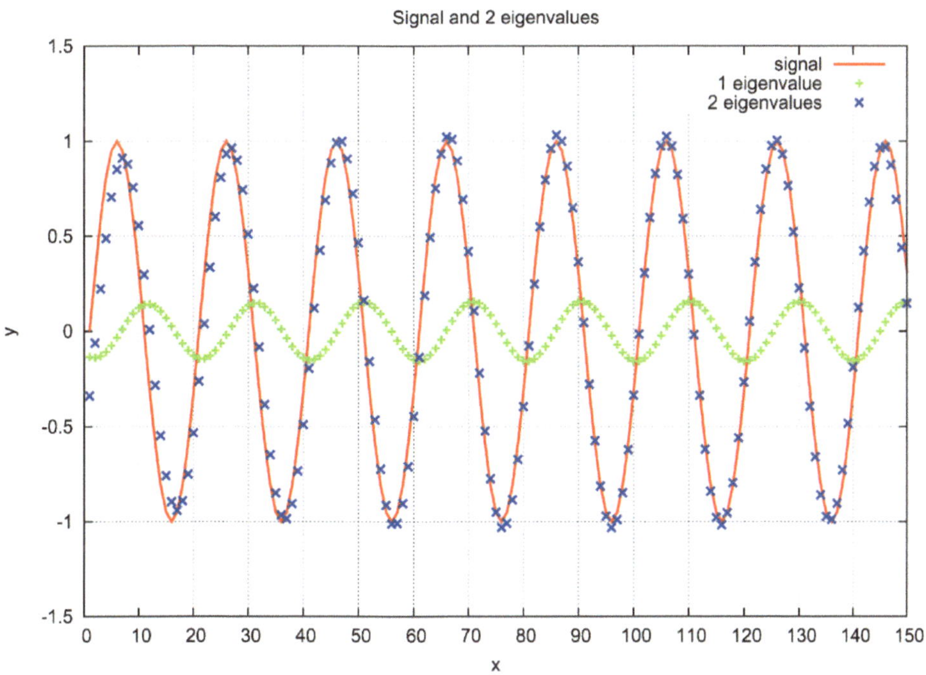

Figure 28.4. Illustration of data compression of $\sin(x)$. The whole curve can be approximated using just two eigenvalues.

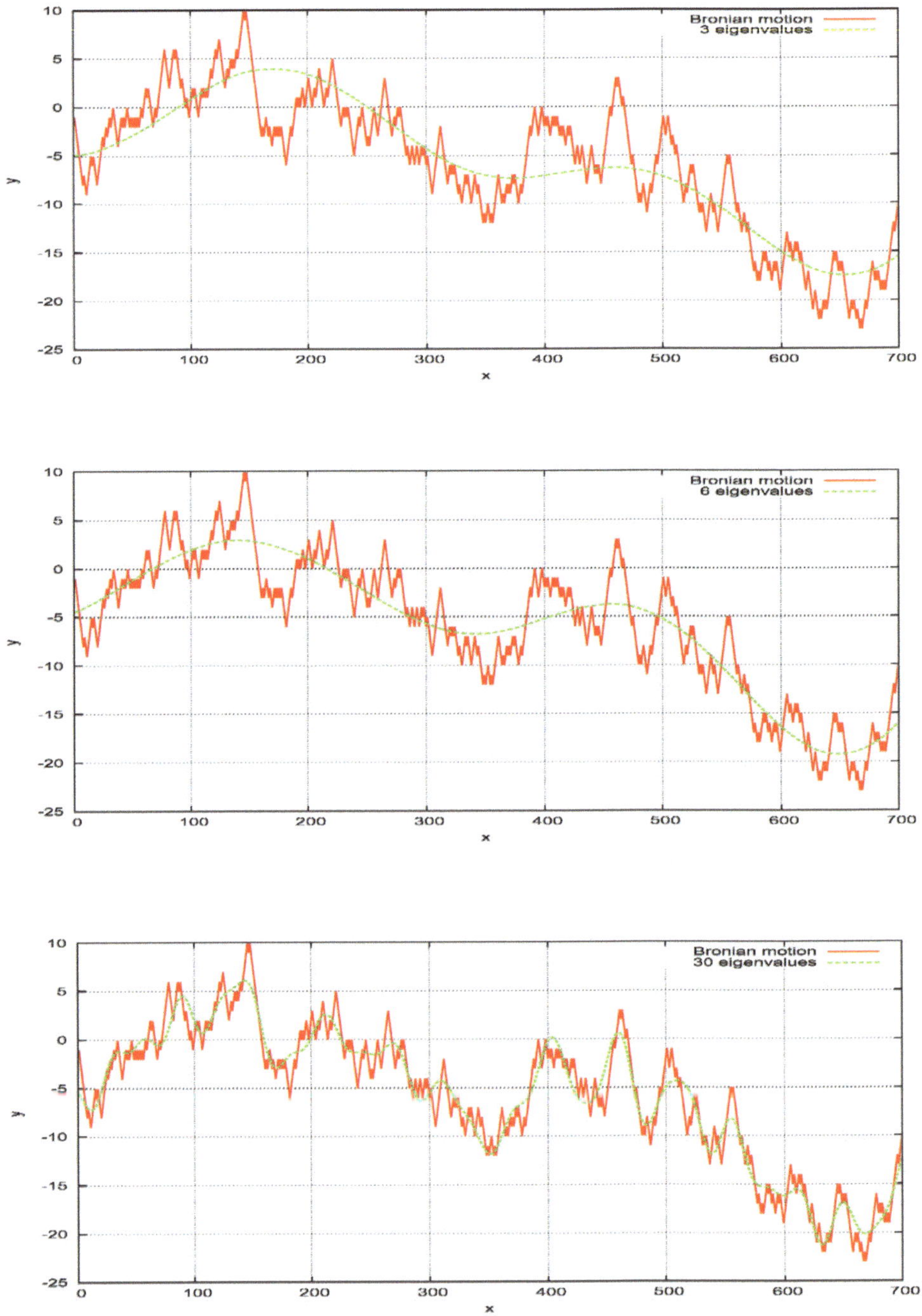

Figure 28.5. Illustration of data compression of a Brownian motion. The top figure shows how three eigenvalues can roughly approximate the data. The middle figure uses six eigenvalues and the bottom 30 out of 700.

28.3 CLASSIFICATION OF SPECTRA

The search for life is the core subject of exobiology. The infrared spectroscopy of samples (e.g., rock, soil, comet, exoplanet atmosphere) is the principal tool used to search for the presence of biomarkers. These are biological molecules present in the environment of a life form. They can be either from a food source or the by-product of bacteria. Identification of the sample composition is difficult without chemical analysis. There is a method involving KLT, or more precisely SVD, that can help.

Singular Value Decomposition (SVD) can be summarized as follows: given an arbitrary matrix **D** of dimension $(n \times p)$, which is a matrix of n observations of p variables measured about their means, then **D** can be written as equation (28.8). **U** is the matrix containing observations represented in the eigenspace. **W** is a square matrix that has eigenvalues on its diagonal, and **V** is the matrix containing the eigenvectors.

$$\mathbf{D} = \mathbf{U}\mathbf{W}\mathbf{V}^T \tag{28.8}$$

Infrared spectra are essentially vectors of wavenumbers.[5] Matrix **D** is formed by joining n spectra of p wavenumbers into a single matrix. The mean value of all wavenumbers is subtracted from each spectra. The data must respect the stationary criterion (at least in a weak sense) used in SVD (see Figure 28.6 for an example).

The SVD algorithm solves the eigenproblem for the covariance matrix **D**. The resulting matrix **U** contains a series of n vectors of p variables (which are not wavenumbers, but rather projections on eigenvectors). By virtue of the data compression nature of eigenvectors, it is possible to use shorter versions of the new vectors by considering only the most important eigenvalues. Instead of n vectors of p elements, the new dataset has n vectors of size m $(m < p)$. It is a form of data compression since the spectra can be represented with shorter vectors than the original dataset.

Cluster analysis[6] techniques can be used on the reduced dataset to group similar vectors. By adding known spectra (i.e., mineral and biological) to matrix **D**, the result of SVD and clustering can be used to classify unknown spectra and identify them. This technique was used in [4] to classify the spectral signatures of rocks and identify possible biomarkers.

Figure 28.7 illustrates the method using infrared spectra taken on several samples (e.g., known and unknown). The spectra *Nunavut* have been taken from rock samples from the Nunavut region in northern Canada. These are compared with spectra of known origin (i.e., mineral or biological). The relative distances from each Nunavut point in relation to the other groups give an indication of the relative importance of the composition of the rock. Note that only two principal components

[5] A wavenumber is the reciprocal of the wavelength. It is a measure used in spectroscopy. It is given in cm^{-1}.

[6] Cluster analysis is a technique used to group a set of objects by using some of their properties. For a set of points in three dimensions, the grouping could simply be by distance.

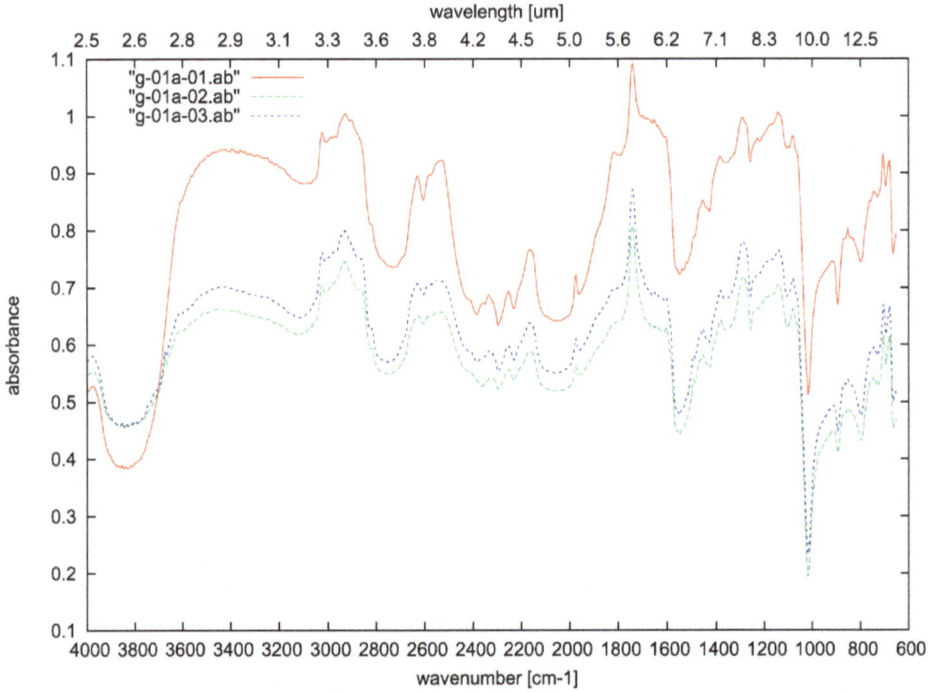

Figure 28.6. Illustration of infrared spectra taken on a rock containing endoliths.

are required to classify the spectra while the original spectra have more than 800 wavenumber values.

It is possible to convert the spectrum to the eigenspace without solving the whole system each time. The matrices WV^T are similar to a rotation matrix and can be used to transfer from one space to another once they have been properly computed with known spectra.

28.4 SIGNAL PROCESSING

Digital signal processing (DSP) is devoted to the analysis of signals in either discrete or continuous time. Signals of interest include any time-varying (i.e., a time series) measurements such as sounds, images, voltages, electrocardiograms, temperatures, etc.

The usage of DSP covers a vast field of activities in communications, medical imaging, radar, sonar, high-fidelity music reproduction, and oil prospecting, to name a few.

This section deals only with the discrete form and considers each observation to be measured at regular intervals of time (i.e., Δt is constant).

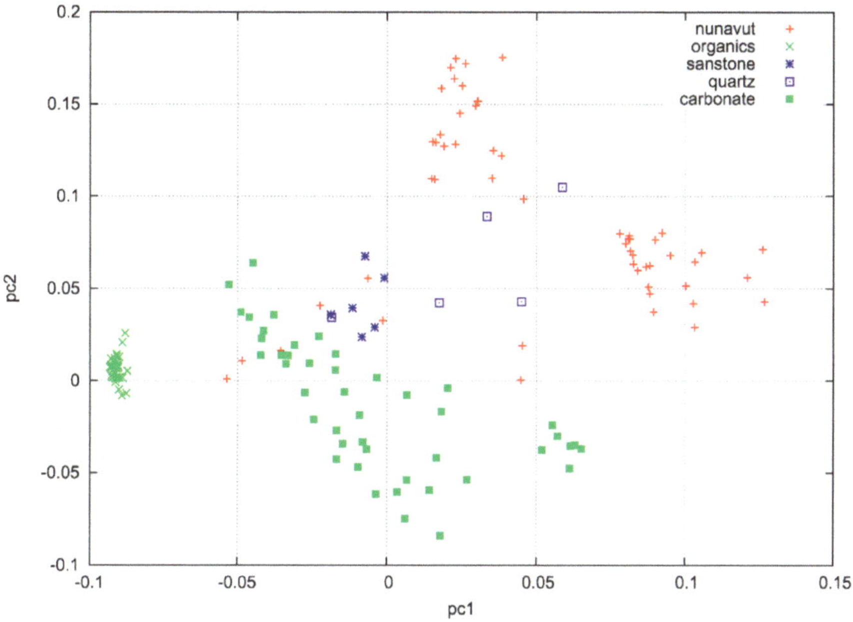

Figure 28.7. Spectral classification of biomarkers and other samples using SVD. The plot shows how the corresponding spectra could be represented by just two eigenvalues (or principal components, pc1 and pc2). The grouping indicates the main composition of each unknown spectra (e.g., Nunavut) relative to those known.

28.4.1 Noise

In signal processing, noise is defined as random fluctuation. Its source may be diverse: instrumentation (e.g., the temperature of the electronics may cause digital noise), the media in which the signal is passing (e.g., interstellar medium for a radiowave), or an unknown phenomenon.

The whole signal is composed of the desired information, $S(t)$, with a superposition of noise, $N_i(t)$, which can be summarized by equation (28.9).

$$X(t) = S(t) + \sum N_i(t) \tag{28.9}$$

The purpose of signal analysis is to characterize the data and remove the N_i (as much as possible) from $X(t)$.

28.4.2 Fourier domain analysis

The most common approach to analyzing a signal is to work in the frequency domain (i.e., the Fourier domain). The signal is represented by a series of sine functions (equation 28.10):

$$X(t) = a_0 + \sum_{n=1}^{N-1} (a_n \cos(\omega_n x) + b_n \sin(\omega_n x)) \tag{28.10}$$

Computer implementation of the Fourier transform (aka the Fast Fourier Transform, FFT) must use a discrete representation of the real signal, instead of a continuous one. The problem related to data sampling may introduce artifacts that must be processed using windowing techniques.

Noise may then be removed by filtering out high frequencies. This implies performing a convolution between the original signal and the filtering function. Such functions depend greatly on the type of desired filtering (e.g., removing high frequencies, low frequencies, or a band of frequencies).

28.4.3 Karhunen–Loève analysis

The KLT can be used to carry out similar analysis (i.e., removing noise). The eigenvectors (Φ_i) are computed from the autocorrelation matrix of $X(t)$, defined by equation (28.11). There is no windowing and the discretization of the data does not bring any problems:

$$R(i,j) = \frac{E[(X_i - \mu_i)(X_j - \mu_j)]}{\sigma_i \sigma_j} \tag{28.11}$$

Projection of the $X(t)$ into the eigenspace is computed using equation (28.12) where $\Phi_n(t)$ is the n-th eigenvector and T is the number of elements in $X(t)$ (i.e., the KL expansion). $X(t)$ can than be reconstructed using the Z_n vectors by simply adding a sufficient number of them (equation 28.13, where K is the number of eigenvectors to sum) until the approximation is good enough:

$$Z_n = \sum_{t=1}^{T} X(t)\Phi_n(t) \tag{28.12}$$

$$X(t) = \sum_{i=1}^{K} Z_i \Phi_i(t) \tag{28.13}$$

The number of eigenvectors required to reconstruct a signal without noise depends on its complexity. A single-frequency signal (e.g., a single sine) will require only one function $\Phi_1(t)$. An estimate of the number of eigenvectors to add can be given by looking at the distribution of λ_i.

28.4.4 Comparison between the FFT and the KLT

Several implementations of the Fast Fourier Transform can be made using software or hardware. While the FFT is faster than the KLT, the KLT performs better when the SNR[7] is very low.

Figure 28.8 demonstrates a noisy sine wave (i.e., with a single frequency $f = 10\,\text{Hz}$ and an SNR of $-23\,\text{dB}$). The Fourier transform cannot isolate the main frequency, while the KLT does show the main frequency as the strongest peak.

[7] Signal-to-noise ratio: defined, in decibels, as $\text{SNR} = 10\log(A/A_0)$ for amplitudes.

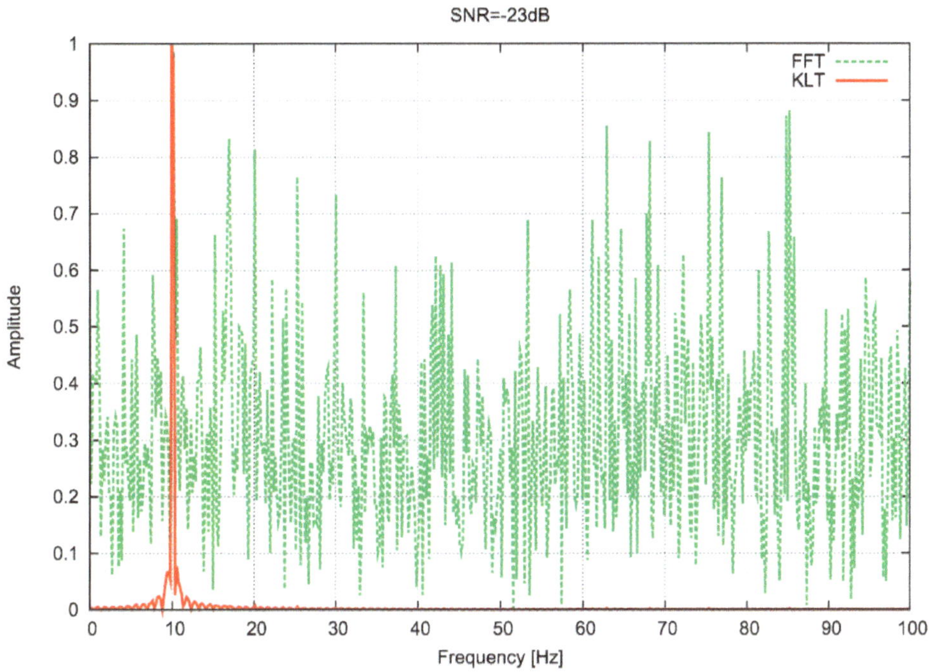

Figure 28.8. Comparison between the FFT and the KLT in the case of a very noisy signal (SNR $= -23$ dB). The main frequency of the signal is located at $f = 10$ on the frequency unit axis.

The KLT method, as implemented by the author for this example, is stable for an SNR ≥ -19 dB, meaning that the algorithm succeeds in finding the main frequency in the noise. For an SNR < -19 dB the algorithm is less stable. The error rate is around 30% at an SNR $= -23$ dB.

28.4.5 Multi-frequency signal

Applying the FFT to a signal, $X(t)$, produces a series of functions, $F_i(\nu)$, describing the signal. Each of these F_i represents a single frequency ν. A signal composed of several frequencies will need several F_i functions to be represented in the Fourier domain.

The eigenvectors of KLT are not directly associated with frequency. They are orthogonal functions defining the eigenspace. They could be sine functions depending on the nature of the original signal.

For a complex signal $X(t)$, several eigenvectors could be required to represent the original information.

Figure 28.9 shows the KLT filtering a signal composed of four sine signals of different frequencies (i.e., 10, 20, 30, and 40 Hz) with an SNR $= -19$dB. Each

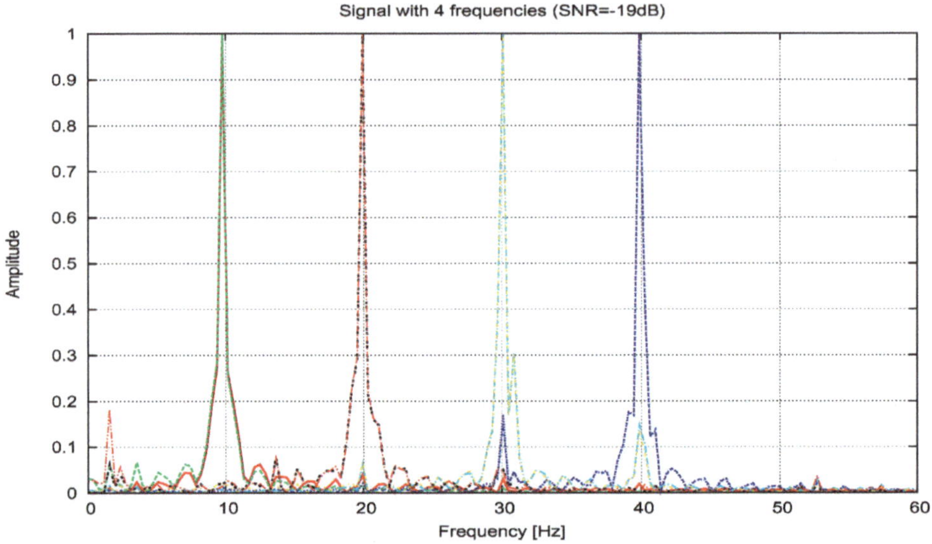

Figure 28.9. Example of KLT applied to a signal with four frequencies (SNR = −19 dB). Each frequency comes out as a strong peak.

frequency is associated with a different λ. Figures 28.10 and 28.11 show the result when the FFT method is used.

28.4.6 BPSK

Phase shift keying (PSK) is a type of digital modulation used in telecommunications. Binary phase shift keying (BPSK) is the simplest form of PSK. It uses two phases which are separated by 180°. This modulation is the most robust of all the PSKs since it would take the highest level of noise or distortion to make the demodulator reach an incorrect decision. The signal is defined by equation (28.14):

$$X(t) = \sqrt{\frac{2E_b}{T_b}} \cos(2\pi f_c t + \pi(1 - n)) \tag{28.14}$$

where n is the bit of information sent (i.e., 0 or 1), E_b is the energy per bit, T_b is the bit duration, and f_c is the frequency. Figure 28.12 shows the KLT vs. FFT analysis when confronted with an SNR = −19 dB. This is another example of the power of the KLT to extract the signal from a sea of noise.

28.5 EIGENIMAGES

Eigenimages can be seen as two-dimensional versions of a time series. They are used to classify hyper-spectral images (i.e., detection of mining resources, analysis of

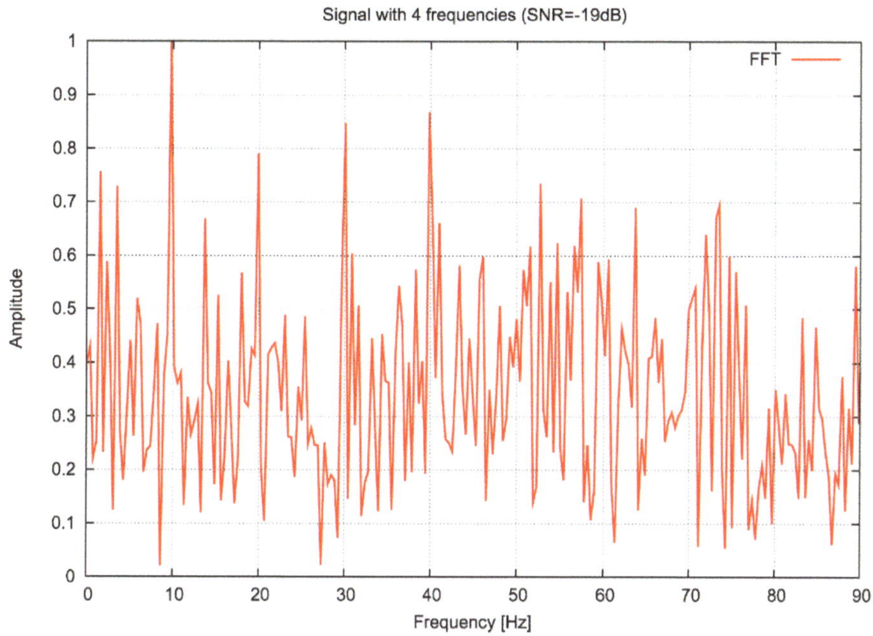

Figure 28.10. FFT of the signal from Figure 28.9. There are too many peaks to be able to isolate the right frequencies.

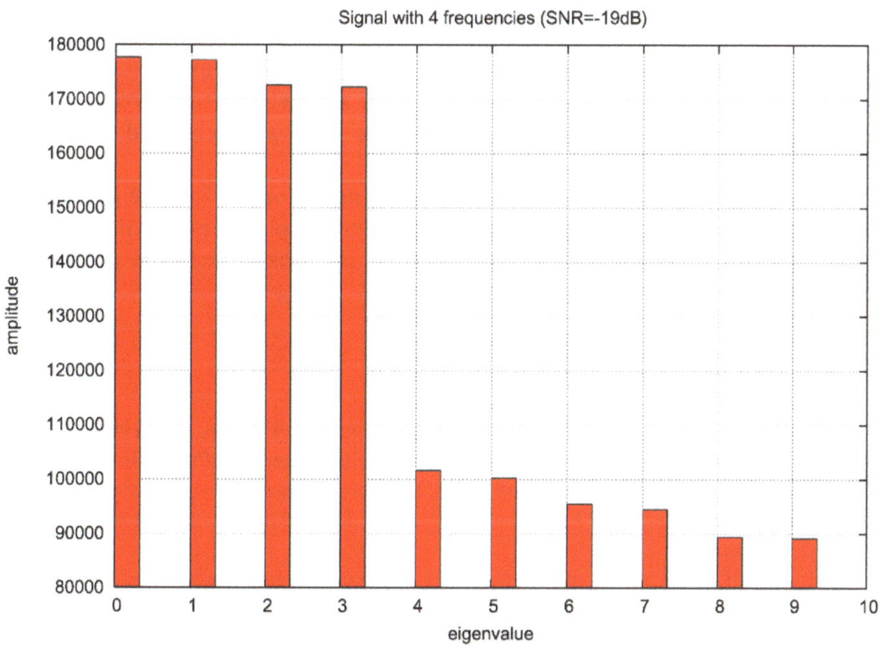

Figure 28.11. Eigenvalues from the signal shown in Figure 28.10. The first four λ are far larger than the rest, indicating the signal can be represented by only 4Φ.

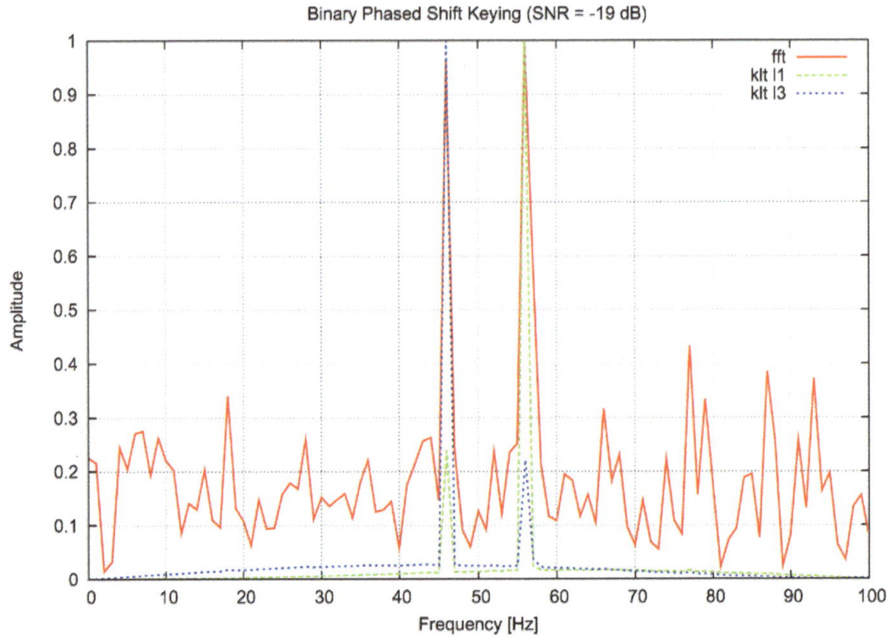

Figure 28.12. Example of a BPSK signal with an SNR = −19 dB. The message is "0101010101". The KLT and the FFT are able to present the two dominant frequencies. However, the KLT shows them as strong peaks while the FFT plot is almost buried in noise.

satellite images) and to adapt computers for human face recognition (i.e., biometric security). Instead of having eigenvectors that are orthonormal vectors in the eigenspace, we work with eigenimages and the data are described as a set of projections on those bases.

Figure 28.13 illustrates the result of superposing 80 (out of 7,040) eigenimages of 80×88 pixels to create a single image. The calculation was performed using the modified Lanczos algorithm (described earlier) and took 21 seconds on a normal workstation.

28.6 REFERENCES

[1] W. H. Press, S. A. Teukolsky, W. T. Vetterling, and B. P. Flannery (1992) *Numerical Recipes: The Art of Scientific Computing*, Cambridge University Press, Boston, MA.

[2] J. K. Cullum and R. A. Willoughby (2002) *Lanczos Algorithms for Large Symmetric Eigenvalue Computations, Vol. 1: Theory* and *Vol. 2: Documentation and Listing*, SIAM, Boston, MA.

[3] G. H. Golub and C. F. van Loan (1996) *Matrix Computations*, The Johns Hopkins University Press, Baltimore, MD.

[4] S. Dumas, Y. Dutil, and G. Joncas (2007) "Detection of endoliths using infrared spectroscopy," *Bioastronomy 2007: Molecules, Microbes and Extraterrestrial Life*, **420**, 377–380.

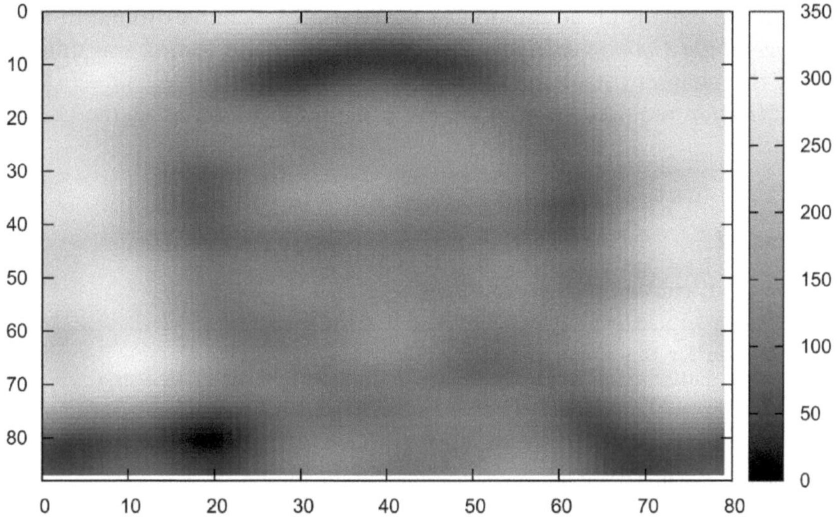

Figure 28.13. Example of an image generated by the superposition of 80 (out of 7,040) eigenimages.

28.7 BIBLIOGRAPHY

Elsner, J. B. and Tsonis, A. A. (2010) *Singular Spectrum Analysis*, Plenum Press, London.

Jolliffe, I. T. (2004) *Principal Component Analysis*, Springer, New York.

Golyandina, N., Nekrutkin, V., and Zhigljavsky, A. (2001) *Analysis of Time Series Structure: SSA and Related Techniques*, Chapman & Hall/CRC, Boca Raton, FL.

Lindgren, B. W. (1968) *Statistical Theory*, The Macmillian Company, Toronto, Canada

Maccone, C. (2009) *Deep Space Flight Communications*, Springer/Praxis, Heidelberg, Germany/Chichester, U.K.

Papoulis, A. and Pillai, S. U. (2002) *Probability, Random Variables and Stochastic Processes*, McGraw-Hill, New York.

Smith, S. W. (1997) *The Scientist and Engineer's Guide to Digital Signal Processing*, California Technical Publishing, Poway, CA.

APPENDIX 28.A FORTRAN CODE FOR KLT SIMULATIONS

28.A.1 Introduction

After so much mathematics, it is natural to think of some computer code capable of simulating the KLTs derived analytically in this book.

The following source code has been written in Fortran 77 to be largely portable. The compiler used to create this example is the GNU Fortran (e.g., *gFORTRAN*). It is freely available for most operating systems. The development of this program was done under Linux.

The source code is an adaptation of the Matlab® examples found in the book *Deep Space Flight Communications*. The programming and testing was done by the author of this chapter (Stephane Dumas).

The *Makefile* required to compile the program is provided at the end of this appendix.

28.A.2 main.f

This is the main routine of the program.

```
c FORTRAN code for the simulation of the KLT.
c based on the matlab code from drs. Nicolo Antonietti & Claudio Maccone.
c FORTRAN version by Stephane Dumas

      program brownian

      implicit none include 'brownian.inc'

c locals
      integer n,neigs,flag
      integer i,j
      real*8 t(max_size),input_data(max_size)
      real*8 amat(max_size,max_size)
      real*8 phi(max_size,max_size)
      real*8 lambda(max_size)
      real*8 klexp(max_eig,max_size)
      real*8 z(max_size)
      real*8 advector(max_size) ! analytic data vector
      real*8 edvector(max_size) ! empiric data vector

c do you prefer to run a new simulation (creating a new and different
c realization of the stochastic process x(t)) or do you wish to load an
c existing data file (produced by a previous simulation) ?

      call input(t,n,input_data,flag)

c decide how many eigenfuctions in the KL expansion you wish to take into
c account for the reconstruction of the process X(t) in the time interval
c between 0 and T. clearly, the number of eigenfunctions taken into account
c is at most equal to the number of instants considered in the simulation.
c in practice, however, you may wish to use fewer eigenfunctions, or even
c just very few eigenfuctions. the reconstruction of X(t) will thus be
c rougher and rougher, but the computation burden will still be affordable
c by your machine. this is the trade-off that the KLT offers to you as
c a lossy compression algorithm.

      write(*,*) 'how many eigenfunctions ?'
      read(*,*) neigs

c computation of the analytic autocorrelation matrix of the brownian
c motion, defined as min(t1, t2). this autocorrelation matrix is fed into
c the code only if you previously selected to run an entirely new
c simulation. if you previously selected to load a pre-existing data file
c (as happens in all experimental applications of the KLT), then the
c data file of the values of x(t), for t ranging between 0 and t, is fed
c into the code.
c the autocorrelation of the brownian motion of size n is defined as
c min(t1, t2) by the function (i.e. by the subroutine)
```

```
c brownian_autocorrelation(n), hereby called by the main code.
c (was) autocorrelation_matrix = brownian_autocorrelation(t);

      call autocorr(input_data,n,amat)

c the next step is the most important step in this main code.
c by virtue of the "eigs" subroutine of matlab, we avoid getting entangled
c in the computation of the eigenvalues lambda and of the eigenfunctions phi
c of the KLT. quite simply, we feed in the autocorrelation matrix (whether
c it was analytic or numeric = experimental) and "eigs" returns both lambda
c and phi! clearly, in non-matlab simulations, this "eigs" routine must be
c very carefully written!

      call eigenvalue(amat,n,phi,lambda)

c we now compute the empiric KLT (as opposed to the analytic KLT derived in
c the book analytically) for the simulation of X(t) under consideration.
c
c this empiric KLT we obtain in the following loop by:
c
c 1) projecting the vector of the input_process_data (i.e. the vector
c representing the stochastic process X(t) to be KL-expanded) onto the
c relevant i-th eigenvector phi(i). this projection is the random variable
c Z(i) of the KL expansion (as follows by inverting the KL expansion,
c just as one does for the fourier series).
c
c 2) defining the ith term of the kl expansion as the product of Z(i) times
c phi(i).

      do i=1,neigs
        z(i) = 0.
        do j=1,n
          z(i) = z(i) + input_data(j)*phi(j,i)
        enddo

        do j=1,n
          klexp(i,j)=z(i)*phi(j,i)
        enddo
      enddo

c we now create the data vector of the empiric reconstruction of x(t)
c achieved by the KLT numerically. this is simply the sum of all the
c kl_expansions_ith_term obtained in the previous step of this main code.

      do i=1,n
        edvector(i)=0.0
      enddo

      do i=1,neigs
        do j=1,n
          edvector(j)=edvector(j)+klexp(i,j)
        enddo
      enddo

c next we create the data vector of the analytic reconstruction of X(t) as
c given by the formulae mathematically demonstrated in the book.
c this requires a separate routine (named hereafter analytic_klt) to be
c called up by this main code. the text of this routine clearly changed
c according to which formula in the book we refer to.

      if (flag.eq.1) then
        call analytic_klt(input_data,n,neigs,t,lambda,advector)

      else if (flag.eq.2) then
```

```
        call analytic_decel(input_data,n,neigs,t,lambda,advector)

      else if (flag.eq.3) then
        call analytic_brow(input_data,n,neigs,t,lambda,advector)

      else if (flag.eq.4) then
        call analytic_brow_dec(input_data,n,neigs,t,lambda,advector)

      else if (flag.eq.5) then
        call analytic_uniform(input_data,n,neigs,t,lambda,advector)
      endif

c Plot the EIGENVALUES of the EMPIRIC RECONSTRUCTION of X(t).
c Plots of:
c 1) The ORIGINAL REALIZATION of X(t)
c 2) The EMPIRIC RECONSTRUCTION OF X(t) by the KLT.
c 3) The ANALYTIC RECONSTRUCTION OF X(t) by the KLT.

      open(unit=10,file='plot_final')
      do i=1,n
        write(10,1000) i,input_data(i),sqrt(1.*i),-sqrt(1.*i),
     &                 advector(i),edvector(i)
      enddo
      close(10)

1000 format(i5,5e15.4)
      end
```

28.A.3 input.f

This subroutine is used to ask the user about the type of analysis and the input data.

```
      subroutine input(time_range,final_t,process_vector,flag)
      implicit none
c this subroutine allows you to either
c 1) create a brand-new realization of the input stochastic process
c    x(t) or
c 2) load an existing file with all the numeric data of the input
c stochastic process. clearly, this arises when you do experimental
c work, such as getting the input of a radiotelescope, etc.

      include 'brownian.inc'

c arguments
      integer flag,final_t
      real*8 time_range(max_size),process_vector(max_size)

c internal
      integer i,j,case_number
      character*256 name

 10   write(*,*) 'what process is going to be analized?'
      write(*,*) ' 1. a standard brownian motion from a still source'
      write(*,*) ' 2. a standard brownian motion from a source in a
     & decelerated motion'
      write(*,*) ' 3. a square standard brownian motion from a still
     & source'
      write(*,*) ' 4. a square standard brownian motion from a source
     & in decelerated motion'
      write(*,*) ' 5. a uniform relativistic motion'
      read(*,*) flag
```

```
      if (flag.lt.1 .or. flag.gt.5) goto 10

 20   write(*,*) 'enter 1 to create a new realization of the brownian
     & motion x(t).'
      write(*,*) 'enter 2 to load an existing brownian motion data file.'
      read(*,*) case_number if (case_number.lt.1 .or. case_number.gt.2) goto 20

c creating the new realization of the stochastic process X(t)
      if (case_number.eq.1) then
        write(*,*) 'please, type the final time unit. (suggested: no
     & more than 1000)'
        read(*,*) final_t
        do i=1,final_t
          time_range(i)=1.*i
        enddo

        call process(final_t,process_vector)

c plot the original realization of X(t) to be later expanded and
c reconstructed by virtue of the KLT.

        open(unit=10,file='plot_input')
        do i=1,final_t
          write(10,1000) i,process_vector(i),sqrt(1.*i),-sqrt(1.*i)
        enddo
        close(10)

c loading an existing experimentally obtained input stochastic process.
      else if (case_number.eq.2) then
        write(*,*) 'please type the full path and file name.'
        read(*,*) name

        open(unit=10,file=name)
        i=1
 30     read(10,*,end=35) j,process_vector(i)
        i=i+1
        goto 30
 35     close(10)

        final_t = i-1

        do i=1,final_t
          time_range(i)=1.*i
        enddo

        open(unit=10,file='plot_input')
        do i=1,final_t
          write(10,1000) i,process_vector(i),sqrt(1.*i),-sqrt(1.*i)
        enddo
        close(10)
      endif

      if (flag .eq. 3) then
        do i=1,final_t
          process_vector(i)=process_vector(i)**2-time_range(i)
        enddo
        else if (flag .eq. 4) then
          do i=1,final_t
            process_vector(i)=process_vector(i)**2
          enddo
        endif

 1000 format(i5,4e15.4)
```

```
      return
      end
```

28.A.4 process.f

This subroutine generates a Brownian motion and stores it as the input vector.

```
      subroutine process(t,x)
      implicit none
c subroutine creating the random walk path of the brownian motion b(t).

      include 'brownian.inc'

c arguments
      integer t
      real*8 x(max_size)

c locals
      integer i
      real*8 rnd

c set to zero all the initial values of the t-element vector b that will
c contain the (random) values of the X(t) process when the new realization
c of x(t) will have been computed.
      x(1)=0.0
      do i=2,t
 10     rnd=rand()
        if (rnd.eq.0.5) goto 10
        if (rnd.gt.0.5) then
          x(i)=x(i-1)-1.0
        else
          x(i)=x(i-1)+1.0
        endif
      enddo

      return
      end
```

28.A.5 autocorr.f

This subroutine computes the autocorrelation of a Brownian motion.

```
      subroutine autocorr(t,n,c)
      implicit none
c this subroutine computes the autocorrelation of the brownian motion
c b(t) by translating its analytical definition min(t1, t2) into a
c numeric matrix. the entries of such a matrix are each the minimum
c between the relevant row and column numbers.

      include 'brownian.inc'

c arguments
      integer n
      real*8 t(max_size),c(max_size,max_size)

c locals
      integer row,col

      do row=1,n
```

```
        do col=1,n
          c(row,col)=min(row,col)
        enddo
      enddo

      return
      end
```

28.A.6 analytic.f

This file contains a series of subroutines responsible for computing the analytical solution of the problem.

```
c ----------------------------------------------------------------------------------
      subroutine analytic_klt(input_data,n,neigs,t,lambda,dvector)
      implicit none
c This subroutine computes the ANALYTIC RECONSTRUCTION of X(t)
c according to the analytic (either exact or approximated formulae
c given in the book.)

      include 'brownian.inc'

c arguments
      integer n,neigs
      real*8 input_data(n),t(n),lambda(n),dvector(n)

c locals
      integer i,j
      real*8 arg(n),c1,zed,phi(n),nn,lda(n)

      do i=1,n
        dvector(i)=0.
      enddo

c This is the KEY SUBROUTINE YIELDING THE ANALYTIC RECONSTRUCTION
c of X(t).
      do i=1,neigs
        do j=1,n
          arg(j)=t(j)*pi*(2*i-1)/(2*(n+1))
        enddo

        nn = sqrt(2/(n+1.))
        lda(i) = 4*(n+1)**2./(pi**2*(2*i-1)**2)

        do j=1,n
          phi(j)=nn*sin(arg(j))
        enddo

        zed=0.
        do j=1,n
          zed=zed+input_data(j)*phi(j)
        enddo

        do j=1,n
          dvector(j)=dvector(j)+zed*phi(j)
        enddo
      enddo

c Plot the EIGENVALUES of the EMPIRIC AND ANALYTIC RECONSTRUCTIONS
c of X(t).
      open(unit=10,file='plot_lambda')
```

```
        do i=1,neigs
          write(10,1000) i,lda(i),lambda(i) ! TO BE CHECKED
        enddo
        close(10)

 1000   format(i5,2e15.5)

        return
        end

c ----------------------------------------------------------------------------------------------------------------------
        subroutine analytic_decel(input_data,n,neigs,t,lambda,dvector)
        implicit none
c This subroutine computes the ANALYTIC RECONSTRUCTION of X(t)
c according to the analytic (either exact or approximated formulae
c given in the book.)

        include 'brownian.inc'

c arguments
        integer n,neigs
        real*8 input_data(n),t(n),lambda(n),dvector(n)

c locals
        integer i,j
        real*8 nn(n),lda(n),arg(n),gam(neigs),phi(n),tau(n)
        real*8 h,c,nu,zeta

c external
        real*8 besselj
        external besselj

        do i=1,n
          dvector(i)=0.
        enddo

c H is the so-called "Hurst parameter" that, in this case of the
c DECELARATED spaceship motion, specifies whether the deceleration is
c "sudden" (i.e. it lasts for a short time only) or "soft" (i.e. it
c takes a longer time to stop).
        write(*,*) 'Value of Hurst parameter H>0.75 ? (Suggested: 0.8)'
        read(*,*) h

c Analytic expression of the constant C as given in the book,
c eq. (17.47).
        c = 1./(2.*h*n**(2*h-1))

c Proper time for the DECELERATED motion case ("Independence Day")
c of a relativistic spaceship approaching the Earth at the speed
c of light.
        do i=1,n
          tau(i)=c*t(i)**(2*h)
        enddo

c This is the KEY SUBROUTINE YIELDING THE ANALYTIC RECONSTRUCTION
c of X(t).
        do i=1,neigs
          gam(i) = i*pi - pi/4. - pi/(2*(2*h+1))
          nu     = 2*h/(2*h+1)

        do j=1,n
          arg(j)=gam(i)*t(j)**(h+0.5)/(n+1)**(h+0.5)
          nn(j)=sqrt(2*h+1)*t(j)**h/(n+1)**(h+0.5)
        enddo
```

```
          lda(i)=(n+1)*(2*h+1)/(gamma(h+1./2.)**2)/(gam(i)**2)

          zeta = 0.
          do j=1,n
            phi(j) = nn(j)*besselj(nu,arg(j))/abs(besselj(nu,gam(i)))
            zeta=zeta+input_data(j)*phi(j)
          enddo

          do j=1,n
            dvector(j)=dvector(j)+zeta*phi(j)
          enddo
        enddo

lc Plot the EIGENVALUES of the EMPIRIC AND ANALYTIC RECONSTRUCTION of X(t).
        open(unit=10,file='plot_lambda')
        do i=1,neigs
          write(10,1000) i,lda(i),lambda(i)
        enddo
        close(10)

 1000 format(i5,2e15.5)

        return
        end

c ------------------------------------------------------------------------------------------------------------------
        subroutine analytic_brow(input_data,n,neigs,t,lambda,dvector)
        implicit none
c This subroutine computes the ANALYTIC RECONSTRUCTION of X(t)
c according to the analytic (either exact or approximated formulae
c given in the book.)

        include 'brownian.inc'

c arguments
        integer n,neigs
        real*8 input_data(n),t(n),lambda(n),dvector(n)

c locals
        integer i,j
        real*8 nn(n),vector(n),lda(n),phi(n),gam(neigs),arg(n)
        real*8 zeta,order

c external
        real*8 besselj
        external besselj

        do i=1,n
          dvector(i)=0.0
        enddo

        order = 2./3.
        do i=1,neigs
          gam(i) = pi * (i-5./12.)

          do j=1,n
            arg(j)=gam(i)*t(j)**(3./2.)/(n+1)**(3./2.)
            nn(j)=sqrt(3.)*t(j)/(n+1)**(3./2.)/abs(besselj(order,gam(i)))
          enddo

          lda(i) = 16./9.*(n+1)**3./gam(i)**2.

          zeta=0.
          do j=1,n
```

```
          phi(j)=nn(j)*besselj(order,arg(j))
          zeta=zeta+input_data(j)*phi(j)
        enddo

        do j=1,n
          dvector(j)=dvector(j)+zeta*phi(j)
        enddo
      enddo

c Plot the EIGENVALUES of the EMPIRIC RECONSTRUCTION of X(t).
      open(unit=10,file='plot_lambda')
      do i=1,neigs
        write(10,1000) i,lda(i),lambda(i)
      enddo
      close(10)

 1000 format(i5,2e15.5)

      return
      end

c ----------------------------------------------------------------------------------------------------------------
      subroutine analytic_brow_dec(input_data,n,neigs,t,lambda,dvector)
      implicit none
c This subroutine computes the ANALYTIC RECONSTRUCTION of X^2(t)
c according to the analytic (either exact or approximated formulae
c given in the book.

      include 'brownian.inc'

c arguments
      integer n,neigs
      real*8 input_data(n),t(n),lambda(n),dvector(n)

c locals
      integer i,j
      real*8 tau(n),gam(neigs),phi(n),arg(n),nn(n),lda(n)
      real*8 h,c,nu,zeta

      do i=1,n
        dvector(i)=0.
      enddo

c H is the so-called "Hurst parameter" that, in this case of the
c DECELERATED spaceship motion, specifies whether the deceleration is
c "sudden" (i.e. it lasts for a short time only) or "soft" (i.e. it
c takes a longer time to stop).
      write(*,*) 'Value of Hurst parameter H > 0.75 ? (Suggested: 0.8)'
      read(*,*) h

c Analytic expression of the constant C as given in the book,
c eq. (17.47).
      c = 1./(2.*H*n**(2.*H-1))

c Proper time for the DECELERATED motion case ("Independence Day") of a
c relativistic spaceship approaching the Earth at the speed of light.

        do j=1,n
          tau(j)=c*t(j)**(2.*h)
        enddo

        do i=1,neigs
          gam(i) = i*pi - pi/4. - pi/(2.*(2*h+1))
          nu = 2.*h/(2.*h+1./2.)
```

```
          do j=1,n
            arg(j)=gam(i)*t(j)**(h+1./2.)/(n+1)**(h+1./2.)
     &         -2.*h*pi/(2.*(2.*h+1))-pi/4.
            nn(j)=sqrt(2*h+1)*t(j)**(h/2.-0.25)/(n+1)**(h/2.+0.25)
          enddo

          lda(i) = (n+1)**2/((h+0.5)**2 / gam(i)**2)

          zeta = 0.
          do j=1,n
            phi(j)=nn(j)*sin(arg(j))
            zeta=zeta+input_data(j)*phi(j)
          enddo

          do j=1,n
            dvector(j)=dvector(j)+zeta*phi(j)
          enddo
        enddo

c Plot the EIGENVALUES of the EMPIRIC RECONSTRUCTION of X(t).
        open(unit=10,file='plot_lambda')
        do i=1,neigs
          write(10,1000) i,lda(i),lambda(i)
        enddo
        close(10)

 1000 format(i5,2e15.5)

        return
        end

c -----------------------------------------------------------------------
        subroutine analytic_uniform(input_data,n,neigs,t,lambda,dvector)
        implicit none
c This subroutine computes the ANALYTIC RECONSTRUCTION of X(t)
c according to the analytic (either exact or approximated formulae
c given in the book.)

        include 'brownian.inc'

c arguments
        integer n,neigs
        real*8 input_data(n),t(n),lambda(n),dvector(n)

c locals
        integer i,j
        real*8 ratio,k,gam(neigs),zed,lda(n),phi(n),arg(n),nn(n)

        write(*,*) 'What ratio of the speed of light is the uniform
       & velocity? (For instance 0.2)'
        read(*,*) ratio

        k = (1.0-ratio**2.0)**(0.25)

        do i=1,n
          dvector(i)=0.
        enddo

        do i=1,neigs
          gam(i) = i*pi - pi/2.0

          do j=1,n
            arg(j)=gam(i)*t(j)/(n+1)
            nn(j)=k*sqrt(2./(n+1))
```

```
          enddo

          lda(i) = k**2.0 * n**2.0 / gam(i)**2

          zed = 0.
          do j=1,n
            phi(j)=nn(j)*sin(arg(j))
            zed=zed+input_data(j)*phi(j)
          enddo

          do j=1,n
            dvector(j)=dvector(j)+zed*phi(j)
          enddo
        enddo

c Plot the EIGENVALUES of the EMPIRIC RECONSTRUCTION of X(t).
        open(unit=10,file='plot_lambda')
        do i=1,neigs
          write(10,1000) i,lda(i),lambda(i)
        enddo
        close(10)

 1000 format(i5,2e15.5)

        return
        end
```

28.A.7 bessel.f

This subroutine computes the bessel of the first kind with real order.

```
        real*8 function besselj(n,x)
        implicit none
c compute the bessel of the first kind, with real order

c arguments
        real*8 n,x

c locals
        integer i
        real*8 m,s,s1,s2,s3

        s=0.
        do i=0,100
          m=1.*i
          s1 = (-1.)**m / gamma(m+1.)
          s2 = gamma(m+n+1)
          s3 = (x/2.)**(2*m+n)
          s=s+s1/s2*s3
        enddo

        besselj = s

        return
        end
```

28.A.8 eigenvalue.f

This is the subroutine responsible for computing eigenvectors and eigenvalues. It is a wrap-around mechanism that uses other subroutines contained in the mathematical library called Numerical Recipes. Licensing forbids us from redistributing their code but it is available on their web page (*http://www.nr.com/*).

Other mathematical libraries offer similar functionality. If you choose to link with one of them, be careful about the arguments.

```
      subroutine eigenvalue(a,n,phi,lambda)
      implicit none
c compute the eigenvalues and eigenvectors of matrix a

      include 'brownian.inc'

c argument
      integer n
      real*8 a(max_size,max_size)
      real*8 phi(max_size,max_size)
      real*8 lambda(max_size)

c locals
      integer nrot

c JACOBI and EIGSRT are routines taken from Numerical Recipes

c Licensing forbids the redistribution of their code
c and it is not possible to give it here.
c The user must get the appropriate subroutines, compile them
c and link them.
c Other mathematical libraries offers such subroutines. They can
c be used instead of those here.
c please refer to the documentation of Numerical Recipes for
c more details.

c phi() contains the eigenvectors
c phi(i,j) refers to the i-th element of the j-th eigenvector

c JACOBI is the subroutine responsible for computing eigenvalues
c and eigenvectors.
      call jacobi(a,n,max_size,lambda,phi,nrot)

c EIGSRT sorts the output of JACOBI so that the eigenvalues
c are from the largest to the smallest. The eigenvectors are also
c sorted accordingly.
      call eigsrt(lambda,phi,n,max_size)

      return
      end
```

28.A.9 brownian.inc

This is an include file that contains some of the definitions all subroutines need.

```
      integer max_size,max_eig
      parameter(max_size=1000,max_eig=20)

      real*8 pi
      parameter(pi=3.14159265358979)
```

28.A.10 Makefile

This is the control file required to compile the whole program.

```
G77 = gFORTRAN

G77FLAGS = -O3 -fbounds-check -ffixed-line-length-none

all : brownian

OBJS = main.o \
       input.o \
       process.o \
       autocorr.o \
       eigenvalue.o \
       analytic.o \
       bessel.o

# Numerical Recipes subroutines
NR_OBJS = eigsrt.o \
       jacobi.o

brownian : $(OBJS) $(NR_OBJS)
       $(G77) -o brownian $(OBJS) $(NR_OBJS)

.f.o:
       $(G77) -c $(G77FLAGS) -o $@ $<

clean :
       rm -f *.o *~ brownian
```

28.A.11 Output files

There are three output files created by the program. They are in plain ASCII, contain several columns of numbers, and can be read by any text editor. The data can be plotted using *Excel*, *gnuplot*, or any other graphical tool.

file **plot_input**	contains the input vector, the file has four columns t X(t) $\sqrt{t}$ $-\sqrt{t}$ t: time X(t)
file **plot_lambda**	contains the eigenvalues, the file has three columns i λ_i Λ_i i: index λ_i: analytical eigenvalues Λ_i: empirical eigenvalues
file **plot_final**	contains a comparison between empirical and analytical solutions t X(t) $\sqrt{t}$ $-\sqrt{t}$ $B_A(t)$ $B_E(t)$ t: time X(t): Brownian motion $B_A(t)$: analytical solution $B_E(t)$: empirical solution

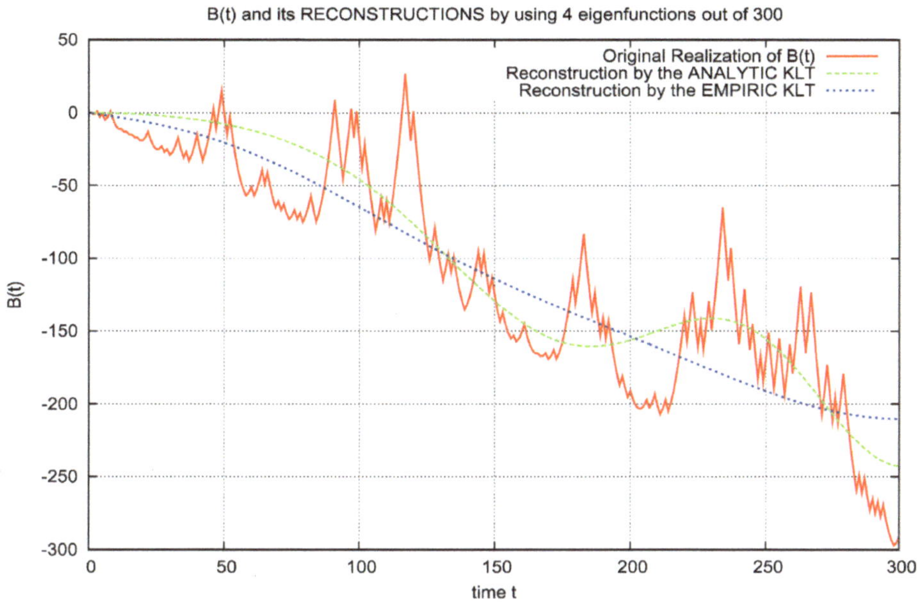

Figure 28.A.1. Examples of the content of the file *plot_final*. The figure was created using *gnuplot*. The analytic KLT gives the value of $B_A(t)$ and the empirical KLT gives the value of $B_E(t)$.

28.A.12 Linking FORTRAN with C

It is possible to use this FORTRAN code with a program written in C without translating the whole code into C. The following is an example of a C program calling the FORTRAN subroutine *bessel*.

```
#include <stdio.h>

extern "C"
{
      // note the "_" at the end of the name.
      double besselj_(double *n, double *x);
}

int main(int argc, char *argv[1])
{
      double n, x;

      n = 1.2;
      x = 2.5;

      // when calling a FORTRAN aubroutine from C, always pass the arguments
      // as pointer, never as reference or plain values.
      printf('BesselJ(%.21f, %.31f) = %.41e\n', n, x, besselj_(&n, &x));

      return 0;
}
```

Finally, this is the *Makefile* to compile and link the program.

```
G77 = gFORTRAN
GXX = g++

all : bessel

bessel : c_bessel.o bessel.o
	$(GXX) -o bessel c_bessel.o bessel.o -lgFORTRAN -lstdc++  -lm

.f.o:
	$(G77) -c $(G77FLAGS) -o $@ $<

.cpp.o:
	$(GXX) -c $(CXXFLAGS) $(INCPATH) -o $@ $<
```

Part IV

The United Nations and protection of the Moon's farside

29

The United Nations and protection of the Moon's farside

29.1 INTRODUCTION: DEFINING THE PAC (PROTECTED ANTIPODE CIRCLE)

Since the 1990s the international scientific community—especially the IAA (International Academy of Astronautics)—have been discussing the need to keep the farside of the Moon free from man-made radio frequency interference (RFI). In fact, the center of the farside, specifically the Daedalus crater, is an ideal site to set up a future radio telescope (or, more realistically, a phased array) to detect radiowaves of all kinds, which are impossible to detect on Earth because of the ever-growing RFI.

Nobody, however, seems to have established a precise border for the circular region around the antipode of the Earth (i.e., 0° latitude and 180° longitude both east and west) that should be protected from uncontrolled human exploitation when several nations will have reached the capability of easy travel to the Moon.

In this chapter we propose creation of the Protected Antipode Circle (PAC), a large circular piece of land about 1,820 km in diameter, centered around the antipode on the farside, spanning an angle of 30° at the center of the Moon in both longitude and latitude and, on the Moon surface in all radial directions from the antipode. It is depicted in Figure 29.1 as the thick red circle tangent to the two parallels at ±30° in latitude.

Sound scientific reasons exist for defining the PAC in this way:

(1) the PAC is the most shielded area of the farside, with an expected attenuation of man-made RFI of 100 dB or higher (as we shall show in Section 29.7);
(2) the PAC is the only area of the farside that will never be reached by radiation emitted by future human space bases located at either of the L4 and L5 Lagrangian points of the Earth–Moon system;
(3) the PAC does not overlap with other areas of interest to human activity except

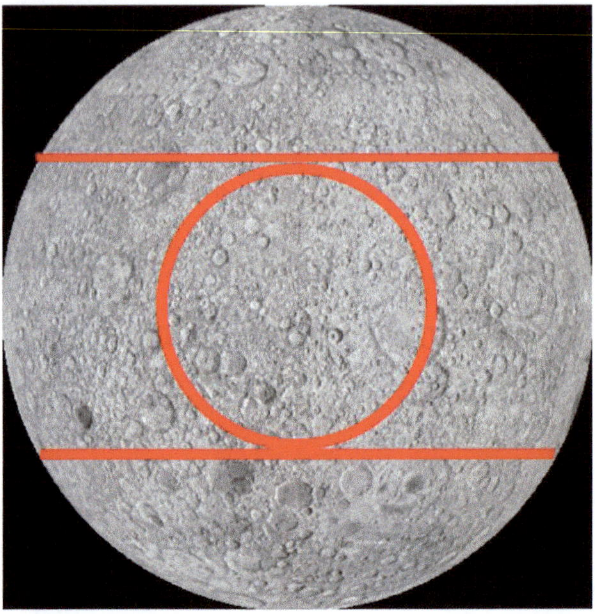

Figure 29.1. A photo of the farside of the Moon, the two parallels at $\pm 30^{\circ}$ drawn by solid red lines, and the PAC shown as the red solid circle centered at the antipode and tangent to the above two parallels at $\pm 30^{\circ}$. In view of these unique features, we propose the PAC to be officially recognized by the United Nations as an International Protected Area, free from radio contamination by humans both now and in the future. This will be an achievement for the benefit of all humankind.

for a minor common area with the Aitken Basin, the southern depression supposed to have been created 3.8 billion years ago during the "Big Wham" between the Earth and the Moon.

In view of these unique features, we propose the PAC to be officially recognized by the United Nations as an International Protected Area, an area prohibiting radio contamination by humans either now or in the future and carried out for the benefit of all humankind.

29.2 THE URGENT NEED FOR RFI-FREE RADIO ASTRONOMY

In order to detect radio signals of all kinds, the stock in trade of radio astronomers, it is mandatory first of all to reject all RFI. But RFI is produced in ever-increasing amounts in line with the technological growth of civilization, and has now reached the point where large bands of the spectrum are blinded by both legal or illegal transmitters of all kinds. In 1994, the late French radio astronomer Jean Heidmann (1920–2000) pointed out that radio astronomy from the surface of the Earth is doomed to die in a few decades if uncontrolled growth of RFI continues. Heidmann also made it clear, however, that advances in modern space technology could breathe new life into radio astronomy if it is done from the farside of the Moon, clearly shielded by the Moon's spherical body from all RFI produced by humans on Earth.

As a result of the developments discussed in this chapter, we present a short review on the five Lagrangian points of the Earth–Moon system (shown in Figure 29.2).

29.3 TERMINAL LONGITUDE λ ON THE MOON FARSIDE FOR RADIOWAVES EMITTED BY TELECOM SATELLITES IN ORBIT AROUND THE EARTH

In this section we prove an important mathematical formula, one that is vital to the selection of any RFI-free base on the Moon's farside.

We want to compute the small angle α beyond the limb (the limb is the meridian having longitude 90° E on the Moon) where radiowaves from telecom satellites in circular orbit around the Earth still reach (i.e., they become tangent to the Moon's spherical body). The new angle $\lambda = \alpha + 90^{\circ}$ we call the "terminal longitude" of these radiowaves. In practice, no radiowave from telecom satellites can hit the Moon surface at longitudes higher than this terminal longitude λ (though, diffraction is not taken into account).

To find α (see Figure 29.3) we draw a straight line tangent to the Moon's sphere from G, the point tangent to the circular orbit having radius R. This straight line forms a right-angled triangle with the Earth–Moon axis, EM, with the right angle at G. Next, consider a straight line parallel to the one above but from the Moon center M, intersecting the EG segment at point P. Once again, the triangle EPM forms a right-angled triangle at P. It is similar to the previous triangle. So, angle α is now equal to the EMP angle. The latter can be found, since:

(1) The Earth–Moon distance $\overline{EM} = D_{Earth-Moon}$ is known and we assume it at perigee: Earth–Moon distance equal to 356,410 km.
(2) The $\overline{EP}$ segment equals the $\overline{EG} = R$ segment minus the Moon radius R_{Moon}.
(3) Using Pythagoras' theorem one finds $\overline{PM} = \sqrt{(\overline{EM})^2 - (\overline{EP})^2}$.
(4) The tangent of the requested angle α is then given by

$$\tan\alpha = \frac{\overline{EP}}{\overline{PM}} = \frac{\overline{EP}}{\sqrt{(\overline{EM})^2 - (\overline{EP})^2}}.$$

Inverting the last equation and making the substitutions described at points (1), (2), and (4), one gets the terminal longitude λ of radiowaves on the Moon farside (between 90° E and 180° E) emitted by a telecom satellite circling around the

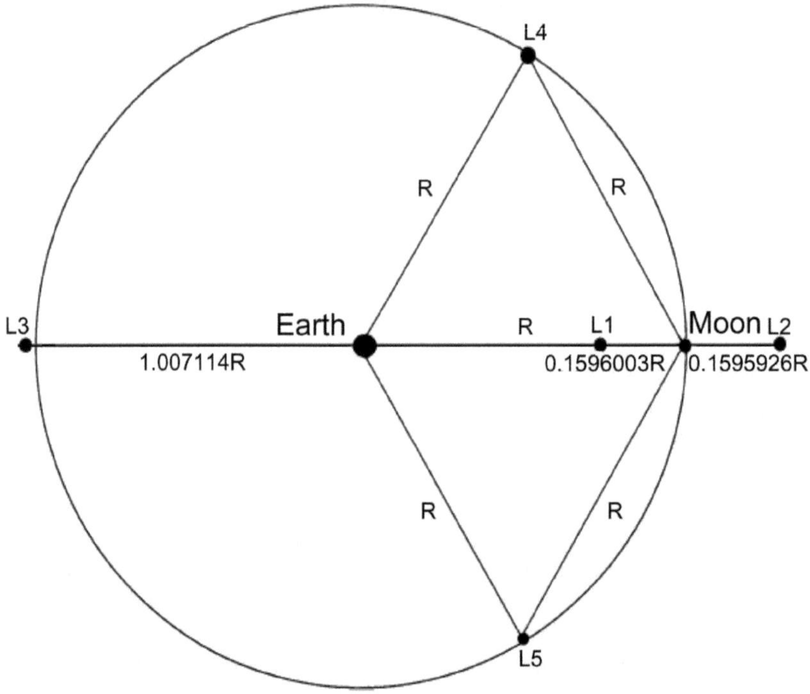

Figure 29.2. The five Earth–Moon Lagrangian points (i.e., the points at which the gravitational pull of the Earth and that of the Moon on a spacecraft cancel out, even allowing for the motion of the Moon around the Earth):

1. Let R denote the Earth–Moon distance (i.e., 384,400 km), and let us suppose the orbit of the Moon to be circular, just to simplify things. Then, the distance between the Moon and Lagrangian point L1 equals $0.1596003 * R$ (i.e., 61,350 km). Consequently the Earth–L1 distance equals $0.8403997 * R$ (i.e., 323,050 km). It can be shown that the L1 point is gravitationally unstable, meaning simply that a spacecraft at L1 is in perfect gravitational balance between the Earth or the Moon! Of course, if it is unable to maintain that balance it will fall towards one or the other until its final destruction on the surface of either the Earth or the Moon. Thus, a spacecraft located at the L1 point must be kept there by the continuous emission of hydrazine, which is of course an expensive and unpleasant technique of control. Plans to establish an international space station (and even a space hotel) have been made since the beginning of the Space Age in the 1960s.
2. The distance between the Moon and Lagrangian point L2 equals $0.1595926 * R$ (i.e., 61,347 km). Like L1, the L2 point is gravitationally unstable. However, a spacecraft put in a "halo orbit" (say, a circular orbit around L2 in the plane perpendicular to the Earth–Moon axis) will have the advantage of being visible from Earth at all times, which is ideal for Earth–spacecraft telecommunications. The drawback of spacecraft in such halo orbits around L2 (or just staying still at the very L2 point) is that their radio transmitters will flood the farside of the Moon, which is exactly what we do *not* want. Yet, the *intensity* of such radio emissions will certainly be much lower than the RFI coming from the Earth, and the notch filter in our radio telescope in Daedalus crater will be enough to prevent these radiowaves from polluting our radio "vision" of the sky from Daedalus crater to any great degree.
3. The distance between the Earth and Lagrangian point L3 equals $1.007114 * R$ (i.e., 387,135 km). No useful way of exploiting the L3 point seems to have been discovered yet, except one for planetary defense (as suggested in [1]).
4. The two triangular Lagrangian points L4 and L5 are equidistant R from the Earth and Moon. They form two equilateral triangles with the Earth and the Moon, and are located on the circular orbit of the Moon around the Earth. It can be proven that a gravitationally stable region exists around each of them. This is the reason the L5 Society was created in 1975 to look into the possibility of establishing at L5 one or more future human space colonies (see *http://en.wikipedia.org/wiki/L5_Society*).

Earth at a distance R:

$$\lambda = \operatorname{atan}\left(\frac{R - R_{Moon}}{\sqrt{D_{Earth-Moon}^2 - (R - R_{Moon})^2}}\right) + \frac{\pi}{2}, \tag{29.1}$$

where the independent variable R can only range between 0 and the maximum value that does not make the above radical become negative (i.e., $0 \le R \le D_{Earth-Moon} + R_{Moon}$). The above equation shows that the $\lambda(R)$ curve becomes vertical for $R \to (D_{Earth-Moon} + R_{Moon})$ and $\lambda = 180^{\circ}$.

29.4 THE QUIET CONE OVERHEAD THE FARSIDE OF THE MOON

The spherical body of the Moon prevents terrestrial electromagnetic waves from reaching the central part of the farside (as described in the previous sections). But this also applies to a certain region of space above the farside itself, which we have come to call the "Quiet Cone". This definition is appropriate inasmuch as radiowaves tangent to the Moon body clearly define a cone, and the apex of such a cone gets closer to the Moon farside the higher the radius R of the telecom satellite around the Earth. To calculate how far into space the Quiet Cone extends, consider Figure 29.3. The similarity between the rectangular triangles EGA and MLA (rectangular at G and L, respectively) immediately shows that $(EM + MA) : EG = MA : ML$; that is, $(D_{Earth-Moon} + H_{Apex})/R = H_{Apex}/R_{Moon}$. Solving for H_{Apex} yields the height H_{Apex} of the apex of the Quiet Cone overhead the Moon farside with respect to the center of the Moon:

$$H_{Apex}(R) = \frac{D_{Earth-Moon} \cdot R_{Moon}}{R - R_{Moon}}. \tag{29.2}$$

A much more elaborate paper about the Quiet Cone was published by this author in 2003 [2]. Two numerical cases are important:

(1) For geostationary orbits of 42,165 km, the corresponding apex of the Quiet Cone is located at about 16,000 km from the center of the Moon's farside. In other words, this is the distance the Quiet Cone extends into space nowadays, but the actual value may already be lower (i.e., as of 2012) for we do not know about military satellites.

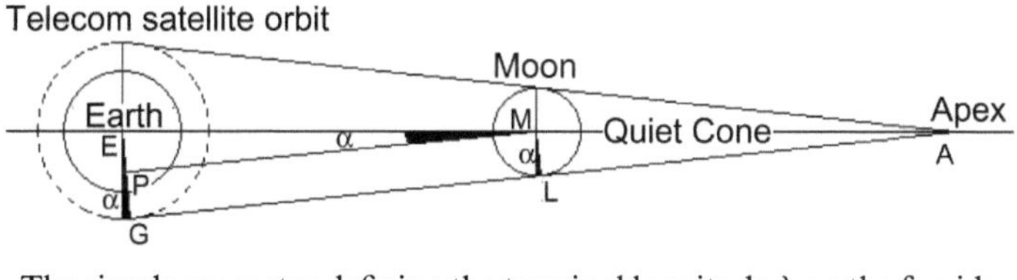

Figure 29.3. The simple geometry defining the terminal longitude λ on the farside of the Moon: radiowaves emitted by telecom satellites circling the Earth at a radius R graze the Moon surface.

(2) The lowest possible value of the apex height corresponds to a satellite circling the Earth at the maximum possible distance (i.e., at the distance of Lagrangian point L1). Thus, the minimal value of the apex is just 2,079 km.

29.5 SELECTING THE DAEDALUS CRATER NEAR THE FARSIDE CENTER

This author claims that the time will come when commercial wars among the large industrial trusts running the telecommunications business will lead them to grab increasingly more space around the Earth, pushing their satellites into orbits with apogees much higher than geostationary ones. A "safe" crater must thus be selected east along the Moon equator. How much farther east? The answer is given by the diagram in Figure 29.4, based on equation (29.1) for λ. The vertical trait predicted by our equation for λ shows up in Figure 29.4 as the upward-going right branch. This shows that, if we only take the equation for λ into account, the maximum distance from the Earth's center for these telecom satellites is about 8.479 times the geostationary radius, corresponding to a circular orbital radius of 358,148 km. If a telecom satellite were put into such a circular orbit around the Earth, its radiowaves would flood Moon longitudes as high as about $< 175°$ (maybe even more).

However, we have not taken the Lagrangian points into consideration yet!

It will never be possible to put a satellite into a circular orbit around the Earth at a distance of 358,148 km simply because this distance already exceeds the distance from Lagrangian point L1 nearest to the Earth, which is located at 323,050 km (Lagrangian points are, by definition, points of zero orbital velocity in the two-body problem).

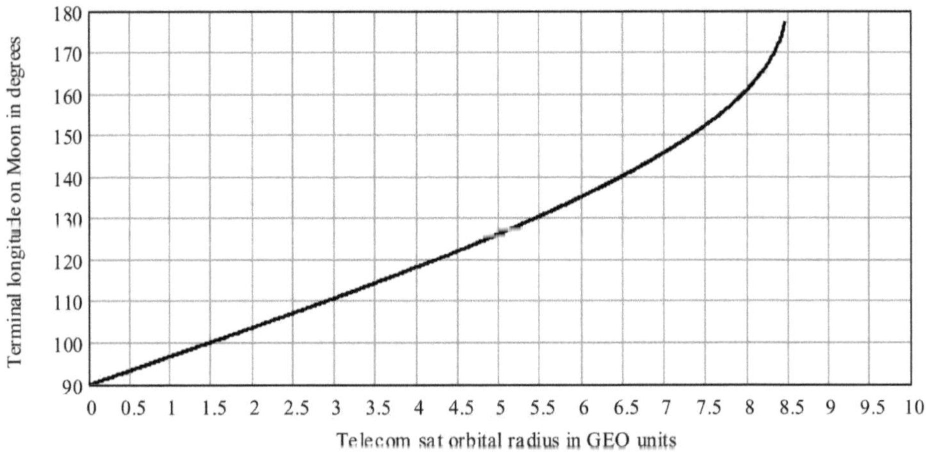

Figure 29.4. Terminal longitude λ (vertical axis) on the Moon farside vs. telecom satellites' orbital radius R around the Earth (horizontal axis) expressed in units of the Earth's geostationary radius (42,241.096 km).

So, this leads us to wonder which is the Moon farside terminal longitude corresponding to the distance of the nearest Lagrangian point, L1? The answer is given by the above equation for λ upon replacing $R = 323{,}050$ km, and the result is $\lambda = 154.359°$. In words, this means the Moon farside sector between 154.359° E and 154.359° W will never be blinded by RFI coming from satellites orbiting just the Earth.

In other words, the limit of the blind longitude as a function of the satellite's orbital radius around the Earth is 180° (E and W longitudes coincide at this meridian, corresponding to the Date Line on Earth). But this is the antipode to Earth on the Moon surface (i.e., the point exactly opposite to the Earth direction on the other side of the Moon). Our theorem simply proves that the antipode is the most shielded point on the Moon surface from radiowaves coming from the Earth—an intuitive and obvious result, really.

So, where are we going to locate our SETI base on the farside of the Moon? Just take a map of the Moon's farside and study it for a moment. One notices that the region of the antipode (where the central meridian crosses the top parallel in Figure 29.5) is too rugged to establish a Moon base. Just about 5° S along the 180° meridian, however, one finds a large crater about 80 km in diameter (just like Saha). This crater is called Daedalus. This author proposes establishing the first RFI-free base on the Moon just inside the Daedalus crater, the most shielded crater of all on the Moon from Earth-made radio pollution!

29.6 OUR VISION OF THE MOON'S FARSIDE FOR RFI-FREE SCIENCE

Let us replace the value of $\lambda = 154.359°$ with the simpler value of $\lambda = 150°$. This fits perfectly with the need for borders of the pristine sector making angles orthogonal to the directions of L4 and L5. The result is this author's vision of the farside of the Moon (shown in Figure 29.6). Figure 29.6 shows a diagram of the Moon as seen from above its north pole with the different colonization regimes proposed by this author. One sees that:

(1) The nearside of the Moon is left totally free for activities of all kinds: scientific, commercial, and industrial.
(2) The farside of the Moon is divided into three sectors (covering 60° in longitude each):
   (a) The Eastern Sector, between 90° E and 150° E, can be used to install radio devices, but only under the aegis of the International Telecommunications Union (i.e., the ITU regime).
   (b) The Central Sector, between 150° E and 150° W, must be kept totally free from human exploitation (i.e., in its pristine radio environment totally free from man-made RFI). This sector is home to the Daedalus crater, a <100 km crater located between 177° E and 179° W and around 5° S. At the moment,

Figure 29.5. An oblique view of the Daedalus crater on the lunar farside as seen from the Apollo 11 spacecraft in lunar orbit. The view looks southwest. Daedalus (formerly referred to as IAU Crater No. 308) is located at 179° E and 5.5° S. Daedalus has a diameter of about 50 statute miles (<80 km). This is a typical scene showing the rugged terrain on the farside of the Moon (from *http://spaceflight.nasa.gov/gallery/images/apollo/apollo11/html/as11_44_6609. html*).

this author is not aware of the height of the circular rim surrounding Daedalus.

(c) The Western Sector, between 90° W and 150° W, can be used to install radio devices, but only as part of the ITU regime.

(d) The Eastern Sector is exactly opposite to the direction of Lagrangian point L4, and so the body of the Moon completely shields the Eastern Sector from RFI produced at L4. Thus, L4 is fully colonizable.

(e) The Western Sector is exactly opposite to the direction of Lagrangian point

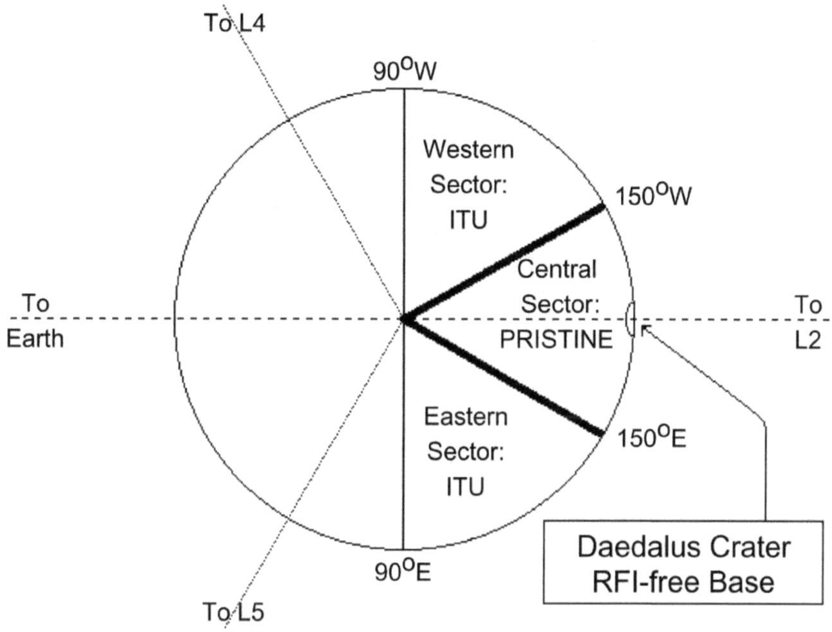

Figure 29.6. This author's vision for the Moon's farside and its protection from RFI. The Central Sector of the farside extends from the north to the south pole between the two meridians at 150° E and 150° W. The Protected Antipode Circle (PAC) is the circular piece of land tangent to these two meridians at the Moon's equator. It is also tangent to the two parallels at +30° and −30°. We claim that the PAC should be kept free from man-made electromagnetic pollution at all times in the future (i.e., it should become an *internationally protected land*, just as Antarctica is on Earth). Also shown is the location for the "Daedalus Crater Base for RFI-free Radio Astronomy, Astrobiology and SETI Science". Future international space stations (ISSs) might be located at both the L4 and L5 Earth–Moon points in the decades to come, but radiation emitted from them will never reach the Central Sector of the farside, nor the PAC. Only point L2, directly overhead the farside, would be able to flood the Central Sector and the PAC with future radio emissions.

L5, and so the body of the Moon completely shields the Western Sector from RFI produced at L5. Thus, L5 is fully colonizable in this author's vision. In other words, this author's vision entails full bilateral symmetry around the plane passing through the Earth–Moon axis and orthogonal to the Moon's orbital plane.

(f) Of course, L2 may not be utilized at all, since it faces the Daedalus crater just at the latter's zenith. Any RFI-producing device located at L2 would flood the whole of the farside and must be ruled out. However, L2 is the only Lagrangian point of the five located in the Earth–Moon system to be kept free. Finally, as L2 is not directly visible from the Earth, it should be left alone!

29.7 TWO FURTHER LAGRANGIAN POINTS (L1 AND L2) OF THE SUN–EARTH SYSTEM AND HOW THEY "POLLUTE" THE FARSIDE OF THE MOON

There remains an unavoidable drawback. This comes from two of the Lagrangian points (L1 and L2) of the Sun–Earth system, located along the Sun–Earth axis and outside the sphere of influence of the Earth, which has a radius of about 924,646 km around the Earth. Precisely, the Sun–Earth L1 point is located at a distance of 1,496,557.035 km from the Earth toward the Sun, and the L2 point at the (virtually identical) distance of 1,496,557.034 km from the Earth in the direction away from the Sun, which is toward the outer solar system. These two points have the convenient property of moving around the Sun at the same angular velocity as the Earth, while keeping at the same distance from the Earth at all times. Thus, they are ideal places for scientific satellites. Actually, the Sun–Earth L1 point has already been in use for scientific satellite location since the NASA ISEE III spacecraft was launched on August 12, 1978, reaching the Sun–Earth L1 region in about a month. On December 2, 1995, the ESA–NASA Soho spacecraft was launched with the aim of investigating the Solar Corona. On February 14, 1996, Soho was inserted into a halo orbit around the Sun–Earth L1 point, where it is still librating in 2012 (Figure 29.7).

As for the Sun–Earth L2 point, there are plans to place NASA's SIM (Space Interferometry Mission) satellite and ESA's GAIA astrometric satellite there. So, all these satellites really will *pollute* the otherwise RFI-free farside of the Moon when the farside is facing them. Unfortunately, the Moon farside faces the Sun–Earth L1

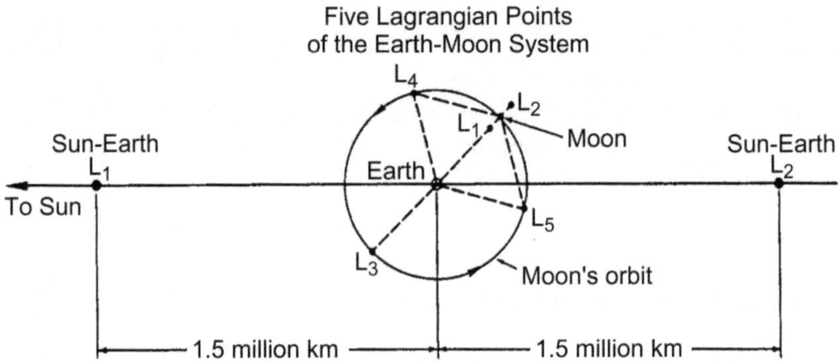

Figure 29.7. In addition to the five Lagrangian points of the Earth–Moon system (already described in Figure 29.1) the next two closest Lagrangian points to the Earth are L1 and L2 of the Sun–Earth system. These are located along the Sun–Earth axis at distances of about 1.5 million kilometers from the Earth toward the Sun (L1) and outward to the outer solar system (L2). Unfortunately, spacecraft located in the neighborhood of these send electromagnetic waves to the farside of the Moon. Examples are the ISEE-III and Soho spacecraft, already orbiting around L1. More spacecraft will do so in the future around both L1 and L2 (courtesy of Dr. Robert "Bob" Farquhar, Johns Hopkins University Applied Physics Laboratory, Laurel, MD).

point for half of the Moon's synodic period, about 14.75 days, and faces the Sun–Earth L2 point for the next 14.75 days—in other words, all the time! Such radio pollution of the Moon farside by scientific satellites located at Lagrangian points L1 and L2 of the Sun–Earth system is unfortunately *unavoidable*. We can only hope that telecom satellites will never be put there. As for the scientific satellites already there or on the way, the radio frequencies they use are well known and usually narrow band. This should help the Fourier transform of future spectrum analyzers to be located on the Moon farside to get rid of these transmissions completely.

29.8 ATTENUATION OF MAN-MADE RFI ON THE MOON'S FARSIDE

In a paper presented by this author at the International Astronautical Congress held in Valencia, Spain in October 2006, his co-worker Salvo Pluchino succeeded in computing RFI attenuation on the farside of the Moon [3]. The basic results proven are shown in Table 29.1.

The precise line frequencies of high scientific importance in Table 29.2 (again taken from [3]) are perhaps even more important than the "generic" frequency values listed in Table 29.1.

Table 29.1. Radiowave attenuation in the lunar equatorial plane and at lunar longitude $\lambda = 180°$ (i.e., near the Daedalus crater) for radio sources emitting at 100 kHz, 100 MHz, and 100 GHz, respectively. All attenuation values are in decibels.

Frequency of radiowaves	$f = 100$ kHz	$f = 100$ MHz	$f = 100$ GHz
Source in GEO	−42.62 dB	−72.62 dB	−102.62 dB
Source in an orbit passing through the L1 point	−30.32 dB	−60.32 dB	−90.32 dB
Source still at the L4 or L5 Lagrangian points	−29.15 dB	−59.15 dB	−89.15 dB

29.9 LEGAL ISSUES: A POSSIBLE STRATEGY TO HAVE THE PAC APPROVED BY THE UNITED NATIONS (COPUOS)

Having thus made clear what the PAC is and why its creation is urgently needed for the benefit of all humankind, our next issue is legal. How can we get the PAC legally approved by the United Nations?

To answer this question, let us first review the function of the United Nations Committee on the Peaceful Uses of Outer Space (COPUOS).

As described at the official website (*http://www.unoosa.org/oosa/COPUOS/copuos.html*) COPUOS was set up by the United Nations General Assembly in

Table 29.2. Attenuation in the lunar equatorial plane and at lunar longitude at $\lambda = 180°$ (near the Daedalus crater) for radiowaves having some of the most important frequencies used by radioastronomers to explore the Universe. It should also be stated that these are the attenuation values on the assumption the Moon is not surrounded by a very thin ionosphere. However, since a very tiny lunar ionosphere might exist, the values below might be slightly out.

Origin of radiowaves	*Radio frequency* f	*Source in GEO* (dB)	*Source in orbit at L1 distance* (dB)	*Source still at L4 or L5* (dB)
ELF	0.003 MHz	−27.39	−15.57	−14.61
VLF	0.030 MHz	−37.39	−25.10	−23.94
Jupiter's storm	20 MHz	−65.63	−53.33	−52.16
Deuterium	327.384 MHz	−77.77	−65.48	−64.30
Hydrogen	1420.406 MHz	−84.14	−71.85	−70.68
Hydroxyl radical	1612.231 MHz	−84.69	−72.40	−71.23
Formaldehyde	4829.660 MHz	−89.46	−77.17	−75.99
Methanol	6668.518 MHz	−90.86	−78.56	−77.39
Water vapor	22.235 GHz	−96.09	−83.79	−82.62
Silicon monoxide	42.519 GHz	−98.90	−86.61	−85.44
Carbon monoxide	109.782 GHz	−103.02	−90.73	−89.56
Water vapor	183.310 GHz	−105.25	−92.95	−91.78

1959 (resolution 1472 (XIV)) to review the scope of COPUOS: to devise programs for the peaceful use of outer space to be undertaken under United Nations auspices, to encourage continued research and the dissemination of information on outer space matters, and to study legal problems arising from the exploration of outer space. As of 2012, the number of member states of COPUOS is 70 (the list can be found at the above website).

The Committee has two standing Subcommittees:

- the Scientific and Technical Subcommittee; and
- the Legal Subcommittee.

There are two important points that may lead to the approval of the PAC by COPUOS:

(1) The Committee and its two Subcommittees meet annually to consider questions put before them by the General Assembly, reports submitted to them, and issues raised by member states.

(2) The Committee and the Subcommittees, working on the basis of consensus, make recommendations to the General Assembly. Detailed information on the work of the Committee and the Subcommittees is contained in their annual reports.

Point (1) makes it clear that there must be at least one member state of the United Nations raising the issue of PAC creation so that COPUOS can later make a recommendation to the General Assembly. Now, as the IAA is only an observer at the United Nations level, it cannot raise issues. Issues can only be raised at the United Nations by member states, and so the steps to follow seem to be:

Step 1. Let at least one member state of the United Nations raise the issue of PAC creation at the United Nations. After Step 1 has been accomplished, COPUOS may well ask the IAA and the IISL (the International Institute of Space Law) about technical clarifications regarding the PAC. Thus, to prepare this request for information by COPUOS from the IAA and the IISL, Step 2 may be started now.

Step 2. The IAA and the IISL will provide COPUOS with all technical information and clarifications about the PAC. In view of this, the author, as an IAA member, has sought permission to lead an IAA study group to study Moon farside protection and the PAC. When this phase has also been accomplished, COPUOS may then take Step 3.

Step 3. COPUOS may finally either (a) submit the PAC issue to the General Assembly of the United Nations for a vote or (b) include the PAC issue in a new Moon Treaty, something long envisaged by some COPUOS members as a good replacement of the "bad Moon Treaty of 1979".

In the following section we look at the already existing Outer Space Treaty of 1967 (sometimes called "the good Moon Treaty") and the (failed, in practice) Moon Treaty of 1979 (sometimes called "the bad Moon Treaty").

29.10 THE EXISTING OUTER SPACE TREATY OF 1967 AND THE (FAILED) MOON TREATY OF 1979

To get a feel for the background to and the substance of the two key space treaties of 1967 and 1979, the reader may wish to consult two Wikipedia sites:

(1) For the Outer Space Treaty of 1967 see *http://en.wikipedia.org/wiki/Outer_Space_Treaty*
(2) For the 1979 Moon Treaty (proposed, but only ratified by some of the major spacefaring nations) see *http://en.wikipedia.org/wiki/Moon_Treaty*

Figures 29.8 and 29.9 are reproduced from the above two websites, respectively.

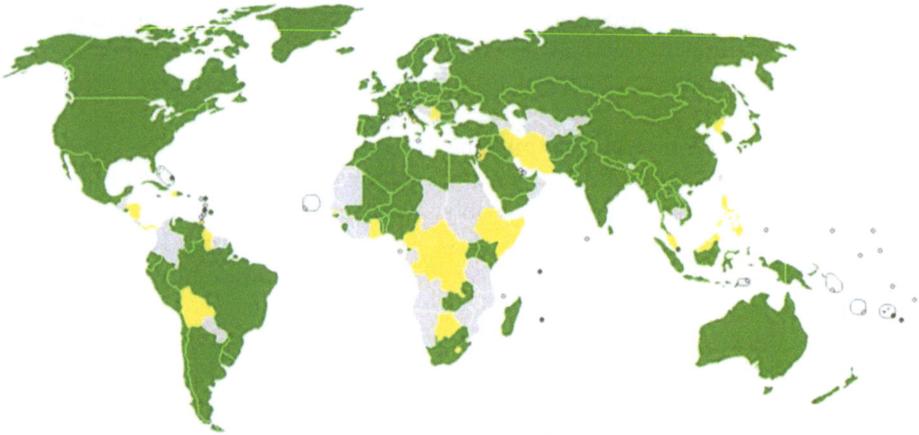

Figure 29.8. Countries that signed and ratified the Outer Space Treaty of 1967 (green) and countries that signed but did not ratify it (yellow) (from *http://en.wikipedia.org/wiki/File: Outer_Space_Treaty.png*).

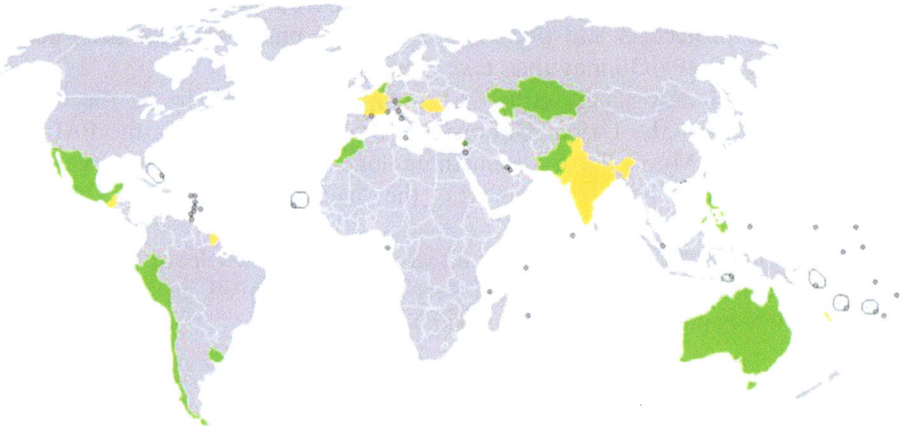

Figure 29.9. Countries that signed and ratified the Moon Treaty of 1979 (green) and countries that signed but did not ratify it (yellow) (from *http://en.wikipedia.org/wiki/Moon_Treaty*).

This author's opinion regarding these two treaties may of course be debatable but points out the great state of uncertainty about "who owns what" on the Moon surface today and tomorrow:

(1) The Outer Space Treaty signed in 1967 by the U.S.A., the U.K., and the Soviet Union was "wise" in that it forbade any country from colonizing the Moon, planets, and asteroids in the name of one country only. As such, there was little surprise that it was later signed and ratified by virtually all the countries in the world. Back in 1967, however, it was hard to envisage a time when private

investors could reach the Moon and other bodies by virtue of their own private means only. By 1980 the situation had already evolved so much in this regard.

(2) When the Moon Treaty of 1979 was finalized and proposed, the United States Senate finally failed to approve it because it forbade both private property and the capability of terraforming (i.e., altering the atmosphere so as to "make it breathable"). In this author's view, the terraforming of the Moon (if at all feasible!) and Mars will be necessary in the long run. So, it was good that the U.S.A. and the rest of the world did not sign the 1979 Treaty. By doing so, however, they left the door open to those who claim that private property on the Moon would be a good thing.

(3) Herein lies the rub: Can pieces of land on the Moon be owned by individuals (whether they be companies or private investors)? The vast majority of legal experts would probably answer "no" to this question, but the matter is not resolved, as is evident from a website on "extraterrestrial real estate" (*http://en.wikipedia.org/wiki/Extraterrestrial_real_estate*).

In the U.S.A. there are, unfortunately, private companies like The Lunar Embassy selling real estate on the Moon who have made a lot of money from this (see *http://www.lunarlandowner.com/*).

Yet, nobody has done anything in the U.S.A. to prevent this illegal sale. Only in China has The Lunar Embassy been recently prevented from doing so by the government (see *http://www.moondaily.com/reports/China_Bans_Firm_From_Selling_Land_On_The_Moon_999.html*).

This author does not know whether The Lunar Embassy has already sold any piece of land within the borders of the PAC, but the issue of having the PAC recognized as a sort of "natural preserve" on the farside of the Moon is urgent indeed.

29.11 THE COMING "NEW MOON RUSH": TRYING TO PROTECT THE FARSIDE FROM COMMERCIAL, ASTRONAUTICAL, AND MILITARY EXPLOITATION

The farside of the Moon should be a unique natural preserve of humankind. It still is uncorrupted by man-made RFI. We have a moral duty to keep it so for the centuries to come by preventing any uncontrolled commercial, astronautical, and military exploitation of the Moon's farside taking place.

How can we bring about legal permanent protection of the farside? The United Nations appears to be the natural international forum for this matter to be discussed. The International Institute of Space Law (IISL), as the leading international mentor institution, stands the best chance of approaching the United Nations with the highest hope of success.

Legal protection of the farside, however, might well be just a dream! Reality could be much tougher, as neatly described, for instance, by the Los Angeles reporter Bill Carswell in his April 18, 2002, *SpaceDaily* article entitled "The Outer Space and

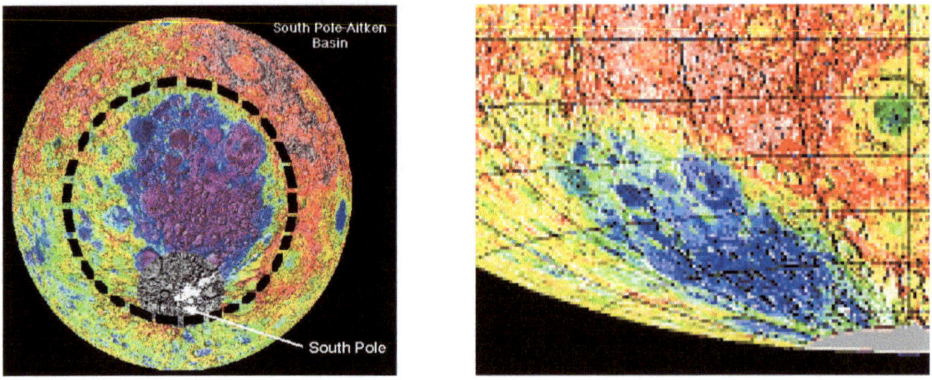

Figure 29.10. Much of the area around the south pole is within the South Pole–Aitken Basin (shown at right in blue on a lunar topography image), a giant impact crater 2,500 km (1,550 miles) in diameter and 12 km deep at its lowest point. Many smaller craters exist on the floor of this basin. Any water ice at the bottom of the crater could probably exist for billions of years at very low temperatures (courtesy NASA).

Moon Treaties and the Coming Moon Rush" (see *http://www.spacedaily.com/news/oped-02c.html*). In essence, any powerful enough spacefaring country could start the Moon Rush right now, paying little attention to international laws.

Even if discussions are confined to purely scientific matters, like geology, astrophysics and astrodynamics, there could still be conflicting exploration issues, such as exploration of the so-called Aitken Basin. On March 5, 1998 it was announced that data returned by Lunar Prospector indicated that water ice is present at both the north and south lunar poles, in agreement with Clementine results for the south pole reported in November 1996. Where is this water ice located? Yes, you've guessed it, within the largest known impact crater in the solar system: the South Pole–Aitken Basin on the farside of Moon, over 1,300 miles across (as shown in Figure 29.7).

Figure 29.10 clearly shows that the southern part of the Central Sector of the farside is of exceptional interest. Hence, there is little chance of its protection by the United Nations or by an international treaty.

Just when it seemed that at least part of the farside (outside the Aitken Basin) could well be protected legally, came the news that the Earth–Moon Lagrangian point L2 would be a wonderful starting point for spaceships to explore the asteroids and outer planets. This remark is no less than the subject matter of the IAA cosmic study S1.1 which has as a "twin" another IAA cosmic study called "The Next Steps in Exploring Deep Space", led by Drs. Wesley T. Huntress Jr. and Robert W. Farquhar. IAA academicians alone may view their PowerPoint pages as a result of restricted IAA access (*http://www.iaanet.org/restrict/presentation.pdf*). The (hopefully apparent) conflict between the two IAA cosmic studies S1.1 and S1.2 lies in the question: "What shall we put in the far future at the Earth–Moon Lagrangian point L2?" Jean Heidmann's initial answer to this question was simply "nothing", clearly to prevent RFI made by this L2 space station flooding the whole of the farside, while Huntress and Farquhar pointed out the gravitational advantages (i.e., less fuel needed) to depart from the Earth–Moon L2. Yet, in recent years, Huntress and Farquhar have come to realize that the Earth–Sun point L2 would also offer similar advantages of fuel saving. Thus, the solution to this apparent contradiction between these two cosmic studies could be "leave the Earth–Moon L2 alone and let the Earth–Sun L2 become the space base for exploration of the outer solar system." The future will tell whether such a "fair solution" will be feasible in the interest of the whole of humankind.

29.12 THIS AUTHOR'S SPEECH AT THE UNITED NATIONS COPUOS (JUNE 10, 2010)

On June 10, 2010 this author was finally allowed (by the IAA and by the United Nations COPUOS) to raise the issue of Moon farside protection in front of the United Nations COPUOS in Vienna. The presentation he made on that occasion can be downloaded free from the U.N. archives (*http://www.unoosa.org/pdf/pres/copuos2010/tech-06E.pdf*) and is reprinted at the end of this chapter as Appendix 29.A. Figure 29.11 was taken while this author was making that speech. He is sitting in the seat granted to IAA at the COPUOS (the IAA is an NGO recognized by the United Nations). This 2010 speech at the United Nations will hopefully start a new era of international collaboration to resolve the issue of protecting the Moon's farside for the benefit of humankind. Developments are foreseen in the years to come.

Figure 29.11. This author's speech at the United Nations COPUOS in Vienna, Austria, on June 10, 2010. This was the first presentation ever made at the United Nations about the need to keep the farside of the Moon free from man-made RFI for all generations to come (for more details see *http://iaaweb.org/iaa/Scientific%20Activity/lunarcopuos.pdf*).

29.13 CONCLUSIONS

The goal of this chapter was to make the reader sensitive to the importance of protecting the central part of the farside of the Moon against any future uncontrolled and anti-scientific exploitation.

In particular, we gave sound scientific reasons for the PAC (Protected Antipode Circle) being declared international land under the protection of the United Nations, or, in absence of that institution, by direct agreement among the governments of the spacefaring nations.

The farside of the Moon is the most unique place for us in the whole universe: it is close to Earth, but well protected from the radio pollution that we ourselves are creating in ever-increasing amounts, making our radio telescopes increasingly blind.

The farside cannot be left in the hands of real estate agents! The urgency with which the issue needs to be addressed cannot be overstressed

29.14 ACKNOWLEDGMENT

The author would like to thank the International Academy of Astronautics (IAA) for allowing him to serve as Coordinator of the "Lunar Farside Radio Lab" IAA cosmic study in the years 2000–2005, and after 2005 for keeping the question alive. Interest in the international role that the United Nations might play in the creation of the PAC also came from several members of the International Institute of Space Law (IISL), in particular Prof. Vladimir Kopal, Chair of the IAA Scientific–Legal Liaison Committee. Hopefully, these and other far-sighted minds will unite their efforts to save the farside of the Moon from newly man-made RFI pollution in the years to come.

29.15 REFERENCES

[1] C. Maccone, "Planetary defense and RFI-free radioastronomy from the farside of the Moon: A unified vision," *Acta Astronautica*, **50** (2002), 185–199.

[2] C. Maccone, "The Quiet Cone above the farside of the Moon," *Acta Astronautica*, **53** (2003), 65–70.

[3] S. Pluchino, N. Antonietti, and C. Maccone, "Protecting the moon farside radio-telescopes from RFI produced at future Lagrangian points space stations," paper IAC-06-D4.1.01 presented at the *International Astronautical Congress held in Valencia, Spain, October 26, 2006*. Later published in the *Journal of the British Interplanetary Society*, **60** (2007), 162–168.

29.16 BIBLIOGRAPHY

J. Heidmann, "Saha Crater: A candidate for a SETI lunar base," *Acta Astronautica*, **32** (1994), 471–472 (Academy Transactions Note).

J. Heidmann, "Sharing the Moon by thirds: An extended Saha crater proposal," *Advances in Space Research*, **26** (2000), 371–375.

J. Heidmann, "Recent progress on the lunar farside Crater Saha proposal," *Acta Astronautica*, **46** (2000), 661–665.

C. Maccone, "Laydown of a tether from Earth visible location to far side for lunar SETI," *Advances in Space Research*, **26** (2000), 359–370.

C. Maccone, "Searching for bioastronomical signals from the farside of the Moon," a paper presented at the *First European Workshop on Exo/Astrobiology held at ESRIN, Frascati, Italy, May 21–23, 2001*. Later this paper was published by ESA in the relevant Proceedings called ESA SP-496 (published in August 2001), pp. 277–280.

C. Maccone, "The lunar SETI cosmic study of IAA: Current status and perspectives," paper IAA-01-IAA.9.1.05 presented at the *52nd International Astronautical Congress held in Toulouse, France, October 1–5, 2001.*

C. Maccone, "The lunar farside radio lab study of IAA," paper IAA-02-IAA.9.1.4 presented at the *53rd International Astronautical Congress (World Space Congress—2002) held in Houston, Texas, October 10–19, 2002.*

C. Maccone, "Lunar farside radio lab: A "cosmic study" by the International Academy of Astronautics," paper IAA.8.3.03 presented at the *14th Space Activity and Society Symposium during the 2003 International Astronautical Congress held in Bremen, Germany, September 29–October 3, 2003.*

C. Maccone, "Moon farside radio lab," *Acta Astronautica*, **56** (2005), 629–639.

C. Maccone, "Protected antipode circle on the farside of the Moon," *Acta Astronautica*, **63** (2008), 110–118.

APPENDIX 29.A

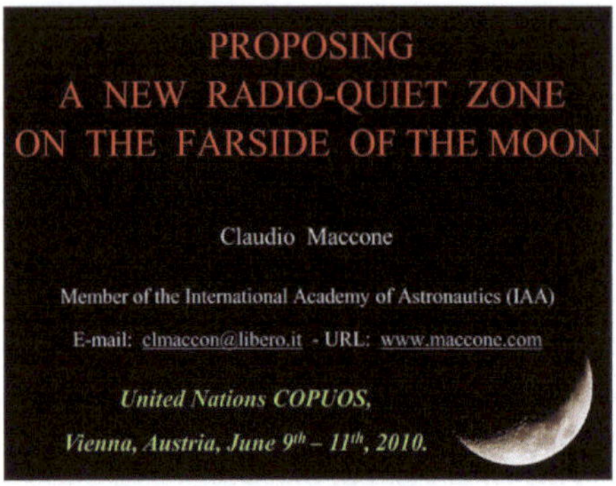

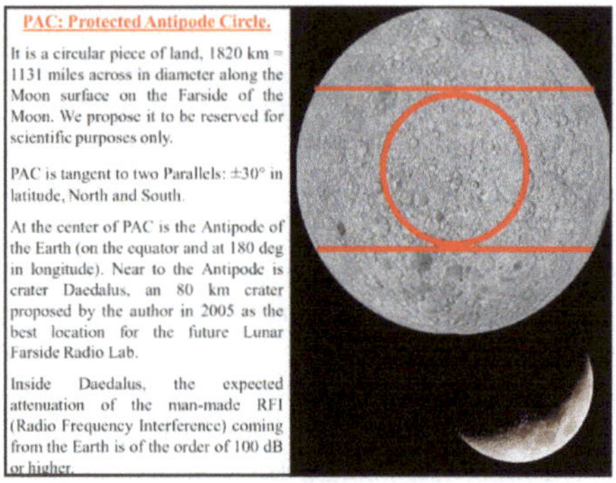

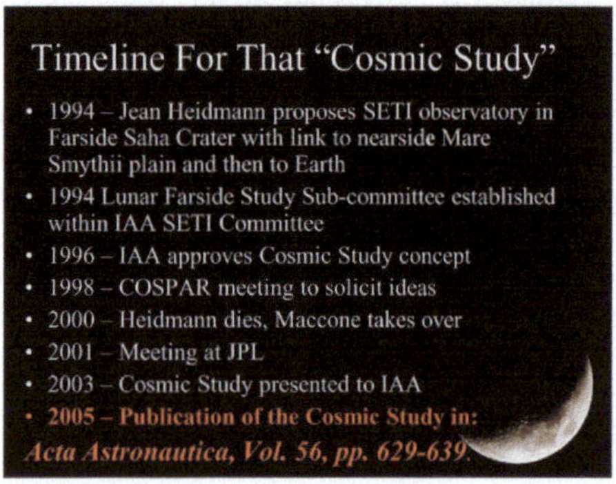

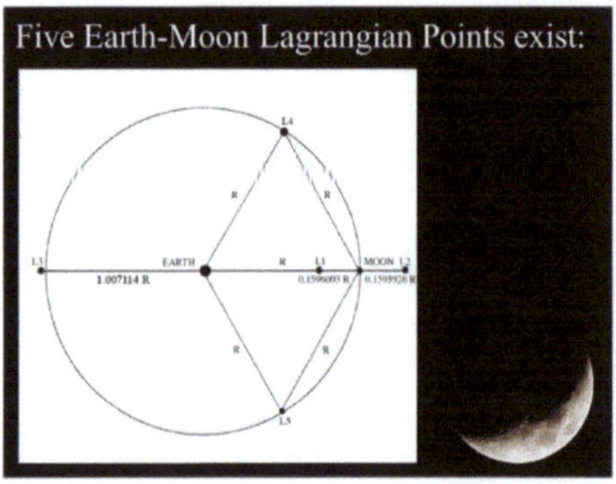

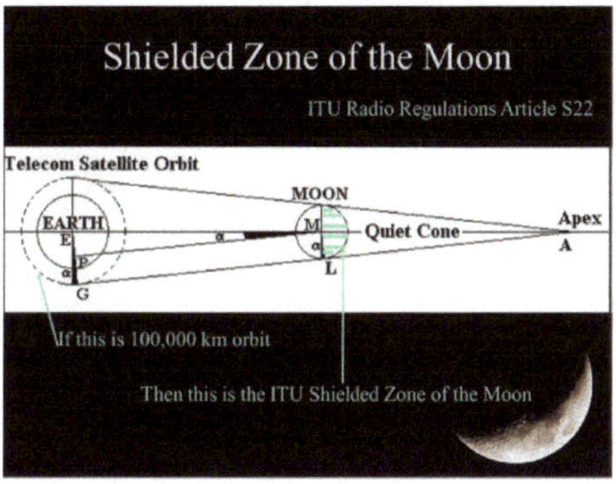

ARTICLE S22

Space Services [1]

Section I. Cessation of Emissions
Section II. Control of Interference to Geostationary-Satellite Systems
Section III. Station Keeping of Space Stations
Section IV. Pointing Accuracy of Antennae on Geostationary Satellites
Section V. Radio Astronomy in the Shielded Zone of the Moon
Section VI. Earth Station Off-Axis Power Limitations in the Fixed-Satellite Service

Section V. Radio Astronomy in the Shielded Zone of the Moon

S22.22
§ 8 (1) In the shielded zone of the Moon [2] emissions causing harmful interference to radio astronomy observations [3] and to other users of passive services shall be prohibited in the entire frequency spectrum except in the following bands:

S22.23
a) the frequency bands allocated to the space research service using active sensors;

S22.24
b) the frequency bands allocated to the space operation service, the earth exploration-satellite service using active sensors, and the radiolocation service using stations on spaceborne platforms, which are required for the support of space research, as well as for radiocommunications and space research transmissions within the lunar shielded zone.

S22.25
(2) In frequency bands in which emissions are not prohibited by Nos. S22.22 to S22.24, radio astronomy observations and passive space research in the shielded zone of the Moon may be protected from harmful interference by agreement between administrations concerned.

S22.22.1
9 - The shielded zone of the Moon comprises the area of the Moon's surface and an adjacent volume of space which are shielded from emissions originating within a distance of 100 000 km from the centre of the Earth.

S22.22.2
10 - The level of harmful interference is determined by agreement between the administrations concerned, with the guidance of the relevant ITU-R Recommendations.

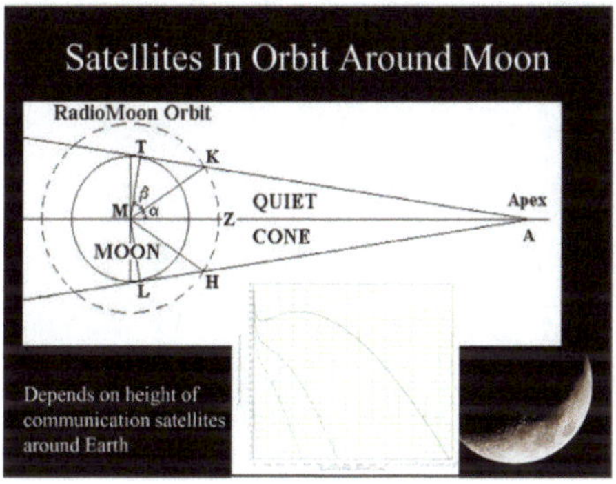

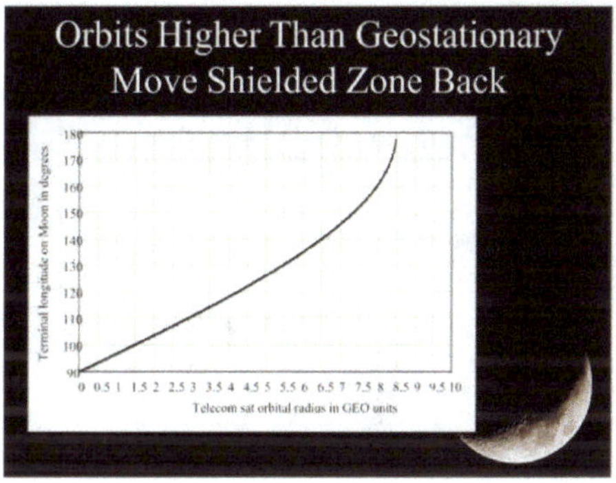

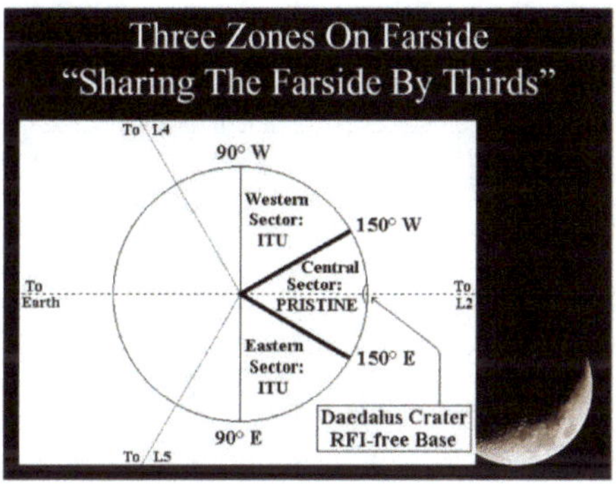

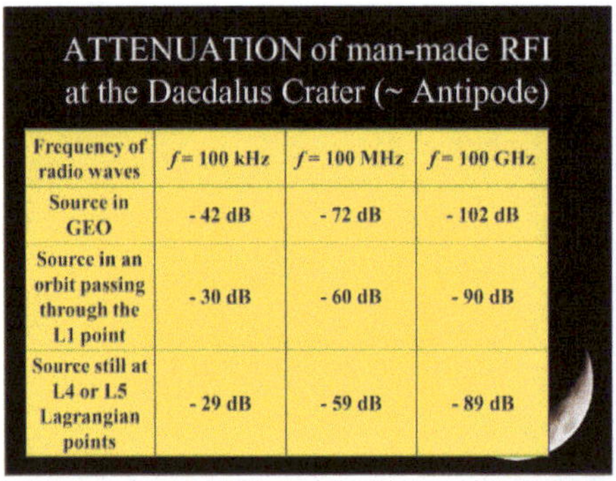

ATTENUATION of man-made RFI at the Daedalus Crater (~ Antipode)

Frequency of radio waves	f = 100 kHz	f = 100 MHz	f = 100 GHz
Source in GEO	- 42 dB	- 72 dB	- 102 dB
Source in an orbit passing through the L1 point	- 30 dB	- 60 dB	- 90 dB
Source still at L4 or L5 Lagrangian points	- 29 dB	- 59 dB	- 89 dB

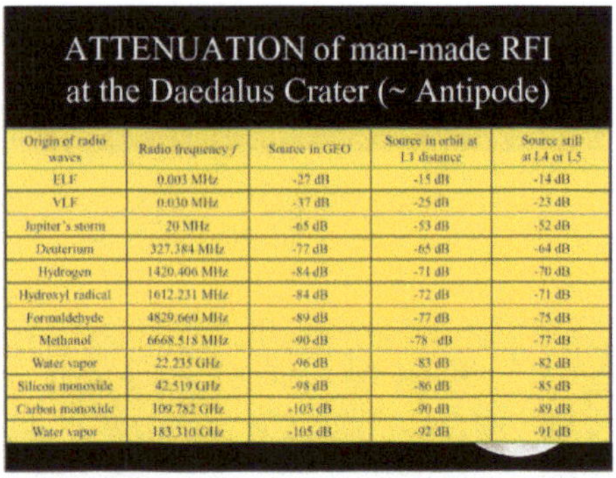

ATTENUATION of man-made RFI at the Daedalus Crater (~ Antipode)

Origin of radio waves	Radio frequency f	Source in GEO	Source in orbit at L1 distance	Source still at L4 or L5
ELF	0.003 MHz	-27 dB	-15 dB	-14 dB
VLF	0.030 MHz	-37 dB	-25 dB	-23 dB
Jupiter's storm	20 MHz	-65 dB	-53 dB	-52 dB
Deuterium	327.384 MHz	-77 dB	-65 dB	-64 dB
Hydrogen	1420.406 MHz	-84 dB	-71 dB	-70 dB
Hydroxyl radical	1612.231 MHz	-84 dB	-72 dB	-71 dB
Formaldehyde	4829.660 MHz	-89 dB	-77 dB	-75 dB
Methanol	6668.518 MHz	-90 dB	-78 dB	-77 dB
Water vapor	22.235 GHz	-96 dB	-83 dB	-82 dB
Silicon monoxide	42.519 GHz	-98 dB	-86 dB	-85 dB
Carbon monoxide	109.782 GHz	-103 dB	-90 dB	-89 dB
Water vapor	183.310 GHz	-105 dB	-92 dB	-91 dB

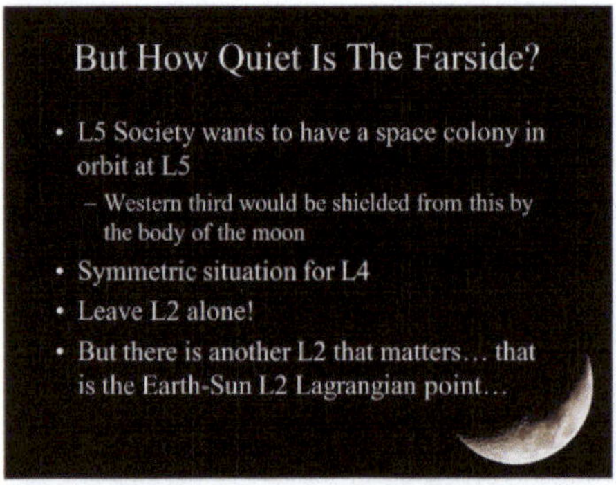

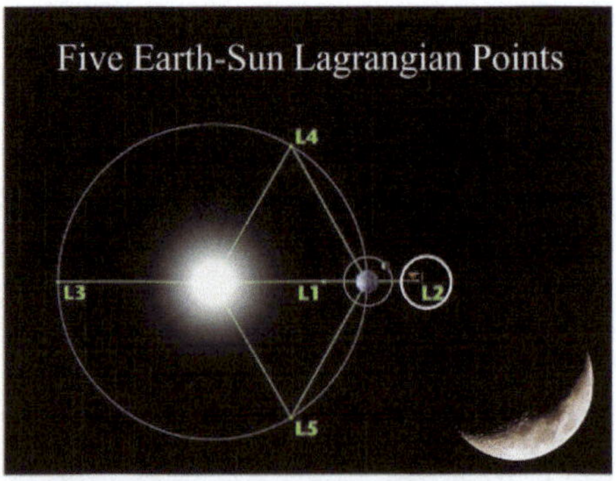

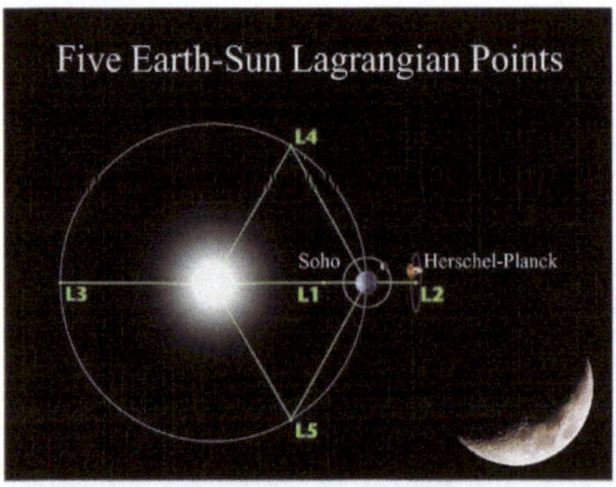

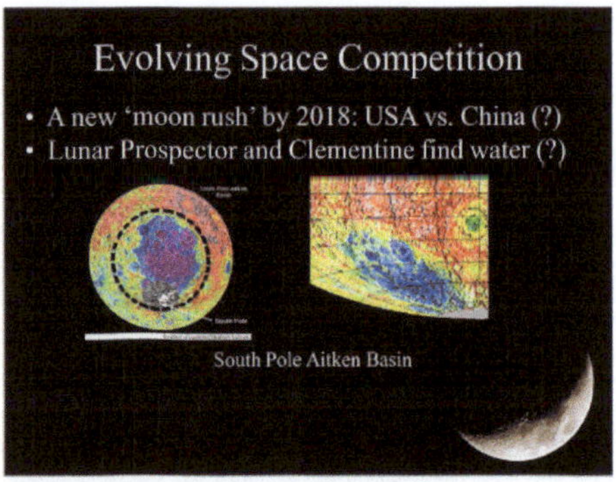

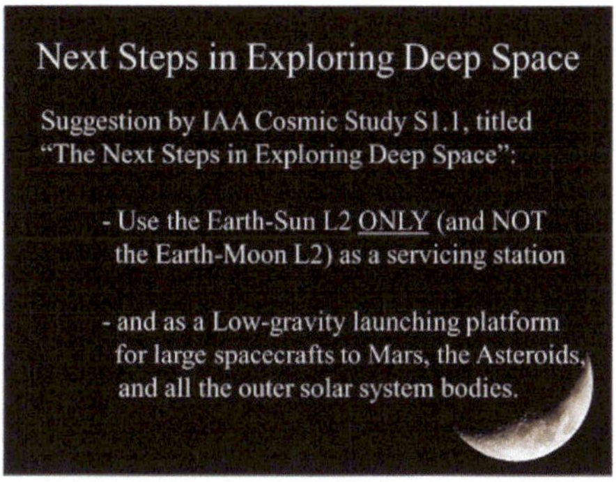
Next Steps in Exploring Deep Space
Suggestion by IAA Cosmic Study S1.1, titled "The Next Steps in Exploring Deep Space":
- Use the Earth-Sun L2 ONLY (and NOT the Earth-Moon L2) as a servicing station
- and as a Low-gravity launching platform for large spacecrafts to Mars, the Asteroids, and all the outer solar system bodies.

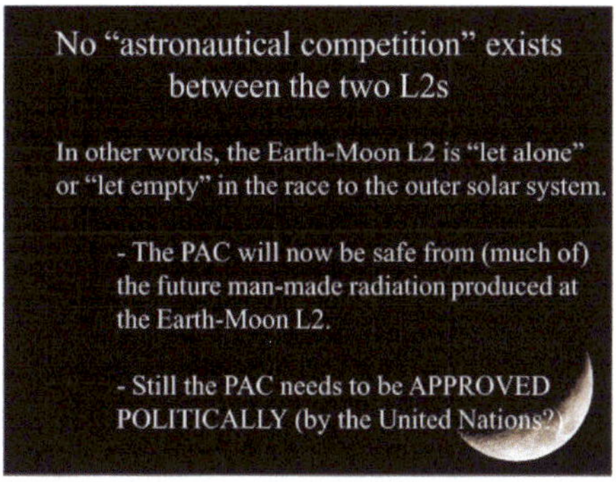
No "astronautical competition" exists between the two L2s
In other words, the Earth-Moon L2 is "let alone" or "let empty" in the race to the outer solar system.
- The PAC will now be safe from (much of) the future man-made radiation produced at the Earth-Moon L2.
- Still the PAC needs to be APPROVED POLITICALLY (by the United Nations?)

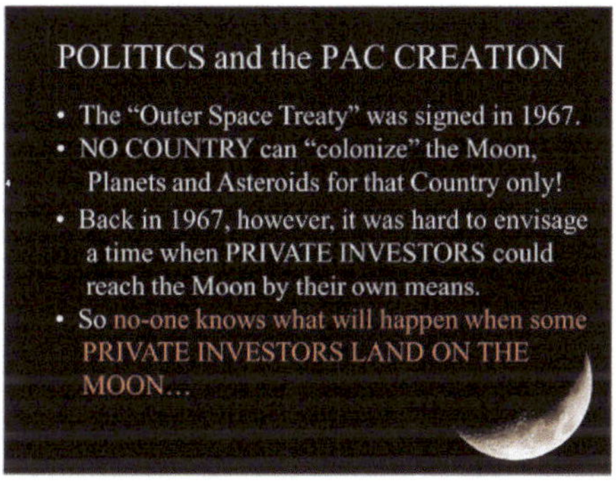
POLITICS and the PAC CREATION
• The "Outer Space Treaty" was signed in 1967.
• NO COUNTRY can "colonize" the Moon, Planets and Asteroids for that Country only!
• Back in 1967, however, it was hard to envisage a time when PRIVATE INVESTORS could reach the Moon by their own means.
• So no-one knows what will happen when some PRIVATE INVESTORS LAND ON THE MOON…

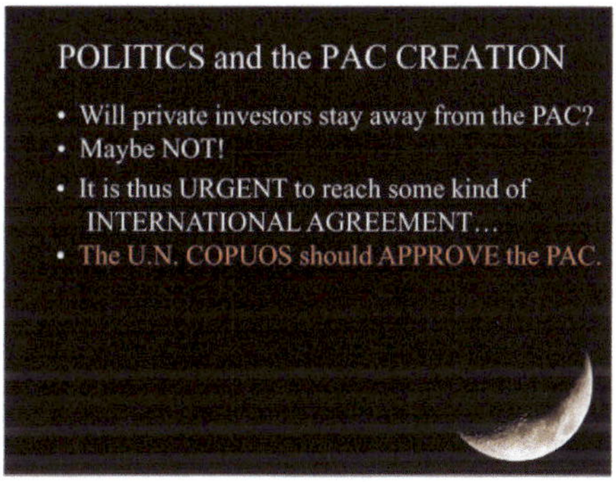
POLITICS and the PAC CREATION
• Will private investors stay away from the PAC?
• Maybe NOT!
• It is thus URGENT to reach some kind of INTERNATIONAL AGREEMENT…
• The U.N. COPUOS should APPROVE the PAC.

POLITICS and the PAC CREATION
• One Country must raise the issue at the U.N.
• Then the U.N. turns to COPUOS…
• Then COPUOS turns to the IAA and IISL…
• Then PAC IS APPROVED BY THE U.N.
• These matters are LEGALLY UNCERTAIN…
• THE POLITICIANS WILL DECIDE…

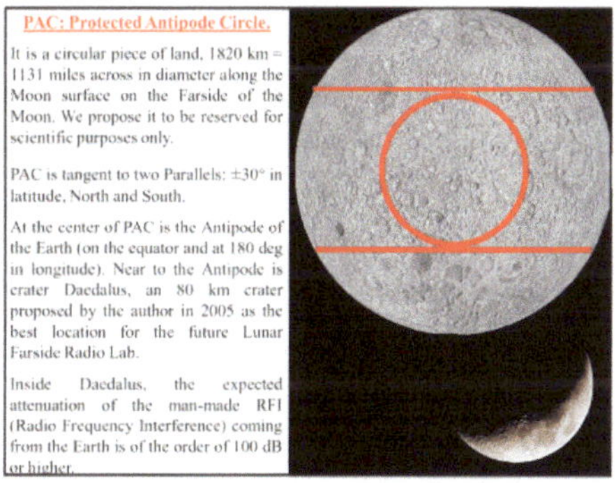
PAC: Protected Antipode Circle.
It is a circular piece of land, 1820 km = 1131 miles across in diameter along the Moon surface on the Farside of the Moon. We propose it to be reserved for scientific purposes only.
PAC is tangent to two Parallels: ±30° in latitude, North and South.
At the center of PAC is the Antipode of the Earth (on the equator and at 180 deg in longitude). Near to the Antipode is crater Daedalus, an 80 km crater proposed by the author in 2005 as the best location for the future Lunar Farside Radio Lab.
Inside Daedalus, the expected attenuation of the man-made RFI (Radio Frequency Interference) coming from the Earth is of the order of 100 dB or higher.

Part V

Epilogue

30

Epilogue: Evolution, progress, and SETI

30.1 INTRODUCTION: JUMPING FROM PHYSICS TO INFORMATION THEORY

Physics, astronomy, and space missions: that is what we have explored so far in this book.

In order to understand evolution, progress, and SETI, however, more profound mathematical tools are required. Information theory (*http://en.wikipedia.org/wiki/Information_theory*), first put forward by Claude Shannon (1916–2001) in 1948, is one such tool.

In particular, we will need the notion of the differential entropy H of an assigned probability density $f_X(x)$, defined by the integral

$$H = -\int_{-\infty}^{\infty} f_X(x) \cdot \ln f_X(x)\, dx. \tag{30.1}$$

Essentially, this is a measure of "how peaked" a pdf is (small entropy) in contrast to a "largely spread" pdf (high entropy), and so a pdf entropy is also called its "uncertainty" (no relationship to Heisenberg's Uncertainty Principle). For more details see *http://en.wikipedia.org/wiki/Entropy_(information_theory)* and, especially, *http://en.wikipedia.org/wiki/Differential_entropy*.

In particular, we need an expression for the differential entropy of a b-lognormal. Let us thus start by finding the differential entropy of the ordinary lognormal (i.e., starting at zero), that is:

$$H_{\text{lognormal}} = -\int_0^\infty \frac{1}{\sqrt{2\pi}\sigma x} e^{-\frac{(\ln x-\mu)^2}{2\sigma^2}} \cdot \ln\left[\frac{1}{\sqrt{2\pi}\sigma x} e^{-\frac{(\ln x-\mu)^2}{2\sigma^2}}\right] dx$$

$$= -\int_0^\infty \frac{1}{\sqrt{2\pi}\sigma x} e^{-\frac{(\ln x-\mu)^2}{2\sigma^2}} \cdot \left[-\ln\left(\sqrt{2\pi}\sigma\right) - \ln x - \frac{(\ln x-\mu)^2}{2\sigma^2}\right] dx$$

$$= \ln\left(\sqrt{2\pi}\sigma\right)\int_0^\infty \frac{1}{\sqrt{2\pi}\sigma x} e^{-\frac{(\ln x-\mu)^2}{2\sigma^2}} dx + \int_0^\infty \ln x \frac{1}{\sqrt{2\pi}\sigma x} e^{-\frac{(\ln x-\mu)^2}{2\sigma^2}} dx$$

$$+ \int_0^\infty \frac{(\ln x-\mu)^2}{2\sigma^2} \cdot \frac{1}{\sqrt{2\pi}\sigma x} e^{-\frac{(\ln x-\mu)^2}{2\sigma^2}} dx =$$

now the subtitution $\ln x = z$ changes all three integrals into well-known integrals of the normal distrubution, equal to 1 (normalization condition), μ (mean value definition), and σ^2 (variance definition), respectively:

$$= \ln\left(\sqrt{2\pi}\sigma\right)\int_{-\infty}^\infty \frac{1}{\sqrt{2\pi}\sigma} e^{-\frac{(z-\mu)^2}{2\sigma^2}} dz + \int_{-\infty}^\infty z \frac{1}{\sqrt{2\pi}\sigma} e^{-\frac{(z-\mu)^2}{2\sigma^2}} dz$$

$$+ \int_{-\infty}^\infty \frac{(z-\mu)^2}{2\sigma^2} \cdot \frac{1}{\sqrt{2\pi}\sigma} e^{-\frac{(z-\mu)^2}{2\sigma^2}} dx$$

$$= \ln\left(\sqrt{2\pi}\sigma\right) + \mu + \frac{1}{2\sigma^2} \cdot \sigma^2 = \ln\left(\sqrt{2\pi}\sigma\right) + \mu + \frac{1}{2}. \tag{30.2}$$

As for the differential entropy of the b-lognormal, it is just the same as (30.2), since the b-lognormal is simply a lognormal shifted to a new origin b along the x-axis, and so all infinite support integrals in (30.2) remain unchanged, and the proof is just the same as (30.2). Thus, in conclusion, the differential entropy of both the ordinary lognormal and the b-lognormal is given by

$$H_{\text{lognormal}} = H_{b\text{-lognormal}} = \ln\left(\sqrt{2\pi}\sigma\right) + \mu + \frac{1}{2}. \tag{30.3}$$

Notice also that (30.3) yields the b-lognormal differential entropy in nats (i.e., in natural logarithms, *http://en.wikipedia.org/wiki/Nat_(information)*). If one wants to express (30.3) in bits (*http://en.wikipedia.org/wiki/Bit*), then one must divide (30.3) by $\ln 2 = 0.693...$. In other words, one has

$$H_{\text{lognormal\_in\_bits}} = H_{b\text{-lognormal\_in\_bits}} = \frac{1}{\ln 2} \cdot \left[\ln\left(\sqrt{2\pi}\sigma\right) + \mu + \frac{1}{2}\right]$$

$$\approx 1.443... \cdot \left[\ln\left(\sqrt{2\pi}\sigma\right) + \mu + \frac{1}{2}\right]. \tag{30.4}$$

The proof of (30.4) from (30.3) simply follows from the well-known change-of-base formula for logarithms

$$\log_2 N = \frac{\ln N}{\ln 2} \approx \frac{\ln N}{0.693} \approx 1.443 \cdot \ln N \tag{30.5}$$

applied to the definition (30.1) of differential entropy with ln(...) replaced by $\log_2(...)$.

30.2 EXPONENTIAL CURVE IN TIME DETERMINED BY TWO POINTS ONLY

Readers will recall from Chapters 7 and 8 that the key point in those chapters is the exponential curve in time

$$E(t) = A\, e^{B\,t}. \tag{30.6}$$

Again in Chapters 7 and 8, the exponential curve (30.6) passes through two (and only two) assigned points of coordinates (p_1, P_1) and (p_2, P_2), respectively. In other words, we assumed that the two simultaneous equations hold

$$\begin{cases} P_1 = A\, e^{B\,p_1} \\ P_2 = A\, e^{B\,p_2}. \end{cases} \tag{30.7}$$

This system (30.7) of two equations may be solved with respect to the two constants A and B in a few simple steps that we omit here but are found in Appendix 30.A (see equations %o2 through %o7). The result is:

$$\begin{cases} A = \dfrac{P_1^{\frac{p_2}{p_2-p_1}}}{P_2^{\frac{p_1}{p_2-p_1}}} \\[2ex] B = \dfrac{\ln\left(\frac{P_2}{P_1}\right)}{p_2 - p_1}. \end{cases} \tag{30.8}$$

Thus, the two equations (30.8) completely solve the problem of finding the exponential curve in time (30.6) passing through the two assigned points of coordinates (p_1, P_1) and (p_2, P_2).

30.3 THE ASSUMPTION THAT THE EXPONENTIAL CURVE IN TIME IS THE GBM MEAN VALUE CURVE

Readers will recall from Chapter 11 that the geometric Brownian motion (GBM) is the natural generalization in time of the statistical Drake equation studied in Chapter 1. The same remark also applies to the statistical Dole equation studied in Chapter 3.

But now, we are going to make a breakthrough in our stochastic representation of Darwinian theory (evolution, as described in Chapter 8) and of historical progress (mathematical history, as described in Chapter 7) if we assume that the exponential curve in time (30.6) is the same thing as the exponential mean value of GBM in time, given by (11.1), that is

$$\langle N(t)\rangle = N_0\, e^{\mu\,t} = N_0\, e^{\mu_{aGBM}\,t}. \tag{30.9}$$

We call this a breakthrough inasmuch as evolution and human progress now cease to be deterministic chains of events, instead they become a stochastic (random) sequence of events, as indeed they are in reality, with all their ups and downs, although constrained to have a deterministic exponential overall average increase. This "mathematical representation of Darwinian evolution and historic human progress as geometric Brownian motion" is thus the #1 message put forward in this book, and we know that they reduce to the statistical Drake and Dole equations (and more similar unnamed equations) in the static case (as described in Chapter 11).

In equation (30.9) we have set

$$\mu = \mu_{aGBM} \tag{30.10}$$

to remind us that the drift parameter μ of the GBM is now the drift parameter of the average GBM, denoted by μ_{aGBM}. Having so said, let us now check (30.6) against (30.9). One then gets

$$\begin{cases} A = N_0 \\ B = \mu_{aGBM}\,. \end{cases} \tag{30.11}$$

On the other hand, it will be remembered from (11.12) that the relationship between N_0 and σ^2_{aGBM} is

$$N_0 = e^{\frac{\sigma^2_{aGBM}}{2}}. \tag{30.12}$$

Thus, upon inserting (30.12) into the upper equation of (30.11), and then solving for both μ_{aGBM} and σ^2_{aGBM} we get

$$\begin{cases} \mu_{aGBM} = B \\ \sigma^2_{aGBM} = 2\ln A\,. \end{cases} \tag{30.13}$$

These two equations may be rewritten in terms of the two points of coordinates (p_1, P_1) and (p_2, P_2) by inserting (30.8) into (30.13). With a little rearranging, the result is

$$\begin{cases} \mu_{aGBM} = \dfrac{\ln\left(\dfrac{P_2}{P_1}\right)}{p_2 - p_1} \\[2ex] \sigma^2_{aGBM} = \dfrac{2\ln\left(\dfrac{P_1^{p_2}}{P_2^{p_1}}\right)}{p_2 - p_1}\,. \end{cases} \tag{30.14}$$

In conclusion, we have

(1) Determined the exponential curve in time passing through the two assigned points of coordinates (p_1, P_1) and (p_2, P_2). These two points will later be identified as the two peaks of the initial b-lognormal (p_1, P_1) and final b-lognormal (p_2, P_2), respectively.
(2) Assumed that the above exponential curve in time is the mean value of a certain GBM. This allows for the representation of Darwinian evolution and later human history as a GBM, with all its ups and downs, but constrained by

definition to have an exponential increase in the "level of evolution", as 3.5 billion years of evolution and progress on Earth plainly show.

(3) Found the two equations (30.14) fully expressing this GBM at least in terms of the two initial and final peaks only. This is in full agreement with the statistical Drake equation static case (as described in Chapters 1, 3, 6, 7, 8, and 11).

The way is thus paved for a full mathematical (statistical, actually) theory of Darwinian evolution and human history, later to be extended to alien civilizations when SETI succeeds, even if these will be post-biological.

30.4 THE "NO-EVOLUTION" STATIONARY STOCHASTIC PROCESS

This short section is devoted to the special case when the two peak ordinates, P_1 and P_2, are equal to each other:

$$P_2 = P_1 . \tag{30.15}$$

Then, (30.8) and (30.15), with a little rearrangement, yield

$$\begin{cases} A = P_1 = P_2 \\ B = 0 . \end{cases} \tag{30.16}$$

This means that the "former exponential" is now a straight line parallel to the time axis and located at a certain positive ordinate $A = P_1 = P_2$ above it. In other words, the former GBM stochastic process has now become a stationary stochastic process, showing that "no-evolution" occurs between the two times p_1 and p_2: for millions or billions of years, living beings are born, reproduce, and die generation after generation with no evolution at all.

Also, the drift parameter μ_{aGBM} vanishes and the σ^2_{aGBM} is a constant in time equal to twice the natural log of the constant peak ordinate $P_1 = P_2$, as one immediately may see by inserting (30.15) into (30.14):

$$\begin{cases} \mu_{aGBM} = 0 \\ \sigma^2_{aGBM} = 2\ln(P_1) = 2\ln(P_2) . \end{cases} \tag{30.17}$$

30.5 DIFFERENTIAL ENTROPY OF THE "RUNNING *b*-LOGNORMAL" PEAKED AT THE GBM EXPONENTIAL AVERAGE

Much more interesting than the no-evolution case just considered is of course the truly exponential evolution case, summarized at a glance in Figure 8.2, which we reproduce at the top of the next page for convenience.

Figure 30.1 intuitively shows the basic mechanism that we propose in this book to unify the notions of Darwinian evolution, human historical progress, and SETI. But all this is just qualitative. If we want to go quantitative, the only way is to resort to Shannon's Information Theory and then resort to the notion of *b*-lognormal

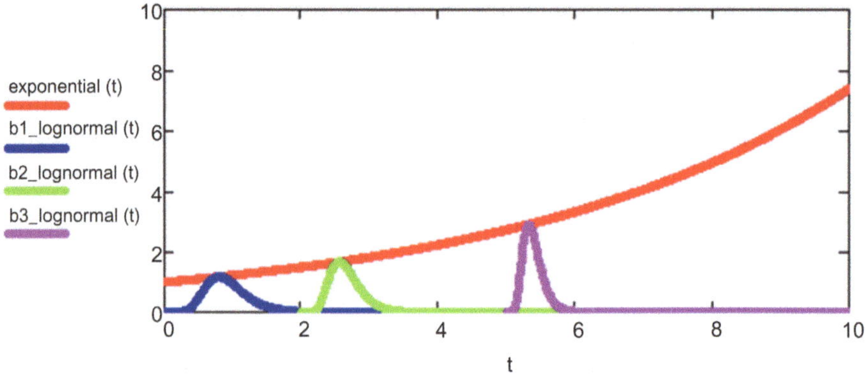

Figure 30.1. Exponential curve in time as the "envelope" of b-lognormals or, more correctly, as the geometric locus of all peaks (i.e., of all ordinates) of the b-lognormals having different peaks in time. This mechanism automatically ensures that "later" b-lognormals are "less spread" than "earlier" b-lognormals or, if one prefers, that later forms of life (or later civilizations) are more neatly peaked (i.e., focused) than the earlier ones.

differential entropy described in Section 30.1, as we describe in the remaining part of this epilogue chapter.

To start with, let us call the generic b-lognormal peaked at the generic positive instant $t = p$ the "running b-lognormal".

It will be remembered from Section 8.4 that the key equations describing the running b-lognormal are the two equations (8.12), which we reproduce here conveniently rewritten in this chapter's notation (the subscript $aRbL$ means "average running b-lognormal"):

$$\begin{cases} \mu_{aRbL} = \dfrac{1}{4\pi A^2} - p\,B \\ \sigma_{aRbL} = \dfrac{1}{\sqrt{2\pi}A}. \end{cases} \tag{30.18}$$

Therefore, the differential entropy of the running b-lognormal in bits is found by inserting (30.18) into (30.4) with the immediate result

$$H_{\text{running\_}b\text{-lognormal\_in\_bits}} = \frac{\ln\left(\sqrt{2\pi}\sigma_{aRbL}\right) + \mu_{aRbL} + \dfrac{1}{2}}{\ln 2} = \frac{-\ln(A) + \dfrac{1}{4\pi A^2} - p\,B + \dfrac{1}{2}}{\ln 2}. \tag{30.19}$$

This equation expresses the running b-lognormal's differential entropy in terms of the two constants A and B of the exponential curve (30.6) and of the abscissa in time (i.e., p) at which the running b-lognormal's peak is positioned. So, in reality, the only "free parameter" in (30.19) is indeed this "free" b-lognormal's peak abscissa p. Let us write this fact neatly as follows

$$H_{\text{running\_}b\text{-lognormal\_in\_bits}}(p) = \frac{-\ln(A) + \dfrac{1}{4\pi A^2} - p\,B + \dfrac{1}{2}}{\ln 2}. \tag{30.20}$$

We may now rewrite the last equation in terms of the coordinates of the two points (p_1, P_1) and (p_2, P_2) by inserting (30.8) into (30.20). The result of this straight substitution is

$$H_{\text{running\_}b\text{-lognormal\_in\_bits}}(p) = \frac{1}{\ln 2} \cdot \left[-\ln\left(\frac{P_1^{\frac{p_2}{p_2-p_1}}}{P_2^{\frac{p_1}{p_2-p_1}}} \right) + \frac{1}{4\pi \left(\frac{P_1^{\frac{p_2}{p_2-p_1}}}{P_2^{\frac{p_1}{p_2-p_1}}} \right)^2} - p \frac{\ln\left(\frac{P_2}{P_1}\right)}{p_2 - p_1} + \frac{1}{2} \right]. \tag{30.21}$$

This formula may only slightly be simplified as follows:

$$H_{\text{running\_}b\text{-lognormal\_in\_bits}}(p) = \frac{1}{\ln 2} \cdot \left[-\frac{\ln\left(\frac{P_1^{p_2}}{P_2^{p_1}} \right)}{p_2 - p_1} + \frac{P_2^{\frac{2p_1}{p_2-p_1}}}{4\pi P_1^{\frac{2p_2}{p_2-p_1}}} - p \frac{\ln\left(\frac{P_2}{P_1}\right)}{p_2 - p_1} + \frac{1}{2} \right]. \tag{30.22}$$

This is the differential entropy of the running b-lognormal peaked at p. It may more conveniently be rewritten as

$$H_{\text{running\_}b\text{-lognormal\_in\_bits}}(p) = -p \frac{\ln\left(\frac{P_2}{P_1}\right)}{(p_2 - p_1)\ln 2} + \text{Part\_not\_depending\_on\_}p. \tag{30.23}$$

We will use (30.23) in a moment!

30.6 DECREASING ENTROPY FOR AN EXPONENTIALLY INCREASING EVOLUTION: PROGRESS!

We now reach the conclusion of this chapter as well as the epilogue of this book on "Mathematical SETI".

Young students at university courses on thermodynamics are still made to learn that "entropy always increases", meaning that the second law of thermodynamics rules so (see, for instance, the Wikipedia site *http://en.wikipedia.org/wiki/Second_law_of_thermodynamics*). But thermodynamics was born in the 1700s amd 1800s upon the discovery of the laws of gases, and quite a few scientists later (in the 1900s and 2000s) came to realize that the sentence "entropy always increases" hardly applies to the evolution of intelligent living species as has occurred on Earth over the last 3.5 billion years. This author belongs to the second category of scientists and is proud to claim that his theory, outlined in this chapter as well as in Chapters 6, 7, and 8, neatly shows that entropy decreases (rather than always increases) when it comes to describing the evolution of life up to current times (i.e., when it comes to describing progress!).

Our proof is as follows.

Consider (30.23), which relates to the entropy of the running b-lognormal.

If we consider the entropy change between the initial b-lognormal peaked at (p_1, P_1) and the final one peaked at (p_2, P_2), this entropy change is, by definition, given by

$$\Delta_{H\_in\_bits} = H_{running\_b\text{-}lognormal\_in\_bits}(p_2) - H_{running\_b\text{-}lognormal\_in\_bits}(p_1). \quad (30.24)$$

When rewritten in terms of (30.23), this entropy change is simpler than (30.23) since the two "Parts_not_depending_on_p" cancel each other, and the result is:

$$\begin{aligned}\Delta_{H\_in\_bits} &= H_{running\_b\text{-}lognormal\_in\_bits}(p_2) - H_{running\_b\text{-}lognormal\_in\_bits}(p_1)\\ &= -p_2\frac{\ln\left(\frac{P_2}{P_1}\right)}{(p_2 - p_1)\ln 2} + p_1\frac{\ln\left(\frac{P_2}{P_1}\right)}{(p_2 - p_1)\ln 2}.\end{aligned} \quad (30.25)$$

But the last equation may be further simplified as follows

$$\begin{aligned}\Delta_{H\_in\_bits} &= -p_2\frac{\ln\left(\frac{P_2}{P_1}\right)}{(p_2 - p_1)\ln 2} + p_1\frac{\ln\left(\frac{P_2}{P_1}\right)}{(p_2 - p_1)\ln 2}\\ &= -(p_2 - p_1)\frac{\ln\left(\frac{P_2}{P_1}\right)}{(p_2 - p_1)\ln 2} = \frac{-\ln\left(\frac{P_2}{P_1}\right)}{\ln 2} = \frac{\ln\left(\frac{P_1}{P_2}\right)}{\ln 2}.\end{aligned} \quad (30.26)$$

Thus, we have reached a result of adamantine beauty: entropy change, when passing from a lower civilization to a higher civilization, is simply given by the log of the ratio between the lower civilization peak and the higher civilization peak (apart from the factor ln 2 at the denominator, necessary to measure entropy in bits):

$$\Delta_{H\_in\_bits} = \frac{\ln\left(\frac{P_1}{P_2}\right)}{\ln 2}. \quad (30.27)$$

Since in the actual unfolding of evolution we always had $P_1 < P_2$, the entropy change (30.27) in passing from the lower civilization to the higher civilization is the log of a number smaller than 1 (i.e., it is a negative number!). Thus, entropy decreases in passing from a lower civilization to a higher civilization, and that proves that *progress amounts to a decrease in entropy*.

One more important result derived from (30.19) is found upon rewriting it explicitly as a function of p, as follows:

$$H_{\text{running\_}b\text{-lognormal\_in\_bits}}(p) = \frac{-\ln(A) + \dfrac{1}{4\pi A^2} - p\,B + \dfrac{1}{2}}{\ln 2}$$

$$= \frac{-B}{\ln 2} \cdot p + \text{Part\_not\_depending\_on\_}p. \qquad (30.28)$$

Then, (30.24) and (30.28) yield immediately

$$\Delta_{H\_\text{in\_bits}} = -\frac{B}{\ln 2} \cdot (p_2 - p_1). \qquad (30.29)$$

This also is a new result of adamantine beauty, since it yields the entropy change in bits by virtue of the difference in time between the two peak abscissas $(p_2 - p_1)$ and the constant B expressing the exponential increase (i.e., the average exponential drift in the GBM) of (30.6). Just to distinguish the two adamantine results (30.27) and (30.29) from each other, we might call (30.27) the "entropy change in evolution by virtue of the peak ordinates only", and (30.29) the "entropy change in evolution by virtue of the peak abscissas and the average GBM exponential drift, B, only".

30.7 SIX EXAMPLES: ENTROPY CHANGES IN DARWINIAN EVOLUTION, HUMAN HISTORY BETWEEN ANCIENT GREECE AND NOW, AND AZTECS AND INCAS VS. SPANIARDS

Appendix 30.A not only provides the full mathematical proofs of all the results we have derived so far in this epilogue: it also offers six examples showing how entropy actually decreased in six different cases of past life on Earth. These six cases are, respectively:

(1) Darwinian evolution on Earth over the last 3.5 billion years. This is shown to correspond to an entropy decrease of 25.57 bits per each living being, if today's number of living species is assumed to be 50 million. Were there more than 50 million species living on Earth right now, our equations (30.27) and (30.29) would yield the corresponding entropy decrease accordingly. As for the proofs, please refer to equations (%i45) through (%o52) of Appendix 30.A.

(2) Entropy changes in human history from Ancient Greece to the end of the British Empire (i.e., to nearly nowadays). According to the b-lognormal theory that we described in Section 7.3, the year 434 BC (i.e., $p_1 = -434 \cdot \text{year}$) corresponds to the peak of the age of Pericles in Athens, while the year 1868 (i.e., $p_2 = -1868 \cdot \text{year}$) corresponds to the peak of the Victorian Age in Britain (Maxwell equations mastering electromagnetism published just four years earlier, in 1864). The entropy change between these two peaks is computed in

equations (%i53) through (%o59), and amounts to an entropy decrease of 1.76 bits per individual.

(3) Entropy changes in human history from Ancient Greece to the end of the American Empire, assumed to yield to China about the year 2050. Again, according to the *b*-lognormal theory that we described in Chapter 7, the year 434 BC (i.e., $p_1 = -434 \cdot \text{year}$) corresponds to the peak of the age of Pericles in Athens, while the year 1973 (i.e., $p_2 = 1973 \cdot \text{year}$) corresponds to the peak of the American Empire (the Americans first landed on the Moon on July 21, 1969 and last landed on December 14, 1972). Equations (%i60) through (%o66), show that the corresponding entropy decrease between the two peaks amounts to 2.38 bits per individual, thus higher than for the Greece-to-Britain GBM exponential average, as is intuitively obvious.

(4) Computing the entropy difference between the Aztecs and the Spaniards at the time they suddenly encountered each other (with no previous contact) in 1519, when Cortez landed in Vera Cruz and began his conquest of the Aztec Empire. This example is particularly important for SETI inasmuch as it might resemble what could happen in case humanity comes across an alien civilization all of a sudden (as in the movie *Independence Day*). Well, we found a way to compute how backward the Aztecs were by exploiting the fact that they were just on the verge of discovering the wheel. Wheels were used by Aztec children in their toys, but not by adults in warfare, and this is just what happened in Mesopotamia, the northern Caucasus, and central Europe about 3,500 years before Christ. Thus, we set $p_1 = -3{,}500 \cdot \text{year}$ to describe the technological backwardness of the Aztecs with respect to the Spaniards, and of course we set $p_2 = 1{,}521 \cdot \text{year}$ to mean that Cortez completed the conquest of the Aztec Empire in 1521. Again, we assumed the Greece-to-Britain exponential to be the right one (since it refers to events that have actually occurred, rather than future events also, as in the Greece-to-U.S.A. case). In conclusion, (30.27) and (30.29) therefore yield an entropy difference of 3.84 bits per individual between the Aztecs and the Spaniards. This is of course more than the 1.76-bit difference between the Victorian Britons and Pericles' Greeks: nearly twice as much, owing to the huge 5,000 years difference in evolution between the Aztecs and the Spaniards, vs. the just about 2,300 years in evolution between Pericles and the Victorian Age.

(5) Computing the entropy difference between the Incas and the Spaniards at the time they suddenly encountered each other (with no previous contact) in 1532, when Pizarro landed in Tumbes and began his conquest of the Inca Empire. Again, the wheel was unknown to the Incas, and so we assumed $p_1 = -3{,}500 \cdot \text{year}$. Naturally, we now assume $p_2 = 1{,}532 \cdot \text{year}$, and the result given by (30.27) and (30.29) is an entropy difference of 3.85 bits per individual. Just very slightly higher than for the Aztec–Spaniard case, meaning that ...

(6) The Aztecs were very slightly more technologically advanced than the Incas when they were both subdued by the Spaniards. In fact, if you assume that the Aztec Empire began around 850 AD (roughly at the peak of the Maya Empire, so that, in some sense, the Aztecs "inherited" part of the Maya civilization), and then you assume that the Inca Empire began around 1250 (when the Incas reached Cuzco),

then, assuming again the Greece-to-Britain exponential as the true exponential of technological development, (30.27) and (30.29) yield an entropy difference of 0.3 bits per individual in favor of the Aztecs (more technologically advanced) over the Incas. A subtle quantitative statement that current historical knowledge about both peoples is unable to prove because it is based on words rather than on mathematics.

So, with these final six entropy measurements, we close our book, hoping to have been the first author to have given a quantitative description of both Darwinian evolution and human history, based on our new discoveries about the mathematical properties of finite and infinite *b*-lognormals. This we did in order to quantitatively estimate the extent to which an alien civilization might be more advanced than us ... when SETI succeeds!

30.8 CONCLUSION

Let us be optimistic!

We all know that the current world on Earth (2012 AD) is much better that any previous "past world" on Earth.

Whether this is true in the Galaxy as well will be apparent only when SETI succeeds.

Let us thus support SETI as the most far-sighted scientific and technological effort ever attempted by humanity.

APPENDIX 30.A

EPILOGUE: Two-Peak GBM.

ENTROPY DECREASES WHEN PROGRESS (= EVOLUTION) IS MADE!

by Claudio Maccone, April 29, 2012.

```
(%i1) kill(all);
(%o0) done
```

```
(%i1) assume(sigma>0);
```
(%o1) $[\sigma>0]$

```
(%i2) first_peak_eq:P1=A*%e^(B*p1);
```
(%o2) $P1=A\ \%e^{p1\ B}$

```
(%i3) second_peak_eq:P2=A*%e^(B*p2);
```
(%o3) $P2=A\ \%e^{p2\ B}$

```
(%i4) without_A:second_peak_eq/first_peak_eq;
```
(%o4) $\frac{P2}{P1}=\%e^{p2\ B-p1\ B}$

```
(%i5) def_B:first(solve(without_A, B));
```
(%o5) $B=\frac{\log\left(\frac{P2}{P1}\right)}{p2-p1}$

```
(%i6) towards_A:first_peak_eq,def_B;
```
(%o6) $P1=A\ \%e^{\frac{p1\ \log\left(\frac{P2}{P1}\right)}{p2-p1}}$

```
(%i7) def_A:first(radcan(solve(towards_A, A)));
```
(%o7) $A=\frac{P1^{\frac{p2}{p2-p1}}}{P2^{\frac{p1}{p2-p1}}}$

```
(%i8) GBM_notation:[A=N0, B=mu[aGBM]];
```
(%o8) $[A=N0,B=\mu_{aGBM}]$

(%i9) def_N0:def_A,GBM_notation;

(%o9) $N0=\frac{P1^{\frac{p2}{p2-p1}}}{P2^{\frac{p1}{p2-p1}}}$

(%i10) def_mu:def_B,GBM_notation;

(%o10) $\mu_{aGBM}=\frac{\log\left(\frac{P2}{P1}\right)}{p2-p1}$

(%i11) def_N0_vs_sigma_square:N0=%e^(sigma[aGBM]^2/2);

(%o11) $N0=\%e^{\frac{\sigma_{aGBM}^2}{2}}$

(%i12) towards_sigma_square:def_N0,def_N0_vs_sigma;

(%o12) $N0=\frac{P1^{\frac{p2}{p2-p1}}}{P2^{\frac{p1}{p2-p1}}}$

(%i13) towards_sigma_square_2:towards_sigma_square,def_N0_vs_sigma_square;

(%o13) $\%e^{\frac{\sigma_{aGBM}^2}{2}}=\frac{P1^{\frac{p2}{p2-p1}}}{P2^{\frac{p1}{p2-p1}}}$

(%i14) obtained_def_sigma_square:first(factor(radcan(solve(towards_sigma_square_2, sigma[aGBM]^2))));

(%o14) $\sigma_{aGBM}^2=-\frac{2\,(p1\,\log(P2)-p2\,\log(P1))}{p2-p1}$

(%i15) easier_def_of_sigma_square:sigma[aGBM]^2=2*(log(P1^p2/P2^p1))/(p2-p1);

(%o15) $\sigma_{aGBM}^2=\frac{2\,\log\left(\frac{P1^{p2}}{P2^{p1}}\right)}{p2-p1}$

(%i16) radcan(easier_def_of_sigma_square-obtained_def_sigma_square);

(%o16) $0=0$

NO-EVOLUTION CASE :

NoEV as a STATIONARY STOCHASTIC PROCESS.

Meaning that the living members of a certain species
are born, reproduce and die for millions or billions of years
WITHOUT any evolution at all.

```
(%i17) def_NoEv:[P2=P1];
```
(%o17) $[P2=P1]$

```
(%i18) zero_mu:def_mu,def_NoEv;
```
(%o18) $\mu_{aGBM}=0$

```
(%i19) constant_sigma_square:easier_def_of_sigma_square, def_NoEv;
```
(%o19) $\sigma_{aGBM}^2=2\log(P1)$

```
(%i20) constant_P1:first(solve(constant_sigma_square, P1));
```
(%o20) $P1=\%e^{\frac{\sigma_{aGBM}^2}{2}}$

RUNNING b-LOGNORMAL.
Its mu and sigma are defined as follows
(see (8.12) and (30.18) of text) :

```
(%i21) def_mu_aRbL:mu[aRbL]=-B*p+1/(4*%pi*A^2);
```
(%o21) $\mu_{aRbL}=\frac{1}{4\pi A^2}-p\,B$

```
(%i22) def_sigma_aRbL:sigma[aRbL]=1/(sqrt(2*%pi)*A);
```
(%o22) $\sigma_{aRbL}=\frac{1}{\sqrt{2}\sqrt{\pi}\,A}$

```
(%i23) mu_sigma_for_aRbL:[def_mu_aRbL, def_sigma_aRbL];
```
(%o23) $[\mu_{aRbL}=\frac{1}{4\pi A^2}-p\,B,\sigma_{aRbL}=\frac{1}{\sqrt{2}\sqrt{\pi}\,A}]$

ENTROPY of the Running b-Lognormal H_aRbL

(%i24) def_aRbL_Entropy:H_aRbL=1/2+(log(2*%pi*sigma[aRbL]^2))/2+mu[aRbL];

(%o24) $H\_aRbL=\frac{\log(2\,\pi\,\sigma_{aRbL}^{2})}{2}+\mu_{aRbL}+\frac{1}{2}$

(%i25) def_H_aRbL_vs_A_B:def_aRbL_Entropy,mu_sigma_for_aRbL;

(%o25) $H\_aRbL=-p\,B-\log(A)+\frac{1}{4\,\pi\,A^{2}}+\frac{1}{2}$

The last equation yields thus the ENTROPY of the RUNNING b-LOGNORMAL.
In other words, the last equation yields the ENTROPY just in terms of the Running b-lognormal's peak abscissa p (the free parameter locating where the peak of the b-lognormal is in time) and of course the two EXPONENTIAL CONSTANTS A and B.

This equation may now be rewritten in two different forms:
1) The Running b-lognormal Entropy as a function of all the four data (p1, P1, p2, P2) of the two peaks, as we now prove immediately by inserting the expressions %o5 and %07 of A and B in terms of the four parameters (p1, P1, p2, P2);
2) The Running b-lognormal Entropy as a function of (p2-p1) ONLY, that, quite surprisingly, turns out to be A LINEAR FUNCTION, namely a DECRASING STRAIGHT LINE as a function of (p2-p1). This is proven in equations %o38 and following.

(%i26) def_H_aRbL_vs_ps_Ps:def_H_aRbL_vs_A_B,def_A,def_B;

(%o26) $H\_aRbL=-\log\left(\frac{P1^{\frac{p2}{p2-p1}}}{P2^{\frac{p1}{p2-p1}}}\right)-\frac{p\,\log\left(\frac{P2}{P1}\right)}{p2-p1}+\frac{P2^{\frac{2\,p1}{p2-p1}}}{4\,\pi\,P1^{\frac{2\,p2}{p2-p1}}}+\frac{1}{2}$

Re-defining the Entropy of the Running b-Lognormal
AS A FUNCTION OF THE PEAK ABSCISSA (p) ONLY.

(%i27) H_aRbL(p):=-(p*log(P2/P1))/(p2-p1)+Part_not_depending_on_p;

(%o27) $H\_aRbL(p):=\frac{-p\,\log\left(\frac{P2}{P1}\right)}{p2-p1}+Part\_not\_depending\_on\_p$

Defining the ENTROPY CHANGE between the
1) Entropy of the b-lognormal with peak at p sub 2 and
2) Entropy of the b-lognormal with peak at p sub 1.

(%i28) def_H_CHANGE:Delta[H]=H_aRbL(p2)-H_aRbL(p1);

$$(\%o28)\ \Delta_H=\frac{p1\log\left(\frac{P2}{P1}\right)}{p2-p1}-\frac{p2\log\left(\frac{P2}{P1}\right)}{p2-p1}$$

(%i29) logcontract(distrib(radcan(%)));

$$(\%o29)\ \Delta_H=\log\left(\frac{P1}{P2}\right)$$

(%i30) def_H_CHANGE_bits:def_H_CHANGE/log(2);

$$(\%o30)\ \frac{\Delta_H}{\log(2)}=\frac{\frac{p1\log\left(\frac{P2}{P1}\right)}{p2-p1}-\frac{p2\log\left(\frac{P2}{P1}\right)}{p2-p1}}{\log(2)}$$

(%i31) def_Delta_sub_H_bits:Delta[H_bits_P1_P2]=logcontract(distrib(radcan(rhs(def_H_CHANGE_bits))));

$$(\%o31)\ \Delta_{H\_bits\_P1\_P2}=\frac{\log\left(\frac{P1}{P2}\right)}{\log(2)}$$

(%i32) log(2)=ev(log(2), numer);

$$(\%o32)\ \log(2)=0.69314718055995$$

CONCLUSION: if P1<P2 (as it is in EVOLUTION) then Delta sub H is NEGATIVE,
i.e. the ENTROPY DECREASED !
Exactly the OPPOSITE of the Second Law of Thermodynamics, as it would have had to be expected !

In other words: EVOLUTION MEANS ENTROPY DECREASE, and NOT increase !

In the NoEv case, of course P2=P1, and the ENTROPY DOES NOT CHANGE !

We may even express the Delta sub H as a function of the two b-lognormals' two peak ordinates P1 and P2 as funtions of the 4 mu1, sigma1, and mu2 sigma2.

(%i33) def_P1:P1=(%e^(sigma[1]^2/2-mu[1]))/(sqrt(2*%pi)*sigma[1]);

$$(\%o33)\ P1=\frac{\%e^{\frac{\sigma_1^2}{2}-\mu_1}}{\sqrt{2}\sqrt{\pi}\,\sigma_1}$$

(%i34) def_P2:P2=(%e^(sigma[2]^2/2-mu[2]))/(sqrt(2*%pi)*sigma[2]);

(%o34) $P2=\frac{\%e^{\frac{\sigma_2^2}{2}-\mu_2}}{\sqrt{2}\sqrt{\pi}\,\sigma_2}$

(%i35) def_2peaks:[def_P1, def_P2] ;

(%o35) $[P1=\frac{\%e^{\frac{\sigma_1^2}{2}-\mu_1}}{\sqrt{2}\sqrt{\pi}\,\sigma_1},P2=\frac{\%e^{\frac{\sigma_2^2}{2}-\mu_2}}{\sqrt{2}\sqrt{\pi}\,\sigma_2}]$

(%i36) def_Delta_sub_H_bits_mus_sigmas:Delta[H_bits_mus_sigmas]=rhs(def_Delta_sub_H_bits),def_2peaks;

(%o36) $\Delta_{H\_bits\_mus\_sigmas}=\frac{\log\left(\frac{\%e^{-\frac{\sigma_2^2}{2}+\mu_2+\frac{\sigma_1^2}{2}-\mu_1}\sigma_2}{\sigma_1}\right)}{\log(2)}$

(%i37) Delta_sub_H_bits_mus_sigmas:logcontract(radcan(def_Delta_sub_H_bits_mus_sigmas));

(%o37) $\Delta_{H\_bits\_mus\_sigmas}=-\frac{\log\left(\frac{\sigma_1^2}{\sigma_2^2}\right)+\sigma_2^2-2\,\mu_2-\sigma_1^2+2\,\mu_1}{2\log(2)}$

APRIL 27, 2012.
Discovered the new formula for Delta H sub in bits, in terms of the Running b-lognormal's (p2-p1) ONLY. In other words, this is a much easier expression of the Running b-lognormal's Entropy, inasmuch as it turns out to ba a LINEAR function of (p2-p1), i.e. a STRAIGHT LINE.

(%i38) def_H_aRbL_vs_A_B:def_aRbL_Entropy,mu_sigma_for_aRbL;

(%o38) $H\_aRbL=-p\,B-\log(A)+\frac{1}{4\,\pi\,A^2}+\frac{1}{2}$

Defining the ENTROPY of the Running b-lognormal IN BITS
as a function of three independent variables:
1) A, B of the exponential and
2) the Running b-lognormal peak abscissa in time, p.

(%i39) H_aRbL_in_bits(p, A, B):=(-p*B-log(A)+1/(4*%pi*A^2)+1/2)/log(2);

(%o39) $H\_aRbL\_in\_bits(p,A,B):=\frac{(-p)\,B-\log(A)+\frac{1}{4\,\pi\,A^2}+\frac{1}{2}}{\log(2)}$

The first derivative of the Running b-Lognormal Entropy with respect to p is a CONSTANT, meaning that Entropy(p) is a STRAIGHT LINE:

(%i40) 'diff('H_aRbL_in_bits(p, A, B), p)=diff(H_aRbL_in_bits(p, A, B), p);

(%o40) $\frac{d}{d\,p}\,\mathrm{H\_aRbL\_in\_bits}(p,A,B)=-\frac{B}{\log(2)}$

So, let us define the equation of this DECREASING STRAIGHT LINE neatly:

(%i41) straight_line(p):=-(B/log(2))*p+(-log(A)+1/(4*%pi*A^2)+1/2)/log(2);

(%o41) $\mathrm{straight\_line}(p):=\left(-\frac{B}{\log(2)}\right)p+\frac{-\log(A)+\frac{1}{4\,\pi\,A^2}+\frac{1}{2}}{\log(2)}$

Again, let us define the ENTROPY DECREASE as the difference between the Entropy at peak p2 minus the Entropy at peak p1:

(%i42) linear_eq:Delta[H_bits_p2_minus_p1]=factor(straight_line(p2)-straight_line(p1));

(%o42) $\Delta_{\mathrm{H\_bits\_p2\_minus\_p1}}=-\frac{(p2-p1)\,B}{\log(2)}$

Finding the INTERCEPT of the straight line with the vertical axis:

(%i43) vertical_axis_intercept=H_aRbL_in_bits(0, A, B);

(%o43) $\mathrm{vertical\_axis\_intercept}=\frac{-\log(A)+\frac{1}{4\,\pi\,A^2}+\frac{1}{2}}{\log(2)}$

Finding the INTERCEPT of the straight line with the horizontal axis:

(%i44) horizontal_axis_intercept=rhs(factor(distrib(first(solve(H_aRbL_in_bits(p, A, B)=0, p)))));

(%o44) $\mathrm{horizontal\_axis\_intercept}=-\frac{4\,\pi\,A^2\log(A)-2\,\pi\,A^2-1}{4\,\pi\,A^2\,B}$

==

==

==

And finally SIX EXAMPLES:
1) DARWINIAN EVOLUTION.
2) MATHEMATICAL HISTORY: Ancient Greece to Britain Exponential.
3) MATHEMATICAL HISTORY: Ancient Greece to USA Exponential.
4) AZTEC-SPANIARDS DIFFERENCE IN CIVILIZATION AMOUNT
at the time of the Cortez conquest of the Aztec Empire (1519-1521).
5) INCAS-SPANIARDS DIFFERENCE IN CIVILIZATION AMOUNT
at the time of the Pizarro conquest of the Inca Empire (1532-1533).
6) INCAS-AZTEC DIFFERENCE IN CIVILIZATION AMOUNT
around 1500 A.D. when they both were about to be subdued by the Spaniards.

==

#1 Example.
DARWINIAN EVOLUTION:
ENTROPY LOSS OVER THE LAST 3.5 BILLION YEARS
IF TODAY'S NUMBER OF SPECIES IS 50 MILLIONS.

```
(%i45) Darwinian_Evolution:[p1=-3.5*10^9*year, P1=1, p2=0*year, P2=50*10^6];
```

(%o45) $[p1=-3.5\ 10^{9}\ year, P1=1, p2=0, P2=50000000]$

```
(%i46) def_Delta_sub_H_bits,Darwinian_Evolution;
```

(%o46) $\Delta_{H\_bits\_P1\_P2}=-\frac{\log(50000000)}{\log(2)}$

```
(%i47) def_Delta_H_bits:ev(%, numer);
```

(%o47) $\Delta_{H\_bits\_P1\_P2}=-25.5754247590989$

So, the entropy decrease all over the 3.5 billion years of Evolution on Earth, namely the entropy decrease between Man (= now) and the first living being 3.5 billion years ago, has been of about 25 bits (per individual).

Now let us re-compute the same result by virtue of the LINEAR expression of the Running b-lognormal Entropy as a function of (p2-p1).
First of all, we must find the numeric value of B for Evolution.

```
(%i48) B_from_%o5:B=log(P2/P1)/(p2-p1);
```

(%o48) $B=\frac{\log\left(\frac{P2}{P1}\right)}{p2-p1}$

(%i49) B_for_Darwin:B_from_%o5,Darwinian_Evolution;

(%o49) $B=\frac{2.857142857142857\ 10^{-10}\ \log(50000000)}{year}$

(%i50) numerical_Darwinian_B:ev(B_for_Darwin,numer);

(%o50) $B=\frac{5.0650095895406915\ 10^{-9}}{year}$

Next, we can find the Entropy by virtue of the straight-line formula %o42.

(%i51) Entropy_linear_decrease:linear_eq,Darwinian_Evolution,numerical_Darwinian_B;

(%o51) $\Delta_{H\_bits\_p2\_minus\_p1}=-\frac{17.72753356339242}{\log(2)}$

(%i52) ev(%, numer);

(%o52) $\Delta_{H\_bits\_p2\_minus\_p1}=-25.5754247590989$

and so we have found the same numerical Entropy as %o47, of course.

==

#2 Example.
MATHEMATICAL HISTORY:
Ancient Greece to Britain Exponential.

(%i53) Greece_to_Britain_EXP:[p1=-434.333*year, P1=2.488*10^(-3), p2=1868*year, P2=8.447*10^-3];

(%o53) [p1=-434.333 year,P1=0.002488,p2=1868 year,P2=0.008447]

(%i54) def_Delta_sub_H_bits,Greece_to_Britain_EXP;

(%o54) $\Delta_{H\_bits\_P1\_P2}=-\frac{1.222332173868427}{\log(2)}$

(%i55) Greece_to_Britain_Delta_H_bits:ev(%, numer);

(%o55) $\Delta_{H\_bits\_P1\_P2}=-1.763452565559005$

(%i56) B_from_%o5:B=log(P2/P1)/(p2-p1);

(%o56) $B=\frac{\log\left(\frac{P2}{P1}\right)}{p2-p1}$

(%i57) B_for_Greece_to_Britain:B_from_%o5,Greece_to_Britain_EXP;

(%o57) $B=\frac{5.3091024359570339\ 10^{-4}}{year}$

Next, we can find the Entropy by virtue of the straight-line formula %o42.

(%i58) Entropy_linear_decrease:linear_eq,Greece_to_Britain_EXP,B_for_Greece_to_Britain;

(%o58) $\Delta_{H\_bits\_p2\_minus\_p1} = -\dfrac{1.222332173868427}{\log(2)}$

(%i59) ev(%, numer);

(%o59) $\Delta_{H\_bits\_p2\_minus\_p1} = -1.763452565559005$

and so we have found the same numerical Entropy as %o55, of course.

==

#3 Example.
MATHEMATICAL HISTORY:
Ancient Greece to USA Exponential.

(%i60) Greece_to_USA_EXP:[p1=-434.333*year, P1=2.488*10^(-3), p2=1973*year, P2=0.013];

(%o60) $[p1 = -434.333\ year, P1 = 0.002488, p2 = 1973\ year, P2 = 0.013]$

(%i61) def_Delta_sub_H_bits,Greece_to_USA_EXP;

(%o61) $\Delta_{H\_bits\_P1\_P2} = -\dfrac{1.653470182584604}{\log(2)}$

(%i62) Greece_to_USA_Delta_H_bits:ev(%, numer);

(%o62) $\Delta_{H\_bits\_P1\_P2} = -2.385453232672576$

(%i63) B_from_%o5:B=log(P2/P1)/(p2-p1);

(%o63) $B = \dfrac{\log\left(\dfrac{P2}{P1}\right)}{p2 - p1}$

(%i64) B_for_Greece_to_USA:B_from_%o5,Greece_to_USA_EXP;

(%o64) $B = \dfrac{6.8684730470799167\ 10^{-4}}{year}$

Next, we can find the Entropy by virtue of the straight-line formula %o42.

(%i65) Entropy_linear_decrease:linear_eq,Greece_to_USA_EXP,B_for_Greece_to_USA;

(%o65) $\Delta_{H\_bits\_p2\_minus\_p1} = -\dfrac{1.653470182584604}{\log(2)}$

```
(%i66) ev(%, numer);
```

(%o66) $\Delta_{H\_bits\_p2\_minus\_p1} = -2.385453232672577$

and so we have found the same numerical Entropy as %o73, of course
(except for a numerical rounding error in the last digit, now a 7 instead of a 6).

Warning: if you think that these ENTROPY LOSSES due to progress are "TOO SMALL",
then you should multiply them by the overall population of all countries involved.
But we shall not tackle this problem now!

==

#4 Example.
AZTEC-SPANIARDS DIFFERENCE IN CIVILIZATION AMOUNT
at the time of the Cortez conquest of the Aztec Empire (1519-1521).

Mountains of words have been used to describe the conquest of the Aztec Empire
of some 20 million people by Cortez and his 600 Spaniards, 16 horses and 10 cannons.
How was that possible ?
Clearly, the Spaniards capitalized on their HIGHER TECHNOLOGICAL LEVEL
with respect to the Aztecs, that made the Aztecs regard the Spaniards as SEMI-GODS.

But while a number of descriptions of the historical events of "the Conquest" exist,
to the best of our knowledge no QUANTITATIVE ESTIMATE of the DIFFERENCE
IN TECHNOLOGICAL EVOLUTION AMOUNT was ever given by any author.
We try to do so in the lines to follow.

Our criterion for estimating numerically the difference in technological evolution between the Aztecs and the Spaniards is the cosideration of the WHEEL.

According to the Wikipedia site http://en.wikipedia.org/wiki/Wheel the wheel was invented almost simultaneously in Mesopotamia, Northern Caucasus and Central Europe roughly about the year 3500 B.C.

Now, when Cortez subdued the Aztecs in 1521, the Aztecs did not know about the wheel, except for CHILDREN'S TOYS ! Thus, it may be said that the Aztecs were just "on the verge" of discovering the wheel, i.e that they had reached the technological level already reached around 3500 B. C. in Mesopotamia, Northern Caucasus and Central Europe.

This we now write in the following "Aztec_to_Cortez" list of four numbers, where we also assume that the GREECE-TO-BRITAIN EXPONENTIAL derived in Appendix 7.A is correct. Indeed, that exponential is based on past events, rather than the Greece-to-USA exponential based on the supposed "end" of the American world supremacy around 2050. So, we assume the Greece-to-Britain Exponential to be the geometric locus of both the Aztec and Spaniards b-lognormal PEAKS.

One thus gets immediately:

```
(%i67) Aztec_to_Cortez_EXP:[p1=-3500*year, p2=1521*year, B_for_Greece_to_Britain];
```

(%o67) $[p1=-3500\ year, p2=1521\ year, B=\frac{5.3091024359570339\ 10^{-4}}{year}]$

```
(%i68) Entropy_linear_decrease:linear_eq,Aztec_to_Cortez_EXP;
```

(%o68) $\Delta_{H\_bits\_p2\_minus\_p1}=-\frac{2.665700333094027}{\log(2)}$

```
(%i69) ev(Entropy_linear_decrease, numer);
```

(%o69) $\Delta_{H\_bits\_p2\_minus\_p1}=-3.845792651050811$

Thus, the difference in technological level between the Aztecs and the Spaniards is quantified as 3.84 bits (per person).
This we claim to be a new, important result in quantifying things much better than mountains of words.

==

#5 Example.
INCAS-SPANIARDS DIFFERENCE IN CIVILIZATION AMOUNT
at the time of the Pizarro conquest of the Inca Empire (1533).

Again, we write the Incas-to-Pizarro list of three numbers:
1) the date of the wheel (that the Incas did not have yet, but were close to have)
2) the date of the Pizarro conquest (Execution of Hatahualpa, August 28, 1533)
3) Greece-to-Britain "best civilization" exponential curve, assumed and extended into the past.

```
(%i70) Incas_to_Pizarro_EXP:[p1=-3500*year, p2=1533*year, B_for_Greece_to_Britain];
```

(%o70) $[p1=-3500\ year, p2=1533\ year, B=\frac{5.3091024359570339\ 10^{-4}}{year}]$

```
(%i71) Entropy_linear_decrease:linear_eq, Incas_to_Pizarro_EXP;
```

(%o71) $\Delta_{H\_bits\_p2\_minus\_p1}=-\frac{2.672071256017175}{\log(2)}$

```
(%i72) ev(Entropy_linear_decrease, numer);
```

(%o72) $\Delta_{H\_bits\_p2\_minus\_p1}=-3.854983949957922$

Thus, the entropy difference between Pizarro and the Incas turns out to be about the same as between the Aztecs and Cortez, actually a little bit higher. This is to say that the Aztecs were a little bit technologically more advanced than the Incas at the time of the Spanish conquest. As we now explore...

==

#6 Example.
INCAS-AZTEC DIFFERENCE IN CIVILIZATION AMOUNT
around 1500 A.D., when they both were about to be subdued by the Spaniards.

We assume that:
1) The Aztec Empire "had started" around 850 A.D.
2) The Inca Empire "had started" around 1250 A.D.
(Incidentally, this is the same year when the Holy Roman Emperor Frederick II died in southern Italy; in practice he ended the Middle Ages in Italy and paved the way to the later Italian Renaissance, that, later still, become the European Renaissance. That "top civilization" would defeat both the Aztecs and Incas in just 250 years).

Of course the above EXACT dates about the Aztecs and Incas may be questionable. We just assume them here to show how the Entropy Difference among the Aztecs and the Incas must be a very smaller number, much smaller than all previously found Entropy Differences.

Actually, with these numbers one finds:

```
(%i73) Aztecs_to_Incas_EXP:[p1=850*year, p2=1250*year, B_for_Greece_to_Britain];
```

(%o73) $[p1=850\ year, p2=1250\ year, B=\frac{5.3091024359570339\ 10^{-4}}{year}]$

```
(%i74) Entropy_linear_decrease:linear_eq, Aztecs_to_Incas_EXP;
```

(%o74) $\Delta_{H\_bits\_p2\_minus\_p1}=-\frac{0.21236409743828}{\log(2)}$

```
(%i75) ev(Entropy_linear_decrease, numer);
```

(%o75) $\Delta_{H\_bits\_p2\_minus\_p1}=-0.30637663023707$

This means that the Aztecs were 0.3 bits more advanced than the Incas (per person) at the time of the Spanish conquest, but, of course, this is just a very small number, and it might even be that the Incas were "more civilized" than the Aztecs, as only more advanced historical investigations may reveal in the future.

==

And, with this final Entropy measurements, we close our book, hoping to have been the first author to give a QUANTITATIVE description of both Darwinian Evolution and Human History, based on our new discoveries about the mathematical properties of the FINITE b-LOGNORMALS.

Should the readers wonder why we did "so much mathematics" just find out how much the Spaniard's technogical level was higher than the Aztec and Incas ones, oh! well, our answer is obvious:

"To be able to QUANTITATIVELY ESTIMATE how much an Alien Civilization might be HIGHER THAN US...

... WHEN SETI SUCCEDS !

Index

MIX
Papier aus verantwortungsvollen Quellen
Paper from responsible sources
FSC® C105338

If you have any concerns about our products,
you can contact us on
ProductSafety@springernature.com

In case Publisher is established outside the EU,
the EU authorized representative is:
Springer Nature Customer Service Center GmbH
Europaplatz 3, 69115 Heidelberg, Germany

Printed by Libri Plureos GmbH
in Hamburg, Germany